W9-CQM-315

FOR STUDENTS

- Career Opportunities
- Career Fitness Program
- Becoming an Electronics Technician
- Free On-Line Study Guides (companion web sites)

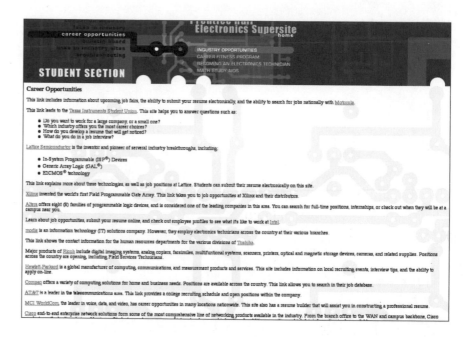

FOR FACULTY*

- Supplements
- On-Line Product Catalog
- Electronics Technology Journal
- Prentice Hall Book Advisor

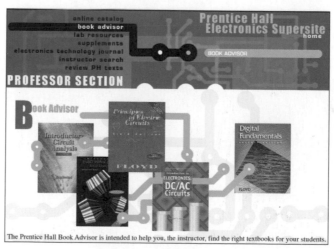

Please contact your Prentice Hall representative for passcode

ELECTRONIC DEVICES

SIXTH EDITION

ELECTRONIC DEVICES

Thomas L. Floyd

Prentice
Hall

Upper Saddle River, New Jersey
Columbus, Ohio

Library of Congress Cataloging-in-Publication Data

Floyd, Thomas L.
    Electronic devices / Thomas L. Floyd.— 6th ed.
        p.   cm.
    Includes index.
    ISBN 0-13-028484-X (alk. paper)
    1. Electronic apparatus and appliances.   2. Solid state electronics.   I. Title.

    TK7870.F52   2002
    627.3815—dc21

                                                    2001031163

Editor in Chief: Stephen Helba
Product Manager: Scott J. Sambucci
Associate Editor: Kate Linsner
Production Editor: Rex Davidson
Design Coordinator: Karrie Converse-Jones
Text Designer: Seventeenth Street Studios
Cover Designer: Thomas Mack
Cover Photo: Image Bank
Production Manager: Pat Tonneman
Illustrations: Jane Lopez and Steve Botts

The book was set in Times Roman by The Clarinda Company and was printed and bound by Von Hoffmann Press, Inc. The cover was printed by Phoenix Color Corp.

Chapter opening photos, System Application photo, and Troubleshooting photo by EyeWire, Inc.

Prentice-Hall International (UK) Limited, *London*
Prentice-Hall of Australia Pty. Limited, *Sydney*
Prentice-Hall Canada, Inc., *Toronto*
Prentice-Hall Hispanoamericana, S. A., *Mexico*
Prentice-Hall of India Private Limited, *New Delhi*
Prentice-Hall of Japan, Inc., *Tokyo*
Prentice-Hall Singapore Pte. Ltd.
Editora Prentice-Hall do Brasil, Ltda, *Rio de Janeiro*

Copyright © 2002, 1999, 1996, 1992, 1988, 1984 by Pearson Education, Inc., Upper Saddle River, New Jersey 07458.
All rights reserved. Printed in the United States of America. This publication is protected by Copyright and permission should be obtained from the publisher prior to any prohibited reproduction, storage in a retrieval system, or transmission in any form or by any means, electronic, mechanical, photocopying, recording, or likewise. For information regarding permission(s), write to: Rights and Permissions Department.

10 9 8 7 6 5 4 3
ISBN: 0-13-028484-X

To Sheila and Taylor,
with love.

Preface

Electronic Devices, Sixth Edition, has been carefully revised. Many suggestions and comments from reviewers and current users have been included to make this edition even more effective. A comprehensive coverage of electronic devices and circuits, including troubleshooting and practical applications, is provided. Chapters 1 through 11 are essentially devoted to discrete devices and circuits, whereas Chapters 12 through 18 primarily cover linear integrated circuits. The majority of chapters include a troubleshooting section, and most chapters have a System Application feature. The use of data sheets provides an important link between theory and the real world. Extensive exercises and problems using Electronics Workbench/Multisim circuit simulations help the student to verify circuit theory and develop troubleshooting and measurement skills. Referenced circuit files in both EWB Version 5 and Multisim are on the CD-ROM packaged with this book.

New to This Edition

New Chapter on Communications Circuits   Chapter 17 includes coverage of basic receivers, multipliers, amplitude and frequency modulation, demodulation, mixers, IF and audio amplifiers, and phase-locked loops.

Two Chapters Combined   Chapter 12 and Chapter 13 have been combined by eliminating certain topics that have become nonessential or less important in the context of modern linear ICs.

Chapters Reordered   Chapters 6 through 9, which cover amplifiers, have been reordered so that the chapter on power amplifiers is now the last in that sequence and includes both BJTs and FETs.

Key Terms   A key terms list is now part of the chapter opener. These key terms are highlighted in color and indicated by a margin icon where they appear in the text. Definitions are given at the end of the chapter as well as at the end of the book in a comprehensive glossary.

System Applications Feature   The System Applications, which were formerly numbered sections, are now presented as a special feature rather than a section. This approach is more appropriate to the optional nature of these topics.

EWB and Multisim Troubleshooting Problems   A set of problems that reference EWB and Multisim files on the CD-ROM have been added to most chapters. Each reference circuit file contains a simulated circuit with an inserted fault. The student must troubleshoot each circuit to determine the faulty component.

General Improvements   Numerous topics throughout the book have been revised either by being completely rewitten, expanded, or condensed to improve and clarify the presentation. Devices that have become obsolete have been replaced by newer, similar devices.

Features

- New, attractive text design and layout provide visual interest and ease of use.
- Full-color format
- Two-page chapter openers with a chapter outline, chapter objectives, introduction, list of key terms, and system application preview

■ An introduction and list of objectives at the beginning of each section within a chapter

■ Numerous worked examples, each with a related problem that provides a student exercise similar to that illustrated in the example. Answers to related problems are at the end of the chapter.

■ An EWB/Multisim simulation on the CD-ROM for selected examples

■ Section Reviews with answers at the end of the chapter

■ A Troubleshooting section in the majority of chapters

■ A System Application at the end of most chapters

■ Chapter Summary, Key Term glossary, Key Formulas, and Self-Test at the end of each chapter. Answers to the self-test are also at the end of the chapter.

■ Problem set at the end of each chapter divided by chapter sections and organized into basic and advanced categories. Selected problem circuits are available in EWB and Multisim on the CD-ROM. Additional problem categories of troubleshooting, data sheet, and EWB/Multisim Troubleshooting are found in many chapters.

■ Table of standard resistor values, derivation of selected equations, data sheets, answers to odd-numbered problems, a comprehensive glossary, and the index are at the end of the book.

Student Resources

Companion Website (*www.prenhall.com/floyd*)   This website offers students a free online study guide that they can check for conceptual understanding of key topics.

Electronics Workbench®/Multisim® CD-ROM   Packaged with each textbook, this software includes simulation circuits for selected examples, troubleshooting sections, and selected problems with and without inserted faults. Each simulated circuit is found in both EWB Version 5 and Multisim. Electronics Workbench and Multisim application software can be obtained through your local bookstore, or by contacting Electronics Workbench at 800-263-5552 or through their website at *www.electronicsworkbench.com.*

Laboratory Exercises for Electronic Devices, Sixth Edition, by Dave Buchla. ISBN: 0-13-092275-7

Experiments in Electronic Devices, Sixth Edition, by Howard Berlin, et al. ISBN: 0-13-092256-0

Electronics Supersite (*www.prenhall.com/electronics*)   Students will find additional troubleshooting exercises, links to industry sites, an interview with an electronics professional, and more.

Instructor Resources

Companion Website (*www.prenhall.com/floyd*)   For the professor, this website offers the ability to post your syllabus online with our Syllabus Builder. This is a great solution for classes taught online, self-paced, or in any computer-assisted manner.

Electronics Workbench/Multisim CD-ROM   Although the CD-ROM accompanying the textbook is primarily for the benefit of the student, solution and fault information is provided on the disk for the instructor's use. Refer to the CD-ROM organization diagram, which shows the folder hierarchy and file naming convention. Circuits containing faults are password-protected so that only the instructor can identify the faults. Solution files are available for each student circuit and are also password-protected and accessible only to the instructor.

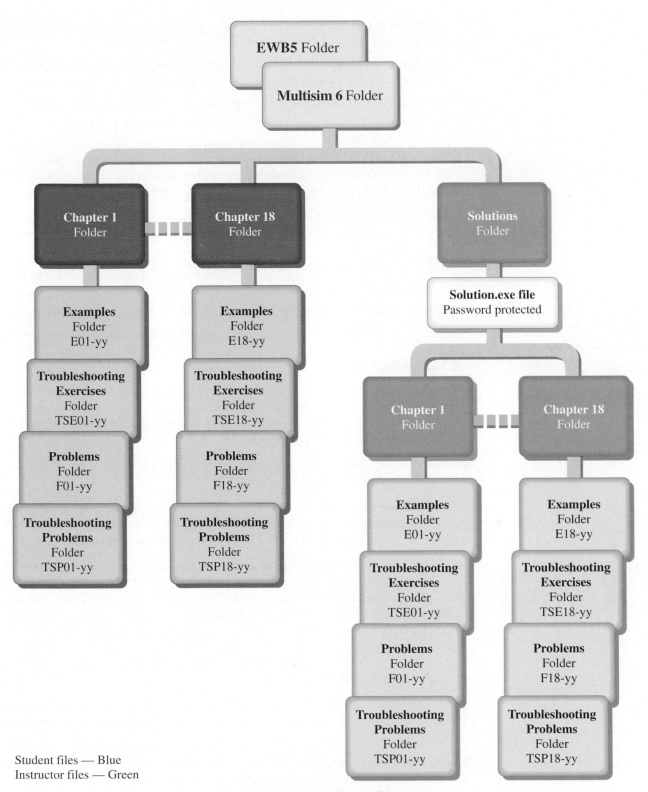

Student files — Blue
Instructor files — Green

CD-ROM Organizational Diagram

Instructor's Resource Manual   Includes solutions to chapter problems, System Application results, and Test Item File. ISBN: 0-13-092257-9

Lab Solutions Manual for *Laboratory Exercises for Electronic Devices* by Buchla   Includes worked-out lab results. ISBN: 0-13-092274-9

Lab Solutions Manual for *Experiments in Electronic Devices* by Berlin et al.   Includes worked-out lab results. ISBN: 0-13-092254-4

Electronics Supersite (*www.prenhall.com/electronics*)   Instructors will find the *Prentice Hall Electronics Technology Journal,* extra classroom resources, and all of the supplements for this text available online for easy access. Contact your local Prentice Hall sales representative for your "User Name" and "Passcode."

Online Course Support   If your program is offering your electronics course in a distance learning format, please contact your local Prentice Hall sales representative for a list of product solutions.

PowerPoint® CD-ROM   Contains slides featuring all figures from the text, as well as text highlights for use in lecture presentations. ISBN: 0-13-092263-3

Prentice Hall Test Manager   This is a CD-ROM version of the Test Item File. ISBN: 0-13-092259-5

Chapter Features

Chapter Opener   Each chapter begins with a two-page spread, as shown in Figure P–1. The chapter opener includes the chapter number and title, a chapter introduction, a list of chapter sections, chapter objectives, key terms, a System Application preview, and a website reference for associated study aids.

Section Opener   Each section in a chapter begins with a brief introduction and section objectives. An example is shown in Figure P–2.

Section Review   Each section in a chapter ends with a review consisting of questions that highlight the main concepts presented in the section. This feature is also illustrated in Figure P–2. The answers to the Section Reviews are at the end of the chapter.

Worked Examples, Related Problems, and EWB/Multisim Exercise   Numerous worked examples throughout each chapter illustrate and clarify basic concepts or specific procedures. Each example ends with a Related Problem that reinforces or expands on the example by requiring the student to work through a problem similar to the example. Selected examples feature an EWB/Multisim exercise keyed to a file on the CD-ROM which contains the circuit illustrated in the example. A typical example with a Related Problem and an EWB/Multisim exercise is shown in Figure P–3. Answers to Related Problems are at the end of the chapter.

Troubleshooting Sections   Many chapters include a troubleshooting section that relates to the topics covered in the chapter and that illustrates troubleshooting procedures and techniques.

System Application   System Applications follow the last section in each chapter (except Chapter 1) and are identified by a special photographic logo and colored background design. A practical application of devices or circuits covered in the chapter is presented. The student learns how the specific device or circuit is used and is asked to compare a schematic to a printed circuit board, develop a test procedure, and troubleshoot specific faults. A typical System Application is shown in Figure P–4. The System Applications are optional and skipping any of them does not affect any other coverage.

Although they are not intended or designed for use as a laboratory project, most System Applications use realistic graphics for printed circuit boards and instruments. Results for the System Applications are provided in the Instructor's Resource Manual.

List of performance-based chapter objectives   System Application Preview

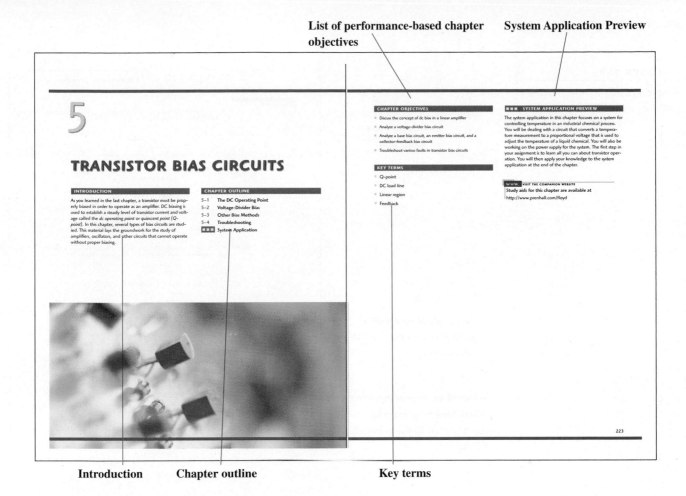

Introduction   Chapter outline   Key terms

▲ **FIGURE P–1**

A typical chapter opener.

Section review questions
end each section.

◀ **FIGURE P–2**

A typical section opener and section review.

Introductory paragraph
begins each section.

Performance-based
section objectives

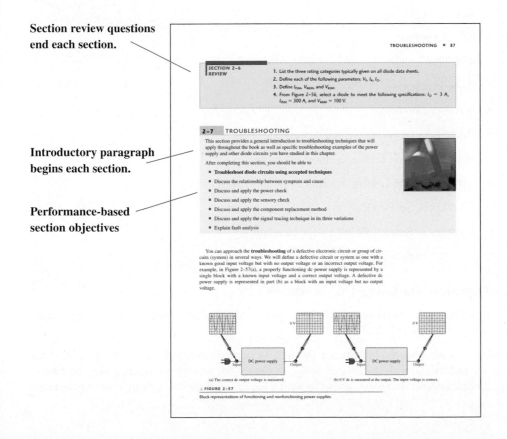

A typical example with a related problem and EWB exercise.

Examples are set off from text.

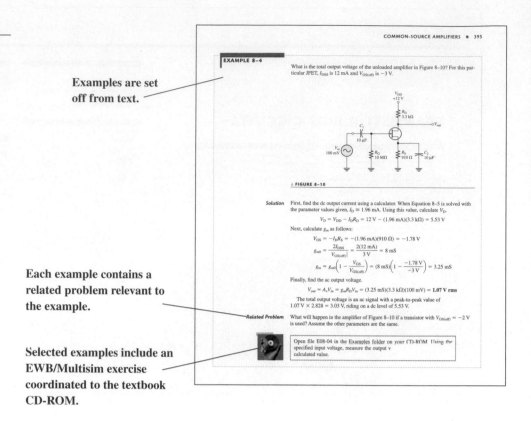

Each example contains a related problem relevant to the example.

Selected examples include an EWB/Multisim exercise coordinated to the textbook CD-ROM.

Chapter End Matter The following pedagogical features are found at the end of each chapter:

- Summary
- Key Term glossary
- Key Formulas
- Self-Test
- Basic Problems
- Advanced Problems
- Data Sheet Problems (selected chapters)
- Troubleshooting Problems (most chapters)
- EWB/Multisim Troubleshooting Problems (most chapters)
- Answers to Section Reviews
- Answers to Related Problems for Examples
- Answers to Self-Test

Suggestions for Using This Textbook

As mentioned, this book covers discrete devices in Chapters 1 through 11 and integrated circuits in Chapters 12 through 18.

Option 1 (two terms) Chapters 1 through 11 can be covered in the first term. Depending on individual preferences and program emphasis, selective coverage may be necessary. For example, you may choose to omit Chapter 11 if the topic of thyristors is covered

System Applications are set off from text.

A series of activities is provided, which simulate "on-the-job" experiences.

An overall introduction to the system application is provided.

Most system applications include realistic PC board graphics.

▲ FIGURE P–4

Portion of a typical system application section.

in a later industrial electronics course. Chapters 12 through 18 can be covered in the second term. Again, selective coverage may be necessary.

Option 2 (one term)   By omitting certain topics and by maintaining a rigorous schedule, this book can be used in one-term courses. For example, a course covering only discrete devices and circuits would cover Chapters 1 through 11 with, perhaps, some selectivity.

Similarly, a course requiring only linear integrated circuit coverage would cover Chapters 12 through 18. Another approach is a very selective coverage of discrete devices and circuits topics followed by a limited coverage of integrated circuits (only op-amps for example).

To the Student

There is an old saying that is very applicable to the study of this textbook as well as many other endeavors in life. It goes like this—*Do one thing at a time, do it very well, and then move on.*

When studying a particular chapter, study one section until you understand it and only then move on to the next one. Read each section and study the related illustrations carefully, think about the material, work through each example step-by-step, work its Related Problem and check the answer, and then answer each question in the section review, checking your answers at the end of the chapter. Don't expect each concept to be crystal clear after a single reading; you may have to read the material two or even three times. Once you believe that you understand the material, review the chapter summary, key formula list, and

Thomas Alva Edison 1847–1931 Born in Milan, Ohio, Thomas Edison was the most prolific inventor of all time. He is credited with 1093 patents and is the only person to ever have at least one patent every year for 65 consecutive years. Mr. Edison's inventions and enterprises encompass many different areas. One of his most famous inventions, the light bulb, was introduced in 1879. Mr. Edison is credited with discovering the diode effect while working with vacuum tubes for the light bulb. Most of his work was done in his laboratory in West Orange, NJ. He also maintained a laboratory at his winter home in Fort Myers, Florida, which was devoted principally to the development of a synthetic rubber using the goldenrod plant. (Photo credit: Library of Congress)

key term definitions at the end of the chapter. Take the multiple-choice self-test. Finally, work the assigned problems at the end of the chapter. Working through these problems is perhaps the most important way to check and reinforce your comprehension of the chapter. By working problems, you acquire an additional level of insight and understanding that reading or classroom lectures alone do not provide.

Generally, you cannot fully understand a concept or procedure by simply watching or listening to someone else. Only hard work and critical thinking will produce the results you expect and deserve.

Milestones in Electronics

Before you begin your study of electronic devices, let's briefly look at some of the important developments that led to the electronics technology we have today. The names of many of the early pioneers in electricity and electromagnetics still live on in terms of familiar units and quantities. Names such as Ohm, Ampere, Volta, Farad, Henry, Coulomb, Oersted, and Hertz are some of the better known examples with which you are already familiar. More widely known names such as Franklin and Edison are also significant in the history of electricity and electronics because of their tremendous contributions. Biographies of a few important figures in the history of electronics are shown.

The Beginning of Electronics Early experiments with electronics involved electric currents in vacuum tubes. Heinrich Geissler (1814–1879) removed most of the air from a glass tube and found that the tube glowed when there was current through it. Later, Sir William Crookes (1832–1919) found the current in vacuum tubes seemed to consist of particles. Thomas Edison (1847–1931) experimented with carbon filament bulbs with plates and discovered that there was a current from the hot filament to a positively charged plate. He patented the idea but never used it.

Other early experimenters measured the properties of the particles that flowed in vacuum tubes. Sir Joseph Thompson (1856–1940) measured properties of these particles, later called *electrons*.

Although wireless telegraphic communication dates back to 1844, electronics is basically a 20th century concept that began with the invention of the vacuum tube amplifier. An early vacuum tube that allowed current in only one direction was constructed by John A. Fleming in 1904. Called the Fleming valve, it was the forerunner of vacuum tube diodes. In 1907, Lee DeForest added a grid to the vacuum tube. The new device, called the audiotron, could amplify a weak signal. By adding the control element, DeForest ushered in the electronics revolution. It was with an improved version of his device that made transcontinental telephone service and radios possible. In 1912, a radio amateur in San Jose, California, was regularly broadcasting music!

In 1921, the secretary of commerce, Herbert Hoover, issued the first license to a broadcast radio station; within two years over 600 licenses were issued. By the end of the 1920s radios were in many homes. A new type of radio, the superheterodyne radio, invented by Edwin Armstrong, solved problems with high-frequency communication. In 1923, Vladimir Zworykin, an American researcher, invented the first television picture tube, and in 1927 Philo T. Farnsworth applied for a patent for a complete television system.

The 1930s saw many developments in radio, including metal tubes, automatic gain control, "midget sets," directional antennas, and more. Also started in this decade was the development of the first electronic computers. Modern computers trace their origins to the work of John Atanasoff at Iowa State University. Beginning in 1937, he envisioned a binary machine that could do complex mathematical work. By 1939, he and graduate student Clifford Berry had constructed a binary machine called ABC, (for Atanasoff-Berry Computer) that used vacuum tubes for logic and condensers (capacitors) for memory. In 1939, the magnetron, a microwave oscillator, was invented in Britain by Henry Boot and John Randall. In the same year, the klystron microwave tube was invented in America by Russell and Sigurd Varian.

Lee DeForest 1873–1961 Born in Iowa, Lee DeForest became an inventor while in college to help defray expenses. He graduated from Yale in 1899 with a PhD. His doctoral thesis, "Reflection of Hertzian Waves from the Ends of Parallel Wires," began his long career in radio. His invention of the vacuum tube triode for use in amplification (audion amplifier) was the most important of his more than 300 inventions. (Photo credit: The National Cyclopedia of American Biography, courtesy AIP Emilio Segrè Visual Archives, T. J. J. See Collection)

During World War II, electronics developed rapidly. Radar and very high-frequency communication were made possible by the magnetron and klystron. Cathode ray tubes were improved for use in radar. Computer work continued during the war. By 1946, John von Neumann had developed the first stored program computer, the Eniac, at the University of Pennsylvania. The decade ended with one of the most important inventions ever, the transistor.

Solid-State Electronics The crystal detectors used in early radios were the forerunners of modern solid-state devices. However, the era of solid-state electronics began with the invention of the transistor in 1947 at Bell Labs. The inventors were Walter Brattain, John Bardeen, and William Shockley. PC (printed circuit) boards were introduced in 1947, the year the transistor was invented. Commercial manufacturing of transistors began in Allentown, Pennsylvania, in 1951.

The most important invention of the 1950s was the integrated circuit. On September 12, 1958, Jack Kilby, at Texas Instruments, made the first integrated circuit (Figure P-5), for which he was awarded a Nobel prize in the fall of 2000. This invention literally created the modern computer age and brought about sweeping changes in medicine, communication, manufacturing, and the entertainment industry. Many billions of "chips"—as integrated circuits came to be called—have since been manufactured.

The 1960s saw the space race begin and spurred work on miniaturization and computers. The space race was the driving force behind the rapid changes in electronics that followed. The first successful "op-amp" was designed by Bob Widlar at Fairchild Semiconductor in 1965. Called the μA709, it was very successful but suffered from "latch-up" and other problems. Later, the most popular op-amp ever, the 741, was taking shape at Fairchild. This op-amp became the industry standard and influenced design of op-amps for years to come. Precursors to the Internet began in the 1960s with remote networked

John Bardeen 1908–1991 An electrical engineer and physicist born in Madison, Wisconsin, Dr. Bardeen was on the faculty of the University of Minnesota from 1938 to 1941 and a physicist at the Naval Ordnance Lab from 1941 to 1945. He then joined Bell Labs and remained there until 1951. Some of his fields of interest were conduction in semiconductors and metals, surface properties of semiconductors, and superconductivity. While at Bell Labs he jointly invented the transistor with colleagues Walter Brattain and William Shockley. After leaving Bell Labs in 1951, Dr. Bardeen joined the faculty at the University of Illinois. (Photo credit: AIP Emilio Segrè Visual Archives, W. F. Meggers Collection)

William Shockley 1910–1989 An American born in London, England, Dr. Shockley obtained his PhD in 1936 from M.I.T. He joined Bell Labs upon graduation and remained there until 1955. His research emphasis included areas of energy bands in solids, theory of vacuum tubes, photoelectrons, ferromagnetic domains, and transistor physics. While at Bell Labs, Dr. Shockley joined John Bardeen and Walter Brattain in the invention of the transistor in 1947. After leaving Bell Labs, Dr. Shockley spent time at Beckman Instruments and at Stanford University. (Photo credit: AIP Emilio Segrè Visual Archives, Physics Today Collection)

Walter H. Brattain 1902–1987 An American born in China, Dr. Brattain joined Bell Telephone Laboratories in 1929. One of his main areas of research was the surface properties of semiconductive materials. His chief contributions were the discovery of the photo effect at the surface of a semiconductor and the invention of the point-contact transistor in 1947, which he jointly invented with John Bardeen and William Shockley. (Photo credit: AIP Emilio Segrè Visual Archives, W. F. Meggers Collection)

▶ **FIGURE P–5**

The first integrated circuit. (Courtesy of Texas Instruments)

Jack S. Kilby 1923–

Jack Kilby was born in Missouri and earned degrees in electrical engineering from the University of Illinois and the University of Wisconsin. From 1947 to 1958, he worked at the Centralab Division of Globe Union, Inc. in Milwaukee. In 1958, he joined Texas Instruments in Dallas where he was responsible for integrated circuit development and applications. Within a year after joining TI he invented the monolithic integrated circuit and the rest is history. Mr. Kilby left TI in 1970. (Photo credit: Courtesy of Texas Instruments)

computers. Systems were in place within Lawrence Livermore National Laboratory that connected over 100 terminals to a computer system (colorfully called the "Octopus system"). In an experiment in 1969 with very remote computers, an exchange took place between researchers at UCLA and Stanford. The UCLA group hoped to connect to a Stanford computer and began by typing the word "login" on its terminal. A separate telephone connection was set up and the following conversation occurred.

The UCLA group asked over the phone, "Do you see the letter L?"

"Yes, we see the L."

The UCLA group typed an O. "Do you see the letter O?"

"Yes, we see the O."

The UCLA group typed a G. At this point the system crashed. Such was technology, but a revolution was in the making.

By 1971, a new company that had been formed by a group from Fairchild introduced the first microprocessor. The company was Intel and the product was the 4004 chip, which had the same processing power as the Eniac computer. Later in that same year, Intel announced the first 8-bit processor, the 8008. In 1975, the first personal computer was introduced by Altair, and *Popular Science* magazine featured it on the cover of the January, 1975, issue. The 1970s also saw the introduction of the pocket calculator and new developments in optical integrated circuits.

By the 1980s, half of all U.S. homes were using cable hookups instead of television antennas. The reliability, speed, and miniaturization of electronics continued throughout the 1980s, including automated testing and calibrating of PC boards. The computer became a part of instrumentation and the virtual instrument was created. Computers became a standard tool on the workbench.

The 1990s saw a widespread application of the Internet. In 1993, there were 130 websites; by the start of the new century (in 2001) there were over 24 million. In the 1990s, companies scrambled to establish a home page and many of the early developments of radio broadcasting had parallels with the Internet. The exchange of information and e-commerce fueled the tremendous economic growth of the 1900s. The Internet became especially important to scientists and engineers, becoming one of the most important scientific communication tools ever.

In 1995, the FCC allocated spectrum space for a new service called Digital Audio Radio Service. Digital television standards were adopted in 1996 by the FCC for the nation's next generation of broadcast television. As the 20th century drew toward a close, historians could only breathe a sign of relief. As one wag put it, "I'm all for new technologies, but I wish they'd let the old ones wear out first."

The 21st century dawned in January 2001 (although most people celebrated the new century the previous year, known as "Y2K"). The major technology story was the continued explosive growth of the Internet. Traffic on the Internet doubles every 100 days with no end in sight. The future of technology looks brighter than ever.

Acknowledgments

Many capable people have been part of this revision for the sixth edition of *Electronic Devices*. It has been thoroughly reviewed and checked for both content and accuracy. Those at Prentice Hall who have contributed greatly to this project throughout the many phases of development and production include Rex Davidson, Kate Linsner, and Scott Sambucci. Lois Porter, whose attention to details is unbelievable, has once more done an outstanding job editing the manuscript. Jane Lopez and Steve Botts have once again provided the excellent illustrations and beautiful graphics work used in the text. As with the previous edition, Gary Snyder has created the circuit files for the Electronics Workbench and Multisim features in this edition. A line-by-line review was done by Dave Hodder.

I wish to express my appreciation to those already mentioned as well as the reviewers who provided many valuable suggestions and constructive criticism that greatly influenced this edition. These reviewers are Roger Harlow, Mesa Community College; James L. Havens, Hallmark Institute of Technology; David Hodder, Yuba Community College; Ron Ingle, Midlands Technical College; Steven D. Johnson, DeVry Institute of Technology; Byron Paul, Bismarck State College; James F. Pechloff, Waukesha County Technical College; and Robert A. Powell, Oakland Community College.

Tom Floyd

Brief Contents

Contents

1

INTRODUCTION

INTRODUCTION

Electronic devices such as diodes, transistors, and integrated circuits are made of a semiconductive material. To understand how these devices work, you need a basic knowledge of the structure of atoms and the interaction of atomic particles. An important concept introduced in this chapter is that of the *pn* junction that is formed when two different types of semiconductive material are joined. The *pn* junction is fundamental to the operation of devices such as the diode and certain types of transistors. Also, the function of the *pn* junction is an essential factor in making electronic circuits operate properly.

CHAPTER OBJECTIVES

- Discuss the basic structure of atoms
- Discuss semiconductors, conductors, and insulators and how they basically differ
- Discuss covalent bonding in silicon
- Describe how current is produced in a semiconductor
- Describe the properties of *n*-type and *p*-type semiconductors
- Describe a diode and how a *pn* junction is formed
- Discuss the bias of a diode
- Analyze the voltage–current (*V–I*) characteristic curve of a diode
- Discuss the operation of diodes and explain the three diode models
- Test a diode using a digital multimeter

KEY TERMS

- Atom
- Proton
- Electron
- Shells
- Valance
- Ionization
- Free electron
- Conductor
- Insulator
- Semiconductor
- Silicon
- Crystal
- Hole
- Doping
- Diode
- *PN* junction
- Barrier potential
- Bias
- Forward bias
- Reverse bias
- *V–I* characteristic
- Cathode
- Anode

WWW. VISIT THE COMPANION WEBSITE
Study aids for this chapter are available at
http://www.prenhall.com/floyd

1–1 ATOMIC STRUCTURE

All matter is made of atoms; and all atoms consist of electrons, protons, and neutrons. In this section, you will learn about the structure of the atom, electron orbits and shells, valence electrons, ions, and two semiconductive materials—silicon and germanium. Semiconductive material is important because the configuration of certain electrons in an atom is the key factor in determining how a given material conducts electrical current.

After completing this section, you should be able to

- **Discuss the basic structure of atoms**
- Define *nucleus, proton, neutron,* and *electron*
- Describe an element's atomic number
- Explain electron shells
- Explain what a valence electron is
- Describe ionization
- Explain what a free electron is

An **atom*** is the smallest particle of an element that retains the characteristics of that element. Each of the known 109 elements has atoms that are different from the atoms of all other elements. This gives each element a unique atomic structure. According to the classical Bohr model, atoms have a planetary type of structure that consists of a central nucleus surrounded by orbiting electrons, as illustrated in Figure 1–1. The **nucleus** consists of positively charged particles called **protons** and uncharged particles called **neutrons**. The basic particles of negative charge are called **electrons**.

Each type of atom has a certain number of electrons and protons that distinguishes it from the atoms of all other elements. For example, the simplest atom is that of hydrogen, which has one proton and one electron, as shown in Figure 1–2(a). As another example, the helium atom, shown in Figure 1–2(b), has two protons and two neutrons in the nucleus and two electrons orbiting the nucleus.

---

*All bold terms are in the end-of-book glossary. The bold terms in color are key terms and are also defined at the end of the chapter.

▶ **FIGURE 1–1**

The Bohr model of an atom showing electrons in orbits around the nucleus, which consists of protons and neutrons. The "tails" on the electrons indicate motion.

● Electron ● Proton ● Neutron

(a) Hydrogen atom   (b) Helium atom

▲ **FIGURE 1–2**

The two simplest atoms, hydrogen and helium.

Atomic Number

All elements are arranged in the periodic table of the elements in order according to their **atomic number**. The atomic number equals the number of protons in the nucleus, which is the same as the number of electrons in an electrically balanced (neutral) atom. For example, hydrogen has an atomic number of 1 and helium has an atomic number of 2. In their normal (or neutral) state, all atoms of a given element have the same number of electrons as protons; the positive charges cancel the negative charges, and the atom has a net charge of zero.

Electron Shells and Orbits

Electrons orbit the nucleus of an atom at certain distances from the nucleus. Electrons near the nucleus have less energy than those in more distant orbits. It is known that only discrete (separate and distinct) values of electron energies exist within atomic structures. Therefore, electrons must orbit only at discrete distances from the nucleus.

Energy Levels   Each discrete distance (**orbit**) from the nucleus corresponds to a certain energy level. In an atom, the orbits are grouped into energy bands known as **shells**. A given atom has a fixed number of shells. Each shell has a fixed maximum number of electrons at permissible energy levels (orbits). The differences in energy levels within a shell are much smaller than the difference in energy between shells. The shells are designated 1, 2, 3, and so on, with 1 being closest to the nucleus. This energy band concept is illustrated in Figure 1–3, which shows the 1st shell with one energy level and the 2nd shell with two energy levels. Additional shells may exist in other types of atoms, depending on the element.

Valence Electrons

Electrons that are in orbits farther from the nucleus have higher energy and are less tightly bound to the atom than those closer to the nucleus. This is because the force of attraction between the positively charged nucleus and the negatively charged electron decreases with increasing distance from the nucleus. Electrons with the highest energy levels exist in the outermost shell of an atom and are relatively loosely bound to the atom. This outermost shell is known as the **valence** shell and electrons in this shell are called *valence electrons*. These valence electrons contribute to chemical reactions and bonding within the structure of a material and determine its electrical properties.

Ionization

When an atom absorbs energy from a heat source or from light, for example, the energy levels of the electrons are raised. The valence electrons possess more energy and are more

▶ **FIGURE 1–3**

Energy levels increase as the distance from the nucleus increases.

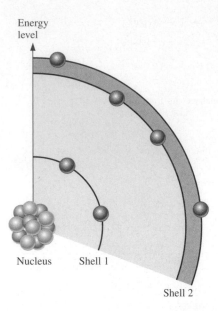

loosely bound to the atom than inner electrons, so they can easily jump to higher orbits within the valence shell when external energy is absorbed.

If a valence electron acquires a sufficient amount of energy, it can actually escape from the outer shell and the atom's influence. The departure of a valence electron leaves a previously neutral atom with an excess of positive charge (more protons than electrons). The process of losing a valence electron is known as **ionization**, and the resulting positively charged atom is called a *positive ion*. For example, the chemical symbol for hydrogen is H. When a neutral hydrogen atom loses its valence electron and becomes a positive ion, it is designated H^+. The escaped valence electron is called a **free electron**. When a free electron loses energy and falls into the outer shell of a neutral hydrogen atom, the atom becomes negatively charged (more electrons than protons) and is called a *negative ion,* designated H^-.

The Number of Electrons in Each Shell

The maximum number of electrons (N_e) that can exist in each shell of an atom is a fact of nature and can be calculated by the formula,

Equation 1–1

$$N_e = 2n^2$$

where n is the number of the shell. The innermost shell is number 1, the next shell is number 2, and so on. The maximum number of electrons that can exist in the innermost shell (shell 1) is

$$N_e = 2n^2 = 2(1)^2 = 2$$

The maximum number of electrons that can exist in the second shell is

$$N_e = 2n^2 = 2(2)^2 = 2(4) = 8$$

The maximum number of electrons that can exist in the third shell is

$$N_e = 2n^2 = 2(3)^2 = 2(9) = 18$$

The maximum number of electrons that can exist in the fourth shell is

$$N_e = 2n^2 = 2(4)^2 = 2(16) = 32$$

All shells in a given atom must be completely filled with electrons except the outer (valence) shell.

**SECTION 1–1
REVIEW**

Answers are at the end of the
chapter.

1. Describe an atom.
2. What is an electron?
3. What is a valence electron?
4. What is a free electron?
5. How are ions formed?

1–2   SEMICONDUCTORS, CONDUCTORS, AND INSULATORS

In terms of their electrical properties, materials can be classified into three groups:
conductors, semiconductors, and insulators. In this section, we will examine the properties of semiconductors and compare them to conductors and insulators.

After completing this section, you should be able to

- **Discuss semiconductors, conductors, and insulators and how they basically
  differ**

- Define the core of an atom

- Describe the atomic structure of copper, silicon, germanium, and carbon

- List the four best conductors

- List four semiconductors

- Discuss the difference between conductors and semiconductors

- Discuss the difference between silicon and germanium semiconductors

- Explain why silicon is much more widely used than germanium

All materials are made up of atoms. These atoms contribute to the electrical properties
of a material, including its ability to conduct electrical current.

For purposes of discussing electrical properties, an atom can be represented by the
valence shell and a **core** that consists of all the inner shells and the nucleus. This concept
is illustrated in Figure 1–4 for a carbon atom. Carbon is used in some types of electrical
resistors. Notice that the carbon atom has four electrons in the valence shell and two electrons in the inner shell. The nucleus consists of six protons and six neutrons so the +6
indicates the positive charge of the six protons. The core has a net charge of +4 (+6 for
the nucleus and −2 for the two inner-shell electrons).

▶ **FIGURE 1–4**

Diagram of a carbon atom.

Conductors

A **conductor** is a material that easily conducts electrical current. The best conductors are single-element materials, such as copper, silver, gold, and aluminum, which are character- ized by atoms with only one valence electron very loosely bound to the atom. These loosely bound valence electrons can easily break away from their atoms and become free electrons. Therefore, a conductive material has many free electrons that, when moving in the same direction, make up the **current**.

Insulators

An **insulator** is a material that does not conduct electrical current under normal conditions. Most good insulators are compounds rather than single-element materials. Valence electrons are tightly bound to the atoms; therefore, there are very few free electrons in an insulator.

Semiconductors

A **semiconductor** is a material that is between conductors and insulators in its ability to conduct electrical current. A semiconductor in its pure (intrinsic) state is neither a good conductor nor a good insulator. The most common single-element semiconductors are **silicon**, **germanium**, and **carbon**. Compound semiconductors such as gallium arsenide are also commonly used. The single-element semiconductors are characterized by atoms with four valence electrons.

Energy Bands

Recall that the valence shell of an atom represents a band of energy levels and that the valence electrons are confined to that band. When an electron acquires enough additional energy, it can leave the valence shell, become a *free electron*, and exist in what is known as the *conduction band.*

The difference in energy between the valence band and the conduction band is called an *energy gap.* This is the amount of energy that a valence electron must have in order to jump from the valence band to the conduction band. Once in the conduction band, the electron is free to move throughout the material and is not tied to any given atom.

Figure 1–5 shows energy diagrams for insulators, semiconductors, and conductors. Notice in part (a) that insulators have a very wide energy gap. Valence electrons do not

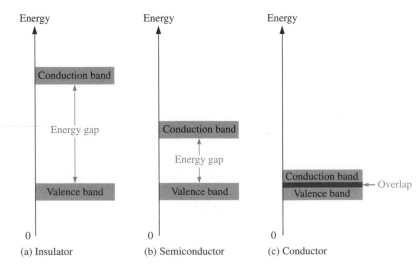

▲ **FIGURE 1–5**

Energy diagrams for the three types of materials.

jump into the conduction band except under breakdown conditions where extremely high voltages are applied across the material. As you can see in part (b), semiconductors have a much narrower energy gap. This gap permits some valence electrons to jump into the conduction band and become free electrons. By contrast, as part (c) illustrates, the energy bands in conductors overlap. In a conductive material there is always a large number of free electrons.

Comparison of a Semiconductor Atom to a Conductor Atom

Now let's examine some basic reasons why silicon is a semiconductor and copper is a conductor. Diagrams of the silicon atom and the copper atom are shown in Figure 1–6. Notice that the core of the silicon atom has a net charge of +4 (14 protons − 10 electrons) and the core of the copper atom has a net charge of +1 (29 protons − 28 electrons).

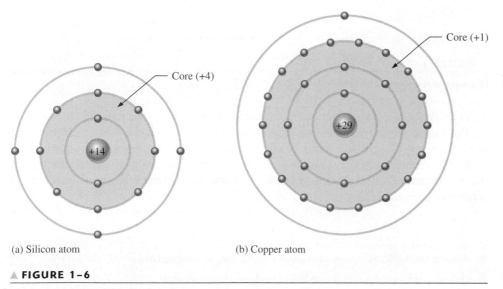

(a) Silicon atom   (b) Copper atom

▲ **FIGURE 1–6**

Diagrams of the silicon and copper atoms.

The valence electron in the copper atom "feels" an attractive force of +1 compared to a valence electron in the silicon atom which "feels" an attractive force of +4. So, there is four times more force trying to hold a valence electron to the atom in silicon than in copper. The copper's valence electron is in the fourth shell, which is a greater distance from its nucleus than the silicon's valence electron in the third shell. Recall that electrons farthest from the nucleus have the most energy.

Therefore, the valence electron in copper has less force holding it to the atom than does the valence electron in silicon. Also, the valence electron in copper has more energy than the valence electron in silicon. This means that it is easier for valence electrons in copper to acquire enough additional energy to escape from their atoms and become free electrons in the conduction band than it is in silicon. In fact, large numbers of valence electrons in copper already have sufficient energy to be free electrons at normal room temperature.

Silicon and Germanium

The atomic structures of silicon and germanium are shown in Figure 1–7. Silicon is the most widely used material in diodes, transistors, integrated circuits, and other semiconductor devices. Notice that both silicon and germanium have the characteristic four valence electrons.

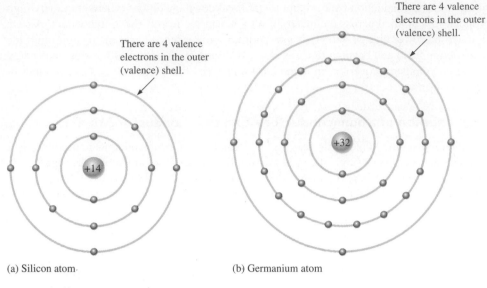

(a) Silicon atom

(b) Germanium atom

▲ FIGURE 1–7

Diagrams of the silicon and germanium atoms.

The valence electrons in germanium are in the fourth shell while those in silicon are in the third shell, closer to the nucleus. This means that the germanium valence electrons are at higher energy levels than those in silicon and, therefore, require a smaller amount of additional energy to escape from the atom. This property makes germanium more unstable at high temperatures, and this is a basic reason why silicon is the most widely used semiconductive material.

SECTION 1–2 REVIEW

1. What is the basic difference between conductors and insulators?
2. How do semiconductors differ from conductors and insulators?
3. How many valence electrons does a conductor such as copper have?
4. How many valence electrons does a semiconductor have?
5. Name three of the best conductive materials.
6. What is the most widely used semiconductive material?
7. Why does a semiconductor have fewer free electrons than a conductor?

1–3 COVALENT BONDS

When atoms combine to form a solid, crystalline material, they arrange themselves in a symmetrical pattern. The atoms within the crystal structure are held together by covalent bonds, which are created by the interaction of the valence electrons of the atoms. Silicon is a crystalline material.

After completing this section, you should be able to

■ **Discuss covalent bonding in silicon**

■ Define a covalent bond

■ Explain what a covalent bond consists of

■ Explain how a silicon crystal is formed

Figure 1–8 shows how each silicon atom positions itself with four adjacent silicon atoms to form a silicon **crystal**. A silicon atom with its four valence electrons shares an electron with each of its four neighbors. This effectively creates eight valence electrons for each atom and produces a state of chemical stability. Also, this sharing of valence electrons produces the **covalent** bonds that hold the atoms together; each shared electron is attracted equally by two adjacent atoms which share it. Covalent bonding in an intrinsic silicon crystal is shown in Figure 1–9. An **intrinsic** crystal is one that has no impurities. Covalent bonding for germanium is similar because it also has four valence electrons.

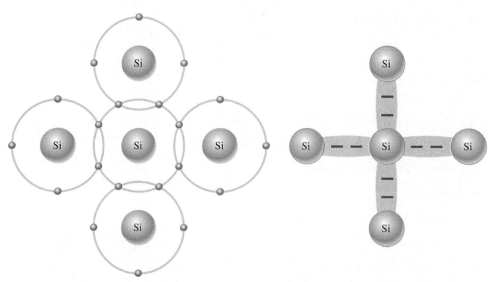

(a) The center atom shares an electron with each of the four surrounding atoms, creating a covalent bond with each. The surrounding atoms are in turn bonded to other atoms, and so on.

(b) Bonding diagram. The red negative signs represent the shared valence electrons.

▲ **FIGURE 1–8**

Illustration of covalent bonds in silicon.

▲ **FIGURE 1–9**

Covalent bonds in a three-dimensional silicon crystal.

SECTION 1–3 REVIEW

1. How are covalent bonds formed?
2. What is meant by the term *intrinsic*?
3. What is a crystal?
4. Effectively, how many valence electrons are there in each atom within a silicon crystal?

1–4 CONDUCTION IN SEMICONDUCTORS

The way a material conducts electrical current is important in understanding how electronic devices operate. You can't really understand the operation of a device such as a diode or transistor without knowing something about the basic current mechanisms. In this section, you will see how conduction occurs in semiconductive material.

After completing this section, you should be able to

- **Describe how current is produced in a semiconductor**
- Describe a conduction electron
- Define *hole*
- Explain what an electron-hole pair is
- Discuss recombination
- Explain the difference between electron current and hole current

As you have learned, the electrons of an atom can exist only within prescribed energy bands. Each shell around the nucleus corresponds to a certain energy band and is separated from adjacent shells by energy gaps, in which no electrons can exist. This is shown in Figure 1–10 for a silicon crystal with only unexcited (no external energy such as heat) silicon atoms. This condition occurs *only* at a temperature of absolute 0 Kelvin.

Conduction Electrons and Holes

An intrinsic (pure) silicon crystal at room temperature has sufficient heat (thermal) energy for some valence electrons to jump the gap from the valence band into the conduction band, becoming free electrons. Free electrons are also called **conduction electrons**. This is illustrated in the energy diagram of Figure 1–11(a) and in the bonding diagram of Figure 1–11(b).

When an electron jumps to the conduction band, a vacancy is left in the valence band within the crystal. This vacancy is called a **hole**. For every electron raised to the conduction band by external energy, there is one hole left in the valence band, creating what is called an **electron-hole pair**. **Recombination** occurs when a conduction-band electron loses energy and falls back into a hole in the valence band.

To summarize, a piece of intrinsic silicon at room temperature has, at any instant, a number of conduction-band (free) electrons that are unattached to any atom and are essentially drifting randomly throughout the material. There is also an equal number of holes in the valence band created when these electrons jump into the conduction band. This is illustrated in Figure 1–12.

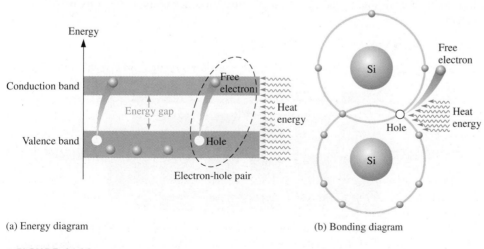

▲ FIGURE 1–10

Energy band diagram for a pure (intrinsic) silicon crystal with unexcited atoms. There are no electrons in the conduction band.

(a) Energy diagram   (b) Bonding diagram

▲ FIGURE 1–11

Creation of electron–hole pairs in a silicon crystal. Electrons in the conduction band are free electrons.

▲ **FIGURE 1–12**

Electron-hole pairs in a silicon crystal. Free electrons are being generated continuously while some recombine with holes.

Electron and Hole Current

When a voltage is applied across a piece of intrinsic silicon, as shown in Figure 1–13, the thermally generated free electrons in the conduction band, which are free to move randomly in the crystal structure, are now easily attracted toward the positive end. This movement of free electrons is one type of current in a semiconductive material and is called *electron current.*

Another type of current occurs at the valence level, where the holes created by the free electrons exist. Electrons remaining in the valence band are still attached to their atoms and are not free to move randomly in the crystal structure as are the free electrons. However, a valence electron can move into a nearby hole with little change in its energy level, thus leaving another hole where it came from. Effectively the hole has moved from one place to another in the crystal structure, as illustrated in Figure 1–14. This is called *hole current.*

▲ **FIGURE 1–13**

Electron current in intrinsic silicon is produced by the movement of thermally generated free electrons.

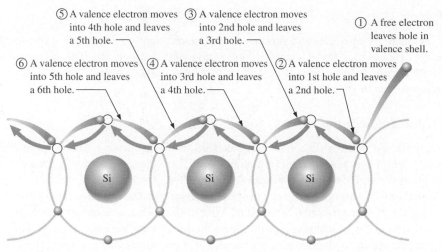

⑤ A valence electron moves into 4th hole and leaves a 5th hole.

③ A valence electron moves into 2nd hole and leaves a 3rd hole.

① A free electron leaves hole in valence shell.

⑥ A valence electron moves into 5th hole and leaves a 6th hole.

④ A valence electron moves into 3rd hole and leaves a 4th hole.

② A valence electron moves into 1st hole and leaves a 2nd hole.

When a valence electron moves left to right to fill a hole while leaving another hole behind, the hole has effectively moved from right to left. Gray arrows indicate effective movement of a hole.

▲ **FIGURE 1–14**

Hole current in intrinsic silicon.

SECTION 1–4 REVIEW	1. Are free electrons in the valence band or in the conduction band?
	2. Which electrons are responsible for current in a material?
	3. What is a hole?
	4. At what energy level does hole current occur?

1–5   N-TYPE AND P-TYPE SEMICONDUCTORS

Semiconductive materials do not conduct current well and are of limited value in their intrinsic state. This is because of the limited number of free electrons in the conduction band and holes in the valence band. Intrinsic silicon (or germanium) must be modified by increasing the number of free electrons or holes to increase its conductivity and make it useful in electronic devices. This is done by adding impurities to the intrinsic material as you will learn in this section. Two types of extrinsic (impure) semiconductive materials, *n*-type and *p*-type, are the key building blocks for most types of electronic devices.

After completing this section, you should be able to

- **Describe the properties of *n*-type and *p*-type semiconductors**
- Define *doping*
- Explain how *n*-type semiconductors are formed
- Explain how *p*-type semiconductors are formed
- Describe a majority carrier and minority carrier

Doping

The conductivity of silicon and germanium can be drastically increased by the controlled addition of impurities to the intrinsic (pure) semiconductive material. This process, called **doping**, increases the number of current carriers (electrons or holes). The two categories of impurities are *n*-type and *p*-type.

N-Type Semiconductor

To increase the number of conduction-band electrons in intrinsic silicon, **pentavalent** impurity atoms are added. These are atoms with five valence electrons such as arsenic (As), phosphorus (P), bismuth (Bi), and antimony (Sb).

As illustrated in Figure 1–15, each pentavalent atom (antimony, in this case) forms covalent bonds with four adjacent silicon atoms. Four of the antimony atom's valence electrons are used to form the covalent bonds with silicon atoms, leaving one extra electron. This extra electron becomes a conduction electron because it is not attached to any atom. Because the pentavalent atom gives up an electron, it is often called a *donor atom*. The number of conduction electrons can be carefully controlled by the number of impurity atoms added to the silicon. A conduction electron created by this doping process does not leave a hole in the valence band because it is in excess of the number required to fill the valence band.

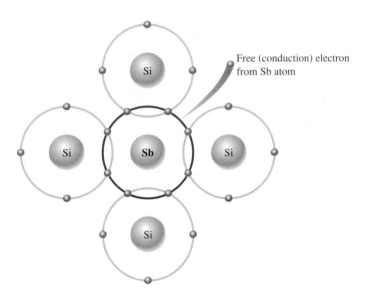

▲ FIGURE 1–15

Pentavalent impurity atom in a silicon crystal structure. An antimony (Sb) impurity atom is shown in the center. The extra electron from the Sb atom becomes a free electron.

Majority and Minority Carriers   Since most of the current carriers are electrons, silicon (or germanium) doped with pentavalent atoms is an *n*-type semiconductor (the *n* stands for the negative charge on an electron). The electrons are called the **majority carriers** in *n*-type material. Although the majority of current carriers in *n*-type material are electrons, there are also a few holes that are created when electron-hole pairs are thermally generated. These holes are *not* produced by the addition of the pentavalent impurity atoms. Holes in an *n*-type material are called **minority carriers**.

P-Type Semiconductor

To increase the number of holes in intrinsic silicon, **trivalent** impurity atoms are added. These are atoms with three valence electrons such as boron (B), indium (In), and gallium

(Ga). As illustrated in Figure 1–16, each trivalent atom (boron, in this case) forms covalent bonds with four adjacent silicon atoms. All three of the boron atom's valence electrons are used in the covalent bonds; and, since four electrons are required, a hole results when each trivalent atom is added. Because the trivalent atom can take an electron, it is often referred to as an *acceptor atom.* The number of holes can be carefully controlled by the number of trivalent impurity atoms added to the silicon. A hole created by this doping process is *not* accompanied by a conduction (free) electron.

▶ **FIGURE 1–16**

Trivalent impurity atom in a silicon crystal structure. A boron (B) impurity atom is shown in the center.

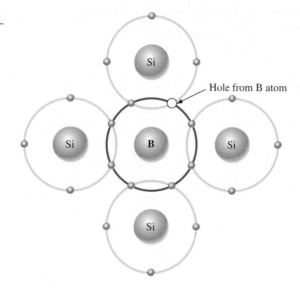

Hole from B atom

Majority and Minority Carriers   Since most of the current carriers are holes, silicon (or germanium) doped with trivalent atoms is called a *p*-type semiconductor. Holes can be thought of as positive charges because the absence of an electron leaves a net positive charge on the atom. The holes are the majority carriers in *p*-type material. Although the majority of current carriers in *p*-type material are holes, there are also a few free electrons that are created when electron-hole pairs are thermally generated. These free electrons are *not* produced by the addition of the trivalent impurity atoms. Electrons in *p*-type material are the minority carriers.

SECTION 1–5 REVIEW	1. Define *doping.*
	2. What is the difference between a pentavalent atom and a trivalent atom? What are other names for these atoms?
	3. How is an *n*-type semiconductor formed?
	4. How is a *p*-type semiconductor formed?
	5. What is the majority carrier in an *n*-type semiconductor?
	6. What is the majority carrier in a *p*-type semiconductor?
	7. By what process are the majority carriers produced?
	8. By what process are the minority carriers produced?
	9. What is the difference between intrinsic and extrinsic semiconductors?

1–6 THE DIODE

If you take a block of silicon and dope part of it with a trivalent impurity and the other part with a pentavalent impurity, a boundary called the *pn* junction is formed between the resulting *p*-type and *n*-type portions and a basic **diode** is created. The *pn* junction is the feature that allows diodes, certain transistors, and other devices to work.

After completing this section, you should be able to

- **Describe a diode and how a *pn* junction is formed**
- Discuss diffusion across a *pn* junction
- Explain the formation of the depletion region
- Define *barrier potential* and discuss its significance
- State the values of barrier potential in silicon and germanium

A *p*-type material consists of silicon atoms and trivalent impurity atoms such as boron. The boron atom adds a hole when it bonds with the silicon atoms. However, since the number of protons and the number of electrons are equal throughout the material, there is no net charge in the material and so it is neutral.

An *n*-type silicon material consists of silicon atoms and pentavalent impurity atoms such as antimony. As you have seen, an impurity atom releases an electron when it bonds with four silicon atoms. Since there is still an equal number of protons and electrons (including the free electrons) throughout the material, there is no net charge in the material and so it is neutral.

If a piece of intrinsic silicon is doped so that part is *n*-type and the other part is *p*-type, a **pn junction** forms at the boundary between the two regions and a diode is created, as indicated in Figure 1–17. The *p* region has many holes (majority carriers) from the impurity atoms and only a few thermally generated free electrons (minority carriers). The *n* region has many free electrons (majority carriers) from the impurity atoms and only a few thermally generated holes (minority carriers).

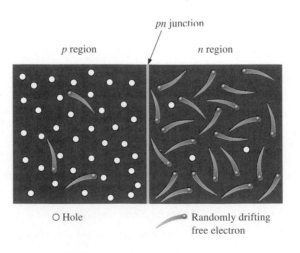

▲ **FIGURE 1–17**

The basic diode structure at the instant of junction formation showing only the majority and minority carriers.

Formation of the Depletion Region

As you have seen, the free electrons in the *n* region are randomly drifting in all directions. At the instant of the *pn* junction formation, the free electrons near the junction in the *n* region begin to diffuse across the junction into the *p* region where they combine with holes near the junction, as shown in Figure 1–18(a).

(a) At the instant of junction formation, free electrons in the *n* region near the *pn* junction begin to diffuse across the junction and fall into holes near the junction in the *p* region.

(b) For every electron that diffuses across the junction and combines with a hole, a positive charge is left in the *n* region and a negative charge is created in the *p* region, forming a barrier potential. This action continues until the voltage of the barrier repels further diffusion.

▲ **FIGURE 1–18**

Formation of the depletion region. The width of the depletion region is exaggerated for illustration purposes.

Before the *pn* junction is formed, recall that there are as many electrons as protons in the *n*-type material making the material neutral in terms of net charge. The same is true for the *p*-type material.

When the *pn* junction is formed, the *n* region loses free electrons as they diffuse across the junction. This creates a layer of positive charges (pentavalent ions) near the junction. As the electrons move across the junction, the *p* region loses holes as the electrons and holes combine. This creates a layer of negative charges (trivalent ions) near the junction. These two layers of positive and negative charges form the **depletion region**, as shown in Figure 1–18(b). The term *depletion* refers to the fact that the region near the *pn* junction is depleted of *charge carriers* (electrons and holes) due to diffusion across the junction. Keep in mind that the depletion region is formed very quickly and is very thin compared to the *n* region and *p* region.

After the initial surge of free electrons across the *pn* junction, the depletion region has expanded to a point where equilibrium is established and there is no further diffusion of electrons across the junction. This occurs as follows. As electrons continue to diffuse across the junction, more and more positive and negative charges are created near the junction as the depletion region is formed. A point is reached where the total negative charge in the depletion region repels any further diffusion of electrons (negatively charged particles) into the *p* region (like charges repel) and the diffusion stops. In other words, the depletion region acts as a barrier to the further movement of electrons across the junction.

Barrier Potential Any time there is a positive charge and a negative charge near each other, there is a force acting on the charges as described by Coulomb's law. In the depletion region there are many positive charges and many negative charges on opposite sides of the *pn* junction. The forces between the opposite charges form a "field of forces" called an *electric field,* as illustrated in Figure 1–18(b) by the red arrows between the positive charges and the negative charges. This electric field is a barrier to the free electrons in the *n* region, and energy must be expended to move an electron through the electric field. That is, external energy must be applied to get the electrons to move across the barrier of the electric field in the depletion region.

The potential difference of the electric field across the depletion region is the amount of voltage required to move electrons through the electric field. This potential difference is called the **barrier potential** and is expressed in *volts.* Stated another way, a certain amount of voltage equal to the barrier potential and with the proper polarity must be applied across a *pn* junction before electrons will begin to flow across the junction. You will learn more about this when we discuss *biasing* in Section 1–7.

The barrier potential of a *pn* junction depends on several factors, including the type of semiconductive material, the amount of doping, and the temperature. The typical barrier potential is approximately 0.7 V for silicon and 0.3 V for germanium at 25°C. Throughout the rest of the book, silicon will be used unless otherwise stated.

Energy Diagrams of the *PN* Junction and Depletion Region

The valence and conduction bands in an *n*-type material are at slightly lower energy levels than the valence and conduction bands in a *p*-type material. This is due to differences in the atomic characteristics of the pentavalent and the trivalent impurity atoms.

An energy diagram for a *pn* junction at the instant of formation is shown in Figure 1–19(a). As you can see, the valence and conduction bands in the *n* region are at lower energy levels than those in the *p* region, but there is a significant amount of overlapping.

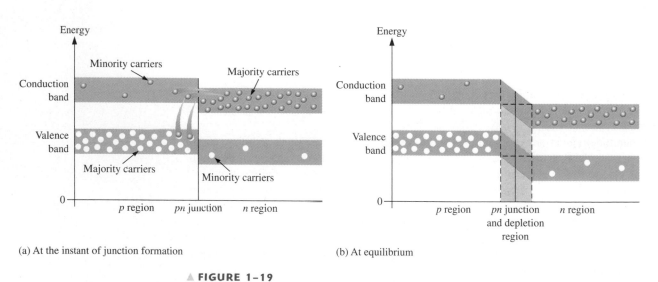

(a) At the instant of junction formation

(b) At equilibrium

▲ **FIGURE 1–19**

Energy diagrams illustrating the formation of the *pn* junction and depletion region.

The free electrons in the *n* region that occupy the upper part of the conduction band in terms of their energy can easily diffuse across the junction (they do not have to gain additional energy) and temporarily become free electrons in the lower part of the *p*-region conduction band. After crossing the junction, the electrons quickly lose energy and fall into the holes in the *p*-region valence band as indicated in Figure 1–19(a).

As the diffusion continues, the depletion region begins to form and the energy level of the *n*-region conduction band decreases. The decrease in the energy level of the conduction band in the *n* region is due to the loss of the higher-energy electrons that have diffused across the junction to the *p* region. Soon, there are no electrons left in the *n*-region conduction band with enough energy to get across the junction to the *p*-region conduction band, as indicated by the alignment of the top of the *n*-region conduction band and the bottom of the *p*-region conduction band in Figure 1–19(b). At this point, the junction is at equilibrium; and the depletion region is complete because diffusion has ceased. There is an energy gradiant across the depletion region which acts as an energy "hill" that an *n*-region electron must climb to get to the *p* region.

Notice that as the energy level of the *n*-region conduction band has shifted downward, the energy level of the valence band has also shifted downward. It still takes the same amount of energy for a valence electron to become a free electron. In other words, the energy gap between the valence band and the conduction band remains the same.

**SECTION 1–6
REVIEW**

1. What is a *pn* junction?
2. Explain what diffusion is.
3. Describe the depletion region.
4. Explain what the barrier potential is and how it is created.
5. What is the typical value of the barrier potential for a silicon diode?
6. What is the typical value of the barrier potential for a germanium diode?

1–7   BIASING THE DIODE

As you have learned, no electrons move through the *pn* junction at equilibrium. Generally the term *bias* refers to the use of a dc voltage to establish certain operating conditions for an electronic device. In relation to a diode, there are two bias conditions: forward and reverse. Either of these bias conditions is established by connecting a sufficient dc voltage of the proper polarity across the *pn* junction.

After completing this section, you should be able to

- **Discuss the bias of a diode**
- Define *forward bias* and state the required conditions
- Define *reverse bias* and state the required conditions
- Discuss the effect of barrier potential on forward bias
- Explain how current is produced in forward bias
- Explain reverse current
- Describe reverse breakdown of a diode
- Explain forward bias and reverse bias in terms of energy diagrams

Forward Bias

To **bias** a diode, you apply a dc voltage across it. **Forward bias** is the condition that allows current through the *pn* junction. Figure 1–20 shows a dc voltage source connected by conductive material (contacts and wire) across a diode in the direction to produce forward bias. This external bias voltage is designated as V_{BIAS}. The resistor, *R,* limits the current to a value that will not damage the *pn* structure.

► **FIGURE 1–20**

A diode connected for forward bias.

► **FIGURE 1–20**

A diode connected for forward bias.

Notice that the negative side of V_{BIAS} is connected to the n region of the diode and the positive side is connected to the p region. This is one requirement for forward bias. A second requirement is that the bias voltage, V_{BIAS}, must be greater than the barrier potential.

A fundamental picture of what happens when a diode is forward-biased is shown in Figure 1–21. Because like charges repel, the negative side of the bias-voltage source "pushes" the free electrons, which are the majority carriers in the n region, toward the pn junction. This flow of free electrons is called *electron current*. The negative side of the source also provides a continuous flow of electrons through the external connection (conductor) and into the n region as shown.

▲ **FIGURE 1–21**

A forward-biased diode showing the flow of majority carriers and the voltage due to the barrier potential across the depletion region.

The bias-voltage source imparts sufficient energy to the free electrons for them to overcome the barrier potential of the depletion region and move on through into the p region. Once in the p region, these conduction electrons have lost enough energy to immediately combine with holes in the valence band.

Now, the electrons are in the valence band in the p region, simply because they have lost too much energy overcoming the barrier potential to remain in the conduction band. Since unlike charges attract, the positive side of the bias-voltage source attracts the valence electrons toward the left end of the p region. The holes in the p region provide the medium or "pathway" for these valence electrons to move through the p region. The electrons move from one hole to the next toward the left. The holes, which are the majority carriers in the p region, effectively (not actually) move to the right toward the junction, as you can see in Figure 1–21. This *effective* flow of holes is called the *hole current*. You can also view the hole current as the flow of valence electrons through the p region, with the holes providing the only means for these electrons to flow.

As the electrons flow out of the p region through the external connection (conductor) and to the positive side of the bias-voltage source, they leave holes behind in the p region;

at the same time, these electrons become conduction electrons in the metal conductor. Recall that the conduction band in a conductor overlaps the valence band so that it takes much less energy for an electron to be a free electron in a conductor than in a semiconductor. So, there is a continuous availability of holes *effectively* moving toward the *pn* junction to combine with the continuous stream of electrons as they come across the junction into the *p* region.

The Effect of Forward Bias on the Depletion Region As more electrons flow into the depletion region, the number of positive ions is reduced. As more holes effectively flow into the depletion region on the other side of the *pn* junction, the number of negative ions is reduced. This reduction in positive and negative ions during forward bias causes the depletion region to narrow, as indicated in Figure 1–22.

(a) At equilibrium (no bias)

(b) Forward bias narrows the depletion region and produces a voltage drop across the *pn* junction equal to the barrier potential.

▲ **FIGURE 1–22**

The depletion region narrows and a voltage drop is produced across the *pn* junction when the diode is forward-biased.

The Effect of the Barrier Potential During Forward Bias Recall that the electric field between the positive and negative ions in the depletion region on either side of the junction creates an "energy hill" that prevents free electrons from diffusing across the junction at equilibrium (see Figure 1–19(b)). This is known as the *barrier potential.*

When forward bias is applied, the free electrons are provided with enough energy from the bias-voltage source to overcome the barrier potential and effectively "climb" the energy hill and cross the depletion region. The energy that the electrons require in order to pass through the depletion region is equal to the barrier potential. In other words, the electrons give up an amount of energy equivalent to the barrier potential when they cross the depletion region. This energy loss results in a voltage drop across the *pn* junction equal to the barrier potential (0.7 V), as indicated in Figure 1–22(b). An additional small voltage drop occurs across the *p* and *n* regions due to the internal resistance of the material. For doped semiconductive material, this resistance, called the **dynamic resistance**, is very small and can usually be neglected. This is discussed in more detail in Section 1–8.

Reverse Bias

Reverse bias is the condition that essentially prevents current through the diode. Figure 1–23 shows a dc voltage source connected across a diode in the direction to produce reverse bias. This external bias voltage is designated as V_{BIAS} just as it was for forward bias. Notice that the positive side of V_{BIAS} is connected to the *n* region of the diode and the negative side is connected to the *p* region. Also note that the depletion region is shown much wider than in forward bias or equilibrium.

▶ **FIGURE 1–23**

A diode connected for reverse bias.

An illustration of what happens when a diode is reverse-biased is shown in Figure 1–24. Because unlike charges attract, the positive side of the bias-voltage source "pulls" the free electrons, which are the majority carriers in the n region, away from the pn junction. As the electrons flow toward the positive side of the voltage source, additional positive ions are created. This results in a widening of the depletion region and a depletion of majority carriers.

▲ **FIGURE 1–24**

The diode during the short transition time immediately after reverse–bias voltage is applied.

In the p region, electrons from the negative side of the voltage source enter as valence electrons and move from hole to hole toward the depletion region where they create additional negative ions. This results in a widening of the depletion region and a depletion of majority carriers. The flow of valence electrons can be viewed as holes being "pulled" toward the positive side.

The initial flow of charge carriers is transitional and lasts for only a very short time after the reverse-bias voltage is applied. As the depletion region widens, the availability of majority carriers decreases. As more of the n and p regions become depleted of majority carriers, the electric field between the positive and negative ions increases in strength until the potential across the depletion region equals the bias voltage, V_{BIAS}. At this point, the transition current essentially ceases except for a very small reverse current that can usually be neglected.

Reverse Current   The extremely small current that exists in reverse bias after the transition current dies out is caused by the minority carriers in the n and p regions that are produced by thermally generated electron-hole pairs. The small number of free minority electrons in the p region are "pushed" toward the pn junction by the negative bias voltage. When these electrons reach the wide depletion region, they "fall down the energy hill" and combine with the minority holes in the n region as valence electrons and flow toward the positive bias voltage, creating a small hole current.

The conduction band in the *p* region is at a higher energy level than the conduction band in the *n* region. Therefore, the minority electrons easily pass through the depletion region because they require no additional energy. Reverse current is illustrated in Figure 1–25.

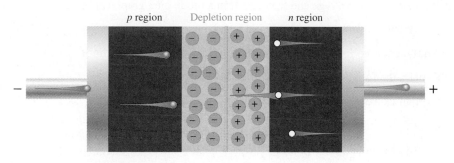

▲ **FIGURE 1–25**

The extremely small reverse current in a reverse-biased diode is due to the minority carriers from thermally generated electron-hole pairs.

Reverse Breakdown Normally, the reverse current is so small that it can be neglected. However, if the external reverse-bias voltage is increased to a value called the *breakdown voltage,* the reverse current will drastically increase.

This is what happens. The high reverse-bias voltage imparts energy to the free minority electrons so that as they speed through the *p* region, they collide with atoms with enough energy to knock valence electrons out of orbit and into the conduction band. The newly created conduction electrons are also high in energy and repeat the process. If one electron knocks only two others out of their valence orbit during its travel through the *p* region, the numbers quickly multiply. As these high-energy electrons go through the depletion region, they have enough energy to go through the *n* region as conduction electrons, rather than combining with holes.

The multiplication of conduction electrons just discussed is known as **avalanche** and results in a very high reverse current that can damage the diode because of excessive heat dissipation.

**SECTION 1–7
REVIEW**

1. Describe forward bias of a diode.
2. Explain how to forward-bias a diode.
3. Describe reverse bias of a diode.
4. Explain how to reverse-bias a diode.
5. Compare the depletion regions in forward bias and reverse bias.
6. Which bias condition produces majority carrier current?
7. How is reverse current in a diode produced?
8. When does reverse breakdown occur in a diode?
9. Define *avalanche* as applied to diodes.

1–8 VOLTAGE-CURRENT CHARACTERISTIC OF A DIODE

As you have learned, forward bias produces current through a diode and reverse bias essentially prevents current, except for a negligible reverse current. Reverse bias prevents current as long as the reverse-bias voltage does not equal or exceed the breakdown voltage of the junction. In this section, we will examine more closely the relationship between the voltage and the current in a diode on a graphical basis.

After completing this section, you should be able to

- ■ **Analyze the voltage-current (*V-I*) characteristic curve of a diode**
  - ■ Explain the forward-bias portion of the *V-I* characteristic curve
  - ■ Explain the reverse-bias portion of the *V-I* characteristic curve
  - ■ Identify the barrier potential
  - ■ Identify the breakdown voltage
  - ■ Discuss temperature effects on a diode

V–I Characteristic for Forward Bias

As you have learned, when a forward-bias voltage is applied across a diode, there is current. This current is called the *forward current* and is designated I_F. Figure 1–26 illustrates what happens as the forward-bias voltage is increased positively from 0 V. The resistor is used to limit the forward current to a value that will not overheat the diode and cause damage.

With 0 V across the diode, there is no forward current. As you gradually increase the forward-bias voltage, the forward current *and* the voltage across the diode gradually increase, as shown in Figure 1–26(a). A portion of the forward-bias voltage is dropped across the limiting resistor. When the forward-bias voltage is increased to a value where

(a) Small forward-bias voltage ($V_F < 0.7$ V), very small forward current.

(b) Forward voltage reaches and remains at approximately 0.7 V. Forward current continues to increase as the bias voltage is increased.

▲ **FIGURE 1–26**

Forward–bias measurements show general changes in V_F and I_F as V_{BIAS} is increased.

the voltage across the diode reaches approximately 0.7 V (barrier potential), the forward current begins to increase rapidly.

As you continue to increase the forward-bias voltage, the current continues to increase very rapidly, but the voltage across the diode increases only gradually above 0.7 V, as illustrated in Figure 1–26(b). This small increase in the diode voltage above the barrier potential is due to the voltage drop across the internal dynamic resistance of the semiconductive material.

Graphing the V-I Curve If you plot the results of the type of measurements shown in Figure 1–26 on a graph, you get the **V-I characteristic** curve for a forward-biased diode, as shown in Figure 1–27(a). The diode forward voltage (V_F) increases to the right along the horizontal axis, and the forward current (I_F) increases upward along the vertical axis.

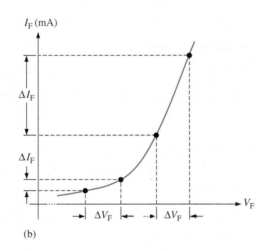

(a)   (b)

▲ **FIGURE 1–27**

V-I characteristic curve for forward bias. Part (b) illustrates how the dynamic resistance r_d' decreases as you move up the curve $(r_d' = \Delta V_F/\Delta I_F)$.

As you can see in Figure 1–27(a), the forward current increases very little until the forward voltage across the *pn* junction reaches approximately 0.7 V at the knee of the curve. After this point, the forward voltage remains at approximately 0.7 V, but I_F increases rapidly. As previously mentioned, there is a slight increase in V_F above 0.7 V as the current increases due mainly to the voltage drop across the dynamic resistance. *Normal operation for a forward-biased diode is above the knee of the curve.* The I_F scale is typically in mA, as indicated.

Three points *A, B,* and *C* are shown on the curve in Figure 1–27(a). Point *A* corresponds to a zero-bias condition. Point *B* corresponds to Figure 1–26(a) where the forward voltage is less than the barrier potential of 0.7 V. Point *C* corresponds to Figure 1–26(b) where the forward voltage *approximately* equals the barrier potential. As the external bias voltage and forward current continue to increase above the knee, the forward voltage will increase slightly above 0.7 V. In reality, the forward voltage can be as much as approximately 0.90 V, depending on the forward current.

Dynamic Resistance Unlike a linear resistance, the resistance of the forward-biased diode is not constant over the entire curve. Because the resistance changes as you move along the V-I curve, it is called *dynamic* or *ac resistance*. Internal resistances of electronic devices are usually designated by lowercase italic *r* with a prime, instead of the standard *R*. The dynamic resistance of a diode is designated r_d'.

Below the knee of the curve the resistance is greatest because the current increases very little for a given change in voltage $(r_d' = \Delta V_F/\Delta I_F)$. The resistance begins to decrease in the region of the knee of the curve and becomes smallest above the knee where there is a large change in current for a given change in voltage. This characteristic is illustrated in Figure 1–27(b) for equal changes in V_F (ΔV_F) on a magnified segment of the V-I curve below and above the knee.

V–I Characteristic for Reverse Bias

When a reverse-bias voltage is applied across a diode, there is only an extremely small reverse current (I_R) through the *pn* junction. With 0 V across the diode, there is no reverse current. As you gradually increase the reverse-bias voltage, there is a very small reverse current and the voltage across the diode increases. When the applied bias voltage is increased to a value where the reverse voltage across the diode (V_R) reaches the breakdown value (V_{BR}), the reverse current begins to increase rapidly.

As you continue to increase the bias voltage, the current continues to increase very rapidly, but the voltage across the diode increases very little above V_{BR}. *Breakdown, with exceptions, is not a normal mode of operation for most pn junction devices.*

Graphing the V–I Curve   If you plot the results of reverse-bias measurements on a graph, you get the V-I characteristic curve for a reverse-biased diode. A typical curve is shown in Figure 1–28. The diode reverse voltage (V_R) increases to the left along the horizontal axis, and the reverse current (I_R) increases downward along the vertical axis.

▶ FIGURE 1–28

V-I characteristic curve for reverse-biased diode.

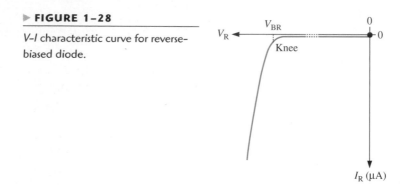

There is very little reverse current (usually μA or nA) until the reverse voltage across the diode reaches approximately the breakdown value (V_{BR}) at the knee of the curve. After this point, the reverse voltage remains at approximately V_{BR}, but I_R increases very rapidly resulting in overheating and possible damage. The breakdown voltage for a typical silicon diode can vary, but a minimum value of 50 V is not unusual.

The Complete V–I Characteristic Curve

Combine the curves for both forward bias and reverse bias, and you have the complete V-I characteristic curve for a diode, as shown in Figure 1–29. Notice that the I_F scale is in mA compared to the I_R scale in μA.

Temperature Effects on the Diode V–I Characteristic

For a forward-biased diode, as temperature is increased, the forward current increases for a given value of forward voltage. Also, for a given value of forward current, the forward voltage decreases. This is shown with the V-I characteristic curves in Figure 1–30. The blue curve is at room temperature (25°C) and the red curve is at an elevated temperature (25°C + ΔT). Notice that the barrier potential decreases as temperature increases.

For a reverse-biased diode, as temperature is increased, the reverse current increases. The difference in the two curves is exaggerated on the graph in Figure 1–30 for illustra-

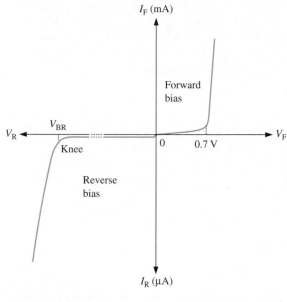

▲ FIGURE 1–29

The complete V-I characteristic curve for a diode.

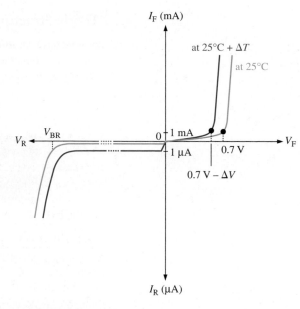

▲ FIGURE 1–30

Temperature effect on the diode V-I characteristic. The 1 mA and 1 μA marks on the vertical axis are given as a basis for a relative comparison of the current scales.

tion. Keep in mind that the reverse current below breakdown remains extremely small and can usually be neglected.

SECTION 1–8 REVIEW

1. Discuss the significance of the knee of the characteristic curve in forward bias.
2. On what part of the curve is a forward-biased diode normally operated?
3. Which is greater, the breakdown voltage or the barrier potential?
4. On what part of the curve is a reverse-biased diode normally operated?
5. What happens to the barrier potential when the temperature increases?

1–9 DIODE MODELS

You have learned that the diode is a *pn* junction device. In this section, you will learn the electrical symbol for a diode and how the diode can be modeled for circuit analysis using three levels of complexity. Also, diode packaging and terminal identification are introduced.

After completing this section, you should be able to

- **Discuss the operation of diodes and explain the three diode models**
- Recognize a diode symbol and indentify the diode terminals
- Recognize diodes in various physical configurations
- Explain the ideal, the practical, and the complete diode models

Diode Structure and Symbol

As you have learned, a diode is a single *pn* junction device with conductive contacts and wire leads connected to each region, as shown in Figure 1–31(a). Part of the diode is an *n*-type semiconductor and the other part is a *p*-type semiconductor.

(a) Basic diode structure (b) Schematic symbol

▲ **FIGURE 1–31**

Diode structure and schematic symbol.

There are several types of diodes, but the schematic symbol for a general-purpose or rectifier diode, such as introduced in this chapter, is shown in Figure 1–31(b). The *n* region is called the **cathode** and the *p* region is called the **anode**. The "arrow" in the symbol points in the direction of conventional current (opposite to electron flow).

Forward-Bias Connection   A diode is forward-biased when a voltage source is connected as shown in Figure 1–32(a). The positive terminal of the source is connected to the anode through a current-limiting resistor. The negative terminal of the source is connected to the cathode. The forward current (I_F) is from anode to cathode as indicated. The forward voltage drop (V_F) due to the barrier potential is from positive at the anode to negative at the cathode.

▶ **FIGURE 1–32**

Forward–bias and reverse–bias connections showing the diode symbol.

(a) Forward bias (b) Reverse bias

Reverse-Bias Connection   A diode is reverse-biased when a voltage source is connected as shown in Figure 1–32(b). The negative terminal is connected to the anode side of the circuit, and the positive terminal is connected to the cathode side. A resistor is not necessary in reverse bias but it is shown for circuit consistency. The reverse current is negligible. Notice that the entire bias voltage (V_{BIAS}) appears across the diode.

The Ideal Diode Model

The ideal model of a diode is a simple switch. When the diode is forward-biased, it acts like a closed (on) switch, as shown in Figure 1–33(a). When the diode is reverse-biased, it acts like an open (off) switch, as shown in part (b). The barrier potential, the forward dynamic resistance, and the reverse current are all neglected.

In Figure 1–33(c), the ideal *V-I* characteristic curve graphically depicts the ideal diode operation. Since the barrier potential and the forward dynamic resistance are neglected,

(a) Forward bias (b) Reverse bias

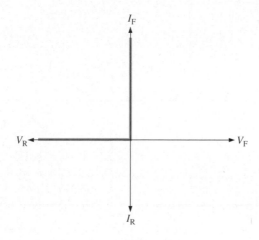

(c) Ideal characteristic curve (blue)

▲ **FIGURE 1–33**

The ideal model of a diode.

the diode is assumed to have a zero voltage across it when forward-biased, as indicated by the portion of the curve on the positive vertical axis.

$$V_F = 0 \text{ V}$$

The forward current is determined by the bias voltage and the limiting resistor using Ohm's law.

$$I_F = \frac{V_{BIAS}}{R_{LIMIT}}$$

Equation 1–2

Since the reverse current is neglected, its value is assumed to be zero, as indicated in Figure 1–33(c) by the portion of the curve on the negative horizontal axis.

$$I_R = 0 \text{ A}$$

The reverse voltage equals the bias voltage.

$$V_R = V_{BIAS}$$

You may want to use the ideal model when you are troubleshooting or trying to figure out the operation of a circuit and are not concerned with more exact values of voltage or current.

The Practical Diode Model

The practical model adds the barrier potential to the ideal switch model. When the diode is forward-biased, it is equivalent to a closed switch in series with a small equivalent voltage source equal to the barrier potential (0.7 V) with the positive side toward the anode,

(a) Forward bias   (b) Reverse bias

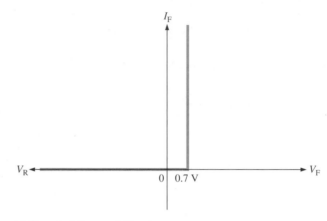

(c) Characteristic curve (silicon)

▲ **FIGURE 1–34**

The practical model of a diode.

as indicated in Figure 1–34(a). This equivalent voltage source represents the fixed voltage drop (V_F) produced across the forward-biased *pn* junction of the diode and is not an active source of voltage.

When the diode is reverse-biased, it is equivalent to an open switch just as in the ideal model, as shown in Figure 1–34(b). The barrier potential does not affect reverse bias, so it is not a factor.

The characteristic curve for the practical diode model is shown in Figure 1–34(c). Since the barrier potential is included and the dynamic resistance is neglected, the diode is assumed to have a voltage across it when forward-biased, as indicated by the portion of the curve to the right of the origin.

$$V_F = 0.7 \text{ V} \quad \text{(silicon)}$$
$$V_F = 0.3 \text{ V} \quad \text{(germanium)}$$

The forward current is determined as follows by first applying Kirchhoff's voltage law to Figure 1–34(a):

$$V_{\text{BIAS}} - V_F - V_{R_{\text{LIMIT}}} = 0$$
$$V_{R_{\text{LIMIT}}} = I_F R_{\text{LIMIT}}$$

Substituting and solving for I_F,

Equation 1–3

$$I_F = \frac{V_{\text{BIAS}} - V_F}{R_{\text{LIMIT}}}$$

The diode is assumed to have zero reverse current, as indicated by the portion of the curve on the negative horizontal axis.

$$I_R = 0 \text{ A}$$
$$V_R = V_{BIAS}$$

The Complete Diode Model

The complete model of a diode consists of the barrier potential, the small forward dynamic resistance (r'_d), and the large internal reverse resistance (r'_R). The reverse resistance is taken into account because it provides a path for the reverse current which is included in this diode model.

When the diode is forward-biased, it acts as a closed switch in series with the barrier potential voltage and the small forward dynamic resistance (r'_d), as indicated in Figure 1–35(a). When the diode is reverse-biased, it acts as an open switch in parallel with the large internal reverse resistance (r'_R), as shown in Figure 1–35(b). The barrier potential does not affect reverse bias, so it is not a factor.

The characteristic curve for the complete diode model is shown in Figure 1–35(c). Since the barrier potential and the forward dynamic resistance are included, the diode is

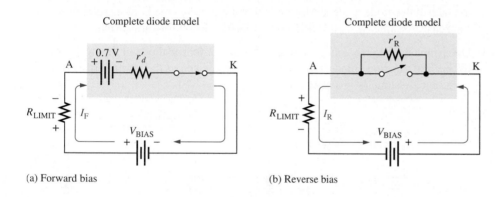

(a) Forward bias   (b) Reverse bias

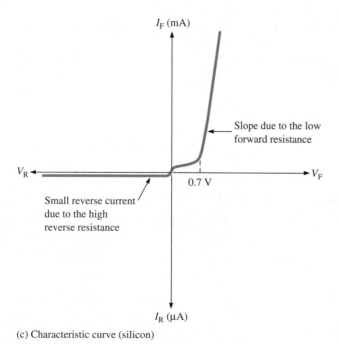

(c) Characteristic curve (silicon)

▲ FIGURE 1–35

The complete model of a diode.

assumed to have a voltage across it when forward-biased. This voltage (V_F) consists of the barrier potential voltage plus the small voltage drop across the dynamic resistance, as indicated by the portion of the curve to the right of the origin. The curve slopes because the voltage drop due to dynamic resistance increases as the current increases. For the complete model of a silicon diode, the following formulas apply:

Equation 1–4

$$V_F = 0.7 \text{ V} + I_F r'_d$$

Equation 1–5

$$I_F = \frac{V_{BIAS} - 0.7 \text{ V}}{R_{LIMIT} + r'_d}$$

The reverse current is taken into account with the parallel resistance and is indicated by the portion of the curve to the left of the origin. The breakdown portion of the curve is not shown because breakdown is not a normal mode of operation for most diodes.

Although only the ideal and practical models will be used in this textbook, the following example illustrates the differences in all three diode models in the analysis of a simple circuit.

EXAMPLE 1–1

(a) Determine the forward voltage and forward current for the diode in Figure 1–36(a) for each of the diode models. Also find the voltage across the limiting resistor in each case. Assume $r'_d = 10 \text{ }\Omega$ at the determined value of forward current.

(b) Determine the reverse voltage and reverse current for the diode in Figure 1–36(b) for each of the diode models. Also find the voltage across the limiting resistor in each case. Assume $I_R = 1 \text{ }\mu\text{A}$.

(a)   (b)

▲ **FIGURE 1–36**

Solution **(a)** Ideal model: $V_F = \textbf{0 V}$

$$I_F = \frac{V_{BIAS}}{R_{LIMIT}} = \frac{10 \text{ V}}{1.0 \text{ k}\Omega} = \textbf{10 mA}$$

$$V_{R_{LIMIT}} = I_F R_{LIMIT} = (10 \text{ mA})(1.0 \text{ k}\Omega) = \textbf{10 V}$$

Practical model: $V_F = \textbf{0.7 V}$

$$I_F = \frac{V_{BIAS} - V_F}{R_{LIMIT}} = \frac{10 \text{ V} - 0.7 \text{ V}}{1.0 \text{ k}\Omega} = \frac{9.3 \text{ V}}{1.0 \text{ k}\Omega} = \textbf{9.3 mA}$$

$$V_{R_{LIMIT}} = I_F R_{LIMIT} = (9.3 \text{ mA})(1.0 \text{ k}\Omega) = \textbf{9.3 V}$$

Complete model: $I_F = \dfrac{V_{BIAS} - 0.7 \text{ V}}{R_{LIMIT} + r'_d} = \dfrac{10 \text{ V} - 0.7 \text{ V}}{1.0 \text{ k}\Omega + 10 \text{ }\Omega} = \dfrac{9.3 \text{ V}}{1010 \text{ }\Omega} = \textbf{9.21 mA}$

$$V_F = 0.7 \text{ V} + I_F r'_d = 0.7 \text{ V} + (9.21 \text{ mA})(10 \text{ }\Omega) = \textbf{792 mV}$$

$$V_{R_{LIMIT}} = I_F R_{LIMIT} = (9.21 \text{ mA})(1.0 \text{ k}\Omega) = \textbf{9.21 V}$$

(b) Ideal model:
$$I_R = \mathbf{0\,A}$$
$$V_R = V_{BIAS} = \mathbf{5\,V}$$
$$V_{R_{LIMIT}} = \mathbf{0\,V}$$

Practical model:
$$I_R = \mathbf{0\,A}$$
$$V_R = V_{BIAS} = \mathbf{5\,V}$$
$$V_{R_{LIMIT}} = \mathbf{0\,V}$$

Complete model:
$$I_R = \mathbf{1\,\mu A}$$
$$V_{R_{LIMIT}} = I_R R_{LIMIT} = (1\,\mu A)(1.0\,k\Omega) = \mathbf{1\,mV}$$
$$V_R = V_{BIAS} - V_{R_{LIMIT}} = 5\,V - 1\,mV = \mathbf{4.999\,V}$$

*Related Problem** Assume that the diode in Figure 1–36(a) fails open. What is the voltage across the diode and the voltage across the limiting resistor?

*Answers are at the end of the chapter.

Open file E01-01 in the Examples folder on your CD-ROM. Measure the voltages across the diode and the resistor in both circuits and compare with the calculated results in this example.

Typical Diodes

Several common physical configurations of diodes are illustrated in Figure 1–37. The anode and cathode are indicated on a diode in several ways, depending on the type of package. The cathode is usually marked by a band, a tab, or some other feature. On those packages where one lead is connected to the case, the case is the cathode. Always check the data sheet, which will be introduced in Chapter 2, for the pin configuration if there is uncertainty.

▲ **FIGURE 1–37**

Typical diode packages with terminal identification.

**SECTION 1–9
REVIEW**

1. What are the two conditions under which the diode is operated?
2. Under what condition is the diode never intentionally operated?
3. What is the simplest way to visualize a diode?
4. To more accurately represent a diode, what factors must be included?
5. Which diode models will be used in this book?

1–10 TESTING A DIODE

A multimeter can be used as a fast and simple way to check a diode. As you know, a good diode will show an extremely high resistance (ideally an open) with reverse bias and a very low resistance with forward bias. A defective open diode will show an extremely high resistance (or open) for both forward and reverse bias. A defective shorted or resistive diode will show zero or a low resistance for both forward and reverse bias. An open diode is the most common type of failure.

After completing this section, you should be able to

- **Test a diode using a digital multimeter**
- Identify a properly functioning diode
- Identify a faulty diode

The DMM Diode Test Position   Many digital multimeters (DMMs) have a diode test position that provides a convenient way to test a diode. A typical DMM, as shown in Figure 1–38, has a small diode symbol to mark the position of the function switch. When set

(a) Forward-bias test

(b) Reverse-bias test

▲ **FIGURE 1–38**

DMM diode test on a properly functioning diode.

to *diode test,* the meter provides an internal voltage sufficient to forward-bias and reverse-bias a diode. This internal voltage may vary among different makes of DMM, but 2.5 V to 3.5 V is a typical range of values. The meter provides a voltage reading or other indication to show the condition of the diode under test.

When the Diode Is Working   In Figure 1–38(a), the red (positive) lead of the meter is connected to the anode and the black (negative) lead is connected to the cathode to forward-bias the diode. If the diode is good, you will get a reading of between approximately 0.5 V and 0.9 V, with 0.7 V being typical for forward bias.

In Figure 1–38(b), the diode is turned around to reverse-bias the diode as shown. If the diode is working properly, you will get a voltage reading based on the meter's internal voltage source. The 2.6 V shown in the figure represents a typical value and indicates that the diode has an extremely high reverse resistance with essentially all of the internal voltage appearing across it.

When the Diode Is Defective   When a diode has failed open, you get an open circuit voltage reading (2.6 V is typical) or "OL" indication for both the forward-bias and the reverse-bias condition, as illustrated in Figure 1–39(a). If a diode is shorted, the meter reads 0 V in both forward- and reverse-bias tests, as indicated in part (b). Sometimes, a failed diode may exhibit a small resistance for both bias conditions rather than a pure short. In this case, the meter will show a small voltage much less than the correct open voltage. For example, a resistive diode may result in a reading of 1.1 V in both directions rather than the correct readings of 0.7 V for forward bias and 2.6 V for reverse bias.

(a) Forward- and reverse-bias tests for an open diode give same indication. Some meters will display "OL."

(b) Forward- and reverse-bias tests for a shorted diode give same 0 V reading. If the diode is resistive, the reading is less than 2.6 V.

▲ **FIGURE 1–39**

Testing a defective diode.

Checking a Diode with the OHMs Function   DMMs that do not have a diode test position can be used to check a diode by setting the function switch on an OHMs range. For a forward-bias check of a good diode, you will get a resistance reading that can vary depending on the meter's internal battery. Many meters do not have sufficient voltage on the OHMs setting to fully forward-bias a diode and you may get a reading of from several hundred to several thousand ohms. For the reverse-bias check of a good diode, you will get some type of out-of-range indication such as "OL" on most DMMs because the reverse resistance is too high for the meter to measure.

Even though you may not get accurate forward- and reverse-resistance readings on a DMM, the relative readings indicate that a diode is functioning properly, and that is usually all you need to know. The out-of-range indication shows that the reverse resistance is extremely high, as you expect. The reading of a few hundred to a few thousand ohms for forward bias is relatively small compared to the reverse resistance, indicating that the diode is working properly. The actual resistance of a forward-biased diode is typically much less than 100 Ω.

SECTION 1–10 REVIEW

1. A properly functioning diode will produce a reading in what range when forward-biased?

2. What reading might a DMM produce when a diode is reverse-biased?

SUMMARY OF DIODE BIAS

FORWARD BIAS: PERMITS MAJORITY-CARRIER CURRENT

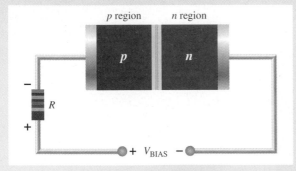

- Bias voltage connections: positive to *p* region; negative to *n* region.
- The bias voltage must be greater than the barrier potential.
- Barrier potential: 0.7 V for silicon.
- Majority carriers flow toward the *pn* junction.
- Majority carriers provide the forward current.
- The depletion region narrows.

REVERSE BIAS: PREVENTS MAJORITY-CARRIER CURRENT

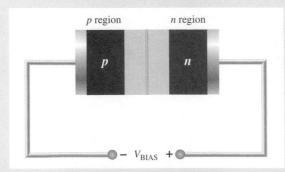

- Bias voltage connections: positive to *n* region; negative to *p* region.
- The bias voltage must be less than the breakdown voltage.
- Majority carriers flow away from the *pn* junction during short transition time.
- Minority carriers provide the extremely small reverse current.
- There is no majority carrier current after transition time.
- The depletion region widens.

CHAPTER SUMMARY

- According to the classical Bohr model, the atom is viewed as having a planetary-type structure with electrons orbiting at various distances around the central nucleus.

- The nucleus of an atom consists of protons and neutrons. The protons have a positive charge and the neutrons are uncharged. The number of protons is the atomic number of the atom.

- Electrons have a negative charge and orbit around the nucleus at distances that depend on their energy level. An atom has discrete bands of energy called *shells* in which the electrons orbit. Atomic structure allows a certain maximum number of electrons in each shell. These shells are designated 1, 2, 3, and so on. In their natural state, all atoms are neutral because they have an equal number of protons and electrons.

- The outermost shell or band of an atom is called the *valence band,* and electrons that orbit in this band are called *valence electrons.* These electrons have the highest energy of all those in the atom. If a valence electron acquires enough energy from an outside source such as heat, it can jump out of the valence band and break away from its atom.

- Semiconductor atoms have four valence electrons. Silicon is the most widely used semiconductive material.

- Materials that are conductors have a large number of free electrons and conduct current very well. Insulating materials have very few free electrons and do not conduct current at all under normal circumstances. Semiconductive materials fall in between conductors and insulators in their ability to conduct current.

- Semiconductor atoms bond together in a symmetrical pattern to form a solid material called a *crystal.* The bonds that hold a crystal together are called *covalent bonds.* Within the crystal structure, the valence electrons that manage to escape from their parent atom are called *conduction electrons* or *free electrons.* They have more energy than the electrons in the valence band and are free to drift throughout the material. When an electron breaks away to become free, it leaves a hole in the valence band creating what is called an *electron-hole pair.* These electron-hole pairs are thermally produced because the electron has acquired enough energy from external heat to break away from its atom.

- A free electron will eventually lose energy and fall back into a hole. This is called *recombination.* But, electron-hole pairs are continuously being thermally generated so there are always free electrons in the material.

- When a voltage is applied across the semiconductor, the thermally produced free electrons move in a net direction and form the current. This is one type of current in an intrinsic (pure) semiconductor.

- Another type of current is the hole current. This occurs as valence electrons move from hole to hole creating, in effect, a movement of holes in the opposite direction.

- An *n*-type semiconductive material is created by adding impurity atoms that have five valence electrons. These impurities are *pentavalent* atoms. A *p*-type semiconductor is created by adding impurity atoms with only three valence electrons. These impurities are *trivalent* atoms.

- The process of adding pentavalent or trivalent impurities to a semiconductor is called *doping.*

- The majority carriers in an *n*-type semiconductor are free electrons acquired by the doping process, and the minority carriers are holes produced by thermally generated electron-hole pairs. The majority carriers in a *p*-type semiconductor are holes acquired by the doping process, and the minority carriers are free electrons produced by thermally generated electron-hole pairs.

- A *pn* junction is formed when part of a material is doped *n*-type and part of it is doped *p*-type. A depletion region forms starting at the junction that is devoid of any majority carriers. The depletion region is formed by ionization.

- There is current through a diode only when it is forward-biased. Ideally, there is no current when there is no bias nor when there is reverse bias. Actually, there is a very small current in reverse bias due to the thermally generated minority carriers, but this can usually be neglected.

- Avalanche occurs in a reverse-biased diode if the bias voltage equals or exceeds the breakdown voltage.

- A diode conducts current when forward-biased and blocks current when reversed-biased.

- The forward-biased barrier potential is typically 0.7 V for a silicon diode and 0.3 V for a germanium diode. These values increase slightly with forward current.

■ Reverse breakdown voltage for a diode is typically greater than 50 V.

■ An ideal diode presents an open when reversed-biased and a short when forward-biased.

KEY TERMS

Key terms and other bold terms are defined in the end-of-book glossary.

Anode   The *p* region of a diode.

Atom   The smallest particle of an element that possesses the unique characteristics of that element.

Barrier potential   The amount of energy required to produce full conduction across the *pn* junction in forward bias.

Bias   The application of a dc voltage to a diode to make it either conduct or block current.

Cathode   The *n* region of a diode.

Conductor   A material that conducts electrical current very well.

Crystal   A solid material in which the atoms are arranged in a symmetrical pattern.

Diode   A semiconductor device with a single *pn* junction that conducts current in only one direction.

Doping   The process of imparting impurities to an intrinsic semiconductive material in order to control its conduction characteristics.

Electron   The basic particle of negative electrical charge.

Forward bias   The condition in which a diode conducts current.

Free electron   An electron that has acquired enough energy to break away from the valence band of the parent atom; also called a *conduction electron*.

Hole   The absence of an electron in the valence band of an atom.

Insulator   A material that does not normally conduct current.

Ionization   The removal or addition of an electron from or to a neutral atom so that the resulting atom (called an ion) has a net positive or negative charge.

***PN* junction**   The boundary between two different types of semiconductive materials.

Proton   The basic particle of positive charge.

Reverse bias   The condition in which a diode prevents current.

Semiconductor   A material that lies between conductors and insulators in its conductive properties. Silicon, germanium, and carbon are examples.

Shell   An energy band in which electrons orbit the nucleus of an atom.

Silicon   A semiconductive material.

Valence   Related to the outer shell of an atom.

***V-I* characteristic**   A curve showing the relationship of diode voltage and current.

KEY FORMULAS

1–1	$N_e = 2n^2$	Maximum number of electrons in any shell
1–2	$I_F = \dfrac{V_{BIAS}}{R_{LIMIT}}$	Forward current, ideal diode model
1–3	$I_F = \dfrac{V_{BIAS} - V_F}{R_{LIMIT}}$	Forward current, practical diode model
1–4	$V_F = 0.7\,V + I_F r'_d$	Forward voltage, complete diode model
1–5	$I_F = \dfrac{V_{BIAS} - 0.7\,V}{R_{LIMIT} + r'_d}$	Forward current, complete diode model

Answers are at the end of the chapter.

1. Every known element has
   (a) the same type of atoms   (b) the same number of atoms
   (c) a unique type of atom   (d) several different types of atoms

2. An atom consists of
   (a) one nucleus and only one electron   (b) one nucleus and one or more electrons
   (c) protons, electrons, and neutrons   (d) answers (b) and (c)

3. The nucleus of an atom is made up of
   (a) protons and neutrons   (b) electrons
   (c) electrons and protons   (d) electrons and neutrons

4. The atomic number of silicon is
   (a) 8   (b) 2   (c) 4   (d) 14

5. The atomic number of germanium is
   (a) 8   (b) 2   (c) 4   (d) 32

6. The valence shell in a silicon atom has the number designation of
   (a) 0   (b) 1   (c) 2   (d) 3

7. Valence electrons are
   (a) in the closest orbit to the nucleus   (b) in the most distant orbit from the nucleus
   (c) in various orbits around the nucleus   (d) not associated with a particular atom

8. A positive ion is formed when
   (a) a valence electron breaks away from the atom
   (b) there are more holes than electrons in the outer orbit
   (c) two atoms bond together
   (d) an atom gains an extra valence electron

9. The most widely used semiconductive material in electronic devices is
   (a) germanium   (b) carbon   (c) copper   (d) silicon

10. The energy band in which free electrons exist is the
   (a) first band   (b) second band   (c) conduction band   (d) valence band

11. Electron-hole pairs are produced by
   (a) recombination   (b) thermal energy   (c) ionization   (d) doping

12. Recombination is when
   (a) an electron falls into a hole
   (b) a positive and a negative ion bond together
   (c) a valence electron becomes a conduction electron
   (d) a crystal is formed

13. In a semiconductor crystal, the atoms are held together by
   (a) the interaction of valence electrons   (b) forces of attraction
   (c) covalent bonds   (d) answers (a), (b), and (c)

14. Each atom in a silicon crystal has
   (a) four valence electrons
   (b) four conduction electrons
   (c) eight valence electrons, four of its own and four shared
   (d) no valence electrons because all are shared with other atoms

15. The current in a semiconductor is produced by
   (a) electrons only   (b) holes only   (c) negative ions   (d) both electrons and holes

16. In an intrinsic semiconductor,
    (a) there are no free electrons   (b) the free electrons are thermally produced
    (c) there are only holes   (d) there are as many electrons as there are holes
    (e) answers (b) and (d)

17. The difference between an insulator and a semiconductor is
    (a) a wider energy gap between the valence band and the conduction band
    (b) the number of free electrons
    (c) the atomic structure
    (d) answers (a), (b), and (c)

18. The process of adding an impurity to an intrinsic semiconductor is called
    (a) doping   (b) recombination   (c) atomic modification   (d) ionization

19. A trivalent impurity is added to silicon to create
    (a) germanium   (b) a *p*-type semiconductor
    (c) an *n*-type semiconductor   (d) a depletion region

20. The purpose of a pentavalent impurity is to
    (a) reduce the conductivity of silicon   (b) increase the number of holes
    (c) increase the number of free electrons   (d) create minority carriers

21. The majority carriers in an *n*-type semiconductor are
    (a) holes   (b) valence electrons   (c) conduction electrons   (d) protons

22. Holes in an *n*-type semiconductor are
    (a) minority carriers that are thermally produced
    (b) minority carriers that are produced by doping
    (c) majority carriers that are thermally produced
    (d) majority carriers that are produced by doping

23. A *pn* junction is formed by
    (a) the recombination of electrons and holes
    (b) ionization
    (c) the boundary of a *p*-type and an *n*-type material
    (d) the collision of a proton and a neutron

24. The depletion region is created by
    (a) ionization   (b) diffusion   (c) recombination   (d) answers (a), (b), and (c)

25. The depletion region consists of
    (a) nothing but minority carriers   (b) positive and negative ions
    (c) no majority carriers   (d) answers (b) and (c)

26. The term *bias* means
    (a) the ratio of majority carriers to minority carriers
    (b) the amount of current across a diode
    (c) a dc voltage is applied to control the operation of a device
    (d) neither (a), (b), nor (c)

27. To forward-bias a diode,
    (a) an external voltage is applied that is positive at the anode and negative at the cathode
    (b) an external voltage is applied that is negative at the anode and positive at the cathode
    (c) an external voltage is applied that is positive at the *p* region and negative at the *n* region
    (d) answers (a) and (c)

28. When a diode is forward-biased,

    (a) the only current is hole current

    (b) the only current is electron current

    (c) the only current is produced by majority carriers

    (d) the current is produced by both holes and electrons

29. Although current is blocked in reverse bias,

    (a) there is some current due to majority carriers

    (b) there is a very small current due to minority carriers

    (c) there is an avalanche current

30. For a silicon diode, the value of the forward-bias voltage typically

    (a) must be greater than 0.3 V

    (b) must be greater than 0.7 V

    (c) depends on the width of the depletion region

    (d) depends on the concentration of majority carriers

31. When forward-biased, a diode

    (a) blocks current   **(b)** conducts current

    (c) has a high resistance   **(d)** drops a large voltage

32. When a voltmeter is placed across a forward-biased diode, it will read a voltage approximately equal to

    (a) the bias battery voltage   **(b)** 0 V

    (c) the diode barrier potential   **(d)** the total circuit voltage

33. A silicon diode is in series with a 1.0 kΩ resistor and a 5 V battery. If the anode is connected to the positive battery terminal, the cathode voltage with respect to the negative battery terminal is

    (a) 0.7 V   **(b)** 0.3 V   **(c)** 5.7 V   **(d)** 4.3 V

34. The positive lead of an ohmmeter is connected to the anode of a diode and the negative lead is connected to the cathode. The diode is

    (a) reversed-biased   **(b)** open   **(c)** forward-biased

    (d) faulty   **(e)** answers (b) and (d)

PROBLEMS

Answers to all odd-numbered problems are at the end of the book.

BASIC PROBLEMS

SECTION 1–1   **Atomic Structure**

1. If the atomic number of a neutral atom is 6, how many electrons does the atom have? How many protons?

2. What is the maximum number of electrons that can exist in the 3rd shell of an atom?

SECTION 1–2   **Semiconductors, Conductors, and Insulators**

3. For each of the energy diagrams in Figure 1–40, determine the class of material based on relative comparisons.

4. A certain atom has four valence electrons. What type of atom is it?

SECTION 1–3   **Covalent Bonds**

5. In a silicon crystal, how many covalent bonds does a single atom form?

SECTION 1–4   **Conduction in Semiconductors**

6. What happens when heat is added to silicon?

7. Name the two energy levels at which current is produced in silicon.

▶ FIGURE 1–40

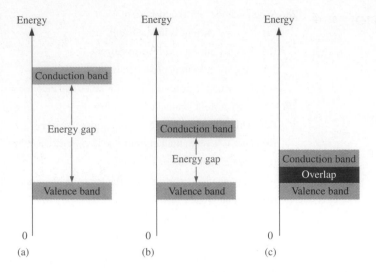

(a)   (b)   (c)

SECTION 1–5   *N*-Type and *P*-Type Semiconductors

8. Describe the process of doping and explain how it alters the atomic structure of silicon.

9. What is antimony? What is boron?

SECTION 1–6   The Diode

10. How is the electric field across the *pn* junction created?

11. Because of its barrier potential, can a diode be used as a voltage source? Explain.

SECTION 1–7   Biasing the Diode

12. To forward-bias a diode, to which region must the positive terminal of a voltage source be connected?

13. Explain why a series resistor is necessary when a diode is forward-biased.

SECTION 1–8   Voltage-Current Characteristic of a Diode

14. Explain how to generate the forward-bias portion of the characteristic curve.

15. What would cause the barrier potential to decrease from 0.7 V to 0.6 V?

▶ FIGURE 1–41

File circuits are identified with a CD logo and are in the Problems folder on your CD-ROM. Filenames correspond to figure numbers (e.g., F01-41).

(a)

(b)

(c)

(d)

SECTION 1–9 Diode Models

16. Determine whether each diode in Figure 1–41 is forward-biased or reverse-biased.

17. Determine the voltage across each diode in Figure 1–41, assuming the practical model.

TROUBLESHOOTING PROBLEMS

SECTION 1–10 Testing A Diode

18. Consider the meter indications in each circuit of Figure 1–42, and determine whether the diode is functioning properly, or whether it is open or shorted. Assume the ideal model.

► FIGURE 1–42

(a)   (b)

(c)   (d)

19. Determine the voltage with respect to ground at each point in Figure 1–43. Assume the practical model.

► FIGURE 1–43

EWB/MULTISIM TROUBLESHOOTING PROBLEMS

These file circuits are in the Troubleshooting Problems folder on your CD-ROM.

20. Open file TSP01-20 and determine the fault.

21. Open file TSP01-21 and determine the fault.

22. Open file TSP01-22 and determine the fault.
23. Open file TSP01-23 and determine the fault.
24. Open file TSP01-24 and determine the fault.
25. Open file TSP01-25 and determine the fault.
26. Open file TSP01-26 and determine the fault.
27. Open file TSP01-27 and determine the fault.
28. Open file TSP01-28 and determine the fault.

ANSWERS

SECTION REVIEWS

SECTION 1–1 Atomic Structure

1. An atom is the smallest particle that retains the characteristics of its element.
2. An electron is the basic particle of negative electrical charge.
3. A valence electron is an electron in the outermost shell of an atom.
4. A free electron is one that has acquired enough energy to break away from the valence band of the parent atom.
5. When a neutral atom loses an electron, the atom becomes a positive ion. When a neutral atom gains an electron, the atom becomes a negative ion.

SECTION 1–2 Semiconductors, Conductors, and Insulators

1. Conductors have many free electrons and conduct current well. Insulators have essentially no free electrons and do not conduct current.
2. Semiconductors do not conduct current as well as conductors do. In terms of conductivity, they are between conductors and insulators.
3. Conductors such as copper have one valence electron.
4. Semiconductors have four valence electrons.
5. Gold, silver, and copper are the best conductors.
6. Silicon is the most widely used semiconductor.
7. The valence electrons of a semiconductor are more tightly bound to the atom than those of conductors.

SECTION 1–3 Covalent Bonds

1. Covalent bonds are formed by the sharing of valence electrons with neighboring atoms.
2. An intrinsic material is one that is in a pure state.
3. A crystal is a solid material formed by atoms bonding together in a fixed pattern.
4. There are eight shared valence electrons in each atom of a silicon crystal.

SECTION 1–4 Conduction in Semiconductors

1. Free electrons are in the conduction band.
2. Free (conduction) electrons are responsible for current in a material.
3. A hole is the absence of an electron in the valence band.
4. Hole current occurs at the valence level.

SECTION 1–5 N-Type and P-Type Semiconductors

1. Doping is the process of adding impurity atoms to a semiconductor in order to modify its conductive properties.
2. A pentavalent atom (donor) has five valence electrons and a trivalent atom (acceptor) has three valence electrons.

3. An *n*-type material is formed by the addition of pentavalent impurity atoms to the intrinsic semi-conductive material.

4. A *p*-type material is formed by the addition of trivalent impurity atoms to the intrinsic semiconductive material.

5. The majority carrier in an *n*-type semiconductor is the free electron.

6. The majority carrier in a *p*-type semiconductor is the hole.

7. Majority carriers are produced by doping.

8. Minority carriers are thermally produced when electron-hole pairs are generated.

9. A pure semiconductor is intrinsic. A doped (impure) semiconductor is extrinsic.

SECTION 1–6   The Diode

1. A *pn* junction is the boundary between *p*-type and *n*-type semiconductors in a diode.

2. Diffusion is the movement of the free electrons (majority carriers) in the *n*-region across the *pn* junction and into the *p* region.

3. The depletion region is the thin layers of positive and negative ions that exist on both sides of the *pn* junction.

4. The barrier potential is the potential difference of the electric field in the depletion region and is the amount of energy required to move electrons through the depletion region.

5. The barrier potential for a silicon diode is approximately 0.7 V.

6. The barrier potential for a germanium diode is approximately 0.3 V.

SECTION 1–7   Biasing the Diode

1. When forward-biased, a diode conducts current. The free electrons in the *n* region move across the *pn* junction and combine with the holes in the *p* region.

2. To forward-bias a diode, the positive side of an external bias voltage is applied to the *p* region and the negative side to the *n* region.

3. When reverse-biased, a diode does not conduct current except for an extremely small reverse current.

4. To reverse-bias a diode, the positive side of an external bias voltage is applied to the *n* region and the negative side to the *p* region.

5. The depletion region for forward bias is much narrower than for reverse bias.

6. Majority carrier current is produced by forward bias.

7. Reverse current is produced by the minority carriers.

8. Reverse breakdown occurs when the reverse-bias voltage equals or exceeds the breakdown voltage of the *pn* junction of a diode.

9. Avalanche is the rapid multiplication of current carriers in reverse breakdown.

SECTION 1–8   Voltage-Current Characteristic of a Diode

1. The knee of the characteristic curve in forward bias is the point at which the barrier potential is overcome and the current increases drastically.

2. A forward-biased diode is normally operated above the knee of the curve.

3. Breakdown voltage is always much greater than the barrier potential.

4. A reverse-biased diode is operated below the breakdown point on the knee of the curve.

5. Barrier potential decreases as temperature increases.

SECTION 1–9   Diode Models

1. The diode is operated in forward bias and reverse bias.

2. The diode should never be operated in reverse breakdown.

3. The diode can be ideally viewed as a switch.

4. A diode includes barrier potential, dynamic resistance, and reverse resistance in the complete model.

5. The ideal and practical diode models (barrier potential) are used.

SECTION 1–10 **Testing a Diode**

1. 0.5 V to 0.9 V
2. 2.60 V

RELATED PROBLEM FOR EXAMPLE

1–1 $V_D = 5$ V; $V_{\text{LIMIT}} = 0$ V

SELF-TEST

1. (c)	**2.** (d)	**3.** (a)	**4.** (d)	**5.** (d)	**6.** (d)	**7.** (b)	**8.** (a)
9. (d)	**10.** (c)	**11.** (b)	**12.** (a)	**13.** (d)	**14.** (c)	**15.** (d)	**16.** (e)
17. (d)	**18.** (a)	**19.** (b)	**20.** (c)	**21.** (c)	**22.** (a)	**23.** (c)	**24.** (d)
25. (d)	**26.** (c)	**27.** (d)	**28.** (d)	**29.** (b)	**30.** (b)	**31.** (b)	**32.** (c)
33. (d)	**34.** (c)						

2

DIODE APPLICATIONS

INTRODUCTION

In Chapter 1, you learned that a semiconductor diode is a device with a single *pn* junction. The importance of the diode in electronic circuits cannot be overemphasized. Its ability to conduct current in one direction while blocking current in the other direction is essential to the operation of many types of circuits. One circuit in particular is the ac rectifier, which is covered in this chapter. Other important applications are circuits such as diode limiters, diode clampers, and diode voltage multipliers. A data sheet is discussed for specific diodes.

CHAPTER OUTLINE

CHAPTER OBJECTIVES

- Explain and analyze the operation of half-wave rectifiers
- Explain and analyze the operation of full-wave rectifiers
- Explain and analyze the operation and characteristics of power supply filters and regulators
- Explain and analyze the operation of diode limiting and clamping circuits
- Explain and analyze the operation of diode voltage multipliers
- Interpret and use a diode data sheet
- Troubleshoot diode circuits using accepted techniques

KEY TERMS

- Power supply
- Filter
- Regulator
- Half-wave rectifier
- Peak inverse voltage (PIV)
- Full-wave rectifier
- Ripple voltage
- Line regulation
- Load regulation
- Limiter
- Clamper

■■■ SYSTEM APPLICATION PREVIEW

As a technician for an electronics company, you are given the responsibility for the final design and testing of a power supply circuit board that your company plans to use in several of its products. Your assignment is to learn all you can about diode circuits by studying this chapter and then apply your knowledge to the system application at the end of the chapter.

WWW. VISIT THE COMPANION WEBSITE
Study aids for this chapter are available at
http://www.prenhall.com/floyd

2–1 HALF-WAVE RECTIFIERS

Because of their ability to conduct current in one direction and block current in the other direction, diodes are used in circuits called rectifiers that convert ac voltage into dc voltage. Rectifiers are found in all dc power supplies that operate from an ac voltage source. A power supply is an essential part of each electronic system from the simplest to the most complex. In this section, you will study the most basic type of rectifier, the half-wave rectifier.

After completing this section, you should be able to

- ■ **Explain and analyze the operation of half-wave rectifiers**
- ■ Describe a basic dc power supply and half-wave rectification
- ■ Determine the average value of a half-wave rectified voltage
- ■ Discuss the effect of barrier potential on a half-wave rectifier output
- ■ Define *peak inverse voltage (PIV)*
- ■ Describe the tranformer-coupled half-wave rectifier

The Basic DC Power Supply

The dc **power supply** converts the standard 110 V, 60 Hz ac available at wall outlets into a constant dc voltage. It is one of the most common electronic circuits that you will find. The dc voltage produced by a power supply is used to power all types of electronic circuits, such as television receivers, stereo systems, VCRs, CD players, and most laboratory equipment.

Basic block diagrams for a rectifier and complete power supply are shown in Figure 2–1. The **rectifier** can be either a half-wave rectifier or a full-wave rectifier (covered in Section 2–2). The rectifier converts the ac input voltage to a pulsating dc voltage, which is half-wave rectified as shown in Figure 2–1(a). A block diagram for a complete power supply is shown in part (b). The **filter** eliminates the fluctuations in the rectified voltage and produces a relatively smooth dc voltage. The power supply filter is covered in Section 2–3. The **regulator** is a circuit that maintains a constant dc voltage for variations in the input line voltage or in the load. Regulators vary from a single device to more complex integrated circuits. The load block is a circuit or device for which the power supply is producing the dc voltage and load current.

The Half-Wave Rectifier

Figure 2–2 illustrates the process called *half-wave rectification*. A diode is connected to an ac source and to a load resistor, R_L, forming a **half-wave rectifier**. Keep in mind that all ground symbols represent the same point electrically. Let's examine what happens during one cycle of the input voltage using the ideal model for the diode. When the sinusoidal input voltage (V_{in}) goes positive, the diode is forward-biased and conducts current through the load resistor, as shown in part (a). The current produces an output voltage across the load R_L, which has the same shape as the positive half-cycle of the input voltage.

When the input voltage goes negative during the second half of its cycle, the diode is reverse-biased. There is no current, so the voltage across the load resistor is 0 V, as shown in Figure 2–2(b). The net result is that only the positive half-cycles of the ac input voltage appear across the load. Since the output does not change polarity, it is a *pulsating dc voltage* with a frequency of 60 Hz, as shown in part (c).

Average Value of the Half-Wave Output Voltage The average value of the half-wave rectified output voltage is the value you would measure on a dc voltmeter. Mathematically, it is determined by finding the area under the curve over a full cycle, as illustrated

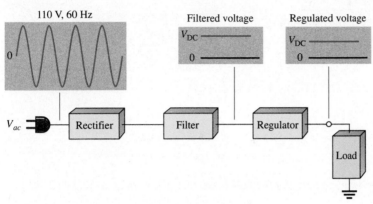

(a) Half-wave rectifier

(b) Complete power supply with rectifier, filter, and regulator

▲ FIGURE 2–1

Block diagram of a rectifier and a dc power supply with a load.

(a) During the positive alternation of the 60 Hz input voltage, the output voltage looks like the positive half of the input voltage. The current path is through ground back to the source.

(b) During the negative alternation of the input voltage, the current is 0, so the output voltage is also 0.

(c) 60 Hz half-wave output voltage for three input cycles

▲ FIGURE 2–2

Half-wave rectifier operation. The diode is considered to be ideal.

in Figure 2–3, and then dividing by 2π, the number of radians in a full cycle. The result of this is expressed in Equation 2–1, where V_p is the peak value of the voltage. This equation shows that V_{AVG} is approximately 31.8% of V_p for a half-wave rectified voltage. See Appendix B for a detailed derivation.

Equation 2–1
$$V_{AVG} = \frac{V_p}{\pi}$$

▶ **FIGURE 2–3**

Average value of the half-wave rectified signal.

<hr>

EXAMPLE 2–1

What is the average value of the half-wave rectified voltage in Figure 2–4?

▲ **FIGURE 2–4**

Solution
$$V_{AVG} = \frac{V_p}{\pi} = \frac{100\ \text{V}}{\pi} = \textbf{31.8 V}$$

*Related Problem** Determine the average value of the half-wave voltage if its peak amplitude is 12 V.

*Answers are at the end of the chapter.

<hr>

Effect of the Barrier Potential on the Half-Wave Rectifier Output

In the previous discussion, the diode was considered ideal. When the practical diode model is used with the barrier potential of 0.7 V taken into account, this is what happens. During the positive half-cycle, the input voltage must overcome the barrier potential before the diode becomes forward-biased. This results in a half-wave output with a peak value that is 0.7 V less than the peak value of the input, as shown in Figure 2–5. The expression for the peak output voltage is

Equation 2–2
$$V_{p(out)} = V_{p(in)} - 0.7\ \text{V}$$

It is usually acceptable to use the ideal diode model, which neglects the effect of the barrier potential, when the peak value of the applied voltage is much greater than the barrier potential (at least 10 V, as a rule of thumb). However, for consistency, we will use the practical model of a diode, taking the 0.7 V barrier potential into account unless stated otherwise.

▲ FIGURE 2–5

The effect of the barrier potential on the half-wave rectified output voltage is to reduce the peak value of the input by about 0.7 V.

EXAMPLE 2–2

Sketch the output voltages of each rectifier for the indicated input voltages, as shown in Figure 2–6. The 1N4001 and 1N4003 are specific rectifier diodes.

▲ FIGURE 2–6

Solution The peak output voltage for circuit (a) is

$$V_{p(out)} = V_{p(in)} - 0.7\,\text{V} = 5\,\text{V} - 0.7\,\text{V} = \textbf{4.30 V}$$

The peak output voltage for circuit (b) is

$$V_{p(out)} = V_{p(in)} - 0.7\,\text{V} = 100\,\text{V} - 0.7\,\text{V} = \textbf{99.3 V}$$

These output voltage waveforms are shown in Figure 2–7. Note that the barrier potential could have been neglected in circuit (b) with very little error (0.7 percent); but, if it is neglected in circuit (a), a significant error results (14 percent).

▲ FIGURE 2–7

Output voltages for the circuits in Figure 2–6. Obviously, they are not shown on the same scale.

Related Problem   Determine the peak output voltages for the rectifiers in Figure 2–6 if the peak input in part (a) is 3 V and the peak input in part (b) is 210 V.

> Open file E02-02 in the Examples folder on your CD-ROM. For the inputs specified in the example, measure the resulting output voltage waveforms. Compare your measured results with those shown in the example.

Peak Inverse Voltage (PIV)

The **peak inverse voltage (PIV)** equals the peak value of the input voltage, and the diode must be capable of withstanding this amount of repetitive reverse voltage. For the diode in Figure 2–8, the maximum value of reverse voltage, designated as PIV, occurs at the peak of each negative alternation of the input voltage when the diode is reverse-biased.

Equation 2–3

$$PIV = V_{p(in)}$$

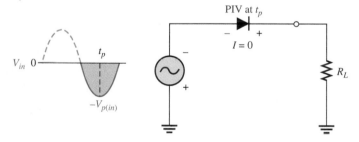

▲ FIGURE 2–8

The PIV occurs at the peak of each half-cycle of the input voltage when the diode is reverse-biased. In this circuit, the PIV occurs at the peak of each negative half-cycle.

Half-Wave Rectifier with Transformer-Coupled Input Voltage

A transformer is often used to couple the ac input voltage from the source to the rectifier, as shown in Figure 2–9. Transformer coupling provides two advantages. First, it allows the source voltage to be stepped up or stepped down as needed. Second, the ac source is electrically isolated from the rectifier, thus preventing a shock hazard in the secondary circuit.

From basic ac circuits recall that the secondary voltage of a transformer equals the turns ratio, n, times the primary voltage, as expressed in Equation 2–4. We will define the turns ratio as the ratio of secondary turns, N_{sec}, to the primary turns, N_{pri}: $n = N_{sec}/N_{pri}$.

Equation 2–4

$$V_{sec} = nV_{pri}$$

▲ **FIGURE 2–9**

Half-wave rectifier with transformer-coupled input voltage.

If $n > 1$, the secondary voltage is greater than the primary voltage. If $n < 1$, the secondary voltage is less than the primary voltage. If $n = 1$, then $V_{sec} = V_{pri}$.

The peak secondary voltage, $V_{p(sec)}$, in a transformer-coupled half-wave rectifier is the same as $V_{p(in)}$ in Equation 2–2. Therefore, Equation 2–2 written in terms of $V_{p(sec)}$ is

$$V_{p(out)} = V_{p(sec)} - 0.7 \text{ V}$$

and Equation 2–3 in terms of $V_{p(sec)}$ is

$$\text{PIV} = V_{p(sec)}$$

EXAMPLE 2–3

Determine the peak value of the output voltage for Figure 2–10 if the turns ratio is 0.5.

▲ **FIGURE 2–10**

Solution
$$V_{p(pri)} = V_{p(in)} = 156 \text{ V}$$

The peak secondary voltage is

$$V_{p(sec)} = nV_{p(pri)} = 0.5(156 \text{ V}) = 78 \text{ V}$$

The rectified peak output voltage is

$$V_{p(out)} = V_{p(sec)} - 0.7 \text{ V} = 78 \text{ V} - 0.7 \text{ V} = \mathbf{77.3 \text{ V}}$$

where $V_{p(sec)}$ is the input to the rectifier.

Related Problem   **(a)** Determine the peak value of the output voltage for Figure 2–10 if $n = 2$ and $V_{p(in)} = 312$ V.

(b) What is the PIV across the diode?

(c) Describe the output voltage if the diode is turned around.

Open file E02-03 in the Examples folder on your CD-ROM. For the specified input, measure the peak output voltage. Compare your measured result with the calculated value.

SECTION 2–1 REVIEW

Answers are at the end of the chapter.

1. At what point on the input cycle does the PIV occur?
2. For a half-wave rectifier, there is current through the load for approximately what percentage of the input cycle?
3. What is the average of a half-wave rectified voltage with a peak value of 10 V?
4. What is the peak value of the output voltage of a half-wave rectifier with a peak sine wave input of 25 V?
5. What PIV rating must a diode have to be used in a rectifier with a peak output voltage of 50 V?

2–2 FULL-WAVE RECTIFIERS

Although half-wave rectifiers have some applications, the full-wave rectifier is the most commonly used type in dc power supplies. In this section, you will use what you learned about half-wave rectification and expand it to full-wave rectifiers. You will learn about two types of full-wave rectifiers: center-tapped and bridge.

After completing this section, you should be able to

■ **Explain and analyze the operation of full-wave rectifiers**

■ Discuss how full-wave rectification differs from half-wave rectification

■ Determine the average value of a full-wave rectified voltage

■ Describe the operation of a center-tapped full-wave rectifier

■ Explain how the transformer turns ratio affects the rectified output voltage

■ Determine the peak inverse voltage (PIV)

■ Describe the operation of a bridge full-wave rectifier

■ Compare the center-tapped rectifier and the bridge rectifier

The difference between full-wave and half-wave rectification is that a **full-wave rectifier** allows unidirectional (one-way) current through the load during the entire 360° of the input cycle, and the half-wave rectifier allows this only during one-half of the cycle. The result of full-wave rectification is an output voltage with a frequency twice the input frequency that pulsates every half-cycle of the input, as shown in Figure 2–11.

The number of positive alternations that make up the full-wave rectified voltage is twice that of the half-wave voltage for the same time interval. The average value, which is

▲ FIGURE 2–11

Full-wave rectification.

the value measured on a dc voltmeter, for a full-wave rectified sinusoidal voltage is twice that of the half-wave, as shown in the following formula:

$$V_{AVG} = \frac{2V_p}{\pi}$$

Equation 2–5

V_{AVG} is approximately 63.7% of V_p for a full-wave rectified voltage.

EXAMPLE 2–4

Find the average value of the full-wave rectified voltage in Figure 2–12.

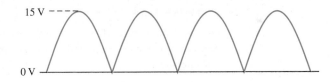

▲ FIGURE 2–12

Solution

$$V_{AVG} = \frac{2V_p}{\pi} = \frac{2(15\text{ V})}{\pi} = \textbf{9.55 V}$$

Related Problem Find the average value of the full-wave rectified voltage if its peak is 155 V.

The Center-Tapped Full-Wave Rectifier

A **center-tapped rectifier** is a type of full-wave rectifier that uses two diodes connected to the secondary of a center-tapped transformer, as shown in Figure 2–13. The input voltage is coupled through the transformer to the center-tapped secondary. Half of the total

▲ FIGURE 2–13

A center-tapped full-wave rectifier.

(a) During positive half-cycles, D_1 is forward-biased and D_2 is reverse-biased.

(b) During negative half-cycles, D_2 is forward-biased and D_1 is reverse-biased.

▲ **FIGURE 2–14**

Basic operation of a center-tapped full-wave rectifier. Note that the current through the load resistor is in the same direction during the entire input cycle, so the output voltage always has the same polarity.

secondary voltage appears between the center tap and each end of the secondary winding as shown.

For a positive half-cycle of the input voltage, the polarities of the secondary voltages are as shown in Figure 2–14(a). This condition forward-biases diode D_1 and reverse-biases diode D_2. The current path is through D_1 and the load resistor R_L, as indicated. For a negative half-cycle of the input voltage, the voltage polarities on the secondary are as shown in Figure 2–14(b). This condition reverse-biases D_1 and forward-biases D_2. The current path is through D_2 and R_L, as indicated. Because the output current during both the positive and negative portions of the input cycle is in the same direction through the load, the output voltage developed across the load resistor is a full-wave rectified dc voltage, as shown.

Effect of the Turns Ratio on the Output Voltage If the transformer's turns ratio is 1, the peak value of the rectified output voltage equals half the peak value of the primary input voltage less the barrier potential, as illustrated in Figure 2–15. This is because half of the primary voltage appears across each half of the secondary winding ($V_{p(sec)} = V_{p(pri)}$). We will begin referring to the forward voltage due to the barrier potential as the **diode drop**.

In order to obtain an output voltage with a peak equal to the input peak (less the diode drop), a step-up transformer with a turns ratio of $n = 2$ must be used, as shown in Figure 2–16. In this case, the total secondary voltage (V_{sec}) is twice the primary voltage ($2V_{pri}$), so the voltage across each half of the secondary is equal to V_{pri}.

In any case, the output voltage of a center-tapped full-wave rectifier is always one-half of the total secondary voltage less the diode drop, no matter what the turns ratio.

Equation 2–6

$$V_{out} = \frac{V_{sec}}{2} - 0.7 \text{ V}$$

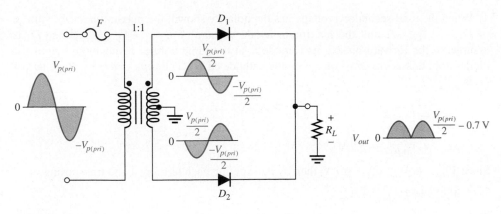

▲ **FIGURE 2–15**

Center-tapped full-wave rectifier with a transformer turns ratio of 1. $V_{p(pri)}$ is the peak value of the primary voltage.

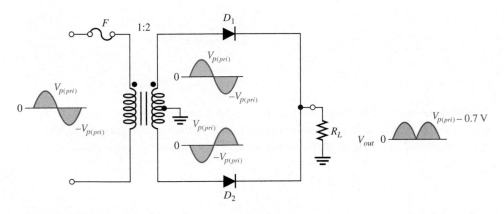

▲ **FIGURE 2–16**

Center-tapped full-wave rectifier with a transformer turns ratio of 2.

Peak Inverse Voltage Each diode in the full-wave rectifier is alternately forward-biased and then reverse-biased. The maximum reverse voltage that each diode must withstand is the peak secondary voltage $V_{p(sec)}$. This is shown in Figure 2–17 where D_2 is assumed to be reverse-biased.

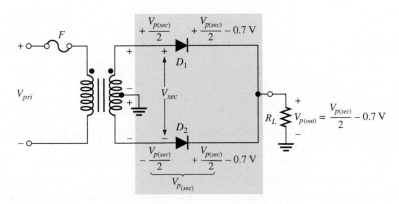

▲ **FIGURE 2–17**

Diode reverse voltage (D_2 shown reverse-biased). The PIV is twice the peak value of the output voltage plus a diode drop.

When the total secondary voltage has the polarity shown, the maximum anode voltage of D_1 is $+V_{p(sec)}/2$ and the maximum anode voltage of D_2 is $-V_{p(sec)}/2$. Since D_1 is assumed to be forward-biased, its cathode is at the same voltage as its anode minus the diode drop; this is also the voltage on the cathode of D_2. The peak inverse voltage across D_2 is

$$\text{PIV} = \left(\frac{V_{p(sec)}}{2} - 0.7 \text{ V} \right) - \left(\frac{-V_{p(sec)}}{2} \right) = \frac{V_{p(sec)}}{2} + \frac{V_{p(sec)}}{2} - 0.7 \text{ V}$$

$$= V_{p(sec)} - 0.7 \text{ V}$$

Since $V_{p(out)} = V_{p(sec)}/2 - 0.7 \text{ V}$, then by multiplying each term by 2 and transposing,

$$V_{p(sec)} = 2V_{p(out)} + 1.4 \text{ V}$$

Therefore, by substitution, the peak inverse voltage across either diode in a full-wave center-tapped rectifier is

Equation 2–7

$$\text{PIV} = 2V_{p(out)} + 0.7 \text{ V}$$

EXAMPLE 2–5

Show the voltage waveforms across each half of the secondary winding and across R_L when a 100 V peak sine wave is applied to the primary winding in Figure 2–18. Also, what minimum PIV rating must the diodes have?

▲ **FIGURE 2–18**

Solution The transformer turns ratio $n = 0.5$. The total peak secondary voltage is

$$V_{p(sec)} = nV_{p(pri)} = 0.5(100 \text{ V}) = 50 \text{ V}$$

There is a 25 V peak across each half of the secondary with respect to ground. The output load voltage has a peak value of 25 V, less the 0.7 V drop across the diode. Each diode must have a minimum PIV rating of

$$\text{PIV} = V_{p(sec)} - 0.7 \text{ V} = 50 \text{ V} - 0.7 \text{ V} = \textbf{49.3 V}$$

The waveforms are shown in Figure 2–19.

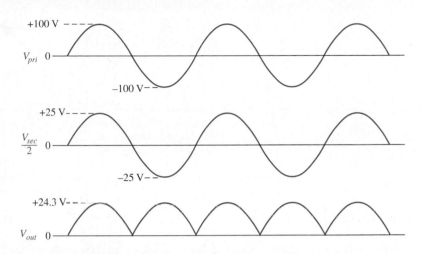

▲ **FIGURE 2–19**

Related Problem What diode PIV rating is required to handle a peak input of 160 V in Figure 2–18?

 Open file E02-05 in the Examples folder on your CD-ROM. For the specified input voltage, measure the voltage waveforms across each half of the secondary and across the load resistor. Compare with the results shown in the example.

The Bridge Full-Wave Rectifier

The **bridge rectifier** uses four diodes connected as shown in Figure 2–20. When the input cycle is positive as in part (a), diodes D_1 and D_2 are forward-biased and conduct current in the direction shown. A voltage is developed across R_L which looks like the positive half of the input cycle. During this time, diodes D_3 and D_4 are reverse-biased.

When the input cycle is negative as in Figure 2–20(b), diodes D_3 and D_4 are forward-biased and conduct current in the same direction through R_L as during the positive half-cycle. During the negative half-cycle, D_1 and D_2 are reverse-biased. A full-wave rectified output voltage appears across R_L as a result of this action.

Bridge Output Voltage A bridge rectifier with a transformer-coupled input is shown in Figure 2–21(a). During the positive half-cycle of the total secondary voltage, diodes D_1 and D_2 are forward-biased. Neglecting the diode drops, the secondary voltage appears across the load resistor. The same is true when D_3 and D_4 are forward-biased during the negative half-cycle.

$$V_{p(out)} = V_{p(sec)}$$

As you can see in Figure 2–21(b), two diodes are always in series with the load resistor during both the positive and negative half-cycles. If these diode drops are taken into account, the output voltage is

$$V_{p(out)} = V_{p(sec)} - 1.4 \text{ V}$$

Equation 2–8

Peak Inverse Voltage Let's assume that D_1 and D_2 are forward-biased and examine the reverse voltage across D_3 and D_4. Visualizing D_1 and D_2 as shorts (ideal model), as in

(a) During positive half-cycle of the input, D_1 and D_2 are forward-biased and conduct current. D_3 and D_4 are reverse-biased.

(b) During negative half-cycle of the input, D_3 and D_4 are forward-biased and conduct current. D_1 and D_2 are reverse-biased.

▲ FIGURE 2–20

Operation of a bridge rectifier.

(a) Ideal diodes

(b) Practical diodes (Diode drops included)

▲ FIGURE 2–21

Bridge operation during a positive half-cycle of the primary and secondary voltages.

Figure 2–22(a), you can see that D_3 and D_4 have a peak inverse voltage equal to the peak secondary voltage. Since the output voltage is *ideally* equal to the secondary voltage,

$$PIV = V_{p(out)}$$

If the diode drops of the forward-biased diodes are included as shown in Figure 2–22(b), the peak inverse voltage across each reverse-biased diode in terms of $V_{p(out)}$ is

$$PIV = V_{p(out)} + 0.7 \text{ V}$$

Equation 2–9

The PIV rating of the bridge diodes is less than that required for the center-tapped configuration. *If the diode drop is neglected, the bridge rectifier requires diodes with half the PIV rating of those in a center-tapped rectifier for the same output voltage.*

(a) For the ideal diode model (forward-biased diodes D_1 and D_2 are shown in blue). PIV = $V_{p(out)}$

(b) For the practical diode model (forward-biased diodes D_1 and D_2 are shown in blue). PIV = $V_{p(out)} + 0.7$ V

▲ **FIGURE 2–22**

Peak inverse voltages across diodes D_3 and D_4 in a bridge rectifier during the positive half-cycle of the secondary voltage.

EXAMPLE 2–6

Determine the peak output voltage for the bridge rectifier in Figure 2–23. Assuming the practical model, what PIV rating is required for the diodes? The transformer is specified to have a 12 V rms secondary voltage for the standard 110 V across the primary.

▲ **FIGURE 2–23**

Solution The peak output voltage (taking into account the two diode drops) is

$$V_{p(sec)} = 1.414V_{rms} = 1.414(12 \text{ V}) \cong 17 \text{ V}$$
$$V_{p(out)} = V_{p(sec)} - 1.4 \text{ V} = 17 \text{ V} - 1.4 \text{ V} = \mathbf{15.6 \text{ V}}$$

The PIV for each diode is

$$PIV = V_{p(out)} + 0.7\,V = 15.6\,V + 0.7\,V = \mathbf{16.3\,V}$$

Related Problem   Determine the peak output voltage for the bridge rectifier in Figure 2–23 if the transformer produces an rms secondary voltage of 30 V. What is the PIV rating for the diodes?

> Open file E02-06 in the Examples folder on your CD-ROM. Measure the output voltage and compare to the calculated value.

SECTION 2–2 REVIEW

1. How does a full-wave voltage differ from a half-wave voltage?
2. What is the average value of a full-wave rectified voltage with a peak value of 60 V?
3. Which type of full-wave rectifier has the greater output voltage for the same input voltage and transformer turns ratio?
4. For a peak output voltage of 45 V, in which type of rectifier would you use diodes with a PIV rating of 50 V?
5. What PIV rating is required for diodes used in the type of rectifier that was not selected in Question 4?

2–3 POWER SUPPLY FILTERS AND REGULATORS

A power supply filter ideally eliminates the fluctuations in the output voltage of a half-wave or full-wave rectifier and produces a constant-level dc voltage. Filtering is necessary because electronic circuits require a constant source of dc voltage and current to provide power and biasing for proper operation. Filters are implemented with capacitors, as you will see in this section. Voltage regulation in power supplies is usually done with integrated circuit voltage regulators. A voltage regulator prevents changes in the filtered dc voltage due to variations in input voltage or load.

After completing this section, you should be able to

- **Explain and analyze the operation and characteristics of power supply filters and regulators**
- Explain the purpose of a filter
- Describe the capacitor-input filter
- Define *ripple voltage* and calculate the ripple factor
- Discuss surge current in a capacitor-input filter
- Discuss voltage regulation
- Describe a typical integrated circuit regulator

In most power supply applications, the standard 60 Hz ac power line voltage must be converted to an approximately constant dc voltage as discussed earlier. The 60 Hz pulsating dc output of a half-wave rectifier or the 120 Hz pulsating output of a full-wave rectifier must be filtered to reduce the large voltage variations. Figure 2–24 illustrates the filtering concept showing a nearly smooth dc output voltage from the filter. The small amount of fluctuation in the filter output voltage is called *ripple*.

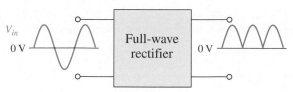

(a) Rectifier without a filter

(b) Rectifier with a filter (output ripple is exaggerated)

▲ **FIGURE 2–24**

Power supply filtering.

Capacitor-Input Filter

A half-wave rectifier with a capacitor-input filter is shown in Figure 2–25. The filter is simply a capacitor connected from the rectifier output to ground. R_L represents the equivalent resistance of a load. We will use the half-wave rectifier to illustrate the basic principle, and then expand the concept to full-wave rectification.

During the positive first quarter-cycle of the input, the diode is forward-biased, allowing the capacitor to charge to within 0.7 V of the input peak, as illustrated in Figure 2–25(a). When the input begins to decrease below its peak, as shown in part (b), the capacitor retains its charge and the diode becomes reverse-biased because the cathode is more positive than the anode. During the remaining part of the cycle, the capacitor can discharge only through the load resistance at a rate determined by the $R_L C$ time constant, which is normally long compared to the period of the input. The larger the time constant, the less the capacitor will discharge. During the first quarter of the next cycle, as illustrated in part (c), the diode will again become forward-biased when the input voltage exceeds the capacitor voltage by approximately 0.7 V.

Ripple Voltage   As you have seen, the capacitor quickly charges at the beginning of a cycle and slowly discharges through R_L after the positive peak of the input voltage (when the diode is reverse-biased). The variation in the capacitor voltage due to the charging and discharging is called the **ripple voltage**. Generally, ripple is undesirable; thus, the smaller the ripple, the better the filtering action, as illustrated in Figure 2–26.

For a given input frequency, the output frequency of a full-wave rectifier is twice that of a half-wave rectifier, as illustrated in Figure 2–27. This makes a full-wave rectifier easier to filter because of the shorter time between peaks. When filtered, the full-wave rectified voltage has a smaller ripple than does a half-wave voltage for the *same* load resistance and capacitor values. The capacitor discharges less during the shorter interval between full-wave pulses, as shown in Figure 2–28.

(a) Initial charging of capacitor (diode is forward-biased) happens only once when power is turned on.

(b) The capacitor discharges through R_L after peak of positive alternation when the diode is reverse-biased. This discharging occurs during the portion of the input voltage indicated by the solid blue curve.

(c) The capacitor charges back to peak of input when the diode becomes forward-biased. This charging occurs during the portion of the input voltage indicated by the solid blue curve.

▲ **FIGURE 2–25**

Operation of a half-wave rectifier with a capacitor-input filter.

(a) Larger ripple means less effective filtering.

(b) Smaller ripple means more effective filtering. Generally, the larger the capacitor value, the smaller the ripple for the same input and load.

▲ **FIGURE 2–26**

Half-wave ripple voltage (blue line).

(a) Half-wave (*hw*)

$$f_{hw} = \frac{1}{T_{hw}} = \frac{1}{16.7 \text{ ms}} = 60 \text{ Hz}$$

$$T_{fw} = \frac{T_{hw}}{2}$$

$$f_{fw} = \frac{1}{T_{fw}} = 2\left(\frac{1}{T_{hw}}\right) = 2f_{hw}$$
$$= 2(60 \text{ Hz}) = 120 \text{ Hz}$$

(b) Full-wave (*fw*)

▲ **FIGURE 2–27**

Frequencies of half-wave and full-wave rectified voltages derived from a 60 Hz sine wave.

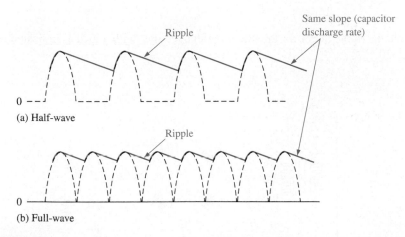

(a) Half-wave

(b) Full-wave

▲ **FIGURE 2–28**

Comparison of ripple voltages for half-wave and full-wave rectified voltages with the same filter capacitor and load and derived from the same sinusoidal input voltage.

Ripple Factor   The **ripple factor** (*r*) is an indication of the effectiveness of the filter and is defined as

$$r = \frac{V_{r(pp)}}{V_{DC}}$$

Equation 2–10

where $V_{r(pp)}$ is the peak-to-peak ripple voltage and V_{DC} is the dc (average) value of the filter's output voltage, as illustrated in Figure 2–29. The lower the ripple factor, the better the filter. *The ripple factor can be lowered by increasing the value of the filter capacitor or increasing the load resistance.*

▲ **FIGURE 2–29**

V_r and V_{DC} determine the ripple factor.

For a full-wave rectifier with a capacitor-input filter, approximations for the peak-to-peak ripple voltage, $V_{r(pp)}$, and the dc value of the filter output voltage, V_{DC}, are given in the following expressions. $V_{p(rect)}$ is the unfiltered peak rectified voltage.

Equation 2–11

$$V_{r(pp)} \cong \left(\frac{1}{fR_LC}\right)V_{p(rect)}$$

Equation 2–12

$$V_{DC} \cong \left(1 - \frac{1}{2fR_LC}\right)V_{p(rect)}$$

These approximate expressions are derived in Appendix B and are based on certain simplifying assumptions.

EXAMPLE 2–7

Determine the ripple factor for the filtered bridge rectifier with a load as indicated in Figure 2–30.

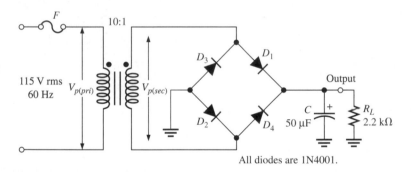

All diodes are 1N4001.

▲ **FIGURE 2–30**

Solution   The transformer turns ratio is $n = 0.1$. The peak primary voltage is

$$V_{p(pri)} = 1.414V_{rms} = 1.414(115 \text{ V}) = 163 \text{ V}$$

The peak secondary voltage is

$$V_{p(sec)} = nV_{p(pri)} = 0.1(163 \text{ V}) = 16.3 \text{ V}$$

The unfiltered peak full-wave rectified voltage is

$$V_{p(rect)} = V_{p(sec)} - 1.4 \text{ V} = 16.3 \text{ V} - 1.4 \text{ V} = 14.9 \text{ V}$$

The frequency of a full-wave rectified voltage is 120 Hz. The approximate peak-to-peak ripple voltage at the output is

$$V_{r(pp)} \cong \left(\frac{1}{fR_LC}\right)V_{p(rect)} = \left(\frac{1}{(120 \text{ Hz})(2.2 \text{ k}\Omega)(50 \text{ }\mu\text{F})}\right)14.9 \text{ V} = 1.13 \text{ V}$$

The approximate dc value of the output voltage is determined as follows:

$$V_{DC} = \left(1 - \frac{1}{2fR_LC}\right)V_{p(rect)} = \left(1 - \frac{1}{(240 \text{ Hz})(2.2 \text{ k}\Omega)(50 \text{ }\mu\text{F})}\right)14.9 \text{ V} = 14.3 \text{ V}$$

The resulting ripple factor is

$$r = \frac{V_{r(pp)}}{V_{DC}} = \frac{1.13 \text{ V}}{14.3 \text{ V}} = \mathbf{0.079}$$

The percent ripple is 7.9%.

Related Problem Determine the peak-to-peak ripple voltage if the filter capacitor in Figure 2–30 is increased to 100 μF and the load resistance changes to 12 kΩ.

Open file E02-07 in the Examples folder on your CD-ROM. For the specified input voltage, measure the peak-to-peak ripple voltage and the dc value at the output. Do the results agree closely with the calculated values? If not, can you explain why?

Surge Current in the Capacitor–Input Filter Before the switch in Figure 2–31(a) is closed, the filter capacitor is uncharged. At the instant the switch is closed, voltage is connected to the bridge and the uncharged capacitor appears as a short, as shown. This produces an initial surge of current, I_{surge}, through the two forward-biased diodes D_1 and D_2. The worst-case situation occurs when the switch is closed at a peak of the secondary voltage and a maximum surge current, $I_{surge(max)}$, is produced, as illustrated in the figure.

It is possible that the surge current could destroy the diodes, and for this reason a surge-limiting resistor is sometimes connected, as shown in Figure 2–31(b). The value of this resistor must be small compared to R_L. Also, the diodes must have a maximum

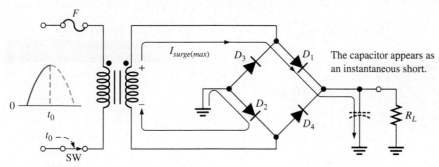

(a) Maximum surge current occurs when switch is closed at peak of an input cycle.

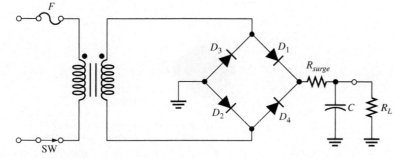

(b) A series resistor (R_{surge}) limits the surge current.

▲ **FIGURE 2–31**

Surge current in a capacitor-input filter.

forward surge current rating such that they can withstand the momentary surge of current. This rating is specified on diode data sheets as I_{FSM}. The minimum surge resistor value can be calculated as follows:

Equation 2–13

$$R_{surge} = \frac{V_{p(sec)} - 1.4 \text{ V}}{I_{FSM}}$$

IC Regulators

While filters can reduce the ripple from power supplies to a low value, the most effective approach is a combination of a capacitor-input filter used with an integrated circuit (IC) voltage regulator. An integrated circuit voltage regulator is connected to the output of a filtered rectifier and maintains a constant output voltage (or current) despite changes in the input, the load current, or the temperature. The capacitor-input filter reduces the input ripple to the regulator to an acceptable level. The combination of a large capacitor and an IC regulator is inexpensive and helps produce an excellent small power supply.

The most popular IC regulators have three terminals—an input terminal, an output terminal, and a reference (or adjust) terminal. The input to the regulator is first filtered with a capacitor to reduce the ripple to <10%. The regulator reduces the ripple to a negligible amount. In addition, most regulators have an internal voltage reference, short-circuit protection, and thermal shutdown circuitry. They are available in a variety of voltages, including positive and negative outputs, and can be designed for variable outputs with a minimum of external components. Typically, IC regulators can furnish a constant output of one or more amps of current with high ripple rejection. IC regulators are available that can supply load currents of over 5 A.

Three-terminal regulators designed for a fixed output voltage require only external capacitors to complete the regulation portion of the power supply, as shown in Figure 2–32. Filtering is accomplished by a large-value capacitor between the input voltage and ground. Sometimes a second smaller-value input capacitor is connected in parallel, especially if the filter capacitor is not close to the IC, to prevent oscillation. This capacitor

(a) Standard configuration

Type number	Output voltage
7805	+5.0 V
7806	+6.0 V
7808	+8.0 V
7809	+9.0 V
7812	+12.0 V
7815	+15.0 V
7818	+18.0 V
7824	+24.0 V

(b) The 7800 series

TO–220
T SUFFIX
CASE 221A

Pin 1. Input
2. Ground
3. Output

Heatsink surface
connected to Pin 2.

D²PAK
D2T SUFFIX
CASE 936

Heatsink surface (shown as terminal 4 in
case outline drawing) is connected to Pin 2.

(c) Typical packages

▲ FIGURE 2–32

The 7800 series three-terminal fixed positive voltage regulators.

needs to be located close to the IC. Finally, an output capacitor (typically 0.1 μF to 1.0 μF) is placed in parallel with the output to improve the transient response.

Examples of fixed three-terminal regulators are the 78XX series of regulators that are available with various output voltages and can supply up to 1 A of load current (with adequate heat sinking). The last two digits of the number stand for the output voltage; thus, the 7812 has a +12 V output. The negative output versions of the same regulator are numbered as the 79XX series; the 7912 has a −12 V output. The output voltage from these regulators is within 3% of the nominal value but will hold a nearly constant output despite changes in the input voltage or output load. A basic fixed +5 V power supply with a 7805 regulator is shown in Figure 2–33.

▲ **FIGURE 2–33**

A basic +5.0 V regulated power supply.

Another type of three-terminal regulator is an adjustable regulator. Figure 2–34 shows a power supply circuit with an adjustable output, controlled by the variable resistor, R_2. Note that R_2 is adjustable from zero to 1.0 kΩ. The LM317 regulator keeps a constant 1.25 V between the output and adjust terminals. This produces a constant current in R_1 of 1.25 V/240 Ω = 52 mA. Neglecting the very small current through the adjust terminal, the current in R_2 is the same as the current in R_1. The output is taken across both R_1 and R_2 and is found from the equation,

$$V_{out} = 1.25 \text{ V}\left(\frac{R_1 + R_2}{R_1}\right)$$

Notice that the output voltage from the power supply is the regulator's 1.25 V multiplied by a ratio of resistances. For the case shown in Figure 2–34, when R_2 is set to the minimum (zero) resistance, the output is 1.25 V. When R_2 is set to the maximum, the output is nearly 6.5 V.

▲ **FIGURE 2–34**

A basic power supply with a variable output voltage (from 1.25 V to 6.5 V).

Percent Regulation

The regulation expressed as a percentage is a figure of merit used to specify the performance of a voltage regulator. It can be in terms of input (line) regulation or load regulation. **Line regulation** specifies how much change occurs in the output voltage for a given change in the input voltage. It is typically defined as a ratio of a change in output voltage for a corresponding change in the input voltage expressed as a percentage.

Equation 2–14

$$\text{Line regulation} = \left(\frac{\Delta V_{\text{OUT}}}{\Delta V_{\text{IN}}} \right) 100\%$$

Load regulation specifies how much change occurs in the output voltage over a certain range of load current values, usually from minimum current (no load, NL) to maximum current (full load, FL). It is normally expressed as a percentage and can be calculated with the following formula:

Equation 2–15

$$\text{Load regulation} = \left(\frac{V_{\text{NL}} - V_{\text{FL}}}{V_{\text{FL}}} \right) 100\%$$

where V_{NL} is the output voltage with no load and V_{FL} is the output voltage with full (maximum) load.

EXAMPLE 2–8

Assume a certain 7805 regulator has a measured no-load output voltage of 5.185 V and a full-load output of 5.152 V. What is the load regulation expressed as a precentage?

Solution

$$\text{Load regulation} = \left(\frac{V_{\text{NL}} - V_{\text{FL}}}{V_{\text{FL}}} \right) 100\% = \left(\frac{5.185 \text{ V} - 5.152 \text{ V}}{5.152 \text{ V}} \right) 100\% = \mathbf{0.64\%}$$

Related Problem

If the no-load output voltage of a regulator is 24.8 V and the full-load output is 23.9 V, what is the load regulation expressed as a percentage?

SECTION 2–3 REVIEW

1. When a 60 Hz sinusoidal voltage is applied to the input of a half-wave rectifier, what is the output frequency?

2. When a 60 Hz sinusoidal voltage is applied to the input of a full-wave rectifier, what is the output frequency?

3. What causes the ripple voltage on the output of a capacitor-input filter?

4. If the load resistance connected to a filtered power supply is decreased, what happens to the ripple voltage?

5. Define *ripple factor*.

6. What advantages are offered by a three-terminal regulator?

7. What is the difference between input (line) regulation and load regulation?

2–4 DIODE LIMITING AND CLAMPING CIRCUITS

Diode circuits, called limiters or clippers, are sometimes used to clip off portions of signal voltages above or below certain levels. Another type of diode circuit, called a clamper, is used to add or restore a dc level to an electrical signal. Both limiter and clamper diode circuits will be examined in this section.

After completing this section, you should be able to

- **Explain and analyze the operation of diode limiting and clamping circuits**
- Explain the operation of diode limiters
- Determine the output voltage of a biased limiter
- Use voltage-divider bias to set the limiting level
- Explain the operation of diode clampers

Diode Limiters

Figure 2–35(a) shows a diode **limiter** (also called **clipper**) that limits or clips the positive part of the input voltage. As the input voltage goes positive, the diode becomes forward-biased. Because the cathode is at ground potential (0 V), the anode cannot exceed 0.7 V (assuming silicon). So point A is limited to $+0.7$ V when the input voltage exceeds this value. When the input voltage goes back below 0.7 V, the diode is reverse-biased and appears as an open. The output voltage looks like the negative part of the input voltage, but with a magnitude determined by the voltage divider formed by R_1 and the load resistor, R_L, as follows:

$$V_{out} = \left(\frac{R_L}{R_1 + R_L}\right)V_{in}$$

If R_1 is small compared to R_L, then $V_{out} = V_{in}$.

Turn the diode around, as in Figure 2–35(b), and the negative part of the input voltage is clipped off. When the diode is forward-biased during the negative part of the input voltage, point A is held at -0.7 V by the diode drop. When the input voltage goes above

(a) Limiting of the positive alternation. The diode is forward-biased during the positive alternation (above 0.7 V) and reverse-biased during the negative alternation.

(b) Limiting of the negative alternation. The diode is forward-biased during the negative alternation (below −0.7 V) and reverse-biased during the positive alternation.

▲ **FIGURE 2–35**

Examples of diode limiters (clippers).

−0.7 V, the diode is no longer forward-biased; and a voltage appears across R_L proportional to the input voltage.

EXAMPLE 2–9

What would you expect to see displayed on an oscilloscope connected across R_L in the limiter shown in Figure 2–36?

▲ **FIGURE 2–36**

Solution The diode is forward-biased and conducts when the input voltage goes below −0.7 V. So, for the negative limiter, the peak output voltage across R_L can be determined by the following equation:

$$V_{p(out)} = \left(\frac{R_L}{R_1 + R_L}\right)V_{p(in)} = \left(\frac{1.0 \text{ k}\Omega}{1.1 \text{ k}\Omega}\right)10 \text{ V} = 9.09 \text{ V}$$

The scope will display an output waveform as shown in Figure 2–37.

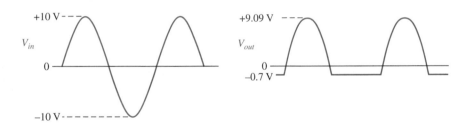

▲ **FIGURE 2–37**

Waveforms for Figure 2–36.

Related Problem Describe the output waveform for Figure 2–36 if the diode is germanium and R_L is changed to 680 Ω.

Open file E02-09 in the Examples folder on your CD-ROM. For the specified input, measure the resulting output waveform. Compare with the waveform shown in the example.

Biased Limiters The level to which an ac voltage is limited can be adjusted by adding a bias voltage, V_{BIAS}, in series with the diode, as shown in Figure 2–38. The voltage at point A must equal $V_{BIAS} + 0.7$ V before the diode will become forward-biased and conduct. Once the diode begins to conduct, the voltage at point A is limited to $V_{BIAS} + 0.7$ V so that all input voltage above this level is clipped off.

▲ **FIGURE 2–38**

A positive limiter.

▲ **FIGURE 2–39**

A negative limiter.

To limit a voltage to a specified negative level, the diode and bias voltage must be connected as in Figure 2–39. In this case, the voltage at point A must go below $-V_{BIAS} - 0.7$ V to forward-bias the diode and initiate limiting action as shown.

By turning the diode around, the positive limiter can be modified to limit the output voltage to the portion of the input voltage waveform above $V_{BIAS} - 0.7$ V, as shown by the output waveform in Figure 2–40(a). Similarly, the negative limiter can be modified to limit the output voltage to the portion of the input voltage waveform below $-V_{BIAS} + 0.7$ V, as shown by the output waveform in part (b).

(a)

(b)

▲ **FIGURE 2–40**

EXAMPLE 2–10

Figure 2–41 shows a circuit combining a positive limiter with a negative limiter. Determine the output voltage waveform.

▲ **FIGURE 2–41**

Solution
When the voltage at point A reaches $+5.7$ V, diode D_1 conducts and limits the waveform to $+5.7$ V. Diode D_2 does not conduct until the voltage reaches -5.7 V. Therefore, positive voltages above $+5.7$ V and negative voltages below -5.7 V are clipped off. The resulting output voltage waveform is shown in Figure 2–42.

▶ **FIGURE 2–42**

Output voltage waveform for Figure 2–41.

Related Problem
Determine the output voltage waveform in Figure 2–41 if both dc sources are 10 V and the input voltage has a peak value of 20 V.

Open file E02-10 in the Examples folder on your CD-ROM. For the specified input, measure the resulting output waveform. Compare with the waveform shown in the example.

Voltage-Divider Bias The bias voltage sources that have been used to illustrate the basic operation of diode limiters can be replaced by a resistive voltage divider that derives the desired bias voltage from the dc supply voltage, as shown in Figure 2–43. The bias voltage is set by the resistor values according to the voltage-divider formula.

$$V_{\text{BIAS}} = \left(\frac{R_3}{R_2 + R_3}\right)V_{\text{SUPPLY}}$$

A positively biased limiter is shown in Figure 2–43(a), a negatively biased limiter is shown in part (b), and a variable positive bias circuit using a potentiometer voltage divider is shown in part (c). The bias resistors must be small compared to R_1 so that the forward current through the diode will not affect the bias voltage.

(a) Positive limiter

(b) Negative limiter

(c) Variable positive limiter

▲ **FIGURE 2–43**

Diode limiters implemented with voltage-divider bias.

EXAMPLE 2–11

Describe the output voltage waveform for the diode limiter in Figure 2–44.

▲ **FIGURE 2–44**

Solution The circuit is a positive limiter and the bias voltage is determined using the voltage-divider formula.

$$V_{BIAS} = \left(\frac{R_3}{R_2 + R_3}\right)V_{SUPPLY} = \left(\frac{220\ \Omega}{100\ \Omega + 220\ \Omega}\right)12\ V = 8.25\ V$$

The output voltage waveform is shown in Figure 2–45. The positive part of the output voltage waveform is limited to $V_{BIAS} + 0.7$ V.

▶ **FIGURE 2–45**

Related Problem Change the voltage divider in Figure 2–44 to limit the output voltage to +6.7 V.

Open file E02-11 in the Examples folder on your CD-ROM. Observe the output voltage on the oscilloscope and compare to the calculated result.

Diode Clampers

A clamper adds a dc level to an ac voltage. **Clampers** are sometimes known as *dc restorers*. Figure 2–46 shows a diode clamper that inserts a positive dc level in the output waveform. The operation of this circuit can be seen by considering the first negative half-cycle of the input voltage. When the input voltage initially goes negative, the diode is forward-biased, allowing the capacitor to charge to near the peak of the input ($V_{p(in)} - 0.7$ V), as shown in Figure 2–46(a). Just after the negative peak, the diode is reverse-biased. This is because the cathode is held near $V_{p(in)} - 0.7$ V by the charge on the capacitor. The capacitor can only discharge through the high resistance of R_L. So, from the peak of one negative half-cycle to the next, the capacitor discharges very little. The amount that is discharged, of course, depends on the value of R_L. For good clamping action, the *RC* time constant should be at least ten times the period of the input frequency.

(a)

(b)

▲ **FIGURE 2–46**

Positive clamper operation.

The net effect of the clamping action is that the capacitor retains a charge approximately equal to the peak value of the input less the diode drop. The capacitor voltage acts essentially as a battery in series with the input voltage. The dc voltage of the capacitor adds to the input voltage by superposition, as in Figure 2–46(b). If the diode is turned around, a negative dc voltage is added to the input voltage to produce the output voltage as shown in Figure 2–47.

▲ **FIGURE 2–47**

Negative clamper.

A Clamper Application   A clamping circuit is often used in television receivers as a dc restorer. The incoming composite video signal is normally processed through capacitively coupled amplifiers that eliminate the dc component, thus losing the black and white reference levels and the blanking level. Before being applied to the picture tube, these reference levels must be restored.

EXAMPLE 2–12

What is the output voltage that you would expect to observe across R_L in the clamping circuit of Figure 2–48? Assume that RC is large enough to prevent significant capacitor discharge.

▲ **FIGURE 2–48**

Solution   Ideally, a negative dc value equal to the input peak less the diode drop is inserted by the clamping circuit.

$$V_{DC} \cong -(V_{p(in)} - 0.7 \text{ V}) = -(24 \text{ V} - 0.7 \text{ V}) = -23.3 \text{ V}$$

Actually, the capacitor will discharge slightly between peaks, and, as a result, the output voltage will have an average value of slightly less than that calculated above. The output waveform goes to approximately +0.7 V, as shown in Figure 2–49.

▶ **FIGURE 2–49**

Output waveform across R_L for Figure 2–48.

Related Problem   What is the output voltage that you would observe across R_L in Figure 2–48 for $C = 22 \ \mu\text{F}$ and $R_L = 18 \text{ k}\Omega$?

Open file E02-12 in the Examples folder on your CD-ROM. For the specified input, measure the output waveform. Compare with the waveform shown in the example.

**SECTION 2–4
REVIEW**

1. Discuss how diode limiters and diode clampers differ in terms of their function.
2. What is the difference between a positive limiter and a negative limiter?
3. What is the maximum voltage across an unbiased positive silicon diode limiter during the positive alternation of the input voltage?
4. To limit the output voltage of a positive limiter to 5 V when a 10 V peak input is applied, what value must the bias voltage be?
5. What component in a clamping circuit effectively acts as a battery?

2–5 VOLTAGE MULTIPLIERS

Voltage multipliers use clamping action to increase peak rectified voltages without the necessity of increasing the input transformer's voltage rating. Multiplication factors of two, three, and four are common. Voltage multipliers are used in high-voltage, low-current applications such as TV receivers.

After completing this section, you should be able to

■ **Explain and analyze the operation of diode voltage multipliers**

■ Discuss voltage doublers

■ Discuss voltage triplers

■ Discuss voltage quadruplers

Voltage Doubler

Half-Wave Voltage Doubler A voltage doubler is a **voltage multiplier** with a multiplication factor of two. A half-wave voltage doubler is shown in Figure 2–50. During the positive half-cycle of the secondary voltage, diode D_1 is forward-biased and D_2 is reverse-biased. Capacitor C_1 is charged to the peak of the secondary voltage (V_p) less the diode drop with the polarity shown in part (a). During the negative half-cycle, diode D_2 is forward-biased and D_1 is reverse-biased, as shown in part (b). Since C_1 can't discharge, the peak voltage on C_1 adds to the secondary voltage to charge C_2 to approximately $2V_p$. Applying Kirchhoff's law around the loop as shown in part (b), the voltage across C_2 is

$$V_{C1} - V_{C2} + V_p = 0$$
$$V_{C2} = V_p + V_{C1}$$

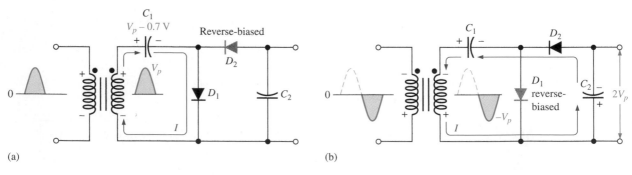

(a)   (b)

▲ **FIGURE 2–50**

Half-wave voltage doubler operation. V_p is the peak secondary voltage.

Neglecting the diode drop of D_2, $V_{C1} = V_p$. Therefore,

$$V_{C2} = V_p + V_p = 2V_p$$

Under a no-load condition, C_2 remains charged to approximately $2V_p$. If a load resistance is connected across the output, C_2 discharges slightly through the load on the next positive half-cycle and is again recharged to $2V_p$ on the following negative half-cycle. The resulting output is a half-wave, capacitor-filtered voltage. The peak inverse voltage across each diode is $2V_p$.

Full–Wave Voltage Doubler A full-wave doubler is shown in Figure 2–51. When the secondary voltage is positive, D_1 is forward-biased and C_1 charges to approximately V_p, as shown in part (a). During the negative half-cycle, D_2 is forward-biased and C_2 charges to approximately V_p, as shown in part (b). The output voltage, $2V_p$, is taken across the two capacitors in series.

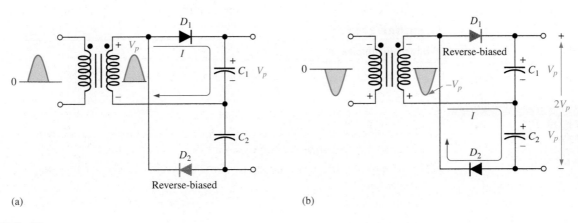

(a)   (b)

▲ FIGURE 2–51

Full-wave voltage doubler operation.

Voltage Tripler

The addition of another diode-capacitor section to the half-wave voltage doubler creates a voltage tripler, as shown in Figure 2–52. The operation is as follows: On the positive half-cycle of the secondary voltage, C_1 charges to V_p through D_1. During the negative half-cycle, C_2 charges to $2V_p$ through D_2, as described for the doubler. During the next positive half-cycle, C_3 charges to $2V_p$ through D_3. The tripler output is taken across C_1 and C_3, as shown in the figure.

▶ FIGURE 2–52

Voltage tripler.

Voltage Quadrupler

The addition of still another diode-capacitor section, as shown in Figure 2–53, produces an output four times the peak secondary voltage. C_4 charges to $2V_p$ through D_4 on a negative half-cycle. The $4V_p$ output is taken across C_2 and C_4, as shown. In both the tripler and quadrupler circuits, the PIV of each diode is $2V_p$.

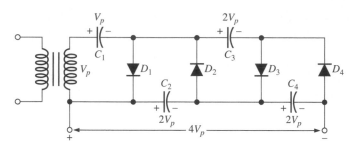

▲ **FIGURE 2–53**

Voltage quadrupler.

SECTION 2–5 REVIEW

1. What must be the peak voltage rating of the transformer secondary for a voltage doubler that produces an output of 200 V?

2. The output voltage of a quadrupler is 620 V. What minimum PIV rating must each diode have?

2–6 THE DIODE DATA SHEET

A manufacturer's data sheet gives detailed information on a device so that it can be used properly in a given application. A typical data sheet provides maximum ratings, electrical characteristics, mechanical data, and graphs of various parameters. In this section, we use a specific example to illustrate a typical data sheet.

After completing this section, you should be able to

- **Interpret and use a diode data sheet**
- Identify maximum voltage and current ratings
- Determine the electrical characteristics of a diode
- Analyze graphical data
- Select an appropriate diode for a given set of specifications

Table 2–1 shows the maximum ratings for a certain series of rectifier diodes (1N4001 through 1N4007). These are the absolute maximum values under which the diode can be operated without damage to the device. For greatest reliability and longer life, the diode should always be operated well under these maximums. Generally, the maximum ratings are specified at 25°C and must be adjusted downward for higher temperatures.

▼ TABLE 2–1

Maximum ratings.

RATING	SYMBOL	1N4001	1N4002	1N4003	1N4004	1N4005	1N4006	1N4007	UNIT
Peak repetitive reverse voltage	V_{RRM}								
Working peak reverse voltage	V_{RWM}	50	100	200	400	600	800	1000	V
DC blocking voltage	V_R								
Nonrepetitive peak reverse voltage	V_{RSM}	60	120	240	480	720	1000	1200	V
rms reverse voltage	$V_{R(rms)}$	35	70	140	280	420	560	700	V
Average rectified forward current (single-phase, resistive load, 60 Hz, $T_A = 75°C$)	I_O	1.0							A
Nonrepetitive peak surge current (surge applied at rated load conditions)	I_{FSM}	30 (for 1 cycle)							A
Operating and storage junction temperature range	T_J, T_{stg}	−65 to +175							°C

An explanation of the parameters from Table 2–1 follows.

V_{RRM} The maximum reverse peak voltage that can be applied repetitively across the diode. Notice that in this case, it is 50 V for the 1N4001 and 1 kV for the 1N4007. This is the same as PIV rating.

V_R The maximum reverse dc voltage that can be applied across the diode.

V_{RSM} The maximum reverse peak value of nonrepetitive voltage that can be applied across the diode.

I_O The maximum average value of a 60 Hz rectified forward current.

I_{FSM} The maximum peak value of nonrepetitive (one cycle) forward surge current. The graph in Figure 2–54 expands on this parameter to show values for more than one cycle at temperatures of 25°C and 175°C. The dashed lines represent values where typical failures occur.

T_A Ambient temperature (temperature of surrounding air).

T_J The operating junction temperature.

T_{stg} The storage junction temperature.

Table 2–2 lists typical and maximum values of certain electrical characteristics. These items differ from the maximum ratings in that they are not selected by design but are the

▲ **FIGURE 2–54**

Nonrepetitive forward surge current capability.

▼ **TABLE 2–2**

Electrical characteristics.

CHARACTERISTICS AND CONDITIONS	SYMBOL	TYPICAL	MAXIMUM	UNIT
Maximum instantaneous forward voltage drop ($I_F = 1$ A, $T_J = 25°C$	v_F	0.93	1.1	V
Maximum full-cycle average forward voltage drop ($I_O = 1$ A, $T_L = 75°C$, 1 inch leads)	$V_{F(avg)}$	—	0.8	V
Maximum reverse current (rated dc voltage) $T_J = 25°C$ $T_J = 100°C$	I_R	 0.05 1.0	 10.0 50.0	μA
Maximum full-cycle average reverse current ($I_O = 1$ A, $T_L = 75°C$, 1 inch leads)	$I_{R(avg)}$	—	30.0	μA

result of operating the diode under specified conditions. A brief explanation of these parameters follows.

v_F The instantaneous voltage across the forward-biased diode when the forward current is 1 A at 25°C. Figure 2–55 shows how the forward voltages vary with forward current.

$V_{F(avg)}$ The maximum forward voltage drop averaged over a full cycle.

I_R The maximum current when the diode is reverse-biased with a dc voltage.

$I_{R(avg)}$ The maximum reverse current averaged over one cycle (when reverse-biased with an ac voltage).

T_L The lead temperature.

Figure 2–56 shows a selection of rectifier diodes arranged in order of increasing I_O, I_{FSM}, and V_{RRM} ratings.

▲ **FIGURE 2–55**

Forward voltage (V_F) versus forward current (I_F).

V_{RRM} (Volts)	I_O, Average Rectified Forward Current (Amperes)					
	1.0	**1.5**	**3.0**			**6.0**
	59-03 (DO-41) Plastic	59-04 Plastic	60-01 Metal	267-03 Plastic	267-02 Plastic	194-04 Plastic
50	1N4001	1N5391	1N4719	MR500	1N5400	MR750
100	1N4002	1N5392	1N4720	MR501	1N5401	MR751
200	1N4003	1N5393 MR5059	1N4721	MR502	1N5402	MR752
400	1N4004	1N5395 MR5060	1N4722	MR504	1N5404	MR754
600	1N4005	1N5397 MR5061	1N4723	MR506	1N5406	MR756
800	1N4006	1N5398	1N4724	MR508		MR758
1000	1N4007	1N5399	1N4725	MR510		MR760
I_{FSM} (Amps)	30	50	300	100	200	400
T_A @ Rated I_O (°C)	75	$T_L = 70$	75	95	$T_L = 105$	60
T_C @ Rated I_O (°C)						
T_J (Max) (°C)	175	175	175	175	175	175

V_{RRM} (Volts)	I_O, Average Rectified Forward Current (Amperes)										
	12	**20**	**24**	**25**	**30**		**40**	**50**	**25**	**35**	**40**
	245A-02 (DO-203AA) Metal		339-02 Plastic	193-04 Plastic	43-02 (DO-21) Metal		42A-01 (DO-203AB) Metal	43-04 Metal	309A-03	309A-02	
50	MR1120 1N1199,A,B	MR2000	MR2400	MR2500	1N3491	1N3659	1N1183A	MR5005	MDA2500	MDA3500	
100	MR1121 1N1200,A,B	MR2001	MR2401	MR2501	1N3492	1N3660	1N1184A	MR5010	MDA2501	MDA3501	
200	MR1122 1N1202,A,B	MR2002	MR2402	MR2502	1N3493	1N3661	1N1186A	MR5020	MDA2502	MDA3502	MDA4002
400	MR1124 1N1204,A,B	MR2004	MR2404	MR2504	1N3495	1N3663	1N1188A	MR5040	MDA2504	MDA3504	MDA4004
600	MR1126 1N1206,A,B	MR2006	MR2406	MR2506			1N1190A		MDA2506	MDA3506	MDA4006
800	MR1128	MR2008		MR2508					MDA2508	MDA3508	MDA4008
1000	MR1130	MR2010		MR2510					MDA2510	MDA3510	
I_{FSM} (Amps)	300	400	400	400	300	400	800	600	400	400	800
T_A @ Rated I_O (°C)											
T_C @ Rated I_O (°C)	150	150	125	150	130	100	150	150	55	55	35
T_J (Max) (°C)	190	175	175	175	175	175	190	195	175	175	175

▲ **FIGURE 2–56**

A selection of rectifier diodes based on maximum ratings of I_O, I_{FSM}, and V_{RRM}.

SECTION 2–6 REVIEW

1. List the three rating categories typically given on all diode data sheets.
2. Define each of the following parameters: V_F, I_R, I_O.
3. Define I_{FSM}, V_{RRM}, and V_{RSM}.
4. From Figure 2–56, select a diode to meet the following specifications: $I_O = 3$ A, $I_{FSM} = 300$ A, and $V_{RRM} = 100$ V.

2–7   TROUBLESHOOTING

This section provides a general introduction to troubleshooting techniques that will apply throughout the book as well as specific troubleshooting examples of the power supply and other diode circuits you have studied in this chapter.

After completing this section, you should be able to

- **Troubleshoot diode circuits using accepted techniques**
- Discuss the relationship between symptom and cause
- Discuss and apply the power check
- Discuss and apply the sensory check
- Discuss and apply the component replacement method
- Discuss and apply the signal tracing technique in its three variations
- Explain fault analysis

You can approach the **troubleshooting** of a defective electronic circuit or group of circuits (system) in several ways. We will define a defective circuit or system as one with a known good input voltage but with no output voltage or an incorrect output voltage. For example, in Figure 2–57(a), a properly functioning dc power supply is represented by a single block with a known input voltage and a correct output voltage. A defective dc power supply is represented in part (b) as a block with an input voltage but no output voltage.

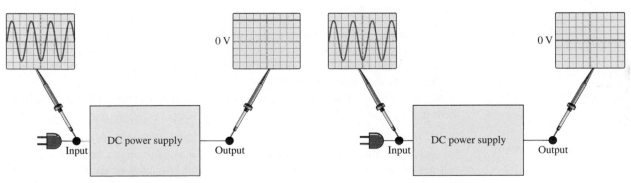

(a) The correct dc output voltage is measured.   (b) 0 V dc is measured at the output. The input voltage is correct.

▲ **FIGURE 2–57**

Block representations of functioning and nonfunctioning power supplies.

The Symptom and the Cause

The first step in troubleshooting a defective circuit or system is to identify the symptom. In the case of our power supply example, this is easy. The symptom is that there is no output voltage. In other situations, symptom identification may be a more crucial step.

The absence of an output voltage from the power supply does not tell much about what the specific cause may be. In some cases, however, a particular symptom may point to a given area where a fault is most likely to have occurred and give you an idea of where to begin to find the cause of the problem.

Troubleshooting Techniques

Several methods can be used to troubleshoot a defective circuit or system. The one to use depends on the type and complexity of the circuit or system, the nature of the problem, and the preference of the individual technician.

Power Check   The first thing you should do in checking out a defective circuit is to make sure the power cord is plugged in and the fuse is not burned out. In the case of battery-powered systems, make sure the battery is good. Something this simple is often the cause of the problem.

Sensory Check   Beyond the power check, the simplest troubleshooting method relies on your senses of observation to check for obvious defects. For example, a burned resistor is often visible as are broken wires, poor solder connections, and burned out fuses. Also, when certain types of components fail, you may be able to detect a smell of smoke if you happen to be there when it fails or shortly afterward.

Since some failures are temperature dependent, you can sometimes unplug the circuit and immediately use your sense of touch to detect an overheated component. This is an intermittent type of failure because the circuit works properly for a while and then fails when it heats up. As a rule, always perform a sensory check before proceeding with more sophisticated troubleshooting methods. Never touch an operating circuit because there may be a risk of burn or electrical shock.

Component Replacement   This method relies largely on educated guesswork based on the symptom or symptoms. Based on your knowledge of the circuit operation, a certain symptom may indicate that a particular component in the circuit is defective. Using this method, you would change the suspected component and test the circuit for proper operation. If you were wrong, you then have to select the next most likely component.

This process can be very quick if you get it right the first time or it can be very time consuming and expensive if you don't. Obviously, component replacement is guaranteed to fix the problem because if you replace enough components, you will eventually find the defective one. This is not a recommended method unless the symptom/cause relationship is glaringly obvious.

Signal Tracing   This troubleshooting technique is very popular and effective. Basically, in signal tracing you look for a point in a circuit or system where you first lose the signal or an incorrect signal first occurs. A point in a circuit where a known voltage can be measured is called a *test point* (TP).

Now, we will discuss three variations of the signal tracing technique and illustrate them in relation to finding a fault in the dc power supply.

Signal tracing method 1:   Start at the input of a circuit where there is a known input signal and work toward the output. Check the signal voltage at successive test points until you get an *incorrect* measurement. When you find no signal or an incorrect signal, you have isolated the problem to the portion of the circuit between the last test point and the present test point. Of course, you must know how the signal is supposed to change as it goes through the circuit or have a schematic with correct voltages marked at the test points. This approach is illustrated in Figure 2–58 for the power supply, which is shown

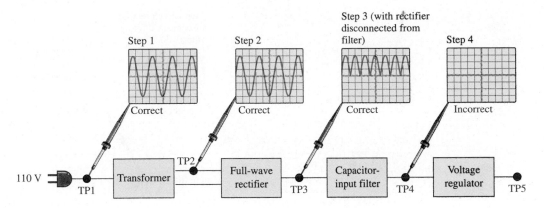

▲ FIGURE 2–58

Example of signal tracing: input to output.

as four functional blocks. The measurements indicate that the fault is in the filter circuit because the input to the filter (TP3) is the correct full-wave rectified voltage but there is no voltage on the output (TP4).

Signal tracing method 2: Start at the output of a circuit and work toward the input. Check for voltage at each test point until you get a *correct* measurement. At this point, you have isolated the problem to the portion of the circuit between the last test point and the current test point. This approach is illustrated in Figure 2–59. Again, the measurements indicate a faulty filter circuit because there is no output voltage from the filter (TP4) but the input is correct (TP3).

Signal tracing method 3: This method is called *half-splitting*. Start in the middle of the circuit. If the beginning test point has a correct signal, you know that the circuit is working properly from the input to that test point. This means the fault is somewhere between the test point and the output; therefore, begin signal tracing from the test point toward the output. If the beginning test point has no signal or an incorrect signal, you know that the fault is between the input and the test point; therefore, begin signal tracing from the test point toward the input. Figure 2–60 illustrates this approach for the power supply. The correct voltage at TP3 indicates that the transformer and rectifier are functioning properly. The measurement of 0 V at TP4 isolates the fault to the filter.

For this example, the half-splitting method took two steps compared to four for method 1 and three for method 2. There is some advantage in method 3 especially for more complex systems because, generally, you will find the problem more quickly.

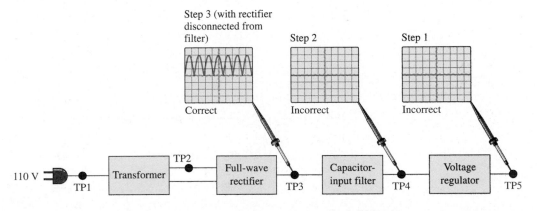

▲ FIGURE 2–59

Example of signal tracing: output to input.

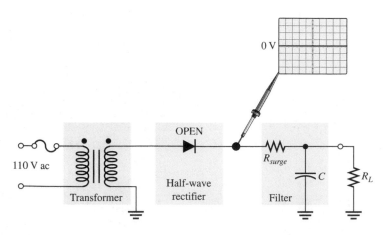

▲ **FIGURE 2–60**

Example of signal tracing: half-splitting.

Signal tracing has isolated the problem in our defective power supply to the filter circuit. The next step is to determine the defective component in the filter. The filter in this example consists of a capacitor and a small-value surge resistor.

In this simple circuit the component replacement method is appropriate. Since we have found that the output voltage is zero for a correct input voltage, the surge resistor must be open. It is possible that the filter capacitor is shorted, but that is less likely to be the case.

Fault Analysis

Fault analysis is an area of troubleshooting that can best be described by this question: *If component X fails in circuit Y, what will be the symptom(s)?*

Fault analysis can be applied when you measure an incorrect voltage at a test point using signal tracing and isolate the fault to a specific circuit. You can perform fault analysis on a circuit using your knowledge of circuit operation and the effects of component failures on that operation to determine which component, if defective, would cause the incorrect voltage at the test point. We will discuss fault analysis by using the power supply circuits and common component failures as examples.

Effect of an Open Diode in a Half-Wave Rectifier A half-wave filtered rectifier with an open diode is shown in Figure 2–61. The resulting symptom is zero output voltage as

▲ **FIGURE 2–61**

The effect of an open diode in a half-wave rectifier is an output of 0 V.

indicated. This is obvious because the open diode breaks the current path from the transformer secondary winding to the filter and load resistor and there is no load current.

Other faults that will cause the same symptom in this circuit are an open transformer winding, an open fuse, or no input voltage.

Effect of an Open Diode in a Full-Wave Rectifier A center-tapped full-wave filtered rectifier is shown in Figure 2–62. If either of the two diodes is open, the output voltage will have a larger than normal ripple voltage at 60 Hz rather than at 120 Hz, as indicated.

▲ **FIGURE 2–62**

The effect of an open diode in a center-tapped rectifier is half-wave rectification and a larger ripple voltage at 60 Hz.

Another fault that will cause the same symptom is an open in one of the halves of the transformer secondary winding.

The reason for the increased ripple at 60 Hz rather than 120 Hz is as follows. If one of the diodes in Figure 2–62 is open, there is current through R_L only during one-half of a cycle of the input voltage. During the other half-cycle of the input, the open path caused by the open diode prevents current through R_L. The result is half-wave rectification, as shown in Figure 2–62, which produces the larger ripple voltage with a frequency of 60 Hz.

An open diode in a bridge full-wave rectifier will produce the same symptom as in the center-tapped circuit, as shown in Figure 2–63. The open diode prevents current through R_L during half of the input voltage cycle. The result is half-wave rectification, which produces an increase in ripple voltage at 60 Hz.

Effect of a Shorted Diode in a Full-Wave Rectifier A shorted diode is one that has failed such that it has a very low resistance in both directions. If a diode suddenly shorts in a bridge rectifier, there will be an excessively high current during one-half of the input voltage cycle, as illustrated in Figure 2–64 with diode D_1 shorted. This condition will most likely burn open one of the diodes if the circuit is not properly fused. During the positive half-cycle of the input voltage as shown in Figure 2–64(a), the path of the load current is through the shorted diode D_1 just the same as if it were forward-biased. During the negative half-cycle of the input, the current is shorted through D_1 and D_4 as shown in

▲ **FIGURE 2–63**

Effect of an open diode in a bridge rectifier.

(a) Positive half-cycle: The shorted diode acts as a forward-biased diode, so the load current is normal.

(b) Negative half-cycle: The shorted diode produces a short circuit across the source. As a result the fuse should blow. If not properly fused, D_1, D_4, or the transformer secondary will probably burn open.

▲ **FIGURE 2–64**

Effect of a shorted diode in a bridge rectifier.

part (b). This is a very low resistance path because there is a forward-biased diode and a shorted diode in series. It is likely that the excessive current will burn either or both of the diodes open if the circuit is not properly fused.

If only one of the diodes, D_1 or D_4, should burn open, the circuit will operate as a half-wave rectifier. If both diodes should burn open or if the excessive current causes the **fuse** to blow (which it should), the output voltage is zero.

Effects of a Faulty Filter Capacitor   Three types of defects of a filter capacitor are illustrated in Figure 2–65.

■ *Open*   If the filter capacitor for a full-wave rectifier opens, the output is a full-wave rectified voltage, as shown in the figure.

■ *Shorted*   If the filter capacitor shorts, the output is 0 V. A shorted capacitor may cause some or all of the diodes in the rectifier to burn open due to excessive current if there is no surge resistor and if the rectifier is not properly fused. If the rectifier is properly fused, the fuse should open and prevent damage to the circuit. In any event, the output is 0 V, as shown in the figure.

■ *Leaky*   A leaky capacitor is equivalent to a capacitor with a parallel leakage resistance. The effect of the leakage resistance is to reduce the time constant and allow the capacitor to discharge more rapidly than normal. This results in an increase in the ripple voltage on the output, as shown in the figure.

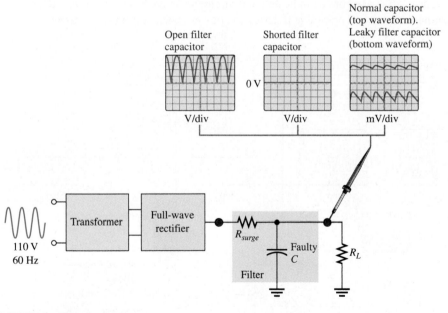

▲ **FIGURE 2–65**

Effects of a faulty filter capacitor.

Effects of a Faulty Transformer   An open primary or secondary winding of a power supply transformer results in an output of 0 V, as mentioned before.

A partially shorted primary winding (which is less likely than an open) results in an increased rectifier output voltage because the turns ratio of the transformer is effectively increased. A partially shorted secondary winding results in a decreased rectifier output voltage because the turns ratio is effectively decreased.

The Complete Troubleshooting Process

Beginning with a malfunctioning or nonfunctioning circuit or system, a typical troubleshooting procedure may go as follows:

1. Identify the symptom(s).

2. Perform a power check.

3. Perform a sensory check.

4. Apply a signal tracing technique to isolate the fault to a single circuit.

5. Apply fault analysis to isolate the fault further to a single component or group of components.

6. Use component replacement to fix the problem.

The following example illustrates the steps and thought process for troubleshooting the power supply.

EXAMPLE 2–13

Determine the faulty component in the power supply represented by the block diagram in Figure 2–66 based on the indicated measurements in each step of the procedure. Assume that you find no problem during the power check and sensory check.

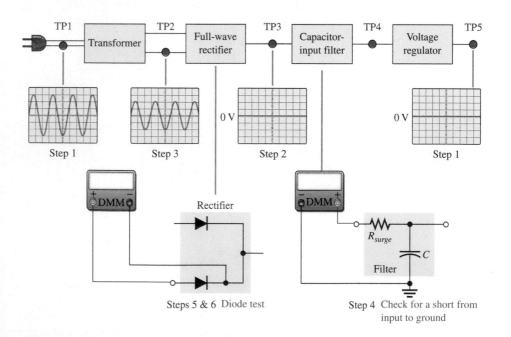

▲ **FIGURE 2–66**

Solution　The step-by-step procedure in reference to Figure 2–66 is as follows:

Step 1: Check the input and output voltages. The measurements indicate that the input voltage is correct but there is no output voltage.

Step 2: Apply the half-splitting method of signal tracing. There is no voltage at TP3. This indicates that the fault is between the input and TP3. It has to be the transformer or the rectifier, or does it? Since the voltage at TP3 is 0 V, there may be a short from the input of the filter to ground.

Step 3: Check the voltage at TP2. If there is no voltage, then the transformer is defective. If the correct voltage appears at TP2, the problem has to be in the rectifier or a shorted filter input. The voltage at TP2 is correct. This narrows it down to a faulty rectifier or a shorted filter input.

Step 4: With the power turned off, use a DMM to perform a check for a short from the filter input to ground. Assume that the DMM indicates no short. The fault is now isolated to the rectifier circuit.

Step 5: Apply fault analysis. Determine the component failures in the rectifier that will produce an output of 0 V. If only one of the diodes in the rectifier is open, there should be a half-wave rectified output voltage, so this is not the problem. In order to have zero volts output, **both of the diodes must be open.**

Step 6: With the power turned off, use a DMM in the diode test mode to check each diode. Replace the defective diodes, turn on the power, and check for proper operation.

Related Problem If you had found a short in Step 4, what would have been the logical next step?

EWB/Multisim Troubleshooting Exercises

These file circuits are in the Troubleshooting Exercises folder on your CD-ROM.

1. Open file TSE02-01. Determine if the circuit is working properly and, if not, determine the fault.

2. Open file TSE02-02. Determine if the circuit is working properly and, if not, determine the fault.

3. Open file TSE02-03 (EWB only). Determine if the circuit is working properly and, if not, determine the fault.

SECTION 2–7 REVIEW

1. What effect does an open diode have on the output voltage of a half-wave rectifier?
2. What effect does an open diode have on the output voltage of a full-wave rectifier?
3. If one of the diodes in a bridge rectifier shorts, what are some possible consequences?
4. What happens to the output voltage of a rectifier if the filter capacitor becomes very leaky?
5. The primary winding of the transformer in a power supply opens. What will you observe on the rectifier output?
6. The dc output voltage of a filtered rectifier is less than it should be. What may be the problem?

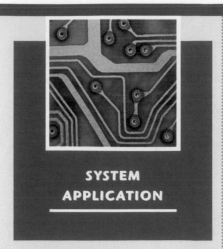

SYSTEM APPLICATION

Assume you are working for an electronics manufacturing company as an electronics technician. Your company designs, manufactures, and tests a wide variety of products. Your first on-the-job assignment is to develop and test a dc power supply that will be used in several different products such as an industrial counting system, a security alarm system, a voice intercom system, and a public address system. You will apply the knowledge you have gained to this point in completing your assignment.

The Power Supply Circuit

The dc power supply that will be produced by your company is shown in Figure 2–67. This is to be a universal design that can be used in several different systems with some modifications. In its basic form, this power supply is to be unregulated. Later modifications will include a regulator. The basic specifications are as follows:

1. Input voltage: 115 V rms at 60 Hz

2. Output voltage (unregulated): 12 V dc ± 10%

3. Maximum ripple factor: 3%

4. Maximum load current: 250 mA

The Components

■ *Transformer* Transformers are usually specified according to their rms output (secondary) voltage. Based on the power supply specifications, select a power transformer from the following list: 24 V, 18 V, 12.6 V, 12 V, 9 V, and 6.3 V.

■ *Diodes* Select a diode type to be used in the rectifier bridge with minimum I_O, I_{FSM}, and V_{RRM} required to do the job. Refer to the diode selection chart in Figure 2–56.

■ *Surge resistor* Determine a value of surge resistor based on the I_{FSM} of the selected diodes. Use the next higher standard value but keep it as small as possible to minimize voltage drop. Refer to the table of standard values in Appendix A.

■ *Primary fuse* Determine the smallest fuse rating that will handle the normal current but will blow if the load current exceeds its normal maximum or if the rectifier output is shorted. You can determine the turns ratio from the voltage ratings and use the basic formula learned in ac circuits to find the primary current. Select from the following standard fuse values: 250 mA, 500 mA, 1 A, 2 A, 3 A, 5 A, 7.5 A, and 10 A.

■ *Filter capacitor* Select a minimum value of filter capacitor to meet the ripple factor requirements with a

maximum load current (minimum load resistance). Choose from the following standard value electrolytics: 1 µF, 4.7 µF, 10 µF, 33 µF, 47 µF, 100 µF, 220 µF, 470 µF, 1000 µF, 1500 µF, 3300 µF, 4700 µF, 6800 µF, and 10,000 µF.

The Schematic

Complete the preliminary schematic in Figure 2–67 by adding component values and/or part numbers.

The Printed Circuit Board

■ Check out the printed circuit board in Figure 2–68 to verify that it is correct according to the schematic. All black resistor bands indicate an unspecified value that you have determined.

■ Label a copy of the board with the component and input/output designations in agreement with the schematic.

A Test Procedure

Develop a step-by-step set of instructions on how to completely check the power supply board for proper operation using the test points (circled numbers) indicated in the test bench setup of Figure 2–69. Specify voltage values and appropriate waveforms for all the measurements to be made. Provide a fault analysis for all possible component failures.

Troubleshooting

A preliminary manufacturing run of the assembled power supply boards are coming off the assembly line. Three boards have been found to be defective. Troubleshooting techniques must be used to determine the problems.

The test bench setup is shown in Figure 2–69. Based on the sequence of measurements for each board indicated in Figure 2–70, determine the most likely fault(s) in each case.

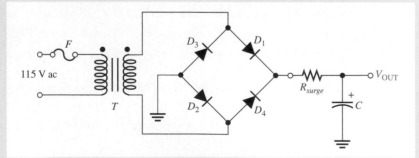

▲ **FIGURE 2–67**

Power supply preliminary schematic.

▲ **FIGURE 2–68**

Power supply printed circuit board.

V⯑ indicates dc/ac function

⯈⯇ indicates diode test

115 V

▲ **FIGURE 2–69**

Power supply test bench.

▲ FIGURE 2–70

Signal tracing sequences for three faulty power supply boards.

The circled numbers indicate test point connections to the circuit board. The DMM function setting is indicated below the display, and the volt/div and sec/div for the oscilloscope are shown on the screen in each case.

Final Report (Optional)

Submit a final written report on the power supply circuit board using an organized format that includes the following:

1. A physical description of the power supply circuit.

2. A discussion of the operation of the power supply.

3. A list of the specifications.

4. A list of parts with part numbers if available.

5. A list of the types of problems on the three faulty circuit boards.

6. A complete description of how you determined the problem on each of the three faulty circuit boards.

SUMMARY OF POWER SUPPLY RECTIFIERS

HALF-WAVE RECTIFIER

Output voltage waveform

- Peak value of output:

$$V_{p(out)} = V_{p(sec)} - 0.7\text{ V}$$

- Average value of output:

$$V_{AVG} = \frac{V_{p(out)}}{\pi}$$

- Diode peak inverse voltage:

$$\text{PIV} = V_{p(sec)}$$

CENTER-TAPPED FULL-WAVE RECTIFIER

Output voltage waveform

- Peak value of output:

$$V_{p(out)} = \frac{V_{p(sec)}}{2} - 0.7\text{ V}$$

- Average value of output:

$$V_{AVG} = \frac{2V_{p(out)}}{\pi}$$

- Diode peak inverse voltage:

$$\text{PIV} = 2V_{p(out)} + 0.7\text{ V}$$

BRIDGE FULL-WAVE RECTIFIER

Output voltage waveform

- Peak value of output:

$$V_{p(out)} = V_{p(sec)} - 1.4\text{ V}$$

- Average value of output:

$$V_{AVG} = \frac{2V_{p(out)}}{\pi}$$

- Diode peak inverse voltage:

$$\text{PIV} = V_{p(out)} + 0.7\text{ V}$$

CHAPTER SUMMARY

- The single diode in a half-wave rectifier is forward-biased and conducts for 180° of the input cycle.
- The output frequency of a half-wave rectifier equals the input frequency.
- PIV (peak inverse voltage) is the maximum voltage appearing across the diode in reverse bias.
- Each diode in a full-wave rectifier is forward-biased and conducts for 180° of the input cycle.
- The output frequency of a full-wave rectifier is twice the input frequency.
- The two basic types of full-wave rectifier are center-tapped and bridge.
- The peak output voltage of a center-tapped full-wave rectifier is approximately one-half of the total peak secondary voltage less one diode drop.
- The PIV for each diode in a center-tapped full-wave rectifier is twice the peak output voltage plus one diode drop.
- The peak output voltage of a bridge rectifier equals the total peak secondary voltage less two diode drops.
- The PIV for each diode in a bridge rectifier is approximately half that required for an equivalent center-tapped configuration and is equal to the peak output voltage plus one diode drop.
- A capacitor-input filter provides a dc output approximately equal to the peak of its rectified input voltage.
- Ripple voltage is caused by the charging and discharging of the filter capacitor.
- The smaller the ripple voltage, the better the filter.
- Regulation of output voltage over a range of input voltages is called *input* or *line regulation*.
- Regulation of output voltage over a range of load currents is called *load regulation*.
- Diode limiters cut off voltage above or below specified levels. Limiters are also called *clippers*.
- Diode clampers add a dc level to an ac voltage.
- A dc power supply typically consists of an input transformer, a diode rectifier, a filter, and a regulator.

KEY TERMS

Key terms and other bold terms in the chapter are defined in the end-of-book glossary.

Clamper   A circuit that adds a dc level to an ac voltage using a diode and a capacitor.

Filter   A capacitor in a power supply used to reduce the variation of the output voltage from a rectifier.

Full-wave rectifier   A circuit that converts an ac sinusoidal input voltage into a pulsating dc voltage with two output pulses occurring for each input cycle.

Half-wave rectifier   A circuit that converts an ac sinusoidal input voltage into a pulsating dc voltage with one output pulse occurring for each input cycle.

Limiter   A diode circuit that clips off or removes part of a waveform above and/or below a specified level.

Line regulation   The change in output voltage of a regulator for a given change in input voltage, normally expressed as a percentage.

Load regulation   The change in output voltage of a regulator for a given range of load currents, normally expressed as a percentage.

Peak inverse voltage (PIV)   The maximum value of reverse voltage which occurs at the peak of the input cycle when the diode is reverse-biased.

Power supply   A circuit that converts ac line voltage to dc voltage and supplies constant power to operate a circuit or system.

Regulator   An electronic device or circuit that maintains an essentially constant output voltage for a range of input voltage or load values; one part of a power supply.

Ripple voltage   The small variation in the dc output voltage of a filtered rectifier caused by the charging and discharging of the filter capacitor.

KEY FORMULAS

2–1	$V_{AVG} = \dfrac{V_p}{\pi}$	Half-wave average value
2–2	$V_{p(out)} = V_{p(in)} - 0.7\text{ V}$	Peak half-wave rectifier output (silicon)
2–3	$PIV = V_{p(in)}$	Peak inverse voltage, half-wave rectifier
2–4	$V_{sec} = n\dot{V}_{pri}$	Transformer secondary voltage
2–5	$V_{AVG} = \dfrac{2V_p}{\pi}$	Full-wave average value
2–6	$V_{out} = \dfrac{V_{sec}}{2} - 0.7\text{ V}$	Center-tapped full-wave output
2–7	$PIV = 2V_{p(out)} + 0.7\text{ V}$	Peak inverse voltage, center-tapped rectifier
2–8	$V_{p(out)} = V_{p(sec)} - 1.4\text{ V}$	Bridge full-wave output
2–9	$PIV = V_{p(out)} + 0.7\text{ V}$	Peak inverse voltage, bridge rectifier
2–10	$r = \dfrac{V_{r(pp)}}{V_{DC}}$	Ripple factor
2–11	$V_{r(pp)} \cong \left(\dfrac{1}{fR_LC}\right)V_{p(rect)}$	Peak-to-peak ripple voltage, capacitor-input filter
2–12	$V_{DC} = \left(1 - \dfrac{1}{2fR_LC}\right)V_{p(rect)}$	DC output voltage, capacitor-input filter
2–13	$R_{surge} = \dfrac{V_{p(sec)} - 1.4\text{ V}}{I_{FSM}}$	Surge resistance
2–14	$\text{Line regulation} = \left(\dfrac{\Delta V_{OUT}}{\Delta V_{IN}}\right)100\%$	
2–15	$\text{Load regulation} = \left(\dfrac{V_{NL} - V_{FL}}{V_{FL}}\right)100\%$	

SELF-TEST

Answers are at the end of the chapter.

1. The average value of a half-wave rectified voltage with a peak value of 200 V is
   (a) 63.7 V   (b) 127.3 V   (c) 141 V   (d) 0 V

2. When a 60 Hz sinusoidal voltage is applied to the input of a half-wave rectifier, the output frequency is
   (a) 120 Hz   (b) 30 Hz   (c) 60 Hz   (d) 0 Hz

3. The peak value of the input to a half-wave rectifier is 10 V. The approximate peak value of the output is
   (a) 10 V   (b) 3.18 V   (c) 10.7 V   (d) 9.3 V

4. For the circuit in Question 3, the diode must be able to withstand a reverse voltage of
   (a) 10 V   (b) 5 V   (c) 20 V   (d) 3.18 V

5. The average value of a full-wave rectified voltage with a peak value of 75 V is
   (a) 53 V   (b) 47.8 V   (c) 37.5 V   (d) 23.9 V

6. When a 60 Hz sinusoidal voltage is applied to the input of a full-wave rectifier, the output frequency is
   (a) 120 Hz   (b) 60 Hz   (c) 240 Hz   (d) 0 Hz

7. The total secondary voltage in a center-tapped full-wave rectifier is 125 V rms. Neglecting the diode drop, the rms output voltage is
   (a) 125 V   (b) 177 V   (c) 100 V   (d) 62.5 V

8. When the peak output voltage is 100 V, the PIV for each diode in a center-tapped full-wave rectifier is (neglecting the diode drop)

   (a) 100 V   (b) 200 V   (c) 141 V   (d) 50 V

9. When the rms output voltage of a bridge full-wave rectifier is 20 V, the peak inverse voltage across the diodes is (neglecting the diode drop)

   (a) 20 V   (b) 40 V   (c) 28.3 V   (d) 56.6 V

10. The ideal dc output voltage of a capacitor-input filter is equal to

   (a) the peak value of the rectified voltage

   (b) the average value of the rectified voltage

   (c) the rms value of the rectified voltage

11. A certain power-supply filter produces an output with a ripple of 100 mV peak-to-peak and a dc value of 20 V. The ripple factor is

   (a) 0.05   (b) 0.005   (c) 0.00005   (d) 0.02

12. A 60 V peak full-wave rectified voltage is applied to a capacitor-input filter. If $f = 120$ Hz, $R_L = 10$ kΩ and $C = 10$ μF, the ripple voltage is

   (a) 0.6 V   (b) 6 mV   (c) 5.0 V   (d) 2.88 V

13. If the load resistance of a capacitor-filtered full-wave rectifier is reduced, the ripple voltage

   (a) increases   (b) decreases

   (c) is not affected   (d) has a different frequency

14. Line regulation is determined by

   (a) load current

   (b) zener current and load current

   (c) changes in load resistance and output voltage

   (d) changes in output voltage and input voltage

15. Load regulation is determined by

   (a) changes in load current and input voltage

   (b) changes in load current and output voltage

   (c) changes in load resistance and input voltage

   (d) changes in zener current and load current

16. A 10 V peak-to-peak sinusoidal voltage is applied across a silicon diode and series resistor. The maximum voltage across the diode is

   (a) 9.3 V   (b) 5 V   (c) 0.7 V   (d) 10 V   (e) 4.3 V

17. If the input voltage to a voltage tripler has an rms value of 12 V, the dc output voltage is approximately

   (a) 36 V   (b) 50.9 V   (c) 33.9 V   (d) 32.4 V

18. If one of the diodes in a bridge full-wave rectifier opens, the output is

   (a) 0 V   (b) one-fourth the amplitude of the input voltage

   (c) a half-wave rectified voltage   (d) a 120 Hz voltage

19. If you are checking a 60 Hz full-wave bridge rectifier and observe that the output has a 60 Hz ripple,

   (a) the circuit is working properly   (b) there is an open diode

   (c) the transformer secondary is shorted   (d) the filter capacitor is leaky

PROBLEMS

Answers to all odd-numbered problems are at the end of the book.

BASIC PROBLEMS

SECTION 2–1   Half-Wave Rectifiers

1. Draw the output voltage waveform for each circuit in Figure 2–71 and include the voltage values.

(a)   (b)

▲ **FIGURE 2–71**

File circuits are identified with a CD logo and are in the Problems folder on your CD-ROM. Filenames correspond to figure numbers (e.g., F02-71).

2. What is the peak forward current through each diode in Figure 2–71?

3. A power-supply transformer has a turns ratio of 5:1. What is the secondary voltage if the primary is connected to a 115 V rms source?

4. Determine the peak and average power delivered to R_L in Figure 2–72.

▶ **FIGURE 2–72**

SECTION 2–2   Full-Wave Rectifiers

5. Find the average value of each voltage in Figure 2–73.

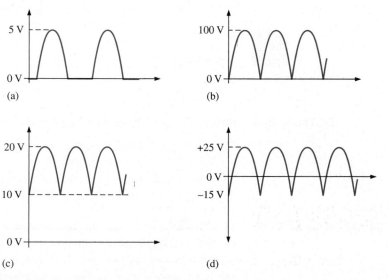

(a)   (b)

(c)   (d)

▲ **FIGURE 2–73**

6. Consider the circuit in Figure 2–74.

   (a) What type of circuit is this?

   (b) What is the total peak secondary voltage?

   (c) Find the peak voltage across each half of the secondary.

   (d) Sketch the voltage waveform across R_L.

   (e) What is the peak current through each diode?

   (f) What is the PIV for each diode?

▶ FIGURE 2–74

7. Calculate the peak voltage across each half of a center-tapped transformer used in a full-wave rectifier that has an average output voltage of 110 V.

8. Show how to connect the diodes in a center-tapped rectifier in order to produce a negative-going full-wave voltage across the load resistor.

9. What PIV rating is required for the diodes in a bridge rectifier that produces an average output voltage of 50 V?

10. The rms output voltage of a bridge rectifier is 20 V. What is the peak inverse voltage across the diodes?

11. Sketch the output voltage of the bridge rectifier in Figure 2–75. Notice that all the diodes are reversed from previous circuits.

▲ FIGURE 2–75

SECTION 2–3   **Power Supply Filters and Regulators**

12. A certain rectifier filter produces a dc output voltage of 75 V with a peak-to-peak ripple voltage of 0.5 V. Calculate the ripple factor.

13. A certain full-wave rectifier has a peak output voltage of 30 V. A 50 μF capacitor-input filter is connected to the rectifier. Calculate the peak-to-peak ripple and the dc output voltage developed across a 600 Ω load resistance.

14. What is the percentage of ripple for the rectifier filter in Problem 13?

15. What value of filter capacitor is required to produce a 1% ripple factor for a full-wave rectifier having a load resistance of 1.5 kΩ? Assume the rectifier produces a peak output of 18 V.

16. A full-wave rectifier produces an 80 V peak rectified voltage from a 60 Hz ac source. If a 10 μF filter capacitor is used, determine the ripple factor for a load resistance of 10 kΩ.

17. Determine the peak-to-peak ripple and dc output voltages in Figure 2–76. The transformer has a 36 V rms secondary voltage rating, and the line voltage has a frequency of 60 Hz.

▲ FIGURE 2–76

18. Refer to Figure 2–76 and sketch the following voltage waveforms in relationship to the input waveforms: V_{AB}, V_{AD}, and V_{CD}. A double letter subscript indicates a voltage from one point to another.

19. If the no-load output voltage of a regulator is 15.5 V and the full-load output is 14.9 V, what is the percent load regulation?

20. Assume a regulator has a percent load regulation of 0.5%. What is output voltage at full-load if the unloaded output is 12.0 V?

SECTION 2–4   Diode Limiting and Clamping Circuits

21. Determine the output waveform for the circuit of Figure 2–77.

▶ FIGURE 2–77

22. Determine the output voltage for the circuit in Figure 2–78(a) for each input voltage in (b), (c), and (d).

(a)

(b)

(c)

(d)

▲ FIGURE 2–78

23. Determine the output voltage waveform for each circuit in Figure 2–79.

▲ **FIGURE 2–79**

24. Determine the R_L voltage waveform for each circuit in Figure 2–80.

▲ **FIGURE 2–80**

25. Sketch the output voltage waveform for each circuit in Figure 2–81.

26. Determine the output voltage waveform for each circuit in Figure 2–82.

27. Describe the output waveform of each circuit in Figure 2–83. Assume the RC time constant is much greater than the period of the input.

28. Repeat Problem 27 with the diodes turned around.

▲ **FIGURE 2–81**

▲ FIGURE 2–82

▲ FIGURE 2–83

SECTION 2–5 Voltage Multipliers

29. A certain voltage doubler has 20 V rms on its input. What is the output voltage? Sketch the circuit, indicating the output terminals and PIV rating for the diode.

30. Repeat Problem 29 for a voltage tripler and quadrupler.

SECTION 2–6 The Diode Data Sheet

31. From the data sheet in Figure 2–56, determine how much peak inverse voltage that a 1N1183A diode can withstand.

32. Repeat Problem 31 for a 1N1188A.

33. If the peak output voltage of a bridge full-wave rectifier is 50 V, determine the minimum value of the surge-limiting resistor required when 1N1183A diodes are used.

TROUBLESHOOTING PROBLEMS

SECTION 2–7 Troubleshooting

34. If one of the diodes in a bridge rectifier opens, what happens to the output?

35. From the meter readings in Figure 2–84, determine if the rectifier is functioning properly. If it is not, determine the most likely failure(s).

36. Each part of Figure 2–85 shows oscilloscope displays of various rectifier output voltages. In each case, determine whether or not the rectifier is functioning properly and if it is not, determine the most likely failure(s).

37. Based on the values given, would you expect the circuit in Figure 2–86 to fail? If so, why?

▲ **FIGURE 2–84**

(a) Output of a half-wave unfiltered rectifier

(b) Output of a full-wave unfiltered rectifier

(c) Output of a full-wave filter

(d) Output of same full-wave filter as part (c)

▲ **FIGURE 2–85**

▲ **FIGURE 2–86**

SYSTEM APPLICATION PROBLEMS

38. Determine the most likely failure in the circuit board of Figure 2–87 for each of the following symptoms. State the corrective action you would take in each case. The transformer has a rated output of 36 V.

(a) No voltage measured from test point 1 to test point 2.

(b) No voltage from test point 3 to test point 4, 110 V rms from test point 1 to test point 2.

(c) 50 V rms from test point 3 to test point 4. Input is correct at 110 V rms.

(d) 25 V rms from test point 3 to test point 4. Input is correct at 110 V rms.

(e) A full-wave rectified voltage with a peak of approximately 50 V at test point 7 with respect to ground.

(f) Excessive 120 Hz ripple voltage at test point 7.

(g) The ripple voltage has a frequency of 60 Hz at test point 7.

(h) No voltage at test point 7.

39. In testing the power supply board in Figure 2–87 with a 10 kΩ load resistor connected, you find the voltage at the positive side of the filter capacitor to have a 60 Hz ripple voltage. You replace all the diodes, plug in the board, and check the point again to verify proper operation and it still has the 60 Hz ripple voltage. What now?

40. If the top diode on the circuit board in Figure 2–87 is incorrectly installed backwards, what voltage would you measure at test point 8?

▲ **FIGURE 2–87**

ADVANCED PROBLEMS

41. A full-wave rectifier with a capacitor-input filter provides a dc output voltage of 35 V to a 3.3 kΩ load. Determine the minimum value of filter capacitor if the maximum peak-to-peak ripple voltage is to be 0.5 V.

42. A certain unfiltered full-wave rectifier with 115 V, 60 Hz input produces an output with a peak of 15 V. When a capacitor-input filter and a 1.0 kΩ load are connected, the dc output voltage is 14 V. What is the peak-to-peak ripple voltage?

43. For a certain full-wave rectifier, the measured surge current in the capacitor filter is 50 A. The transformer is rated for a secondary voltage of 24 V with a 110 V, 60 Hz input. Determine the value of the surge resistor in this circuit.

44. Design a full-wave rectifier using an 18 V center-tapped transformer. The output ripple is not to exceed 5% of the output voltage with a load resistance of 680 Ω. Specify the I_O and PIV ratings of the diodes and select an appropriate diode from Figure 2–56.

45. Design a filtered power supply that can produce dc output voltages of $+9\text{ V} \pm 10\%$ and $-9\text{ V} \pm 10\%$ with a maximum load current of 100 mA. The voltages are to be switch selectable across one set of output terminals. The ripple voltage must not exceed 0.25 V rms.

46. Design a circuit to limit a 20 V rms sinusoidal voltage to a maximum positive amplitude of 18 V and a maximum negative amplitude of 10 V using a single 24 V dc voltage source.

▶ **FIGURE 2–88**

47. Determine the voltage across each capacitor in the circuit of Figure 2–88.

EWB/MULTISIM TROUBLESHOOTING PROBLEMS

These file circuits are in the Troubleshooting Problems folder on your CD-ROM.

48. Open file TSP02-48 and determine the fault.

49. Open file TSP02-49 and determine the fault.

50. Open file TSP02-50 and determine the fault.

51. Open file TSP02-51 and determine the fault.

52. Open file TSP02-52 and determine the fault.

53. Open file TSP02-53 and determine the fault.

54. Open file TSP02-54 and determine the fault.

55. Open file TSP02-55 and determine the fault.

56. Open file TSP02-56 and determine the fault.

ANSWERS

SECTION REVIEWS

SECTION 2–1   Half-Wave Rectifiers

1. PIV across the diode occurs at the peak of the input.

2. There is current through the load for approximately half (50%) of the input cycle.

3. The average value is $10 \text{ V}/\pi = 3.18$ V.

4. The peak output voltage is $25 \text{ V} - 0.7 \text{ V} = 24.3$ V.

5. The PIV must be at least 50 V.

SECTION 2–2   Full-Wave Rectifiers

1. A full-wave voltage occurs on each half of the input cycle and has a frequency of twice the input frequency. A half-wave voltage occurs once each input cycle and has a frequency equal to the input frequency.

2. The average value of $2(60 \text{ V})/\pi = 38.12$ V

3. The bridge rectifier has the greater output voltage.

4. The 50 V diodes must be used in the bridge rectifier.

5. In the center-tapped rectifier, diodes with a PIV rating of at least 90 V would be required.

SECTION 2–3   Power Supply Filters and Regulators

1. The output frequency is 60 Hz.

2. The output frequency is 120 Hz.

3. The ripple voltage is caused by the slight charging and discharging of the capacitor through the load resistor.

4. The ripple voltage amplitude increases when the load resistance decreases.

5. Ripple factor is the ratio of the ripple voltage to the average or dc voltage.

6. Simple, reliable, and inexpensive

7. Input regulation means that the output is constant over a range of input voltages. Load regulation means that the output voltage is constant over a range of load current values.

SECTION 2–4 Diode Limiting and Clamping Circuits

1. Limiters clip off or remove portions of a waveform. Clampers insert a dc level.
2. A positive limiter clips off positive voltages. A negative limiter clips off negative voltages.
3. 0.7 V appears across the diode.
4. The bias voltage must be 5 V − 0.7 V = 4.3 V.
5. The capacitor acts as a battery.

SECTION 2–5 Voltage Multipliers

1. The peak voltage rating must be 100 V.
2. The PIV rating must be at least 310 V.

SECTION 2–6 The Diode Data Sheet

1. The three rating categories on a diode data sheet are maximum ratings, electrical characteristics, and mechanical data.
2. V_F is forward voltage, I_R is reverse current, and I_O is peak average forward current.
3. I_{FSM} is maximum forward surge current, V_{RRM} is maximum reverse peak repetitive voltage, and V_{RSM} is maximum reverse peak nonrepetitive voltage.
4. The 1N4720 has an I_O = 3.0 A, I_{FSM} = 300 A, and V_{RRM} = 100 V.

SECTION 2–7 Troubleshooting

1. An open diode results in no output voltage.
2. An open diode produces a half-wave output voltage.
3. The shorted diode may burn open. Transformer will be damaged. Fuse will blow.
4. The amplitude of the ripple voltage increases with a leaky filter capacitor.
5. There will be no output voltage when the primary opens.
6. The problem may be a partially shorted secondary winding.

RELATED PROBLEMS FOR EXAMPLES

2–1   3.82 V
2–2   (a) 2.3 V   (b) 209.3 V
2–3   (a) 623.3 V   (b) 624 V   (c) negative half-cycles rather than positive half cycles
2–4   98.7 V
2–5   79.3 V including diode drop
2–6   41.0 V; 41.7 V
2–7   103 mV
2–8   3.7%
2–9   A positive peak of 8.72 V and clipped at −0.3 V
2–10 Limited at +10.7 V and −10.7 V
2–11 Change R_3 to 1.0 kΩ or R_2 to 2.2 kΩ.
2–12 Same voltage waveform as Figure 2–49
2–13 Check for a physical short in the circuit or check for shorted diodes in the rectifier.

SELF-TEST

1. (a)   2. (c)   3. (d)   4. (a)   5. (b)   6. (a)
7. (d)   8. (b)   9. (c)   10. (a)   11. (b)   12. (c)
13. (a)   14. (d)   15. (b)   16. (d)   17. (b)   18. (c)   19. (b)

3

SPECIAL-PURPOSE DIODES

INTRODUCTION

Chapter 2 was devoted to general-purpose and rectifier diodes, which are the most widely used types. In this chapter, we will cover several other types of diodes that are designed for specific applications, including the zener, varactor (variable-capacitance), light-emitting, photodiode, current regulator, Schottky, tunnel, *pin*, step-recovery, and laser diodes.

CHAPTER OBJECTIVES

■ Describe the characteristics of a zener diode and analyze its operation

■ Explain how a zener is used in voltage regulation and limiting and analyze zener circuits

■ Describe the variable-capacitance characteristics of a varactor diode and analyze its operation in a typical circuit

■ Discuss the operation and characteristics of LEDs and photo-diodes

■ Discuss the basic characteristics of the current regulator diode, the Schottky diode, the *pin* diode, the step-recovery diode, the tunnel diode, and the laser diode

■ Troubleshoot zener diode regulators

KEY TERMS

■ Zener diode

■ Zener breakdown

■ Varactor

■ Light-emitting diode (LED)

■ Electroluminescence

■ Photodiode

■ Laser

■■■ SYSTEM APPLICATION PREVIEW

You are given the assignment to analyze and test a new system for counting and controlling items for packaging and shipment. The first system is to be installed in a sporting goods manufacturing plant to control the packaging of baseballs for shipment. The first step in your assignment is to learn all you can about various special-purpose diodes. You will then apply your knowledge to the system application at the end of the chapter.

WWW. **VISIT THE COMPANION WEBSITE**
Study aids for this chapter are available at
http://www.prenhall.com/floyd

3–1 ZENER DIODES

A major application for zener diodes is as a type of voltage regulator for providing stable reference voltages for use in power supplies, voltmeters, and other instruments. In this section, you will see how the zener diode maintains a nearly constant dc voltage under the proper operating conditions. You will learn the conditions and limitations for properly using the zener diode and the factors that affect its performance.

After completing this section, you should be able to

- **Describe the characteristics of a zener diode and analyze its operation**
- Identify a zener diode by its symbol
- Discuss avalanche and zener breakdown
- Analyze the characteristic (*V-I*) curve of a zener diode
- Discuss the zener equivalent circuit
- Define temperature coefficient and apply it to zener analysis
- Discuss power dissipation in a zener and apply derating
- Interpret a zener diode data sheet

Cathode (K)

Anode (A)

▲ **FIGURE 3–1**

Zener diode symbol.

The symbol for a zener diode is shown in Figure 3–1. The **zener diode** is a silicon *pn* junction device that differs from rectifier diodes because *it is designed for operation in the reverse-breakdown region.* The breakdown voltage of a zener diode is set by carefully controlling the doping level during manufacture. Recall, from the discussion of the diode characteristic curve in Chapter 1, that when a diode reaches reverse breakdown, its voltage remains almost constant even though the current changes drastically. This volt-ampere characteristic is shown again in Figure 3–2 with normal operating regions for rectifier diodes and for zener diodes shown as shaded areas. If a zener diode is forward-biased, it operates the same as a rectifier diode.

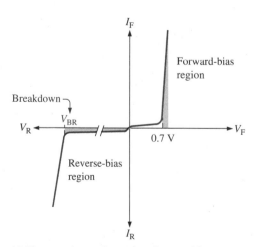

(a) The normal operating regions for a rectifier diode are shown as shaded areas.

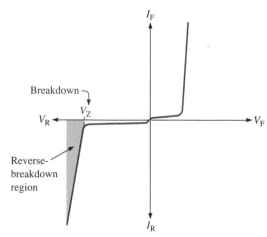

(b) The normal operating region for a zener diode is shaded.

▲ **FIGURE 3–2**

General diode *V-I* characteristic.

Zener Breakdown

Zener diodes are designed to operate in reverse breakdown. Two types of reverse break-down in a zener diode are *avalanche* and *zener*. The avalanche breakdown, discussed in Chapter 1, occurs in both rectifier and zener diodes at a sufficiently high reverse voltage. **Zener breakdown** occurs in a zener diode at low reverse voltages. A zener diode is heavily doped to reduce the breakdown voltage. This causes a very thin depletion region. As a result, an intense electric field exists within the depletion region. Near the zener break-down voltage (V_Z), the field is intense enough to pull electrons from their valence bands and create current.

Zener diodes with breakdown voltages of less than approximately 5 V operate predominately in zener breakdown. Those with breakdown voltages greater than approximately 5 V operate predominately in **avalanche breakdown**. Both types, however, are called *zener diodes*. Zeners are commercially available with breakdown voltages of 1.8 V to 200 V with specified tolerances from 1% to 20%.

Breakdown Characteristics

Figure 3–3 shows the reverse portion of a zener diode's characteristic curve. Notice that as the reverse voltage (V_R) is increased, the reverse current (I_R) remains extremely small up to the "knee" of the curve. The reverse current is also called the zener current, I_Z. At this point, the breakdown effect begins; the internal zener resistance, also called zener impedance (Z_Z), begins to decrease as the reverse current increases rapidly. From the bottom of the knee, the zener breakdown voltage (V_Z) remains essentially constant although it increases slightly as the zener current, I_Z, increases.

▶ **FIGURE 3–3**

Reverse characteristic of a zener diode. V_Z is usually specified at the zener test current, I_{ZT}, and is designated V_{ZT}.

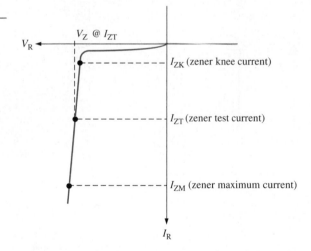

Zener Regulation   The ability to keep the reverse voltage across its terminals essentially constant is the key feature of the zener diode. A zener diode operating in breakdown acts as a voltage regulator because it maintains a nearly constant voltage across its terminals over a specified range of reverse-current values.

A minimum value of reverse current, I_{ZK}, must be maintained in order to keep the diode in breakdown for voltage regulation. You can see on the curve in Figure 3–3 that when the reverse current is reduced below the knee of the curve, the voltage decreases drastically and regulation is lost. Also, there is a maximum current, I_{ZM}, above which the diode may be damaged due to excessive power dissipation. So, basically, the zener diode maintains a nearly constant voltage across its terminals for values of reverse current ranging from I_{ZK} to I_{ZM}. A nominal zener voltage, V_{ZT}, is usually specified on a data sheet at a value of reverse current called the *zener test current, I_{ZT}*.

Zener Equivalent Circuit

Figure 3–4(a) shows the ideal model of a zener diode in reverse breakdown. It has a constant voltage drop equal to the nominal zener voltage. This constant voltage drop is represented by a dc voltage source even though the zener diode does not actually produce an emf voltage. The dc source simply indicates that the effect of reverse breakdown is a constant voltage across the zener terminals.

Figure 3–4(b) represents the practical model of a zener diode, where the zener impedance (Z_Z) is included. Since the actual voltage curve is not ideally vertical, a change in zener current (ΔI_Z) produces a small change in zener voltage (ΔV_Z), as illustrated in Figure 3–4(c). By Ohm's law, the ratio of ΔV_Z to ΔI_Z is the impedance, as expressed in the following equation:

Equation 3–1

$$Z_Z = \frac{\Delta V_Z}{\Delta I_Z}$$

Normally, Z_Z is specified at I_{ZT}, the zener test current, and is designated Z_{ZT}. In most cases, you can assume that Z_Z is constant over the full linear range of zener current values and is purely resistive.

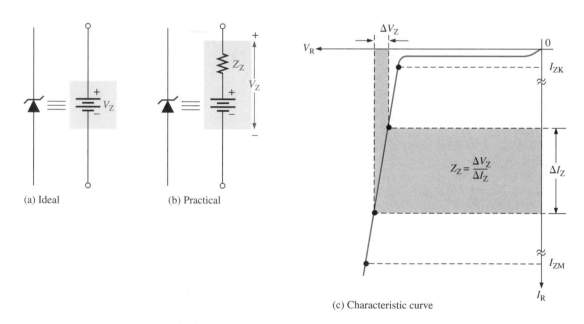

▲ **FIGURE 3–4**

Zener diode equivalent circuit models and the characteristic curve illustrating Z_Z.

EXAMPLE 3–1

A zener diode exhibits a certain change in V_Z for a certain change in I_Z on a portion of the linear characteristic curve between I_{ZK} and I_{ZM} as illustrated in Figure 3–5. What is the zener impedance?

▲ FIGURE 3-5

Solution

$$Z_Z = \frac{\Delta V_Z}{\Delta I_Z} = \frac{50 \text{ mV}}{5 \text{ mA}} = \mathbf{10 \ \Omega}$$

Related Problem* Calculate the zener impedance if the change in zener voltage is 100 mV for a 20 mA change in zener current on the linear portion of the characteristic curve.

*Answers are at the end of the chapter.

EXAMPLE 3–2

A 1N4736 zener diode has a Z_{ZT} of 3.5 Ω. The data sheet gives $V_{ZT} = 6.8$ V at $I_{ZT} = 37$ mA and $I_{ZK} = 1$ mA. What is the voltage across the zener terminals when the current is 50 mA? When the current is 25 mA?

Solution Figure 3–6 represents the zener diode.

▶ FIGURE 3–6

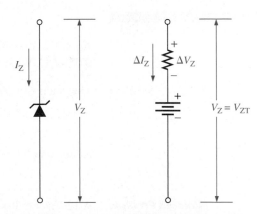

For $I_Z = 50$ mA: The 50 mA current is a 13 mA increase above $I_{ZT} = 37$ mA.

$$\Delta I_Z = I_Z - I_{ZT} = +13 \text{ mA}$$
$$\Delta V_Z = \Delta I_Z Z_{ZT} = (13 \text{ mA})(3.5 \text{ }\Omega) = +45.5 \text{ mV}$$

The change in voltage due to the increase in current above the I_{ZT} value causes the zener terminal voltage to increase. The zener voltage for $I_Z = 50$ mA is

$$V_Z = 6.8 \text{ V} + \Delta V_Z = 6.8 \text{ V} + 45.5 \text{ mV} = \textbf{6.85 V}$$

For $I_Z = 25$ mA: The 25 mA current is a 12 mA decrease below $I_{ZT} = 37$ mA.

$$\Delta I_Z = -12 \text{ mA}$$
$$\Delta V_Z = \Delta I_Z Z_{ZT} = (-12 \text{ mA})(3.5 \text{ }\Omega) = -42 \text{ mV}$$

The change in voltage due to the decrease in current below I_{ZT} causes the zener terminal voltage to decrease. The zener voltage for $I_Z = 25$ mA is

$$V_Z = 6.8 \text{ V} - \Delta V_Z = 6.8 \text{ V} - 42 \text{ mV} = \textbf{6.76 V}$$

Related Problem Repeat the analysis for $I_Z = 10$ mA and for $I_Z = 30$ mA using a 1N4742 zener with $V_{ZT} = 12$ V at $I_{ZT} = 21$ mA and $Z_{ZT} = 9 \text{ }\Omega$.

Temperature Coefficient

The temperature coefficient specifies the percent change in zener voltage for each °C change in temperature. For example, a 12 V zener diode with a positive temperature coefficient of 0.01%/°C will exhibit a 1.2 mV increase in V_Z when the junction temperature increases one Celsius degree. The formula for calculating the change in zener voltage for a given junction temperature change, for a specified temperature coefficient, is

Equation 3–2

$$\Delta V_Z = V_Z \times TC \times \Delta T$$

where V_Z is the nominal zener voltage at 25°C, TC is the temperature coefficient, and ΔT is the change in temperature. A positive TC means that the zener voltage increases with an increase in temperature or decreases with a decrease in temperature. A negative TC means that the zener voltage decreases with an increase in temperature or increases with a decrease in temperature.

In some cases, the temperature coefficient is expressed in mV/°C rather than as %/°C. For these cases, ΔV_Z is calculated as

Equation 3–3

$$\Delta V_Z = TC \times \Delta T$$

EXAMPLE 3–3

An 8.2 V zener diode (8.2 V at 25°C) has a positive temperature coefficient of 0.05%/°C. What is the zener voltage at 60°C?

Solution The change in zener voltage is

$$\Delta V_Z = V_Z \times TC \times \Delta T = (8.2 \text{ V})(0.05\%/°C)(60°C - 25°C)$$
$$= (8.2 \text{ V})(0.0005/°C)(35°C) = 144 \text{ mV}$$

Notice that 0.05%/°C was converted to 0.0005/°C. The zener voltage at 60°C is

$$V_Z + \Delta V_Z = 8.2 \text{ V} + 144 \text{ mV} = \textbf{8.34 V}$$

Related Problem A 12 V zener has a positive temperature coefficient of 0.075%/°C. How much will the zener voltage change when the junction temperature decreases 50°C?

Zener Power Dissipation and Derating

Zener diodes are specified to operate at a maximum power called the maximum dc power dissipation, $P_{D(max)}$. For example, the 1N746 zener is rated at a $P_{D(max)}$ of 500 mW and the 1N3305A is rated at a $P_{D(max)}$ of 50 W. The dc power dissipation is determined by the formula,

$$P_D = V_Z I_Z$$

Power Derating The maximum power dissipation of a zener diode is typically specified for temperatures at or below a certain value (50°C, for example). Above the specified temperature, the maximum power dissipation is reduced according to a derating factor. The derating factor is expressed in mW/°C. The maximum derated power can be determined with the following formula:

$$P_{D(derated)} = P_{D(max)} - (mW/°C)\Delta T$$

Equation 3–4

EXAMPLE 3–4

A certain zener diode has a maximum power rating of 400 mW at 50°C and a derating factor of 3.2 mW/°C. Determine the maximum power the zener can dissipate at a temperature of 90°C.

Solution

$$\begin{aligned} P_{D(derated)} &= P_{D(max)} - (mW/°C)\Delta T \\ &= 400 \text{ mW} - (3.2 \text{ mW/°C})(90°C - 50°C) \\ &= 400 \text{ mW} - 128 \text{ mW} = \mathbf{272 \text{ mW}} \end{aligned}$$

Related Problem A certain 50 W zener diode must be derated with a derating factor of 0.5 W/°C above 75°C. Determine the maximum power it can dissipate at 160°C.

Zener Diode Data Sheet Information

The amount and type of information found on data sheets for zener diodes (or any category of electronic device) varies from one type of diode to the next. The data sheet for some zeners contains more information than for others. Figure 3–7 gives an example of the type of information that you have studied that can be found on a typical data sheet but does not represent the complete data sheet. This particular information is for a popular zener series, the 1N4728–1N4764.

Electrical Characteristics The electrical characteristics are listed in a tabular form in Figure 3–7(a) with the zener type numbers in the first column. This feature is common to most device data sheets.

Zener voltage For each zener type number, the nominal zener voltage, V_Z, for a specified value of zener test current, I_{ZT}, is listed in the second column. The nominal value of V_Z can vary depending on the tolerance. For example, the 1N4738 has a nominal V_Z of 8.2 V. For 10% tolerance, this value can range from 7.38 V to 9.02 V.

Zener test current The value of zener current, I_{ZT}, in mA at which the nominal zener voltage is specified is listed in the third column of the table in Figure 3–7(a).

Maximum Ratings

Rating	Symbol	Value	Unit
DC power dissipation @ T_A = 50°C	P_D	1.0	Watt
Derate above 50°C		6.67	mW/°C
Operating and storage junction Temperature range	T_J, T_{stg}	−65 to +200	°C

Electrical Characteristics (T_A = 25°C unless otherwise noted) V_F = 1.2 V max, I_F = 200 mA for all types.

JEDEC Type No. (Note 1)	Nominal Zener Voltage V_Z @ I_{ZT} Volts	Test Current I_{ZT} mA	Maximum Zener Impedance			Leakage Current	
			Z_{ZT} @ I_{ZT} Ohms	Z_{ZK} @ I_{ZK} Ohms	I_{ZK} mA	I_R μA Max	V_R Volts
1N4728	3.3	76	10	400	1.0	100	1.0
1N4729	3.6	69	10	400	1.0	100	1.0
1N4730	3.9	64	9.0	400	1.0	50	1.0
1N4731	4.3	58	9.0	400	1.0	10	1.0
1N4732	4.7	53	8.0	500	1.0	10	1.0
1N4733	5.1	49	7.0	550	1.0	10	1.0
1N4734	5.6	45	5.0	600	1.0	10	2.0
1N4735	6.2	41	2.0	700	1.0	10	3.0
1N4736	6.8	37	3.5	700	1.0	10	4.0
1N4737	7.5	34	4.0	700	0.5	10	5.0
1N4738	8.2	31	4.5	700	0.5	10	6.0
1N4739	9.1	28	5.0	700	0.5	10	7.0
1N4740	10	25	7.0	700	0.25	10	7.6
1N4741	11	23	8.0	700	0.25	5.0	8.4
1N4742	12	21	9.0	700	0.25	5.0	9.1
1N4743	13	19	10	700	0.25	5.0	9.9
1N4744	15	17	14	700	0.25	5.0	11.4
1N4745	16	15.5	16	700	0.25	5.0	12.2
1N4746	18	14	20	750	0.25	5.0	13.7
1N4747	20	12.5	22	750	0.25	5.0	15.2
1N4748	22	11.5	23	750	0.25	5.0	16.7
1N4749	24	10.5	25	750	0.25	5.0	18.2
1N4750	27	9.5	35	750	0.25	5.0	20.6
1N4751	30	8.5	40	1000	0.25	5.0	22.8
1N4752	33	7.5	45	1000	0.25	5.0	25.1
1N4753	36	7.0	50	1000	0.25	5.0	27.4
1N4754	39	6.5	60	1000	0.25	5.0	29.7
1N4755	43	6.0	70	1500	0.25	5.0	32.7
1N4756	47	5.5	80	1500	0.25	5.0	35.8
1N4757	51	5.0	95	1500	0.25	5.0	38.8
1N4758	56	4.5	110	2000	0.25	5.0	42.6
1N4759	62	4.0	125	2000	0.25	5.0	47.1
1N4760	68	3.7	150	2000	0.25	5.0	51.7
1N4761	75	3.3	175	2000	0.25	5.0	56.0
1N4762	82	3.0	200	3000	0.25	5.0	62.2
1N4763	91	2.8	250	3000	0.25	5.0	69.2
1N4764	100	2.5	350	3000	0.25	5.0	76.0

NOTE 1 — Tolerance and Type Number Designation. The JEDEC type numbers listed have a standard tolerance on the nominal zener voltage of ±10%. A standard tolerance of ±5% on individual units is also available and is indicated by suffixing "A" to the standard type number. C for ±2.0%, D for ±1.0%.

(a) Electrical characteristics

(b) Power derating

(c) Temperature coefficient

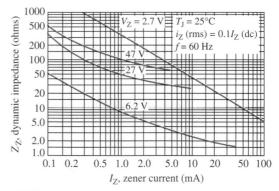

(d) Effect of zener current on zener impedance

▲ **FIGURE 3–7**

Partial data sheet for the 1N4728–1N4764 series 1 W zener diodes.

Zener impedance Z_{ZT} is the value of dynamic impedance in ohms measured at the test current. The values of Z_{ZT} for each zener type are listed in the fourth column. The term *dynamic* means that it is measured as an ac quantity; that is, the *change* in voltage for a specified *change* in current ($Z_{ZT} = \Delta V_Z/\Delta I_Z$). You cannot get Z_{ZT} using V_Z and I_{ZT}, which are dc values. The table also includes Z_{ZK}, which is the impedance measured at the zener knee current, I_{ZK}.

Reverse leakage current   The values of leakage current are listed in the fifth column of the table. The leakage current is the current through the reverse-biased zener diode for values of reverse voltage less than the value at the knee of the characteristic curve. Notice that the values are extremely small as was the case for rectifier diodes.

Maximum zener current   The maximum dc current, I_{ZM}, is not specified on this particular data sheet. However, it is worth mentioning because you will find it on some data sheets. The value of I_{ZM} is specified based on the power rating, the zener voltage at I_{ZM}, and the zener voltage tolerance. An approximate value for I_{ZM} can be calculated using the maximum power dissipation, $P_{D(max)}$ and V_Z at I_{ZT} as follows:

$$I_{ZM} = \frac{P_{D(max)}}{V_Z}$$

Equation 3–5

Graphical Data   Some data sheets provide various types of data in the form of graphs while others do not. Figure 3–7 includes graphs for data related to concepts covered in this section.

Power derating   Figure 3–7(b) shows a power derating curve for this particular series of zener diodes. Notice that the zeners are rated for a maximum power dissipation of 1 W for temperatures of 50°C and below. Above 50°C the power rating decreases linearly as shown. For example, at 140°C the power rating is approximately 400 mW.

Temperature coefficients   Figure 3–7(c) shows the temperature coefficient in mV/°C versus zener voltage for zener voltages up to 12 V. The two curves define a range for the temperature coefficient. For example, a 6 V zener diode exhibits a temperature coefficient that can range from about 1.5 mV/°C to about 3 mV/°C.

Effect of zener current on zener impedance   Figure 3–7(d) shows how the zener impedance, Z_Z, varies with current for selected values of nominal zener voltage: 2.7 V, 6.2 V, 27 V, and 47 V. Notice that Z_Z decreases with increasing current.

SECTION 3–1 REVIEW

Answers are at the end of the chapter.

1. In what region of their characteristic curve are zener diodes operated?
2. At what value of zener current is the zener voltage normally specified?
3. How does the zener impedance affect the voltage across the terminals of the device?
4. For a certain zener diode, $V_Z = 10\ V$ at $I_{ZT} = 30\ mA$. If $Z_Z = 8\ \Omega$, what is the terminal voltage at $I_Z = 50\ mA$?
5. What does a positive temperature coefficient of 0.05%/°C mean?
6. Explain power derating.

3-2 ZENER DIODE APPLICATIONS

The zener diode is often used as a type of voltage regulator for providing stable reference voltages. In this section, the concept of zener voltage regulation is introduced. Also, you will see how zeners can be used as simple limiters or clippers.

After completing this section, you should be able to

- **Explain how a zener is used in voltage regulation and limiting and analyze zener circuits**

- Analyze zener diode regulators under both varying input and varying load conditions

- Analyze zener waveform-limiting circuits

Zener Regulation with a Varying Input Voltage

Zener diodes are widely used for voltage regulation. Figure 3–8 illustrates how a zener diode can be used to regulate a varying dc voltage. As the input voltage varies (within limits), the zener diode maintains a nearly constant output voltage across its terminals. However, as V_{IN} changes, I_Z will change proportionally so that the limitations on the input voltage variation are set by the minimum and maximum current values (I_{ZK} and I_{ZM}) with

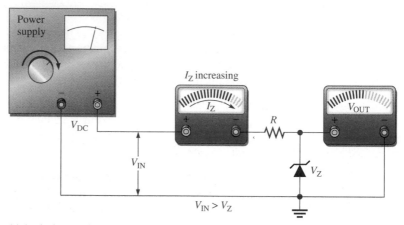

(a) As the input voltage increases, the output voltage remains constant ($I_{ZK} < I_Z < I_{ZM}$).

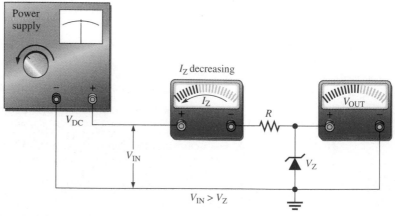

(b) As the input voltage decreases, the output voltage remains constant ($I_{ZK} < I_Z < I_{ZM}$).

▲ **FIGURE 3–8**

Zener regulation of a varying input voltage.

which the zener can operate. Resistor R is the series current-limiting resistor. The DMM displays indicate the relative values and trends.

For example, suppose that the 1N4740 10 V zener diode in Figure 3–9 can maintain regulation over a range of zener current values from $I_{ZK} = 0.25$ mA to $I_{ZM} = 100$ mA ($I_{ZM} = P_{D(max)}/V_Z = 1$ W/10 V $= 100$ mA). For the minimum zener current, the voltage across the 220 Ω resistor is

$$V_R = I_{ZK}R = (0.25 \text{ mA})(220 \text{ Ω}) = 55 \text{ mV}$$

Since $V_R = V_{IN} - V_Z$,

$$V_{IN(min)} \cong V_R + V_Z = 55 \text{ mV} + 10 \text{ V} = 10.055 \text{ V}$$

For the maximum zener current, the voltage across the 220 Ω resistor is

$$V_R = I_{ZM}R = (100 \text{ mA})(220 \text{ Ω}) = 22 \text{ V}$$

Therefore,

$$V_{IN(max)} \cong 22 \text{ V} + 10 \text{ V} = 32 \text{ V}$$

This shows that this zener diode can regulate an input voltage from 10.055 V to 32 V and maintain an approximate 10 V output. The output will vary slightly because of the zener impedance, which has been neglected in these calculations.

▶ **FIGURE 3–9**

EXAMPLE 3–5

Determine the minimum and the maximum input voltages that can be regulated by the zener diode in Figure 3–10.

▶ **FIGURE 3–10**

Solution　From the data sheet in Figure 3–7, the following information for the 1N4733 is obtained: $V_Z = 5.1$ V at $I_{ZT} = 49$ mA, $I_{ZK} = 1$ mA, and $Z_Z = 7$ Ω at I_{ZT}. For simplicity, assume this value of Z_Z over the range of current values. The equivalent circuit is shown in Figure 3–11. At $I_{ZK} = 1$ mA, the output voltage is

$$V_{OUT} \cong 5.1 \text{ V} - \Delta V_Z = 5.1 \text{ V} - (I_{ZT} - I_{ZK})Z_Z$$
$$= 5.1 \text{ V} - (48 \text{ mA})(7 \text{ Ω}) = 5.1 \text{ V} - 0.336 \text{ V} = 4.76 \text{ V}$$

Therefore,

$$V_{IN(min)} = I_{ZK}R + V_{OUT} = (1 \text{ mA})(100 \text{ Ω}) + 4.76 \text{ V} = \textbf{4.86 V}$$

To find the maximum input voltage, first calculate the maximum zener current. Assume the temperature is 50°C or below, so from the graph in Figure 3–7(b), the power dissipation is 1 W.

$$I_{ZM} = \frac{P_{D(max)}}{V_Z} = \frac{1 \text{ W}}{5.1 \text{ V}} = 196 \text{ mA}$$

At I_{ZM}, the output voltage is

$$V_{OUT} \cong 5.1 \text{ V} + \Delta V_Z = 5.1 \text{ V} + (I_{ZM} - I_{ZT})Z_Z$$
$$= 5.1 \text{ V} + (147 \text{ mA})(7 \text{ }\Omega) = 5.1 \text{ V} + 1.03 \text{ V} = 6.13 \text{ V}$$

Therefore,

$$V_{IN(max)} = I_{ZM}R + V_{OUT} = (196 \text{ mA})(100 \text{ }\Omega) + 6.13 \text{ V} = \textbf{25.7 V}$$

▲ **FIGURE 3–11**

Equivalent of circuit in Figure 3–10.

Related Problem Determine the minimum and maximum input voltages that can be regulated if a 1N4736 zener diode is used in Figure 3–10.

> Open file E03-05 in the Examples folder on your CD-ROM. For the calculated minimum and maximum dc input voltages, measure the resulting output voltages. Compare with the calculated values.

Zener Regulation with a Variable Load

Figure 3–12 shows a zener voltage regulator with a variable load resistor across the terminals. The zener diode maintains a nearly constant voltage across R_L as long as the zener current is greater than I_{ZK} and less than I_{ZM}.

▶ **FIGURE 3–12**

Zener regulation with a variable load.

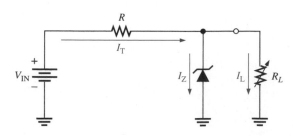

From No Load to Full Load

When the output terminals of the zener regulator are open ($R_L = \infty$), the load current is zero and *all* of the current is through the zener. When a load resistor (R_L) is connected, part of the total current is through the zener and part through R_L. As R_L is decreased, the load current, I_L increases and I_Z decreases. The zener diode continues to regulate the voltage until I_Z reaches its minimum value, I_{ZK}. At this point the load current is maximum. The total current through R remains essentially constant. The following example will illustrate this.

EXAMPLE 3–6

Determine the minimum and the maximum load currents for which the zener diode in Figure 3–13 will maintain regulation. What is the minimum R_L that can be used? $V_Z = 12$ V, $I_{ZK} = 1$ mA, and $I_{ZM} = 50$ mA. Assume $Z_Z = 0\ \Omega$ and V_Z remains a constant 12 V over the range of current values, for simplicity.

▶ **FIGURE 3–13**

Solution   When $I_L = 0$ A ($R_L = \infty$), I_Z is maximum and equal to the total circuit current I_T.

$$I_{Z(max)} = I_T = \frac{V_{IN} - V_Z}{R} = \frac{24\text{ V} - 12\text{ V}}{470\ \Omega} = 25.5\text{ mA}$$

Since $I_{Z(max)}$ is less than I_{ZM}, 0 A is an acceptable minimum value for I_L because the zener can handle all of the 25.5 mA. This means that R_L can be removed from the circuit and regulation will be maintained.

$$I_{L(min)} = \mathbf{0\ A}$$

The maximum value of I_L occurs when I_Z is minimum ($I_Z = I_{ZK}$), so solve for $I_{L(max)}$.

$$I_{L(max)} = I_T - I_{ZK} = 25.5\text{ mA} - 1\text{ mA} = \mathbf{24.5\ mA}$$

The minimum value of R_L is

$$R_{L(min)} = \frac{V_Z}{I_{L(max)}} = \frac{12\text{ V}}{24.5\text{ mA}} = \mathbf{490\ \Omega}$$

Therefore, if R_L is less than 490 Ω, R_L will draw more current away from the zener and I_Z will be reduced below I_{ZK}. This will cause the zener to lose regulation. In summary, regulation is maintained for any value of R_L between 490 Ω and infinity.

Related Problem   Find the minimum and maximum load currents for which the circuit in Figure 3–13 will maintain regulation. Determine the minimum R_L that can be used. $V_Z = 3.3$ V (constant), $I_{ZK} = 1$ mA, $I_{ZM} = 150$ mA. Assume $Z_Z = 0\ \Omega$ for simplicity.

Open file E03-06 in the Examples folder on your CD-ROM. For the calculated minimum value of load resistance, verify that regulation occurs.

In the last example, we assumed that Z_Z was zero and, therefore, the zener voltage remained constant over the range of currents. This assumption was made in order to more easily demonstrate the concept of how the regulator works with a varying load. Such a simplifying assumption is often acceptable and in some cases produces results that are accurate enough. In Example 3–7, we will take the zener impedance into account.

EXAMPLE 3–7

For the circuit in Figure 3–14:

(a) Determine V_{OUT} at I_{ZK} and at I_{ZM}.

(b) Calculate the value of R that should be used.

(c) Determine the minimum value of R_L that can be used.

▶ **FIGURE 3–14**

Solution   The 1N4744 zener used in the regulator circuit of Figure 3–14 is a 15 V diode. The data sheet in Figure 3–7(a) gives the following information: $V_Z = 15$ V @ I_{ZT}, $I_{ZK} = 0.25$ mA, $I_{ZT} = 17$ mA, and $Z_{ZT} = 14$ Ω.

(a) For I_{ZK}:

$$V_{OUT} = V_Z = 15 \text{ V} - \Delta I_Z Z_{ZT} = 15 \text{ V} - (I_{ZT} - I_{ZK})Z_{ZT}$$
$$= 15 \text{ V} - (16.75 \text{ mA})(14 \text{ }\Omega) = 15 \text{ V} - 0.235 \text{ V} = \textbf{14.76 V}$$

Calculate the zener maximum current. The power dissipation is 1 W.

$$I_{ZM} = \frac{P_{D(max)}}{V_Z} = \frac{1 \text{ W}}{15 \text{ V}} = 66.7 \text{ mA}$$

For I_{ZM}:

$$V_{OUT} = V_Z = 15 \text{ V} + \Delta I_Z Z_{ZT}$$
$$= 15 \text{ V} + (I_{ZM} - I_{ZT})Z_{ZT} = 15 \text{ V} + (49.7 \text{ mA})(14 \text{ }\Omega) = \textbf{15.7 V}$$

(b) The value of R is calculated for the maximum zener current that occurs when there is no load as shown in Figure 3–15(a).

$$R = \frac{V_{IN} - V_Z}{I_{ZM}} = \frac{24 \text{ V} - 15.7 \text{ V}}{66.7 \text{ mA}} = 124 \text{ }\Omega$$

$R = \textbf{130 Ω}$ (nearest larger standard value).

▶ **FIGURE 3–15**

(a)

(b)

(c) For the minimum load resistance (maximum load current), the zener current is minimum ($I_{ZK} = 0.25$ mA) as shown in Figure 3–15(b).

$$I_T = \frac{V_{IN} - V_{OUT}}{R} = \frac{24 \text{ V} - 14.76 \text{ V}}{130 \text{ }\Omega} = 71.0 \text{ mA}$$

$$I_L = I_T - I_{ZK} = 71.0 \text{ mA} - 0.25 \text{ mA} = 70.75 \text{ mA}$$

$$R_{L(min)} = \frac{V_{OUT}}{I_L} = \frac{14.76 \text{ V}}{70.75 \text{ mA}} = \mathbf{209 \text{ }\Omega}$$

Related Problem Repeat each part of the preceding analysis if the zener is changed to a 12 V device (1N4742).

Zener Limiting

In addition to voltage regulation applications, zener diodes can be used in ac applications to limit voltage swings to desired levels. Figure 3–16 shows three basic ways the limiting action of a zener diode can be used. Part (a) shows a zener used to limit the positive peak of a signal voltage to the selected zener voltage. During the negative alternation, the zener acts as a forward-biased diode and limits the negative voltage to −0.7 V. When the zener is turned around, as in part (b), the negative peak is limited by zener action and the positive voltage is limited to +0.7 V. Two back-to-back zeners limit both peaks to the zener voltage plus or minus 0.7 V, as shown in part (c). During the positive alternation, D_2 is functioning as the zener limiter and D_1 is functioning as a forward-biased diode. During the negative alternation, the roles are reversed.

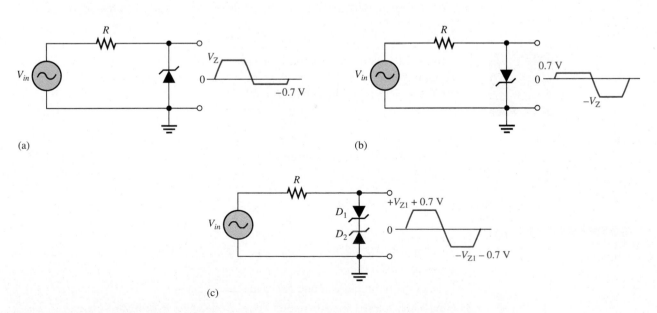

▲ FIGURE 3–16

Basic zener limiting action with a sinusoidal input voltage.

EXAMPLE 3–8

Determine the output voltage for each zener limiting circuit in Figure 3–17.

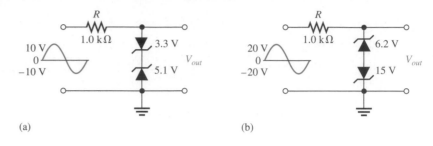

(a)   (b)

▲ **FIGURE 3–17**

Solution   See Figure 3–18 for the resulting output voltages. Remember, when one zener is operating in breakdown, the other one is forward-biased with approximately 0.7 V across it.

(a)   (b)

▲ **FIGURE 3–18**

Related Problem   **(a)** What is the output in Figure 3–17(a) if the input voltage is increased to a peak value of 20 V?

(b) What is the output in Figure 3–17(b) if the input voltage is decreased to a peak value of 5 V?

Open file E03-08 in the Examples folder on your CD-ROM. For the specified input voltages, measure the resulting output waveforms. Compare with the waveforms shown in the example.

SECTION 3–2 REVIEW

1. In a zener diode regulator, what value of load resistance results in the maximum zener current?

2. Explain the terms *no-load* and *full-load*.

3. How much voltage appears across a zener diode when it is forward-biased?

3-3 VARACTOR DIODES

Varactor diodes are also known as variable-capacitance diodes because the junction capacitance varies with the amount of reverse-bias voltage. Varactors are specifically designed to take advantage of this variable-capacitance characteristic. These devices are commonly used in electronic tuning circuits used in communications systems.

After completing this section, you should be able to

- ■ **Describe the variable-capacitance characteristics of a varactor diode and analyze its operation in a typical circuit**
- ■ Identify a varactor diode symbol
- ■ Explain why a reverse-biased varactor exhibits capacitance
- ■ Discuss how the capacitance varies with reverse-bias voltage
- ■ Interpret a varactor data sheet
- ■ Define *tuning ratio*
- ■ Define *quality factor, Q*
- ■ Discuss varactor temperature coefficients
- ■ Analyze a varactor-tuned band-pass filter

A **varactor** is a diode that always operates in reverse-bias and is doped to maximize the inherent capacitance of the depletion region. The depletion region, widened by the reverse bias, acts as a capacitor dielectric because of its nonconductive characteristic. The *p* and *n* regions are conductive and act as the capacitor plates, as illustrated in Figure 3–19.

▶ **FIGURE 3–19**

The reverse-biased varactor diode acts as a variable capacitor.

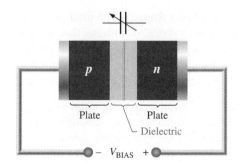

Plate Plate
Dielectric
– V_{BIAS} +

Basic Operation

As the reverse-bias voltage increases, the depletion region widens, effectively increasing the plate separation and the dielectric thickness and thus decreasing the capacitance. When the reverse-bias voltage decreases, the depletion region narrows, thus increasing the capacitance. This action is shown in Figure 3–20(a) and (b). A graph of diode capacitance (C_T) versus reverse voltage for a certain varactor is shown in Figure 3–20(c). For this particular device, C_T varies from 40 pF to slightly greater than 4 pF as V_R varies from 1 V to 40 V.

(a) Greater reverse bias, less capacitance   (b) Less reverse bias, greater capacitance   (c) Graph of diode capacitance versus reverse voltage

▲ **FIGURE 3–20**

Varactor diode capacitance varies with reverse voltage.

▲ **FIGURE 3–21**

Varactor diode symbol.

Recall that capacitance is determined by the parameters of plate area (A), dielectric constant (ϵ), and dielectric thickness (d), as expressed in the following formula:

$$C = \frac{A\epsilon}{d}$$

In a varactor diode, these capacitance parameters are controlled by the method of doping near the pn junction and the size and geometry of the diode's construction. Nominal varactor capacitances are typically available from a few picofarads to several hundred picofarads. Figure 3–21 shows a common symbol for a varactor.

Varactor Data Sheet Information

A partial data sheet for a specific series of varactor diodes (1N5139–1N5148) is shown in Figure 3–22. The values of nominal diode capacitance, C_T, are measured at a reverse voltage of 4 V dc and range from 6.8 pF to 47 pF for this particular series.

Capacitance Tolerance Range The minimum and maximum values of C_T are based on 10% tolerance. For example, this means that when reverse-biased at 4 V, the 1N5139 can exhibit a capacitance anywhere between 6.1 pF and 7.5 pF. This tolerance range should not be confused with the range of capacitance values that result from varying the reverse bias as determined by the tuning ratio, which we will discuss next.

Tuning Ratio The varactor **tuning ratio** is also called the *capacitance ratio*. It is the ratio of the diode capacitance at a minimum reverse voltage to the diode capacitance at a maximum reverse voltage. For the varactor diodes represented in Figure 3–22, the tuning ratio is the ratio of C_T measured at a V_R of 4 V divided by C_T measured at a V_R of 60 V. The tuning ratio is designated as C_4/C_{60} in this case.

For the 1N5139, the typical tuning ratio is 2.9. This means that the capacitance value decreases by a factor of 2.9 as V_R is increased from 4 V to 60 V. The following calculation illustrates how to use the tuning ratio (TR) to find the capacitance range for the 1N5139. From the data table in Figure 3–22(a), $C_4 = 6.8$ pF, and the typical $TR = C_4/C_{60} = 2.9$. Therefore,

$$C_{60} = \frac{C_4}{TR} = \frac{6.8 \text{ pF}}{2.9} = 2.3 \text{ pF}$$

The diode capacitance varies from 6.8 pF to 2.3 pF when V_R is increased from 4 V to 60 V.

The capacitance range can also be determined from the graph in Figure 3–22(b), which shows how the varactor capacitance varies for reverse voltages from 1 V to 60 V.

Maximum Ratings ($T_C = 25°C$ unless otherwise noted)

Rating	Symbol	Value	Unit
Reverse voltage	V_R	60	Volts
Forward current	I_F	250	mA
RF power input*	P_{in}	5.0	Watts
Device dissipation @ $T_A = 25°C$ Derate above 25°C	P_D	400 2.67	mW mW/°C
Device dissipation @ $T_C = 25°C$ Derate above 25°C	P_C	2.0 13.3	Watts mW/°C
Junction temperature	T_J	+175	°C
Storage temperature range	T_{stg}	−65 to +200	°C

*The RF power input rating assumes that an adequate heat sink is provided.

Electrical Characteristics ($T_A = 25°C$ unless otherwise noted)

Characteristic	Symbol	Min	Typ	Max	Unit
Reverse breakdown voltage ($I_R = 10$ μA dc)	$V_{(BR)R}$	60	70	–	V dc
Reverse voltage leakage current ($V_R = 55$ V dc, $T_A = 25°C$) ($V_R = 55$ V dc, $T_A = 150°C$)	I_R	– –	– –	0.02 20	μA dc
Series inductance ($f = 250$ MHz, $L ≈ 1/16''$)	I_S	–	5.0	–	nH
Case capacitance ($f = 1.0$ MHz, $L ≈ 1/16''$)	C_C	–	0.25	–	pF
Diode capacitance temperature coefficient ($V_R = 4.0$ V dc, $f = 1.0$ MHz)	TC_C	–	200	300	ppm/°C

	C_T, Diode Capacitance $V_R = 4.0$ V dc, $f = 1.0$ MHz pF			Q, Figure of Merit $V_R = 4.0$ V dc $f = 50$ MHz	TR, Tuning Ratio C_4/C_{60} $f = 1.0$ MHz	
Device	Min	Typ	Max	Min	Min	Typ
1N5139	6.1	6.8	7.5	350	2.7	2.9
1N5140	9.0	10	11	300	2.8	3.0
1N5141	10.8	12	13.2	300	2.8	3.0
1N5142	13.5	15	16.5	250	2.8	3.0
1N5143	16.2	18	19.8	250	2.8	3.0
1N5144	19.8	22	24.2	200	3.2	3.4
1N5145	24.3	27	29.7	200	3.2	3.4
1N5146	29.7	33	36.3	200	3.2	3.4
1N5147	36.1	39	42.9	200	3.2	3.4
1N5148	42.3	47	51.7	200	3.2	3.4

(a) Electrical characteristics

(b) Diode capacitance

(c) Figure of merit

▲ **FIGURE 3–22**

Partial data sheet for the 1N5139–1N5148 varactor diodes.

On the graph, you can see that the capacitance for the 1N5139 is approximately 10.5 pF at $V_R = 1$ V and approximately 2.3 pF at $V_R = 60$ V.

The 1N51xx series of varactors are abrupt junction devices. The doping in the n and p regions is made uniform so that at the pn junction there is a relatively abrupt change from n to p instead of the more gradual change found in the rectifier diodes. The abruptness of the pn junction determines the tuning ratio. Other types of varactors such as the MV1401 are hyper-abrupt devices in which the doping pattern results in an even more abrupt junction. Many hyper-abrupt varactors exhibit tuning ratios from 10 to 15.

Figure of Merit The figure of merit or **quality factor (*Q*)** of a reactive component is the ratio of energy stored and then returned by a capacitor (or inductor) to the energy dissipated in the resistance. The 1N5139 has a minimum *Q* of 350 at $V_R = 4$ V, which indicates that the energy stored and returned by the diode capacitance is 350 times greater than the energy lost in the resistance of the device. High values of *Q* are desirable. Figure 3–22(c) is a graph showing how the typical figure of merit increases with increasing reverse voltage for three varactors in the series.

Temperature Coefficients The diode capacitance has a positive temperature coefficient so C_T increases a small amount as the temperature increases. The figure of merit has a negative temperature coefficient, so *Q* decreases as the temperature increases.

An Application

A major application of varactors is in tuning circuits. For example, electronic tuners in TV and other commercial receivers utilize varactors. When used in a resonant circuit, the varactor acts as a variable capacitor, thus allowing the resonant frequency to be adjusted by a variable voltage level, as illustrated in Figure 3–23 where the varactor diode provides the total variable capacitance in the parallel resonant band-pass filter.

▲ **FIGURE 3–23**

A resonant band-pass filter using a varactor diode for adjusting the resonant frequency over a specified range.

The varactor diode and the inductor form a parallel resonant circuit from the output to ac ground. Capacitors C_1, C_2, C_3, and C_4 are coupling capacitors to prevent the dc bias circuit from being loaded by the filter circuit. These capacitors have no effect on the filter's frequency response because their reactances are negligible at the resonant frequencies. C_1 prevents a dc path from the potentiometer wiper back to the ac source through the inductor and R_1. C_2 prevents a dc path from the cathode to the anode of the varactor through the inductor. C_3 prevents a dc path from the wiper to a load on the output through the inductor. C_4 prevents a dc path from the wiper to ground.

Resistors R_2, R_3, R_5, and potentiometer R_4 form a variable dc voltage divider for biasing the varactor. The reverse-bias voltage across the varactor can be varied with the potentiometer.

Recall that the parallel resonant frequency is

$$f_r \cong \frac{1}{2\pi\sqrt{LC}}$$

EXAMPLE 3–9

For the varactor-tuned band-pass filter in Figure 3–24, determine the range of resonant frequencies over which it can be adjusted. The values of the bias resistors are selected to prevent significant ac loading on the filter.

▲ **FIGURE 3–24**

Solution   From the data sheet information in Figure 3–22(a), the 1N5148 varactor has a nominal capacitance of 47 pF at a reverse bias of 4 V.

First, determine the range of reverse-bias voltages for the filter circuit. The dc voltage at the cathode (V_K) of the varactor is fixed at

$$V_K = \left(\frac{R_3 + R_4 + R_5}{R_2 + R_3 + R_4 + R_5} \right) V_{BIAS} = \left(\frac{2.44\ M\Omega}{4.64\ M\Omega} \right) 60\ V = 31.6\ V$$

The dc voltage at the anode (V_A) of the varactor can be varied from a minimum to a maximum with the potentiometer R_4.

$$V_{A(min)} = \left(\frac{R_5}{R_2 + R_3 + R_4 + R_5} \right) V_{BIAS} = \left(\frac{220\ k\Omega}{4.64\ M\Omega} \right) 60\ V = 2.85\ V$$

$$V_{A(max)} = \left(\frac{R_4 + R_5}{R_2 + R_3 + R_4 + R_5} \right) V_{BIAS} = \left(\frac{2.22\ M\Omega}{4.64\ M\Omega} \right) 60\ V = 28.7\ V$$

The minimum and maximum values for the reverse voltage, V_R, are determined as follows:

$$V_{R(min)} = V_K - V_{A(max)} = 31.6\ V - 28.7\ V = 2.9\ V$$

$$V_{R(max)} = V_K - V_{A(min)} = 31.6\ V - 2.85\ V = 29\ V$$

Although it is difficult to get exact figures from the graph in Figure 3–22(b), the approximate capacitance values of the varactor at 2.9 V and 29 V are $C_{2.9} \cong 55$ pF and $C_{29} \cong 17$ pF. The minimum resonant frequency for the filter is

$$f_{r(min)} \cong \frac{1}{2\pi \sqrt{LC}} = \frac{1}{2\pi \sqrt{(1\ mH)(55\ pF)}} = \textbf{679 kHz}$$

The maximum resonant frequency for the filter is

$$f_{r(max)} \cong \frac{1}{2\pi\sqrt{LC}} = \frac{1}{2\pi\sqrt{(1\ mH)(17\ pF)}} = \textbf{1.22 MHz}$$

Related Problem   If the bias voltage source in Figure 3–24 is reduced to 30 V, determine the range of the reverse voltage across the varactor.

SECTION 3–3 REVIEW

1. What is the key feature of a varactor diode?
2. Under what bias condition is a varactor operated?
3. What part of the varactor produces the capacitance?
4. Based on the graph in Figure 3–22(b), what happens to the diode capacitance when the reverse voltage is increased?
5. Define *tuning ratio*.

3–4   OPTICAL DIODES

In this section, two types of optoelectronic devices—the light-emitting diode (LED) and the photodiode—are introduced. As the name implies, the LED is a light emitter. The photodiode, on the other hand, is a light detector. We will examine the characteristics of both devices, and you will see an example of their use in a system application at the end of the chapter.

After completing this section, you should be able to

- **Discuss the operation and characteristics of LEDs and photodiodes**
- Identify LED and photodiode symbols
- Explain basically how an LED emits light
- Analyze the spectral output curves and radiation patterns of LEDs
- Interpret an LED data sheet
- Define *radiant intensity* and *irradiance*
- Use an LED seven-segment display
- Explain how a photodiode detects light
- Analyze the response curve of a photodiode
- Interpret a photodiode data sheet
- Discuss photodiode sensitivity

The Light-Emitting Diode (LED)

The symbol for an LED is shown in Figure 3–25.

▶ **FIGURE 3–25**

Symbol for an LED. When forward-biased, it emits light.

The basic operation of the **light-emitting diode (LED)** is as follows. When the device is forward-biased, electrons cross the *pn* junction from the *n*-type material and recombine with holes in the *p*-type material. Recall from Chapter 1 that these free electrons are in the conduction band and at a higher energy level than the holes in the valence band. When recombination takes place, the recombining electrons release energy in the form of heat and light. A large exposed surface area on one layer of the semiconductive material permits the **photons** to be emitted as visible light. This process, called **electroluminescence**, is illustrated in Figure 3–26. Various impurities are added during the doping process to establish the wavelength of the emitted light. The wavelength determines the color of the light and if it is visible or invisible (infrared).

▶ **FIGURE 3–26**

Electroluminescence in a forward–biased LED.

Light energy

p region

n region

Semiconductive Materials LEDs are made of gallium arsenide (GaAs), gallium arsenide phosphide (GaAsP), or gallium phosphide (GaP). Silicon and germanium are not used because they are essentially heat-producing materials and are very poor at producing light. GaAs LEDs emit **infrared (IR)** radiation, which is nonvisible, GaAsP produces either red or yellow visible light, and GaP emits red or green visible light. LEDs that emit blue light are also available. Red is the most common.

LED Biasing The forward voltage across an LED is considerably greater than for a silicon diode. Typically the maximum V_F for LEDs is between 1.2 V and 3.2 V, depending on the device. Reverse breakdown for an LED is much less than for a silicon rectifier diode (3 V to 10 V is typical).

The LED emits light in response to a sufficient forward current, as shown in Figure 3–27(a). The amount of power output translated into light is directly proportional to the forward current, as indicated in Figure 3–27(b). An increase in I_F corresponds proportionally to an increase in light output.

(a) Forward-biased operation

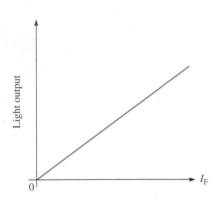

(b) General light output versus forward current

▲ **FIGURE 3–27**

Basic operation of an LED.

Light Emission   The **wavelength** of light determines whether it is visible or infrared. An LED emits light over a specified range of wavelengths as indicated by the **spectral** output curves in Figure 3–28. The curves in part (a) represent the light output versus wavelength for typical visible LEDs, and the curve in part (b) is for a typical infrared LED. The wavelength (λ) is expressed in nanometers (nm). The normalized output of the visible red LED peaks at 660 nm, the yellow at 590 nm, green at 540 nm, and blue at 460 nm. The output for the infrared LED peaks at 940 nm.

(a) Visible light

(b) Nonvisible infrared (IR)

▲ **FIGURE 3–28**

Examples of typical spectral output curves for LEDs.

The graph in Figure 3–29 is the **radiation** pattern for a typical LED. It shows how directional the emitted light is. The radiation pattern depends on the type of lens structure of the LED. The narrower the radiation pattern, the more the light is concentrated in a particular direction. Also, colored lenses are used to enhance the color.

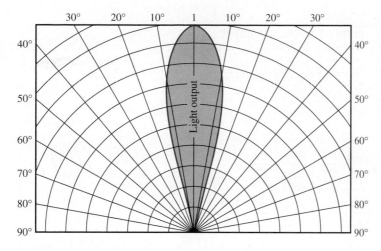

▲ **FIGURE 3–29**

General radiation pattern of a typical LED.

Typical LEDs are shown in Figure 3–30. Photodiodes, to be studied next, generally have a similar appearance.

▶ **FIGURE 3–30**

Typical LEDs.

Cathode
(lead on right
looking from front)

Anode
(longer lead)

Anode
(lead near tab)

LED Data Sheet Information

A partial data sheet for an MLED81 infrared light-emitting diode is shown in Figure 3–31. Notice that the maximum reverse voltage is only 5 V, the maximum continuous forward current is 100 mA, and the forward voltage drop is 1.35 V for $I_F = 100$ mA.

From the graph in part (c), you can see that the peak power output for this device occurs at a wavelength of 940 nm; its radiation pattern is shown in part (d). At 30° on either side of the maximum orientation, the output power drops to approximately 60% of maximum.

Radiant Intensity and Irradiance   In Figure 3–31(a), the axial **radiant intensity**, I_e (symbol not to be confused with current), is the output power per steradian and is specified as 15 mW/sr. The steradian (sr) is the unit of solid angular measurement. **Irradiance, H,** is the power per unit area at a given distance from an LED source expressed in mW/cm^2 and can be calculated using radiant intensity and the distance in centimeters (cm) using the following formula:

$$H = \frac{I_e}{d^2}$$

Equation 3–6

Irradiance is important because the response of a detector (photodiode) used in conjunction with an LED depends on the irradiance of the light it receives. We will discuss this further in relation to photodiodes.

Maximum Ratings

Rating	Symbol	Value	Unit
Reverse voltage	V_R	5	Volts
Forward current — continuous	I_F	100	mA
Forward current — peak pulse	I_F	1	A
Total power dissipation @ $T_A = 25°C$	P_D	100	mW
Derate above 25°C		2.2	mW/°C
Ambient operating temperature range	T_A	−30 to +70	°C
Storage temperature	T_{stg}	−30 to +80	°C
Lead soldering temperature, 5 seconds max, 1/16 inch from case	—	260	°C

Electrical Characteristics ($T_A = 25°C$ unless otherwise noted)

Characteristic	Symbol	Min	Typ	Max	Unit
Reverse leakage current ($V_R = 3$ V)	I_R	—	10	—	nA
Reverse leakage current ($V_R = 5$ V)	I_R	—	1	10	μA
Forward voltage ($I_F = 100$ mA)	V_F	—	1.35	1.7	V
Temperature coefficient of forward voltage	ΔV_F	—	− 1.6	—	mV/K
Capacitance ($f = 1$ MHz)	C	—	25	—	pF

Optical Characteristics ($T_A = 25°C$ unless otherwise noted)

Characteristic	Symbol	Min	Typ	Max	Unit
Peak wavelength ($I_F = 100$ mA)	λp	—	940	—	nm
Spectral half-power bandwidth	$\Delta\lambda$	—	50	—	nm
Total power output ($I_F = 100$ mA)	ϕe	—	16	—	mW
Temperature coefficient of total power output	$\Delta\phi e$	—	− 0.25	—	%/K
Axial radiant intensity ($I_F = 100$ mA)	I_e	10	15	—	mW/sr
Temperature coefficient of axial radiant intensity	ΔI_e	—	− 0.25	—	%/K
Power half-angle	ϕ	—	±30	—	°

MLED 81

(a) Ratings and characteristics

(b) LED forward voltage versus forward current

(c) Relative spectral emission

(d) Spatial radiation pattern

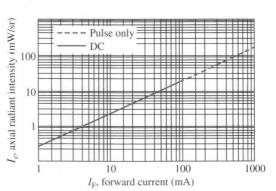

(e) Intensity versus forward current

▲ **FIGURE 3–31**

Partial data sheet for an MLED81 IR light–emitting diode.

EXAMPLE 3–10

From the LED data sheet in Figure 3–31 determine the following:

(a) The radiant intensity at 900 nm if the maximum output is 15 mW/sr.

(b) The forward voltage drop for $I_F = 20$ mA.

(c) The radiant intensity for $I_F = 30$ mA.

(d) The maximum irradiance at a distance of 10 cm from the LED source.

Solution (a) From the relative spectral emission graph in Figure 3–31(c), the relative radiant intensity at 900 nm is approximately 0.75. The radiant intensity is, therefore,

$$I_e = 0.75(15 \text{ mW/sr}) = \textbf{11.3 mW/sr}$$

(b) From the graph in part (b), $V_F = \textbf{1.23 V}$ for $I_F = 20$ mA.

(c) From the graph in part (e), $I_e = \textbf{5 mW/sr}$ for $I_F = 30$ mA.

(d) $H = \dfrac{I_e}{d^2} = \dfrac{15 \text{ mW/sr}}{(10 \text{ cm})^2} = \textbf{0.15 mW/cm}^2$

Related Problem If $I_e = 12$ mW/sr, at a wavelength of 940 nm, determine the radiant intensity at 1000 nm.

Applications LEDs are used for indicator lamps and readout displays on a wide variety of instruments, ranging from consumer appliances to scientific apparatus. A common type of display device using LEDs is the seven-segment display. Combinations of the segments form the ten decimal digits as illustrated in Figure 3–32. Each segment in the display is an LED. By forward-biasing selected combinations of segments, any decimal digit and a decimal point can be formed. Two types of LED circuit arrangements are the common anode and common cathode as shown.

(a) LED segment arrangement and typical device

(b) Common anode   (c) Common cathode

▲ **FIGURE 3–32**

The 7-segment LED display.

IR-emitting diodes are used in optical coupling applications, often in conjunction with fiber optics. Areas of application include industrial processing and control, position encoders, bar graph readers, and optical switching.

The Photodiode

The **photodiode** is a device that operates in reverse bias, as shown in Figure 3–33(a), where I_λ is the reverse current. The photodiode has a small transparent window that allows light to strike the *pn* junction. Some typical photodiodes are shown in Figure 3–33(b). An alternate photodiode symbol is shown in Figure 3–33(c).

(a) Reverse-bias operation   (b) Typical devices   (c) Alternate symbol

▲ **FIGURE 3–33**

Photodiode.

Recall that when reverse-biased, a rectifier diode has a very small reverse leakage current. The same is true for a photodiode. The reverse-biased current is produced by thermally generated electron-hole pairs in the depletion region, which are swept across the *pn* junction by the electric field created by the reverse voltage. In a rectifier diode, the reverse leakage current increases with temperature due to an increase in the number of electron-hole pairs.

A photodiode differs from a rectifier diode in that when its *pn* junction is exposed to light, the reverse current increases with the light intensity. When there is no incident light, the reverse current, I_λ, is almost negligible and is called the **dark current**. An increase in the amount of light intensity, expressed as irradiance (mW/cm^2), produces an increase in the reverse current, as shown by the graph in Figure 3–34(a).

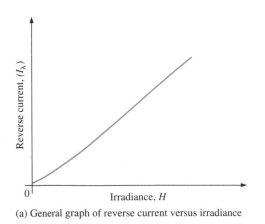

(a) General graph of reverse current versus irradiance

(b) Example of a graph of reverse current versus reverse voltage for several values of irradiance

▲ **FIGURE 3–34**

Typical photodiode characteristics.

From the graph in Figure 3–34(b), you can see that the reverse current for this particular device is approximately 1.4 μA at a reverse-bias voltage of 10 V with an irradiance of 0.5 mW/cm². Therefore, the resistance of the device is

$$R_R = \frac{V_R}{I_\lambda} = \frac{10 \text{ V}}{1.4 \ \mu\text{A}} = 7.14 \text{ M}\Omega$$

At 20 mW/cm², the current is approximately 55 μA at V_R = 10 V. The resistance under this condition is

$$R_R = \frac{V_R}{I_\lambda} = \frac{10 \text{ V}}{55 \ \mu\text{A}} = 182 \text{ k}\Omega$$

These calculations show that the photodiode can be used as a variable-resistance device controlled by light intensity.

Figure 3–35 illustrates that the photodiode allows essentially no reverse current (except for a very small dark current) when there is no incident light. When a light beam strikes the photodiode, it conducts an amount of reverse current that is proportional to the light intensity (irradiance).

(a) No light, no current except negligible dark current

(b) Where there is incident light, resistance decreases and there is reverse current.

▲ **FIGURE 3–35**

Operation of a photodiode.

Photodiode Data Sheet Information

A partial data sheet for an MRD821 photodiode is shown in Figure 3–36. Notice that the maximum reverse voltage is 35 V and the dark current (reverse current with no light) is typically 3 nA for a reverse voltage of 10 V. As the graphs in parts (b) and (c) show, the dark current (leakage current) increases with an increase in reverse voltage and also with an increase in temperature.

Sensitivity   From the graph in part (d), you can see that the maximum sensitivity for this device occurs at a wavelength of 940 nm. The angular response graph in part (e) shows a broad area of response measured as relative sensitivity. At 50° on either side of the maximum orientation, the sensitivity drops to approximately 80% of maximum.

Maximum Ratings

Rating	Symbol	Value	Unit
Reverse voltage	V_R	35	Volts
Forward current — continuous	I_F	100	mA
Total power dissipation @ $T_A = 25°C$ Derate above 25°C	P_D	150 3.3	mW mW/°C
Ambient operating temperature range	T_A	−30 to +70	°C
Storage temperature	T_{stg}	−40 to +80	°C
Lead soldering temperature, 5 seconds max, 1/16 inch from case	—	260	°C

Electrical Characteristics ($T_A = 25°C$ unless otherwise noted)

Characteristic	Symbol	Min	Typ	Max	Unit
Dark current ($V_R = 10$ V)	I_D	—	3	30	nA
Capacitance ($f = 1$ MHz, $V = 0$)	C_J	—	175	—	pF

Optical Characteristics ($T_A = 25°C$ unless otherwise noted)

Characteristic	Symbol	Min	Typ	Max	Unit
Wavelength of maximum sensitivity	λmax	—	940	—	nm
Spectral range	$\Delta\lambda$	—	170	—	nm
Sensitivity ($\lambda = 940$ nm, $V_R = 20$ V)	S	—	50	—	$\mu A/mW/cm^2$
Temperature coefficient of sensitivity	ΔS	—	0.18	—	%/K
Acceptance half-angle	φ	—	±70	—	°
Short circuit current (Ev = 1000 lux)	I_S	—	50	—	μA
Open circuit voltage (Ev = 1000 lux)	V_L	—	0.3	—	V

(a) Ratings and characteristics

MRD 821

(b) Dark current versus reverse voltage

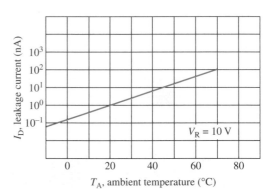

(c) Dark current versus temperature

(d) Relative spectral sensitivity

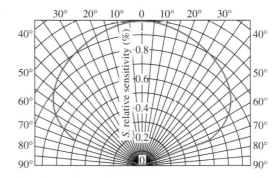

(e) Angular response

▲ **FIGURE 3–36**

Partial data sheet for the MRD821 photodiode.

In Figure 3–36(a), the typical sensitivity is specified as 50 μA/mW/cm^2 for a wavelength of 940 nm and a reverse voltage of 20 V. This means, for example, that if the irradiance is 1 mW/cm^2, there are 50 μA of reverse current and if the irradiance is 0.5 mW/cm^2, there are 25 μA of reverse current.

EXAMPLE 3–11

An MRD821 photodiode is exposed to a 1000 nm infrared light with an irradiance (H) of 2.5 mW/cm^2. The angle at which the light strikes the photodiode is 35°. Determine the response of the photodiode in terms of the reverse current (I_λ) through the device.

Solution   From the photodiode data sheet in Figure 3–36, the sensitivity of the photodiode is 50 μA/mW/cm^2 at 940 nm. The light on the photodiode is at a wavelength of 1000 nm. From the data sheet graph in part (d), the sensitivity (S) at 1000 nm is approximately 83% of the sensitivity at 940 nm.

$$S_{1000} = 0.83S_{940} = 0.83(50~\mu\text{A/mW/cm}^2) = 41.5~\mu\text{A/mW/cm}^2$$

Also, the angle at which the light strikes the photodiode reduces the sensitivity further. From the graph in Figure 3–36(e), at an angle of 35° from the maximum orientation (0°), the relative sensitivity is approximately 90%.

$$S = 0.9(41.5~\mu\text{A/mW/cm}^2) = 37.4~\mu\text{A/mW/cm}^2$$

For an irradiance, H, of 2.5 mW/cm^2, the reverse current is

$$I_\lambda = S \times H = (37.4~\mu\text{A/mW/cm}^2)(2.5~\text{mW/cm}^2) = \mathbf{93.5~\mu A}$$

Related Problem   Determine the MRD821 response (reverse current) to an irradiance of 1 mW/cm^2 for a wavelength of 900 nm at an angle of 60° from maximum orientation.

**SECTION 3–4
REVIEW**

1. Name two types of LEDs in terms of their light-emission spectrum.
2. Which has the greater wavelength, visible light or infrared?
3. In what bias condition is an LED normally operated?
4. What happens to the light emission of an LED as the forward current increases?
5. The forward voltage drop of an LED is 0.7 V. (true or false)
6. In what bias condition is a photodiode normally operated?
7. When the intensity of the incident light (irradiance) on a photodiode increases, what happens to its internal reverse resistance?
8. What is *dark current*?

3–5 OTHER TYPES OF DIODES

In this section, several types of diodes that you are less likely to encounter as a technician but are nevertheless important are introduced. Among these are the current regulator, the Schottky diode, the *pin* diode, the step-recovery diode, the tunnel diode, and the laser diode.

After completing this section, you should be able to

- **Discuss the basic characteristics of the current regulator diode, the Schottky diode, the *pin* diode, the step-recovery diode, the tunnel diode, and the laser diode**

- Identify the various diode symbols

- Discuss how the current regulator diode maintains a constant forward current

- Describe the characteristics of the Schottky diode

- Describe the characteristics of the *pin* diode

- Describe the characteristics of the step-recovery diode

- Describe the characteristics of the tunnel diode and explain its negative resistance

- Describe the laser diode and how it differs from an LED

Current Regulator Diode

The current regulator diode is often referred to as a constant-current diode. Rather than maintaining a constant voltage, as the zener diode does, this diode maintains a constant current. The symbol is shown in Figure 3–37.

A typical characteristic curve is shown in Figure 3–38. The current regulator diode operates in forward bias (shaded region) and the forward current becomes a specified constant value at forward voltages ranging from about 1.5 V to about 6 V, depending on the diode type. The constant forward current is called the *regulator current* and is designated I_P. For example, the 1N5283–1N5314 series have nominal regulator currents ranging from 220 μA to 4.7 mA. These diodes may be used in parallel to obtain higher currents. This diode does not have a sharply defined reverse breakdown, so the reverse current begins to increase for V_{AK} values of less than 0 V (unshaded region). This device should never be operated in reverse bias.

Anode ——————○——|———— Cathode

▲ **FIGURE 3–37**

Symbol for a current regulator diode.

▲ **FIGURE 3–38**

Typical characteristic curve for a current regulator diode.

In forward bias, the diode regulation begins at the limiting voltage, V_L, and extends up to the POV (peak operating voltage). Notice that between V_K and POV, the current is essentially constant. V_T is the test voltage at which I_P and the diode impedance, Z_T, are specified on a data sheet. The impedance Z_T has very high values ranging from 235 kΩ to 25 MΩ for the diode series mentioned before.

The Schottky Diode

Schottky diodes are used primarily in high-frequency and fast-switching applications. They are also known as *hot-carrier diodes*. A Schottky diode symbol is shown in Figure 3–39. A Schottky diode is formed by joining a doped semiconductor region (usually *n*-type) with a metal such as gold, silver, or platinum. So, rather than a *pn* junction, there is a metal-to-semiconductor junction, as shown in Figure 3–40. The forward voltage drop is typically around 0.3 V.

▲ **FIGURE 3–39**

Schottky diode symbol.

▶ **FIGURE 3–40**

Basic internal construction of a Schottky diode.

The Schottky diode operates only with majority carriers. There are no minority carriers and thus no reverse leakage current as in other types of diodes. The metal region is heavily occupied with conduction-band electrons, and the *n*-type semiconductor region is lightly doped. When forward-biased, the higher energy electrons in the *n* region are injected into the metal region where they give up their excess energy very rapidly. Since there are no minority carriers, as in a conventional rectifier diode, there is a very rapid response to a change in bias. The Schottky is a very fast-switching diode, and most of its applications make use of this property. It can be used in high-frequency applications and in many digital circuits to decrease switching times.

The *PIN* Diode

The *pin* diode consists of heavily doped *p* and *n* regions separated by an intrinsic (*i*) region, as shown in Figure 3–41(a). When reverse-biased, the *pin* diode acts like a nearly constant capacitance. When forward-biased, it acts like a current-controlled variable resistance. This is shown in Figure 3–41(b) and (c). The low forward resistance of the intrinsic region decreases with increasing current.

(a) Construction   (b) Reverse-biased   (c) Forward-biased

▲ **FIGURE 3–41**

PIN diode.

▲ **FIGURE 3–42**

PIN diode characteristics.

The forward series resistance characteristic and the reverse capacitance characteristic are shown graphically in Figure 3–42 for a typical *pin* diode.

The *pin* diode is used as a dc-controlled microwave switch operated by rapid changes in bias or as a modulating device that takes advantage of the variable forward-resistance characteristic. Since no rectification occurs at the *pn* junction, a high-frequency signal can be modulated (varied) by a lower-frequency bias variation. A *pin* diode can also be used in attenuator applications because its resistance can be controlled by the amount of current. Certain types of *pin* diodes are used as photodetectors in fiber-optic systems.

The Step-Recovery Diode

The step-recovery diode employs graded doping where the doping level of the semiconductive materials is reduced as the *pn* junction is approached. This produces an abrupt turn-off time by allowing a very fast release of stored charge when switching from forward to reverse bias. It also allows a rapid re-establishment of forward current when switching from reverse to forward bias. This diode is used in very high frequency (VHF) and fast-switching applications.

The Tunnel Diode

The **tunnel diode** exhibits a special characteristic known as *negative resistance*. This feature makes it useful in oscillator and microwave amplifier applications. Two alternate symbols are shown in Figure 3–43. Tunnel diodes are constructed with germanium or gallium arsenide by doping the *p* and *n* regions much more heavily than in a conventional rectifier diode. This heavy doping results in a very narrow depletion region. The heavy doping allows conduction for all reverse voltages so that there is no breakdown effect as with the conventional rectifier diode. This is shown in Figure 3–44.

Also, the extremely narrow depletion region permits electrons to "tunnel" through the *pn* junction at very low forward-bias voltages, and the diode acts as a conductor. This is shown in Figure 3–44 between points *A* and *B*. At point *B*, the forward voltage begins to develop a barrier, and the current begins to decrease as the forward voltage continues to increase. This is the *negative-resistance region*.

$$R_F = \frac{\Delta V_F}{\Delta I_F}$$

This effect is opposite to that described in Ohm's law, where an increase in voltage results in an increase in current. At point *C*, the diode begins to act as a conventional forward-biased diode.

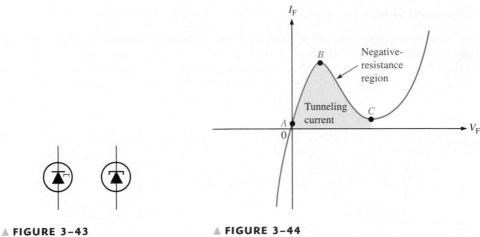

▲ FIGURE 3–43

Tunnel diode symbols.

▲ FIGURE 3–44

Tunnel diode characteristic curve.

An Application   A parallel resonant circuit can be represented by a capacitance, inductance, and resistance in parallel, as in Figure 3–45(a). R_P is the parallel equivalent of the series winding resistance of the coil. When the tank circuit is "shocked" into oscillation by an application of voltage as in Figure 3–45(b), a damped sinusoidal output results. The damping is due to the resistance of the tank, which prevents a sustained oscillation because energy is lost when there is current through the resistance.

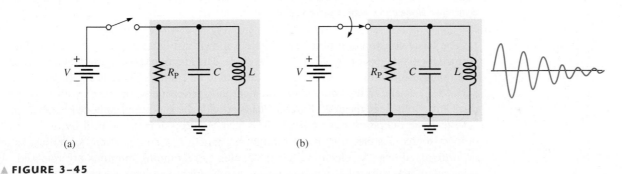

(a)   (b)

▲ FIGURE 3–45

Parallel resonant circuit.

If a tunnel diode is placed in series with the tank circuit and biased at the center of the negative-resistance portion of its characteristic curve, as shown in Figure 3–46, a sustained oscillation (constant sinusoidal voltage) will result on the output. This is because the negative-resistance characteristic of the tunnel diode counteracts the positive-resistance characteristic of the tank resistance.

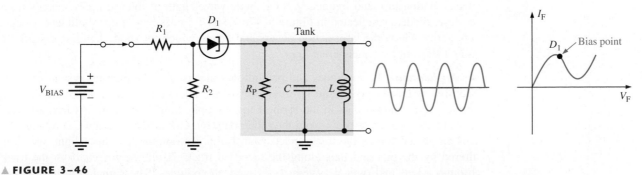

▲ FIGURE 3–46

Basic tunnel diode oscillator.

The Laser Diode

The term **laser** stands for *l*ight *a*mplification by *s*timulated *e*mission of *r*adiation. Laser light is **monochromatic**, which means that it consists of a single color and not a mixture of colors. Laser light is also called **coherent light**, a single wavelength, as compared to incoherent light, which consists of a wide band of wavelengths. The laser diode normally emits coherent light, whereas the LED emits incoherent light. The symbols are the same as shown in Figure 3–47(a).

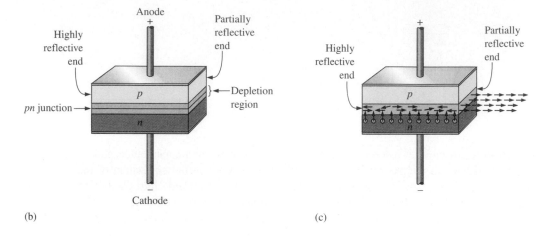

(a) Symbol   (b)   (c)

▲ FIGURE 3–47

Basic laser diode construction and operation.

The basic construction of a laser diode is shown in Figure 3–47(b). A *pn* junction is formed by two layers of doped gallium arsenide. The length of the *pn* junction bears a precise relationship with the wavelength of the light to be emitted. There is a highly reflective surface at one end of the *pn* junction and a partially reflective surface at the other end produced by "polishing" the ends. External leads provide the anode and cathode connections.

The basic operation is as follows. The laser diode is forward-biased by an external voltage source. As electrons move through the junction, recombination occurs just as in an ordinary diode. As electrons fall into holes to recombine, photons are released. A released photon can strike an atom, causing another photon to be released. As the forward current is increased, more electrons enter the depletion region and cause more photons to be emitted. Eventually some of the photons that are randomly drifting within the depletion region strike the reflected surfaces perpendicularly. These reflected photons move along the depletion region, striking atoms and releasing additional photons due to the avalanche effect. This back-and-forth movement of photons increases as the generation of photons "snowballs" until a very intense beam of laser light is formed by the photons that pass through the partially reflective end of the *pn* junction.

Each photon produced in this process is identical to the other photons in energy level, phase relationship, and frequency. So a single wavelength of intense light emerges from the laser diode, as indicated in Figure 3–47(c). Laser diodes have a threshold level of current above which the laser action occurs and below which the diode behaves essentially as an LED, emitting incoherent light.

An Application   Laser diodes and photodiodes are used in the pick-up system of compact disk (CD) players. Audio information (sound) is digitally recorded in stereo on the surface of a compact disk in the form of microscopic "pits" and "flats." A lens arrangement focuses the laser beam from the diode onto the CD surface. As the CD rotates, the lens and beam follow the track under control of a servomotor. The laser light, which is altered by the pits and flats along the recorded track, is reflected back from the track through a lens and optical system to infrared photodiodes. The signal from the photodiodes is then used to reproduce the digitally recorded sound.

**SECTION 3–5
REVIEW**

1. Between what two voltages does a current regulator diode operate?
2. What are the primary application areas for Schottky diodes?
3. What is a hot-carrier diode?
4. What is the key characteristic of a tunnel diode?
5. What is one application for a tunnel diode?
6. Name the three regions of a *pin* diode.
7. What does *laser* mean?
8. What is the difference between incoherent and coherent light and which is produced by a laser diode?

3–6   TROUBLESHOOTING

In this section, you will see how a faulty zener diode can affect the output of a regulated dc power supply. Although IC regulators are generally used for power supply outputs, the zener is occasionally used when less precise regulation is required. Like other diodes, the zener can fail open, it can exhibit degraded performance in which its internal resistance increases significantly, or it can short out.

After completing this section, you should be able to

- ■ **Troubleshoot zener diode regulators**
- ■ Recognize the effects of an open zener
- ■ Recognize the effects of a zener with excessive impedance

A Zener-Regulated DC Power Supply

Figure 3–48 shows a filtered dc power supply that produces a constant 24 V before it is regulated down to 15 V by the zener regulator. The 1N4744 zener diode is the same as the one in Example 3–7. A no-load check of the regulated output voltage shows 15.5 V as indicated in part (a). The voltage expected at maximum zener current (I_{ZM}) for this particular diode is 15.7 V. In part (b), a potentiometer is connected to provide a variable load resistance. It is adjusted to a minimum value for a full-load test as determined by the following calculations. The full-load test is at minimum zener current (I_{ZK}). The meter reading of 14.8 V indicates approximately the expected output voltage of 14.76 V.

$$I_T = \frac{24 \text{ V} - 14.76 \text{ V}}{180 \text{ }\Omega} = 51.3 \text{ mA}$$

$$I_L = 51.3 \text{ mA} - 0.25 \text{ mA} = 51.1 \text{ mA}$$

$$R_{L(min)} = \frac{14.76 \text{ V}}{51.1 \text{ mA}} = 289 \text{ }\Omega$$

Case 1: Zener Diode Open   If the zener diode fails open, the power supply test gives the approximate results indicated in Figure 3–49. In the no-load check shown in part (a), the output voltage is 24 V because there is no voltage dropped between the filtered output of the power supply and the output terminal. This definitely indicates an open between the output terminal and ground. In the full-load check, the voltage of 14.8 V results from the voltage-divider action of the 180 Ω series resistor and the 289 Ω load. In this case, the result is too close to the normal reading to be a reliable fault indication but the no-load

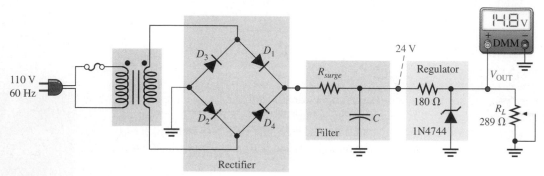

(a) Correct output voltage with no load

(b) Correct output voltage with full load

▲ FIGURE 3–48

Zener–regulated power supply test.

(a) Open zener diode with no load

(b) Open zener diode cannot be detected by full-load measurement in this case.

▲ FIGURE 3–49

Indications of an open zener.

check will verify the problem. Also, if R_L is varied, V_{OUT} will vary if the zener diode is open.

Case 2: Excessive Zener Impedance   As indicated in Figure 3–50, a no-load check that results in an output voltage greater than the maximum zener voltage but less than the power supply output voltage indicates that the zener has failed such that its internal impedance is more than it should be. The 20 V output in this case is 4.5 V higher than the expected value of 15.5 V. That additional voltage is caused by the drop across the excessive internal impedance of the zener.

▲ **FIGURE 3–50**

Indication of excessive zener impedance.

EWB/Multisim Troubleshooting Exercises

These file circuits are in the Troubleshooting Exercises folder on your CD-ROM.

1. Open file TSE03-01. Determine if the circuit is working properly and, if not, determine the fault.

2. Open file TSE03-02. Determine if the circuit is working properly and, if not, determine the fault.

3. Open file TSE03-03. Determine if the circuit is working properly and, if not, determine the fault.

4. Open file TSE03-04. Determine if the circuit is working properly and, if not, determine the fault.

SECTION 3–6 REVIEW

1. In a zener regulator, what are the symptoms of an open zener diode?

2. If a zener regulator fails so that the zener impedance is greater than the specified value, is the output voltage more or less than it should be?

3. If you measure 0 V at the output of a zener-regulated power supply, what is the most likely fault(s)?

4. The zener diode regulator in a power supply is open. What will you observe on the output with a voltmeter if the load resistance is varied within its specified range?

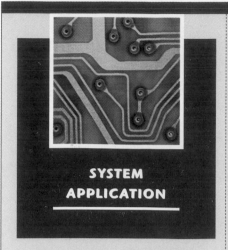

SYSTEM APPLICATION

You have been assigned to modify the power supply circuit board from the system application of Chapter 2. You are to incorporate voltage regulation and light emission and detection circuits to be used in a new system your company is developing. This system will be used in a sporting goods manufacturing plant for counting and controlling the number of baseballs going into various sizes of boxes for shipment. You will apply the knowledge you have gained in this chapter in completing your assignment.

The Counting and Control System

This particular system is used to count baseballs as they are fed down a chute into a box for shipping. It can also be applied to inventory and shipping control for many other types of products. The portion of the system for which you are responsible consists of the regulated power supply, an infrared emitter circuit, and an infrared detector circuit that are all on the same board.

The complete system also includes a threshold circuit that senses the output of the infrared detector and provides a pulse output to a digital counter. The output of the counter goes to a display and a control mechanism for stopping the baseballs when a box is full. The system concept and block diagram are shown in Figure 3–51.

The Power Supply Circuit

The dc power supply circuit is the same as the one developed in the system application of Chapter 2 except that it is modified by the addition of a zener voltage regulator as shown in the schematic of Figure 3–52. Also, the new circuit board will include the IR emitter and IR detector circuits. The basic power supply specifications are

1. Input voltage: 115 V rms at 60 Hz
2. Unregulated output voltage: 12 V dc ± 10%
3. Regulated output voltage: 5.1 V ± 10%
4. Maximum ripple factor: 3%
5. Maximum load current: 100 mA

The Power Supply Components

■ *The unregulated portion of the power supply* This portion of the circuit is the same as the one developed in Chapter 2.

■ *The regulator* Select a zener diode to be used as the regulator. Refer to the data sheet in Figure 3–7(a).

■ *The limiting resistor* Determine a value of limiting resistor to be used in the regulator.

■ *The fuse* Determine a rating for the fuse to be used in the power supply.

The IR Emitter and IR Detector Circuits

The MLED81 light-emitting diode is used as the IR emitter, and the MRD821 photodiode is used as the IR detector. These devices are located on either side of the tube through which the baseballs are routed. The diameter of the tube is 1.5 cm greater than the diameter of a baseball. The LED emits a constant beam of infrared light directly toward the photo-

▲ **FIGURE 3–51**

Basic system concept and block diagram of the counting and control system.

▲ **FIGURE 3–52**

Zener regulated power supply preliminary schematic.

diode; this beam of light is interrupted as a baseball passes in the tube.

The IR detector senses the interruption in the LED emission and produces a minimum positive-going output transition of 3 V for the threshold circuit that generates a pulse to advance the digital counter. The counter is advanced by one count for each baseball that passes the IR beam. When a preset number of baseballs has been packed into a box, the control circuit produces a signal to activate the stop mechanism. The system is reset for the next box.

Both the LED and photodiode are connected to their series-limiting resistors and the voltage source on the circuit board with a four-wire cable. Figure 3–53 shows the basic IR emitter and IR detector circuits.

The IR Emitter and IR Detector Components

- *The distance between emitter and detector*   Determine the distance from the LED to the photodiode (diameter of the tube) based on the diameter of a baseball (7.3 cm) plus a clearance of 1.5 cm.

- *The IR emitter series-limiting resistor*   Determine a value for the current-limiting series resistor to achieve the maximum possible irradiance (H) at the detector. Assume the emitter and detector are aligned for maximum angular response. Refer to the data sheet in Figure 3–31.

- *The IR detector resistor*   Determine a value of resistor in the IR detector to produce minimum voltage transitions of 3 V on the threshold output when the photodiode turns on and off. Refer to the data sheet in Figure 3–36.

The Schematic

Produce a complete schematic that includes the regulated power supply, the IR emitter, and the IR detector.

The Printed Circuit Board

- Check out the printed circuit board in Figure 3–54 to verify that it is correct according to the schematic.

- Label a copy of the board with the component and input/output designations in agreement with the schematic.

(a) IR emitter   (b) IR detector

▲ **FIGURE 3–53**

Basic schematics of the IR emitter and IR detector circuits.

A Test Procedure

■ Develop a step-by-step set of instructions on how to completely check the power supply and IR emitter/detector for proper operation using the test points (circled numbers) indicated in the test bench setup of Figure 3–55. There is a test fixture that simulates the operation of the LED and photodiode by momentarily inserting a blocking plate for the IR beam in a tube-type mounting device. The IR beam is shown as a red beam for illustration.

■ Specify voltage values and appropriate waveforms for all the measurements to be made.

■ Provide a fault analysis for all possible component failures.

■ Expand the test procedure you developed for the unregulated portion of the power supply circuit in Chapter 2 to include the additional circuits.

Troubleshooting

Several prototype boards have been assembled and are ready for testing. The test bench setup is shown in Figure 3–55. Based on the sequence of measurements for each board indicated in Figure 3–56, determine the most likely fault in each case.

The circled numbers indicate test point connections to the circuit board. The DMM function setting is indicated below the display and the volt/div and sec/div settings for the oscilloscope are shown on the screen in each case.

Final Report (Optional)

Submit a final written report on the power supply and IR emitter/detector circuit board using an organized format that includes the following:

1. A physical description of the circuits.

2. A discussion of the operation of each circuit.

3. A list of the specifications.

4. A list of parts with part numbers if available.

5. A list of the types of problems on the three circuit boards.

6. A complete description of how you determined the problem on each of the three circuit boards.

▲ **FIGURE 3–54**

Power supply and IR emitter/detector printed circuit board. All black resistor bands indicate values to be determined.

V≈ indicates dc/ac function

▸▸ indicates diode test

Tube fixture for testing IR emitter and detector.
Beam (red) is shown unblocked.

▲ **FIGURE 3–55**

Power supply and IR emitter/detector test bench.

▲ **FIGURE 3–56**

Test results of three prototype circuit boards.

CHAPTER SUMMARY

- The zener diode operates in reverse breakdown.
- There are two breakdown mechanisms in a zener diode: avalanche breakdown and zener breakdown.
- When $V_Z < 5$ V, zener breakdown is predominant.
- When $V_Z > 5$ V, avalanche breakdown is predominant.
- A zener diode maintains a nearly constant voltage across its terminals over a specified range of zener currents.
- Zener diodes are used as voltage regulators and limiters.
- Zener diodes are available in many voltage ratings ranging from 1.8 V to 200 V.
- A varactor diode acts as a variable capacitor under reverse-bias conditions.
- The capacitance of a varactor varies inversely with reverse-bias voltage.
- The current regulator diode keeps its forward current at a constant specified value.
- The Schottky diode has a metal-to-semiconductor junction. It is used in fast-switching applications.
- The tunnel diode is used in oscillator circuits.
- An LED emits light when forward-biased.
- LEDs are available for either infrared or visible light.
- The photodiode exhibits an increase in reverse current with light intensity.
- The *pin* diode has a *p* region, an *n* region, and an intrinsic (*i*) region and displays a variable resistance characteristic when forward-biased and a constant capacitance when reverse-biased.
- A laser diode is similar to an LED except that it emits coherent (single wavelength) light when the forward current exceeds a threshold value.
- A summary of special-purpose diode symbols is given in Figure 3–57.

(a) Zener   (b) Light-emitting   (c) Photo   (d) Varactor   (e) Current-regulator

(f) Schottky   (g) *PIN*   (h) Tunnel   (i) Laser

▲ **FIGURE 3–57**

Diode symbols.

KEY TERMS

Key terms and other bold terms in the chapter are defined in the end–of–book glossary.

Electroluminescence   The process of releasing light energy by the recombination of electrons in a semiconductor.

Laser   *L*ight *a*mplification by *s*timulated *e*mission of *r*adiation.

Light-emitting diode (LED)   A type of diode that emits light when there is forward current.

Photodiode   A diode in which the reverse current varies directly with the amount of light.

Varactor   A variable capacitance diode.

Zener breakdown   The lower voltage breakdown in a zener diode.

Zener diode   A diode designed for limiting the voltage across its terminals in reverse bias.

KEY FORMULAS

3–1	$Z_Z = \dfrac{\Delta V_Z}{\Delta I_Z}$	Zener impedance
3–2	$\Delta V_Z = V_Z \times TC \times \Delta T$	V_Z temperature change when TC is %/°C
3–3	$\Delta V_Z = TC \times \Delta T$	V_Z temperature change when TC is mV/°C
3–4	$P_{D(derated)} = P_{D(max)} - (mW/°C)\Delta T$	Derated power dissipation
3–5	$I_{ZM} = \dfrac{P_{D(max)}}{V_Z}$	Maximum zener current
3–6	$H = \dfrac{I_e}{d^2}$	Irradiance

SELF-TEST

Answers are at the end of the chapter.

1. The cathode of a zener diode in a voltage regulator is normally
   (a) more positive than the anode   (b) more negative than the anode
   (c) at +0.7 V   (d) grounded

2. If a certain zener diode has a zener voltage of 3.6 V, it operates in
   (a) regulated breakdown   (b) zener breakdown
   (c) forward conduction   (d) avalanche breakdown

3. For a certain 12 V zener diode, a 10 mA change in zener current produces a 0.1 V change in zener voltage. The zener impedance for this current range is
   (a) 1 Ω   (b) 100 Ω   (c) 10 Ω   (d) 0.1 Ω

4. The data sheet for a particular zener gives V_Z = 10 V at I_{ZT} = 500 mA. Z_Z for these conditions is
   (a) 50 Ω   (b) 20 Ω   (c) 10 Ω   (d) unknown

5. A no-load condition means that
   (a) the load has infinite resistance   (b) the load has zero resistance
   (c) the output terminals are open   (d) answers (a) and (c)

6. A varactor diode exhibits
   (a) a variable capacitance that depends on reverse voltage
   (b) a variable resistance that depends on reverse voltage
   (c) a variable capacitance that depends on forward current
   (d) a constant capacitance over a range of reverse voltages

7. An LED
   (a) emits light when reverse-biased   (b) senses light when reverse-biased
   (c) emits light when forward-biased   (d) acts as a variable resistance

8. Compared to a visible red LED, an infrared LED
   (a) produces light with shorter wavelengths   (b) produces light of all wavelengths
   (c) produces only one color of light   (d) produces light with longer wavelengths

9. The internal resistance of a photodiode
   (a) increases with light intensity when reverse-biased
   (b) decreases with light intensity when reverse-biased
   (c) increases with light intensity when forward-biased
   (d) decreases with light intensity when forward-biased

10. A diode that has a negative resistance characteristic is the
    (a) Schottky diode   (b) tunnel diode   (c) laser diode   (d) hot-carrier diode

11. An infrared LED is optically coupled to a photodiode. When the LED is turned off, the reading on an ammeter in series with the reverse-biased photodiode will

   (a) not change   **(b)** decrease   **(c)** increase   **(d)** fluctuate

12. In order for a system to function properly, the various types of circuits that make up the system must be

   (a) properly biased   **(b)** properly connected   **(c)** properly interfaced

   (d) all of the above   **(e)** answers (a) and (b)

PROBLEMS

Answers to all odd-numbered problems are at the end of the book.

BASIC PROBLEMS

SECTION 3–1   Zener Diodes

1. A certain zener diode has a $V_Z = 7.5$ V and an $Z_Z = 5 \ \Omega$ at a certain current. Sketch the equivalent circuit.

2. From the characteristic curve in Figure 3–58, what is the approximate minimum zener current (I_{ZK}) and the approximate zener voltage at I_{ZK}?

3. When the reverse current in a particular zener diode increases from 20 mA to 30 mA, the zener voltage changes from 5.6 V to 5.65 V. What is the impedance of this device?

4. A zener has an impedance of 15 Ω. What is its terminal voltage at 50 mA if $V_{ZT} = 4.7$ V at $I_{ZT} = 25$ mA?

5. A certain zener diode has the following specifications: $V_Z = 6.8$ V at 25°C and $TC = +0.04\%/°C$. Determine the zener voltage at 70°C.

▲ FIGURE 3–58   ▲ FIGURE 3–59

SECTION 3–2   Zener Diode Applications

6. Determine the minimum input voltage required for regulation to be established in Figure 3–59. Assume an ideal zener diode with $I_{ZK} = 1.5$ mA and $V_Z = 14$ V.

7. Repeat Problem 6 with $Z_Z = 20 \ \Omega$ and $V_{ZT} = 14$ V at 30 mA.

8. To what value must R be adjusted in Figure 3–60 to make $I_Z = 40$ mA? Assume $V_Z = 12$ V at 30 mA and $Z_Z = 30$ Ω.

9. A 20 V peak sinusoidal voltage is applied to the circuit in Figure 3–60 in place of the dc source. Sketch the output waveform. Use the parameter values established in Problem 8.

▶ FIGURE 3–60

10. A loaded zener regulator is shown in Figure 3–61. $V_Z = 5.1$ V at $I_{ZT} = 49$ mA, $I_{ZK} = 1$ mA, $Z_Z = 7$ Ω, and $I_{ZM} = 70$ mA. Determine the minimum and maximum permissible load currents.

▶ FIGURE 3–61

File circuits are identified with a CD logo and are in the Problems folder on your CD-ROM. Filenames correspond to figure numbers (e.g., F03-61).

11. Find the load regulation expressed as a percentage in Problem 10. Refer to Chapter 2, Equation 2–15.

12. Analyze the circuit in Figure 3–61 for percent line regulation using an input voltage from 6 V to 12 V with no load. Refer to Chapter 2, Equation 2–14.

13. The no-load output voltage of a certain zener regulator is 8.23 V, and the full-load output is 7.98 V. Calculate the load regulation expressed as a percentage. Refer to Chapter 2, Equation 2–15.

14. In a certain zener regulator, the output voltage changes 0.2 V when the input voltage goes from 5 V to 10 V. What is the input regulation expressed as a percentage? Refer to Chapter 2, Equation 2–14.

15. The output voltage of a zener regulator is 3.6 V at no load and 3.4 V at full load. Determine the load regulation expressed as a percentage. Refer to Chapter 2, Equation 2–15.

SECTION 3–3   Varactor Diodes

16. Figure 3–62 is a curve of reverse voltage versus capacitance for a certain varactor. Determine the change in capacitance if V_R varies from 5 V to 20 V.

17. Refer to Figure 3–62 and determine the value of V_R that produces 25 pF.

18. What capacitance value is required for each of the varactors in Figure 3–63 to produce a resonant frequency of 1 MHz?

19. At what value must the voltage V_R be set in Problem 18 if the varactors have the characteristic curve in Figure 3–62?

▲ FIGURE 3–62

▲ FIGURE 3–63

SECTION 3–4   Optical Diodes

20. The LED in Figure 3–64(a) has a light-producing characteristic as shown in part (b). Neglecting the forward voltage drop of the LED, determine the amount of radiant (light) power produced in mW.

(a)

(b)

▲ FIGURE 3–64

21. Determine how to connect the seven-segment display in Figure 3–65 to display "5." The maximum continuous forward current for each LED is 30 mA and a +5 V dc source is to be used.

22. For a certain photodiode at a given irradiance, the reverse resistance is 200 kΩ and the reverse voltage is 10 V. What is the current through the device?

▲ FIGURE 3–65

23. What is the resistance of each photodiode in Figure 3–66?

24. When the switch in Figure 3–67 is closed, will the microammeter reading increase or decrease? Assume D_1 and D_2 are optically coupled.

(a)   (b)   (c)

▲ **FIGURE 3–66**

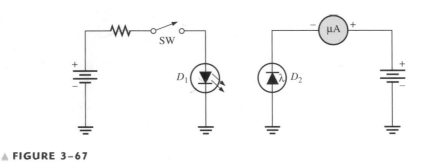

▲ **FIGURE 3–67**

SECTION 3–5 Other Types of Diodes

25. The *V-I* characteristic of a certain tunnel diode shows that the current changes from 0.25 mA to 0.15 mA when the voltage changes from 125 mV to 200 mV. What is the resistance?

26. In what type of circuit are tunnel diodes commonly used?

27. What purpose do the reflective surfaces in the laser diode serve? Why is one end only partially reflective?

TROUBLESHOOTING PROBLEMS

SECTION 3–6 Troubleshooting

28. For each set of measured voltages at the points (1, 2, and 3) indicated in Figure 3–68, determine if they are correct and if not, identify the most likely fault(s). State what you would do to correct the problem once it is isolated. The zener is rated at 12 V.

    (a) $V_1 = 110$ V rms, $V_2 = 30$ V dc, $V_3 = 12$ V dc

    (b) $V_1 = 100$ V rms, $V_2 = 30$ V dc, $V_3 = 30$ V dc

    (c) $V_1 = 0$ V, $V_2 = 0$ V, $V_3 = 0$ V

    (d) $V_1 = 110$ V rms, $V_2 = 30$ V peak full-wave 120 Hz, $V_3 = 12$ V 120 Hz pulsating voltage

    (e) $V_1 = 110$ V rms, $V_2 = 9$ V, $V_3 = 0$ V

29. What is the output voltage in Figure 3–68 for each of the following faults?

    (a) D_5 open   **(b)** R open   **(c)** C leaky   **(d)** C open   **(e)** D_3 open

    (f) D_2 open   **(g)** T open   **(h)** F open

▲ FIGURE 3–68

SYSTEM APPLICATION PROBLEMS

30. The counting and control system has been installed at a customer's location. The system is performing erratically, so you decide to first check the power supply and IR emitter/detector board. Based on Figure 3–69, determine the problem.

▲ FIGURE 3–69

31. Another problem has developed with the counter and control system. This time, the system completely quits working and again you decide to first check the power supply board. Based on Figure 3–70, determine the problem.

32. List the possible reasons for the LED in Figure 3–55 not emitting infrared light when the power supply is plugged in.

▲ FIGURE 3–70

33. List the possible reasons for the photodiode in Figure 3–55 not responding to the infrared light from the LED. List the steps in sequence that you would take to isolate the problem.

DATA SHEET PROBLEMS

34. Refer to the zener diode data sheet in Figure 3–7.

   (a) What is the maximum dc power dissipation at 25°C for a 1N4738?

   (b) Determine the maximum power dissipation at 70°C and at 100°C for a 1N4751.

   (c) What is the minimum current required by the 1N4738 for regulation?

   (d) What is the maximum current for the 1N4750 at 25°C?

   (e) The current through a 1N4740 changes from 25 mA to 0.25 mA. How much does the zener impedance change?

   (f) What is the maximum zener voltage of a 1N4736 at 50°C?

   (g) What is the minimum zener voltage for a 1N4747 at 75°C?

35. Refer to the varactor diode data sheet in Figure 3–22.

   (a) What is the maximum reverse voltage for the 1N5139?

   (b) Determine the maximum power dissipation for a 1N5141 at an ambient temperature of 60°C.

   (c) Determine the maximum power dissipation for a 1N5148 at a case temperature of 80°C.

   (d) What is the capacitance of a 1N5148 at a reverse voltage of 20 V?

   (e) If figure of merit were the only criteria, which varactor diode would you select?

   (f) What is the typical capacitance at $V_R = 60$ V for a 1N5142?

36. Refer to the LED data sheet in Figure 3–31.

   (a) Can 9 V be applied in reverse across an MLED81?

   (b) Determine the minimum value of series resistor for the MLED81 when a voltage of 5.1 V is used to forward-bias the diode.

   (c) Assume the forward current is 50 mA and the forward voltage drop is 1.5 V at an ambient temperature of 45°C. Is the maximum power rating exceeded?

   (d) Determine the axial radiant intensity for a forward current of 30 mA.

   (e) What is the radiant intensity at an angle of 20° from the axis if the forward current is 100 mA?

37. Refer to the photodiode data sheet in Figure 3–36.

   (a) An MRD821 is connected in series with a 10 kΩ resistor and a reverse-bias voltage source. There is no incident light on the diode. What is the voltage drop across the resistor?

   (b) At what wavelength will the reverse current be the greatest for a given irradiance?

   (c) What is the dark current at an ambient temperature of 60°C?

   (d) At what wavelength is sensitivity of the MRD821 at a maximum?

   (e) If the maximum sensitivity is 50 μA/mW/cm^2, what is the sensitivity at 900 nm?

   (f) An infrared light with a wavelength of 900 nm strikes an MRD821 with an irradiance of 3 mW/cm^2 and an angle of 40° from the maximum axis. Determine the reverse current.

ADVANCED PROBLEMS

38. Develop the schematic for the circuit board in Figure 3–71 and determine what type of circuit it is.

39. If a 110 V rms, 60 Hz input voltage is connected to the ac inputs, determine the output voltages on the circuit board in Figure 3–71.

40. If each output of the board in Figure 3–71 is loaded with 1.0 kΩ, what fuse rating should be used?

41. Design a zener voltage regulator to meet the following specifications:
The input voltage is 24 V dc, the load current is 35 mA, and the load voltage is 8.2 V.

42. The varactor-tuned band-pass filter in Figure 3–24 is to be redesigned to produce a bandwidth of from 350 kHz to 850 kHz within a 10% tolerance. Using the basic circuit in Figure 3–72, determine all components necessary to meet the specification. Use the nearest standard values.

Rectifier diodes: 1N4001
Zener diodes: D1-1N4736, D2-1N4749
Filter capacitors: 100 μF

▲ **FIGURE 3–71**

▲ **FIGURE 3–72**

43. Design a seven-segment LED display circuit in which any of the ten digits can be displayed using a set of switches. Each LED segment is to have a current of 20 mA ± 10% from a 12 V source and the circuit must be designed with a minimum number of switches.

44. If you used a common-anode seven-segment display in Problem 43, redesign it for a common-cathode display or vice versa.

EWB/MULTISIM TROUBLESHOOTING PROBLEMS

These file circuits are in the Troubleshooting Problems folder on your CD-ROM.

45. Open file TSP03-45 and determine the fault.

46. Open file TSP03-46 and determine the fault.

47. Open file TSP03-47 and determine the fault.

48. Open file TSP03-48 and determine the fault.

ANSWERS

SECTION REVIEWS

SECTION 3–1 Zener Diodes

1. Zener diodes are operated in the reverse-breakdown region.
2. The test current, I_{ZT}
3. The zener impedance causes the voltage to vary slightly with current.
4. $V_Z = 10\,\text{V} + (20\,\text{mA})(8\,\Omega) = 10.16\,\text{V}$
5. The zener voltage increases (or decreases) 0.05% for each degree centigrade increase (or decrease).
6. Power derating is the reduction in the power rating of a device as a result of an increase in temperature.

SECTION 3–2 Zener Diode Applications

1. An infinite resistance (open)
2. With no load, there is no current to a load. With full load, there is maximum current to the load.
3. Approximately 0.7 V, just like a rectifier diode

SECTION 3–3 Varactor Diodes

1. A varactor exhibits variable capacitance.
2. A varactor is operated in reverse bias.
3. The depletion region
4. Capacitance decreases with more reverse bias.
5. The tuning ratio is the ratio of a varactor's capacitance at a specified minimum voltage to the capacitance at a specified maximum voltage.

SECTION 3–4 Optical Diodes

1. Infrared and visible light
2. Infrared has the greater wavelength.
3. An LED operates in forward bias.
4. Light emission increases with forward current.
5. False, V_F of an LED is usually greater than 1.2 V.
6. A photodiode operates in reverse bias.
7. The internal resistance decreases.
8. Dark current is the reverse photodiode current when there is no light.

SECTION 3–5 Other Types of Diodes

1. A current regulator operates between V_L (limiting voltage) and POV (peak operating voltage).
2. High-frequency and fast-switching circuits
3. *Hot carrier* is another name for Schottky diodes.
4. Tunnel diodes have negative resistance.
5. Oscillators
6. *p* region, *n* region, and intrinsic (*i*) region
7. *l*ight *a*mplification by *s*timulated *e*mission of *r*adiation
8. Coherent light has only a single wavelength, but incoherent light has a wide band of wavelengths. A laser diode produces coherent light.

SECTION 3–6   **Troubleshooting**

1. The output voltage is too high and equal to the rectifier output.

2. More

3. Series limiting resistor open, fuse blown

4. The output voltage changes as the load resistance changes.

RELATED PROBLEMS FOR EXAMPLES

3–1 $5 \, \Omega$

3–2 $V_Z = -11.9$ V at 10 mA; $V_Z = 12.08$ V at 30 mA

3–3   0.45 V to 11.55 V

3–4   7.5 W

3–5 $V_{IN(min)} = 6.77$ V; $V_{IN(max)} = 21.9$ V

3–6 $I_{L(min)} = 0$ A; $I_{L(max)} = 43$ mA; $R_{L(min)} = 76.7 \, \Omega$

3–7   **(a)** 11.8 V at I_{ZK}; 12.6 V at I_{ZM} **(b)** $144 \, \Omega$ **(c)** $140 \, \Omega$

3–8   **(a)** A waveform identical to Figure 3–18(a)

  (b) A sine wave with a peak value of 5 V

3–9 $V_{R(min)} = 1.43$ V; $V_{R(max)} = 14.4$ V

3–10   2.4 mW/sr

3–11   29.8 μA

SELF-TEST

1. (a)   2. (b)   3. (c)   4. (b)   5. (d)   6. (a)

7. (c)   8. (d)   9. (b)   10. (b)   11. (b)   12. (d)

4

BIPOLAR JUNCTION TRANSISTORS (BJTs)

INTRODUCTION

The transistor was invented by a team of three men at Bell Laboratories in 1947. Although this first transistor was not a bipolar junction device, it was the beginning of a technological revolution that is still continuing. All of the complex electronic devices and systems today are an outgrowth of early developments in semiconductor transistors.

Two basic types of transistors are the bipolar junction transistor (BJT), which we will begin to study in this chapter, and the field-effect transistor (FET), which we will cover in later chapters. The BJT is used in two broad areas—as a linear amplifier to boost or amplify an electrical signal and as an electronic switch. Both of these applications are introduced in this chapter.

CHAPTER OUTLINE

- Describe the basic structure of the BJT (bipolar junction transistor)

- Explain how a transistor is biased and discuss the transistor currents and their relationships

- Discuss transistor parameters and characteristics and use these to analyze a transistor circuit

- Discuss how a transistor is used as a voltage amplifier

- Discuss how a transistor is used as an electronic switch

- Identify various types of transistor package configurations

- Troubleshoot various faults in transistor circuits

KEY TERMS

- BJT (bipolar junction transistor)

- Emitter

- Base

- Collector

- Bias

- Beta

- Gain

- Saturation

- Linear

- Cutoff

- Amplification

■■■ SYSTEM APPLICATION PREVIEW

Suppose you work for a company that makes a security alarm system for protecting homes and places of business against illegal entry. You are given the responsibility for final development and for testing each system before it goes to the customer's location. The first step in your assignment is to learn all you can about transistor operation. You will then apply your knowledge to the system application at the end of the chapter.

---

www. VISIT THE COMPANION WEBSITE

Study aids for this chapter are available at

http://www.prenhall.com/floyd

4–1 TRANSISTOR STRUCTURE

The basic structure of the bipolar junction transistor (BJT) determines its operating characteristics. In this section, you will see how semiconductive materials are used to form a transistor, and you will learn the standard transistor symbols.

After completing this section, you should be able to

- **Describe the basic structure of the BJT (bipolar junction transistor)**
- Explain the difference between the structure of an *npn* and a *pnp* transistor
- Identify the symbols for *npn* and *pnp* transistors
- Name the three regions of a BJT and their labels

The **BJT (bipolar junction transistor)** is constructed with three doped semiconductor regions separated by two *pn* junctions, as shown in the epitaxial planar structure in Figure 4–1(a). The three regions are called **emitter**, **base**, and **collector**. Physical representations of the two types of BJTs are shown in Figure 4–1(b) and (c). One type consists of two *n* regions separated by a *p* region (*npn*), and the other type consists of two *p* regions separated by an *n* region (*pnp*).

The *pn* junction joining the base region and the emitter region is called the *base-emitter junction*. The *pn* junction joining the base region and the collector region is called the *base-collector junction*, as indicated in Figure 4–1(b). A wire lead connects to each of the three regions, as shown. These leads are labeled E, B, and C for emitter, base, and collector, respectively. The base region is lightly doped and very thin compared to the heavily doped emitter and the moderately doped collector regions. (The reason for this is discussed in the next section.)

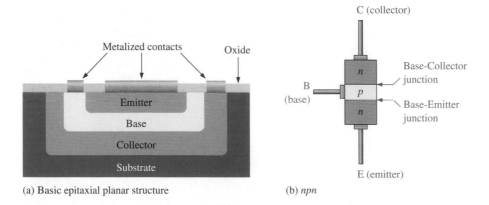

(a) Basic epitaxial planar structure

(b) *npn*

(c) *pnp*

▲ **FIGURE 4–1**

Basic BJT construction.

Figure 4–2 shows the schematic symbols for the *npn* and *pnp* bipolar junction transistors. The term **bipolar** refers to the use of both holes and electrons as carriers in the transistor structure.

► FIGURE 4-2

Standard BJT (bipolar junction transistor) symbols.

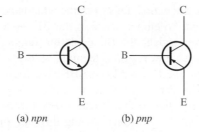

(a) *npn*   (b) *pnp*

SECTION 4-1 REVIEW

Answers are at the end of the chapter.

1. Name the two types of BJTs according to their structure.
2. The BJT is a three-terminal device. Name the three terminals.
3. What separates the three regions in a BJT?

4-2   BASIC TRANSISTOR OPERATION

In order for the transistor to operate properly as an amplifier, the two *pn* junctions must be correctly biased with external dc voltages. In this section, we use the *npn* transistor for illustration. The operation of the *pnp* is the same as for the *npn* except that the roles of the electrons and holes, the bias voltage polarities, and the current directions are all reversed.

After completing this section, you should be able to

- **Explain how a transistor is biased and discuss the transistor currents and their relationships**

- Describe forward-reverse bias

- Show how to connect a transistor to the bias-voltage sources

- Describe the basic internal operation of a transistor

- State the formula relating the collector, emitter, and base currents in a transistor

Figure 4–3 shows the proper **bias** arrangement for both *npn* and *pnp* **transistors** for active operation as an **amplifier**. Notice that in both cases the base-emitter (BE) junction is forward-biased and the base-collector (BC) junction is reverse-biased.

▲ FIGURE 4-3

Forward–reverse bias of a BJT.

To illustrate transistor action, let's examine what happens inside the *npn* transistor. The forward bias from base to emitter narrows the BE depletion region, and the reverse bias from base to collector widens the BC depletion region, as depicted in Figure 4–4. The heavily doped *n*-type emitter region is teeming with conduction-band (free) electrons that easily diffuse through the forward-biased BE junction into the *p*-type base region where they become minority carriers, just as in a forward-biased diode. The base region is lightly doped and very thin so that it has a very limited number of holes. Thus, only a small percentage of all the electrons flowing through the BE junction can combine with the available holes in the base. These relatively few recombined electrons flow out of the base lead as valence electrons, forming the small base electron current, as shown in Figure 4–4.

▲ **FIGURE 4–4**

Illustration of BJT action.

Most of the electrons flowing from the emitter into the thin, lightly doped base region do not recombine but diffuse into the BC depletion region. Once in this region they are pulled through the reverse-biased BC junction by the electric field set up by the force of attraction between the positive and negative ions. Actually, you can think of the electrons as being pulled across the reverse-biased BC junction by the attraction of the collector supply voltage. The electrons now move through the collector region, out through the collector lead, and into the positive terminal of the collector voltage source. This forms the collector electron current, as shown in Figure 4–4. The collector current is much larger than the base current. This is the reason transistors exhibit current gain.

Transistor Currents

The directions of the currents in an *npn* transistor and its schematic symbol are as shown in Figure 4–5(a); those for a *pnp* transistor are shown in Figure 4–5(b). Notice that the arrow on the emitter of the transistor symbols points in the direction of conventional current. These diagrams show that the emitter current (I_E) is the sum of the collector current (I_C) and the base current (I_B), expressed as follows:

$$I_E = I_C + I_B$$

Equation 4–1

As mentioned before, I_B is very small compared to I_E or I_C. The capital-letter subscripts indicate dc values.

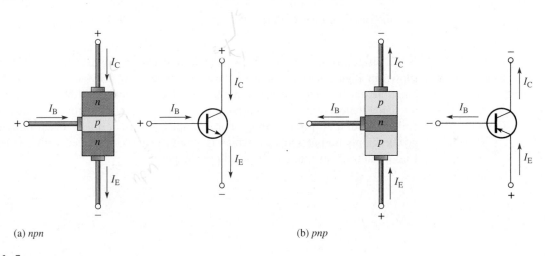

(a) *npn*   (b) *pnp*

▲ **FIGURE 4–5**

Transistor currents.

SECTION 4–2 REVIEW

1. What are the bias conditions of the base-emitter and base-collector junctions for a transistor to operate as an amplifier?

2. Which is the largest of the three transistor currents?

3. Is the base current smaller or larger than the emitter current?

4. Is the base region much thinner or much wider than the collector and emitter regions?

5. If the collector current is 1 mA and the base current is 10 μA, what is the emitter current?

4–3 TRANSISTOR CHARACTERISTICS AND PARAMETERS

Two important parameters, β_{DC} (dc current gain) and α_{DC} are introduced and used to analyze a transistor circuit. Also, transistor characteristic curves are covered, and you will learn how a transistor's operation can be determined from these curves. Finally, maximum ratings of a transistor are discussed.

After completing this section, you should be able to

- ■ **Discuss transistor parameters and characteristics and use these to analyze a transistor circuit**
- ■ Define dc *beta* (β_{DC})
- ■ Define dc *alpha* (α_{DC})
- ■ Identify all currents and voltages in a transistor circuit
- ■ Analyze a basic transistor dc circuit
- ■ Interpret collector characteristic curves and use a dc load line
- ■ Describe how β_{DC} varies with temperature and collector current
- ■ Discuss and apply maximum transistor ratings
- ■ Derate a transistor for power dissipation
- ■ Interpret a transistor data sheet

As discussed in the last section, when a transistor is connected to dc bias voltages, as shown in Figure 4–6 for both *npn* and *pnp* types, V_{BB} forward-biases the base-emitter junction, and V_{CC} reverse-biases the base-collector junction. Although in this chapter we are using battery symbols to represent the bias voltages, in practice the voltages are often derived from a dc power supply. For example, V_{CC} is normally taken directly from the power supply output and V_{BB} (which is smaller) can be produced with a voltage divider. Bias circuits are examined thoroughly in Chapter 5.

(a) *npn*   (b) *pnp*

▲ **FIGURE 4–6**

Transistor dc bias circuits.

DC Beta (β_{DC}) and DC Alpha (α_{DC})

The ratio of the dc collector current (I_C) to the dc base current (I_B) is the dc **beta** (β_{DC}), which is the dc current **gain** of a transistor.

Equation 4–2

$$\beta_{DC} = \frac{I_C}{I_B}$$

Typical values of β_{DC} range from less than 20 to 200 or higher. β_{DC} is usually designated as an equivalent hybrid (h) parameter, h_{FE}, on transistor data sheets. h-parameters are covered in Chapter 6. All you need to know now is that

$$h_{FE} = \beta_{DC}$$

The ratio of the dc collector current (I_C) to the dc emitter current (I_E) is the dc **alpha** (α_{DC}). The alpha is a less-used parameter than beta in transistor circuits.

$$\alpha_{DC} = \frac{I_C}{I_E}$$

Typically, values of α_{DC} range from 0.95 to 0.99 or greater, but α_{DC} is always less than 1. The reason is that I_C is always slightly less than I_E by the amount of I_B. For example, if $I_E = 100$ mA and $I_B = 1$ mA, then $I_C = 99$ mA and $\alpha_{DC} = 0.99$.

EXAMPLE 4–1

Determine β_{DC} and I_E for a transistor where $I_B = 50$ μA and $I_C = 3.65$ mA.

Solution

$$\beta_{DC} = \frac{I_C}{I_B} = \frac{3.65 \text{ mA}}{50 \text{ }\mu\text{A}} = \mathbf{73}$$

$$I_E = I_C + I_B = 3.65 \text{ mA} + 50 \text{ }\mu\text{A} = \mathbf{3.70 \text{ mA}}$$

Related Problem* A certain transistor has a β_{DC} of 200. When the base current is 50 μA, determine the collector current.

*Answers are at the end of the chapter.

Current and Voltage Analysis

Consider the basic transistor bias circuit configuration in Figure 4–7. Three transistor dc currents and three dc voltages can be identified.

I_B: dc base current

I_E: dc emitter current

I_C: dc collector current

V_{BE}: dc voltage at base with respect to emitter

V_{CB}: dc voltage at collector with respect to base

V_{CE}: dc voltage at collector with respect to emitter

▶ **FIGURE 4–7**

Transistor currents and voltages.

V_{BB} forward-biases the base-emitter junction and V_{CC} reverse-biases the base-collector junction. When the base-emitter junction is forward-biased, it is like a forward-biased diode and has a nominal forward voltage drop of

Equation 4–3

$$V_{BE} \cong 0.7 \text{ V}$$

Although in an actual transistor V_{BE} can be as high as 0.9 V and is dependent on current, we will use 0.7 V throughout this text in order to simplify analysis for the purpose of presenting the basic concepts.

Since the emitter is at ground (0 V), by Kirchhoff's voltage law, the voltage across R_B is

$$V_{R_B} = V_{BB} - V_{BE}$$

Also, by Ohm's law,

$$V_{R_B} = I_B R_B$$

Substituting for V_{R_B} yields

$$I_B R_B = V_{BB} - V_{BE}$$

Solving for I_B,

Equation 4–4

$$I_B = \frac{V_{DD} - V_{BE}}{R_B}$$

The voltage at the collector with respect to the grounded emitter is

$$V_{CE} = V_{CC} - V_{R_C}$$

Since the drop across R_C is

$$V_{R_C} = I_C R_C$$

the voltage at the collector can be written as

Equation 4–5

$$V_{CE} = V_{CC} - I_C R_C$$

where $I_C = \beta_{DC} I_B$. The voltage across the reverse-biased collector-base junction is

Equation 4–6

$$V_{CB} = V_{CE} - V_{BE}$$

EXAMPLE 4–2

Determine I_B, I_C, I_E, V_{BE}, V_{CE}, and V_{CB} in the circuit of Figure 4–8. The transistor has a $\beta_{DC} = 150$.

▶ **FIGURE 4–8**

Solution   From Equation 4–3, $V_{BE} \cong$ **0.7 V**. Calculate the base, collector, and emitter currents as follows:

$$I_B = \frac{V_{BB} - V_{BE}}{R_B} = \frac{5\,V - 0.7\,V}{10\,k\Omega} = \textbf{430 } \boldsymbol{\mu}\textbf{A}$$

$$I_C = \beta_{DC}I_B = (150)(430\,\mu A) = \textbf{64.5 mA}$$

$$I_E = I_C + I_B = 64.5\,mA + 430\,\mu A = \textbf{64.9 mA}$$

Solve for V_{CE} and V_{CB}.

$$V_{CE} = V_{CC} - I_C R_C = 10\,V - (64.5\,mA)(100\,\Omega) = 10\,V - 6.45\,V = \textbf{3.55 V}$$

$$V_{CB} = V_{CE} - V_{BE} = 3.55\,V - 0.7\,V = \textbf{2.85 V}$$

Since the collector is at a higher voltage than the base, the collector-base junction is reverse-biased.

Related Problem   Determine I_B, I_C, I_E, V_{CE}, and V_{CB} in Figure 4–8 for the following values: $R_B = 22\,k\Omega$, $R_C = 220\,\Omega$, $V_{BB} = 6\,V$, $V_{CC} = 9\,V$, and $\beta_{DC} = 90$.

Open file E04-02 in the Examples folder on your CD-ROM. Measure each current and voltage and compare with the calculated values.

Collector Characteristic Curves

Using a circuit like that shown in Figure 4–9(a), you can generate a set of *collector characteristic curves* that show how the collector current, I_C, varies with the collector-to-emitter voltage, V_{CE}, for specified values of base current, I_B. Notice in the circuit diagram that both V_{BB} and V_{CC} are variable sources of voltage.

Assume that V_{BB} is set to produce a certain value of I_B and V_{CC} is zero. For this condition, both the base-emitter junction and the base-collector junction are forward-biased because the base is at approximately 0.7 V while the emitter and the collector are at 0 V. The base current is through the base-emitter junction because of the low impedance path to ground and, therefore, I_C is zero. When both junctions are forward-biased, the transistor is in the **saturation** region of its operation.

As V_{CC} is increased, V_{CE} increases gradually as the collector current increases. This is indicated by the portion of the characteristic curve between points A and B in Figure 4–9(b). I_C increases as V_{CC} is increased because V_{CE} remains less than 0.7 V due to the forward-biased base-collector junction.

Ideally, when V_{CE} exceeds 0.7 V, the base-collector junction becomes reverse-biased and the transistor goes into the *active* or **linear** region of its operation. Once the base-collector junction is reverse-biased, I_C levels off and remains essentially constant for a given value of I_B as V_{CE} continues to increase. Actually, I_C increases very slightly as V_{CE} increases due to widening of the base-collector depletion region. This results in fewer holes for recombination in the base region which effectively causes a slight increase in β_{DC}. This is shown by the portion of the characteristic curve between points B and C in Figure 4–9(b). For this portion of the characteristic curve, the value of I_C is determined only by the relationship expressed as $I_C = \beta_{DC}I_B$.

When V_{CE} reaches a sufficiently high voltage, the reverse-biased base-collector junction goes into breakdown; and the collector current increases rapidly as indicated by the part of the curve to the right of point C in Figure 4–9(b). A transistor should never be operated in this breakdown region.

(a) Circuit

(b) I_C versus V_{CE} curve for one value of I_B

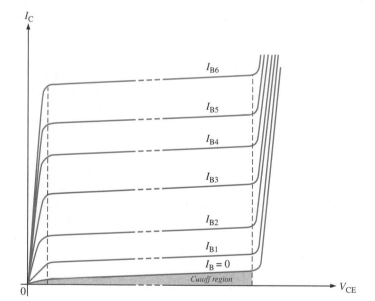

(c) Family of I_C versus V_{CE} curves for several values of I_B
($I_{B1} < I_{B2} < I_{B3}$, etc.)

▲ **FIGURE 4–9**

Collector characteristic curves.

A family of collector characteristic curves is produced when I_C versus V_{CE} is plotted for several values of I_B, as illustrated in Figure 4–9(c). When $I_B = 0$, the transistor is in the **cutoff** region although there is a very small collector leakage current as indicated. The amount of collector leakage current for $I_B = 0$ is exaggerated on the graph for purposes of illustration.

EXAMPLE 4–3

Sketch an ideal family of collector curves for the circuit in Figure 4–10 for $I_B = 5\ \mu A$ to $25\ \mu A$ in $5\ \mu A$ increments. Assume $\beta_{DC} = 100$ and that V_{CE} does not exceed breakdown.

▲ **FIGURE 4–10**

Solution   Using the relationship $I_C = \beta_{DC}I_B$, values of I_C are calculated and tabulated in Table 4–1. The resulting curves are plotted in Figure 4–11. These are ideal curves because the slight increase in I_C for a given value of I_B as V_{CE} increases in the active region is neglected.

▶ **TABLE 4–1**

I_B	I_C
$5\ \mu A$	$0.5\ mA$
$10\ \mu A$	$1.0\ mA$
$15\ \mu A$	$1.5\ mA$
$20\ \mu A$	$2.0\ mA$
$25\ \mu A$	$2.5\ mA$

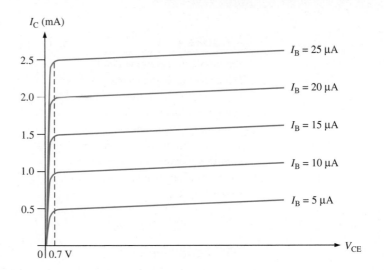

▲ **FIGURE 4–11**

Related Problem   Where would the curve for $I_B = 0$ appear on the graph in Figure 4–11, neglecting collector leakage current?

Cutoff

As previously mentioned, when $I_B = 0$, the transistor is in the cutoff region of its operation. This is shown in Figure 4–12 with the base lead open, resulting in a base current of zero. Under this condition, there is a very small amount of collector leakage current, I_{CEO}, due mainly to thermally produced carriers. Because I_{CEO} is extremely small, it will usually be neglected in circuit analysis so that $V_{CE} = V_{CC}$. In cutoff, both the base-emitter and the base-collector junctions are reverse-biased.

▶ **FIGURE 4–12**

Cutoff: Collector leakage current (I_{CEO}) is extremely small and is usually neglected. Base–emitter and base-collector junctions are reverse-biased.

Saturation

When the base-emitter junction becomes forward-biased and the base current is increased, the collector current also increases ($I_C = \beta_{DC}I_B$) and V_{CE} decreases as a result of more drop across the collector resistor ($V_{CE} = V_{CC} - I_C R_C$). This is illustrated in Figure 4–13. When V_{CE} reaches its saturation value, $V_{CE(sat)}$, the base-collector junction becomes forward-biased and I_C can increase no further even with a continued increase in I_B. At the point of saturation, the relation $I_C = \beta_{DC}I_B$ is no longer valid. $V_{CE(sat)}$ for a transistor occurs somewhere below the knee of the collector curves, and it is usually only a few tenths of a volt for silicon transistors.

▶ **FIGURE 4–13**

Saturation: As I_B increases due to increasing V_{BB}, I_C also increases and V_{CE} decreases due to the increased voltage drop across R_C. When the transistor reaches saturation, I_C can increase no further regardless of further increase in I_B. Base-emitter and base-collector junctions are forward-biased.

DC Load Line

Cutoff and saturation can be illustrated in relation to the collector characteristic curves by the use of a load line. Figure 4–14 shows a dc load line drawn on a family of curves connecting the cutoff point and the saturation point. The bottom of the load line is at ideal cutoff where $I_C = 0$ and $V_{CE} = V_{CC}$. The top of the load line is at saturation where $I_C = I_{C(sat)}$ and $V_{CE} = V_{CE(sat)}$. In between cutoff and saturation along the load line is the *active region* of the transistor's operation. Load line operation is discussed more in Chapter 5.

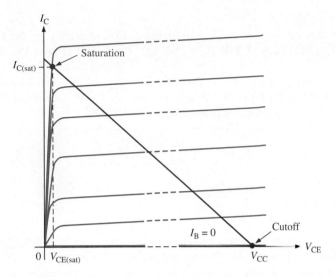

▲ **FIGURE 4–14**

DC load line on a family of collector characteristic curves illustrating the cutoff and saturation conditions.

EXAMPLE 4–4

Determine whether or not the transistor in Figure 4–15 is in saturation. Assume $V_{CE(sat)} = 0.2$ V.

▶ **FIGURE 4–15**

Solution   First, determine $I_{C(sat)}$.

$$I_{C(sat)} = \frac{V_{CC} - V_{CE(sat)}}{R_C} = \frac{10 \text{ V} - 0.2 \text{ V}}{1.0 \text{ k}\Omega} = \frac{9.8 \text{ V}}{1.0 \text{ k}\Omega} = 9.8 \text{ mA}$$

Now, see if I_B is large enough to produce $I_{C(sat)}$.

$$I_B = \frac{V_{BB} - V_{BE}}{R_B} = \frac{3 \text{ V} - 0.7 \text{ V}}{10 \text{ k}\Omega} = \frac{2.3 \text{ V}}{10 \text{ k}\Omega} = 0.23 \text{ mA}$$

$$I_C = \beta_{DC} I_B = (50)(0.23 \text{ mA}) = 11.5 \text{ mA}$$

This shows that with the specified β_{DC}, this base current is capable of producing an I_C greater than $I_{C(sat)}$. Therefore, the **transistor is saturated**, and the collector current value of 11.5 mA is never reached. If you further increase I_B, the collector current remains at its saturation value.

Related Problem   Determine whether or not the transistor in Figure 4–15 is saturated for the following values: $\beta_{DC} = 125$, $V_{BB} = 1.5$ V, $R_B = 6.8$ kΩ, $R_C = 180$ Ω, and $V_{CC} = 12$ V.

Open file E04-04 in the Examples folder on your CD-ROM. Determine if the transistor is in saturation and explain how you did this.

More About β_{DC}

The β_{DC} or h_{FE} is a very important bipolar junction transistor parameter that we need to examine further. β_{DC} is not truly constant but varies with both collector current and with temperature. Keeping the junction temperature constant and increasing I_C causes β_{DC} to increase to a maximum. A further increase in I_C beyond this maximum point causes β_{DC} to decrease. If I_C is held constant and the temperature is varied, β_{DC} changes directly with the temperature. If the temperature goes up, β_{DC} goes up and vice versa. Figure 4–16 shows the variation of β_{DC} with I_C and junction temperature (T_J) for a typical transistor.

▲ **FIGURE 4–16**

Variation of β_{DC} with I_C for several temperatures.

A transistor data sheet usually specifies β_{DC} (h_{FE}) at specific I_C values. Even at fixed values of I_C and temperature, β_{DC} varies from device to device for a given transistor due to inconsistencies in the manufacturing process that are unavoidable. The β_{DC} specified at a certain value of I_C is usually the minimum value, $\beta_{DC(min)}$, although the maximum and typical values are also sometimes specified.

Maximum Transistor Ratings

A transistor, like any other electronic device, has limitations on its operation. These limitations are stated in the form of maximum ratings and are normally specified on the manufacturer's data sheet. Typically, maximum ratings are given for collector-to-base voltage, collector-to-emitter voltage, emitter-to-base voltage, collector current, and power dissipation.

The product of V_{CE} and I_C must not exceed the maximum power dissipation. Both V_{CE} and I_C cannot be maximum at the same time. If V_{CE} is maximum, I_C can be calculated as

$$I_C = \frac{P_{D(max)}}{V_{CE}}$$

Equation 4–7

If I_C is maximum, V_{CE} can be calculated by rearranging Equation 4–7 as follows:

$$V_{CE} = \frac{P_{D(max)}}{I_C}$$

For any given transistor, a maximum power dissipation curve can be plotted on the collector characteristic curves, as shown in Figure 4–17(a). These values are tabulated in Figure 4–17(b). Assume $P_{D(max)}$ is 500 mW, $V_{CE(max)}$ is 20 V, and $I_{C(max)}$ is 50 mA. The curve shows that this particular transistor cannot be operated in the shaded portion of the graph. $I_{C(max)}$ is the limiting rating between points A and B, $P_{D(max)}$ is the limiting rating between points B and C, and $V_{CE(max)}$ is the limiting rating between points C and D.

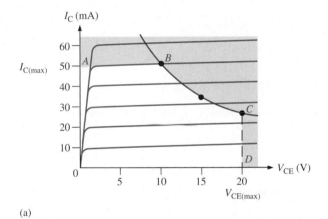

$P_{D(max)}$	V_{CE}	I_C
500 mW	5 V	100 mA
500 mW	10 V	50 mA
500 mW	15 V	33 mA
500 mW	20 V	25 mA

(a) (b)

▲ FIGURE 4–17

Maximum power dissipation curve.

EXAMPLE 4–5

A certain transistor is to be operated with $V_{CE} = 6$ V. If its maximum power rating is 250 mW, what is the most collector current that it can handle?

Solution

$$I_C = \frac{P_{D(max)}}{V_{CE}} = \frac{250\ mW}{6\ V} = \textbf{41.7 mA}$$

Remember that this is not necessarily the maximum I_C. The transistor can handle more collector current if V_{CE} is reduced, as long as $P_{D(max)}$ is not exceeded.

Related Problem If $P_{D(max)} = 1$ W, how much voltage is allowed from collector to emitter if the transistor is operating with $I_C = 100$ mA?

EXAMPLE 4–6

The transistor in Figure 4–18 has the following maximum ratings: $P_{D(max)} = 800$ mW, $V_{CE(max)} = 15$ V, and $I_{C(max)} = 100$ mA. Determine the maximum value to which V_{CC} can be adjusted without exceeding a rating. Which rating would be exceeded first?

▶ **FIGURE 4–18**

Solution First, find I_B so that you can determine I_C.

$$I_B = \frac{V_{BB} - V_{BE}}{R_B} = \frac{5\text{ V} - 0.7\text{ V}}{22\text{ k}\Omega} = 195\ \mu\text{A}$$

$$I_C = \beta_{DC}I_B = (100)(195\ \mu\text{A}) = 19.5\text{ mA}$$

I_C is much less than $I_{C(max)}$ and will not change with V_{CC}. It is determined only by I_B and β_{DC}.

The voltage drop across R_C is

$$V_{R_C} = I_C R_C = (19.5\text{ mA})(1.0\text{ k}\Omega) = 19.5\text{ V}$$

Now you can determine the value of V_{CC} when $V_{CE} = V_{CE(max)} = 15$ V.

$$V_{R_C} = V_{CC} - V_{CE}$$

So,

$$V_{CC(max)} = V_{CE(max)} + V_{R_C} = 15\text{ V} + 19.5\text{ V} = \mathbf{34.5\ V}$$

V_{CC} can be increased to 34.5 V, under the existing conditions, before $V_{CE(max)}$ is exceeded. However, at this point it is not known whether or not $P_{D(max)}$ has been exceeded.

$$P_D = V_{CE(max)} I_C = (15\text{ V})(19.5\text{ mA}) = 293\text{ mW}$$

Since $P_{D(max)}$ is 800 mW, it is *not* exceeded when $V_{CC} = 34.5$ V. So, $V_{CE(max)} = 15$ V is the limiting rating in this case. If the base current is removed causing the transistor to turn off, $V_{CE(max)}$ **will be exceeded first** because the entire supply voltage, V_{CC}, will be dropped across the transistor.

Related Problem The transistor in Figure 4–18 has the following maximum ratings: $P_{D(max)} = 500$ mW, $V_{CE(max)} = 25$ V, and $I_{C(max)} = 200$ mA. Determine the maximum value to which V_{CC} can be adjusted without exceeding a rating. Which rating would be exceeded first?

Derating $P_{D(max)}$

$P_{D(max)}$ is usually specified at 25°C. For higher temperatures, $P_{D(max)}$ is less. Data sheets often give derating factors for determining $P_{D(max)}$ at any temperature above 25°C. For example, a derating factor of 2 mW/°C indicates that the maximum power dissipation is reduced 2 mW for each centigrade degree increase in temperature.

EXAMPLE 4–7

A certain transistor has a $P_{D(max)}$ of 1 W at 25°C. The derating factor is 5 mW/°C. What is the $P_{D(max)}$ at a temperature of 70°C?

Solution   The change (reduction) in $P_{D(max)}$ is

$$\Delta P_{D(max)} = (5 \text{ mW/°C})(70°C - 25°C) = (5 \text{ mW/°C})(45°C) = 225 \text{ mW}$$

Therefore, the $P_{D(max)}$ at 70°C is

$$1 \text{ W} - 225 \text{ mW} = \textbf{775 mW}$$

Related Problem   A transistor has a $P_{D(max)} = 5$ W at 25°C. The derating factor is 10 mW/°C. What is the $P_{D(max)}$ at 70°C?

Transistor Data Sheet

A partial data sheet for the 2N3903 and 2N3904 *npn* transistors is shown in Figure 4–19 on the following page. Notice that the maximum collector-emitter voltage (V_{CEO}) is 40 V. The CEO subscript indicates that the voltage is measured from collector (C) to emitter (E) with the base open (O). In the text, we use $V_{CE(max)}$ for clarity. Also notice that the maximum collector current is 200 mA.

The β_{DC} (h_{FE}) is specified for several values of I_C and, as you can see, h_{FE} varies with I_C as we previously discussed.

The collector-emitter saturation voltage, $V_{CE(sat)}$ is 0.2 V maximum for $I_{C(sat)} = 10$ mA and increases with the current.

SECTION 4–3 REVIEW

1. Define β_{DC} and α_{DC}. What is h_{FE}?
2. If the dc current gain of a transistor is 100, determine β_{DC} and α_{DC}.
3. What two variables are plotted on a collector characteristic curve?
4. What bias conditions must exist for a transistor to operate as an amplifier?
5. Does β_{DC} increase or decrease with temperature?
6. For a given type of transistor, can β_{DC} be considered to be a constant?

Maximum Ratings

Rating	Symbol	Value	Unit
Collector-Emitter voltage	V_{CEO}	40	V dc
Collector-Base voltage	V_{CBO}	60	V dc
Emitter-Base voltage	V_{EBO}	6.0	V dc
Collector current — continuous	I_C	200	mA dc
Total device dissipation @ $T_A = 25°C$ Derate above 25°C	P_D	625 5.0	mW mW/°C
Total device dissipation @ $T_C = 25°C$ Derate above 25°C	P_D	1.5 12	Watts mW/°C
Operating and storage junction Temperature range	T_J, T_{stg}	−55 to +150	°C

Thermal Characteristics

Characteristic	Symbol	Max	Unit
Thermal resistance, junction to case	$R_{\theta JC}$	83.3	°C/W
Thermal resistance, junction to ambient	$R_{\theta JA}$	200	°C/W

2N3903
2N3904

3 Collector

2
Base

1 Emitter

General-Purpose
Transistors

NPN Silicon

Electrical Characteristics ($T_A = 25°C$ unless otherwise noted.)

Characteristic		Symbol	Min	Max	Unit
OFF Characteristics					
Collector-Emitter breakdown voltage ($I_C = 1.0$ mA dc, $I_B = 0$)		$V_{(BR)CEO}$	40	–	V dc
Collector-Base breakdown voltage ($I_C = 10$ µA dc, $I_E = 0$)		$V_{(BR)CBO}$	60	–	V dc
Emitter-Base breakdown voltage ($I_E = 10$ µA dc, $I_C = 0$)		$V_{(BR)EBO}$	6.0	–	V dc
Base cutoff current ($V_{CE} = 30$ V dc, $V_{EB} = 3.0$ V dc)		I_{BL}	–	50	nA dc
Collector cutoff current ($V_{CE} = 30$ V dc, $V_{EB} = 3.0$ V dc)		I_{CEX}	–	50	nA dc
ON Characteristics					
DC current gain ($I_C = 0.1$ mA dc, $V_{CE} = 1.0$ V dc)	2N3903 2N3904	h_{FE}	20 40	– –	–
($I_C = 1.0$ mA dc, $V_{CE} = 1.0$ V dc)	2N3903 2N3904		35 70	– –	
($I_C = 10$ mA dc, $V_{CE} = 1.0$ V dc)	2N3903 2N3904		50 100	150 300	
($I_C = 50$ mA dc, $V_{CE} = 1.0$ V dc)	2N3903 2N3904		30 60	– –	
($I_C = 100$ mA dc, $V_{CE} = 1.0$ V dc)	2N3903 2N3904		15 30	– –	
Collector-Emitter saturation voltage ($I_C = 10$ mA dc, $I_B = 1.0$ mA dc) ($I_C = 50$ mA dc, $I_B = 5.0$ mA dc)		$V_{CE(sat)}$	– –	0.2 0.3	V dc
Base-Emitter saturation voltage ($I_C = 10$ mA dc, $I_B = 1.0$ mA dc) ($I_C = 50$ mA dc, $I_B = 5.0$ mA dc)		$V_{BE(sat)}$	0.65 –	0.85 0.95	V dc

▲ **FIGURE 4–19**

Partial transistor data sheet.

4–4   THE TRANSISTOR AS AN AMPLIFIER

Amplification is the process of linearly increasing the amplitude of an electrical signal and is one of the major properties of a transistor. As you learned, the transistor exhibits current gain (called β). When a transistor is biased in the active (or linear) region, as previously described, the BE junction has a low resistance due to forward bias and the BC junction has a high resistance due to reverse bias.

After completing this section, you should be able to

- **Discuss how a transistor is used as a voltage amplifier**

- Describe amplification

- Develop the ac equivalent circuit for a basic transistor amplifier

- Determine the voltage gain of a basic transistor amplifier

DC and AC Quantities

Before introducing the concept of transistor **amplification**, the designations that we will use for the circuit quantities of current, voltage, and resistance must be explained because amplifier circuits have both dc and ac quantities.

In this text, italic capital letters are used for both dc and ac currents (I) and voltages (V). This rule applies to rms, average, peak, and peak-to-peak ac values. AC current and voltage values are always rms unless stated otherwise. Although some texts use lowercase i and v for ac current and voltage, we reserve the use of lowercase i and v only for instantaneous values, as you learned in your dc/ac circuits course. In this text, the distinction between a dc current or voltage and an ac current or voltage is in the subscript.

DC quantities always carry an uppercase roman (nonitalic) subscript. For example, I_B, I_C, and I_E are the dc transistor currents. V_{BE}, V_{CB}, and V_{CE} are the dc voltages from one transistor terminal to another. Single subscripted voltages such as V_B, V_C, and V_E are dc voltages from the transistor terminals to ground.

AC and all time-varying quantities always carry a lowercase italic subscript. For example, I_b, I_c, and I_e are the ac transistor currents. V_{be}, V_{cb}, and V_{ce} are the ac voltages from one transistor terminal to another. Single subscripted voltages such as V_b, V_c, and V_e are ac voltages from the transistor terminals to ground.

The rule is different for *internal* transistor resistances. As you will see later, transistors have internal ac resistances that are designated by lowercase r' with an appropriate subscript. For example, the internal ac emitter resistance is designated as r'_e.

Circuit resistances external to the transistor itself use the standard italic capital R with a subscript that identifies the resistance as dc or ac (when applicable), just as for current and voltage. For example R_E is an external dc emitter resistance and R_e is an external ac emitter resistance.

Transistor Amplification

As you have learned, a transistor amplifies current because the collector current is equal to the base current multiplied by the current gain, β. The base current in a transistor is very small compared to the collector and emitter currents. Because of this, the collector current is approximately equal to the emitter current.

With this in mind, let's look at the circuit in Figure 4–20(a). An ac voltage, V_{in}, is superimposed on the dc bias voltage V_{BB} by connecting them in series with the base resistor, R_B, as shown. The dc bias voltage V_{CC} is connected to the collector through the collector resistor, R_C.

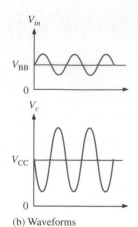

(a) Circuit with ac input voltage V_{in} and dc bias
voltage superimposed

(b) Waveforms

▲ **FIGURE 4–20**

Basic transistor amplifier circuit.

    The ac input voltage produces an ac base current, which results in a much larger ac collector current. The ac collector current produces an ac voltage across R_C, thus producing an amplified, but inverted, reproduction of the ac input voltage in the active region of operation, as illustrated in Figure 4–20(b).

    The forward-biased base-emitter junction presents a very low resistance to the ac signal. This internal ac emitter resistance is designated r'_e. In Figure 4–20(a), the ac emitter current is

$$I_e \cong I_c = \frac{V_b}{r'_e}$$

The ac collector voltage, V_c, equals the ac voltage drop across R_C.

$$V_c = I_c R_C$$

Since $I_c \cong I_e$, the ac collector voltage is

$$V_c \cong I_e R_C$$

V_b can be considered the transistor ac input voltage where $V_b = V_{in} - I_b R_B$. V_c can be considered the transistor ac output voltage. The ratio of V_c to V_b is the ac voltage gain, A_v, of the transistor circuit.

$$A_v = \frac{V_c}{V_b}$$

Substituting $I_e R_C$ for V_c and $I_e r'_e$ for V_b yields

$$A_v = \frac{V_c}{V_b} \cong \frac{I_e R_C}{I_e r'_e}$$

The I_e terms cancel; therefore,

Equation 4–8
$$A_v \cong \frac{R_C}{r'_e}$$

Equation 4–8 shows that the transistor in Figure 4–20 provides amplification in the form of voltage gain, which is dependent on the value of R_C and r'_e.

    Since R_C is always considerably larger in value than r'_e, the output voltage is always greater than the input voltage. Various types of amplifiers are covered in detail in later chapters.

EXAMPLE 4–8

Determine the voltage gain and the ac output voltage in Figure 4–21 if $r'_e = 50\ \Omega$.

▶ **FIGURE 4–21**

Solution The voltage gain is

$$A_v \cong \frac{R_C}{r'_e} = \frac{1.0\ \text{k}\Omega}{50\ \Omega} = \mathbf{20}$$

Therefore, the ac output voltage is

$$V_{out} = A_v V_b = (20)(100\ \text{mV}) = \mathbf{2\ V\ rms}$$

Related Problem What value of R_C in Figure 4–21 will it take to have a voltage gain of 50?

**SECTION 4–4
REVIEW**

1. What is amplification?
2. How is voltage gain defined?
3. Name two factors that determine the voltage gain of an amplifier.
4. What is the voltage gain of a transistor amplifier that has an output of 5 V rms and an input of 250 mV rms?
5. A transistor connected as in Figure 4–21 has an $r'_e = 20\ \Omega$. If R_C is 1200 Ω, what is the voltage gain?

4–5 THE TRANSISTOR AS A SWITCH

In the previous section, you saw how the transistor can be used as a linear amplifier. The second major application area is switching applications. When used as an electronic switch, a transistor is normally operated alternately in cutoff and saturation. Digital circuits make use of the switching characteristics of transistors.

After completing this section, you should be able to

■ **Discuss how a transistor is used as an electronic switch**

■ Analyze a transistor switching circuit for cutoff and saturation

■ Describe the conditions that produce cutoff

■ Describe the conditions that produce saturation

■ Discuss a basic application of a transistor switching circuit

Figure 4–22 illustrates the basic operation of the transistor as a switching device. In part (a), the transistor is in the cutoff region because the base-emitter junction is not forward-biased. In this condition, there is, ideally, an *open* between collector and emitter, as indicated by the switch equivalent. In part (b), the transistor is in the saturation region because the base-emitter junction and the base-collector junction are forward-biased and the base current is made large enough to cause the collector current to reach its saturation value. In this condition, there is, ideally, a *short* between collector and emitter, as indicated by the switch equivalent. Actually, a voltage drop of up to a few tenths of a volt normally occurs, which is the saturation voltage, $V_{CE(sat)}$.

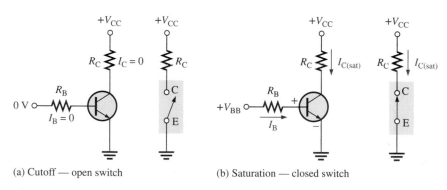

(a) Cutoff — open switch   (b) Saturation — closed switch

▲ **FIGURE 4–22**

Ideal switching action of a transistor.

Conditions in Cutoff

As mentioned before, a transistor is in the cutoff region when the base-emitter junction is *not* forward-biased. Neglecting leakage current, all of the currents are zero, and V_{CE} is equal to V_{CC}.

Equation 4–9

$$V_{CE(cutoff)} = V_{CC}$$

Conditions in Saturation

As you have learned, when the base-emitter junction is forward-biased and there is enough base current to produce a maximum collector current, the transistor is saturated. The formula for collector saturation current is

$$I_{C(sat)} = \frac{V_{CC} - V_{CE(sat)}}{R_C}$$

Equation 4–10

Since $V_{CE(sat)}$ is very small compared to V_{CC}, it can usually be neglected.
The minimum value of base current needed to produce saturation is

$$I_{B(min)} = \frac{I_{C(sat)}}{\beta_{DC}}$$

Equation 4–11

I_B should be significantly greater than $I_{B(min)}$ to keep the transistor well into saturation.

EXAMPLE 4–9

(a) For the transistor circuit in Figure 4–23, what is V_{CE} when $V_{IN} = 0$ V?

(b) What minimum value of I_B is required to saturate this transistor if β_{DC} is 200? Neglect $V_{CE(sat)}$.

(c) Calculate the maximum value of R_B when $V_{IN} = 5$ V.

▶ **FIGURE 4–23**

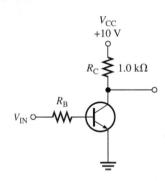

Solution (a) When $V_{IN} = 0$ V, the transistor is in cutoff (acts like an open switch) and

$$V_{CE} = V_{CC} = \mathbf{10\ V}$$

(b) Since $V_{CE(sat)}$ is neglected (assumed to be 0 V),

$$I_{C(sat)} = \frac{V_{CC}}{R_C} = \frac{10\ V}{1.0\ k\Omega} = 10\ mA$$

$$I_{B(min)} = \frac{I_{C(sat)}}{\beta_{DC}} = \frac{10\ mA}{200} = \mathbf{50\ \mu A}$$

This is the value of I_B necessary to drive the transistor to the point of saturation. Any further increase in I_B will drive the transistor deeper into saturation but will not increase I_C.

(c) When the transistor is on, $V_{BE} \cong 0.7$ V. The voltage across R_B is

$$V_{R_B} = V_{IN} - V_{BE} \cong 5\,\text{V} - 0.7\,\text{V} = 4.3\,\text{V}$$

Calculate the maximum value of R_B needed to allow a minimum I_B of 50 μA by Ohm's law as follows:

$$R_{B(max)} = \frac{V_{R_B}}{I_{B(min)}} = \frac{4.3\,\text{V}}{50\,\mu\text{A}} = \textbf{86 k}\boldsymbol{\Omega}$$

Related Problem   Determine the minimum value of I_B required to saturate the transistor in Figure 4–23 if β_{DC} is 125 and $V_{CE(sat)}$ is 0.2 V.

A Simple Application of a Transistor Switch

The transistor in Figure 4–24 is used as a switch to turn the LED on and off. For example, a square wave input voltage with a period of 2 s is applied to the input as indicated. When the square wave is at 0 V, the transistor is in cutoff and, since there is no collector current, the LED does not emit light. When the square wave goes to its high level, the transistor saturates. This forward-biases the LED, and the resulting collector current through the LED causes it to emit light. So, we have a blinking LED that is on for 1 s and off for 1 s.

▶ **FIGURE 4–24**

A transistor used to switch an LED on and off.

EXAMPLE 4–10

The LED in Figure 4–24 requires 30 mA to emit a sufficient level of light. Therefore, the collector current should be approximately 30 mA. For the following circuit values, determine the amplitude of the square wave input voltage necessary to make sure that the transistor saturates. Use double the minimum value of base current as a safety margin to ensure saturation. $V_{CC} = 9$ V, $V_{CE(sat)} = 0.3$ V, $R_C = 270\ \Omega$, $R_B = 3.3$ kΩ, and $\beta_{DC} = 50$.

Solution

$$I_{C(sat)} = \frac{V_{CC} - V_{CE(sat)}}{R_C} = \frac{9\,\text{V} - 0.3\,\text{V}}{270\,\Omega} = 32.2\,\text{mA}$$

$$I_{B(min)} = \frac{I_{C(sat)}}{\beta_{DC}} = \frac{32.2\,\text{mA}}{50} = 644\,\mu\text{A}$$

To ensure saturation, use twice the value of $I_{B(min)}$, which gives 1.29 mA. Then

$$I_B = \frac{V_{R_B}}{R_B} = \frac{V_{in} - V_{BE}}{R_B} = \frac{V_{in} - 0.7 \text{ V}}{3.3 \text{ k}\Omega}$$

Solve for the voltage amplitude of the square wave input V_{in}.

$$V_{in} - 0.7 \text{ V} = 2I_{B(min)}R_B = (1.29 \text{ mA})(3.3 \text{ k}\Omega)$$

$$V_{in} = (1.29 \text{ mA})(3.3 \text{ k}\Omega) + 0.7 \text{ V} = \textbf{4.96 V}$$

Related Problem   If you change the LED in Figure 4–24 to one that requires 50 mA for a specified light emission and you can't increase the input amplitude above 5 V or V_{CC} above 9 V, how would you modify the circuit? Specify the component(s) to be changed and the value(s).

Open file E04-10 in the Examples folder on your CD-ROM. Using a 0.5 Hz square wave input with the calculated amplitude, verify that the transistor is switching between cutoff and saturation and that the LED is alternately turning on and off.

SECTION 4–5 REVIEW

1. When a transistor is used as a switch, in what two states is it operated?
2. When is the collector current maximum?
3. When is the collector current approximately zero?
4. Under what condition is $V_{CE} = V_{CC}$?
5. When is V_{CE} minimum?

4–6 TRANSISTOR PACKAGES AND TERMINAL IDENTIFICATION

Transistors are available in a wide range of package types for various applications. Those with mounting studs or heat sinks are usually power transistors. Low-power and medium-power transistors are usually found in smaller metal or plastic cases. Still another package classification is for high-frequency devices. You should be familiar with common transistor packages and be able to identify the emitter, base, and collector terminals. This section is about transistor packages and terminal identification.

After completing this section, you should be able to

- **Identify various types of transistor package configurations**

- List three broad categories of transistors

- Recognize various types of cases and identify the pin configurations

Transistor Categories

Manufacturers generally classify their bipolar junction transistors into three broad categories: general-purpose/small-signal devices, power devices, and RF (radio frequency/microwave) devices. Although each of these categories, to a large degree, has its own unique package types, you will find certain types of packages used in more than one device category. While keeping in mind there is some overlap, we will look at transistor packages for each of the three categories, so that you will be able to recognize a transistor when you see one on a circuit board and have a good idea of what general category it is in.

General–Purpose/Small–Signal Transistors   General-purpose/small-signal transistors are generally used for low- or medium-power amplifiers or switching circuits. The packages are either plastic or metal cases. Certain types of packages contain multiple transistors. Figure 4–25 illustrates common plastic cases, Figure 4–26 shows packages called *metal cans,* and Figure 4–27 shows multiple-transistor packages. Some of the multiple-transistor packages such as the dual-in-line (DIP) and the small-outline (SO) are the same as those used for many integrated circuits. Typical pin connections are shown so you can identify the emitter, base, and collector.

Power Transistors   Power transistors are used to handle large currents (typically more than 1 A) and/or large voltages. For example, the final audio stage in a stereo system uses a power transistor amplifier to drive the speakers. Figure 4–28 shows some common package configurations. In most applications, the metal tab or the metal case is common

(a) TO-92 or TO-226AA   (b) TO-92 or TO-226AE   (c) SOT-23 or TO-236AB

▲ **FIGURE 4–25**

Plastic cases for general–purpose/small–signal transistors. Both old and new JEDEC TO numbers are given. Pin configurations may vary. Always check the data sheet.

(a) TO-18 or TO-206AA   (b) TO-39 or TO-205AD   (c) TO-46 or TO-206AB

(d) TO-52 or TO-206AC   (e) TO-72 or TO-206AF   (f) Pin configuration (bottom view). Emitter is closest to tab.

▲ **FIGURE 4–26**

Metal cases for general–purpose/small–signal transistors.

(a) Dual metal can

(b) Quad dual-in-line (DIP) and quad
flat-pack. Dot indicates pin 1.

(c) Quad small outline (SO) package for
surface-mount technology

(d) Dual ceramic flat-pack

▲ FIGURE 4–27

Typical multiple-transistor packages.

(a) TO-3 or TO-204AE

(b) TO-218

(c) TO-218AC

(d) TO-220AB

(e) TO-225AA

(f) Surface-mount technology

(g) Greatly enlarged cutaway view of tiny transistor chip mounted in the
encapsulated package

▲ FIGURE 4–28

Typical power transistors.

to the collector and is thermally connected to a heat sink for heat dissipation. Notice in part (g) how the small transistor chip is mounted inside the much larger package.

RF Transistors   RF transistors are designed to operate at extremely high frequencies and are commonly used for various purposes in communications systems and other high-frequency applications. Their unusual shapes and lead configurations are designed to optimize certain high-frequency parameters. Figure 4–29 shows some examples.

▲ FIGURE 4–29

Examples of RF transistors.

SECTION 4–6 REVIEW

1. List the three broad categories of bipolar junction transistors.
2. In a single-transistor metal case, how do you identify the leads?
3. In power transistors, the metal mounting tab or case is connected to which transistor region?

4–7   TROUBLESHOOTING

As you already know, a critical skill in electronics work is the ability to identify a circuit malfunction and to isolate the failure to a single component if necessary. In this section, the basics of troubleshooting transistor bias circuits and testing individual transistors are covered.

After completing this section, you should be able to

- **Troubleshoot various faults in transistor circuits**
- Explain floating point measurement
- Use voltage measurements to identify a fault in a transistor circuit
- Use a DMM to test a transistor
- Explain how a transistor can be viewed in terms of a diode equivalent
- Discuss in-circuit and out-of-circuit testing
- Discuss point-of-measurement in troubleshooting
- Discuss leakage and gain measurements

Troubleshooting a Biased Transistor

Several faults can occur in a simple transistor bias circuit. Possible faults are open bias resistors, open or resistive connections, shorted connections, and opens or shorts internal to the transistor itself. Figure 4–30 is a basic transistor bias circuit with all voltages referenced to ground. The two bias voltages are $V_{BB} = 3$ V and $V_{CC} = 9$ V. The correct voltage measurements at the base and collector are shown. Analytically, these voltages are verified as follows. A $\beta_{DC} = 200$ is taken as midway between the minimum and maximum values of h_{FE} given on the data sheet for the 2N3904 in Figure 4–19. A different h_{FE} (β_{DC}), of course, will produce different results for the given circuit.

$$V_B = V_{BE} = 0.7 \text{ V}$$

$$I_B = \frac{3 \text{ V} - 0.7 \text{ V}}{56 \text{ k}\Omega} = \frac{2.3 \text{ V}}{56 \text{ k}\Omega} = 41.1 \text{ } \mu\text{A}$$

$$I_C = \beta_{DC}I_B = 200(41.1 \text{ } \mu\text{A}) = 8.2 \text{ mA}$$

$$V_C = 9 \text{ V} - I_CR_C = 9 \text{ V} - (8.2 \text{ mA})(560 \text{ } \Omega) = 4.4 \text{ V}$$

▶ **FIGURE 4–30**

A basic transistor bias circuit.

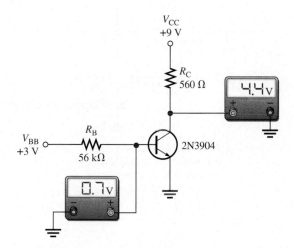

Several faults that can occur in the circuit and the accompanying symptoms are illustrated in Figure 4–31. Symptoms are shown in terms of measured voltages that are incorrect. The term **floating point** refers to a point in the circuit that is not electrically connected to ground or a "solid" voltage. Normally, very small and sometimes fluctuating voltages in the μV to low mV range are generally measured at floating points. The faults in Figure 4–31 are typical but do not represent all possible faults that may occur.

Testing a Transistor with a DMM

A digital multimeter can be used as a fast and simple way to check a transistor for open or shorted junctions. For this test, you can view the transistor as two diodes connected as shown in Figure 4–32 for both *npn* and *pnp* transistors. The base-collector junction is one diode and the base-emitter junction is the other.

Recall that a good diode will show an extremely high resistance (or open) with reverse bias and a very low resistance with forward bias. A defective open diode will show an extremely high resistance (or open) for both forward and reverse bias. A defective shorted or resistive diode will show zero or a very low resistance for both forward and reverse bias. An open diode is the most common type of failure. Since the transistor *pn* junctions are, in effect diodes, the same basic characteristics apply.

(a) **Fault:** Open base resistor.
Symptoms: Readings from μV to a few mV at base due to floating point. 9 V at collector because transistor is in cutoff.

(b) **Fault:** Open collector resistor.
Symptoms: Readings from μV to a few mV at collector due to floating point. 0.5 V – 0.7 V at base due to forward voltage drop across the base-emitter junction.

(c) **Fault:** Base internally open.
Symptoms: 3 V at base lead. 9 V at collector because transistor is in cutoff.

(d) **Fault:** Collector internally open.
Symptoms: 0.5 V – 0.7 V at base lead due to forward voltage drop across base-emitter junction. 9 V at collector because the open prevents collector current.

(e) **Fault:** Emitter internally open.
Symptoms: 3 V at base lead. 9 V at collector because there is no collector current. 0 V at the emitter as normal.

(f) **Fault:** Open ground connection.
Symptoms: 3 V at base lead. 9 V at collector because there is no collector current. 2.5 V or more at the emitter due to the forward voltage drop across the base-emitter junction. The measuring voltmeter provides a forward current path through its internal resistance.

▲ **FIGURE 4–31**

Typical faults and symptoms in the basic transistor bias circuit.

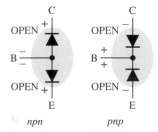

(a) Both junctions should read 0.7 V ± 0.2 V when forward-biased.

(b) Both junctions should ideally read OPEN when reverse-biased.

▲ **FIGURE 4–32**

A transistor viewed as two diodes.

The DMM Diode Test Position Many digital multimeters (DMMs) have a *diode test* position that provides a convenient way to test a transistor. A typical DMM, as shown in Figure 4–33, has a small diode symbol to mark the position of the function switch. When set to diode test, the meter provides an internal voltage sufficient to forward-bias and reverse-bias a transistor junction. This internal voltage may vary among different makes of DMM, but 2.5 V to 3.5 V is a typical range of values. The meter provides a voltage reading to indicate the condition of the transistor junction under test.

(a) Forward-bias test of BE junction

(b) Reverse-bias test of BE junction

(c) Forward-bias test of BC junction

(d) Reverse-bias test of BC junction

▲ **FIGURE 4–33**

Typical DMM test of a properly functioning *npn* transistor. Leads are reversed for a *pnp* transistor.

When the Transistor Is Not Defective In Figure 4–33(a), the red (positive) lead of the meter is connected to the base of an *npn* transistor and the black (negative) lead is connected to the emitter to forward-bias the base-emitter junction. If the junction is good, you will get a reading of between 0.5 V and 0.9 V, with 0.7 V being typical for forward bias.

In Figure 4–33(b), the leads are switched to reverse-bias the base-emitter junction, as shown. If the transistor is working properly, you will get a voltage reading based on the meter's internal voltage source. The 2.6 V shown in the figure represents a typical value and indicates that the junction has an extremely high reverse resistance with essentially all of the internal voltage appearing across it.

The process just described is repeated for the base-collector junction as shown in Figure 4–33(c) and (d). For a *pnp* transistor, the polarity of the meter leads are reversed for each test.

When the Transistor Is Defective When a transistor has failed with an open junction or internal connection, you get an open circuit voltage reading (2.6 V is typical for many DMMs) for both the forward-bias and the reverse-bias conditions for that junction, as illustrated in Figure 4–34(a). If a junction is shorted, the meter reads 0 V in both forward- and reverse-bias tests, as indicated in part (b). Sometimes, a failed junction may exhibit a small resistance for both bias conditions rather than a pure short. In this case, the meter will show a small voltage much less than the correct open voltage. For example, a resistive junction may result in a reading of 1.1 V in both directions rather than the correct readings of 0.7 V for forward bias and 2.6 V for reverse bias.

Some DMMs provide a test socket on their front panel for testing a transistor for the h_{FE} (β_{DC}) value. If the transistor is inserted improperly in the socket or if it is not functioning properly due to a faulty junction or internal connection, a typical meter will flash

(a) Forward-bias test and reverse-bias test give the same reading (2.60 V is typical) for an open BC junction.

(b) Forward- and reverse-bias tests for a shorted junction give the same 0 V reading. If the junction is resistive, the reading is less than 2.6 V.

▲ **FIGURE 4–34**

Testing a defective *npn* transistor. Leads are reversed for a *pnp* transistor.

a 1 or display a 0. If a value of β_{DC} within the normal range for the specific transistor is displayed, the device is functioning properly. The normal range of β_{DC} can be determined from the data sheet.

Checking a Transistor with the OHMs Function   DMMs that do not have a diode test position or an h_{FE} socket can be used to test a transistor for open or shorted junctions by setting the function switch to an OHMs range. For the forward-bias check of a good transistor *pn* junction, you will get a resistance reading that can vary depending on the meter's internal battery. Many DMMs do not have sufficient voltage on the OHMs range to fully forward-bias a junction, and you may get a reading of from several hundred to several thousand ohms.

For the reverse-bias check of a good transistor, you will get an out-of-range indication on most DMMs because the reverse resistance is too high to measure. An out-of-range indication may be a flashing 1 or a display of dashes, depending on the particular DMM.

Even though you may not get accurate forward and reverse resistance readings on a DMM, the relative readings are sufficient to indicate a properly functioning transistor *pn* junction. The out-of-range indication shows that the reverse resistance is very high, as you expect. The reading of a few hundred to a few thousand ohms for forward bias indicates that the forward resistance is small compared to the reverse resistance, as you expect.

Transistor Testers

An individual transistor can be tested either in-circuit or out-of-circuit with a transistor tester. For example, let's say that an amplifier on a particular printed circuit (PC) board has malfunctioned. Good troubleshooting practice dictates that you do not unsolder a component from a circuit board unless you are reasonably sure that it is bad or you simply cannot isolate the problem down to a single component. When components are removed, there is a risk of damage to the PC board contacts and traces.

The first step is to do an in-circuit check of the transistor using a transistor tester similar to the one shown in Figure 4–35. The three clip-leads are connected to the transistor terminals and the tester gives a positive indication if the transistor is good.

Case 1   If the transistor tests defective, it should be carefully removed and replaced with a known good one. An out-of-circuit check of the replacement device is usually a good

▲ FIGURE 4–35

Transistor tester (courtesy of B & K Precision).

idea, just to make sure it is OK. The transistor is plugged into the socket on the transistor tester for out-of-circuit tests.

Case 2 If the transistor tests good in-circuit but the circuit is not working properly, examine the circuit board for a poor connection at the collector pad or for a break in the connecting trace. A poor solder joint often results in an open or a highly resistive contact. The physical point at which you actually measure the voltage is very important in this case. For example, if you measure on the collector lead when there is an external open at the collector pad, you will measure a floating point. If you measure on the connecting trace or on the R_C lead, you will read V_{CC}. This situation is illustrated in Figure 4–36.

▶ FIGURE 4–36

The indication of an open, when it is in the circuit external to the transistor, depends on where you measure.

Meter reads dc
supply voltage.

A few μV to a few
mV indicates
a floating point.

E B C

GND ---

OPEN connection
at pad

$--V_{CC}$

Importance of Point-of-Measurement in Troubleshooting   In case 2, if you had taken the initial measurement on the transistor lead itself and the open were *internal* to the transistor as shown in Figure 4–37, you would have measured V_{CC}. This would have indicated a defective transistor even before the tester was used. This simple concept emphasizes the importance of point-of-measurement in certain troubleshooting situations.

▶ **FIGURE 4–37**

Illustration of an internal open.
Compare with Figure 4–36.

EXAMPLE 4–11

What fault do the measurements in Figure 4–38 indicate?

▲ **FIGURE 4–38**

Solution   The transistor is in cutoff, as indicated by the 10 V measurement on the collector lead. The base bias voltage of 3 V appears on the PC board contact but not on the transistor lead, as indicated by the floating point measurement. This shows that there is an open external to the transistor between the two measured base points. Check the solder joint at the base contact on the PC board. If the open were internal, there would be 3 V on the base lead.

Related Problem   If the meter in Figure 4–38 that now reads 3 V indicates a floating point when touching the circuit board pad, what is the most likely fault?

Leakage Measurement

Very small leakage currents exist in all transistors and in most cases are small enough to neglect (usually nA). When a transistor is connected with the base open ($I_B = 0$), it is in cutoff. Ideally $I_C = 0$; but actually there is a small current from collector to emitter, as mentioned earlier, called I_{CEO} (collector-to-emitter current with base open). This leakage current is usually in the nA range for silicon. A faulty transistor will often have excessive leakage current and can be checked in a transistor tester. Another leakage current in transistors is the reverse collector-to-base current, I_{CBO}. This is measured with the emitter open. If it is excessive, a shorted collector-base junction is likely.

Gain Measurement

In addition to leakage tests, the typical transistor tester also checks the β_{DC}. A known value of I_B is applied, and the resulting I_C is measured. The reading will indicate the value of the I_C/I_B ratio, although in some units only a relative indication is given. Most testers provide for an in-circuit β_{DC} check, so that a suspected device does not have to be removed from the circuit for testing.

Curve Tracers

The *curve tracer* is an oscilloscope type of instrument that can display transistor characteristics such as a family of collector curves. In addition to the measurement and display of various transistor characteristics, diode curves can also be displayed.

EWB/Multisim Troubleshooting Exercises

These file circuits are in the Troubleshooting Exercises folder on your CD-ROM.

1. Open file TSE04-01. Determine if the circuit is working properly and, if not, determine the fault.

2. Open file TSE04-02. Determine if the circuit is working properly and, if not, determine the fault.

3. Open file TSE04-03. Determine if the circuit is working properly and, if not, determine the fault.

SECTION 4–7 REVIEW

1. If a transistor on a circuit board is suspected of being faulty, what should you do?
2. In a transistor bias circuit, such as the one in Figure 4–30, what happens if R_B opens?
3. In a circuit such as the one in Figure 4–30, what are the base and collector voltages if there is an external open between the emitter and ground?

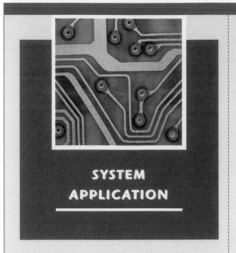

SYSTEM
APPLICATION

You have been assigned to work on an electronic security alarm system. This sys-

tem has several parts, but you will concentrate on the transistor circuits that detect a break (open) in the loops containing remote sensors for windows and doors. In this particular application, the transistors are used as switching devices. You will apply the knowledge you have gained in this chapter in completing your assignment.

The Security Alarm System

The block diagram for a three-zone security alarm system is shown in Figure 4–39. A detail of the circuit board containing the zone-monitoring circuits is included

because this is the part of the system that is the focus of this assignment.

Two of the zones contain conductive loops with several magnetic switches that protect the windows and doors in that particular zone. The third zone protects the main entry door.

When an intrusion occurs, a magnetic switch at that point of entry breaks contact and opens the zone loop. This causes the input to the monitoring circuit for that zone to go to a 0 V level, activating the circuit which, in turn, energizes the relay. The closure of the relay contacts sets off the audible alarm (siren) and/or initi-

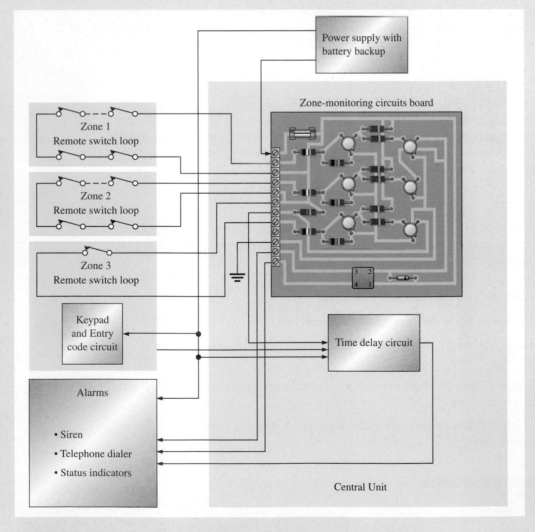

▲ FIGURE 4–39

Basic block diagram of the security alarm system.

ates an automatic telephone dialing sequence. The monitoring circuits for either zone 1 or zone 2 can energize the common relay.

For zone 3, which is the main entry, the output of the monitoring circuit goes to the time-delay circuit to allow time for keying in an entry code that will disarm the system. If, after a preset time interval, no code or an incorrect code has been entered, the time-delay circuit will set off the alarm and/or the telephone dialing sequence.

There are many features in a typical security system, and systems can range from very simple to very complex. For this assignment, you will concentrate on the zone-monitoring circuits board.

The Zone-Monitoring Circuits

There are three identical transistor monitoring circuits, one for each zone. The magnetic switches in the zone loop are normally closed and open only when there is an intrusion (a window or door opened). Because the magnetic switches are in series, the entire loop is open when one switch is open.

The circuit for one zone consists of two transistors and is shown in Figure 4–40. Transistor Q_1 detects when one of the remote magnetic switches is open, and transistor Q_2 drives the relay that activates

the alarm. The transistors operate only in cutoff or saturation.

The zone loop is normally closed, keeping the input to R_1 at 12 V and transistor Q_1 in saturation. Since Q_1 is operating in saturation, its collector voltage (with respect to ground) is no more than 0.2 V. This keeps transistor Q_2 in cutoff; and since there is no Q_2 collector current through the relay coil, the system is unactivated.

When a magnetic switch in the zone loop opens, the 12 V at the input is removed and transistor Q_1 switches to cutoff because R_2 pulls the base to ground. When Q_1 goes into cutoff, Q_2 is biased into saturation by the 12 V and the base current through R_3 and R_4. The

▲ **FIGURE 4–40**

Zone-monitoring circuit. All three circuits are identical with the exception that the zone 3 circuit has a collector resistor instead of the relay.

resulting Q_2 collector current through the relay coil energizes the normally open relay and causes the contact to close; this activates the alarm. Diode D_1 across the relay coil suppresses the induced transient voltage when the relay is deactivated.

The Components

- *The transistors*   The transistors are 2N3947s.

- *The resistors*   Determine the minimum power rating for each of the resistors in the circuit and specify a standard rating.

- *The relay*   Based on the applicable transistor ratings and parameters and on the power supply voltage, select one of the following relays:

  Relay A: coil voltage, 12 V;
  coil resistance, 55 Ω;
  turn-on current, 0.2 A;
  holding current, 0.15 A

  Relay B: coil voltage, 12 V;
  coil resistance, 95 Ω;
  turn-on current, 0.5 A;
  holding current, 0.4 A

  Relay C: coil voltage, 24 V;
  coil resistance, 150 Ω;
  turn-on current, 0.75 A;
  holding current, 0.5 A

- *The diode*   Select a diode with a PIV rating of at least 100 V.

Analysis of the Circuit

Verify that the resistors in Figure 4–40 have values that produce adequate currents for the transistors to operate in cut-off and saturation and to drive the selected relay. Refer to the transistor data sheet in Figure 4–41.

The Printed Circuit Board

- Using the complete zone-monitoring circuits schematic in Figure 4–42(b), check out the printed circuit board in Figure 4–42(a) to verify that it is correct.

- Label a copy of the board with the component and input/output designations in agreement with the schematic.

A Test Procedure

- Develop a step-by-step set of instructions on how to completely check the zone-monitoring circuits board for proper operation using the test points (circled numbers) and connector terminals indicated in the test bench setup of Figure 4–43. The zone loop for each circuit is simulated with a single SPST switch.

- Specify voltage values and/or resistance for all the measurements to be made. Provide a fault analysis for all possible component failures.

Troubleshooting

Three boards in the initial manufacturing run have been found to be defective. Troubleshooting must be used to determine the problems.

The test results are shown in Figure 4–44 on page 210. Based on the sequence of measurements and the loop switch configuration for each defective board, determine the most likely fault(s) in each case.

The red circled numbers indicate test points on the circuit board and the black numbers in a square are for connector terminals on the circuit board. The DMM function setting is indicated below the display.

Final Report (Optional)

Submit a final written report on the zone-monitoring circuits board using an organized format that includes the following:

1. A physical description of the circuits.

2. A discussion of the operation of each circuit.

3. A list of the specifications.

4. A list of parts with part numbers if available.

5. A list of the types of problems on the three defective circuit boards.

6. A complete description of how you determined the problem on each of the three defective boards.

Maximum Ratings

Rating	Symbol	Value	Unit
Collector-Emitter voltage	V_{CEO}	40	V dc
Collector-Base voltage	V_{CBO}	60	V dc
Emitter-Base voltage	V_{EBO}	6.0	V dc
Collector current — continuous	I_C	200	mA dc
Total device dissipation @ $T_A = 25°C$ Derate above 25°C	P_D	0.36 2.06	Watts mW/°C
Total device dissipation @ $T_C = 25°C$ Derate above 25°C	P_D	1.2 6.9	Watts mW/°C
Operating and storage junction Temperature range	T_J, T_{stg}	−65 to +200	°C

Thermal Characteristics

Characteristic	Symbol	Max	Unit
Thermal resistance, junction to case	$R_{\theta JC}$	0.15	°C/mW
Thermal resistance, junction to ambient	$R_{\theta JA}$	0.49	°C/mW

2N3946
2N3947

3 Collector

2 Base

1 Emitter

General-Purpose Transistors

NPN Silicon

Electrical Characteristics (T_A = 25°C unless otherwise noted.)

Characteristic	Symbol	Min	Max	Unit
OFF Characteristics				
Collector-Emitter breakdown voltage ($I_C = 10$ mA dc)	$V_{(BR)CEO}$	40	−	V dc
Collector-Base breakdown voltage ($I_C = 10$ μA dc, $I_E = 0$)	$V_{(BR)CBO}$	60	−	V dc
Emitter-Base breakdown voltage ($I_E = 10$ μA dc, $I_C = 0$)	$V_{(BR)EBO}$	6.0	−	V dc
Collector cutoff current ($V_{CE} = 40$ V dc, $V_{OB} = 3.0$ V dc) ($V_{CE} = 40$ V dc, $V_{OB} = 3.0$ V dc, $T_A = 150°C$)	I_{CEX}	− −	0.010 15	μA dc
Base cutoff current ($V_{CE} = 40$ V dc, $V_{OB} = 3.0$ V dc)	I_{BL}	−	.025	μA dc
ON Characteristics				
DC current gain ($I_C = 0.1$ mA dc, $V_{CE} = 1.0$ V dc) 2N3946 2N3947	h_{FE}	30 60	− −	−
($I_C = 1.0$ mA dc, $V_{CE} = 1.0$ V dc) 2N3946 2N3947		45 90	− −	
($I_C = 10$ mA dc, $V_{CE} = 1.0$ V dc) 2N3946 2N3947		50 100	150 300	
($I_C = 50$ mA dc, $V_{CE} = 1.0$ V dc) 2N3946 2N3947		20 40	− −	
Collector-Emitter saturation voltage ($I_C = 10$ mA dc, $I_B = 1.0$ mA dc) ($I_C = 50$ mA dc, $I_B = 5.0$ mA dc)	$V_{CE(sat)}$	− −	0.2 0.3	V dc
Base-Emitter saturation voltage ($I_C = 10$ mA dc, $I_B = 1.0$ mA dc) ($I_C = 50$ mA dc, $I_B = 5.0$ mA dc)	$V_{BE(sat)}$	0.6 −	0.9 1.0	V dc
Small-Signal Characteristics				
Current gain — Bandwidth product ($I_C = 10$ mA dc, $V_{CE} = 20$ V dc, $f = 100$ MHz) 2N3946 2N3947	f_T	250 300	− −	MHz
Output capacitance ($V_{CB} = 10$ V dc, $I_E = 0$, $f = 100$ kHz)	C_{obo}	−	4.0	pF

▲ **FIGURE 4–41**

Partial data sheet for the 2N3947 *npn* transistor.

(a)

(b)

▲ FIGURE 4–42

Zone-monitoring circuits schematic and printed circuit board.

V≈ indicates dc/ac function

➤◄— indicates diode test

SPST switch

▲ **FIGURE 4–43**

Zone-monitoring circuits board test bench.

▲ **FIGURE 4–44**

Test results for the three defective circuit boards.

SUMMARY OF BIPOLAR JUNCTION TRANSISTORS

SYMBOLS

npn   *pnp*

CURRENTS AND VOLTAGES

$$I_E = I_C + I_B$$

AMPLIFICATION

■ DC current gain

$$I_C = \beta_{DC}I_B$$

■ AC voltage gain

$$A_v = \frac{V_c}{V_b} = \frac{R_C}{r'_e}$$

DC circuit   **Linear:** BE junction forward-biased
BC junction reverse-biased

AC equivalent circuit

CUTOFF AND SATURATION

Cutoff: BE junction reverse-biased
BC junction reverse-biased

Ideal switch
equivalent for
cutoff

Saturation: BE junction forward-biased
BC junction forward-biased

Ideal switch
equivalent for
saturation

CHAPTER SUMMARY

- The BJT (bipolar junction transistor) is constructed with three regions: base, collector, and emitter.
- The BJT has two *pn* junctions, the base-emitter junction and the base-collector junction.
- Current in a BJT consists of both free electrons and holes, thus the term *bipolar*.
- The base region is very thin and lightly doped compared to the collector and emitter regions.
- The two types of bipolar junction transistor are the *npn* and the *pnp*.
- To operate as an amplifier, the base-emitter junction must be forward-biased and the base-collector junction must be reverse-biased. This is called *forward-reverse bias*.
- The three currents in the transistor are the base current (I_B), emitter current (I_E), and collector current (I_C).
- I_B is very small compared to I_C and I_E.
- The dc current gain of a transistor is the ratio of I_C to I_B and is designated β_{DC}. Values typically range from less than 20 to several hundred.
- β_{DC} is usually referred to as h_{FE} on transistor data sheets.
- The ratio of I_C to I_E is called α_{DC}. Values typically range from 0.95 to 0.99.
- When a transistor is forward reverse biased, the voltage gain depends on the internal emitter resistance and the external collector resistance.
- A transistor can be operated as an electronic switch in cutoff and saturation.
- In cutoff, both *pn* junctions are reverse-biased and there is essentially no collector current. The transistor ideally behaves like an open switch between collector and emitter.
- In saturation, both *pn* junctions are forward-biased and the collector current is maximum. The transistor ideally behaves like a closed switch between collector and emitter.
- There is a variation in β_{DC} over temperature and also from one transistor to another of the same type.
- There are many types of transistor packages using plastic, metal, or ceramic.
- It is best to check a transistor in-circuit before removing it.
- Common faults are open junctions, low β_{DC}, excessive leakage currents, and external opens and shorts on the circuit board.

KEY TERMS

Key terms and other bold terms in the chapter are defined in the end-of-book glossary.

Amplification   The process of increasing the power, voltage, or current by electronic means.

Base   One of the semiconductor regions in a BJT. The base is very thin and lightly doped compared to the other regions.

Beta (β)   The ratio of dc collector current to dc base current in a BJT; current gain from base to collector.

Bias   The necessary application of a dc voltage to a transistor or other device to produce a desired mode of operation.

BJT (bipolar junction transistor)   A transistor constructed with three doped semiconductor regions separated by two *pn* junctions.

Collector   The largest of the three semiconductor regions of a BJT.

Cutoff   The nonconducting state of a transistor.

Emitter   The most heavily doped of the three semiconductor regions of a BJT.

Gain   The amount by which an electrical signal is increased or amplified.

Linear   Characterized by a straight-line relationship.

Saturation   The state of a BJT in which the collector current has reached a maximum and is independent of the base current.

KEY FORMULAS

4–1	$I_E = I_C + I_B$	Transistor currents
4–2	$\beta_{DC} = \dfrac{I_C}{I_B}$	DC current gain
4–3	$V_{BE} \cong 0.7\text{ V}$	Base-to-emitter voltage (silicon)
4–4	$I_B = \dfrac{V_{BB} - V_{BE}}{R_B}$	Base current
4–5	$V_{CE} = V_{CC} - I_C R_C$	Collector-to-emitter voltage (common-emitter)
4–6	$V_{CB} = V_{CE} - V_{BE}$	Collector-to-base voltage
4–7	$I_C = \dfrac{P_{D(max)}}{V_{CE}}$	Maximum I_C for given V_{CE}
4–8	$A_v \cong \dfrac{R_C}{r'_e}$	Approximate ac voltage gain
4–9	$V_{CE(cutoff)} = V_{CC}$	Cutoff condition
4–10	$I_{C(sat)} = \dfrac{V_{CC} - V_{CE(sat)}}{R_C}$	Collector saturation current
4–11	$I_{B(min)} = \dfrac{I_{C(sat)}}{\beta_{DC}}$	Minimum base current for saturation

SELF-TEST

Answers are at the end of the chapter.

1. The three terminals of a bipolar junction transistor are called
   (a) *p, n, p*   (b) *n, p, n*
   (c) input, output, ground   (d) base, emitter, collector

2. In a *pnp* transistor, the *p*-regions are
   (a) base and emitter   (b) base and collector   (c) emitter and collector

3. For operation as an amplifier, the base of an *npn* transistor must be
   (a) positive with respect to the emitter   (b) negative with respect to the emitter
   (c) positive with respect to the collector   (d) 0 V

4. The emitter current is always
   (a) greater than the base current   (b) less than the collector current
   (c) greater than the collector current   (d) answers (a) and (c)

5. The β_{DC} of a transistor is its
   (a) current gain   (b) voltage gain   (c) power gain   (d) internal resistance

6. If I_C is 50 times larger than I_B, then β_{DC} is
   (a) 0.02   (b) 100   (c) 50   (d) 500

7. The approximate voltage across the forward-biased base-emitter junction of a silicon BJT is
   (a) 0 V   (b) 0.7 V   (c) 0.3 V   (d) V_{BB}

8. The bias condition for a transistor to be used as a linear amplifier is called
   (a) forward-reverse   (b) forward-forward   (c) reverse-reverse   (d) collector bias

9. If the output of a transistor amplifier is 5 V rms and the input is 100 mV rms, the voltage gain is
   (a) 5   (b) 500   (c) 50   (d) 100

10. When operated in cutoff and saturation, the transistor acts like
    (a) a linear amplifier   (b) a switch   (c) a variable capacitor   (d) a variable resistor

11. In cutoff, V_{CE} is

    (a) 0 V   **(b)** minimum   **(c)** maximum

    (d) equal to V_{CC} **(e)** answers (a) and (b)   **(f)** answers (c) and (d)

12. In saturation, V_{CE} is

    (a) 0.7 V   **(b)** equal to V_{CC} **(c)** minimum   **(d)** maximum

13. To saturate a BJT,

    (a) $I_B = I_{C(sat)}$ **(b)** $I_B > I_{C(sat)}/\beta_{DC}$

    (c) V_{CC} must be at least 10 V   **(d)** the emitter must be grounded

14. Once in saturation, a further increase in base current will

    (a) cause the collector current to increase   **(b)** not affect the collector current

    (c) cause the collector current to decrease   **(d)** turn the transistor off

15. If the base-emitter junction is open, the collector voltage is

    (a) V_{CC} **(b)** 0 V   **(c)** floating   **(d)** 0.2 V

PROBLEMS

Answers to all odd-numbered problems are at the end of the book.

BASIC PROBLEMS

SECTION 4–1   Transistor Structure

1. What are the majority carriers in the base region of an *npn* transistor called?

2. Explain the purpose of a thin, lightly doped base region.

SECTION 4–2   Basic Transistor Operation

3. Why is the base current in a transistor so much less than the collector current?

4. In a certain transistor circuit, the base current is 2 percent of the 30 mA emitter current. Determine the collector current.

5. For normal operation of a *pnp* transistor, the base must be (+ or −) with respect to the emitter, and (+ or −) with respect to the collector.

6. What is the value of I_C for $I_E = 5.34$ mA and $I_B = 475$ μA?

SECTION 4–3   Transistor Characteristics and Parameters

7. What is the α_{DC} when $I_C = 8.23$ mA and $I_E = 8.69$ mA?

8. A certain transistor has an $I_C = 25$ mA and an $I_B = 200$ μA. Determine the β_{DC}.

9. What is the β_{DC} of a transistor if $I_C = 20.5$ mA and $I_E = 20.3$ mA?

10. What is the α_{DC} if $I_C = 5.35$ mA and $I_B = 50$ μA?

11. A certain transistor exhibits an α_{DC} of 0.96. Determine I_C when $I_E = 9.35$ mA.

12. A base current of 50 μA is applied to the transistor in Figure 4–45, and a voltage of 5 V is dropped across R_C. Determine the β_{DC} of the transistor.

13. Calculate α_{DC} for the transistor in Problem 12.

▶ **FIGURE 4–45**

14. Determine each current in Figure 4–46. What is the β_{DC}?

▶ **FIGURE 4–46**

File circuits are identified with a CD logo and are in the Problems folder on your CD-ROM. Filenames correspond to figure numbers (e.g., F04-46).

15. Find V_{CE}, V_{BE}, and V_{CB} in both circuits of Figure 4–47.

▶ **FIGURE 4–47**

(a)   (b)

16. Determine whether or not the transistors in Figure 4–47 are saturated.

17. Find I_B, I_E, and I_C in Figure 4–48. $\alpha_{DC} = 0.98$.

18. Determine the terminal voltages of each transistor with respect to ground for each circuit in Figure 4–49. Also determine V_{CE}, V_{BE}, and V_{BC}.

19. If the β_{DC} in Figure 4–49(a) changes from 100 to 150 due to a temperature increase, what is the change in collector current?

(a)   (b)

▲ **FIGURE 4–48**

▲ **FIGURE 4–49**

20. A certain transistor is to be operated at a collector current of 50 mA. How high can V_{CE} go without exceeding a $P_{D(max)}$ of 1.2 W?

21. The power dissipation derating factor for a certain transistor is 1 mW/°C. The $P_{D(max)}$ is 0.5 W at 25°C. What is $P_{D(max)}$ at 100°C?

SECTION 4–4 The Transistor as an Amplifier

22. A transistor amplifier has a voltage gain of 50. What is the output voltage when the input voltage is 100 mV?

23. To achieve an output of 10 V with an input of 300 mV, what voltage gain is required?

24. A 50 mV signal is applied to the base of a properly biased transistor with $r'_e = 10 \ \Omega$ and $R_C = 560 \ \Omega$. Determine the signal voltage at the collector.

SECTION 4–5 The Transistor as a Switch

25. Determine $I_{C(sat)}$ for the transistor in Figure 4–50. What is the value of I_B necessary to produce saturation? What minimum value of V_{IN} is necessary for saturation? Assume $V_{CE(sat)} = 0$ V.

26. The transistor in Figure 4–51 has a β_{DC} of 50. Determine the value of R_B required to ensure saturation when V_{IN} is 5 V. What must V_{IN} be to cut off the transistor? Assume $V_{CE(sat)} = 0$ V.

▲ FIGURE 4–50   ▲ FIGURE 4–51

SECTION 4–6 Transistor Packages and Terminal Identification

27. Identify the leads on the transistors in Figure 4–52. Bottom views are shown.

▶ FIGURE 4–52

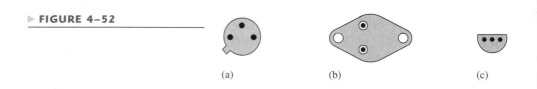

(a)   (b)   (c)

28. What is the most probable category of each transistor in Figure 4–53?

(a)   (b)   (c)   (d)   (e)

▲ FIGURE 4–53

TROUBLESHOOTING PROBLEMS

SECTION 4–7 Troubleshooting

29. In an out-of-circuit test of a good *npn* transistor, what should an analog ohmmeter indicate when its positive probe is touching the emitter and the negative probe is touching the base? When its positive probe is touching the base and the negative probe is touching the collector?

30. What is the most likely problem, if any, in each circuit of Figure 4–54? Assume a β_{DC} of 75.

31. What is the value of the β_{DC} of each transistor in Figure 4–55?

▲ FIGURE 4–54

▲ FIGURE 4–55

SYSTEM APPLICATION PROBLEMS

32. This problem relates to the circuit board and schematic in Figure 4–42. A remote switch loop is connected between pins 2 and 3. When the remote switches are closed, the relay (RY1) contacts between pin 11 and pin 12 are normally open. When a remote switch is opened, the relay contacts do not close. Determine the possible causes of this malfunction.

33. This problem relates to Figure 4–42. The relay contact remains closed between pins 11 and 12 of the circuit board no matter what any of the inputs are. This means that the relay is energized continuously. What are the possible faults?

34. This problem relates to Figure 4–42. Pin 7 stays at approximately 0.1 V, regardless of the input at pin 8. What do you think is wrong? What would you check first?

DATA SHEET PROBLEMS

35. Refer to the partial transistor data sheet in Figure 4–19.

  (a) What is the maximum collector-to-emitter voltage for a 2N3903?

  (b) How much dc collector current can the 2N3904 handle?

  (c) How much power can a 2N3903 dissipate if the surrounding air is at a temperature of 25°C?

  (d) How much power can a 2N3904 dissipate if the case is at a temperature of 25°C?

  (e) What is the minimum h_{FE} of a 2N3903 if the collector current is 1 mA?

36. Refer to the transistor data sheet in Figure 4–19. A 2N3904 is operating in an environment where the ambient temperature is 65°C. What is the most power that it can dissipate?

37. Refer to the transistor data sheet in Figure 4–19. A 2N3903 is operating with a case temperature of 45°C. What is the most power that it can dissipate?

38. Refer to the transistor data sheet in Figure 4–19. Determine if any rating is exceeded in each circuit of Figure 4–56 based on minimum specified values.

(a) $T_A = 50°C$ (b) $T_A = 25°C$

▲ **FIGURE 4–56**

39. Refer to the transistor data sheet in Figure 4–19. Determine whether or not the transistor is saturated in each circuit of Figure 4–57 based on the maximum specified value of h_{FE}.

40. Refer to the partial transistor data sheet in Figure 4–41. Determine the minimum and maximum base currents required to produce a collector current of 10 mA in a 2N3946. Assume that the transistor is not in saturation and $V_{CE} = 1$ V.

41. For each of the circuits in Figure 4–58, determine if there is a problem based on the data sheet information in Figure 4–41. Use the maximum specified h_{FE}.

▲ FIGURE 4–57

▲ FIGURE 4–58

ADVANCED PROBLEMS

42. Derive a formula for α_{DC} in terms of β_{DC}.

43. A certain 2N3904 dc bias circuit with the following values is in saturation. $I_B = 500\ \mu A$, $V_{CC} = 10$ V, and $R_C = 180\ \Omega$, $h_{FE} = 150$. If you increase V_{CC} to 15 V, does the transistor come out of saturation? If so, what is the collector-to-emitter voltage and the collector current?

44. Design a dc bias circuit for a 2N3904 operating from a collector supply voltage of 9 V and a base-bias voltage of 3 V that will supply 150 mA to a resistive load that acts as the collector resistor. The circuit must not be in saturation. Assume the minimum specified β_{DC} from the data sheet.

45. Modify the design in Problem 44 to use a single 9 V dc source rather than two different sources. Other requirements remain the same.

46. Design a dc bias circuit for an amplifier in which the voltage gain is to be a minimum of 50 and the output signal voltage is to be "riding" on a dc level of 5 V. The maximum input signal voltage at the base is 10 mV rms, $V_{CC} = 12$ V, and $V_{BB} = 4$ V. Assume $r'_e = 8\ \Omega$.

EWB/MULTISIM TROUBLESHOOTING PROBLEMS

These file circuits are in the Troubleshooting Problems folder on your CD-ROM.

47. Open file TSP04-47 and determine the fault.

48. Open file TSP04-48 and determine the fault.

49. Open file TSP04-49 and determine the fault.

50. Open file TSP04-50 and determine the fault.

51. Open file TSP04-51 and determine the fault.

52. Open file TSP04-52 and determine the fault.

53. Open file TSP04-53 and determine the fault.

54. Open file TSP04-54 and determine the fault.

ANSWERS

SECTION REVIEWS

SECTION 4–1   Transistor Structure

1. The two types of BJTs are *npn* and *pnp*.

2. The terminals of a BJT are base, collector, and emitter.

3. The three regions of a BJT are separated by two *pn* junctions.

SECTION 4–2   Basic Transistor Operation

1. To operate as an amplifier, the base-emitter is forward-biased and the base-collector is reverse-biased.

2. The emitter current is the largest.

3. The base current is much smaller than the emitter current.

4. The base region is very narrow compared to the other two regions.

5. $I_E = 1$ mA $+ 10 \mu A = 1.01$ mA

SECTION 4–3   Transistor Characteristics and Parameters

1. $\beta_{DC} = I_C/I_B$; $\alpha_{DC} = I_C/I_E$; h_{FE} is β_{DC}.

2. $\beta_{DC} = 100$; $\alpha_{DC} = 100/(100 + 1) = 0.99$

3. I_C is plotted versus V_{CE}.

4. Forward-reverse bias is required for amplifier operation.

5. β_{DC} increases with temperature.

6. β_{DC} generally varies some from one device to the next for a given type.

SECTION 4–4   The Transistor as an Amplifier

1. Amplification is the process where a smaller signal is used to produce a larger identical signal.

2. Voltage gain is the ratio of output voltage to input voltage.

3. R_C and r'_e determine the voltage gain.

4. $A_v = 5$ V/250 mV $= 20$

5. $A_v = 1200 \, \Omega/20 \, \Omega = 60$

SECTION 4–5   The Transistor as a Switch

1. A transistor switch operates in cutoff and saturation.

2. The collector current is maximum in saturation.

3. The collector current is approximately zero in cutoff.

4. $V_{CE} = V_{CC}$ in cutoff.

5. V_{CE} is minimum in saturation.

SECTION 4–6   Transistor Packages and Terminal Identification

1. Three categories of BJTs are small signal/general purpose, power, and RF.

2. Going clockwise from tab: emitter, base, and collector (bottom view).

3. The metal mounting tab or case in power transistors is the collector.

SECTION 4-7 Troubleshooting

1. First, test it in-circuit.
2. If R_B opens, the transistor is in cutoff.
3. The base voltage is +3 V and the collector voltage is +9 V.

RELATED PROBLEMS FOR EXAMPLES

4-1 10 mA

4-2 $I_B = 241\ \mu A$; $I_C = 21.7$ mA; $I_E = 21.94$ mA; $V_{CE} = 4.23$ V; $V_{CB} = 3.53$ V

4-3 Along the horizontal axis

4-4 Not saturated

4-5 10 V

4-6 $V_{CC(max)} = 44.5$ V; $V_{CE(max)}$ is exceeded first.

4-7 4.55 W

4-8 2.5 kΩ

4-9 78.4 μA

4-10 Reduce R_C to 160 Ω and R_B to 2.2 kΩ.

4-11 R_B open

SELF-TEST

1. (d) 2. (c) 3. (a) 4. (d) 5. (a)

6. (c) 7. (b) 8. (a) 9. (c) 10. (b)

11. (f) 12. (c) 13. (b) 14. (b) 15. (a)

5

TRANSISTOR BIAS CIRCUITS

INTRODUCTION

As you learned in the last chapter, a transistor must be properly biased in order to operate as an amplifier. DC biasing is used to establish a steady level of transistor current and voltage called the *dc operating point* or *quiescent point* (*Q-point*). In this chapter, several types of bias circuits are studied. This material lays the groundwork for the study of amplifiers, oscillators, and other circuits that cannot operate without proper biasing.

CHAPTER OBJECTIVES

■ Discuss the concept of dc bias in a linear amplifier

■ Analyze a voltage-divider bias circuit

■ Analyze a base bias circuit, an emitter bias circuit, and a collector-feedback bias circuit

■ Troubleshoot various faults in transistor bias circuits

KEY TERMS

■ Q-point

■ DC load line

■ Linear region

■ Feedback

■ ■ ■ SYSTEM APPLICATION PREVIEW

The system application in this chapter focuses on a system for controlling temperature in an industrial chemical process. You will be dealing with a circuit that converts a temperature measurement to a proportional voltage that is used to adjust the temperature of a liquid chemical. You will also be working on the power supply for the system. The first step in your assignment is to learn all you can about transistor operation. You will then apply your knowledge to the system application at the end of the chapter.

WWW. **VISIT THE COMPANION WEBSITE**

Study aids for this chapter are available at
http://www.prenhall.com/floyd

5–1 THE DC OPERATING POINT

A transistor must be properly biased with a dc voltage in order to operate as an amplifier. A dc operating point must be set so that signal variations at the input terminal are amplified and accurately reproduced at the output terminal. As you learned in Chapter 4, when you bias a transistor, you establish a certain current and voltage condition. This means, for example, that at the dc operating point, I_C and V_{CE} have specified values. The dc operating point is often referred to as the Q-point (quiescent point).

After completing this section, you should be able to

- ■ **Discuss the concept of dc bias in a linear amplifier**
  - ■ Describe how to generate collector characteristic curves for a biased transistor
  - ■ Draw a dc load line for a given biased transistor circuit
  - ■ Explain Q-point
  - ■ Explain the conditions for linear operation
  - ■ Explain the conditions for saturation and cutoff
  - ■ Discuss the reasons for output waveform distortion

DC Bias

Bias establishes the dc operating point for proper linear operation of an amplifier. If an amplifier is not biased with correct dc voltages on the input and output, it can go into saturation or cutoff when an input signal is applied. Figure 5–1 shows the effects of proper and improper dc biasing of an inverting amplifier. In part (a), the output signal is an amplified replica of the input signal except that it is inverted, which means that it is 180° out of phase with the input. The output signal swings equally above and below the dc bias level of the output, $V_{DC(out)}$. Improper biasing can cause distortion in the output signal, as illustrated in parts (b) and (c). Part (b) illustrates limiting of the positive portion of the output voltage as a result of a dc operating point (**Q-point**) being too close to cutoff. Part (c) shows limiting of the negative portion of the output voltage as a result of a dc operating point being too close to saturation.

(a) Linear operation: larger output has same shape as input except that it is inverted

(b) Nonlinear operation: output voltage limited (clipped) by cutoff

(c) Nonlinear operation: output voltage limited (clipped) by saturation

▲ **FIGURE 5–1**

Examples of linear and nonlinear operation of an inverting amplifier (the triangle symbol).

Graphical Analysis The transistor in Figure 5–2(a) is biased with variable voltages V_{CC} and V_{BB} to obtain certain values of I_B, I_C, I_E, and V_{CE}. The collector characteristic curves for this particular transistor are shown in Figure 5–2(b); we will use these curves to graphically illustrate the effects of dc bias.

(a) DC biased circuit   (b) Collector characteristic curves

▲ **FIGURE 5–2**

A dc–biased transistor circuit with variable bias voltages (V_{BB} and V_{CC}) for generating the collector characteristic curves shown in part (b).

In Figure 5–3, we assign three values to I_B and observe what happens to I_C and V_{CE}. First, V_{BB} is adjusted to produce an I_B of 200 μA, as shown in Figure 5–3(a). Since $I_C = \beta_{DC}I_B$, the collector current is 20 mA, as indicated, and

$$V_{CE} = V_{CC} - I_C R_C = 10\,\text{V} - (20\,\text{mA})(220\,\Omega) = 10\,\text{V} - 4.4\,\text{V} = 5.6\,\text{V}$$

This Q-point is shown on the graph of Figure 5–3(a) as Q_1.

Next, as shown in Figure 5–3(b), V_{BB} is increased to produce an I_B of 300 μA and an I_C of 30 mA.

$$V_{CE} = 10\,\text{V} - (30\,\text{mA})(220\,\Omega) = 10\,\text{V} - 6.6\,\text{V} = 3.4\,\text{V}$$

The Q-point for this condition is indicated by Q_2 on the graph.

Finally, as in Figure 5–3(c), V_{BB} is increased to give an I_B of 400 μA and an I_C of 40 mA.

$$V_{CE} = 10\,\text{V} - (40\,\text{mA})(220\,\Omega) = 10\,\text{V} - 8.8\,\text{V} = 1.2\,\text{V}$$

Q_3 is the corresponding Q-point on the graph.

DC Load Line Notice that when I_B increases, I_C increases and V_{CE} decreases. When I_B decreases, I_C decreases and V_{CE} increases. So, as V_{BB} is adjusted up or down, the dc operating point of the transistor moves along a sloping straight line, called the **dc load line**, connecting each separate Q-point. At any point along the line, values of I_B, I_C, and V_{CE} can be picked off the graph, as shown in Figure 5–4.

The dc load line intersects the V_{CE} axis at 10 V, the point where $V_{CE} = V_{CC}$. This is the transistor cutoff point because I_B and I_C are zero (ideally). Actually, there is a small leakage current, I_{CBO}, at cutoff as indicated, and therefore V_{CE} is slightly less than 10 V but normally this can be neglected.

The dc load line intersects the I_C axis at 45.5 mA ideally. This is the transistor saturation point because I_C is maximum at the point where $V_{CE} = 0$ V and $I_C = V_{CC}/R_C$.

(a)

(b)

(c)

▲ FIGURE 5–3

Illustration of Q-point adjustments.

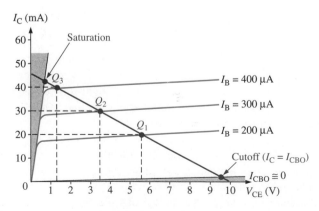

▲ FIGURE 5–4

The dc load line.

Actually, there is a small voltage ($V_{CE(sat)}$) across the transistor, and $I_{C(sat)}$ is slightly less than 45.5 mA, as indicated in Figure 5–4. Note that Kirchhoff's voltage law applied around the collector loop gives

$$V_{CC} - I_C R_C - V_{CE} = 0$$

This results in a straight line equation for the load line of the form $y = mx + b$ as follows:

$$I_C = -\left(\frac{1}{R_C}\right)V_{CE} + \frac{V_{CC}}{R_C}$$

where $-1/R_C$ is the slope and V_{CC}/R_C is the y-axis intercept point.

Linear Operation   The region along the load line including all points between saturation and cutoff is generally known as the **linear region** of the transistor's operation. As long as the transistor is operated in this region, the output voltage is ideally a linear reproduction of the input.

Figure 5–5 shows an example of the linear operation of a transistor. AC quantities are indicated by lowercase italic subscripts. Assume a sinusoidal voltage, V_{in}, is superimposed on V_{BB}, causing the base current to vary sinusoidally 100 μA above and below its Q-point value of 300 μA. This, in turn, causes the collector current to vary 10 mA above and below its Q-point value of 30 mA. As a result of the variation in collector current, the collector-to-emitter voltage varies 2.2 V above and below its Q-point value of 3.4 V. Point A on the load line in Figure 5–5 corresponds to the positive peak of the sinusoidal input voltage. Point B corresponds to the negative peak, and point Q corresponds to the zero value of the sine wave, as indicated. V_{CEQ}, I_{CQ}, and I_{BQ} are dc Q-point values with no input sinusoidal voltage applied.

▲ **FIGURE 5–5**

Variations in collector current and collector-to-emitter voltage as a result of a variation in base current. Notice that ac quantities are indicated by lowercase italic subscripts.

Waveform Distortion As previously mentioned, under certain input signal conditions the location of the Q-point on the load line can cause one peak of the V_{ce} waveform to be limited or clipped, as shown in parts (a) and (b) of Figure 5–6. In each case the input signal is too large for the Q-point location and is driving the transistor into cutoff or saturation during a portion of the input cycle. When both peaks are limited as in Figure 5–6(c), the transistor is being driven into both saturation and cutoff by an excessively large input signal. When only the positive peak is limited, the transistor is being driven into cutoff but not saturation. When only the negative peak is limited, the transistor is being driven into saturation but not cutoff.

(a) Transistor is driven into saturation because the Q-point is too close to saturation for the given input signal.

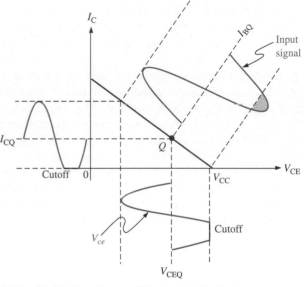

(b) Transistor is driven into cutoff because the Q-point is too close to cutoff for the given input signal.

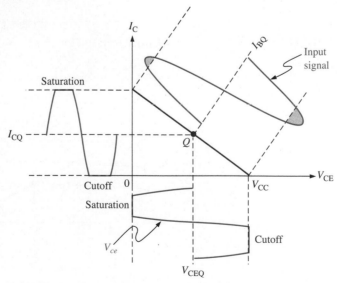

(c) Transistor is driven into both saturation and cutoff because the input signal is too large.

▲ **FIGURE 5–6**

Graphical load line illustration of a transistor being driven into saturation and/or cutoff.

EXAMPLE 5–1

Determine the Q-point in Figure 5–7, and find the maximum peak value of base current for linear operation. Assume $\beta_{DC} = 200$.

◆ **FIGURE 5–7**

Solution

The Q-point is defined by the values of I_C and V_{CE}. Find these values by using formulas you learned in Chapter 4.

$$I_B = \frac{V_{BB} - V_{BE}}{R_B} = \frac{10 \text{ V} - 0.7 \text{ V}}{47 \text{ k}\Omega} = 198 \text{ } \mu A$$

$$I_C = \beta_{DC} I_B = (200)(198 \text{ } \mu A) = 39.6 \text{ mA}$$

$$V_{CE} = V_{CC} - I_C R_C = 20 \text{ V} - 13.07 \text{ V} = 6.93 \text{ V}$$

The Q-point is at $I_C = 39.6$ mA and at $V_{CE} = 6.93$ V. Since $I_{C(cutoff)} = 0$, you need to know $I_{C(sat)}$ to determine how much variation in collector current can occur and still maintain linear operation of the transistor.

$$I_{C(sat)} = \frac{V_{CC}}{R_C} = \frac{20 \text{ V}}{330 \text{ }\Omega} = 60.6 \text{ mA}$$

The dc load line is graphically illustrated in Figure 5–8, showing that before saturation is reached, I_C can increase an amount ideally equal to

$$I_{C(sat)} - I_{CQ} = 60.6 \text{ mA} - 39.6 \text{ mA} = 21 \text{ mA}$$

However, I_C can decrease by 39.6 mA before cutoff ($I_C = 0$ V) is reached. Therefore, the limiting excursion is 21 mA because the *Q-point is closer to saturation than to cutoff*. The 21 mA is the maximum peak variation of the collector current. Actually, it would be slightly less in practice because $V_{CE(sat)}$ is not quite zero. The maximum peak variation of the base current is determined as follows:

$$I_{b(peak)} = \frac{I_{c(peak)}}{\beta_{DC}} = \frac{21 \text{ mA}}{200} = \textbf{105 } \boldsymbol{\mu} \textbf{A}$$

◆ **FIGURE 5–8**

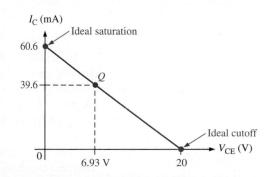

Related Problem*   Find the Q-point in Figure 5–7, and determine the maximum peak value of base current for linear operation for the following circuit values: $\beta_{DC} = 100$, $R_C = 1.0\ k\Omega$, and $V_{CC} = 24$ V.

*Answers are at the end of the chapter.

> Open file E05-01 in the Examples folder on your CD-ROM. Measure I_C and V_{CE} and compare with the calculated values.

SECTION 5–1 REVIEW

Answers are at the end of the chapter.

1. What are the upper and lower limits on a dc load line in terms of V_{CE} and I_C?
2. Define *Q-point*.
3. At what point on the load line does saturation begin? At what point does cutoff occur?
4. For maximum V_{ce}, where should the Q-point be placed?

5–2   VOLTAGE-DIVIDER BIAS

You will now study a method of biasing a transistor for linear operation using a single-source resistive voltage-divider. This is the most widely used biasing method. Three other methods are covered in the next section.

After completing this section, you should be able to

- **Analyze a voltage-divider bias circuit**

- Discuss the effect of the input resistance on the bias circuit

- Discuss the stability of voltage-divider bias

- Explain how to minimize or essentially eliminate the effects of β_{DC} and V_{BE} on the stability of the Q-point

- Discuss voltage-divider bias for a *pnp* transistor

Up to this point a separate dc source, V_{BB}, was used to bias the base-emitter junction strictly for convenience in illustrating transistor operation since it could be varied independently of V_{CC}. A more practical method is to use V_{CC} as the single bias source, as shown in Figure 5–9. To simplify the schematics, the battery symbol can be omitted and replaced by a line termination circle with a voltage indicator (V_{CC}), as shown in Figure 5–9.

A dc bias voltage at the base of the transistor can be developed by a resistive voltage-divider consisting of R_1 and R_2 as shown in Figure 5–9. V_{CC} is the dc collector supply voltage. There are two current paths between point A and ground: one through R_2 and the other through the base-emitter junction of the transistor and R_E.

If the base current is much smaller than the current through R_2, the bias circuit can be viewed as a voltage-divider consisting of R_1 and R_2, as indicated in Figure 5–10(a). If I_B is *not* small enough to neglect compared to I_2, then the dc input resistance, $R_{IN(base)}$, that appears from the base of the transistor to ground must be considered. $R_{IN(base)}$ is in parallel with R_2, as shown in Figure 5–10(b).

▶ FIGURE 5–10

Simplified voltage-divider.

| (a) Unloaded | (b) Loaded |

Input Resistance at the Transistor Base

To develop a formula for the dc input resistance at the base of a transistor, we will use the diagram in Figure 5–11. V_{IN} is applied between base and ground, and I_{IN} is the current into the base as shown.

By Ohm's law,

$$R_{IN(base)} = \frac{V_{IN}}{I_{IN}}$$

Kirchhoff's voltage law applied around the base-emitter circuit yields

$$V_{IN} = V_{BE} + I_E R_E$$

With the assumption that $V_{BE} << I_E R_E$, the equation reduces to

$$V_{IN} \cong I_E R_E$$

Now, since $I_E \cong I_C = \beta_{DC} I_B$,

$$V_{IN} \cong \beta_{DC} I_B R_E$$

► **FIGURE 5–11**

DC input resistance is V_{IN}/I_{IN}.

The input current is the base current:

$$I_{IN} = I_B$$

By substitution,

$$R_{IN(base)} = \frac{V_{IN}}{I_{IN}} \cong \frac{\beta_{DC}I_B R_E}{I_B}$$

Cancelling the I_B terms gives

Equation 5–1

$$R_{IN(base)} \cong \beta_{DC}R_E$$

EXAMPLE 5–2

Determine the dc input resistance looking in at the base of the transistor in Figure 5–12. $\beta_{DC} = 125$.

► **FIGURE 5–12**

Solution $R_{IN(base)} \cong \beta_{DC}R_E = (125)(1.0 \text{ k}\Omega) = \textbf{125 k}\boldsymbol{\Omega}$

Related Problem   What is $R_{IN(base)}$ if $\beta_{DC} = 60$ and $R_E = 910 \ \Omega$ in Figure 5–12?

Analysis of a Voltage-Divider Bias Circuit

A voltage-divider biased *npn* transistor is shown in Figure 5–13(a). Let's begin the analysis by determining the voltage at the base using the voltage-divider formula, which is developed as follows:

$$R_{\text{IN(base)}} \cong \beta_{\text{DC}} R_{\text{E}}$$

The total resistance from base to ground is

$$R_2 \parallel R_{\text{IN(base)}}$$

Substituting $R_{\text{IN(base)}} \cong \beta_{\text{DC}} R_{\text{E}}$,

$$R_2 \parallel \beta_{\text{DC}} R_{\text{E}}$$

A voltage-divider is formed by R_1 and the resistance from base to ground ($\beta_{\text{DC}} R_{\text{E}}$) in parallel with R_2 as shown in Figure 5–13(b). Applying the voltage-divider formula yields

$$V_{\text{B}} = \left(\frac{R_2 \parallel \beta_{\text{DC}} R_{\text{E}}}{R_1 + (R_2 \parallel \beta_{\text{DC}} R_{\text{E}})} \right) V_{\text{CC}}$$

If $\beta_{\text{DC}} R_{\text{E}} \gg R_2$, (at least ten times greater) then the formula simplifies to

$$V_{\text{B}} \cong \left(\frac{R_2}{R_1 + R_2} \right) V_{\text{CC}}$$

Equation 5–2

▶ **FIGURE 5–13**

An *npn* transistor with voltage-divider bias.

(a) (b)

Once you know the base voltage, you can determine the emitter voltage, which equals V_{B} less the value of the base-emitter drop (V_{BE}).

$$V_{\text{E}} = V_{\text{B}} - V_{\text{BE}}$$

Equation 5–3

You can find the emitter current by using Ohm's law.

$$I_{\text{E}} = \frac{V_{\text{E}}}{R_{\text{E}}}$$

Equation 5–4

Once you know I_{E}, you can find all the other circuit values.

$$I_{\text{C}} \cong I_{\text{E}}$$

Equation 5–5

$$V_{\text{C}} = V_{\text{CC}} - I_{\text{C}} R_{\text{C}}$$

Equation 5–6

Once you know V_{C} and V_{E}, you can determine V_{CE}.

$$V_{\text{CE}} = V_{\text{C}} - V_{\text{E}}$$

Also, you can express V_{CE} in terms of I_C by using Kirchhoff's voltage law as follows:

$$V_{CC} - I_C R_C - I_E R_E - V_{CE} = 0$$

Since $I_C \cong I_E$,

$$V_{CE} \cong V_{CC} - I_C R_C - I_C R_E$$

Equation 5–7

$$V_{CE} \cong V_{CC} - I_C(R_C + R_E)$$

EXAMPLE 5–3

Determine V_{CE} and I_C in the voltage-divider biased transistor circuit of Figure 5–14 if $\beta_{DC} = 100$.

▶ **FIGURE 5–14**

Solution First, determine the dc input resistance at the base to see if it can be neglected.

$$R_{IN(base)} \cong \beta_{DC} R_E = (100)(560\ \Omega) = 56\ k\Omega$$

A common rule-of-thumb is that if two resistors are in parallel and one is at least ten times the other, the total resistance is approximately equal to the smaller value. However, in some cases, this may result in unacceptable inaccuracy.

In this case, $R_{IN(base)} = 10R_2$, so neglect $R_{IN(base)}$. In the related exercise, you will rework this example taking $R_{IN(base)}$ into account and compare the difference. Proceed with the analysis by determining the base voltage.

$$V_B \cong \left(\frac{R_2}{R_1 + R_2}\right)V_{CC} = \left(\frac{5.6\ k\Omega}{15.6\ k\Omega}\right)10\ V = 3.59\ V$$

So,

$$V_E = V_B - V_{BE} = 3.59\ V - 0.7\ V = 2.89\ V$$

and

$$I_E = \frac{V_E}{R_E} = \frac{2.89\ V}{560\ \Omega} = 5.16\ mA$$

Therefore,

$$I_C \cong \mathbf{5.16\ mA}$$

and

$$V_{CE} \cong V_{CC} - I_C(R_C + R_E) = 10\ V - 5.16\ mA(1.56\ k\Omega) = \mathbf{1.95\ V}$$

Since $V_{CE} > 0$ V (or greater than a few tenths of a volt), you know that the transistor is *not* in saturation.

Related Problem Rework this example taking into account $R_{IN(base)}$ and compare the results.

Open file E05-03 in the Examples folder on your CD-ROM. Measure I_C and V_{CE}. Your results should agree more closely with those in the Related Problem than with those calculated in the example. Can you explain this?

Stability of Voltage-Divider Bias

Another way to analyze a voltage-divider biased transistor circuit is to apply Thevenin's theorem. We will use this method to evaluate the stability of the circuit. First, let's get an equivalent base-emitter circuit for Figure 5–13 using Thevenin's theorem. Looking out from the base terminal, the bias circuit can be redrawn as shown in Figure 5–15(a). Apply Thevenin's theorem to the circuit left of point A, with V_{CC} replaced by a short to ground and the transistor disconnected from the circuit. The voltage at point A with respect to ground is

$$V_{TH} = \left(\frac{R_2}{R_1 + R_2}\right)V_{CC}$$

and the resistance is

$$R_{TH} = \frac{R_1 R_2}{R_1 + R_2}$$

The Thevenin equivalent of the bias circuit, connected to the transistor base, is shown in the beige box in Figure 5–15(b). Applying Kirchhoff's voltage law around the equivalent base-emitter loop gives

$$V_{TH} - V_{R_{TH}} - V_{BE} - V_{R_E} = 0$$

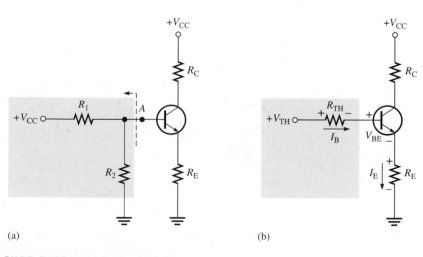

(a)   (b)

▲ **FIGURE 5–15**

Thevenizing the bias circuit.

Substituting, using Ohm's law, and solving for V_{TH},

$$V_{TH} = I_B R_{TH} + V_{BE} + I_E R_E$$

Substituting I_E/β_{DC} for I_B,

$$V_{TH} = I_E(R_E + R_{TH}/\beta_{DC}) + V_{BE}$$

or, solving for I_E,

$$I_E = \frac{V_{TH} - V_{BE}}{R_E + R_{TH}/\beta_{DC}}$$

If $R_E >> R_{TH}/\beta_{DC}$, then

$$I_E \cong \frac{V_{TH} - V_{BE}}{R_E}$$

This last equation shows that I_E, and therefore I_C, is essentially independent of β_{DC} (notice that β_{DC} does not appear in the equation) for the stated condition. This can be achieved in practice by selecting a value for R_E that is at least ten times the resistance of the parallel combination of the voltage-divider resistors (R_{TH}) divided by the minimum β_{DC}. The voltage-divider bias is widely used because *reasonably good stability is achieved with a single supply voltage.*

Voltage-Divider Biased *PNP* Transistor

As you know, a *pnp* transistor requires bias polarities opposite to the *npn*. This can be accomplished with a negative collector supply voltage, as in Figure 5–16(a), or with a positive emitter supply voltage, as in Figure 5–16(b). In a schematic, the *pnp* is often drawn upside down so that the supply voltage line can be drawn across the top of the schematic and ground at the bottom, as in Figure 5–17. The analysis procedure is basically the same as for an *npn* circuit, as demonstrated in the following steps with reference to Figure 5–17. The base voltage is determined by using the voltage-divider formula.

$$V_B = \left(\frac{R_1}{R_1 + R_2 \| \beta_{DC} R_E}\right) V_{EE}$$

and

$$V_E = V_B + V_{BE}$$

(a) Negative collector
supply voltage, V_{CC}

(b) Positive collector
supply voltage, V_{EE}

▲ **FIGURE 5–16**

Voltage-divider biased *pnp* transistor.

▲ **FIGURE 5–17**

By Ohm's law,

$$I_E = \frac{V_{EE} - V_E}{R_E}$$

and

$$V_C = I_C R_C$$
$$V_{EC} = V_E - V_C$$

EXAMPLE 5–4

Find I_C and V_{EC} in Figure 5–18.

▶ **FIGURE 5–18**

Solution   First, check to see if $R_{IN(base)}$ can be neglected.

$$R_{IN(base)} = \beta_{DC} R_E = (150)(1.0\ k\Omega) = 150\ k\Omega$$

Since 150 kΩ is more than ten times R_2, the condition $\beta_{DC} R_E >> R_2$ is met and $R_{IN(base)}$ can be neglected. Now, calculate V_B.

$$V_B \cong \left(\frac{R_1}{R_1 + R_2}\right) V_{EE} = \left(\frac{22\ k\Omega}{32\ k\Omega}\right) 10\ V = 6.88\ V$$

Then

$$V_E = V_B + V_{BE} = 6.88\ V + 0.7\ V = 7.58\ V$$

and

$$I_E = \frac{V_{EE} - V_E}{R_E} = \frac{10\ V - 7.58\ V}{1.0\ k\Omega} = 2.42\ mA$$

From I_E, you can determine I_C and V_{CE} as follows:

$$I_C \cong I_E = \mathbf{2.42\ mA}$$

and

$$V_C = I_C R_C = (2.42\ mA)(2.2\ k\Omega) = 5.32\ V$$

Therefore,

$$V_{EC} = V_E - V_C = 7.58\ V - 5.32\ V = \mathbf{2.26\ V}$$

Related Problem Determine I_C and V_{EC} in Figure 5–18 with $R_{IN(base)}$ taken into account.

> Open file E05-04 in the Examples folder on your CD-ROM. Measure I_C and V_{EC}. Your results should agree more closely with those in the Related Exercise than with those calculated in the example. Can you explain?

EXAMPLE 5–5

Find I_C and V_{CE} in Figure 5–16(a) for $R_1 = 68$ kΩ, $R_2 = 47$ kΩ, $R_C = 1.8$ kΩ, $R_E = 2.2$ kΩ, $V_{CC} = -6$ V, and $\beta_{DC} = 75$.

Solution $R_{IN(base)} = \beta_{DC}R_E = 75(2.2 \text{ kΩ}) = 165 \text{ kΩ}$

Since $R_{IN(base)}$ is not ten times greater than R_2, it must be taken into account. Determine the base voltage as follows:

$$V_B = \left(\frac{R_2 \| R_{IN(base)}}{R_1 + R_2 \| R_{IN(base)}}\right)(V_{CC}) = \left(\frac{47 \text{ kΩ} \| 165 \text{ kΩ}}{68 \text{ kΩ} + 47 \text{ kΩ} \| 165 \text{ kΩ}}\right)(-6 \text{ V})$$

$$= \left(\frac{36.6 \text{ kΩ}}{68 \text{ kΩ} + 36.6 \text{ kΩ}}\right)(-6 \text{ V}) = -2.1 \text{ V}$$

Next, calculate the emitter voltage and current.

$$V_E = V_B + V_{BE} = -2.1 \text{ V} + 0.7 \text{ V} = -1.4 \text{ V}$$

$$I_E = \frac{V_E}{R_E} = \frac{-1.4 \text{ V}}{2.2 \text{ kΩ}} = -636 \text{ μA}$$

From I_E, you can determine I_C and V_{CE} as follows:

$$I_C \cong I_E = \mathbf{-636 \text{ μA}}$$
$$V_C = V_{CC} - I_C R_C = -6 \text{ V} - (-636 \text{ μA})(1.8 \text{ kΩ}) = -4.86 \text{ V}$$
$$V_{CE} = V_C - V_E = -4.86 \text{ V} - (-1.4 \text{ V}) = \mathbf{-3.46 \text{ V}}$$

Related Problem What value of β_{DC} is required in this example in order to neglect $R_{IN(base)}$ in keeping with the basic ten-times rule?

SECTION 5–2 REVIEW

1. If the voltage at the base of a transistor is 5 V and the base current is 5 μA, what is the dc input resistance at the base?

2. If a transistor has a dc beta of 190 and its emitter resistor is 1.0 kΩ, what is the dc input resistance at the base?

3. What bias voltage is developed at the base of a transistor if both resistors in the voltage divider are equal and $V_{CC} = +10$ V? Assume the input resistance at the base is large enough to neglect.

4. What are two advantages of voltage-divider bias?

5-3 OTHER BIAS METHODS

In this section, three additional methods for dc biasing a transistor circuit are discussed. Although these methods are not as common as voltage-divider bias, you should be able to recognize them when you see them.

After completing this section, you should be able to

- **Analyze three additional types of bias circuits**
- Recognize base bias
- Recognize emitter bias
- Recognize collector-feedback bias
- Discuss the stability of each bias circuit and compare with the voltage-dividier bias

Base Bias

Figure 5–19 shows a base-biased transistor. The analysis of this circuit for the linear region is as follows. Starting with Kirchhoff's voltage law around the base circuit,

$$V_{CC} - V_{R_B} - V_{BE} = 0$$

Substituting $I_B R_B$ for V_{R_B}, you get

$$V_{CC} - I_B R_B - V_{BE} = 0$$

Then solving for I_B,

$$I_B = \frac{V_{CC} - V_{BE}}{R_B}$$

▲ **FIGURE 5–19**

Base bias.

Kirchhoff's voltage law applied around the collector circuit in Figure 5–19 gives the following equation:

$$V_{CC} - I_C R_C - V_{CE} = 0$$

Solving for V_{CE},

$$V_{CE} = V_{CC} - I_C R_C$$

Equation 5–8

Substituting this expression for I_B into the formula $I_C = \beta_{DC} I_B$ yields

$$I_C = \beta_{DC} \left(\frac{V_{CC} - V_{BE}}{R_B} \right)$$

Equation 5–9

Q-Point Stability of Base Bias   Notice that Equation 5–9 shows that I_C is dependent on β_{DC}. The disadvantage of this is that a variation in β_{DC} causes I_C and, as a result, V_{CE} to change, thus changing the Q-point of the transistor. This makes the base bias circuit extremely beta-dependent and very unstable.

Recall that β_{DC} varies with temperature and collector current. In addition, there is a large spread of β_{DC} values from one transistor to another of the same type due to manufacturing variations.

EXAMPLE 5–6

Determine how much the Q-point (I_C, V_{CE}) in Figure 5–20 will change over a temperature range where β_{DC} increases from 85 to 100 and V_{BE} decreases from 0.7 V to 0.6 V.

▶ **FIGURE 5–20**

Solution For $\beta_{DC} = 85$ and $V_{BE} = 0.7$ V,

$$I_{C(1)} = \beta_{DC}\left(\frac{V_{CC} - V_{BE}}{R_B}\right) = 85\left(\frac{12\ \text{V} - 0.7\ \text{V}}{100\ \text{k}\Omega}\right) = 9.61\ \text{mA}$$

$$V_{CE(1)} = V_{CC} - I_C R_C = 12\ \text{V} - (9.61\ \text{mA})(560\ \Omega) = 6.62\ \text{V}$$

For $\beta_{DC} = 100$ and $V_{BE} = 0.6$ V,

$$I_{C(2)} = \beta_{DC}\left(\frac{V_{CC} - V_{BE}}{R_B}\right) = 100\left(\frac{12\ \text{V} - 0.6\ \text{V}}{100\ \text{k}\Omega}\right) = 11.4\ \text{mA}$$

$$V_{CE(2)} = V_{CC} = I_C R_C = 12\ \text{V} - (11.4\ \text{mA})(560\ \Omega) = 5.62\ \text{V}$$

The percent change in I_C as β_{DC} changes from 85 to 100 and V_{BE} changes from 0.7 V to 0.6 V is

$$\% \ \Delta I_C = \left(\frac{I_{C(2)} - I_{C(1)}}{I_{C(1)}}\right)100\%$$

$$= \left(\frac{11.4\ \text{mA} - 9.61\ \text{mA}}{9.61\ \text{mA}}\right)100\% = \mathbf{18.6\%} \quad \text{(an increase)}$$

The percent change in V_{CE} is

$$\% \ \Delta V_{CE} = \left(\frac{V_{CE(2)} - V_{CE(1)}}{V_{CE(1)}}\right)100\%$$

$$= \left(\frac{5.62\ \text{V} - 6.62\ \text{V}}{6.62\ \text{V}}\right)100\% = \mathbf{-15.1\%} \quad \text{(a decrease)}$$

As you can see, the Q-point is very dependent on β_{DC} in this circuit and therefore makes the base bias arrangement very unstable. Consequently, base bias is not normally used if linear operation is required. However, it can be used in switching applications.

Related Problem If $\beta_{DC} = 50$ at 0°C and 125 at 100°C for the circuit in Figure 5–20, determine the percent change in the Q-point values over the temperature range. Assume no chnage in V_{BE}.

Open file E05-06 in the Examples folder on your CD-ROM. Set $\beta_{DC} = 85$ and measure I_C and V_{CE}. Next, set $\beta_{DC} = 100$ and measure I_C and V_{CE}. Compare results with the calculated values.

Emitter Bias

Emitter bias uses both a positive and a negative supply voltage, as shown in Figure 5–21. In this circuit, the V_{EE} supply voltage forward-biases the base-emitter junction. Kirchhoff's voltage law applied around the base-emitter circuit in Figure 5–21(a), which has been redrawn in Figure 5–21(b) for analysis, gives the following equation:

$$V_{EE} + V_{R_B} + V_{BE} + V_{R_E} = 0$$

Substituting, using Ohm's law,

$$V_{EE} + I_B R_B + V_{BE} + I_E R_E = 0$$

Solving for V_{EE},

$$I_B R_B + I_E R_E + V_{BE} = -V_{EE}$$

Since $I_C \cong I_E$ and $I_C = \beta_{DC} I_B$,

$$I_B \cong \frac{I_E}{\beta_{DC}}$$

Substituting for I_B,

$$\left(\frac{I_E}{\beta_{DC}}\right) R_B + I_E R_E + V_{BE} = -V_{EE}$$

Factoring out I_E yields

$$I_E\left(\frac{R_B}{\beta_{DC}} + R_E\right) + V_{BE} = -V_{EE}$$

Transposing V_{BE} and then solving for I_E,

$$I_E = \frac{-V_{EE} - V_{BE}}{R_E + R_B/\beta_{DC}}$$

Since $I_C \cong I_E$,

$$I_C \cong \frac{-V_{EE} - V_{BE}}{R_E + R_B/\beta_{DC}}$$

Equation 5–10

(a)

(b)

▲ **FIGURE 5–21**

An *npn* transistor with emitter bias. Polarities are reversed for a *pnp* transistor. Single subscripts indicate voltages with respect to ground.

Voltages with respect to ground are indicated by a single subscript. The emitter voltage with respect to ground is

$$V_E = V_{EE} + I_E R_E$$

The base voltage with respect to ground is

$$V_B = V_E + V_{BE}$$

The collector voltage with respect to ground is

$$V_C = V_{CC} - I_C R_C$$

Q-Point Stability of Emitter Bias The formula for I_E shows that the emitter bias circuit is dependent on V_{BE} and β_{DC}, both of which change with temperature and current.

$$I_E = \frac{-V_{EE} - V_{BE}}{R_E + R_B/\beta_{DC}}$$

If $R_E \gg R_B/\beta_{DC}$, the R_B/β_{DC} term can be dropped and the equation becomes

$$I_E \cong \frac{-V_{EE} - V_{BE}}{R_E}$$

This condition makes I_E essentially independent of β_{DC}.

A further approximation can be made; if $V_{EE} \gg V_{BE}$, the V_{BE} term can be dropped.

$$I_E \cong \frac{V_{EE}}{R_E}$$

This condition makes I_E essentially independent of V_{BE}.

If I_E is independent of β_{DC} and V_{BE}, then the Q-point is not affected by variations in these parameters. Thus, emitter bias can provide a stable Q-point *if properly designed*.

EXAMPLE 5–7

Determine how much the Q-point (I_C, V_{CE}) in Figure 5–22 will change over a temperature range where β_{DC} increases from 85 to 100 and V_{BE} decreases from 0.7 V to 0.6 V.

▶ **FIGURE 5–22**

Solution For $\beta_{DC} = 85$ and $V_{BE} = 0.7$ V,

$$I_{C(1)} \cong I_E = \frac{-V_{EE} - V_{BE}}{R_E + R_B/\beta_{DC}} = \frac{-(-20\text{ V}) - 0.7\text{ V}}{10\text{ k}\Omega + 100\text{ k}\Omega/85} = 1.73\text{ mA}$$

$$V_C = V_{CC} - I_C R_C = 20\text{ V} - (1.73\text{ mA})(4.7\text{ k}\Omega) = 11.9\text{ V}$$

$$V_E = V_{EE} + I_E R_E = -20\text{ V} + (1.73\text{ mA})(10\text{ k}\Omega) = -2.7\text{ V}$$

Therefore,

$$V_{CE(1)} = V_C - V_E = 11.9\text{ V} - (-2.7\text{ V}) = 14.6\text{ V}$$

For $\beta_{DC} = 100$ and $V_{BE} = 0.6$ V,

$$I_{C(2)} \cong I_E = \frac{-V_{EE} - V_{BE}}{R_E + R_B/\beta_{DC}} = \frac{-(-20\text{ V}) - 0.6\text{ V}}{10\text{ k}\Omega + 100\text{ k}\Omega/100} = 1.76\text{ mA}$$

$$V_C = V_{CC} - I_C R_C = 20\text{ V} - (1.76\text{ mA})(4.7\text{ k}\Omega) = 11.7\text{ V}$$

$$V_E = V_{EE} + I_E R_E = -20\text{ V} + (1.76\text{ mA})(10\text{ k}\Omega) = 2.4\text{ V}$$

Therefore,

$$V_{CE(2)} = V_C - V_E = 11.7\text{ V} - (-1.5\text{ V}) = 13.2\text{ V}$$

The percent change in I_C as β_{DC} changes from 85 to 100 and V_{BE} changes from 0.7 V to 0.6 V is

$$\% \Delta I_C = \left(\frac{I_{C(2)} - I_{C(1)}}{I_{C(1)}}\right)100\% = \left(\frac{1.76\text{ mA} - 1.73\text{ mA}}{1.73\text{ mA}}\right)100\% = \mathbf{1.73\%}$$

The percent change in V_{CE} is

$$\% \Delta V_{CE} = \left(\frac{V_{CE(2)} - V_{CE(1)}}{V_{CE(1)}}\right)100\% = \left(\frac{13.2\text{ V} - 14.6\text{ V}}{14.6\text{ V}}\right)100\% = \mathbf{-9.6\%}$$

Related Problem Determine how much the Q-point in Figure 5–22 changes over a temperature range where β_{DC} increases from 65 to 75 and V_{BE} decreases from 0.75 V to 0.59 V. The supply voltages are ± 10 V.

Collector-Feedback Bias

In Figure 5–23, the base resistor R_B is connected to the collector rather than to V_{CC}, as it was in the base bias arrangement discussed earlier. The collector voltage provides the bias for the base-emitter junction. The negative **feedback** creates an "offsetting" effect that tends to keep the Q-point stable. If I_C tries to increase, it drops more voltage across R_C, thereby causing V_C to decrease. When V_C decreases, there is a decrease in voltage across R_B, which decreases I_B. The decrease in I_B produces less I_C which, in turn, drops less voltage across R_C and thus offsets the decrease in V_C.

Analysis of a Collector-Feedback Bias Circuit By Ohm's law, the base current can be expressed as

$$I_B = \frac{V_C - V_{BE}}{R_B}$$

Let's assume that $I_C \gg I_B$. The collector voltage is

$$V_C \cong V_{CC} - I_C R_C$$

▲ **FIGURE 5–23**

Collector-feedback bias.

Also,

$$I_B = \frac{I_C}{\beta_{DC}}$$

Substituting for I_B and V_C in the first equation,

$$\frac{I_C}{\beta_{DC}} = \frac{V_{CC} - I_C R_C - V_{BE}}{R_B}$$

The terms can be arranged so that

$$\frac{I_C R_B}{\beta_{DC}} + I_C R_C = V_{CC} - V_{BE}$$

Then you can solve for I_C as follows:

$$I_C(R_C + R_B/\beta_{DC}) = V_{CC} - V_{BE}$$

Equation 5–11

$$I_C = \frac{V_{CC} - V_{BE}}{R_C + R_B/\beta_{DC}}$$

Since the emitter is ground, $V_{CE} = V_C$.

Equation 5–12

$$V_{CE} = V_{CC} - I_C R_C$$

Q-Point Stability Over Temperature Equation 5–11 shows that the collector current is dependent to some extent on β_{DC} and V_{BE}. This dependency, of course, can be minimized by making $R_C \gg R_B/\beta_{DC}$ and $V_{CC} \gg V_{BE}$. An important feature of collector-feedback bias is that it essentially eliminates the β_{DC} and V_{BE} dependency even if the stated conditions are met.

As you have learned, β_{DC} varies directly with temperature, and V_{BE} varies inversely with temperature. As the temperature goes up in a collector-feedback circuit, β_{DC} goes up and V_{BE} goes down. The increase in β_{DC} acts to increase I_C. The decrease in V_{BE} acts to increase I_B which, in turn also acts to increase I_C. As I_C tries to increase, the voltage drop across R_C also tries to increase. This tends to reduce the collector voltage and therefore the voltage across R_B, thus reducing I_B and offsetting the attempted increase in I_C and the attempted decrease in V_C. The result is that the collector-feedback circuit maintains a relatively stable Q-point. The reverse action occurs when the temperature decreases.

EXAMPLE 5–8

Calculate the Q-point values (I_C and V_{CE}) for the circuit in Figure 5–24.

▶ **FIGURE 5–24**

Solution   Use Equation 5–11.

$$I_C = \frac{V_{CC} - V_{BE}}{R_C + R_B/\beta_{DC}} = \frac{10\text{ V} - 0.7\text{ V}}{10\text{ k}\Omega + 100\text{ k}\Omega/100} = \mathbf{845\ \mu A}$$

Using Equation 5–12, the collector-to-emitter voltage is

$$V_{CE} = V_{CC} - I_C R_C = 10 \text{ V} - (845 \ \mu\text{A})(10 \text{ k}\Omega) = \mathbf{1.55 \ V}$$

Related Problem Calculate the Q-point values in Figure 5–24 for $\beta_{DC} = 85$ and determine the change in the Q-point from $\beta_{DC} = 100$ to $\beta_{DC} = 85$.

Open file E05-08 in the Examples folder on your CD-ROM. Measure I_C and V_{CE}. Compare with the calculated values.

**SECTION 5–3
REVIEW**

1. What is the main disadvantage of the base bias method?
2. Explain why the base bias Q-point changes with temperature.
3. Why is emitter bias more stable than base bias?
4. What is the main disadvantage of emitter bias?
5. Explain how an increase in β_{DC} causes a reduction in base current in a collector-feedback circuit.

5–4   TROUBLESHOOTING

In a biased transistor circuit, the transistor can fail or sometimes a resistor in the bias circuit can fail. We will examine several possibilities in this section using the voltage-divider bias arrangement. Many circuit failures result from open resistors, internally open transistor leads and junctions, or shorted junctions. Often, these failures can produce an apparent cutoff or saturation condition when voltage is measured at the collector.

After completing this section, you should be able to

- **Troubleshoot various faults in transistor bias circuits**
- Use voltage measurements to identify a fault in a transistor bias circuit
- Analyze a transistor bias circuit for several common faults

Troubleshooting a Voltage-Divider Biased Transistor

An example of a transistor with voltage-divider bias is shown in Figure 5–25. For the specific component values shown, you should get the voltage readings approximately as indicated when the circuit is operating properly.

For this type of bias circuit, a particular group of faults will cause the transistor collector to be at V_{CC} when measured with respect to ground. Five faults are shown in Figure 5–26(a). The collector voltage is equal to 10 V with respect to ground for each of the faults as indicated. Also, for each of the faults, the base voltage and the emitter voltage with respect to ground are displayed in part (b).

Fault 1: Resistor R_1 Open This fault removes the bias voltage from the base, thus connecting the base to ground through R_2 and forcing the transistor into cutoff because $V_B = 0$ V and $I_B = 0$ A. The transistor is nonconducting so there is no I_C and, therefore, no

▶ **FIGURE 5–25**

A voltage–divider biased transistor with correct voltages.

▲ **FIGURE 5–26**

Faults for which $V_C = V_{CC}$.

(a) For all indicated faults, $V_C = V_{CC}$

(b) V_B and V_E for each indicated fault in part (a)

voltage drop across R_C. This makes the collector voltage equal to V_{CC} (10 V). Since there is no base current or collector current, there is also no emitter current and $V_E = 0$ V.

Fault 2: Resistor R_E Open This fault prevents base current, emitter current, and collector current except for a very small I_{CBO} that can be neglected. Since $I_C = 0$ A, there is no voltage drop across R_C and, therefore, $V_C = V_{CC} = 10$ V. The voltage divider produces a voltage at the base with respect to ground as follows:

$$V_B = \left(\frac{R_2}{R_1 + R_2} \right) V_{CC} = \left(\frac{4.7 \text{ k}\Omega}{14.7 \text{ k}\Omega} \right) 10 \text{ V} = 3.20 \text{ V}$$

When a voltmeter is connected to the emitter, it provides a current path through its high internal impedance, resulting in a forward-biased base-emitter junction. Therefore, the emitter voltage is $V_E = V_B - V_{BE}$. The amount of the forward voltage drop across the BE junction depends on the current. $V_{BE} = 0.7$ V is assumed for purposes of illustration, but it may be much less. The result is an emitter voltage as follows:

$$V_E = V_B - V_{BE} = 3.2 \text{ V} - 0.7 \text{ V} = 2.5 \text{ V}$$

Fault 3: Base Lead Internally Open An internal transistor fault is more likely to happen than an open resistor. Again, the transistor is nonconducting so $I_C = 0$ A and $V_C = V_{CC} = 10$ V. Just as for the case of the open R_E, the voltage divider produces 3.2 V at the external base connection. The voltage at the external emitter connection is 0 V because there is no emitter current through R_E and, thus, no voltage drop.

Fault 4: BE Junction or Emitter Connection Internally Open Again, the transistor is nonconducting, so $I_C = 0$ A and $V_C = V_{CC} = 10$ V. Just as for the case of the open R_E and the internally open base, the voltage divider produces 3.2 V at the base. The voltage at the external emitter lead is 0 V because that point is open and connected to ground through R_E. Notice that Faults 3 and 4 produce identical symptoms.

Fault 5: BC Junction or Collector Connection Internally Open Since there is an internal open in the transistor collector, there is no I_C and, therefore, $V_C = V_{CC} = 10$ V. In this situation, the voltage divider is loaded by R_E through the forward-biased BE junction, as shown by the approximate equivalent circuit in Figure 5–27. The base voltage and emitter voltage are determined as follows:

$$V_B \cong \left(\frac{R_2\|R_E}{R_1 + R_2\|R_E}\right)V_{CC} + 0.7 \text{ V}$$

$$= \left(\frac{427 \ \Omega}{10.427 \ k\Omega}\right)10 \text{ V} + 0.7 \text{ V} = 0.41 \text{ V} + 0.7 \text{ V} = 1.11 \text{ V}$$

$$V_E = V_B - V_{BE} = 1.11 \text{ V} - 0.7 \text{ V} = 0.41 \text{ V}$$

▶ **FIGURE 5–27**

Equivalent bias circuit for an internally open collector.

There are two possible additional faults for which the transistor is conducting or appears to be conducting, based on the collector voltage measurement. These are indicated in Figure 5–28.

Fault 6: Resistor R_C Open For this fault, which is illustrated in Figure 5–28(a), the collector voltage may lead you to think that the transistor is in saturation, but actually it is nonconducting. Obviously, if R_C is open, there can be no collector current. In this situation,

(a) R_C open

(b) R_2 open

▲ **FIGURE 5–28**

Faults for which the transistor is conducting or appears to be conducting.

the equivalent bias circuit is the same as for Fault 5, as illustrated in Figure 5–27. Therefore, $V_B = 1.11$ V and since the BE junction is forward-biased, $V_E = V_B - V_{BE} = 1.11$ V $- 0.7$ V $= 0.41$ V. When a voltmeter is connected to the collector to measure V_C, a current path is provided through the internal impedance of the meter and the BC junction is forward-biased by V_B. Therefore, $V_C = V_B - V_{BC} = 1.11$ V $- 0.7$ V $= 0.41$ V. Again the forward drops across the internal transistor junctions depend on the current. We are using 0.7 V for illustration, but the forward drops may be much less.

Fault 7: Resistor R_2 Open   When R_2 opens as shown in Figure 5–28(b), the base voltage and base current increase from their normal values because the voltage divider is now formed by R_1 and $\beta_{DC}R_{IN(base)}$. In this case, the base voltage is determined by the emitter voltage ($V_B = V_E + V_{BE}$).

First, verify whether the transistor is in saturation or not. The collector saturation current and the base current *required* to produce saturation are determined as follows (assuming $V_{CE(sat)} = 0.2$ V):

$$I_{C(sat)} = \frac{V_{CC} - V_{CE(sat)}}{R_C + R_E} = \frac{9.8 \text{ V}}{1.47 \text{ k}\Omega} = 6.67 \text{ mA}$$

$$I_{B(sat)} = \frac{I_{C(sat)}}{\beta_{DC}} = \frac{6.67 \text{ mA}}{300} = 22.2 \text{ } \mu\text{A}$$

Assuming the transistor is saturated, the maximum base current is determined.

$$R_{IN(base)} = \beta_{DC}R_E = 300(470 \text{ }\Omega) = 141 \text{ k}\Omega$$

$$I_{B(max)} \cong \frac{V_{CC}}{R_1 + R_{IN(base)}} = \frac{10 \text{ V}}{151 \text{ k}\Omega} = 66.2 \text{ } \mu\text{A}$$

Since this amount of base current is more than enough to produce saturation, the transistor is definitely saturated. Therefore, V_E, V_B, and V_C are as follows:

$$V_E \cong I_{C(sat)} R_E = (6.67 \text{ mA})(470 \text{ }\Omega) = 3.13 \text{ V}$$

$$V_B = V_E + V_{BE} = 3.13 \text{ V} + 0.7 \text{ V} = 3.83 \text{ V}$$

$$V_C = V_{CC} - I_{C(sat)}R_C = 10 \text{ V} - (6.67 \text{ mA})(1.0 \text{ k}\Omega) = 3.33 \text{ V}$$

EWB/Multisim Troubleshooting Exercises

These file circuits are in the Troubleshooting Exercises folder on your CD-ROM.

1. Open file TSE05-01. Determine if the circuit is working properly and, if not, determine the fault.

2. Open file TSE05-02. Determine if the circuit is working properly and, if not, determine the fault.

3. Open file TSE05-03. Determine if the circuit is working properly and, if not, determine the fault.

4. Open file TSE05-04. Determine if the circuit is working properly and, if not, determine the fault.

**SECTION 5–4
REVIEW**

1. How do you determine when a transistor is saturated? When a transistor is in cutoff?

2. In a voltage-divider biased *npn* transistor circuit, you measure V_{CC} at the collector and an emitter voltage 0.7 V less than the base voltage. Is the transistor functioning in cutoff, or is R_E open?

3. What symptoms does an open R_C produce?

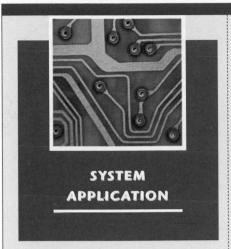

SYSTEM APPLICATION

Assume you are working on an industrial temperature control system to be installed in a chemical plant. Several people will be working on various portions of the system, and you have been assigned responsibility for the temperature-to-voltage conversion circuit and the power supply. You will apply the knowledge you have gained in this chapter in completing your assignment.

The Industrial Temperature Control System

This system is to be used for maintaining a preset sequence of specified temperatures in a liquid chemical substance during the mixing process in a large vat. The system is an example of a process-control system that is based on the principle of closed-loop feedback as shown in block diagram form in Figure 5–29.

The temperature of the chemical is sensed by a temperature-sensitive resistor called a **thermistor.** The resistance varies inversely with temperature; that is, it has a negative temperature coefficient. Essentially, the thermistor converts temperature to resistance.

The thermistor is a part of the temperature-to-voltage conversion circuit, for which you are responsible. The temperature of the chemical mixture determines the resistance of the thermistor. The temperature-to-voltage conversion circuit produces an output voltage that is pro-

portional to the temperature. This output voltage goes to an analog-to-digital converter (ADC), which converts the voltage to digital form. The digital processor adds the appropriate scaling and linearization to the digitized voltage and sends it to the valve control circuit that is programmed to sequence through a series of set-point temperatures during the mixing process. The valve control circuit compares the set-point temperature to the actual temperature as measured by the temperature-to-voltage conversion circuit and makes the proper adjustments to the burner.

If the temperature of the chemical is below its set-point value, the valve control circuit adjusts the valve to allow increased fuel flow to the burner to raise the temperature of the chemical. When the temperature of the chemical reaches the set-point value, the valve control circuit reduces the fuel flow and continuously adjusts it to maintain the constant

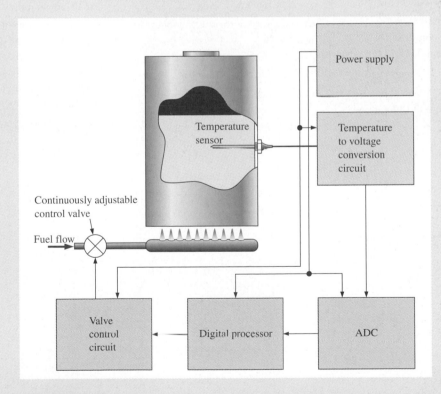

▲ **FIGURE 5–29**

Basic block diagram of the industrial temperature control system.

set-point temperature. A similar process occurs when the temperature of the chemical is above the set-point value.

In addition to the temperature-to-voltage conversion circuit, you are also responsible for the dc power supply that will provide dc voltages and currents to each part of the system. Although you do not need to know how the other blocks work, you must know their voltage and current requirements and the load impedance that the ADC presents to the temperature-to-voltage conversion circuit.

The System Requirements

■ There are three set-point temperatures in the mixing process.

■ Each set-point temperature is maintained for a preprogrammed time interval determined by the valve control circuit.

■ The set-point temperatures are 46°C, 50°C, and 54°C.

■ The dc voltages and currents for each circuit are as follows:

Temperature-to-voltage conversion circuit: 9.1 V regulated, 10 mA

Analog-to-digital converter (ADC): 5.1 V regulated, 50 mA

Digital processor: 5.1 V regulated, 25 mA

Valve control circuit: 9.1 V regulated, 40 mA

■ The input resistance of the ADC is 100 kΩ.

The Temperature-to-Voltage Conversion Circuit

The schematic for the temperature-to-voltage conversion circuit is shown in Figure 5–30. As you can see, it is simply a voltage-divider biased transistor with the thermistor used as one of the voltage-divider resistors.

Basic Operation When the temperature increases, the resistance of the thermistor decreases and reduces the base bias voltage which, in turn increases the collector (output) voltage proportionally. When the temperature decreases, the resistance of the thermistor increases and increases the base bias voltage, which then causes a decrease in the output voltage.

The output voltage of the temperature-to-voltage conversion circuit tracks the changing temperature of the chemical and, therefore, must operate in its linear or active region and not in saturation or cutoff. The output must interface with the ADC. Notice the schematic symbol for the thermistor in Figure 5–30.

The Thermistor Characteristics

■ The response characteristic of the thermistor is nonlinear. This means that the resistance does not vary in a straight-line relationship with temperature.

■ The resistance versus temperature characteristic of the thermistor for the range of 45°C to 55°C is given in Figure 5–31.

Analysis of the Temperature-to-Voltage Conversion Circuit

A 100 kΩ load resistance is connected to the output to simulate the ADC input resistance.

■ Determine the output voltage at each specified set-point temperature from 45°C to 55°C.

■ Determine if the transistor is operating in its linear (active) region over the prescribed temperature range. That is, verify that the transistor does not go into saturation or cutoff. If it does, suggest a modification to the circuit to correct the design flaw.

The Power Supply Circuit

The dc power supply circuit is the same as the one developed in previous system applications except that it needs to be modified for different voltage and current requirements, as shown in the schematic of Figure 5–32. The basic power supply specifications are

1. Input voltage: 115 V rms at 60 Hz

▲ **FIGURE 5–30**

Schematic of the temperature-to-voltage conversion circuit.

▲ **FIGURE 5–31**

Thermistor resistance versus temperature.

▲ FIGURE 5–32

Power supply schematic.

2. Regulated output voltages: 5.1 V dc ± 5%; 9.1 V dc ± 5%

3. Maximum ripple factor: 3%

4. Maximum load currents: 75 mA @ 5.1 V; 50 mA @ 9.1 V

Component Values and the Schematic

▪ Determine the values of all the unspecified components in the power supply. You may wish to refer to previous system application assignments.

▪ You decide to combine the power supply and the temperature-to-voltage conversion circuit on one printed circuit board. Produce a complete schematic that includes both circuits.

▲ FIGURE 5–33

Power supply and temperature detector printed circuit board. All black resistor bands indicate values to be determined.

The Printed Circuit Board

■ Check out the printed circuit board in Figure 5–33 to verify that it is correct according to the schematic.

■ Label a copy of the board with the component and input/output designations in agreement with the schematic.

Test Procedure

■ Develop a step-by-step set of instructions on how to completely check the power supply and temperature-to-voltage conversion circuit for proper operation using the test points (circled numbers) indicated in the test bench setup of Figure 5–34. This particular thermistor is a tubular shaft with a threaded mount for inserting in a ves-sel. The oven-type device is a controlled-temperature chamber.

■ Specify voltage values for all the measurements to be made. Provide a fault analysis for all possible component failures. Utilize and modify the test procedure you developed for the power supply circuit in Chapters 2 and 3 to include the changes.

▲ **FIGURE 5–34**

Test bench setup for the circuit board.

Troubleshooting

Three boards in the initial manufacturing run have been found to be defective. The test results are shown in Figure 5–35. Based on the sequence of measurements for each defective board, determine the most likely fault(s) in each case.

The red circled numbers indicate test points on the circuit board. The DMM function setting is indicated below the display.

Final Report (Optional)

Submit a final written report on the power supply and temperature-to-voltage conversion circuit board using an organized format that includes the following:

1. A physical description of the circuits.
2. A discussion of the operation of each circuit.
3. A list of the specifications.
4. A list of parts with part numbers if available.
5. A list of types of problems on the three circuit boards.
6. A complete description of how you determined the problem on each of the three circuit boards.

▲ **FIGURE 5–35**

Test results for the three faulty circuit boards.

SUMMARY OF TRANSISTOR BIAS CIRCUITS

npn transistors are shown. Supply voltage polarities are reversed for *pnp* transistors.

VOLTAGE-DIVIDER BIAS

- Q-point values ($I_C \cong I_E$)
- Collector current:

$$I_C \cong \frac{V_{TH} - V_{BE}}{R_E + R_{TH}/\beta_{DC}}$$

where $R_{TH} = \dfrac{R_2}{R_1 + R_2}$

and $V_{TH} = R_{TH}V_{CC}$

- Collector-to-emitter voltage:

$$V_{CE} \cong V_{CC} - I_C(R_C + R_E)$$

BASE BIAS

- Q-point values ($I_C \cong I_E$)
- Collector current:

$$I_C = \beta_{DC}\left(\frac{V_{CC} - V_{BE}}{R_B}\right)$$

- Collector-to-emitter voltage:

$$V_{CE} = V_{CC} - I_C R_C$$

EMITTER BIAS

- Q-point values ($I_C \cong I_E$)
- Collector current:

$$I_C \cong \frac{V_{EE} - V_{BE}}{R_E + R_B/\beta_{DC}}$$

- Collector-to-emitter voltage:

$$V_{CE} \cong V_{CC} + V_{EE} - I_C(R_C + R_E)$$

COLLECTOR-FEEDBACK BIAS

- Q-point values ($I_C \cong I_E$)
- Collector current:

$$I_C = \frac{V_{CC} - V_{BE}}{R_C + R_B/\beta_{DC}}$$

- Collector-to-emitter voltage:

$$V_{CE} = V_{CC} - I_C R_C$$

CHAPTER SUMMARY

- The purpose of biasing a circuit is to establish a proper stable dc operating point (Q-point).
- The Q-point of a circuit is defined by specific values for I_C and V_{CE}. These values are called the coordinates of the Q-point.
- A dc load line passes through the Q-point on a transistor's collector curves intersecting the vertical axis at approximately $I_{C(sat)}$ and the horizontal axis at $V_{CE(off)}$.
- The linear (active) operating region of a transistor lies along the load line below saturation and above cutoff.
- The dc input resistance at the base of a BJT is approximately $\beta_{DC}R_E$.
- Voltage-divider bias provides good Q-point stability with a single-polarity supply voltage. It is the most common bias circuit.
- The base bias circuit arrangement has poor stability because its Q-point varies widely with β_{DC}.

- Emitter bias generally provides good Q-point stability but requires both positive and negative supply voltages.
- Collector-feedback bias provides good stability using negative feedback from collector to base.

KEY TERMS

Key terms and other bold terms in the chapter are defined in the end-of-book glossary.

DC load line   A straight line plot of I_C and V_{CE} for a transistor circuit.

Feedback   The process of returning a portion of a circuit's output back to the input in such a way as to oppose or aid a change in the output.

Linear region   The region of operation along the load line between saturation and cutoff.

Q-point   The dc operating (bias) point of an amplifier specified by voltage and current values.

KEY FORMULAS

Voltage-Divider Bias

$$5\text{--}1 \qquad R_{IN(base)} \cong \beta_{DC}R_E$$

$$5\text{--}2 \qquad V_B \cong \left(\frac{R_2}{R_1 + R_2}\right)V_{CC} \qquad \text{for } npn, \text{ assuming, } \beta_{DC}R_E >> R_2$$

$$5\text{--}3 \qquad V_E = V_B - V_{BE}$$

$$5\text{--}4 \qquad I_E = \frac{V_E}{R_E}$$

$$5\text{--}5 \qquad I_C \cong I_E$$

$$5\text{--}6 \qquad V_C = V_{CC} - I_C R_C$$

$$5\text{--}7 \qquad V_{CE} \cong V_{CC} - I_C(R_C + R_E)$$

Base Bias

$$5\text{--}8 \qquad V_{CE} = V_{CC} - I_C R_C$$

$$5\text{--}9 \qquad I_C = \beta_{DC}\left(\frac{V_{CC} - V_{BE}}{R_B}\right)$$

Emitter Bias

$$5\text{--}10 \qquad I_C \cong \frac{-V_{EE} - V_{BE}}{R_E + R_B/\beta_{DC}}$$

Collector-Feedback Bias

$$5\text{--}11 \qquad I_C = \frac{V_{CC} - V_{BE}}{R_C + R_B/\beta_{DC}}$$

$$5\text{--}12 \qquad V_{CE} = V_{CC} - I_C R_C$$

SELF-TEST

Answers are at the end of the chapter.

1. The maximum value of collector current in a biased transistor is
   (a) $\beta_{DC}I_B$ (b) $I_{C(sat)}$ (c) greater than I_E (d) $I_E - I_B$

2. Ideally, a dc load line is a straight line drawn on the collector characteristic curves between
   (a) the Q-point and cutoff   (b) the Q-point and saturation
   (c) $V_{CE(cutoff)}$ and $I_{C(sat)}$ (d) $I_B = 0$ and $I_B = I_C/\beta_{DC}$

3. If a sinusoidal voltage is applied to the base of a biased *npn* transistor and the resulting sinusoidal collector voltage is clipped near zero volts, the transistor is
   (a) being driven into saturation   (b) being driven into cutoff
   (c) operating nonlinearly   (d) answers (a) and (c)   (e) answers (b) and (c)

4. The input resistance at the base of a biased transistor depends mainly on

   (a) β_{DC} (b) R_B (c) R_E (d) β_{DC} and R_E

5. In a voltage-divider biased transistor circuit such as in Figure 5–13, $R_{IN(base)}$ can generally be neglected in calculations when

   (a) $R_{IN(base)} > R_2$ (b) $R_2 > 10R_{IN(base)}$ (c) $R_{IN(base)} > 10R_2$ (d) $R_1 << R_2$

6. In a certain voltage-divider biased *npn* transistor, V_B is 2.95 V. The dc emitter voltage is approximately

   (a) 2.25 V   (b) 2.95 V   (c) 3.65 V   (d) 0.7 V

7. Voltage-divider bias

   (a) cannot be independent of β_{DC}

   (b) can be essentially independent of β_{DC}

   (c) is not widely used

   (d) requires fewer components than all the other methods

8. The disadvantage of base bias is that

   (a) it is very complex   (b) it produces low gain

   (c) it is too beta dependent   (d) it produces high leakage current

9. Emitter bias is

   (a) essentially independent of β_{DC} (b) very dependent on β_{DC}

   (c) provides a stable bias point   (d) answers (a) and (c)

10. In an emitter bias circuit, $R_E = 2.7$ kΩ and $V_{EE} = 15$ V. The emitter current

    (a) is 5.3 mA   (b) is 2.7 mA   (c) is 180 mA   (d) cannot be determined

11. Collector-feedback bias is

    (a) based on the principle of positive feedback   (b) based on beta multiplication

    (c) based on the principle of negative feedback   (d) not very stable

12. In a voltage-divider biased *npn* transistor, if the upper voltage-divider resistor (the one connected to V_{CC}) opens,

    (a) the transistor goes into cutoff   (b) the transistor goes into saturation

    (c) the transistor burns out   (d) the supply voltage is too high

13. In a voltage-divider biased *npn* transistor, if the lower voltage-divider resistor (the one connected to ground) opens,

    (a) the transistor is not affected

    (b) the transistor may be driven into cutoff

    (c) the transistor may be driven into saturation

    (d) the collector current will decrease

14. In a voltage-divider biased *pnp* transistor, there is no base current, but the base voltage is approximately correct. The most likely problem(s) is

    (a) a bias resistor is open   (b) the collector resistor is open

    (c) the base-emitter junction is open   (d) the emitter resistor is open

    (e) answers (a) and (c)   (f) answers (c) and (d)

PROBLEMS

Answers to all odd-numbered problems are at the end of the book.

BASIC PROBLEMS

SECTION 5–1   **The DC Operating Point**

1. The output (collector voltage) of a biased transistor amplifier is shown in Figure 5–36. Is the transistor biased too close to cutoff or too close to saturation?

▶ **FIGURE 5-36**

≈ 0 V ───

2. What is the Q-point for a biased transistor as in Figure 5–2 with $I_B = 150$ μA, $\beta_{DC} = 75$, $V_{CC} = 18$ V, and $R_C = 1.0$ kΩ?

3. What is the saturation value of collector current in Problem 2?

4. What is the cutoff value of V_{CE} in Problem 2?

5. Determine the intercept points of the dc load line on the vertical and horizontal axes of the collector-characteristic curves for the circuit in Figure 5–37.

▶ **FIGURE 5-37**

File circuits are identified with a CD logo and are in the Problems folder on your CD–ROM. Filenames correspond to figure numbers (e.g., F05-37).

6. Assume that you wish to bias the transistor in Figure 5–37 with $I_B = 20$ μA. To what voltage must you change the V_{BB} supply? What are I_C and V_{CE} at the Q-point, given that $\beta_{DC} = 50$?

7. Design a biased-transistor circuit using $V_{BB} = V_{CC} = 10$ V, for a Q-point of $I_C = 5$ mA and $V_{CE} = 4$ V. Assume $\beta_{DC} = 100$. The design involves finding R_B, R_C, and the *minimum* power rating of the transistor. (The actual power rating should be greater.) Sketch the circuit.

8. Determine whether the transistor in Figure 5–38 is biased in cutoff, saturation, or the linear region. Keep in mind that $I_C = \beta_{DC}I_B$ is valid only in the linear region.

▶ **FIGURE 5-38**

SECTION 5-2   Voltage-Divider Bias

9. What is the minimum value of β_{DC} in Figure 5–39 that makes $R_{IN(base)} \cong 10R_2$?

10. The bias resistor R_2 in Figure 5–39 is replaced by a 15 kΩ potentiometer. What minimum resistance setting causes saturation?

11. If the potentiometer described in Problem 10 is set at 2 kΩ, what are the values for I_C and V_{CE}?

12. Determine all transistor terminal voltages with respect to ground in Figure 5–40. Do not neglect the input resistance at the base or V_{BE}.

13. Show the connections required to replace the transistor in Figure 5–40 with a *pnp* device.

$V_B = \left(\dfrac{R_2}{R_1 + R_2}\right) V_{CC}$

$V_E = V_B - V_{BE}$

$I_E = \dfrac{V_E}{R_E}$

▲ FIGURE 5–39

▲ FIGURE 5–40

14. (a) Determine V_B in Figure 5–41.
    (b) If R_E is doubled, what is the value of V_B?

15. (a) Find the Q-point values for Figure 5–41.
    (b) Find the minimum power rating of the transistor in Figure 5–41.

▶ FIGURE 5–41

SECTION 5-3   Other Bias Methods

16. Determine I_B, I_C, and V_{CE} for a base-biased transistor circuit with the following values: $\beta_{DC} = 90$, $V_{CC} = 12$ V, $R_B = 22$ kΩ, and $R_C = 100$ Ω.

17. If β_{DC} in Problem 16 doubles over temperature, what are the Q-point values?

18. You have two base bias circuits connected for testing. They are identical except that one is biased with a separate V_{BB} source and the other is biased with the base resistor connected to V_{CC}. Ammeters are connected to measure collector current in each circuit. You vary the V_{CC} supply voltage and observe that the collector current varies in one circuit, but not in the other. In which circuit does the collector current change? Explain your observation.

19. The data sheet for a particular transistor specifies a minimum β_{DC} of 50 and a maximum β_{DC} of 125. What range of Q-point values can be expected if an attempt is made to mass-produce the circuit in Figure 5–42? Is this range acceptable if the Q-point must remain in the transistor's linear region?

20. The base bias circuit in Figure 5–42 is subjected to a temperature variation from 0°C to 70°C. The β_{DC} decreases by 50 percent at 0°C and increases by 75 percent at 70°C from its nominal value of 110 at 25°C. What are the changes in I_C and V_{CE} over the temperature range of 0°C to 70°C?

▲ FIGURE 5–42

▲ FIGURE 5–43

21. Analyze the circuit in Figure 5–43 to determine the correct voltages at the transistor terminals with respect to ground. Assume $\beta_{DC} = 100$.

22. To what value can R_E in Figure 5–43 be reduced without the transistor going into saturation?

23. Taking V_{BE} into account in Figure 5–43, how much will I_E change with a temperature increase from 25°C to 100°C? The V_{BE} is 0.7 V at 25°C and decreases 2.5 mV per degree Celsius. Neglect β_{DC}.

24. When can the effect of a change in β_{DC} be neglected in the emitter bias circuit?

25. Determine I_C and V_{CE} in the *pnp* emitter bias circuit of Figure 5–44. Assume $\beta_{DC} = 100$.

26. Determine V_B, V_C, and I_C in Figure 5–45.

27. What value of R_C can be used to decrease I_C in Problem 26 by 25 percent?

28. What is the minimum power rating for the transistor in Problem 27?

29. A collector-feedback circuit uses an *npn* transistor with $V_{CC} = 12$ V, $R_C = 1.2$ kΩ, and $R_B = 47$ kΩ. Determine the collector current and the collector voltage if $\beta_{DC} = 200$.

▲ FIGURE 5–44

▲ FIGURE 5–45

TROUBLESHOOTING PROBLEMS

SECTION 5–4 Troubleshooting

30. Assume the emitter becomes shorted to ground in Figure 5–46 by a solder splash or stray wire clipping. What do the meters read? When you correct the problem, what do the meters read?

31. Determine the most probable failures, if any, in each circuit of Figure 5–47, based on the indicated measurements.

32. Determine if the DMM readings 2 through 4 in the breadboard circuit of Figure 5–48 are correct. If they are not, isolate the problem(s). The transistor is a *pnp* device with a specified dc beta range of 35 to 100.

33. Determine each meter reading in Figure 5–48 for each of the following faults:

   (a) the 680 Ω resistor open   **(b)** the 5.6 kΩ resistor open

   (c) the 10 kΩ resistor open   **(d)** the 1.0 kΩ resistor open

   (e) a short from emitter to ground   **(f)** an open base-emitter junction

▶ **FIGURE 5–46**

(a)

(b)

(c)

(d)

▲ **FIGURE 5–47**

▲ **FIGURE 5–48**

SYSTEM APPLICATION PROBLEMS

34. Determine V_B, V_E, and V_C in the temperature-to-voltage conversion circuit in Figure 5–30 if R_1 fails open.

35. What faults will cause the transistor in the temperature-to-voltage conversion circuit to go into cutoff?

36. Assume that a 5.1 V zener diode is inadvertently installed in the place of the 9.1 V zener on the circuit board in Figure 5–33. What voltages will you measure at the base, emitter, and collector of the temperature-to-voltage conversion circuit when the temperature is 45°C?

37. Explain how you would identify an open collector-base junction in the transistor in Figure 5–30.

DATA SHEET PROBLEMS

38. Analyze the temperature-to-voltage conversion circuit in Figure 5–49(a) at the temperature extremes for both minimum and maximum specified data sheet values of h_{FE}. Refer to the partial data sheet in Figure 5–50.

(a)

(b)

▲ **FIGURE 5–49**

Maximum Ratings

Rating	Symbol	Value	Unit
Collector-Emitter voltage	V_{CEO}	40	V dc
Collector-Base voltage	V_{CBO}	60	V dc
Emitter-Base voltage	V_{EBO}	6.0	V dc
Collector current — continuous	I_C	200	mA dc
Total device dissipation @ $T_A = 25°C$ Derate above 25°C	P_D	625 5.0	mW mW/°C
Total device dissipation @ $T_C = 25°C$ Derate above 25°C	P_D	1.5 12	Watts mW/°C
Operating and storage junction Temperature range	T_J, T_{stg}	−55 to +150	°C

Thermal Characteristics

Characteristic	Symbol	Max	Unit
Thermal resistance, junction to case	$R_{\theta JC}$	83.3	°C/W
Thermal resistance, junction to ambient	$R_{\theta JA}$	200	°C/W

Electrical Characteristics ($T_A = 25°C$ unless otherwise noted.)

Characteristic	Symbol	Min	Max	Unit
Off Characteristics				
Collector-Emitter breakdown voltage $I_C = 1.0$ mA dc, $I_B = 0$	$V_{(BR)CEO}$	40	—	V dc
Collector-Base breakdown voltage $I_C = 10$ μA dc, $I_E = 0$	$V_{(BR)CBO}$	60	—	V dc
Emitter-Base breakdown voltage $I_E = 10$ μA dc, $I_C = 0$	$V_{(BR)EBO}$	6.0	—	V dc
Base cutoff current ($V_{CE} = 30$ V dc, $V_{EB} = 3.0$ V dc)	I_{BL}	—	50	nA dc
Collector cutoff current ($V_{CE} = 30$ V dc, $V_{EB} = 3.0$ V dc)	I_{CEX}	—	50	nA dc

On Characteristics

Characteristic		Symbol	Min	Max	Unit
DC current gain		h_{FE}			—
($I_C = 0.1$ mA dc, $V_{CE} = 1.0$ V dc)	2N3903		20	—	
	2N3904		40	—	
($I_C = 1.0$ mA dc, $V_{CE} = 1.0$ V dc)	2N3903		35	—	
	2N3904		70		
($I_C = 10$ mA dc, $V_{CE} = 1.0$ V dc)	2N3903		50	150	
	2N3904		100	300	
($I_C = 50$ mA dc, $V_{CE} = 1.0$ V dc)	2N3903		30	—	
	2N3904		60	—	
($I_C = 100$ mA dc, $V_{CE} = 1.0$ V dc)	2N3903		15	—	
	2N3904		30	—	
Collector-Emitter saturation voltage		$V_{CE(sat)}$			V dc
($I_C = 10$ mA dc, $I_B = 1.0$ mA dc)			—	0.2	
($I_C = 50$ mA dc, $I_B = 5.0$ mA dc)			—	0.3	
Base-Emitter saturation voltage		$V_{BE(sat)}$			V dc
($I_C = 10$ mA dc, $I_B = 1.0$ mA dc)			0.65	0.85	
($I_C = 50$ mA dc, $I_B = 5.0$ mA dc)			—	0.95	

2N3903
2N3904

3 Collector

2
Base

1 Emitter

General Purpose Transistors

NPN Silicon

▲ **FIGURE 5–50**

39. Verify that no maximum ratings are exceeded in the temperature-to-voltage conversion circuit in Figure 5–49. Refer to the partial data sheet in Figure 5–50.

40. Refer to the partial data sheet in Figure 5–51.

   (a) What is the maximum collector current for a 2N2222A?

   (b) What is the maximum reverse base-emitter voltage for a 2N2118?

41. Determine the maximum power dissipation for a 2N2222 at 100°C.

42. When you increase the collector current in a 2N2219 from 1 mA to 500 mA, how much does the minimum β_{DC} (h_{FE}) change?

ADVANCED PROBLEMS

43. Design a circuit using base bias that operates from a 15 V dc voltage and draws a maximum current from the dc source ($I_{CC(max)}$) of 10 mA. The Q-point values are to be $I_C = 5$ mA and $V_{CE} = 5$ V. The transistor is a 2N3903. Assume a midpoint value for β_{DC}.

Maximum Ratings

Rating	Symbol	2N2218 2N2219 2N2221 2N2222	2N2218A 2N2219A 2N2221A 2N2222A	2N5581 2N5582	Unit
Collector-Emitter voltage	V_{CEO}	30	40	40	V dc
Collector-Base voltage	V_{CBO}	60	75	75	V dc
Emitter-Base voltage	V_{EBO}	5.0	6.0	6.0	V dc
Collector current — continuous	I_C	800	800	800	mA dc
		2N2218,A 2N2219,A	2N2221,A 2N2222,A	2N5581 2N5582	
Total device dissipation @ $T_A = 25°C$ Derate above 25°C	P_D	0.8 4.57	0.5 2.28	0.6 3.33	Watt mW/°C
Total device dissipation @ $T_C = 25°C$ Derate above 25°C	P_D	3.0 17.1	1.2 6.85	2.0 11.43	Watt mW/°C
Operating and storage junction Temperature range	T_J, T_{stg}	−65 to +200			°C

2N2218, A/2N2219, A
2N2221, A/2N2222, A
2N5581/82

2N2218, A/2N2219, A
Case 79-04
TO-39 (TO-205AD)
Style 1

2N2221, A/2N2222, A
Case 22-03
TO-18 (TO-206AA)
Style 1

2N5581/2N5582
Case 26-03
TO-46 (TO-206AB)
Style 1

3 Collector
2 Base
1 Emitter

General Purpose Transistors
NPN Silicon

Electrical Characteristics ($T_A = 25°C$ unless otherwise noted.)

Characteristic		Symbol	Min	Max	Unit
Off Characteristics					
Collector-Emitter breakdown voltage ($I_C = 10$ mA dc, $I_B = 0$)	Non-A Suffix	$V_{(BR)CEO}$	30	—	V dc
	A-Suffix, 2N5581, 2N5582		40	—	
Collector-Base breakdown voltage ($I_C = 10$ μA dc, $I_E = 0$)	Non-A Suffix	$V_{(BR)CBO}$	60	—	V dc
	A-Suffix, 2N5581, 2N5582		75	—	
Emitter-Base breakdown voltage ($I_E = 10$ μA dc, $I_C = 0$)	Non-A Suffix	$V_{(BR)EBO}$	5.0	—	V dc
	A-Suffix, 2N5581, 2N5582		6.0	—	
Collector cutoff current ($V_{CE} = 60$ V dc, $V_{EB(off)} = 3.0$ V dc)	A-Suffix, 2N5581, 2N5582	I_{CEX}	—	10	nA dc
Collector cutoff current ($V_{CB} = 50$ V dc, $I_E = 0$) ($V_{CB} = 60$ V dc, $I_E = 0$) ($V_{CB} = 50$ V dc, $I_E = 0$, $T_A = 150°C$) ($V_{CB} = 60$ V dc, $I_E = 0$, $T_A = 150°C$)	Non-A Suffix A-Suffix, 2N5581, 2N5582 Non-A Suffix A-Suffix, 2N5581, 2N5582	I_{CBO}	— — — —	0.01 0.01 10 10	μA dc
Emitter cutoff current ($V_{EB} = 3.0$ V dc, $I_C = 0$)	A-Suffix, 2N5581, 2N5582	I_{EBO}	—	10	μA dc
Base cutoff current ($V_{CE} = 60$ V dc, $V_{EB(off)} = 3.0$ V dc)	A-Suffix	I_{BL}	—	20	nA dc
On Characteristics					
DC current gain ($I_C = 0.1$ mA dc, $V_{CE} = 10$ V dc)	2N2218,A, 2N2221,A, 2N5581(1) 2N2219,A, 2N2222,A, 2N5582(1)	h_{FE}	20 35	— —	—
($I_C = 1.0$ mA dc, $V_{CE} = 10$ V dc)	2N2218,A, 2N2221,A, 2N5581 2N2219,A, 2N2222,A, 2N5582		25 50	— —	
($I_C = 10$ mA dc, $V_{CE} = 10$ V dc)	2N2218,A, 2N2221,A, 2N5581(1) 2N2219,A, 2N2222,A, 2N5582(1)		35 75	— —	
($I_C = 10$ mA dc, $V_{CE} = 10$ V dc, $T_A = -55°C$)	2N2218,A, 2N2221,A, 2N5581 2N2219,A, 2N2222,A, 2N5582		15 35	— —	
($I_C = 150$ mA dc, $V_{CE} = 10$ V dc)	2N2218,A, 2N2221,A, 2N5581 2N2219,A, 2N2222,A, 2N5582		40 100	120 300	
($I_C = 150$ mA dc, $V_{CE} = 1.0$ V dc)	2N2218,A, 2N2221,A, 2N5581 2N2219,A, 2N2222,A, 2N5582		20 50	— —	
($I_C = 500$ mA dc, $V_{CE} = 10$ V dc)	2N2218, 2N2221 2N2219, 2N2222 2N2218A, 2N2221A, 2N5581 2N2219A, 2N2222A, 2N5582		20 30 25 40	— — — —	
Collector-Emitter saturation voltage ($I_C = 150$ mA dc, $I_B = 15$ mA dc)	Non-A Suffix A-Suffix, 2N5581, 2N5582	$V_{CE(sat)}$	— —	0.4 0.3	V dc
($I_C = 500$ mA dc, $I_B = 50$ mA dc)	Non-A Suffix A-Suffix, 2N5581, 2N5582		— —	1.6 1.0	
Base-Emitter saturation voltage ($I_C = 150$ mA dc, $I_B = 15$ mA dc)	Non-A Suffix A-Suffix, 2N5581, 2N5582	$V_{BE(sat)}$	0.6 0.6	1.3 1.2	V dc
($I_C = 500$ mA dc, $I_B = 50$ mA dc)	Non-A Suffix A-Suffix, 2N5581, 2N5582		— —	2.6 2.0	

▲ **FIGURE 5–51**

44. Design a circuit using emitter bias that operates from dc voltages of $+12$ V and -12 V. The maximum I_{CC} is to be 20 mA and the Q-point is at 10 mA and 4 V. The transistor is a 2N3904.

45. Design a circuit using voltage-divider bias for the following specifications: $V_{CC} = 9$ V, $I_{CC(max)} = 5$ mA, $I_C = 1.5$ mA, and $V_{CE} = 3$ V. The transistor is a 2N3904.

46. Design a collector-feedback circuit using a 2N2222 with $V_{CC} = 5$ V, $I_C = 10$ mA, and $V_{CE} = 1.5$ V.

47. Can you replace the 2N3904 in Figure 5–49 with a 2N2222A and maintain the same range of output voltage over a temperature range from 45°C to 55°C?

48. Refer to the data sheet graph in Figure 5–52 and the partial data sheet in Figure 5–51. Determine the minimum dc current gain for a 2N2222 at $-55°C$, 25°C, and 175°C for $V_{CE} = 1$ V.

▲ **FIGURE 5–52**

49. A design change is required in the ADC block of the industrial temperature control system shown in Figure 5–29. The new design will have an ADC input resistance of 10 kΩ. Determine the effect this change has on the temperature-to-voltage conversion circuit.

50. Investigate the feasibility of redesigning the temperature-to-voltage conversion circuit in Figure 5–30 to operate from a dc supply voltage of 5.1 V and produce the same range of output voltages determined in the System Application over the required thermistor temperature range from 46°C to 54°C.

EWB/MULTISIM TROUBLESHOOTING PROBLEMS

These file circuits are in the Troubleshooting Problems folder on your CD-ROM.

51. Open file TSP05-51 and determine the fault.

52. Open file TSP05-52 and determine the fault.

53. Open file TSP05-53 and determine the fault.

54. Open file TSP05-54 and determine the fault.

55. Open file TSP05-55 and determine the fault.

56. Open file TSP05-56 and determine the fault.

ANSWERS

SECTION REVIEWS

SECTION 5–1   The DC Operating Point

1. The upper load line limit is $I_{C(sat)}$ and $V_{CE(sat)}$. The lower limit is $I_C = 0$ and $V_{CE(cutoff)}$.

2. The Q-point is the dc point at which a transistor is biased specified by V_{CE} and I_C.

3. Saturation begins at the intersection of the load line and the vertical portion of the collector curve. Cutoff occurs at the intersection of the load line and the $I_B = 0$ curve.

4. The Q-point must be centered on the load line for maximum V_{ce}.

SECTION 5–2 Voltage-Divider Bias

1. $R_{IN(base)} = V_{IN}/I_{IN} = 5\ \text{V}/5\ \mu\text{A} = 1\ \text{M}\Omega$

2. $R_{IN(base)} = \beta_{DC}R_E = 190(1.0\ \text{k}\Omega) = 190\ \text{k}\Omega$

3. $V_B = 5\ \text{V}$

4. Voltage-divider bias is stable and requires only one supply voltage.

SECTION 5–3 Other Bias Methods

1. Base bias is beta-dependent.

2. The Q-point changes due to changes in β_{DC} and V_{CE} over temperature.

3. Emitter bias is much less dependent on the value of beta than is base bias.

4. Emitter bias requires two separate supply voltages.

5. I_C increases with β_{DC}, causing a reduction in V_C and, therefore, less voltage across R_B, thus less I_B.

SECTION 5–4 Troubleshooting

1. A transistor is saturated when $V_{CE} = 0\ \text{V}$. A transistor is in cutoff when $V_{CE} = V_{CC}$.

2. R_E is open because the BE junction of the transistor is still forward-biased.

3. If R_C is open, V_C is about 0.7 V less than V_B.

RELATED PROBLEMS FOR EXAMPLES

5–1 $I_{CQ} = 19.8\ \text{mA};\ V_{CEQ} = 4.2\ \text{V};\ I_{b(peak)} = 42\ \mu\text{A}$

5–2 54.6 kΩ

5–3 $V_{CE} = 2.56\ \text{V},\ I_C = 4.77\ \text{mA}$

5–4 $I_C \cong 2.29\ \text{mA},\ V_{EC} = 2.67\ \text{V}$

5–5 214

5–6 $\%\ \Delta I_C = 150\%;\ \%\ \Delta V_{CE} = 53.6\%$

5–7 $\%\ \Delta I_C = 2.40\%;\ \%\ \Delta V_{CE} = -3.87\%$

5–8 $I_C = 832\ \mu\text{A},\ V_{CE} = 1.68\ \text{V},\ \Delta I_C = -13\ \mu\text{A},\ \Delta V_{CE} = 0.13\ \text{V}$

SELF-TEST

1. (b) 2. (c) 3. (d) 4. (d) 5. (c)

6. (a) 7. (b) 8. (c) 9. (d) 10. (a)

11. (c) 12. (a) 13. (c) 14. (f)

6

BJT AMPLIFIERS

INTRODUCTION

The things you learned about biasing a transistor in Chapter 5 are now applied in this chapter where bipolar junction transistor (BJT) circuits are used as small-signal amplifiers. The term *small-signal* refers to the use of signals that take up a relatively small percentage of an amplifier's operational range. Additionally, you will learn how to reduce an amplifier to an equivalent dc and ac circuit for easier analysis, and you will learn about multistage amplifiers.

- Understand the amplifier concept

- Identify and apply internal transistor parameters

- Understand and analyze the operation of common-emitter amplifiers

- Understand and analyze the operation of common-collector amplifiers

- Understand and analyze the operation of common-base amplifiers

- Discuss multistage amplifiers and analyze their operation

- Troubleshoot amplifier circuits

KEY TERMS

- *r* parameter

- Common–emitter

- ac ground

- Input resistance

- Output resistance

- Attenuation

- Overall voltage gain

- Bypass capacitor

- Emitter–follower

- Common–base

■■■ SYSTEM APPLICATION PREVIEW

The system application in this chapter involves a preamplifier circuit for a public address and paging system. The complete system includes the preamplifier, a power amplifier, and a dc power supply. You will focus on the preamplifier in this chapter and then on the power amplifier in Chapter 9. The first step in your assignment is to learn all you can about amplifier operation. You will then apply your knowledge to the system application at the end of the chapter.

WWW. **VISIT THE COMPANION WEBSITE**

Study aids for this chapter are available at

http://www.prenhall.com/floyd

6–1 AMPLIFIER OPERATION

The biasing of a transistor is purely a dc operation. The purpose of biasing is to establish a Q-point about which variations in current and voltage can occur in response to an ac input signal. In applications where small signal voltages must be amplified—such as from an antenna, a microphone, etc.—variations about the Q-point are relatively small. Amplifiers designed to handle these small ac signals are often referred to as small-signal amplifiers.

After completing this section, you should be able to

- **Understand the amplifier concept**
- Interpret labels used for dc and ac voltages and currents
- Discuss the general operation of a small-signal amplifier
- Analyze ac load line operation
- Describe phase inversion

AC Quantities

In the previous chapters, dc quantities were identified by nonitalic uppercase (capital) subscripts such as I_C, I_E, V_C, and V_{CE}. Lowercase italic subscripts are used to indicate ac quantities of rms, peak, and peak-to-peak currents and voltages, for example, I_c, I_e, I_b, V_c, and V_{ce} (rms values are assumed unless otherwise stated). Instantaneous quantities are represented by both lowercase letters and subscripts such as i_c, i_e, i_b, and v_{ce}. Figure 6–1 illustrates these quantities for a specific voltage waveform.

In addition to currents and voltages, resistances often have different values when a circuit is analyzed from an ac viewpoint as opposed to a dc viewpoint. Lowercase subscripts are used to identify ac resistance values. For example, R_c is the ac collector resistance, and R_C is the dc collector resistance. You will see the need for this distinction later. Resistance values *internal* to the transistor use a lowercase r'. An example is the internal ac emitter resistance, r'_e.

The Linear Amplifier

A voltage-divider biased transistor with a sinusoidal ac source capacitively coupled to the base through C_1 and a load capacitively coupled to the collector through C_2 is shown in

▶ **FIGURE 6–1**

V_{ce} can represent rms, average, peak, or peak-to-peak, but rms will be assumed unless stated otherwise. v_{ce} can be any instantaneous value on the curve.

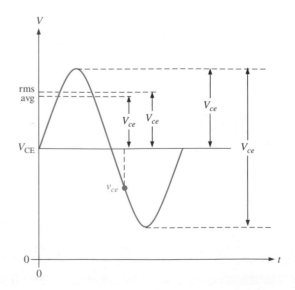

Figure 6–2. The coupling capacitors block dc and thus prevent the internal source resistance, R_s, and the load resistance, R_L, from changing the dc bias voltages at the base and collector. The capacitors appear ideally as shorts to the signal voltage. The sinusoidal source voltage causes the base voltage to vary sinusoidally above and below its dc bias level. The resulting variation in base current produces a larger variation in collector current because of the current gain of the transistor.

▲ **FIGURE 6–2**

An amplifier with voltage-divider bias driven by an ac voltage source with an internal resistance, R_s.

As the sinusoidal collector current increases, the collector voltage decreases. The collector current varies above and below its Q-point value in phase with the base current. The sinusoidal collector-to-emitter voltage varies above and below its Q-point value 180° out of phase with the base voltage, as illustrated in Figure 6–2. A transistor always produces a phase inversion between the base voltage and the collector voltage.

A Graphical Picture   The operation just described can be illustrated graphically on the collector-characteristic curves, as shown in Figure 6–3. The sinusoidal voltage at the base produces a base current that varies above and below the Q-point on the ac load line, as shown by the arrows. Lines projected from the peaks of the base current, across to the I_C axis, and down to the V_{CE} axis, indicate the peak-to-peak variations of the

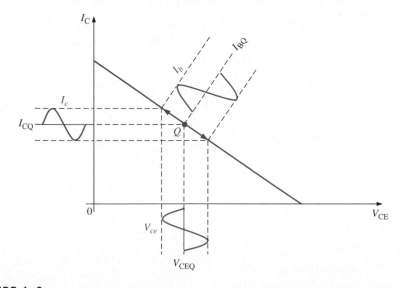

▲ **FIGURE 6–3**

Graphical operation of the amplifier showing the variation of the base current, collector current, and collector-to-emitter voltage about their dc Q-point values. I_b and I_c are on different scales.

collector current and collector-to-emitter voltage, as shown. The ac load line differs from the dc load line because the effective ac collector resistance is R_L in parallel with R_C and is less than the dc collector resistance R_C alone without R_L in parallel. This difference between the dc and the ac load lines is covered in Chapter 9 in relation to power amplifiers.

EXAMPLE 6–1

The ac load line operation of a certain amplifier extends 10 μA above and below the Q-point base current value of 50 μA, as shown in Figure 6–4. Determine the resulting peak-to-peak values of collector current and collector-to-emitter voltage from the graph.

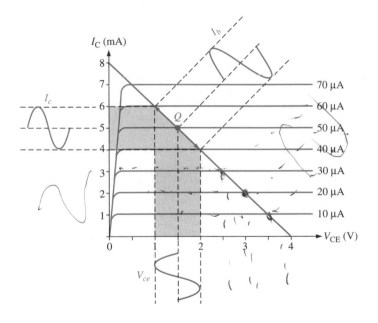

▲ **FIGURE 6–4**

Solution   Projections on the graph of Figure 6–4 show the collector current varying from 6 mA to 4 mA for a peak-to-peak value of **2 mA** and the collector-to-emitter voltage varying from 1 V to 2 V for a peak-to-peak value of **1 V**.

*Related Problem**   What are the Q-point values of I_C and V_{CE} in Figure 6–4?

*Answers are at the end of the chapter.

SECTION 6–1 REVIEW*

Answers are at the end of the chapter.

1. When I_b is at its positive peak, I_c is at its _____ peak, and V_{ce} is at its _____ peak.
2. What is the difference between V_{CE} and V_{ce}?
3. What is the difference between R_e and r'_e?

6–2 TRANSISTOR AC EQUIVALENT CIRCUITS

To visualize the operation of a transistor in an amplifier circuit, it is often useful to represent the device by an equivalent circuit. An equivalent circuit uses various internal transistor parameters to represent the transistor's operation. Equivalent circuits are described in this section based on resistance or r parameters. Another system of parameters, called h parameters, are briefly described.

After completing this section, you should be able to

- **Identify and apply internal transistor parameters**
- Define the r parameters
- Represent a transistor by an r-parameter equivalent circuit
- Distinguish between the dc beta and the ac beta
- Define the h parameters

r Parameters

The resistance, r, parameters are the most commonly used for BJTs. The five r parameters are given in Table 6–1. The italic lowercase letter r with a prime denotes resistances internal to the transistor.

r parameters.

r PARAMETER	DESCRIPTION
α_{ac}	ac alpha (I_c/I_e)
β_{ac}	ac beta (I_c/I_b)
r_e'	ac emitter resistance
r_b'	ac base resistance
r_c'	ac collector resistance

r-Parameter Equivalent Circuits

An **r-parameter** equivalent circuit for a bipolar junction transistor is shown in Figure 6–5(a). For most general analysis work, Figure 6–5(a) can be simplified as follows: The

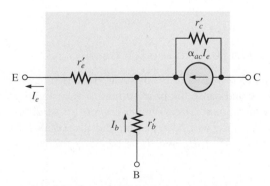

(a) Generalized r-parameter equivalent circuit for a bipolar junction transistor

(b) Simplified r-parameter equivalent circuit for a bipolar junction transistor

▲ FIGURE 6–5

r-parameter equivalent circuits.

▶ FIGURE 6–6

Relation of transistor symbol to
r-parameter equivalent.

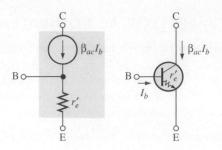

effect of the ac base resistance (r'_b) is usually small enough to neglect, so it can be replaced by a short. The ac collector resistance (r'_c) is usually several hundred kilohms and can be replaced by an open. The resulting simplified r-parameter equivalent circuit is shown in Figure 6–5(b).

The interpretation of this equivalent circuit in terms of a transistor's ac operation is as follows: A resistance (r'_e) appears between the emitter and base terminals. This is the resistance "seen" looking into the emitter of a forward-biased transistor. The collector effectively acts as a current source of $\alpha_{ac}I_e$ or, equivalently, $\beta_{ac}I_b$. These factors are shown with a transistor symbol in Figure 6–6.

Determining r'_e by a Formula

For amplifier analysis, r'_e is the most important of the r parameters. To get r'_e, you can use the simple formula of Equation 6–1 to calculate an approximate value.

Equation 6–1

$$r'_e \cong \frac{25 \text{ mV}}{I_E}$$

Although the formula is simple, its derivation is not and is therefore reserved for Appendix B.

EXAMPLE 6–2

Determine the r'_e of a transistor that is operating with a dc emitter current of 2 mA.

Solution

$$r'_e \cong \frac{25 \text{ mV}}{I_E} = \frac{25 \text{ mV}}{2 \text{ mA}} = \textbf{12.5 } \Omega$$

Related Problem   What is I_E if $r'_e = 8 \ \Omega$?

Comparison of the AC Beta (β_{ac}) to the DC Beta (β_{DC})

For a typical transistor, a graph of I_C versus I_B is nonlinear, as shown in Figure 6–7(a). If you pick a Q-point on the curve and cause the base current to vary an amount ΔI_B, then the collector current will vary an amount ΔI_C as shown in part (b). At different points on the nonlinear curve, the ratio $\Delta I_C/\Delta I_B$ will be different, and it may also differ from the I_C/I_B ratio at the Q-point. Since $\beta_{DC} = I_C/I_B$ and $\beta_{ac} = \Delta I_C/\Delta I_B$, the values of these two quantities can differ slightly.

h Parameters

Because they are typically specified on a manufacturer's data sheet, h (hybrid) parameters (h_i, h_r, h_f, and h_o) are important. These parameters are often specified by manufacturers because they are relatively easy to measure.

▶ FIGURE 6–7

I_C-versus-I_B curve illustrates the difference between $\beta_{DC} = I_C/I_B$ and $\beta_{ac} = \Delta I_C/\Delta I_B$.

(a) $\beta_{DC} = I_C/I_B$ at Q-point   (b) $\beta_{ac} = \Delta I_C/\Delta I_B$

The four basic ac h parameters and their descriptions are given in Table 6–2. Each of the four h parameters carries a second subscript letter to designate the common-emitter (e), common-base (b), or common-collector (c) amplifier configuration, as listed in Table 6–3. The characteristics of each of these three BJT amplifier configurations are covered later in this chapter.

▶ TABLE 6–2

Basic ac h parameters.

h PARAMETER	DESCRIPTION	CONDITION
h_i	Input impedance (resistance)	Output shorted
h_r	Voltage feedback ratio	Input open
h_f	Forward current gain	Output shorted
h_o	Output admittance (conductance)	Input open

▶ TABLE 6–3

Subscripts of h parameters for the three amplifier configurations.

CONFIGURATION	h PARAMETERS
Common-Emitter	$h_{ie}, h_{re}, h_{fe}, h_{oe}$
Common-Base	$h_{ib}, h_{rb}, h_{fb}, h_{ob}$
Common-Collector	$h_{ic}, h_{rc}, h_{fc}, h_{oc}$

Relationships of h Parameters and r Parameters

The ac current ratios, α_{ac} and β_{ac}, convert directly from h parameters as follows:

$$\alpha_{ac} = h_{fb}$$
$$\beta_{ac} = h_{fe}$$

Because data sheets often provide only common-emitter h parameters, the following formulas show how to convert them to r parameters. We will use r parameters throughout the text.

$$r'_e = \frac{h_{re}}{h_{oe}}$$

$$r'_c = \frac{h_{re} + 1}{h_{oe}}$$

$$r'_b = h_{ie} - \frac{h_{re}}{h_{oe}}(1 + h_{fe})$$

| SECTION 6–2 REVIEW | 1. Define each of the parameters: α_{ac}, β_{ac}, r'_e, r'_b, and r'_c.
2. Which h parameter is equivalent to β_{ac}?
3. If $I_E = 15$ mA, what is the approximate value of r'_e? |

6–3 THE COMMON-EMITTER AMPLIFIER

Now that you have an idea of how a transistor can be represented in an ac equivalent circuit, a complete amplifier circuit will be examined. The common-emitter (CE) configuration is covered in this section. CE amplifiers exhibit high voltage and current gains. The common-collector and common-base configurations are covered in the following sections.

After completing this section, you should be able to

- **Understand and analyze the operation of common-emitter amplifiers**
- Represent a CE amplifier by its dc equivalent circuit
- Analyze the dc operation of a CE amplifier
- Represent a CE amplifier by its ac equivalent circuit
- Analyze the ac operation of a CE amplifier
- Determine the input resistance
- Determine the output resistance
- Determine the voltage gain
- Explain the effects of an emitter bypass capacitor
- Describe swamping and discuss its purpose and effects
- Describe the effect of a load resistor on the voltage gain
- Discuss phase inversion in a CE amplifier
- Determine current gain
- Determine power gain

Figure 6–8 shows a **common-emitter** amplifier with voltage-divider bias and coupling capacitors, C_1 and C_3, on the input and output and a bypass capacitor, C_2, from emitter to ground. The circuit has a combination of dc and ac operation, both of which must be considered. The input signal, V_{in}, is capacitively coupled into the base, and the output signal, V_{out}, is capacitively coupled from the collector. The amplified output is 180° out of phase with the input.

DC Analysis

To analyze the amplifier in Figure 6–8, the dc bias values must first be determined. To do this, a dc equivalent circuit is developed by replacing the coupling and bypass capacitors with opens (remember, a capacitor appears open to dc), as shown in Figure 6–9.

▲ FIGURE 6–8

A common–emitter amplifier.

▶ FIGURE 6–9

DC equivalent circuit for the amplifier in Figure 6–8.

Recall from Chapter 5 that the dc input resistance at the base is determined as follows:

$$R_{IN(base)} = \beta_{DC}R_E = (150)(560\ \Omega) = 84\ k\Omega$$

Since $R_{IN(base)}$ is more than ten times R_2 in this case, it will be neglected when calculating the dc base voltage.

$$V_B \cong \left(\frac{R_2}{R_1 + R_2}\right)V_{CC} = \left(\frac{6.8\ k\Omega}{28.8\ k\Omega}\right)12\ V = 2.83\ V$$

and

$$V_E = V_B - V_{BE} = 2.83\ V - 0.7\ V = 2.13\ V$$

Therefore,

$$I_E = \frac{V_E}{R_E} = \frac{2.13\ V}{560\ \Omega} = 3.80\ mA$$

Since $I_C \cong I_E$, then

$$V_C = V_{CC} - I_CR_C = 12\ V - (3.80\ mA)(1.0\ k\Omega) = 12\ V - 3.80\ V = 8.20\ V$$

Finally,

$$V_{CE} = V_C - V_E = 8.20\ V - 2.13\ V = 6.07\ V$$

The AC Equivalent Circuit

To analyze the ac signal operation of an amplifier, an ac equivalent circuit is developed as follows:

The capacitors C_1, C_2, and C_3 are replaced by effective shorts because their values are selected so that $X_C \cong 0\ \Omega$ at the signal frequency.

AC Ground The dc source is replaced by a ground. You assume that the voltage source has an internal resistance of approximately $0\ \Omega$ so that no ac voltage is developed across the source terminals. Therefore, the V_{CC} terminal is at a zero-volt ac potential and is called **ac ground**.

The ac equivalent circuit for the common-emitter amplifier in Figure 6–8 is shown in Figure 6–10(a). Notice that both R_C and R_1 have one end connected to ac ground because, in the actual circuit, they are connected to V_{CC} which is, in effect, ac ground.

In ac analysis, the ac ground and the actual ground are treated as the same point electrically. The amplifier in Figure 6–8 is a common-emitter type because the bypass capacitor C_2 keeps the emitter at ac ground. Ground is the common point in the circuit.

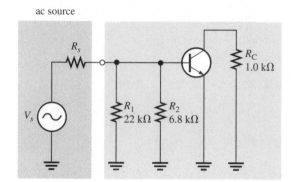

(a) Without an ac voltage source. AC ground is shown in red.

(b) With an ac voltage source

▲ **FIGURE 6–10**

AC equivalent circuit for the amplifier in Figure 6–8.

Signal (AC) Voltage at the Base An ac voltage source is shown connected to the input in Figure 6–10(b). If the internal resistance of the ac source is $0\ \Omega$, then all of the source voltage appears at the base terminal. If, however, the ac source has a nonzero internal resistance, then three factors must be taken into account in determining the actual signal voltage at the base. These are the *source resistance* (R_s), the *bias resistance* $(R_1 \parallel R_2)$, and the *input resistance* $(R_{in(base)})$, (some prefer to use the term *impedance*) at the base. This is

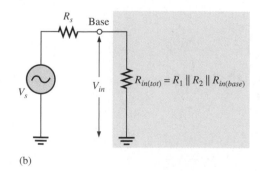

(a)

(b)

▲ **FIGURE 6–11**

AC equivalent base circuit.

illustrated in Figure 6–11(a) and is simplified by combining R_1, R_2, and $R_{in(base)}$ in parallel to get the total input resistance, $R_{in(tot)}$, as shown in Figure 6–11(b). As you can see, the source voltage, V_s, is divided down by R_s (source resistance) and $R_{in(tot)}$ so that the signal voltage at the base of the transistor is found by the voltage-divider formula as follows:

$$V_b = \left(\frac{R_{in(tot)}}{R_s + R_{in(tot)}}\right)V_s$$

If $R_s \ll R_{in(tot)}$, then $V_b \cong V_s$ where V_b is the input voltage, V_{in}, to the amplifier.

Input Resistance To develop an expression for the **input resistance** as seen by an ac source looking in at the base, we will use the simplified *r*-parameter model of the transistor. Figure 6–12 shows the transistor connected with the external collector resistor, R_C. The input resistance looking in at the base is

$$R_{in(base)} = \frac{V_{in}}{I_{in}} = \frac{V_b}{I_b}$$

The base voltage is

$$V_b = I_e r'_e$$

and since $I_e \cong I_c$,

$$I_b \cong \frac{I_e}{\beta_{ac}}$$

Substituting for V_b and I_b,

$$R_{in(base)} = \frac{V_b}{I_b} = \frac{I_e r'_e}{I_e/\beta_{ac}}$$

Cancelling I_e,

$$R_{in(base)} = \beta_{ac} r'_e$$

Equation 6–2

The total input resistance seen by the source is the parallel combination of R_1, R_2, and $R_{in(base)}$.

$$R_{in(tot)} = R_1 \parallel R_2 \parallel R_{in(base)}$$

Equation 6–3

▶ **FIGURE 6–12**

r-parameter transistor model (inside shaded block) connected to external circuit.

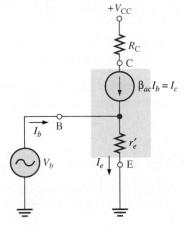

Output Resistance The **output resistance** of the common-emitter amplifier, looking in at the collector, is approximately equal to the collector resistor.

$$R_{out} \cong R_C$$

Equation 6–4

Actually, $R_{out} = R_C \parallel r'_c$, but since the internal ac collector resistance of the transistor, r'_c, is typically much larger than R_C, the approximation is usually valid.

EXAMPLE 6–3

Determine the signal voltage at the base of the transistor in Figure 6–13. This circuit is the ac equivalent of the amplifier in Figure 6–8 with a 10 mV rms, 300 Ω signal source. I_E was previously found to be 3.80 mA.

▲ FIGURE 6–13

Solution First, determine the ac emitter resistance.

$$r'_e \cong \frac{25\text{ mV}}{I_E} = \frac{25\text{ mV}}{3.80\text{ mA}} = 6.58\ \Omega$$

Then,

$$R_{in(base)} = \beta_{ac}r'_e = 160(6.58\ \Omega) = 1.05\text{ k}\Omega$$

Next, determine the total input resistance viewed from the source.

$$R_{in(tot)} = R_1 \| R_2 \| R_{in(base)} = \cfrac{1}{\cfrac{1}{22\text{ k}\Omega} + \cfrac{1}{6.8\text{ k}\Omega} + \cfrac{1}{1.05\text{ k}\Omega}} = 873\ \Omega$$

The source voltage is divided down by R_s and $R_{in(tot)}$, so the signal voltage at the base is the voltage across $R_{in(tot)}$.

$$V_b = \left(\frac{R_{in(tot)}}{R_s + R_{in(tot)}}\right)V_s = \left(\frac{873\ \Omega}{1173\ \Omega}\right)10\text{ mV} = \textbf{7.44 mV}$$

As you can see, there is attenuation (reduction) of the source voltage due to the source resistance and amplifier's input resistance acting as a voltage divider.

Related Problem Determine the signal voltage at the base of Figure 6–13 if the source resistance is 75 Ω and another transistor with an ac beta of 200 is used.

▌Voltage Gain of the CE Amplifier

The ac voltage gain expression is developed using the equivalent circuit in Figure 6–14. The gain is the ratio of ac output voltage at the collector (V_c) to ac input voltage at the base (V_b).

$$A_v = \frac{V_{out}}{V_{in}} = \frac{V_c}{V_b}$$

Notice in the figure that $V_c = \alpha_{ac}I_eR_C \cong I_eR_C$ and $V_b = I_er'_e$. Therefore,

$$A_v = \frac{I_eR_C}{I_er'_e}$$

▶ FIGURE 6–14

Equivalent circuit for obtaining ac voltage gain.

The I_e terms cancel, so

$$A_v = \frac{R_C}{r'_e}$$

Equation 6–5

Equation 6–5 is the voltage gain from base to collector. To get the overall gain of the amplifier from the source voltage to collector, the attenuation of the input circuit must be included. **Attenuation** means that the signal voltage is reduced as it passes through a circuit.

The attenuation from source to base multiplied by the gain from base to collector is the *overall* amplifier gain. Suppose the source produces 10 mV and the source resistance and input resistance is such that the base voltage is 5 mV. The attenuation is therefore 5 mV/10 mV = 0.5. Now assume the amplifier has a voltage gain from base to collector of 20. The output voltage is 5 mV × 20 = 100 mV. Therefore, the overall gain is 100 mV/10 mV = 10 and is equal to the attenuation times the gain (0.5 × 20 = 10). Overall gain is illustrated in Figure 6–15.

▲ FIGURE 6–15

Base circuit attenuation and overall gain.

The expression for the attenuation in the base circuit where R_s and $R_{in(tot)}$ act as a voltage divider is

$$\text{Attenuation} = \frac{V_b}{V_s} = \frac{R_{in(tot)}}{R_s + R_{in(tot)}}$$

The **overall voltage gain,** A'_v, is the product of the attenuation and the gain from base to collector, A_v.

$$A'_v = \left(\frac{V_b}{V_s}\right) A_v$$

Effect of the Emitter Bypass Capacitor on Voltage Gain The emitter **bypass capacitor,** which is C_2 in Figure 6–8, provides an effective short to the ac signal around the emitter resistor, thus keeping the emitter at ac ground, as you have seen. With the bypass capacitor, the gain of a given amplifier is maximum and equal to R_C/r'_e.

The value of the bypass capacitor must be large enough so that its reactance over the frequency range of the amplifier is very small (ideally 0 Ω) compared to R_E. A good rule-of-thumb is that X_C of the bypass capacitor should be at least 10 times smaller than R_E at the minimum frequency for which the amplifier must operate.

$$10X_C \leq R_E$$

EXAMPLE 6–4

Select a minimum value for the emitter bypass capacitor, C_2, in Figure 6–16 if the amplifier must operate over a frequency range from 2 kHz to 10 kHz.

▶ **FIGURE 6–16**

Solution Since $R_E = 560\ \Omega$, X_C of the bypass capacitor, C_2, should be no greater than

$$10X_C = R_E$$

$$X_C = \frac{R_E}{10} = \frac{560\ \Omega}{10} = 56\ \Omega$$

The capacitance value is determined at the minimum frequency of 2 kHz as follows:

$$C_2 = \frac{1}{2\pi f X_C} = \frac{1}{2\pi(2\text{ kHz})(56\ \Omega)} = \textbf{1.42 }\boldsymbol{\mu}\textbf{F}$$

This is the minimum value for the bypass capacitor for this circuit. You can use a larger value, although cost and physical size usually impose limitations.

Related Problem If the minimum frequency is reduced to 1 kHz, what value of bypass capacitor must you use?

Voltage Gain Without the Bypass Capacitor   To see how the bypass capacitor affects ac voltage gain, let's remove it from the circuit in Figure 6–16 and compare voltage gains.

Without the bypass capacitor, the emitter is no longer at ac ground. Instead, R_E is seen by the ac signal between the emitter and ground and effectively adds to r'_e in the voltage gain formula.

$$A_v = \frac{R_C}{r'_e + R_E}$$

Equation 6–6

The effect of R_E is to decrease the ac voltage gain.

EXAMPLE 6–5

Calculate the base-to-collector voltage gain of the amplifier in Figure 6–16 both without and with an emitter bypass capacitor if there is no load resistor.

Solution   From Example 6–3, $r'_e = 6.58\ \Omega$ for this same amplifier. Without C_2, the gain is

$$A_v = \frac{R_C}{r'_e + R_E} = \frac{1.0\ k\Omega}{566.58\ \Omega} = \mathbf{1.76}$$

With C_2, the gain is

$$A_v = \frac{R_C}{r'_e} = \frac{1.0\ k\Omega}{6.58\ \Omega} = \mathbf{152}$$

As you can see, the bypass capacitor makes quite a difference.

Related Problem   Determine the base-to-collector voltage gain in Figure 6–16 with R_E bypassed, for the following circuit values: $R_C = 1.8\ k\Omega$, $R_E = 1.0\ k\Omega$, $R_1 = 33\ k\Omega$, and $R_2 = 6.8\ k\Omega$.

Effect of a Load on the Voltage Gain   When a **load**, R_L, is connected to the output through the coupling capacitor C_3, as shown in Figure 6–17(a), the collector resistance at the signal frequency is effectively R_C in parallel with R_L. Remember, the upper end of R_C is effectively at ac ground. The ac equivalent circuit is shown in Figure 6–17(b). The total ac collector resistance is

$$R_c = \frac{R_C R_L}{R_C + R_L}$$

(a) Complete amplifier

(b) AC equivalent ($X_{C1} = X_{C2} = X_{C3} = 0$)

▲ **FIGURE 6–17**

A common–emitter amplifier with an ac (capacitively) coupled load.

Replacing R_C with R_c in the voltage gain expression gives

Equation 6–7

$$A_v = \frac{R_c}{r'_e}$$

When $R_c < R_C$, the voltage gain is reduced. If $R_L \gg R_C$, then $R_c \cong R_C$ and the load has very little effect on the gain.

EXAMPLE 6–6

Calculate the base-to-collector voltage gain of the amplifier in Figure 6–16 when a load resistance of 5 kΩ is connected to the output. The emitter is effectively bypassed and $r'_e = 6.58$ Ω.

Solution   The ac collector resistance is

$$R_c = \frac{R_C R_L}{R_C + R_L} = \frac{(1.0 \text{ k}\Omega)(5 \text{ k}\Omega)}{6 \text{ k}\Omega} = 833 \text{ }\Omega$$

Therefore,

$$A_v = \frac{R_c}{r'_e} = \frac{833 \text{ }\Omega}{6.58 \text{ }\Omega} = \mathbf{127}$$

The unloaded gain was found to be 152 in Example 6–5.

Related Problem   Determine the base-to-collector voltage gain in Figure 6–16 when a 10 kΩ load resistance is connected from collector to ground. Change the resistance values as follows: $R_C = 1.8$ kΩ, $R_E = 1.0$ kΩ, $R_1 = 33$ kΩ, and $R_2 = 6.8$ kΩ. The emitter resistor is effectively bypassed and $r'_e = 10$ Ω.

Stability of the Voltage Gain

Although bypassing R_E does produce the maximum voltage gain, there is a **stability** problem because the ac voltage gain is dependent on r'_e ($A_v = R_C/r'_e$). Also, r'_e depends on I_E and on temperature. This causes the gain to be unstable over temperature because when r'_e increases, the gain decreases and vice versa.

With no bypass capacitor, the gain is decreased because R_E is now in the ac circuit ($A_v = R_C/(r'_e + R_E)$). However, with R_E unbypassed, the gain is much less dependent on r'_e. If $R_E \gg r'_e$, the gain is essentially independent of r'_e because

$$A_v \cong \frac{R_C}{R_E}$$

Swamping r'_e to Stabilize the Voltage Gain   *Swamping* is a method used to minimize the effect of r'_e without reducing the voltage gain to its minimum value. This method "swamps" out the effect of r'_e on the voltage gain. Swamping is, in effect, a compromise between having a bypass capacitor across R_E and having no bypass capacitor at all.

In a swamped amplifier, R_E is partially bypassed so that a reasonable gain can be achieved, and the effect of r'_e on the gain is greatly reduced or eliminated. The total external emitter resistance, R_E, is formed with two separate emitter resistors, R_{E1} and R_{E2}, as indicated in Figure 6–18. One of the resistors, R_{E2}, is bypassed and the other is not.

Both resistors ($R_{E1} + R_{E2}$) affect the dc bias while only R_{E1} affects the ac voltage gain.

$$A_v = \frac{R_C}{r'_e + R_{E1}}$$

▶ FIGURE 6–18

A swamped amplifier uses a partially bypassed emitter resistance to minimize the affect of r'_e on the gain in order to achieve gain stability.

If R_{E1} is at least ten times larger than r'_e, then the effect of r'_e is minimized and the approximate voltage gain for the swamped amplifier is

$$A_v \cong \frac{R_C}{R_{E1}}$$

Equation 6–8

EXAMPLE 6–7

Determine the voltage gain of the swamped amplifier in Figure 6–19. Assume that the bypass capacitor has a negligible reactance for the frequency at which the amplifier is operated. Assume $r'_e = 20\ \Omega$.

▶ **FIGURE 6–19**

Solution R_{E2} is bypassed by C_2. R_{E1} is more than ten times r'_e so the approximate voltage gain is

$$A_v \cong \frac{R_C}{R_{E1}} = \frac{3.3\ \text{k}\Omega}{330\ \Omega} = 10$$

Related Problem   What would be the voltage gain without C_2? What would be the voltage gain with C_2 bypassing both R_{E1} and R_{E2}?

The Effect of Swamping on the Amplifier's Input Resistance   The ac input resistance, looking in at the base of a common-emitter amplifier with R_E completely bypassed, is $R_{in} = \beta_{ac}r'_e$. When the emitter resistance is partially bypassed, the portion of the resistance that is unbypassed is seen by the ac signal and contributes to the input resistance by appearing in series with r'_e. The formula is

$$R_{in(base)} = \beta_{ac}(r'_e + R_{E1})$$

Equation 6–9

Phase Inversion in a Common-Emitter Amplifier

The output voltage at the collector of a common-emitter amplifier is 180° out of phase with the input voltage at the base. The phase inversion is sometimes indicated by a negative sign in front of voltage gain, $-A_v$.

EXAMPLE 6–8

For the amplifier in Figure 6–20, determine the total collector voltage and the total output voltage (dc and ac). Draw the waveforms.

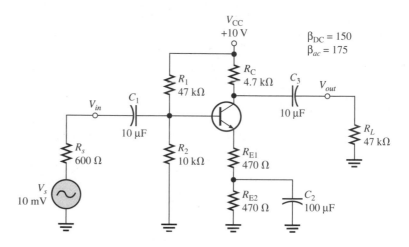

▲ **FIGURE 6–20**

Solution Two sets of calculations are necessary to determine the total collector voltage.

Step 1: DC analysis

Determine the dc bias values. Refer to the dc equivalent circuit in Figure 6–21.

$$R_{IN(base)} = \beta_{DC}(R_{E1} + R_{E2}) = 150(940\ \Omega) = 141\ k\Omega$$

▶ **FIGURE 6–21**

DC equivalent for the circuit in Figure 6–20.

Since $R_{\text{IN(base)}}$ is more than ten times larger than R_2, it can be neglected in the dc base voltage calculation.

$$V_B \cong \left(\frac{R_2}{R_1 + R_2}\right) V_{CC} = \left(\frac{10\ \text{k}\Omega}{47\ \text{k}\Omega + 10\ \text{k}\Omega}\right) 10\ \text{V} = 1.75\ \text{V}$$

$$V_E = V_B - 0.7\ \text{V} = 1.75\ \text{V} - 0.7\ \text{V} = 1.05\ \text{V}$$

$$I_E = \frac{V_E}{R_{E1} + R_{E2}} = \frac{1.05\ \text{V}}{940\ \Omega} = 1.12\ \text{mA}$$

$$V_C = V_{CC} - I_C R_C = 10\ \text{V} - (1.12\ \text{mA})(4.7\ \text{k}\Omega) = \textbf{4.74 V}$$

Step 2: AC analysis

The ac analysis is based on the ac equivalent circuit in Figure 6–22. The first thing to do in the ac analysis is calculate r'_e.

$$r'_e \cong \frac{25\ \text{mV}}{I_E} = \frac{25\ \text{mV}}{1.12\ \text{mA}} = 22\ \Omega$$

Next, determine the attenuation in the base circuit. Looking from the 600 Ω source, the total R_{in} is

$$R_{in(tot)} = R_1 \| R_2 \| R_{in(base)}$$
$$R_{in(base)} = \beta_{ac}(r'_e + R_{E1}) = 175(492\ \Omega) = 86.1\ \text{k}\Omega$$

Therefore,

$$R_{in(tot)} = 47\ \text{k}\Omega \| 10\ \text{k}\Omega \| 86.1\ \text{k}\Omega = 7.53\ \text{k}\Omega$$

The attenuation from source to base is

$$\text{Attenuation} = \frac{V_b}{V_s} = \frac{R_{in(tot)}}{R_s + R_{in(tot)}} = \frac{7.53\ \text{k}\Omega}{600\ \Omega + 7.53\ \text{k}\Omega} = 0.93$$

Before A_v can be determined, you must know the ac collector resistance R_c.

$$R_c = \frac{R_C R_L}{R_C + R_L} = \frac{(4.7\ \text{k}\Omega)(47\ \text{k}\Omega)}{4.7\ \text{k}\Omega + 47\ \text{k}\Omega} = 4.27\ \text{k}\Omega$$

The voltage gain from base to collector is

$$A_v \cong \frac{R_c}{R_{E1}} = \frac{4.27\ \text{k}\Omega}{470\ \Omega} = 9.09$$

The overall voltage gain is the attenuation times the amplifier voltage gain.

$$A'_v = \left(\frac{V_b}{V_s}\right) A_v = (0.93)(9.09) = 8.45$$

▲ **FIGURE 6–22**

AC equivalent for the circuit in Figure 6–20.

The source produces 10 mV rms, so the rms voltage at the collector is

$$V_c = A'_v V_{in} = (8.45)(10 \text{ mV}) = \mathbf{84.5 \text{ mV}}$$

Step 3: Collector voltage and output voltage waveforms

The total collector voltage is the signal voltage of 84.5 mV rms riding on a dc level of 4.74 V, as shown in Figure 6–23(a), where approximate peak values are determined as follows:

$$\text{Max } V_{c(p)} = 4.74 \text{ V} + (84.5 \text{ mV})(1.414) = 4.86 \text{ V}$$
$$\text{Min } V_{c(p)} = 4.74 \text{ V} - (84.5 \text{ mV})(1.414) = 4.62 \text{ V}$$

The coupling capacitor, C_3, keeps the dc level from getting to the output. So, V_{out} is equal to the ac portion of the collector voltage ($V_{out(p)} = 119$ mV), as indicated in Figure 6–23(b). The source voltage is shown to emphasize the phase inversion.

(a) Total collector voltage

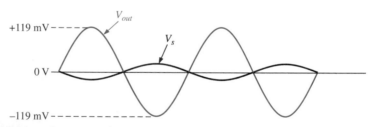

(b) Source and output ac voltages

▲ **FIGURE 6–23**

Voltages for Figure 6–20.

Related Problem   What is A_v in Figure 6–20 with R_L removed?

Open file E06-08 in the Examples folder on your CD-ROM. Measure the dc and the ac values of the collector voltage and compare with the calculated values.

Current Gain

The current gain from base to collector is I_c/I_b or β_{ac}. However, the overall current gain of the amplifier is

$$A_i = \frac{I_c}{I_s}$$

Equation 6–10

I_s is the total signal current produced by the source, part of which is base current and part of which is the signal current in the bias circuit ($R_1 \parallel R_2$), as shown in Figure 6–24. The total current signal produced by the source is

$$I_s = \frac{V_s}{R_{in(tot)} + R_s}$$

▶ **FIGURE 6–24**

Total input signal current (directions shown are for the positive half-cycle of V_s).

Power Gain

The power gain is the product of the overall voltage gain and the current gain.

$$A_p = A_v' A_i$$

Equation 6–11

SECTION 6–3 REVIEW

1. In the dc equivalent circuit of an amplifier, how are the capacitors treated?
2. When the emitter resistor is bypassed with a capacitor, how is the gain of the amplifier affected?
3. Explain swamping.
4. List the elements included in the total input resistance of a common-emitter amplifier.
5. What elements determine the overall voltage gain of a common-emitter amplifier?
6. When a load resistor is capacitively coupled to the collector of a CE amplifier, is the voltage gain increased or decreased?
7. What is the phase relationship of the input and output voltages of a CE amplifier?

6–4 THE COMMON-COLLECTOR AMPLIFIER

The common-collector (CC) amplifier is usually referred to as an emitter-follower (EF). The input is applied to the base through a coupling capacitor, and the output is at the emitter. The voltage gain of a CC amplifier is approximately 1, and its main advantages are its high input resistance and current gain.

After completing this section, you should be able to

- **Understand and analyze the operation of common-collector amplifiers**
  - Represent a CC amplifier by its dc and ac equivalent circuits
  - Analyze the dc and ac operation of a CC amplifier
  - Determine the voltage gain
  - Determine the input resistance
  - Determine the output resistance
  - Determine the current gain
  - Determine the power gain
  - Discuss the darlington pair and describe its main advantage

 An **emitter-follower** circuit with voltage-divider bias is shown in Figure 6–25. Notice that the input signal is capacitively coupled to the base, the output signal is capacitively coupled from the emitter, and the collector is at ac ground. There is no phase inversion, and the output is approximately the same amplitude as the input.

▲ FIGURE 6–25

Emitter-follower with voltage-divider bias.

Voltage Gain

As in all amplifiers, the voltage gain is $A_v = V_{out}/V_{in}$. The capacitive reactances are assumed to be negligible at the frequency of operation. For the emitter-follower, as shown in the ac equivalent model in Figure 6–26,

$$V_{out} = I_e R_e \quad \text{and} \quad V_{in} = I_e(r'_e + R_e)$$

Therefore, the voltage gain is

$$A_v = \frac{I_e R_e}{I_e(r'_e + R_e)}$$

The I_e current terms cancel, and the base-to-emitter voltage gain expression simplifies to

$$A_v = \frac{R_e}{r'_e + R_e}$$

where R_e is the parallel combination of R_E and R_L. If there is no load, then $R_e = R_E$. Notice that the gain is always less than 1. If $R_e >> r'_e$, then a good approximation is

Equation 6–12

$$A_v \cong 1$$

▶ **FIGURE 6–26**

Emitter-follower model for voltage gain derivation.

Since the output voltage is at the emitter, it is in phase with the base voltage, so there is *no inversion* from input to output. Because there is no inversion and because the voltage gain is approximately 1, the output voltage closely *follows* the input voltage in both phase and amplitude; thus the term *emitter-follower*.

Input Resistance

The emitter-follower is characterized by a high input resistance; this is what makes it a useful circuit. Because of the high input resistance, it can be used as a buffer to minimize loading effects when a circuit is driving a low resistance load. The derivation of the input resistance, looking in at the base of the common-collector amplifier, is similar to that for the common-emitter amplifier. In a common-collector circuit, however, the emitter resistor is *never* bypassed because the output is taken across R_e.

$$R_{in(base)} = \frac{V_{in}}{I_{in}} = \frac{V_b}{I_b} = \frac{I_e(r'_e + R_e)}{I_b}$$

Since $I_e \cong I_c = \beta_{ac}I_b$,

$$R_{in(base)} \cong \frac{\beta_{ac}I_b(r'_e + R_e)}{I_b}$$

The I_b terms cancel; therefore,

$$R_{in(base)} \cong \beta_{ac}(r'_e + R_e)$$

If $R_e >> r'_e$, then the input resistance at the base is simplified to

$$R_{in(base)} \cong \beta_{ac}R_e$$

Equation 6–13

The bias resistors in Figure 6–25 appear in parallel with $R_{in(base)}$, looking from the input source; and just as in the common-emitter circuit, the total input resistance is

$$R_{in(tot)} = R_1 \parallel R_2 \parallel R_{in(base)}$$

Output Resistance

With the load removed, the output resistance, looking into the emitter of the emitter-follower, is approximated as follows:

$$R_{out} \cong \left(\frac{R_s}{\beta_{ac}}\right) \parallel R_E$$

Equation 6–14

R_s is the resistance of the input source. The derivation of this expression is relatively involved and several simplifying assumptions have been made, as shown in Appendix B. The output resistance is very low, making the emitter-follower useful for driving low resistance loads.

Current Gain

The overall gain for the emitter-follower in Figure 6–25 is I_e/I_{in}. You can calculate I_{in} as $V_{in}/R_{in(tot)}$. If the resistance of the parallel combination of the voltage-divider bias resistors R_1 and R_2 is much greater than $R_{in(base)}$, then most of the input current goes into the base; thus, the current gain of the amplifier approaches the current gain of the transistor, β_{ac}, which is equal to I_c/I_b. This is because very little signal current is diverted to the bias resistors. Stated concisely, if

$$R_1 \parallel R_2 >> \beta_{ac}R_e$$

then

$$A_i \cong \beta_{ac}$$

Otherwise,

Equation 6–15

$$A_i = \frac{I_e}{I_{in}}$$

β_{ac} is the maximum achievable current gain in both common-collector and common-emitter amplifiers.

Power Gain

The common-collector power gain is the product of the voltage gain and the current gain. For the emitter-follower, the overall power gain is approximately equal to the current gain because the voltage gain is approximately 1.

$$A_p = A_v A_i$$

Since $A_v \cong 1$, the overall power gain is

Equation 6–16

$$A_p \cong A_i$$

EXAMPLE 6–9

Determine the total input resistance of the emitter-follower in Figure 6–27. Also find the voltage gain, current gain, and power gain in terms of power delivered to the load, R_L. Assume $\beta_{ac} = 175$ and that the capacitive reactances are negligible at the frequency of operation.

▲ **FIGURE 6–27**

Solution The ac emitter resistance external to the transistor is

$$R_e = R_E \parallel R_L = 1.0 \text{ k}\Omega \parallel 1.0 \text{ k}\Omega = 500 \text{ }\Omega$$

The approximate resistance, looking in at the base, is

$$R_{in(base)} \cong \beta_{ac}R_e = (175)(500 \text{ }\Omega) = 87.5 \text{ k}\Omega$$

The total input resistance is

$$R_{in(tot)} = R_1 \parallel R_2 \parallel R_{in(base)} = 18 \text{ k}\Omega \parallel 18 \text{ k}\Omega \parallel 87.5 \text{ k}\Omega = \mathbf{8.16 \text{ k}\Omega}$$

The voltage gain is $A_v \cong 1$. By using r'_e, you can determine a more precise value of A_v if necessary.

$$V_E = \left(\frac{R_2}{R_1 + R_2}\right)V_{CC} - V_{BE} = (0.5)(10 \text{ V}) - 0.7 \text{ V} = 4.3 \text{ V}$$

Therefore,

$$I_E = \frac{V_E}{R_E} = \frac{4.3 \text{ V}}{1.0 \text{ k}\Omega} = 4.3 \text{ mA}$$

and

$$r'_e \cong \frac{25 \text{ mV}}{I_E} = \frac{25 \text{ mV}}{4.3 \text{ mA}} = 5.8 \text{ }\Omega$$

So,

$$A_v = \frac{R_e}{r'_e + R_e} = \frac{500 \text{ }\Omega}{505.8 \text{ }\Omega} = \mathbf{0.989}$$

The small difference in A_v as a result of considering r'_e is insignificant in most cases.
   The current gain is $A_i = I_e/I_{in}$. The calculations are as follows:

$$I_e = \frac{V_e}{R_e} = \frac{A_v V_b}{R_e} \cong \frac{1 \text{ V}}{500 \text{ }\Omega} = 2 \text{ mA}$$

$$I_{in} = \frac{V_{in}}{R_{in(tot)}} = \frac{1 \text{ V}}{8.16 \text{ k}\Omega} = 123 \text{ }\mu\text{A}$$

$$A_i = \frac{I_e}{I_{in}} = \frac{2 \text{ mA}}{123 \text{ }\mu\text{A}} = \mathbf{16.3}$$

The overall power gain is

$$A_p \cong A_i = 16.3$$

Since $R_L = R_E$, one-half of the total power is dissipated in R_L. So in terms of power to the load, the power gain is one-half of the overall power gain.

$$A_{p(load)} = \frac{A_p}{2} = \frac{16.3}{2} = \mathbf{8.15}$$

Related Problem If R_L in Figure 6–27 is decreased in value, does power gain to the load increase or decrease?

Open file E06-09 in the Examples folder on your CD-ROM. Measure the voltage gain and compare with the calculated value.

The Darlington Pair

As you have seen, β_{ac} is a major factor in determining the input resistance of an amplifier. The β_{ac} of the transistor limits the maximum achievable input resistance you can get from a given emitter-follower circuit.

One way to boost input resistance is to use a **darlington pair**, as shown in Figure 6–28. The collectors of two transistors are connected, and the emitter of the first drives the base of the second. This configuration achieves β_{ac} multiplication as shown in the following steps. The emitter current of the first transistor is

$$I_{e1} \cong \beta_{ac1}I_{b1}$$

This emitter current becomes the base current for the second transistor, producing a second emitter current of

$$I_{e2} \cong \beta_{ac2}I_{e1} = \beta_{ac1}\beta_{ac2}I_{b1}$$

Therefore, the effective current gain of the darlington pair is

$$\beta_{ac} = \beta_{ac1}\beta_{ac2}$$

Neglecting r'_e by assuming that it is much smaller than R_E, the input resistance is

Equation 6–17

$$R_{in} = \beta_{ac1}\beta_{ac2}R_E$$

▶ **FIGURE 6–28**

A darlington pair multiplies β_{ac}.

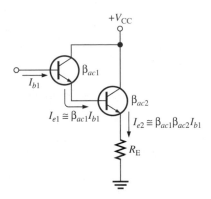

An Application

The emitter-follower is often used as an interface between a circuit with a high output resistance and a low-resistance load. In such an application, the emitter-follower is called a *buffer.*

For example, suppose a common-emitter amplifier with a 1.0 kΩ collector resistance (output resistance) must drive a low-resistance load such as an 8 Ω low-power speaker. If the speaker is capacitively coupled to the output of the amplifier, the 8 Ω load appears— to the ac signal—in parallel with the 1.0 kΩ collector resistor. This results in an ac collector resistance of

$$R_c = R_C \| R_L = 1.0 \text{ k}\Omega \| 8 \text{ }\Omega = 7.94 \text{ }\Omega$$

Obviously, this is not acceptable because most of the voltage gain is lost ($A_v = R_c/r'_e$). For example, if $r'_e = 5 \text{ }\Omega$, the voltage gain is reduced from

$$A_v = \frac{R_C}{r'_e} = \frac{1.0 \text{ k}\Omega}{5 \text{ }\Omega} = 200$$

with no load to

$$A_v = \frac{R_c}{r'_e} = \frac{7.94\ \Omega}{5\ \Omega} = 1.59$$

with an 8 Ω speaker load.

An emitter-follower using a darlington pair can be used to interface the amplifier and the speaker, as shown in Figure 6–29.

Common-emitter amplifier   Darlington emitter-follower   Load

▲ **FIGURE 6–29**

A darlington emitter-follower used as a buffer between a common-emitter amplifier and a low-resistance load such as a speaker.

EXAMPLE 6–10

In Figure 6–29 for the common-emitter amplifier, $V_{CC} = 12$ V, $R_C = 1.0$ kΩ and $r'_e = 5\ \Omega$. For the darlington emitter-follower, $R_1 = 10$ kΩ, $R_2 = 22$ kΩ, $R_E = 22\ \Omega$, $R_L = 8\ \Omega$, $V_{CC} = 12$ V, and $\beta_{DC} = \beta_{ac} = 100$ for each transistor.

(a) Determine the voltage gain of the common-emitter amplifier.

(b) Determine the voltage gain of the darlington emitter-follower.

(c) Determine the overall voltage gain and compare to the gain of the common-emitter amplifier driving the speaker directly without the darlington emitter-follower.

Solution (a) To determine A_v for the common-emitter amplifier, first find r'_e for the darlington emitter-follower.

$$V_B = \left(\frac{R_2 \| \beta^2_{DC}R_E}{R_1 + R_2 \| \beta^2_{DC}R_E}\right)V_{CC} = \left(\frac{20\text{ k}\Omega}{30\text{ k}\Omega}\right)12\text{ V} = 8.0\text{ V}$$

$$I_E = \frac{V_E}{R_E} = \frac{V_B - 2V_{BE}}{R_E} = \frac{8.0\text{ V} - 1.4\text{ V}}{22\ \Omega} = \frac{6.6\text{ V}}{22\ \Omega} = 300\text{ mA}$$

$$r'_e = \frac{25\text{ mV}}{I_E} = \frac{25\text{ mV}}{300\text{ mA}} = 83\text{ m}\Omega$$

Note that R_E must dissipate a power of

$$P_{R_E} = I^2_E R_E = (300\text{ mA})^2\, 22\ \Omega = 1.98\text{ W}$$

and transistor Q_2 must dissipate

$$P_{Q2} = (V_{CC} - V_E)I_E = (5.4 \text{ V})(300 \text{ mA}) = 1.62 \text{ W}$$

Next, the ac emitter resistance of the darlington emitter-follower is

$$R_e = R_E \| R_L = 22 \text{ Ω} \| 8 \text{ Ω} = 5.87 \text{ Ω}$$

The total input resistance of the darlington emitter-follower is

$$R_{in(tot)} = R_1 \| R_2 \| \beta_{ac}^2(r'_e + R_e)$$
$$= 10 \text{ kΩ} \| 22 \text{ kΩ} \| 100^2(83 \text{ mΩ} + 5.87 \text{ Ω}) = 6.16 \text{ kΩ}$$

The effective ac collector resistance of the common-emitter amplifier is

$$R_c = R_C \| R_{in(tot)} = 1.0 \text{ kΩ} \| 6.16 \text{ kΩ} = 860 \text{ Ω}$$

The voltage gain of the common-emitter amplifier is

$$A_v = \frac{R_c}{r'_e} = \frac{860 \text{ Ω}}{5 \text{ Ω}} = \mathbf{172}$$

(b) The effective ac emitter resistance was found in part (a) to be 5.87 Ω. The voltage gain for the darlington emitter-follower is

$$A_v = \frac{R_e}{r'_e + R_e} = \frac{5.87 \text{ Ω}}{83 \text{ mΩ} + 5.87 \text{ Ω}} = \mathbf{0.99}$$

(c) The overall voltage gain is

$$A'_v = A_{v(EF)}A_{v(CE)} = (0.99)(172) = \mathbf{170}$$

If the common-emitter amplifier drives the speaker directly, the gain is 1.59 as we previously calculated.

Related Problem   Using the same circuit values, determine the voltage gain of the common-emitter amplifier in Figure 6–29 if a single transistor is used in the emitter-follower in place of the darlington pair. Assume $\beta_{DC} = \beta_{ac} = 100$. Explain the difference in the voltage gain without the darlington pair.

SECTION 6-4
REVIEW

1. What is a common-collector amplifier called?
2. What is the ideal maximum voltage gain of a common-collector amplifier?
3. What characteristic of the common-collector amplifier makes it a useful circuit?

6-5   THE COMMON-BASE AMPLIFIER

The common-base (CB) amplifier provides high voltage gain with a maximum current gain of 1. Since it has a low input resistance, the CB amplifier is the most appropriate type for certain applications where sources tend to have very low-resistance outputs.

After completing this section, you should be able to

- **Understand and analyze the operation of common-base amplifiers**
- Represent a CB amplifier by its dc and ac equivalent circuits
- Analyze the dc and ac operation of a CB amplifier
- Determine the voltage gain
- Determine the input resistance
- Determine the output resistance
- Determine the current gain
- Determine the power gain

A typical **common-base** amplifier is shown in Figure 6–30. The base is the common terminal and is at ac ground because of capacitor C_2. The input signal is capacitively coupled to the emitter. The output is capacitively coupled from the collector to a load resistor.

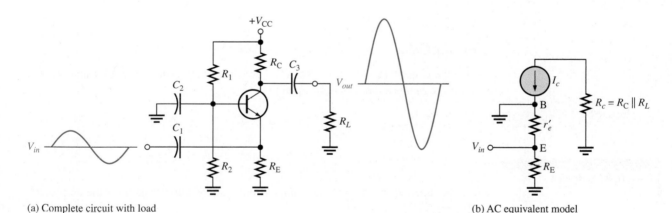

(a) Complete circuit with load

(b) AC equivalent model

▲ **FIGURE 6–30**

Common-base amplifier with voltage-divider bias.

Voltage Gain

The voltage gain from emitter to collector is developed as follows ($V_{in} = V_e$, $V_{out} = V_c$).

$$A_v = \frac{V_{out}}{V_{in}} = \frac{V_c}{V_e} = \frac{I_c R_c}{I_e(r'_e \| R_E)} \cong \frac{I_e R_c}{I_e(r'_e \| R_E)}$$

If $R_E \gg r'_e$, then

$$A_v \cong \frac{R_c}{r'_e}$$

Equation 6–18

where $R_c = R_C \| R_L$. Notice that the gain expression is the same as for the common-emitter amplifier. However, there is no phase inversion from emitter to collector.

Input Resistance

The resistance, looking in at the emitter, is

$$R_{in(emitter)} = \frac{V_{in}}{I_{in}} = \frac{V_e}{I_e} = \frac{I_e(r'_e \| R_E)}{I_e}$$

If $R_E \gg r'_e$, then

Equation 6–19

$$R_{in(emitter)} \cong r'_e$$

R_E is typically much greater than r'_e, so the assumption that $r'_e \| R_E \cong r'_e$ is usually valid.

Output Resistance

Looking into the collector, the ac collector resistance, r'_c, appears in parallel with R_C. As you have previously seen in connection with the CE amplifier, r'_c is typically much larger than R_C, so a good approximation for the output resistance is

Equation 6–20

$$R_{out} \cong R_C$$

Current Gain

The current gain is the output current divided by the input current. I_c is the ac output current, and I_e is the ac input current. Since $I_c \cong I_e$, the current gain is approximately 1.

Equation 6–21

$$A_i \cong 1$$

Power Gain

Since the current gain is approximately 1 for the common-base amplifier and $A_p = A_v A_i$, the power gain is approximately equal to the voltage gain.

Equation 6–22

$$A_p \cong A_v$$

EXAMPLE 6–11

Find the input resistance, voltage gain, current gain, and power gain for the amplifier in Figure 6–31. $\beta_{DC} = 250$.

▲ **FIGURE 6–31**

Solution   First, find I_E so that you can determine r'_e. Then $R_{in} \cong r'_e$. Since $\beta_{DC}R_E >> R_2$, then

$$V_B \cong \left(\frac{R_2}{R_1 + R_2}\right)V_{CC} = \left(\frac{12 \text{ k}\Omega}{68 \text{ k}\Omega}\right)10 \text{ V} = 1.76 \text{ V}$$

$$V_E = V_B - 0.7 \text{ V} = 1.76 \text{ V} - 0.7 \text{ V} = 1.06 \text{ V}$$

$$I_E = \frac{V_E}{R_E} = \frac{1.06 \text{ V}}{1.0 \text{ k}\Omega} = 1.06 \text{ mA}$$

Therefore,

$$R_{in} \cong r'_e = \frac{25 \text{ mV}}{I_E} = \frac{25 \text{ mV}}{1.06 \text{ mA}} = \mathbf{23.6 \ \Omega}$$

The voltage gain is found as follows:

$$R_c = R_C \| R_L = 2.2 \text{ k}\Omega \| 10 \text{ k}\Omega = 1.8 \text{ k}\Omega$$

$$A_v = \frac{R_c}{r'_e} = \frac{1.8 \text{ k}\Omega}{23.6 \ \Omega} = \mathbf{76.3}$$

Also, $A_i \cong \mathbf{1}$ and $A_p \cong A_v = \mathbf{76.3}$.

Related Problem   Find A_v in Figure 6–31 if $\beta_{DC} = 50$.

> Open file E06-11 in the Examples folder on your CD-ROM. Measure the voltage gain and compare with the calculated value.

SECTION 6–5 REVIEW

1. Can the same voltage gain be achieved with a common-base as with a common-emitter amplifier?
2. Does the common-base amplifier have a low or a high input resistance?
3. What is the maximum current gain in a common-base amplifier?

6–6   MULTISTAGE AMPLIFIERS

Two or more amplifiers can be connected in a cascaded arrangement with the output of one amplifier driving the input of the next. Each amplifier in a cascaded arrangement is known as a stage. The basic purpose of a multistage arrangement is to increase the overall voltage gain. Although discrete multistage amplifiers are not as common as they once were, a familiarization with this area provides insight into how some circuits affect each other when they are connected together.

After completing this section, you should be able to

- ■ **Discuss multistage amplifiers and analyze their operation**
- ■ Determine multistage voltage gain
- ■ Express the voltage gain in decibels (dB)
- ■ Determine the loading effects in a multistage amplifier
- ■ Analyze each stage to determine the overall voltage gain
- ■ Discuss capacitive coupling in multistage amplifiers
- ■ Describe a basic direct-coupled multistage amplifier
- ■ Describe a basic transformer-coupled multistage amplifier

Multistage Voltage Gain

The overall voltage gain, A'_v, of **cascaded** amplifiers, as shown in Figure 6–32, is the product of the individual voltage gains.

Equation 6–23

$$A'_v = A_{v1}A_{v2}A_{v3} \cdots A_{vn}$$

where n is the number of **stages**.

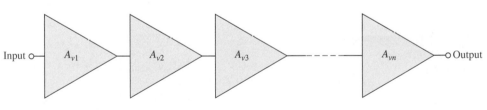

▲ **FIGURE 6–32**

Cascaded amplifiers. Each triangular symbol represents a separate amplifier.

Voltage Gain Expressed in Decibels

Amplifier voltage gain is often expressed in **decibels** (dB) as follows:

Equation 6–24

$$A_{v(dB)} = 20 \log A_v$$

This is particularly useful in **multistage** systems because the overall voltage gain in dB is the *sum* of the individual voltage gains in dB.

Equation 6–25

$$A'_{v(dB)} = A_{v1(dB)} + A_{v2(dB)} + \cdots + A_{vn(dB)}$$

EXAMPLE 6–12

A certain cascaded amplifier arrangement has the following voltage gains: $A_{v1} = 10$, $A_{v2} = 15$, and $A_{v3} = 20$. What is the overall voltage gain? Also express each gain in decibels (dB) and determine the total voltage gain in dB.

Solution

$$A'_v = A_{v1}A_{v2}A_{v3} = (10)(15)(20) = \mathbf{3000}$$
$$A_{v1(dB)} = 20 \log 10 = \mathbf{20.0\ dB}$$
$$A_{v2(dB)} = 20 \log 15 = \mathbf{23.5\ dB}$$
$$A_{v3(dB)} = 20 \log 20 = \mathbf{26.0\ dB}$$
$$A'_{v(dB)} = 20.0\ dB + 23.5\ dB + 26.0\ dB = \mathbf{69.5\ dB}$$

Related Problem

In a certain multistage amplifier, the individual stages have the following voltage gains: $A_{v1} = 25$, $A_{v2} = 5$, and $A_{v3} = 12$. What is the overall gain? Express each gain in dB and determine the total voltage gain in dB.

Multistage Amplifier Analysis

For purposes of illustration, we will use the two-stage capacitively coupled amplifier in Figure 6–33. Notice that both stages are identical common-emitter amplifiers with the output of the first stage capacitively coupled to the input of the second stage. Capacitive coupling prevents the dc bias of one stage from affecting that of the other but allows the ac signal to pass without attenuation because $X_C \cong 0\ \Omega$ at the frequency of operation. Notice, also, that the transistors are labeled Q_1 and Q_2.

$$\beta_{DC} = \beta_{ac} = 150, \text{ for } Q_1 \text{ and } Q_2$$

▲ **FIGURE 6–33**

A two-stage common–emitter amplifier.

Loading Effects In determining the voltage gain of the first stage, you must consider the loading effect of the second stage. Because the coupling capacitor C_3 effectively appears as a short at the signal frequency, the total input resistance of the second stage presents an ac load to the first stage.

Looking from the collector of Q_1, the two biasing resistors in the second stage, R_5 and R_6, appear in parallel with the input resistance at the base of Q_2. In other words, the signal at the collector of Q_1 "sees" R_3, R_5, R_6, and $R_{in(base2)}$ of the second stage all in parallel to ac ground. Thus, the effective ac collector resistance of Q_1 is the total of all these resistances in parallel, as Figure 6–34 illustrates. The voltage gain of the first stage is reduced by the loading of the second stage because the effective ac collector resistance of the first stage is less than the actual value of its collector resistor, R_3. Remember that $A_v = R_c/r'_e$.

▲ **FIGURE 6–34**

AC equivalent of first stage in Figure 6–33, showing loading from second stage input resistance.

Voltage Gain of the First Stage   The ac collector resistance of the first stage is

$$R_{c1} = R_3 \parallel R_5 \parallel R_6 \parallel R_{in(base2)}$$

Remember that lowercase italic subscripts denote ac quantities such as for R_c.

You can verify that $I_E = 1.05$ mA, $r'_e = 23.8\ \Omega$, and $R_{in(base2)} = 3.57$ kΩ. The effective ac collector resistance of the first stage is as follows:

$$R_{c1} = 4.7\ \text{k}\Omega \parallel 47\ \text{k}\Omega \parallel 10\ \text{k}\Omega \parallel 3.57\ \text{k}\Omega = 1.63\ \text{k}\Omega$$

Therefore, the base-to-collector voltage gain of the first stage is

$$A_{v1} = \frac{R_{c1}}{r'_e} = \frac{1.63\ \text{k}\Omega}{23.8\ \Omega} = 68.5$$

Voltage Gain of the Second Stage   The second stage has no load resistor, so the ac collector resistance is R_7, and the gain is

$$A_{v2} = \frac{R_7}{r'_e} = \frac{4.7\ \text{k}\Omega}{23.8\ \Omega} = 197$$

Compare this to the gain of the first stage, and notice how much the loading from the second stage reduced the gain.

Overall Voltage Gain   The overall amplifier gain with no load on the output is

$$A'_v = A_{v1}A_{v2} = (68.5)(197) \cong 13,495$$

If an input signal of 100 μV, for example, is applied to the first stage and if there is no attenuation in the input base circuit due to the source resistance, an output from the second stage of (100 μV)(13,495) $\cong$ 1.35 V will result. The overall voltage gain can be expressed in dB as follows:

$$A'_{v(\text{dB})} = 20 \log (13,495) = 82.6\ \text{dB}$$

DC Voltages in the Capacitively Coupled Multistage Amplifier   Since both stages in Figure 6–33 are identical, the dc voltages for Q_1 and Q_2 are the same. Since $\beta_{DC}R_4 \gg R_2$ and $\beta_{DC}R_8 \gg R_6$, the dc base voltage for Q_1 and Q_2 is

$$V_B \cong \left(\frac{R_2}{R_1 + R_2}\right)V_{CC} = \left(\frac{10\ \text{k}\Omega}{57\ \text{k}\Omega}\right)10\ \text{V} = 1.75\ \text{V}$$

The dc emitter and collector voltages are as follows:

$$V_E = V_B - 0.7 \text{ V} = 1.05 \text{ V}$$

$$I_E = \frac{V_E}{R_4} = \frac{1.05 \text{ V}}{1.0 \text{ k}\Omega} = 1.05 \text{ mA}$$

$$I_C \cong I_E = 1.05 \text{ mA}$$

$$V_C = V_{CC} - I_C R_3 = 10 \text{ V} - (1.05 \text{ mA})(4.7 \text{ k}\Omega) = 5.07 \text{ V}$$

Direct-Coupled Multistage Amplifiers

A basic two-stage, direct-coupled amplifier is shown in Figure 6–35. Notice that there are no coupling or bypass capacitors in this circuit. The dc collector voltage of the first stage provides the base-bias voltage for the second stage. Because of the direct coupling, this type of amplifier has a better low-frequency response than the capacitively coupled type in which the reactance of coupling and bypass capacitors at very low frequencies may become excessive. The increased reactance of capacitors at lower frequencies produces gain reduction in capacitively coupled amplifiers.

▶ **FIGURE 6–35**

A basic two-stage direct-coupled amplifier.

Direct-coupled amplifiers, on the other hand, can be used to amplify low frequencies all the way down to dc (0 Hz) without loss of voltage gain because there are no capacitive reactances in the circuit. The disadvantage of direct-coupled amplifiers is that small changes in the dc bias voltages from temperature effects or power-supply variation are amplified by the succeeding stages, which can result in a significant drift in the dc levels throughout the circuit.

**SECTION 6–6
REVIEW**

1. What does the term *stage* mean?
2. How is the overall voltage gain of a multistage amplifier determined?
3. Express a voltage gain of 500 in dB.
4. Discuss a disadvantage of a capacitively coupled amplifier.

6-7 TROUBLESHOOTING

In working with any circuit, you must first know how it is supposed to work before you can troubleshoot it for a failure. The two-stage capacitively coupled amplifier discussed in Section 6–6 is used to illustrate a typical troubleshooting procedure.

After completing this section, you should be able to

■ **Troubleshoot amplifier circuits**

■ Discuss the complete troubleshooting process

■ Apply the troubleshooting process to a two-stage amplifier

■ Use the signal-tracing method

■ Apply fault analysis

At this point, you should review the general troubleshooting techniques discussed in Chapter 2. The two-stage common-emitter amplifier that was discussed in Section 6–6 is used to illustrate basic multistage amplifier troubleshooting.

When you are faced with having to troubleshoot a circuit, the first thing you need is a schematic with the proper dc and signal voltages labeled. You must know what the correct voltages in the circuit should be before you can identify an incorrect voltage. Schematics of some circuits are available with voltages indicated at certain points. If this is not the case, you must use your knowledge of the circuit operation to determine the correct voltages. Figure 6–36 is the schematic for the two-stage amplifier that was analyzed in the previous section. The correct voltages are indicated at each point.

The Complete Troubleshooting Process

Beginning with a malfunctioning or nonfunctioning circuit or system, a typical troubleshooting procedure may go as follows:

1. Identify the symptom(s).

2. Perform a power check.

3. Perform a sensory check.

4. Apply a signal-tracing technique to isolate the fault to a single circuit.

5. Apply fault analysis to isolate the fault further to a single component or group of components.

6. Use replacement or repair to fix the problem.

Applying the Troubleshooting Process to a Two-Stage Amplifier   To determine the faulty component in a multistage amplifier, use the general six-step troubleshooting procedure which is illustrated as follows.

Step 1: *Check the input and output voltages.* Assume the measurements indicate that the input signal voltage is correct. However, there is no output signal voltage or the output signal voltage is much less than it should be, as shown by the diagram in Figure 6–37.

Step 2: *Perform a power check.* Assume the dc supply voltage is correct as indicated in Figure 6–37.

Step 3: *Perform a sensory check.* Assume there are no detectable signs of a fault.

▲ **FIGURE 6–36**

A two-stage common-emitter amplifier with correct voltages indicated. Both transistors have betas of 150. Different values of β will produce slightly different results.

▲ **FIGURE 6–37**

Initial check of a faulty two-stage amplifiier.

Step 4: *Apply the half-splitting method of signal tracing.* Check the voltages at the output of the first stage. No signal voltage or a much less than normal signal voltage indicates that the problem is in the first stage. An incorrect dc voltage also indicates a first-stage problem. If the signal voltage and the dc voltage are correct at the output of the first stage, the problem is in the second stage. After this check, you have narrowed the problem to one of the two stages. This step is illustrated in Figure 6–38.

Step 5: *Apply fault analysis.* Focus on the faulty stage and determine the component failure that can produce the incorrect output.

Symptom: DC voltages incorrect.
Faults: A failure of any resistor or the transistor will produce an incorrect dc bias voltage. A leaky bypass or coupling capacitor will also affect the dc bias voltages. Further measurements in the stage are necessary to isolate the faulty component.

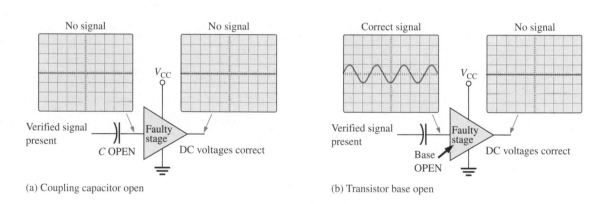

▲ **FIGURE 6–38**

Half-splitting signal tracing isolates the faulty stage.

(a) Coupling capacitor open

(b) Transistor base open

(c) Bypass capacitor open

▲ **FIGURE 6–39**

Troubleshooting a faulty stage.

Incorrect ac voltages and the most likely fault(s) are illustrated in Figure 6–39 as follows:

(a) *Symptom 1:* Signal voltage at output missing; dc voltage correct.
Symptom 2: Signal voltage at base missing; dc voltage correct.
Fault: Input coupling capacitor open. This prevents the signal from getting to the base.

(b) *Symptom:* Correct signal at base but no output signal.
Fault: Transistor base open.

(c) *Symptom:* Signal voltage at output much less than normal; dc voltage correct.
Fault: Bypass capacitor open.

Step 6: *Replace or repair.* With the power turned off, replace the defective component or repair the defective connection. Turn on the power, and check for proper operation.

EXAMPLE 6–13

The two-stage amplifier in Figure 6–36 has malfunctioned. Specify the step-by-step troubleshooting procedure for an assumed fault.

Solution   The troubleshooting procedure for a certain fault scenario is as follows:

Step 1: There is a verified input signal voltage, but no output signal voltage is measured.

Step 2: There is power to the circuit as indicated by a correct V_{CC} measurement.

Step 3: Assume there are no visual or other indications of a problem such as a charred resistor, solder splash, wire clipping, broken connection, or extremely hot component.

Step 4: The signal voltage and the dc voltage at the collector of Q_1 are correct. This means that the problem is in the second stage or the coupling capacitor C_3 between the stages.

Step 5: The correct signal voltage and dc bias voltage are measured at the base of Q_2. This eliminates the possibility of a fault in C_3 or the second stage bias circuit.

The collector of Q_2 is at 10 V and there is no signal voltage. This measurement, made directly on the transistor collector, indicates that either the collector is shorted to V_{CC} or the transistor is internally open.

A visual inspection reveals no physical shorts between V_{CC} and the collector connection. It is unlikely that the collector resistor R_7 is shorted but to verify, turn off the power and use an ohmmeter to check.

The possibility of a short is eliminated by the ohmmeter check. The other possible faults are (a) transistor Q_2 internally open or (b) emitter resistor or connection open. Use a transistor tester and/or ohmmeter to check each of these possible faults with power off.

Step 6: Replace the faulty component or repair open connection and retest the circuit for proper operation.

Related Problem   Determine the possible fault(s) if, in Step 5, you find no signal voltage at the base of Q_2 but the dc voltage is correct.

EWB/Multisim Troubleshooting Exercises

These file circuits are in the Troubleshooting Exercises folder on your CD-ROM.

1. Open file TSE06-01. Determine if the circuit is working properly and, if not, determine the fault.

2. Open file TSE06-02. Determine if the circuit is working properly and, if not, determine the fault.

3. Open file TSE06-03. Determine if the circuit is working properly and, if not, determine the fault.

4. Open file TSE06-04. Determine if the circuit is working properly and, if not, determine the fault.

SECTION 6–7 REVIEW

1. If C_4 in Figure 6–36 were open, how would the output signal be affected? How would the dc level at the collector of Q_2 be affected?

2. If R_5 in Figure 6–36 were open, how would the output signal be affected?

3. If the coupling capacitor C_3 in Figure 6–36 shorted out, would any of the dc voltages in the amplifier be changed? If so, which ones?

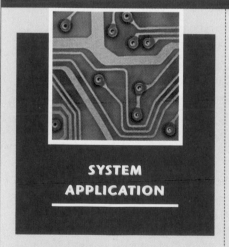

SYSTEM APPLICATION

Let's assume that you work for a company that develops, manufactures, and markets public address and voice paging systems for use in service garages, supermarkets, department stores, and the like. You are responsible for the most basic product in this line, which is a simple preamplifier with a microphone input followed by a power amplifier and a speaker. You will apply the knowledge you have gained in this chapter in completing your assignment.

The Public Address System

This system includes a magnetic microphone, a two-stage preamplifier, a power amplifier, and a speaker. This design is the most basic public address system that your company plans to produce and market and is intended for limited application in small businesses for paging employees or making announcements.

A block diagram of the system is shown in Figure 6–40. The microphone input goes to the preamplifier, which has an adjustable volume control. The output of the preamplifier goes to a power amplifier that drives a horn speaker. A dc power supply provides the required dc voltage for the circuits and is included as part of the system.

The System Requirements

■ The magnetic microphone has an output impedance of 30 Ω and produces an average output voltage of 2 mV rms

in response to a typical speaking voice. The frequency response is from 100 Hz to 8,000 Hz.

■ The speaker has an impedance of 8 Ω and can handle up to 15 W. The frequency response is from 100 Hz to 7500 Hz.

■ The frequency range of the amplifier is to be at least 300 Hz to 5000 Hz.

Basic Operation The schematic of the two-stage preamplifier is shown in Figure 6–41. The first stage is a common-base amplifier, and the second stage is a common-emitter amplifier with a swamping resistor (R_8) and an emitter resistor (R_9) with variable bypass for volume control.

The output of the amplifier will connect to a power amplifier to be added later in the development process (Chapter 9). For now, the assignment is to check out the preamplifier, which is in a preproduction phase.

Analysis of the Preamplifier Circuit

■ Determine the input resistance of the first stage.

■ For each milliwatt of power produced by the microphone, how much is transferred to the amplifier?

■ Determine the dc voltages at the base, collector, and emitter of Q_1 and Q_2.

■ Determine the overall voltage gain of the amplifier.

■ Calculate the current that is drawn from the dc power supply when there is no input signal to the amplifier.

■ Specify the minimum standard power rating for each resistor.

■ *Optional analysis:* Determine the lowest frequency that the circuit can amplify without the gain decreasing more than 3 dB. The coupling capacitors and the input resistances of the amplifier stages determine the low-frequency response.

The Power Supply Circuit

The dc power supply circuit is the same as the one developed in the system

application for Chapter 2 except that it must produce a regulated voltage of approximately +12 V. Specify the changes necessary to adapt the power supply to this system application.

The Printed Circuit Board

■ Check out the printed circuit board in Figure 6–42 to verify that it is correct according to the schematic.

■ Label a copy of the board with component and input/output designations in agreement with the schematic.

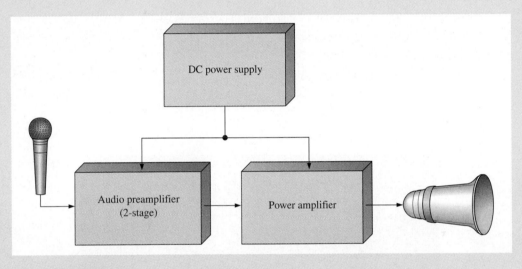

▲ **FIGURE 6–40**

Basic public address/paging system block diagram.

▲ **FIGURE 6–41**

Schematic of the preamplifier circuit. Both transistors are 2N3904. Assume $\beta_{DC} = \beta_{ac} = 100$.

Test Procedure

- Develop a step-by-step set of instructions on how to check the preamplifier circuit board for proper operation at a test frequency of 5 kHz using the test points (circled numbers) indicated in the test bench setup of Figure 6–43. A function generator is the signal source with a potentiometer used as an attenuator, if necessary, to reduce the signal to a value small enough to simulate the microphone input.

- Specify voltage values for all the measurements to be made. Provide a fault analysis for all possible component failures.

Troubleshooting

Several prototype boards have been assembled and are ready for testing. Based on the sequence of test bench measurements for each board indicated in Figure 6–44, determine the most likely fault in each case. The circled numbers indicate test point connections to the circuit board.

Final Report (Optional)

Submit a final written report on the audio preamplifier circuit board using an organized format that includes the following:

1. A physical description of the circuits.

2. A discussion of the operation of the circuit.

3. A list of the specifications.

4. A list of parts with part numbers if available.

5. A list of the types of problems on the three faulty circuit boards.

6. A complete description of how you determined the problem on each of the faulty circuit boards.

▲ **FIGURE 6–42**

Preamplifier circuit board.

Sinusoidal output from
function generator

+12 V from power supply

Power supply ground

▲ FIGURE 6–43

Test bench setup for the preamplifier board.

▲ FIGURE 6–44

Test results for three faulty circuit boards.

SUMMARY OF THE COMMON-EMITTER AMPLIFIER

CIRCUIT WITH VOLTAGE-DIVIDER BIAS

- Input is at the base. Output is at the collector.

- There is a phase inversion from input to output.

- C_1 and C_3 are coupling capacitors for the input and output signals.

- C_2 is the emitter-bypass capacitor.

- All capacitors must have a negligible reactance at the frequency of operation.

- Emitter is at ac ground due to the bypass capacitor.

EQUIVALENT CIRCUITS AND FORMULAS

DC equivalent

- DC formulas:

$$V_B = \left(\frac{R_2 \| \beta_{DC} R_E}{R_1 + R_2 \| \beta_{DC} R_E} \right) V_{CC}$$

$$V_E = V_B - V_{BE}$$

$$I_E = \frac{V_E}{R_E}$$

$$V_C = V_{CC} - I_C R_C$$

AC equivalent

- AC formulas:

$$r'_e = \frac{25\ \text{mV}}{I_E}$$

$$R_{in(base)} = \beta_{ac} r'_e$$

$$R_{out} \cong R_C$$

$$A_v = \frac{R_C}{r'_e} \quad \text{(base to collector)}$$

$$A'_v = \left(\frac{V_b}{V_s} \right) A_v \quad \text{(overall)}$$

$$A_i = \frac{I_c}{I_{in}}$$

$$A_p = A'_v A_i$$

SUMMARY OF THE COMMON-EMITTER AMPLIFIER, *continued*

SWAMPED AMPLIFIER WITH RESISTIVE LOAD

Swamping resistor

AC equivalent

- AC formulas:

$$A_v \cong \frac{R_C \| R_L}{R_{E1}}$$

$$R_{in(base)} = \beta_{ac}(r_e' + R_{E1})$$

- Swamping stabilizes gain by minimizing the effect of r_e'.

- Swamping reduces the voltage gain from its unswamped value.

- Swamping increases input resistance.

- The load resistance reduces the voltage gain. The smaller the load resistance, the less the gain.

SUMMARY OF THE COMMON-COLLECTOR AMPLIFIER

CIRCUIT WITH VOLTAGE-DIVIDER BIAS

- Input is at the base. Output is at the emitter.

- There is no phase inversion from input to output.

- Input resistance is high. Output resistance is low.

- Maximum voltage gain is 1.

- Collector is at ac ground.

- Capacitors must have a negligible reactance at the frequency of operation.

EQUIVALENT CIRCUITS AND FORMULAS

DC equivalent

- DC formulas:

$$V_B = \left(\frac{R_2 \| \beta_{DC} R_E}{R_1 + R_2 \| \beta_{DC} R_E} \right) V_{CC}$$

$$V_E = V_B - V_{BE}$$

$$I_E = \frac{V_E}{R_E}$$

$$V_C = V_{CC}$$

AC equivalent

- AC formulas:

$$r'_e = \frac{25 \text{ mV}}{I_E}$$

$$R_{in(base)} = \beta_{ac}(r'_e + R_e) \cong \beta_{ac} R_e$$

$$R_{out} = \left(\frac{R_s}{\beta_{ac}} \right) \| R_E$$

$$A_v = \frac{R_e}{r'_e + R_e} \cong 1$$

$$A_i = \frac{I_e}{I_{in}}$$

$$A_p = A_i$$

SUMMARY OF COMMON-BASE AMPLIFIER

CIRCUIT WITH VOLTAGE-DIVIDER BIAS

- Input is at the emitter. Output is at the collector.
- There is no phase inversion from input to output.
- Input resistance is low. Output resistance is high.
- Maximum current gain is 1.
- Base is at ac ground.

EQUIVALENT CIRCUITS AND FORMULAS

DC equivalent

- DC formulas:

$$V_B = \left(\frac{R_2 \| \beta_{DC} R_E}{R_1 + R_2 \| \beta_{DC} R_E} \right) V_{CC}$$

$$V_E = V_B - V_{BE}$$

$$I_E = \frac{V_E}{R_E}$$

$$V_C = V_{CC} - I_C R_C$$

AC equivalent

- AC formulas:

$$r_e' = \frac{25 \text{ mV}}{I_E}$$

$$R_{in(emitter)} \cong r_e'$$

$$R_{out} \cong R_C$$

$$A_v \cong \frac{R_c}{r_e'}$$

$$A_i \cong 1$$

$$A_p \cong A_v$$

CHAPTER SUMMARY

- A small-signal amplifier uses only a small portion of its load line under signal conditions.
- *r* parameters are easily identifiable and applicable with a transistor's circuit operation.
- *h* parameters are important to technicians and technologists because manufacturers' data sheets specify transistors using *h* parameters.
- A common-emitter amplifier has good voltage, current, and power gains, but a relatively low input resistance.
- A common-collector amplifier has high input resistance and good current gain, but its voltage gain is approximately 1.
- The common-base amplifier has a good voltage gain, but it has a very low input resistance and its current gain is approximately 1.
- A darlington pair provides beta multiplication for increased input resistance.
- The total gain of a multistage amplifier is the product of the individual gains (sum of dB gains).
- Single-stage amplifiers can be connected in sequence with various coupling methods to form multistage amplifiers.
- Common-emitter, common-collector, and common-base amplifier configurations are summarized in Table 6–4.

▶ **TABLE 6–4**

Relative comparison of amplifier configurations. The current gains and the input and output resistances are the approximate maximum achievable values, with the bias resistors neglected.

	CE	CC	CB
Voltage gain, A_v	High R_C/r'_e	Low $\cong 1$	High R_C/r'_e
Current gain, $A_{i(max)}$	High β_{ac}	High β_{ac}	Low $\cong 1$
Power gain, A_p	Very high A_iA_v	High $\cong A_i$	High $\cong A_v$
Input resistance, $R_{in(max)}$	Low $\beta_{ac}r'_e$	High $\beta_{ac}R_E$	Very low r'_e
Output resistance, R_{out}	High R_C	Very low $(R_s/\beta_{ac}) \parallel R_E$	High R_C

KEY TERMS

Key terms and other bold terms in the chapter are defined in the end–of–book glossary.

ac ground A point in a circuit that appears as ground to ac signals only.

Attenuation The reduction in the level of power, current, or voltage.

Bypass capacitor A capacitor placed across the emitter resistor of an amplifier.

Common-base (CB) A BJT amplifier configuration in which the base is the common terminal to an ac signal or ground.

Common-emitter (CE) A BJT amplifier configuration in which the emitter is the common terminal to an ac signal or ground.

Emitter-follower A popular term for a common-collector amplifier.

Input resistance The resistance looking in at the transistor base.

Output resistance The resistance looking in at the transistor collector.

Overall voltage gain The product of the attenuation and the gain from base to collector of an amplifier.

***r* parameter** One of a set of BJT characteristic parameters that include α_{ac}, β_{ac}, r'_e, r'_b, and r'_c.

KEY FORMULAS

$$6\text{–}1 \qquad r'_e \cong \frac{25\ \text{mV}}{I_E} \qquad\qquad \text{Internal ac emitter resistance}$$

Common-Emitter

6–2	$R_{in(base)} = \beta_{ac}\, r'_e$	Input resistance at base
6–3	$R_{in(tot)} = R_1 \parallel R_2 \parallel R_{in(base)}$	Total amplifier input resistance, voltage-divider bias
6–4	$R_{out} \cong R_C$	Output resistance

$$6\text{–}5 \qquad A_v = \frac{R_C}{r'_e} \qquad\qquad \text{Voltage gain, base-to-collector, unloaded}$$

$$6\text{–}6 \qquad A_v = \frac{R_C}{r'_e + R_E} \qquad\qquad \text{Voltage gain without bypass capacitor}$$

$$6\text{–}7 \qquad A_v = \frac{R_c}{r'_e} \qquad\qquad \text{Voltage gain, base-to-collector, loaded, bypassed } R_E$$

$$6\text{–}8 \qquad A_v \cong \frac{R_C}{R_{E1}} \qquad\qquad \text{Voltage gain, swamped amplifier}$$

6–9	$R_{in(base)} = \beta_{ac}\,(r'_e + R_{E1})$	Input resistance at base, swamped amplifier

$$6\text{–}10 \qquad A_i = \frac{I_c}{I_s} \qquad\qquad \text{Current gain, input source to collector}$$

6–11	$A_p = A'_v A_i$	Power gain

Common-Collector (Emitter-Follower)

6–12	$A_v = 1$	Voltage gain, base-to-emitter
6–13	$R_{in(base)} \cong \beta_{ac} R_e$	Input resistance at base, loaded

$$6\text{–}14 \qquad R_{out} \cong \left(\frac{R_s}{\beta_{ac}}\right) \parallel R_E \qquad\qquad \text{Output resistance}$$

$$6\text{–}15 \qquad A_i = \frac{I_e}{I_{in}} \qquad\qquad \text{Current gain}$$

6–16	$A_p \cong A_i$	Power gain
6–17	$R_{in} = \beta_{ac1}\beta_{ac2} R_E$	Input resistance, darlington pair

Common-Base

$$6\text{–}18 \qquad A_v \cong \frac{R_c}{r'_e} \qquad\qquad \text{Voltage gain, emitter-to-collector}$$

6–19	$R_{in(emitter)} \cong r'_e$	Input resistance at emitter
6–20	$R_{out} \cong R_C$	Output resistance
6–21	$A_i \cong 1$	Current gain
6–22	$A_p \cong A_v$	Power gain

Multistage Amplifier

6–23	$A'_v = A_{v1}A_{v2}A_{v3}\cdots A_{vn}$	Overall voltage gain
6–24	$A_{v(dB)} = 20\log A_v$	Voltage gain expressed in dB
6–25	$A'_{v(dB)} = A_{v1(dB)} + A_{v2(dB)} + \cdots + A_{vn(dB)}$	Overall voltage gain in dB

Answers are at the end of the chapter.

1. A small-signal amplifier

   (a) uses only a small portion of its load line

   (b) always has an output signal in the mV range

   (c) goes into saturation once on each input cycle

   (d) is always a common-emitter amplifier

2. The parameter h_{fe} corresponds to

   (a) β_{DC} (b) β_{ac} (c) r'_e (d) r'_c

3. If the dc emitter current in a certain transistor amplifier is 3 mA, the approximate value of r'_e is

   (a) 3 kΩ   (b) 3 Ω   (c) 8.33 Ω   (d) 0.33 kΩ

4. A certain common-emitter amplifier has a voltage gain of 100. If the emitter bypass capacitor is removed,

   (a) the circuit will become unstable   (b) the voltage gain will decrease

   (c) the voltage gain will increase   (d) the Q-point will shift

5. For a common-collector amplifier, $R_E = 100\ \Omega$, $r'_e = 10\ \Omega$, and $\beta_{ac} = 150$. The ac input resist-ance at the base is

   (a) 1500 Ω   (b) 15 kΩ   (c) 110 Ω   (d) 16.5 kΩ

6. If a 10 mV signal is applied to the base of the emitter-follower circuit in Question 5, the output signal is approximately

   (a) 100 mV   (b) 150 mV   (c) 1.5 V   (d) 10 mV

7. For a common-emitter amplifier, $R_C = 1.0\ k\Omega$, $R_E = 390\ \Omega$, $r'_e = 15\ \Omega$, and $\beta_{ac} = 75$. Assum-ing that R_E is completely bypassed at the operating frequency, the voltage gain is

   (a) 66.7   (b) 2.56   (c) 2.47   (d) 75

8. In the circuit of Question 7, if the frequency is reduced to the point where $X_{C(bypass)} = R_E$, the voltage gain

   (a) remains the same   (b) is less   (c) is greater

9. In a certain emitter-follower circuit, the current gain is 50. The power gain is approximately

   (a) $50A_v$ (b) 50   (c) 1   (d) answers (a) and (b)

10. In a darlington pair configuration, each transistor has an ac beta of 125. If R_E is 560 Ω, the input resistance is

    (a) 560 Ω   (b) 70 kΩ   (c) 8.75 MΩ   (d) 140 kΩ

11. The input resistance of a common-base amplifier is

    (a) very low   (b) very high   (c) the same as a CE   (d) the same as a CC

12. In a common-emitter amplifier with voltage-divider bias, $R_{in(base)} = 68\ k\Omega$, $R_1 = 33\ k\Omega$, and $R_2 = 15\ k\Omega$. The total input resistance is

    (a) 68 kΩ   (b) 8.95 kΩ   (c) 22.2 kΩ   (d) 12.3 kΩ

13. A CE amplifier is driving a 10 kΩ load. If $R_C = 2.2\ k\Omega$ and $r'_e = 10\ \Omega$, the voltage gain is approximately

    (a) 220   (b) 1000   (c) 10   (d) 180

14. Each stage of a four-stage amplifier has a voltage gain of 15. The overall voltage gain is

    (a) 60   (b) 15   (c) 50,625   (d) 3078

15. The overall gain found in Question 14 can be expressed in decibels as

    (a) 94.1 dB   (b) 47.0 dB   (c) 35.6 dB   (d) 69.8 dB

PROBLEMS

Answers to all odd-numbered problems are at the end of the book.

BASIC PROBLEMS

SECTION 6–1 Amplifier Operation

1. What is the lowest value of dc collector current to which a transistor having the characteristic curves in Figure 6–4 can be biased and still retain linear operation with a peak-to-peak base current swing of 20 μA?

2. What is the highest value of I_C under the conditions described in Problem 1?

SECTION 6–2 Transistor AC Equivalent Circuits

3. If the dc emitter current in a transistor is 3 mA, what is the value of r'_e?

4. If the h_{fe} of a transistor is specified as 200, determine β_{ac}.

5. A certain transistor has a dc beta (h_{FE}) of 130. If the dc base current is 10 μA, determine r'_e. $\alpha_{DC} = 0.99$.   *$\beta_{DC} = 130$*   *$I_B = 10\,\mu A$*

6. At the dc bias point of a certain transistor circuit, $I_B = 15$ μA and $I_C = 2$ mA. Also, a variation in I_B of 3 μA about the Q-point produces a variation in I_C of 0.35 mA about the Q-point. Determine β_{DC} and β_{ac}.

SECTION 6–3 The Common-Emitter Amplifier

7. Draw the dc equivalent circuit and the ac equivalent circuit for the unloaded amplifier in Figure 6–45.

8. Determine the following values for the amplifier in Figure 6–45.

   (a) $R_{in(base)}$ **(b)** $R_{in(tot)}$ **(c)** A_v

9. Connect a bypass capacitor across R_E in Figure 6–45, and repeat Problem 8.

10. Connect a 10 kΩ load resistor to the output in Figure 6–45, and repeat Problem 9.

11. Determine the following dc values for the amplifier in Figure 6–46.

    (a) V_B **(b)** V_E **(c)** I_E **(d)** I_C **(e)** V_C **(f)** V_{CE}

12. Determine the following ac values for the amplifier in Figure 6–46.

    (a) $R_{in(base)}$ **(b)** R_{in} **(c)** A_v **(d)** A_i **(e)** A_p

▲ **FIGURE 6–45**

File circuits are identified with a CD logo and are in the Problems folder on your CD-ROM. Filenames correspond to figure numbers (e.g., F06-45).

▲ **FIGURE 6–46**

EWB only.

13. Assume that a 600 Ω, 12 μV rms voltage source is driving the amplifier in Figure 6–46. Determine the overall voltage gain by taking into account the attenuation in the base circuit, and find the *total* output voltage (ac and dc). What is the phase relationship of the collector signal voltage to the base signal voltage?

14. The amplifier in Figure 6–47 has a variable gain control, using a 100 Ω potentiometer for R_E with the wiper ac-grounded. As the potentiometer is adjusted, more or less of R_E is bypassed to ground, thus varying the gain. The total R_E remains constant to dc, keeping the bias fixed. Determine the maximum and minimum gains for this unloaded amplifier.

▶ FIGURE 6–47

15. If a load resistance of 600 Ω is placed on the output of the amplifier in Figure 6–47, what are the maximum and minimum gains?

16. Find the overall maximum voltage gain for the amplifier in Figure 6–47 with a 1.0 kΩ load if it is being driven by a 300 Ω source.

17. Modify the schematic to show how you would "swamp out" the temperature effects of r'_e in Figure 6–46 by making R_e at least ten times larger than r'_e. Keep the same total R_E. How does this affect the voltage gain?

SECTION 6–4 The Common-Collector Amplifier

18. Determine the *exact* voltage gain for the unloaded emitter-follower in Figure 6–48.

19. What is the total input resistance in Figure 6–48? What is the dc output voltage?

20. A load resistance is capacitively coupled to the emitter in Figure 6–48. In terms of signal operation, the load appears in parallel with R_E and reduces the effective emitter resistance. How does this affect the voltage gain?

▶ FIGURE 6–48

21. In Problem 20, what value of R_L will cause the voltage gain to drop to 0.9?

22. For the circuit in Figure 6–49, determine the following:

    (a) Q_1 and Q_2 dc terminal voltages   **(b)** overall β_{ac}

    (c) r'_e for each transistor   **(d)** total input resistance

23. Find the overall current gain A_i in Figure 6–49.

▲ **FIGURE 6–49**

SECTION 6–5   The Common-Base Amplifier

24. What is the main disadvantage of the common-base amplifier compared to the common-emitter and the emitter-follower amplifiers?

25. Find $R_{in(emitter)}$, A_v, A_i, and A_p for the unloaded amplifier in Figure 6–50.

▲ **FIGURE 6–50**

26. Match the following generalized characteristics with the appropriate amplifier configuration.

    (a) Unity current gain, good voltage gain, very low input resistance

    (b) Good current gain, good voltage gain, low input resistance

    (c) Good current gain, unity voltage gain, high input resistance

SECTION 6–6   Multistage Amplifiers

27. Each of two cascaded amplifier stages has an $A_v = 20$. What is the overall gain?

28. Each of three cascaded amplifier stages has a dB voltage gain of 10 dB. What is the overall voltage gain in dB? What is the actual overall voltage gain?

29. For the two-stage, capacitively coupled amplifier in Figure 6–51, find the following values:

    (a) voltage gain of each stage   **(b)** overall voltage gain

    (c) Express the gains found in (a) and (b) in dB.

30. If the multistage amplifier in Figure 6–51 is driven by a 75 Ω, 50 μV source and the second stage is loaded with an $R_L = 18$ kΩ, determine

    (a) voltage gain of each stage   **(b)** overall voltage gain

    (c) Express the gains found in (a) and (b) in dB.

31. Figure 6–52 shows a direct-coupled (that is, with no coupling capacitors between stages) two-stage amplifier. The dc bias of the first stage sets the dc bias of the second. Determine all dc voltages for both stages and the overall ac voltage gain.

▲ **FIGURE 6–51**

EWB only.

▲ **FIGURE 6–52**

32. Express the following voltage gains in dB:

    (a) 12   **(b)** 50   **(c)** 100   **(d)** 2500

33. Express the following voltage gains in dB as standard voltage gains:

    (a) 3 dB   **(b)** 6 dB   **(c)** 10 dB   **(d)** 20 dB   **(e)** 40 dB

TROUBLESHOOTING PROBLEMS

SECTION 6–7 **Troubleshooting**

34. Assume that the coupling capacitor C_3 is shorted in Figure 6–33. What dc voltage will appear at the collector of Q_1?

35. Assume that R_5 opens in Figure 6–33. Will Q_2 be in cutoff or in conduction? What dc voltage will you observe at the Q_2 collector?

36. Refer to Figure 6–51 and determine the general effect of each of the following failures:

    (a) C_2 open   **(b)** C_3 open

    (c) C_4 open   **(d)** C_2 shorted

    (e) base-collector junction of Q_1 open   **(f)** base-emitter junction of Q_2 open

37. Assume that you must troubleshoot the amplifier in Figure 6–51. Set up a table of test point values, input, output, and all transistor terminals that include both dc and rms values that you expect to observe when a 300 Ω test signal source with a 25 μV rms output is used.

SYSTEM APPLICATION PROBLEMS

38. Refer to the public address/paging system block diagram in Figure 6–40. You are asked to repair a system that is not working. After a preliminary check, you find that there is no output from the power amplifier or from the preamplifier. Based on this check and assuming that only one of the blocks is faulty, which block can you eliminate as the faulty one? What would you check next?

39. Determine the dc and ac voltages at the output (collector of Q_2) for each of the following faults in the amplifier of Figure 6–53. Assume a 2 mV rms input signal and $\beta_{ac} = 200$.

    (a) Open C_2 **(b)** Open C_1 **(c)** Open C_3

    (d) Open C_4 **(e)** Q_1 collector internally open   **(f)** Q_2 emitter shorted to ground

40. Suppose a 220 kΩ resistor is incorrectly installed in the R_6 position of the amplifier in Figure 6–53. What effect does this have on the circuit and what is the output voltage (dc and ac) if the input voltage is 3 mV rms?

41. The connection from R_1 to the supply voltage in Figure 6–53 has opened.

    (a) What happens to Q_1?

    (b) What is the voltage at the Q_1 collector?

    (c) What is the voltage at the Q_2 collector?

▲ **FIGURE 6–53**

DATA SHEET PROBLEMS

42. Refer to the 2N3947 data sheet in Appendix C. Determine the minimum value for each of the following r parameters:

    (a) β_{ac} **(b)** r'_e **(c)** r'_c

43. Repeat Problem 42 for maximum values.

44. Should you use a 2N3946 or a 2N3947 transistor in a certain application if the criteria is maximum current gain?

ADVANCED PROBLEMS

45. In an amplifier such as the one in Figure 6–53, explain the general effect that a leaky coupling capacitor would have on circuit performance.

46. Draw the dc and ac equivalent circuits for the amplifier in Figure 6–53.

47. Redesign the 2-stage amplifier in Figure 6–53 using *pnp* transistors such as the 2N3906. Maintain the same voltage gain.

48. Design a single-stage common-emitter amplifier with a voltage gain of 40 dB that operates from a dc supply voltage of $+12$ V. Use a 2N2222 transistor, voltage-divider bias, and a 330 Ω swamping resistor. The maximum input signal is 25 mV rms.

49. Design an emitter-follower with a minimum input resistance of 50 kΩ using a 2N3904 *npn* transistor with a $\beta_{ac} = 100$.

50. Repeat Problem 49 using a 2N3906 with a $\beta_{ac} = 100$.

51. Design a single-stage common-base amplifier for a voltage gain of 75. Use a 2N3904 with emitter bias. The dc supply voltages are to be ±6 V.

52. Refer to the amplifier in Figure 6–53 and determine the minimum value of coupling capacitors necessary for the amplifier to produce the same output voltage at 100 Hz that it does at 5000 Hz.

53. Prove that for any unloaded common-emitter amplifier with a collector resistor R_C and R_E bypassed, the voltage gain is $A_v \cong 40V_{R_C}$.

EWB/MULTISIM TROUBLESHOOTING PROBLEMS

These file circuits are in the Troubleshooting Problems folder on your CD-ROM.

54. Open file TSP06-54 and determine the fault.

55. Open file TSP06-55 and determine the fault.

56. Open file TSP06-56 and determine the fault.

57. Open file TSP06-57 and determine the fault.

58. Open file TSP06-58 and determine the fault.

59. Open file TSP06-59 and determine the fault.

ANSWERS

SECTION REVIEWS

SECTION 6–1   **Amplifier Operation**

1. Positive, negative

2. V_{CE} is a dc quantity and V_{ce} is an ac quantity.

3. R_e is the external emitter ac resistance, r'_e is the internal emitter ac resistance.

SECTION 6–2   **Transistor AC Equivalent Circuits**

1. α_{ac}—ac alpha, I_c/I_e; β_{ac}—ac beta, I_c/I_b; r'_e—ac emitter resistance; r'_b—ac base resistance; r'_c—ac collector resistor.

2. h_{fe} is equivalent to β_{ac}.

3. $r'_e = 25$ mV/15 mA $= 1.67\ \Omega$

SECTION 6–3 The Common-Emitter Amplifier

1. The capacitors are treated as opens.
2. The gain increases with a bypass capacitor.
3. Swamping eliminates the effects of r'_e by partially bypassing R_E.
4. Total input resistance includes the bias resistors, r'_e, and any unbypassed R_E.
5. The gain is determined by R_c, r'_e, and any unbypassed R_E.
6. The voltage gain decreases with a load.
7. The input and output are 180° out of phase.

SECTION 6–4 The Common-Collector Amplifier

1. A common-collector amplifier is an emitter-follower.
2. The maximum voltage gain of a common-collector amplifier is 1.
3. A common-collector amplifier has a high input resistance.

SECTION 6–5 The Common-Base Amplifier

1. Yes
2. The common-base amplifier has a low input resistance.
3. The maximum current gain is 1 in a CB amplifier.

SECTION 6–6 Multistage Amplifiers

1. A stage is one amplifier in a cascaded arrangement.
2. The overall voltage gain is the product of the individual gains.
3. $20 \log(500) = 54.0$ dB
4. At lower frequencies, X_C becomes large enough to affect the gain.

SECTION 6–7 Troubleshooting

1. If C_4 opens, the gain drops. The dc level would not be affected.
2. Q_2 would be biased in cutoff.
3. The collector voltage of Q_1 and the base, emitter, and collector voltages of Q_2 would change.

RELATED PROBLEMS FOR EXAMPLES

6–1 $I_C = 5$ mA; $V_{CE} = 1.5$ V

6–2 3.13 mA

6–3 9.3 mV

6–4 $C_2 = 28.4$ μF

6–5 97.3

6–6 153

6–7 4.85; 165

6–8 9.56

6–9 Increases

6–10 71. A single transistor loads the CE amplifier much more than the darlington pair.

6–11 55.9

6–12 $A_v' = 1500$; $A_{v1(dB)} = 27.96$ dB; $A_{v2(dB)} = 13.98$ dB; $A_{v3(dB)} = 21.58$ dB; $A_{v(dB)}' = 63.52$ dB

6–13 C_3 open

SELF-TEST

1. (a) **2.** (b) **3.** (c) **4.** (b) **5.** (d)

6. (d) **7.** (a) **8.** (b) **9.** (d) **10.** (c)

11. (a) **12.** (b) **13.** (d) **14.** (c) **15.** (a)

7

FIELD-EFFECT TRANSISTORS (FETs)

INTRODUCTION

BJTs (bipolar junction transistors) were covered in previous chapters. Now we will discuss the second major type of transistor, the FET (field-effect transistor). FETs are unipolar devices because, unlike BJTs that use both electron and hole current, they operate only with one type of charge carrier. The two main types of FETs are the junction field–effect transistor (JFET) and the metal oxide semiconductor field–effect transistor (MOSFET).

Recall that a BJT is a current-controlled device; that is, the base current controls the amount of collector current. A FET is different. It is a voltage-controlled device, where the voltage between two of the terminals (gate and source) controls the current through the device. As you will learn, a major feature of FETs is their very high input resistance.

- Explain the operation of JFETs
- Define, discuss, and apply important JFET parameters
- Discuss and analyze JFET bias circuits
- Explain the operation of MOSFETs
- Define, discuss, and apply important MOSFET parameters
- Discuss and analyze MOSFET bias circuits
- Troubleshoot FET circuits

KEY TERMS

- JFET
- Drain
- Source
- Gate
- Pinch-off voltage
- Transconductance
- MOSFET
- Depletion
- Enhancement

■■■ SYSTEM APPLICATION PREVIEW

The system application at the end of the chapter involves the electronic control circuits for a waste water treatment system. In particular, you will focus on the application of field-effect transistors in the sensing circuits for chemical measurements.

WWW. VISIT THE COMPANION WEBSITE
Study aids for this chapter are available at
http://www.prenhall.com/floyd

7–1   THE JFET

The JFET (junction field-effect transistor) is a type of FET that operates with a reverse-biased *pn* junction to control current in a channel. Depending on their structure, JFETs fall into either of two categories, *n* channel or *p* channel.

After completing this section, you should be able to

- **Explain the operation of JFETs**
- Identify the three terminals of a JFET
- Explain what a channel is
- Describe the structural difference between an *n*-channel JFET and a *p*-channel JFET
- Discuss how voltage controls the current in a JFET
- Identify the symbols for *n*-channel and *p*-channel JFETs

Figure 7–1(a) shows the basic structure of an *n*-channel **JFET** (junction field-effect transistor). Wire leads are connected to each end of the *n*-channel; the **drain** is at the upper end, and the **source** is at the lower end. Two *p*-type regions are diffused in the *n*-type material to form a **channel**, and both *p*-type regions are connected to the **gate** lead. For simplicity, the gate lead is shown connected to only one of the *p* regions. A *p*-channel JFET is shown in Figure 7–1(b).

▶ **FIGURE 7–1**

A representation of the basic structure of the two types of JFET.

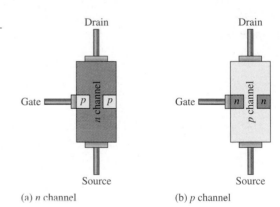

(a) *n* channel   (b) *p* channel

Basic Operation

To illustrate the operation of a JFET, Figure 7–2 shows dc bias voltages applied to an *n*-channel device. V_{DD} provides a drain-to-source voltage and supplies current from drain to source. V_{GG} sets the reverse-bias voltage between the gate and the source, as shown.

The JFET is always operated with the gate-source pn junction reverse-biased. Reverse-biasing of the gate-source junction with a negative gate voltage produces a depletion region along the *pn* junction, which extends into the *n* channel and thus increases its resistance by restricting the channel width.

The channel width and thus the channel resistance can be controlled by varying the gate voltage, thereby controlling the amount of drain current, I_D. Figure 7–3 illustrates this concept. The white areas represent the depletion region created by the reverse bias. It is wider toward the drain end of the channel because the reverse-bias voltage between the gate and the drain is greater than that between the gate and the source. We will discuss JFET characteristic curves and some important parameters in Section 7–2.

▶ **FIGURE 7–2**

A biased *n*-channel JFET.

(a) JFET biased for conduction

(b) Greater V_{GG} narrows the channel (between the white areas) which increases the resistance of the channel and decreases I_D.

(c) Less V_{GG} widens the channel (between the white areas) which decreases the resistance of the channel and increases I_D.

▲ **FIGURE 7–3**

Effects of V_{GS} on channel width, resistance, and drain current ($V_{GG} = V_{GS}$).

JFET Symbols

The schematic symbols for both *n*-channel and *p*-channel JFETs are shown in Figure 7–4. Notice that the arrow on the gate points "in" for *n* channel and "out" for *p* channel.

▶ **FIGURE 7–4**

JFET schematic symbols.

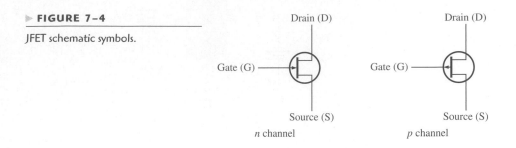

n channel

p channel

SECTION 7–1 REVIEW

Answers are at the end of the chapter.

1. Name the three terminals of a JFET.
2. Does an *n*-channel JFET require a positive or negative value for V_{GS}?
3. How is the drain current controlled in a JFET?

7–2 JFET CHARACTERISTICS AND PARAMETERS

In this section, you will see how the JFET operates as a voltage-controlled, constant-current device. You will also learn about cutoff and pinch-off as well as JFET transfer characteristics.

After completing this section, you should be able to

- **Define, discuss, and apply important JFET parameters**
- Explain ohmic area, constant-current area, and breakdown
- Define *pinch-off voltage*
- Describe how gate-to-source voltage controls the drain current
- Define *cutoff voltage*
- Compare pinch-off and cutoff
- Analyze a JFET transfer characteristic curve
- Use the equation for the transfer characteristic to calculate I_D
- Use a JFET data sheet
- Define *transconductance*
- Explain and determine input resistance and capacitance
- Determine drain-to-source resistance

Consider the case when the gate-to-source voltage is zero ($V_{GS} = 0$ V). This is produced by shorting the gate to the source, as in Figure 7–5(a) where both are grounded. As V_{DD} (and thus V_{DS}) is increased from 0 V, I_D will increase proportionally, as shown in the graph of Figure 7–5(b) between points *A* and *B*. In this area, the channel resistance is essentially constant because the depletion region is not large enough to have significant effect. This is called the ohmic area because V_{DS} and I_D are related by Ohm's law.

At point B in Figure 7–5(b), the curve levels off and I_D becomes essentially constant. As V_{DS} increases from point B to point C, the reverse-bias voltage from gate to drain (V_{GD}) produces a depletion region large enough to offset the increase in V_{DS}, thus keeping I_D relatively constant.

(a) JFET with $V_{GS} - 0$ V and a variable V_{DS} (V_{DD})   (b) Drain characteristic

▲ **FIGURE 7–5**

The drain characteristic curve of a JFET for $V_{GS} = 0$ showing pinch-off.

Pinch-Off Voltage

For $V_{GS} = 0$ V, the value of V_{DS} at which I_D becomes essentially constant (point B on the curve in Figure 7–5(b)) is the **pinch-off voltage**, V_P. For a given JFET, V_P has a fixed value. As you can see, a continued increase in V_{DS} above the pinch-off voltage produces an almost constant drain current. This value of drain current is I_{DSS} (*D*rain to *S*ource current with gate *S*horted) and is always specified on JFET data sheets. I_{DSS} is the *maximum* drain current that a specific JFET can produce regardless of the external circuit, and it is always specified for the condition, $V_{GS} = 0$ V.

As shown in the graph in Figure 7–5(b), **breakdown** occurs at point C when I_D begins to increase very rapidly with any further increase in V_{DS}. Breakdown can result in irreversible damage to the device, so JFETs are always operated below breakdown and within the constant-current area (between points B and C on the graph). The JFET action that produces the drain characteristic curve to the point of breakdown for $V_{GS} = 0$ V is illustrated in Figure 7–6.

V_{GS} Controls I_D

Let's connect a bias voltage, V_{GG}, from gate to source as shown in Figure 7–7(a). As V_{GS} is set to increasingly more negative values by adjusting V_{GG}, a family of drain characteristic curves is produced, as shown in Figure 7–7(b). Notice that I_D decreases as the magnitude of V_{GS} is increased to larger negative values because of the narrowing of the channel. Also notice that, for each increase in V_{GS}, the JFET reaches pinch-off (where constant current begins) at values of V_{DS} less than V_P. So, the amount of drain current is controlled by V_{GS}, as illustrated in Figure 7–8.

(a) When $V_{DS} = 0$, $I_D = 0$.

(b) I_D increases proportionally with V_{DS} in the ohmic area.

(c) When $V_{DS} = V_P$, I_D is consistant and equal to I_{DSS}.

(d) As V_{DS} increases further, I_D remains at I_{DSS} until breakdown occurs.

▲ **FIGURE 7–6**

JFET action that produces the characteristic curve for $V_{GS} = 0$ V.

(a) JFET biased at $V_{GS} = -1$ V

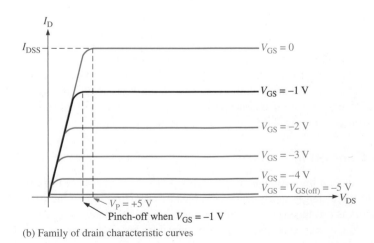

(b) Family of drain characteristic curves

▲ **FIGURE 7–7**

Pinch–off occurs at a lower V_{DS} as V_{GS} is increased to more negative values.

(a) $V_{GS} = 0$ V, $V_{DS} \geq V_P$, $I_D = I_{DSS}$

(b) When V_{GS} is negative, I_D decreases and is constant above pinch-off, which is less than V_P.

(c) As V_{GS} is made more negative, I_D continues to decrease but is constant above pinch-off, which has also decreased.

(d) Until $V_{GS} = -V_{GS(off)}$, I_D continues to decrease. When $V_{GS} \geq -V_{GS(off)}$, $I_D \cong 0$.

▲ **FIGURE 7–8**

V_{GS} controls I_D.

Cutoff Voltage

The value of V_{GS} that makes I_D approximately zero is the **cutoff voltage**, $V_{GS(off)}$. The JFET must be operated between $V_{GS} = 0$ V and $V_{GS(off)}$. For this range of gate-to-source voltages, I_D will vary from a maximum of I_{DSS} to a minimum of almost zero.

As you have seen, for an *n*-channel JFET, the more negative V_{GS} is, the smaller I_D becomes in the constant-current area. When V_{GS} has a sufficiently large negative value, I_D is reduced to zero. This cutoff effect is caused by the widening of the depletion region to a point where it completely closes the channel, as shown in Figure 7–9.

▲ **FIGURE 7–9**

JFET at cutoff.

The basic operation of a *p*-channel JFET is the same as for an *n*-channel device except that a *p*-channel JFET requires a negative V_{DD} and a positive V_{GS}, as illustrated in Figure 7–10.

▶ **FIGURE 7–10**

A biased *p*-channel JFET.

Comparison of Pinch-Off and Cutoff

As you have seen, there is a difference between pinch-off and cutoff. There is also a connection. V_P is the value of V_{DS} at which the drain current becomes constant and is always measured at $V_{GS} = 0$ V. However, pinch-off occurs for V_{DS} values less than V_P when V_{GS} is nonzero. So, although V_P is a constant, the minimum value of V_{DS} at which I_D becomes constant varies with V_{GS}.

$V_{GS(off)}$ and V_P are always equal in magnitude but opposite in sign. A data sheet usually will give either $V_{GS(off)}$ or V_P, but not both. However, when you know one, you have the other. For example, if $V_{GS(off)} = -5$ V, then $V_P = +5$ V, as shown in Figure 7–7(b).

EXAMPLE 7–1

For the JFET in Figure 7–11, $V_{GS(off)} = -4$ V and $I_{DSS} = 12$ mA. Determine the *minimum* value of V_{DD} required to put the device in the constant-current area of operation.

▶ **FIGURE 7–11**

Solution   Since $V_{GS(off)} = -4$ V, $V_P = 4$ V. The minimum value of V_{DS} for the JFET to be in its constant-current area is

$$V_{DS} = V_P = 4 \text{ V}$$

In the constant-current area with $V_{GS} = 0$ V,

$$I_D = I_{DSS} = 12 \text{ mA}$$

The drop across the drain resistor is

$$V_{RD} = I_D R_D = (12 \text{ mA})(560 \ \Omega) = 6.72 \text{ V}$$

Apply Kirchhoff's law around the drain circuit.

$$V_{DD} = V_{DS} + V_{RD} = 4 \text{ V} + 6.72 \text{ V} = \textbf{10.7 V}$$

This is the value of V_{DD} to make $V_{DS} = V_P$ and put the device in the constant-current area.

Related Problem* If V_{DD} is increased to 15 V, what is the drain current?

EXAMPLE 7–2

A particular p-channel JFET has a $V_{GS(off)} = +4$ V. What is I_D when $V_{GS} = +6$ V?

Solution The p-channel JFET requires a positive gate-to-source voltage. The more positive the voltage, the less the drain current. When $V_{GS} = 4$ V, $I_D = 0$. Any further increase in V_{GS} keeps the JFET cut off, so I_D remains **0**.

Related Problem What is V_P for the JFET described in this example?

JFET Transfer Characteristic

You have learned that a range of V_{GS} values from zero to $V_{GS(off)}$ controls the amount of drain current. For an n-channel JFET, $V_{GS(off)}$ is negative, and for a p-channel JFET, $V_{GS(off)}$ is positive. Because V_{GS} does control I_D, the relationship between these two quantities is very important. Figure 7–12 is a general transfer characteristic curve that illustrates graphically the relationship between V_{GS} and I_D.

▶ **FIGURE 7–12**

JFET transfer characteristic curve (*n*-channel).

Notice that the bottom end of the curve is at a point on the V_{GS} axis equal to $V_{GS(off)}$, and the top end of the curve is at a point on the I_D axis equal to I_{DSS}. This curve, of course, shows that the operating limits of a JFET are

$$I_D = 0 \quad \text{when } V_{GS} = V_{GS(off)}$$

and

$$I_D = I_{DSS} \quad \text{when } V_{GS} = 0$$

The transfer characteristic curve can be developed from the drain characteristic curves by plotting values of I_D for the values of V_{GS} taken from the family of drain curves at pinch-off, as illustrated in Figure 7–13 for a specific set of curves. Each point on the transfer characteristic curve corresponds to specific values of V_{GS} and I_D on the drain

▲ FIGURE 7–13

Example of the development of an *n*-channel JFET transfer characteristic curve (blue) from the JFET drain characteristic curves (green).

curves. For example, when $V_{GS} = -2$ V, $I_D = 4.32$ mA. Also, for this specific JFET, $V_{GS(off)} = -5$ V and $I_{DSS} = 12$ mA.

A JFET transfer characteristic curve is nearly parabolic in shape and can be expressed as

Equation 7–1

$$I_D = I_{DSS}\left(1 - \frac{V_{GS}}{V_{GS(off)}}\right)^2$$

With Equation 7–1, I_D can be determined for any V_{GS} if $V_{GS(off)}$ and I_{DSS} are known. These quantities are usually available from the data sheet for a given JFET. Notice the squared term in the equation. Because of its form, a parabolic relationship is known as a *square law,* and therefore, JFETs and MOSFETs are often referred to as *square-law devices.*

The data sheet for a typical JFET series is shown in Figure 7–14.

EXAMPLE 7–3

The data sheet in Figure 7–14 for a 2N5459 JFET indicates that typically $I_{DSS} = 9$ mA and $V_{GS(off)} = -8$ V (maximum). Using these values, determine the drain current for $V_{GS} = 0$ V, -1 V, and -4 V.

Solution For $V_{GS} = 0$ V,

$$I_D = I_{DSS} = \textbf{9 mA}$$

For $V_{GS} = -1$ V, use Equation 7–1.

$$I_D = I_{DSS}\left(1 - \frac{V_{GS}}{V_{GS(off)}}\right)^2 = (9 \text{ mA})\left(1 - \frac{-1 \text{ V}}{-8 \text{ V}}\right)^2$$

$$= (9 \text{ mA})(1 - 0.125)^2 = (9 \text{ mA})(0.766) = \textbf{6.89 mA}$$

For $V_{GS} = -4$ V,

$$I_D = (9 \text{ mA})\left(1 - \frac{-4 \text{ V}}{-8 \text{ V}}\right)^2 = (9 \text{ mA})(1 - 0.5)^2 = (9 \text{ mA})(0.25) = \textbf{2.25 mA}$$

Related Problem   Determine I_D for $V_{GS} = -3$ V for the 2N5459 JFET.

**2N5457
thru
2N5459**

**Case 29-04, Style 5
TO-92 (TO-226AA)**

**JFETs
General Purpose**

N channel — Depletion

Maximum Ratings

Rating	Symbol	Value	Unit
Drain-Source voltage	V_{DS}	25	V dc
Drain-Gate voltage	V_{DG}	25	V dc
Reverse gate-source voltage	V_{GSR}	−25	V dc
Gate current	I_G	10	mA dc
Total device dissipation @ $T_A = 25°C$ Derate above 25°C	P_D	310 2.82	mW mW/°C
Junction temperature	T_J	125	°C
Storage channel temperature range	T_{stg}	−65 to +150	°C

Electrical Characteristics ($T_A = 25°C$ unless otherwise noted.)

Characteristic		Symbol	Min	Typ	Max	Unit
OFF Characteristics						
Gate-Source breakdown voltage ($I_G = -10 \text{ μA dc}, V_{DS} = 0$)		$V_{(BR)GSS}$	−25	–	–	V dc
Gate reverse current ($V_{GS} = -15 \text{ V dc}, V_{DS} = 0$) ($V_{GS} = -15 \text{ V dc}, V_{DS} = 0, T_A = 100°C$)		I_{GSS}	– –	– –	−1.0 −200	nA dc
Gate-Source cutoff voltage ($V_{DS} = 15 \text{ V dc}, I_D = 10 \text{ nA dc}$)	2N5457 2N5458 2N5459	$V_{GS(off)}$	−0.5 −1.0 −2.0	– – –	−6.0 −7.0 −8.0	V dc
Gate-Source voltage ($V_{DS} = 15 \text{ V dc}, I_D = 100 \text{ μA dc}$) ($V_{DS} = 15 \text{ V dc}, I_D = 200 \text{ μA dc}$) ($V_{DS} = 15 \text{ V dc}, I_D = 400 \text{ μA dc}$)	2N5457 2N5458 2N5459	V_{GS}	– – –	−2.5 −3.5 −4.5	– – –	V dc
ON Characteristics						
Zero-Gate-Voltage drain current ($V_{DS} = 15 \text{ V dc}, V_{GS} = 0$)	2N5457 2N5458 2N5459	I_{DSS}	1.0 2.0 4.0	3.0 6.0 9.0	5.0 9.0 16	mA dc
Small-signal Characteristics						
Forward transfer admittance common source ($V_{DS} = 15 \text{ V dc}, V_{GS} = 0, f = 1.0 \text{ kHz}$)	2N5457 2N5458 2N5459	$|y_{fs}|$	1000 1500 2000	– – –	5000 5500 6000	μmhos or μS
Output admittance common source ($V_{DS} = 15 \text{ V dc}, V_{GS} = 0, f = 1.0 \text{ kHz}$)		$|y_{os}|$	–	10	50	μmhos or μS
Input capacitance ($V_{DS} = 15 \text{ V dc}, V_{GS} = 0, f = 1.0 \text{ MHz}$)		C_{iss}	–	4.5	7.0	pF
Reverse transfer capacitance ($V_{DS} = 15 \text{ V dc}, V_{GS} = 0, f = 1.0 \text{ MHz}$)		C_{rss}	–	1.5	3.0	pF

▲ **FIGURE 7–14**

JFET data sheet.

JFET Forward Transconductance

The forward **transconductance** (transfer conductance), g_m, is the change in drain current (ΔI_D) for a given change in gate-to-source voltage (ΔV_{GS}) with the drain-to-source voltage constant. It is expressed as a ratio and has the unit of siemens (S).

$$g_m = \frac{\Delta I_D}{\Delta V_{GS}}$$

Other common designations for this parameter are g_{fs} and y_{fs} (forward transfer admittance). As you will see in Chapter 8, g_m is important in FET amplifiers as a major factor in determining the voltage gain.

Because the transfer characteristic curve for a JFET is nonlinear, g_m varies in value depending on the location on the curve as set by V_{GS}. The value for g_m is greater near the top of the curve (near $V_{GS} = 0$) than it is near the bottom (near $V_{GS(off)}$), as illustrated in Figure 7–15.

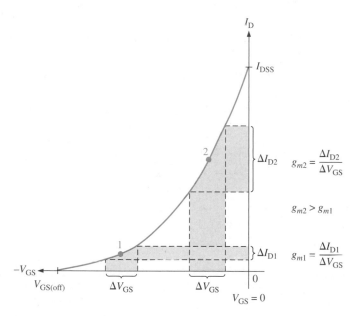

▲ **FIGURE 7–15**

g_m varies depending on the bias point (V_{GS}).

A data sheet normally gives the value of g_m measured at $V_{GS} = 0$ V (g_{m0}). For example, the data sheet for the 2N5457 JFET specifies a minimum g_{m0} (y_{fs}) of 1000 μS with $V_{DS} = 15$ V.

Given g_{m0}, you can calculate an approximate value for g_m at any point on the transfer characteristic curve using the following formula:

Equation 7–2

$$g_m = g_{m0}\left(1 - \frac{V_{GS}}{V_{GS(off)}}\right)$$

When a value of g_{m0} is not available, you can calculate it using values of I_{DSS} and $V_{GS(off)}$. The vertical lines indicate an absolute value (no sign).

Equation 7–3

$$g_{m0} = \frac{2I_{DSS}}{|V_{GS(off)}|}$$

EXAMPLE 7–4

The following information is included on the data sheet in Figure 7–14 for a 2N5457 JFET: typically, $I_{DSS} = 3.0$ mA, $V_{GS(off)} = -6$ V maximum, and $y_{fs(max)} = 5000$ μS. Using these values, determine the forward transconductance for $V_{GS} = -4$ V, and find I_D at this point.

Solution $g_{m0} = y_{fs} = 5000$ μS. Use Equation 7–2 to calculate g_m.

$$g_m = g_{m0}\left(1 - \frac{V_{GS}}{V_{GS(off)}}\right) = (5000 \ \mu S)\left(1 - \frac{-4 \text{ V}}{-6 \text{ V}}\right) = \mathbf{1667 \ \mu S}$$

Next, use Equation 7–1 to calculate I_D at $V_{GS} = -4$ V.

$$I_D = I_{DSS}\left(1 - \frac{V_{GS}}{V_{GS(off)}}\right)^2 = (3.0 \text{ mA})\left(1 - \frac{-4 \text{ V}}{-6 \text{ V}}\right)^2 = \mathbf{333 \ \mu A}$$

Related Problem A given JFET has the following characteristics: $I_{DSS} = 12$ mA, $V_{GS(off)} = -5$ V, and $g_{m0} = 3000$ μS. Find g_m and I_D when $V_{GS} = -2$ V.

Input Resistance and Capacitance

As you know, a JFET operates with its gate-source junction reverse-biased, which makes the input resistance at the gate very high. This high input resistance is one advantage of the JFET over the BJT. (Recall that a bipolar junction transistor operates with a forward-biased base-emitter junction.) JFET data sheets often specify the input resistance by giving a value for the gate reverse current, I_{GSS}, at a certain gate-to-source voltage. The input resistance can then be determined using the following equation, where the vertical lines indicate an absolute value (no sign):

$$R_{IN} = \left|\frac{V_{GS}}{I_{GSS}}\right|$$

For example, the 2N5457 data sheet in Figure 7–14 lists a maximum I_{GSS} of -1.0 nA for $V_{GS} = -15$ V at 25°C. I_{GSS} increases with temperature, so the input resistance decreases.

The input capacitance, C_{iss}, is a result of the JFET operating with a reverse-biased *pn* junction. Recall that a reverse-biased *pn* junction acts as a capacitor whose capacitance depends on the amount of reverse voltage. For example, the 2N5457 has a maximum C_{iss} of 7 pF for $V_{GS} = 0$.

EXAMPLE 7–5

A certain JFET has an I_{GSS} of -2 nA for $V_{GS} = -20$ V. Determine the input resistance.

Solution $$R_{IN} = \left|\frac{V_{GS}}{I_{GSS}}\right| = \frac{20 \text{ V}}{2 \text{ nA}} = \mathbf{10,000 \ M\Omega}$$

Related Problem Determine the minimum input resistance for the 2N5458 from the data sheet in Figure 7–14.

Drain-to-Source Resistance

You learned from the drain characteristic curve that, above pinch-off, the drain current is relatively constant over a range of drain-to-source voltages. Therefore, a large change in V_{DS} produces only a very small change in I_D. The ratio of these changes is the drain-to-source resistance of the device, r'_{ds}.

$$r'_{ds} = \frac{\Delta V_{DS}}{\Delta I_D}$$

Data sheets often specify this parameter in terms of the output conductance, g_{os}, or output admittance, y_{os}.

SECTION 7–2 REVIEW

1. The drain-to-source voltage at the pinch-off point of a particular JFET is 7 V. If the gate-to-source voltage is zero, what is V_P?
2. The V_{GS} of a certain n-channel JFET is increased negatively. Does the drain current increase or decrease?
3. What value must V_{GS} have to produce cutoff in a p-channel JFET with a $V_P = -3$ V?

7–3 JFET BIASING

Using some of the FET parameters discussed in the previous sections, you will now see how to dc-bias JFETs. Just as with the BJT, the purpose of biasing is to select the proper dc gate-to-source voltage to establish a desired value of drain current and, thus, a proper Q-point. You will learn about two types of bias circuits, self-bias and voltage-divider bias.

After completing this section, you should be able to

- **Discuss and analyze JFET bias circuits**
- Describe self-bias
- Analyze a self-biased JFET circuit
- Set the self-biased Q-point
- Analyze a voltage-divider-biased JFET circuit
- Use transfer characteristic curves to analyze JFET bias circuits
- Discuss Q-point stability

Self-Bias

Self-bias is the most common type of JFET bias. Recall that a JFET must be operated such that the gate-source junction is always reverse-biased. This condition requires a negative V_{GS} for an n-channel JFET and a positive V_{GS} for a p-channel JFET. This can be achieved using the self-bias arrangements shown in Figure 7–16. The gate resistor, R_G, does not affect the bias because it has essentially no voltage drop across it; and therefore the gate remains at 0 V. R_G is necessary only to isolate an ac signal from ground in amplifier applications, as you will see later.

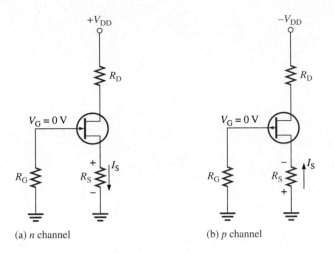

▲ FIGURE 7–16

Self-biased JFETs ($I_S = I_D$ in all FETs).

For the n-channel JFET in Figure 7–16(a), I_S produces a voltage drop across R_S and makes the source positive with respect to ground. Since $I_S = I_D$ and $V_G = 0$, then $V_S = I_D R_S$. The gate-to-source voltage is

$$V_{GS} = V_G - V_S = 0 - I_D R_S = -I_D R_S$$

Thus,

$$V_{GS} = -I_D R_S$$

For the p-channel JFET shown in Figure 7–16(b), the current through R_S produces a negative voltage at the source, making the gate positive with respect to the source. Therefore, since $I_S = I_D$,

$$V_{GS} = +I_D R_S$$

In the following analysis, the n-channel JFET in Figure 7–16(a) is used for illustration. Keep in mind that analysis of the p-channel JFET is the same except for opposite-polarity voltages. The drain voltage with respect to ground is determined as follows:

$$V_D = V_{DD} - I_D R_D$$

Since $V_S = I_D R_S$, the drain-to-source voltage is

$$V_{DS} = V_D - V_S$$
$$= V_{DD} - I_D(R_D + R_S)$$

EXAMPLE 7–6

Find V_{DS} and V_{GS} in Figure 7–17. For the particular JFET in this circuit, the internal parameter values such as g_m, $V_{GS(off)}$, and I_{DSS} are such that a drain current (I_D) of approximately 5 mA is produced. Another JFET, even of the same type, may not produce the same results when connected in this circuit due the variations in parameter values.

▶ **FIGURE 7–17**

Solution

$$V_S = I_D R_S = (5 \text{ mA})(220 \text{ }\Omega) = 1.1 \text{ V}$$

$$V_D = V_{DD} - I_D R_D = 15 \text{ V} - (5 \text{ mA})(1.0 \text{ k}\Omega) = 15 \text{ V} - 5 \text{ V} = 10 \text{ V}$$

Therefore,

$$V_{DS} = V_D - V_S = 10 \text{ V} - 1.1 \text{ V} = \mathbf{8.9 \text{ V}}$$

Since $V_G = 0$ V,

$$V_{GS} = V_G - V_S = 0 \text{ V} - 1.1 \text{ V} = \mathbf{-1.1 \text{ V}}$$

Related Problem Determine V_{DS} and V_{GS} in Figure 7–17 when $I_D = 8$ mA. Assume that $R_D = 860 \text{ }\Omega$, $R_S = 390 \text{ }\Omega$, and $V_{DD} = 12$ V.

> Open file E07-06 in the Examples folder on your CD-ROM. Measure I_D, V_{GS}, and V_{DS} and compare to the calculated values from the Related Problem.

Setting the Q-Point of a Self-Biased JFET

The basic approach to establishing a JFET bias point is to determine I_D for a desired value of V_{GS} or vice versa. Then calculate the required value of R_S using the following relationship. The vertical lines indicate an absolute value.

$$R_S = \left| \frac{V_{GS}}{I_D} \right|$$

For a desired value of V_{GS}, I_D can be determined in either of two ways: from the transfer characteristic curve for the particular JFET or, more practically, from Equation 7–1 using I_{DSS} and $V_{GS(off)}$ from the JFET data sheet. The next two examples illustrate these procedures.

EXAMPLE 7–7

Determine the value of R_S required to self-bias an *n*-channel JFET that has the transfer characteristic curve shown in Figure 7–18 at $V_{GS} = -5$ V.

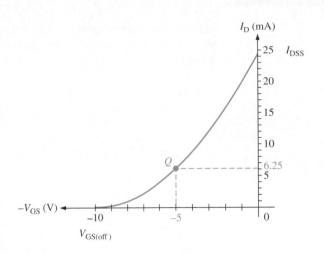

▲ FIGURE 7–18

Solution   From the graph, $I_D = 6.25$ mA when $V_{GS} = -5$ V. Calculate R_S.

$$R_S = \left| \frac{V_{GS}}{I_D} \right| = \frac{5 \text{ V}}{6.25 \text{ mA}} = \mathbf{800 \ \Omega}$$

Related Problem   Find R_S for $V_{GS} = -3$ V.

EXAMPLE 7–8

Determine the value of R_S required to self-bias a p-channel JFET with $I_{DSS} = 25$ mA and $V_{GS(off)} = 15$ V. V_{GS} is to be 5 V.

Solution   Use Equation 7–1 to calculate I_D.

$$I_D = I_{DSS}\left(1 - \frac{V_{GS}}{V_{GS(off)}} \right)^2 = (25 \text{ mA})\left(1 - \frac{5 \text{ V}}{15 \text{ V}} \right)^2 = (25 \text{ mA})(1 - 0.333)^2 = 11.1 \text{ mA}$$

Now, determine R_S.

$$R_S = \left| \frac{V_{GS}}{I_D} \right| = \frac{5 \text{ V}}{11.1 \text{ mA}} = \mathbf{450 \ \Omega}$$

Related Problem   Find the value of R_S required to self-bias a p-channel JFET with $I_{DSS} = 18$ mA and $V_{GS(off)} = 8$ V. $V_{GS} = 4$ V.

Midpoint Bias   It is usually desirable to bias a JFET near the midpoint of its transfer characteristic curve where $I_D = I_{DSS}/2$. Under signal conditions, midpoint bias allows the maximum amount of drain current swing between I_{DSS} and 0. Using Equation 7–1, it is shown in Appendix B that I_D is approximately one-half of I_{DSS} when $V_{GS} = V_{GS(off)}/3.4$.

$$I_D = I_{DSS}\left(1 - \frac{V_{GS}}{V_{GS(off)}} \right)^2 = I_{DSS}\left(1 - \frac{V_{GS(off)}/3.4}{V_{GS(off)}} \right)^2 = 0.5 I_{DSS}$$

So, by selecting $V_{GS} = V_{GS(off)}/3.4$, you should get a midpoint bias in terms of I_D.

To set the drain voltage at midpoint ($V_D = V_{DD}/2$), select a value of R_D to produce the desired voltage drop. Choose R_G arbitrarily large to prevent loading on the driving stage in a cascaded amplifier arrangement. Example 7–9 illustrates these concepts.

EXAMPLE 7–9

Select resistor values for R_D and R_S in Figure 7–19 to set up an approximate midpoint bias. For this particular JFET, the parameters are $I_{DSS} = 12$ mA and $V_{GS(off)} = -3$ V. V_D should be approximately 6 V (one-half of V_{DD}).

▶ **FIGURE 7–19**

Solution   For midpoint bias,

$$I_D \cong \frac{I_{DSS}}{2} = 6 \text{ mA}$$

and

$$V_{GS} \cong \frac{V_{GS(off)}}{3.4} = \frac{-3 \text{ V}}{3.4} = -882 \text{ mV}$$

Then

$$R_S = \left| \frac{V_{GS}}{I_D} \right| = \frac{882 \text{ mV}}{6 \text{ mA}} = \textbf{147 } \boldsymbol{\Omega}$$

$$V_D = V_{DD} - I_D R_D$$

$$I_D R_D = V_{DD} - V_D$$

$$R_D = \frac{V_{DD} - V_D}{I_D} = \frac{12 \text{ V} - 6 \text{ V}}{6 \text{ mA}} = \textbf{1 k}\boldsymbol{\Omega}$$

Related Problem   Select resistor values in Figure 7–19 to set up an approximate midpoint bias. The JFET parameters are $I_{DSS} = 10$ mA and $V_{GS(off)} = -10$ V. $V_{DD} = 15$ V.

Open file E07-09 in the Examples folder on your CD-ROM. The circuit has the calculated values for R_D and R_S from the Related Problem. Verify that an approximate midpoint bias is established by measuring V_D and I_D.

Graphical Analysis of a Self-Biased JFET

You can use the transfer characteristic curve of a JFET and certain parameters to determine the Q-point (I_D and V_{GS}) of a self-biased circuit. A circuit is shown in Figure 7–20(a), and a transfer characteristic curve is shown in Figure 7–20(b). If a curve is not

▲ FIGURE 7–20

A self-biased JFET and its transfer characteristic curve.

available from the data sheet, you can plot it from Equation 7–1 using data sheet values for I_{DSS} and $V_{GS(off)}$.

To determine the Q-point of the circuit in Figure 7–20(a), a self-bias dc load line is established as follows. First, calculate V_{GS} when I_D is zero.

$$V_{GS} = -I_D R_S = (0)(470\ \Omega) = 0\ \text{V}$$

This establishes a point at the origin on the graph ($I_D = 0$, $V_{GS} = 0$). Next, get I_{DSS} from the data sheet and calculate V_{GS} when $I_D = I_{DSS}$. From the curve, $I_{DSS} = 10$ mA for the JFET in Figure 7–20.

$$V_{GS} = -I_D R_S = -(10\ \text{mA})(470\ \Omega) = -4.7\ \text{V}$$

This establishes a second point on the graph ($I_D = 10$ mA, $V_{GS} = -4.7$ V). Now, with two points, the load line can be drawn on the graph of the transfer characteristic curve as shown in Figure 7–21. The point where the line intersects the transfer characteristic curve is the Q-point of the circuit.

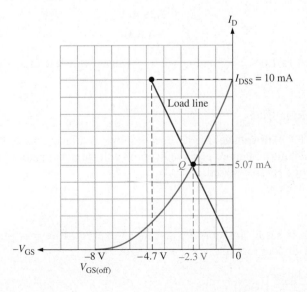

▲ FIGURE 7–21

The intersection of the self-bias dc load line and the transfer characteristic curve is the Q-point.

EXAMPLE 7–10

Determine the Q-point for the JFET circuit in Figure 7–22(a). The transfer characteristic curve is given in Figure 7–22(b).

(a)   (b)

▲ **FIGURE 7–22**

Solution   For $I_D = 0$,

$$V_{GS} = -I_D R_S = (0)(680 \ \Omega) = 0 \ V$$

This gives a point at the origin. From the curve, $I_{DSS} = 4$ mA. So for $I_D = I_{DSS} = 4$ mA.

$$V_{GS} = -I_D R_S = -(4 \ \text{mA})(680 \ \Omega) = -2.72 \ V$$

This gives a second point at 4 mA and -2.72 V. A line is now drawn between the two points, and the values of I_D and V_{GS} at the intersection of the line and the curve are taken from the graph, as illustrated in Figure 7–22(b). The Q-point values from the graph are

$$I_D = \textbf{2.25 mA}$$
$$V_{GS} = \textbf{–1.5 V}$$

Related Problem   If R_S is increased to 1.0 kΩ in Figure 7–22(a), what is the new Q-point?

Voltage-Divider Bias

An *n*-channel JFET with voltage-divider bias is shown in Figure 7–23. The voltage at the source of the JFET must be more positive than the voltage at the gate in order to keep the gate-source junction reverse-biased.

The source voltage is

$$V_S = I_D R_S$$

The gate voltage is set by resistors R_1 and R_2 as expressed by the following equation using the voltage-divider formula:

$$V_G = \left(\frac{R_2}{R_1 + R_2} \right) V_{DD}$$

▶ **FIGURE 7–23**

An *n*-channel JFET with voltage-divider bias ($I_S = I_D$).

The gate-to-source voltage is

$$V_{GS} = V_G - V_S$$

and the source voltage is

$$V_S = V_G - V_{GS}$$

The drain current can be expressed as

$$I_D = \frac{V_S}{R_S}$$

Substituting for V_S,

$$I_D = \frac{V_G - V_{GS}}{R_S}$$

EXAMPLE 7–11

Determine I_D and V_{GS} for the JFET with voltage-divider bias in Figure 7–24, given that for this particular JFET the internal parameter values are such that $V_D \cong 7$ V.

▶ **FIGURE 7–24**

Solution $$I_D = \frac{V_{DD} - V_D}{R_D} = \frac{12\text{ V} - 7\text{ V}}{3.3\text{ k}\Omega} = \frac{5\text{ V}}{3.3\text{ k}\Omega} = \textbf{1.52 mA}$$

Calculate the gate-to-source voltage as follows:

$$V_S = I_D R_S = (1.52 \text{ mA})(2.2 \text{ k}\Omega) = 3.34 \text{ V}$$

$$V_G = \left(\frac{R_2}{R_1 + R_2}\right)V_{DD} = \left(\frac{1.0 \text{ M}\Omega}{7.8 \text{ M}\Omega}\right)12 \text{ V} = 1.54 \text{ V}$$

$$V_{GS} = V_G - V_S = 1.54 \text{ V} - 3.34 \text{ V} = \mathbf{-1.8 \text{ V}}$$

If V_D had not been given in this example, the Q-point values could not have been found without the transfer characteristic curve.

Related Problem   Given that $V_D = 6$ V when another JFET is inserted in the circuit of Figure 7–24, determine the Q-point.

Graphical Analysis of a JFET with Voltage-Divider Bias

An approach similar to the one used for self-bias can be used with voltage-divider bias to graphically determine the Q-point of a circuit on the transfer characteristic curve.

In a JFET with voltage-divider bias when $I_D = 0$, V_{GS} is not zero, as in the self-biased case, because the voltage divider produces a voltage at the gate independent of the drain current. The voltage-divider dc load line is determined as follows.

For $I_D = 0$,

$$V_S = I_D R_S = (0)R_S = 0 \text{ V}$$
$$V_{GS} = V_G - V_S = V_G - 0 \text{ V} = V_G$$

Therefore, one point on the line is at $I_D = 0$ and $V_{GS} = V_G$.

For $V_{GS} = 0$,

$$I_D = \frac{V_G - V_{GS}}{R_S} = \frac{V_G}{R_S}$$

A second point on the line is at $I_D = V_G/R_S$ and $V_{GS} = 0$. The generalized dc load line is shown in Figure 7–25.

▶ **FIGURE 7–25**

Generalized dc load line for a JFET with voltage-divider bias.

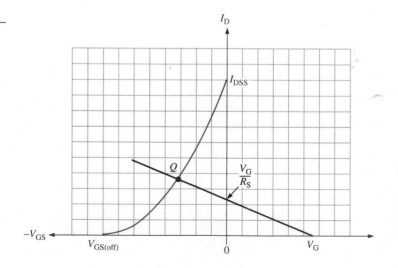

EXAMPLE 7-12

Determine the approximate Q-point for the JFET with voltage-divider bias in Figure 7–26(a), given that this particular device has a transfer characteristic curve as shown in Figure 7–26(b).

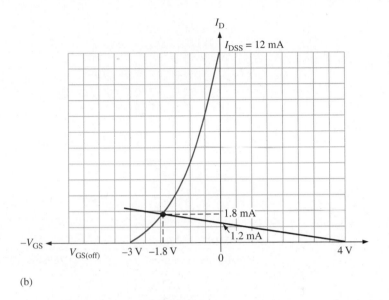

▲ **FIGURE 7-26**

Solution   First, establish the two points for the bias line. For $I_D = 0$,

$$V_{GS} = V_G = \left(\frac{R_2}{R_1 + R_2}\right)V_{DD} = \left(\frac{2.2 \text{ M}\Omega}{4.4 \text{ M}\Omega}\right)8 \text{ V} = 4 \text{ V}$$

The first point is at $I_D = 0$ and $V_{GS} = 4$ V. For $V_{GS} = 0$,

$$I_D = \frac{V_G - V_{GS}}{R_S} = \frac{V_G}{R_S} = \frac{4 \text{ V}}{3.3 \text{ k}\Omega} = 1.2 \text{ mA}$$

The second point is at $I_D = 1.2$ mA and $V_{GS} = 0$.

The load line is drawn in Figure 7–26(b), and the approximate Q-point values of $I_D \cong$ 1.8 mA and $V_{GS} \cong -1.8$ V are picked off the graph, as indicated.

Related Problem   Change R_S to 4.7 kΩ and determine the Q-point for the circuit in Figure 7–26(a).

Open file E07-12 in the Examples folder on your CD-ROM. Measure the Q-point values of I_D and V_{GS} and see how they compare to the graphically determined values from the Related Problem.

Q-Point Stability

Unfortunately, the transfer characteristic of a JFET can differ considerably from one device to another of the same type. If, for example, a 2N5459 JFET is replaced in a given bias circuit with another 2N5459, the transfer characteristic curve can vary greatly, as illustrated in Figure 7–27(a). In this case, the maximum I_{DSS} is 16 mA and the minimum

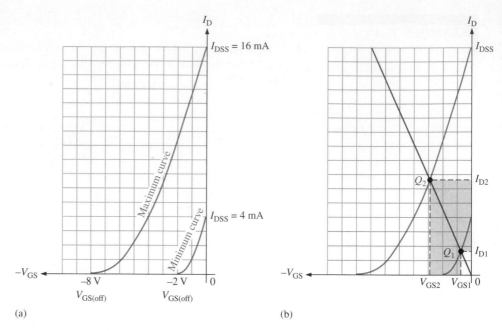

(a)

(b)

▲ **FIGURE 7–27**

Variation in the transfer characteristic of 2N5459 JFETs and the effect on the Q-point.

I_{DSS} is 4 mA. Likewise, the maximum $V_{GS(off)}$ is -8 V and the minimum $V_{GS(off)}$ is -2 V. This means that if you have a selection of 2N5459s and you randomly pick one out, it can have values anywhere within these ranges.

If a self-bias dc load line is drawn as illustrated in Figure 7–27(b), the same circuit using a 2N5459 can have a Q-point anywhere along the line from Q_1, the minimum bias point, to Q_2, the maximum bias point. Accordingly, the drain current can be any value

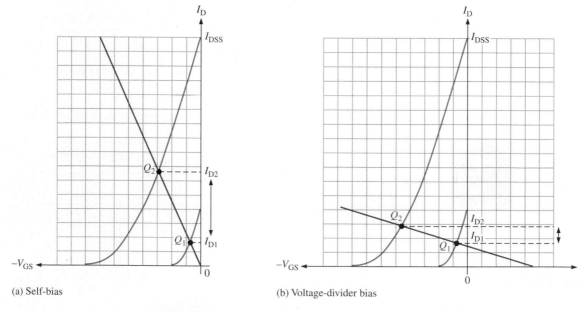

(a) Self-bias

(b) Voltage-divider bias

▲ **FIGURE 7–28**

The change in I_D between the minimum and the maximum Q-points is much less for a JFET with voltage-divider bias than for a self-biased JFET.

between I_{D1} and I_{D2}, as shown. This means that the dc voltage at the drain can have a range of values depending on I_D. Also, the gate-to-source voltage can be any value between V_{GS1} and V_{GS2}, as indicated.

With voltage-divider bias, the dependency of I_D on the range of Q-points is reduced because the slope of the bias line is less than for self-bias for a given JFET. Although V_{GS} varies quite a bit for both self-bias and voltage-divider bias, I_D is much more stable with voltage-divider bias, as illustrated in Figure 7–28.

SECTION 7–3 REVIEW

1. Should a *p*-channel JFET have a positive or a negative V_{GS}?
2. In a certain self-biased *n*-channel JFET circuit, I_D = 8 mA and R_S = 1.0 kΩ. Determine V_{GS}.
3. An *n*-channel JFET with voltage-divider bias has a gate voltage of 3 V and a source voltage of 5 V. Calculate V_{GS}.

7–4 THE MOSFET

The MOSFET (metal oxide semiconductor field-effect transistor) is the second category of field-effect transistor. The MOSFET differs from the JFET in that it has no *pn* junction structure; instead, the gate of the MOSFET is insulated from the channel by a silicon dioxide (SiO_2) layer. The two basic types of MOSFETs are depletion (D) and enhancement (E). Because of the insulated gate, these devices are sometimes called IGFETs.

After completing this section, you should be able to

- **Explain the operation of MOSFETs**
- Describe the structural difference between an *n*-channel and a *p*-channel depletion MOSFET (D-MOSFET)
- Explain the depletion mode
- Explain the enhancement mode
- Identify the symbols for *n*-channel and *p*-channel D-MOSFETs
- Describe the structural difference between an *n*-channel and a *p*-channel enhancement MOSFET (E-MOSFET)
- Identify the symbols for *n*-channel and *p*-channel E-MOSFETs
- Explain how D-MOSFETs and E-MOSFETs differ
- Discuss power MOSFETs
- Discuss dual-gate MOSFETs

Depletion MOSFET (D-MOSFET)

One type of **MOSFET** is the depletion MOSFET (D-MOSFET), and Figure 7–29 illustrates its basic structure. The drain and source are diffused into the substrate material and then connected by a narrow channel adjacent to the insulated gate. Both *n*-channel and *p*-channel devices are shown in the figure. We will use the *n*-channel device to describe the basic operation. The *p*-channel operation is the same, except the voltage polarities are opposite those of the *n*-channel.

(a) *n* channel   (b) *p* channel

▲ **FIGURE 7–29**

Representation of the basic structure of D-MOSFETs.

The D-MOSFET can be operated in either of two modes—the depletion mode or the enhancement mode—and is sometimes called a *depletion/enhancement MOSFET.* Since the gate is insulated from the channel, either a positive or a negative gate voltage can be applied. The *n*-channel MOSFET operates in the **depletion** mode when a negative gate-to-source voltage is applied and in the **enhancement** mode when a positive gate-to-source voltage is applied. These devices are generally operated in the depletion mode.

Depletion Mode Visualize the gate as one plate of a parallel-plate capacitor and the channel as the other plate. The silicon dioxide insulating layer is the dielectric. With a negative gate voltage, the negative charges on the gate repel conduction electrons from the channel, leaving positive ions in their place. Thereby, the *n* channel is depleted of some of its electrons, thus decreasing the channel conductivity. The greater the negative voltage on the gate, the greater the depletion of *n*-channel electrons. At a sufficiently negative gate-to-source voltage, $V_{GS(off)}$, the channel is totally depleted and the drain current is zero. This depletion mode is illustrated in Figure 7–30(a). Like the *n*-channel JFET, the *n*-channel D-MOSFET conducts drain current for gate-to-source voltages between $V_{GS(off)}$ and zero. In addition, the D-MOSFET conducts for values of V_{GS} above zero.

(a) Depletion mode: V_{GS} negative and less than $V_{GS(off)}$ (b) Enhancement mode: V_{GS} positive

▲ **FIGURE 7–30**

Operation of *n*-channel D-MOSFET.

Enhancement Mode   With a positive gate voltage, more conduction electrons are attracted into the channel, thus increasing (enhancing) the channel conductivity, as illustrated in Figure 7–30(b).

D–MOSFET Symbols   The schematic symbols for both the *n*-channel and the *p*-channel depletion MOSFETs are shown in Figure 7–31. The substrate, indicated by the arrow, is normally (but not always) connected internally to the source. Sometimes, there is a separate substrate pin. An inward-pointing substrate arrow is for *n* channel, and an outward-pointing arrow is for *p* channel.

▶ **FIGURE 7–31**

D-MOSFET schematic symbols.

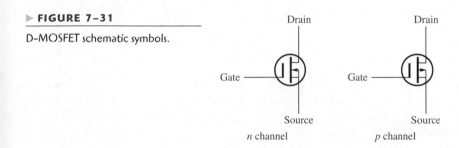

n channel   *p* channel

Enhancement MOSFET (E-MOSFET)

The E-MOSFET operates *only* in the enhancement mode and has no depletion mode. It differs in construction from the D-MOSFET in that it has no structural channel. Notice in Figure 7–32(a) that the substrate extends completely to the SiO_2 layer. For an *n*-channel device, a positive gate voltage above a threshold value *induces* a channel by creating a thin layer of negative charges in the substrate region adjacent to the SiO_2 layer, as shown in Figure 7–32(b). The conductivity of the channel is enhanced by increasing the gate-to-source voltage and thus pulling more electrons into the channel area. For any gate voltage below the threshold value, there is no channel.

(a) Basic construction   (b) Induced channel ($V_{GS} > V_{GS(th)}$)

▲ **FIGURE 7–32**

Representation of the basic E-MOSFET construction and operation (*n*-channel).

The schematic symbols for the *n*-channel and *p*-channel E-MOSFETs are shown in Figure 7–33. The broken lines symbolize the absence of a physical channel. Like the D-MOSFET, some devices have a separate substrate connection.

▲ **FIGURE 7–33**

E-MOSFET schematic symbols.

▲ **FIGURE 7–34**

Cross section of conventional E-MOSFET structure.

Power MOSFETs

The conventional enhancement MOSFETs have a long thin lateral channel as shown in the structural view in Figure 7–34. This results in a relatively high drain-to-source resistance and limits the E-MOSFET to low power applications. When the gate is positive, the channel is formed close to the gate between the source and the drain, as shown.

Lateral Double Diffused MOSFET (LDMOSFET)   The LDMOSFET is a type of enhancement MOSFET designed for power applications. This device has a shorter channel between drain and source than does the conventional E-MOSFET. The shorter channel results in lower resistance, which allows higher current and voltage.

Figure 7–35 shows the basic structure of an LDMOSFET. When the gate is positive, a very short *n* channel is induced in the *p* layer between the lightly doped source and the n^- region. There is current from the drain through the *n* regions and the induced channel to the source, as indicated.

▶ **FIGURE 7–35**

Cross section of LDMOSFET structure.

VMOSFET   The V-groove MOSFET is another variation of the conventional E-MOSFET designed to achieve higher power capability by creating a shorter and wider channel with less resistance between the drain and source. The shorter, wider channels allow for higher currents and, thus, greater power dissipation. Frequency response is also improved.

The VMOSFET has two source connections, the gate connection on top, and the drain connection on the bottom, as shown in Figure 7–36. The channel is induced vertically along both sides of the V-shaped groove between the drain (n^+ substrate where n^+ means a higher doping level than n^-) and the source connections. The channel length is set by the thickness of the layers, which is controlled by doping densities and diffusion time rather than by mask dimensions.

▶ **FIGURE 7–36**

Cross section of VMOSFET structure.

TMOSFET TMOSFET is similar to VMOSFET except that it doesn't use a V-shaped groove and is, therefore, easier to manufacture. The structure of TMOSFET is illustrated in Figure 7–37. The gate structure is embedded in a silicon dioxide layer, and the source contact is continuous over the entire surface area. The drain is on the bottom. TMOSFET achieves greater packing density than VMOSFET, while retaining the short vertical channel advantage.

Dual-Gate MOSFETs

The dual-gate MOSFET can be either a depletion or an enhancement type. The only difference is that it has two gates, as shown in Figure 7–38. As previously mentioned, one drawback of a FET is its high input capacitance, which restricts its use at higher frequencies. By using a dual-gate device, the input capacitance is reduced, thus making the device useful in high-frequency RF amplifier applications. Another advantage of the dual-gate arrangement is that it allows for an automatic gain control (AGC) input in certain RF amplifiers.

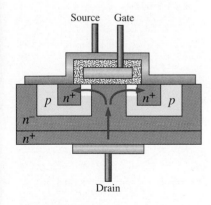

▲ **FIGURE 7–37**

Cross section of TMOSFET structure.

(a) D-MOSFET   (b) E-MOSFET

▲ **FIGURE 7–38**

Dual-gate *n*-channel MOSFET symbols.

SECTION 7–4 REVIEW

1. Name the two basic types of MOSFETs.

2. If the gate-to-source voltage in a depletion MOSFET is zero, what is the current from drain to source?

3. If the gate-to-source voltage in an E-MOSFET is zero, what is the current from drain to source?

7–5   MOSFET CHARACTERISTICS AND PARAMETERS

Much of the discussion concerning JFET characteristics and parameters applies equally to MOSFETs. In this section, the differences are highlighted.

After completing this section, you should be able to

- **Define, discuss, and apply important MOSFET parameters**
- Analyze a D-MOSFET transfer characteristic curve
- Use the equation for the D-MOSFET transfer characteristic to calculate I_D
- Analyze an E-MOSFET transfer characteristic curve
- Use the equation for the E-MOSFET transfer characteristic to calculate I_D
- Use a MOSFET data sheet
- Discuss handling precautions for MOS devices

D-MOSFET Transfer Characteristic

As previously discussed, the D-MOSFET can operate with either positive or negative gate voltages. This is indicated on the general transfer characteristic curves in Figure 7–39 for both n-channel and p-channel MOSFETs. The point on the curves where $V_{GS} = 0$ corresponds to I_{DSS}. The point where $I_D = 0$ corresponds to $V_{GS(off)}$. As with the JFET, $V_{GS(off)} = -V_P$.

The square-law expression in Equation 7–1 for the JFET curve also applies to the D-MOSFET curve, as Example 7–13 demonstrates.

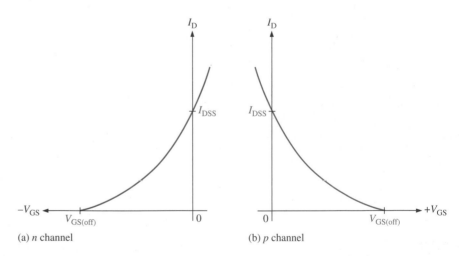

(a) n channel   (b) p channel

▲ **FIGURE 7–39**

D–MOSFET general transfer characteristic curves.

EXAMPLE 7–13

For a certain D-MOSFET, $I_{DSS} = 10$ mA and $V_{GS(off)} = -8$ V.

(a) Is this an n-channel or a p-channel?

(b) Calculate I_D at $V_{GS} = -3$ V.

(c) Calculate I_D at $V_{GS} = +3$ V.

Solution (a) The device has a negative $V_{GS(off)}$; therefore, it is an **n-channel** MOSFET.

(b) $I_D = I_{DSS}\left(1 - \dfrac{V_{GS}}{V_{GS(off)}}\right)^2 = (10\ mA)\left(1 - \dfrac{-3\ V}{-8\ V}\right)^2 = \mathbf{3.91\ mA}$

(c) $I_D = (10\ mA)\left(1 - \dfrac{+3\ V}{-8\ V}\right)^2 = \mathbf{18.9\ mA}$

Related Problem For a certain D-MOSFET, $I_{DSS} = 18$ mA and $V_{GS(off)} = +10$ V.

(a) Is this an *n*-channel or a *p*-channel?

(b) Determine I_D at $V_{GS} = +4$ V.

(c) Determine I_D at $V_{GS} = -4$ V.

E-MOSFET Transfer Characteristic

The E-MOSFET uses only channel enhancement. Therefore, an *n*-channel device requires a positive gate-to-source voltage, and a *p*-channel device requires a negative gate-to-source voltage. Figure 7–40 shows the general transfer characteristic curves for both types of E-MOSFETs. As you can see, there is no drain current when $V_{GS} = 0$. Therefore, the E-MOSFET does not have a significant I_{DSS} parameter, as do the JFET and the D-MOSFET. Notice also that there is ideally no drain current until V_{GS} reaches a certain nonzero value called the *threshold voltage*, $V_{GS(th)}$.

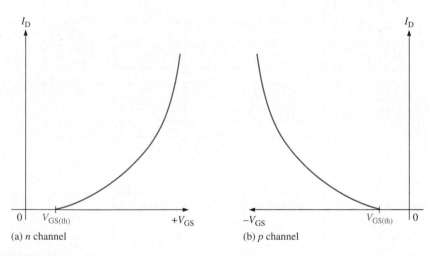

(a) *n* channel (b) *p* channel

▲ **FIGURE 7–40**

E-MOSFET general transfer characteristic curves.

The equation for the parabolic transfer characteristic curve of the E-MOSFET differs from that of the JFET and the D-MOSFET because the curve starts at $V_{GS(th)}$ rather than $V_{GS(off)}$ on the horizontal axis and never intersects the vertical axis. The equation for the E-MOSFET transfer characteristic curve is

$$I_D = K(V_{GS} - V_{GS(th)})^2$$

Equation 7–4

The constant K depends on the particular MOSFET and can be determined from the data sheet by taking the specified value of I_D, called $I_{D(on)}$, at the given value of V_{GS} and substituting the values into Equation 7–4. A typical E-MOSFET data sheet is given in Figure 7–41.

Maximum Ratings

Rating	Symbol	Value	Unit
Drain-Source voltage	V_{DSS}	60	V dc
Drain-Gate voltage ($R_{GS} = 1\ M\Omega$)	V_{DGR}	60	V dc
Gate-Source voltage	V_{GS}	±40	V dc

Drain current
  Continuous
  Pulsed
	I_D
	I_{DM}
		150
		1000
			mA dc

Total power dissipation @ $T_A = 25°C$ Derate above 25°C	P_D	400 3.2	mW mW/°C
Operating and storage temperature range	T_J, T_{stg}	−55 to +150	°C

Thermal Characteristics

Thermal resistance junction to ambient	$R_{\theta JA}$	312.5	°C/W

Maximum lead temperature for
  soldering purposes, 1/16" from case
  for 10 seconds
	T_L	300	°C

2N7008

Case 29-04, Style 22
TO-92 (TO-226AA)

3 Drain

2 Gate

1 Source

**TMOS FET
Transistor**

N channel — Enhancement

Electrical Characteristics ($T_C = 25°C$ unless otherwise noted.)

Characteristic	Symbol	Min	Max	Unit
OFF Characteristics				
Drain-Source breakdown voltage ($V_{GS} = 0, I_D = 100\ \mu A$)	$V_{(BR)DSS}$	60	–	V dc

Zero gate voltage drain current
  ($V_{DS} = 50\ V, V_{GS} = 0$)
  ($V_{DS} = 50\ V, V_{GS} = 0, T_J = 125°C$)
	I_{DSS}	–
		–
			1.0
			500
				μA dc

Gate-Body leakage current, forward ($V_{GSF} = 30\ V\ dc, V_{DS} = 0$)	I_{GSSF}	–	−100	nA dc
ON Characteristics				
Gate threshold voltage ($V_{DS} = V_{GS}, I_D = 250\ \mu A$)	$V_{GS(th)}$	1.0	2.5	V dc

Static drain-source on-resistance
  ($V_{GS} = 5.0\ V\ dc, I_D = 50\ A\ dc$)
  ($V_{GS} = 10\ V\ dc, I_D = 500\ mA\ dc, T_C = 125°C$)
	$r_{DS(on)}$	–
		–
			7.5
			13.5
				Ohm

Drain-Source on-voltage
  ($V_{GS} = 5.0\ V, I_D = 50\ mA$)
  ($V_{GS} = 10\ V, I_D = 500\ mA$)
	$V_{DS(on)}$	–
		–
			1.5
			3.75
				V dc

On-state drain current ($V_{GS} = 10\ V, V_{DS} \geq 2.0 V_{D(on)}$)	$I_{D(on)}$	500	–	mA
Forward transconductance ($V_{DS} \geq 2.0 V_{DS(on)}, I_D = 200\ mA$)	g_{fs}	80	–	μmhos or μS
Dynamic Characteristics				
Input capacitance	C_{iss}	–	50	pF
Output capacitance	C_{oss}	–	25	
Reverse transfer capacitance	C_{rss}	–	5.0	
Switching Characteristics				
Turn-On delay time	t_{on}	–	20	ns
Turn-Off delay time	t_{off}	–	20	

For the Dynamic Characteristics rows the test conditions are ($V_{DS} = 25\ V, V_{GS} - 0$, $f = 1.0\ MHz$). For the Switching Characteristics rows the test conditions are ($V_{DD} = 30\ V, I_D = 200\ mA$, $R_{gen} = 25\ ohms, R_L = 150\ ohms$).

▲ **FIGURE 7–41**

Data sheet for the 2N7008 *n*–channel E-MOSFET (TMOSFET construction).

EXAMPLE 7–14

The data sheet in Figure 7–41 for a 2N7008 E-MOSFET gives $I_{D(on)} = 500$ mA (minimum) at $V_{GS} = 10$ V and $V_{GS(th)} = 1$ V. Determine the drain current for $V_{GS} = 5$ V.

Solution   First, solve for *K* using Equation 7–4.

$$K = \frac{I_{D(on)}}{(V_{GS} - V_{GS(th)})^2} = \frac{500\ mA}{(10\ V - 1\ V)^2} = \frac{500\ mA}{81\ V^2} = 6.17\ mA/V^2$$

Next, using the value of *K*, calculate I_D for $V_{GS} = 5$ V.

$$I_D = K(V_{GS} - V_{GS(th)})^2 = (6.17\ mA/V^2)(5\ V - 1\ V)^2 = \textbf{98.7 mA}$$

Related Problem   The data sheet for an E-MOSFET gives $I_{D(on)} = 100$ mA at $V_{GS} = 8$ V and $V_{GS(th)} = 4$ V. Find I_D when $V_{GS} = 6$ V.

Handling Precautions

All MOS devices are subject to damage from electrostatic discharge (ESD). Because the gate of a MOSFET is insulated from the channel, the input resistance is extremely high (ideally infinite). The gate leakage current, I_{GSS}, for a typical MOSFET is in the pA range, whereas the gate reverse current for a typical JFET is in the nA range. The input capacitance results from the insulated gate structure. Excess static charge can be accumulated because the input capacitance combines with the very high input resistance and can result in damage to the device. To avoid damage from ESD, certain precautions should be taken when handling MOSFETs:

1. MOS devices should be shipped and stored in conductive foam.

2. All instruments and metal benches used in assembly or test should be connected to earth ground (round or third prong of 110 V wall outlets).

3. The assembler's or handler's wrist should be connected to earth ground with a length of wire and a high-value series resistor.

4. Never remove a MOS device (or any other device, for that matter) from the circuit while the power is on.

5. Do not apply signals to a MOS device while the dc power supply is off.

SECTION 7–5 REVIEW

1. What is the major difference in construction of the D–MOSFET and the E–MOSFET?
2. Name two parameters of an E–MOSFET that are not specified for D–MOSFETs?
3. What is ESD?

7–6   MOSFET BIASING

In this section, you will see three ways to bias a MOSFET: zero-bias, voltage-divider bias, and drain-feedback bias. As you know, biasing is very important in amplifier applications, which you will study in the next chapter.

After completing this section, you should be able to

■ **Discuss and analyze MOSFET bias circuits**

■ Describe zero-bias of a D-MOSFET

■ Analyze a zero-biased MOSFET circuit

■ Describe voltage-divider bias of an E-MOSFET

■ Describe drain-feedback bias of an E-MOSFET

D-MOSFET Bias

Recall that D-MOSFETs can be operated with either positive or negative values of V_{GS}. A simple bias method is to set $V_{GS} = 0$ so that an ac signal at the gate varies the gate-to-source voltage above and below this 0 V bias point. A MOSFET with zero bias is shown

▶ **FIGURE 7–42**

A zero-biased D-MOSFET.

(a)   (b)

in Figure 7–42(a). Since $V_{GS} = 0$, $I_D = I_{DSS}$ as indicated. The drain-to-source voltage is expressed as follows:

$$V_{DS} = V_{DD} - I_{DSS}R_D$$

The purpose of R_G is to accommodate an ac signal input by isolating it from ground, as shown in Figure 7–42(b). Since there is no dc gate current, R_G does not affect the zero gate-to-source bias.

EXAMPLE 7–15

Determine the drain-to-source voltage in the circuit of Figure 7–43. The MOSFET data sheet gives $V_{GS(off)} = -8$ V and $I_{DSS} = 12$ mA.

▶ **FIGURE 7–43**

Solution   Since $I_D = I_{DSS} = 12$ mA, the drain-to-source voltage is

$$V_{DS} = V_{DD} - I_{DSS}R_D = 18\text{ V} - (12\text{ mA})(620\text{ }\Omega) = \textbf{10.6 V}$$

Related Problem   Find V_{DS} in Figure 7–43 when $V_{GS(off)} = -10$ V and $I_{DSS} = 20$ mA.

E-MOSFET Bias

Recall that E-MOSFETs must have a V_{GS} greater than the threshold value, $V_{GS(th)}$, so zero bias cannot be used. Figure 7–44 shows two ways to bias an E-MOSFET (D-MOSFETs can also be biased using these methods). An *n*-channel device is used for purposes of illustration. In either the voltage-divider or drain-feedback bias arrangement, the purpose

► FIGURE 7–44

Common E-MOSFET biasing arrangements.

(a) Voltage-divider bias   (b) Drain-feedback bias

is to make the gate voltage more positive than the source by an amount exceeding $V_{GS(th)}$. Equations for the analysis of the voltage-divider bias in Figure 7–44(a) are as follows:

$$V_{GS} = \left(\frac{R_2}{R_1 + R_2}\right)V_{DD}$$

$$V_{DS} = V_{DD} - I_D R_D$$

where $I_D = K(V_{GS} - V_{GS(th)})^2$ from Equation 7–4.

In the drain-feedback bias circuit in Figure 7–44(b), there is negligible gate current and, therefore, no voltage drop across R_G. This makes $V_{GS} = V_{DS}$.

EXAMPLE 7–16

Determine V_{GS} and V_{DS} for the E-MOSFET circuit in Figure 7–45. Assume this particular MOSFET has minimum values of $I_{D(on)} = 200$ mA at $V_{GS} = 4$ V and $V_{GS(th)} = 2$ V.

► FIGURE 7–45

Solution

$$V_{GS} = \left(\frac{R_2}{R_1 + R_2}\right)V_{DD} = \left(\frac{15\ k\Omega}{115\ k\Omega}\right)24\ V = \textbf{3.13 V}$$

Find K using the minimum value of $I_{D(on)}$ in Equation 7–4.

$$K = \frac{I_{D(on)}}{(V_{GS} - V_{GS(th)})^2} = \frac{200\ mA}{(4\ V - 2\ V)^2} = \frac{200\ mA}{4\ V^2} = 50\ mA/V^2$$

Now calculate I_D for $V_{GS} = 3.13$ V.

$$I_D = K(V_{GS} - V_{GS(th)})^2 = (50 \text{ mA} / \text{V}^2)(3.13 \text{ V} - 2 \text{ V})^2$$
$$= (50 \text{ mA} / \text{V}^2)(1.13 \text{ V})^2 = 63.8 \text{ mA}$$

Finally, calculate V_{DS}.

$$V_{DS} = V_{DD} - I_D R_D = 24 \text{ V} - (63.8 \text{ mA})(200 \text{ } \Omega) = \textbf{11.2 V}$$

Related Problem   Determine V_{GS} and V_{DS} for the maximum specified value of $V_{GS(th)}$ if the MOSFET in Figure 7–45 is a 2N7008. Refer to the data sheet in Figure 7–41.

> Open file E07-16 in the Examples folder on your CD-ROM. Measure V_{GS} and V_{DS} and compare to the calculated values.

EXAMPLE 7–17

Determine the amount of drain current in Figure 7–46. The MOSFET has a $V_{GS(th)} = 3$ V.

▶ **FIGURE 7–46**

Solution   The meter indicates $V_{GS} = 8.5$ V. Since this is a drain-feedback configuration, $V_{DS} = V_{GS} = 8.5$ V.

$$I_D = \frac{V_{DD} - V_{DS}}{R_D} = \frac{15 \text{ V} - 8.5 \text{ V}}{4.7 \text{ k}\Omega} = \textbf{1.38 mA}$$

Related Problem   Determine I_D if the meter in Figure 7–46 reads 5 V.

SECTION 7–6 REVIEW

1. For a D-MOSFET biased at $V_{GS} = 0$, is the drain current equal to zero, I_{GSS}, or I_{DSS}?
2. For an *n*-channel E-MOSFET with $V_{GS(th)} = 2$ V, V_{GS} must be in excess of what value in order to conduct?

7–7   TROUBLESHOOTING

In this section, we discuss some common faults that may be encountered in several FET circuits and the probable causes for each fault.

After completing this section, you should be able to

■ **Troubleshoot FET circuits**

■ Troubleshoot self-biased JFET circuits

■ Troubleshoot zero-biased MOSFET circuits

Faults in Self-Biased JFET Circuits

Symptom 1: $V_D = V_{DD}$ For this condition, the drain current must be zero because there is no voltage drop across R_D, as illustrated in Figure 7–47(a). As in any circuit, it is good troubleshooting practice to first check for obvious problems such as open or poor connections, as well as charred resistors. Next, disconnect power and measure suspected resistors for opens. If these are okay, the JFET is probably bad. A list of faults that can produce this symptom is as follows:

1. No ground connection at R_S

2. R_S open

3. Open drain lead connection

4. Open source lead connection

5. FET internally open between drain and source

(a) Symptom 1: Drain voltage equal to supply voltage

(b) Symptom 2: Drain voltage less than normal

▲ **FIGURE 7–47**

Two symptoms in a self-biased JFET circuit.

Symptom 2: V_D Significantly Less Than Normal   For this condition, unless the supply voltage is lower than it should be, the drain current must be larger than normal because the drop across R_D is too much. Figure 7–47(b) indicates this situation. This symptom can be caused by any one of the following:

1. Open R_G

2. Open gate lead

3. FET internally open at gate

Any one of these faults will cause the depletion region in the JFET to disappear and the channel to widen, so that the drain current is limited only by R_D, R_S, and the small channel resistance.

Faults in D-MOSFET and E-MOSFET Circuits

One fault that is difficult to detect is when the gate opens in a zero-biased D-MOSFET or an E-MOSFET with drain-feedback bias. In a zero-biased D-MOSFET, the gate-to-source voltage remains zero when an open occurs in the gate circuit; thus, the drain current doesn't change, and the bias appears normal, as indicated in Figure 7–48.

▶ FIGURE 7–48

An open fault in the gate circuit of a D-MOSFET causes no change in I_D.

(a) Normal operation

(b) Gate circuit open (can be either external or internal)

In an E-MOSFET circuit with voltage-divider bias, an open R_1 makes the gate voltage zero. This causes the transistor to be off and act like an open switch because a gate-to-source threshold voltage greater than zero is required to turn the device on. This condition is illustrated in Figure 7–49(a).

If R_2 opens, the gate is at $+V_{DD}$ and the channel resistance is very low so the device approximates a closed switch. The drain current is limited only by R_D. This condition is illustrated in Figure 7–49(b).

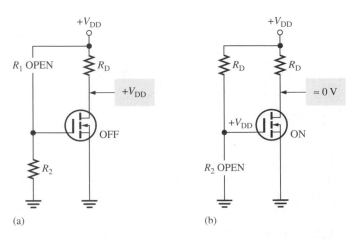

(a)

(b)

▲ FIGURE 7–49

Failures in an E-MOSFET circuit with voltage-divider bias.

EWB/Multisim Troubleshooting Exercises

These file circuits are in the Troubleshooting Exercises folder on your CD-ROM.

1. Open file TSE07-01. Determine if the circuit is working properly and, if not, determine the fault.

2. Open file TSE07-02. Determine if the circuit is working properly and, if not, determine the fault.

3. Open file TSE07-03. Determine if the circuit is working properly and, if not, determine the fault.

SECTION 7–7 REVIEW

1. In a self-biased JFET circuit, the drain voltage equals V_{DD}. If the JFET is okay, what are other possible faults?

2. Why doesn't the drain current change when an open occurs in the gate circuit of a zero-biased D-MOSFET circuit?

3. If the gate of an E-MOSFET becomes shorted to ground in a circuit with voltage-divider bias, what is the drain voltage?

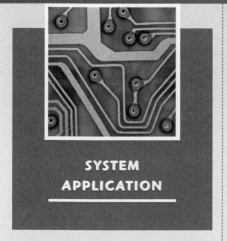

SYSTEM APPLICATION

This application involves electronic instrumentation for waste water treatment facilities. You have been given responsibility for evaluating the circuits used in a waste water neutralization system. Although both digital and analog circuits are used, you will begin by focusing on the pH sensor circuit board and apply the knowledge you have gained in this chapter in completing your assignment.

The Waste Water Neutralization System

Basic Operation The diagram of the waste water neutralization system is shown in Figure 7–50. The system measures and controls the pH of waste water. The pH is a measure of the degree of acidity or alkalinity of a solution. Values of pH range from 0 for the strongest acids, through 7 for neutral solutions, on up to 14 for the strongest bases. Typically waste water is not a strong acid or a strong base, so the range of pH values is typically greater than 2 and less than 11. The pH of the water is measured by sensors at the inlet and outlet of the neutralization basin and at the outlet of the smoothing basin where the pH should be 7, indicating a neutral solution.

The pH sensor produces a small voltage proportional to the pH of the liquid in which it is immersed. The output voltage of each pH sensor is fed to the gate of a MOSFET on the pH sensor circuit board.

The small gate voltage from the sensor controls the drain current which produces an output voltage at the drain that is inversely proportional to the gate voltage but with a larger magnitude. Rheostats are used as drain resistors for calibrating each circuit individually so that, for a given pH, all the output voltages are equal. This is necessary because of variation in the MOSFET characteristics from one device to another.

The MOSFET output voltages go to the analog-to-digital converters and the digital controller. Based on the digitized pH values, the controller determines whether to add sulfuric acid or caustic reagent to the water and the amount that should be added. The digital controller activates the control valves for the correct amount of chemical to properly adjust the pH level. Also, the digitized pH values are sent to a display panel for visual monitoring.

The Printed Circuit Board

■ The system documentation is incomplete and the schematic for the pH sensor circuit board is missing. The transistors are 2N3797.

■ From the circuit board in Figure 7–51, create a schematic and label all components. There are two interconnections on the back side shown as dark traces.

■ Label a copy of the board with component and input/output designations in agreement with the schematic.

Analysis of the pH Sensor Circuits

Refer to the partial data sheet in Figure 7–52 and the pH sensor graph in Figure 7–53.

■ Determine the input resistance of the D-MOSFET in each circuit.

■ Determine the minimum, typical, and maximum resistance values to which the rheostat must be set in each circuit to provide a dc drain voltage of +7 V for a neutral solution (pH = 7). The regulated dc supply voltage is +15 V.

■ Determine the range of output voltage (drain voltage) for a change in the pH sensor voltage of from −500 mV to +500 mV. What is the range of pH values represented? Use typical values from the MOSFET data sheet.

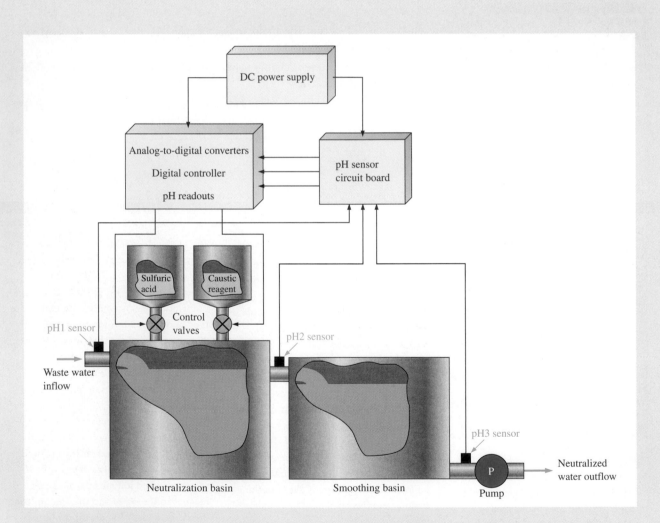

▲ FIGURE 7–50

Diagram of the waste water neutralization system.

Test Procedure

- Develop a step-by-step set of instructions on how to check the pH sensor circuit board for proper operation using the test points (circled numbers) indicated in the test bench setup of Figure 7–54. Assume the test solutions for pH values from 2 to 11 are available.

- Specify voltage values for all the measurements to be made. Take into account the loading effect of the 10 MΩ input resistance of the DMM.

- Provide a fault analysis for all possible component failures.

Troubleshooting

Problems have developed in two boards. Based on the sequence of test bench measurements for each board indicated in Figure 7–55 on page 370, determine the most likely fault in each case. The circled numbers indicate test point connections to the circuit board. Assume typical data sheet values for the MOSFETs. The circuits are supposed to be calibrated, but don't rely on it.

Final Report (Optional)

Submit a final written report on the pH sensor circuit board using an organized format that includes the following:

1. A physical description of the circuits.
2. A discussion of the operation of the circuits.
3. A list of the specifications.
4. A list of parts with part numbers if available.
5. A list of the types of problems on the two faulty circuit boards.
6. A complete description of how you determined the problem on each of the faulty circuit boards.

▲ FIGURE 7–51

pH sensor circuit board.

Maximum Ratings

Rating	Symbol	Value	Unit
Drain-Source voltage 2N3796 2N3797	V_{DS}	 25 20	V dc
Gate-Source voltage	V_{GS}	±10	V dc
Drain current	I_D	20	mA dc
Total device dissipation @ $T_A = 25°C$ Derate above 25°C	P_D	200 1.14	mW mW/°C
Junction temperature range	T_J	+175	°C
Storage channel temperature range	T_{stg}	−65 to +200	°C

2N3796
2N3797

Case 22-03, Style 2
TO-18 (TO-206AA)

3 Drain
Gate 2
1 Source

MOSFETs
Low Power Audio
N channel — Depletion

Electrical Characteristics ($T_A = 25°C$ unless otherwise noted.)

Characteristic		Symbol	Min	Typ	Max	Unit
OFF Characteristics						
Drain-Source breakdown voltage ($V_{GS} = -4.0$ V, $I_D = 5.0$ μA) ($V_{GS} = -7.0$ V, $I_D = 5.0$ μA)	 2N3796 2N3797	$V_{(BR)DSX}$	 25 20	 30 25	 – –	V dc
Gate reverse current ($V_{GS} = -10$ V, $V_{DS} = 0$) ($V_{GS} = -10$ V, $V_{DS} = 0$, $T_A = 150°C$)		I_{GSS}	 – –	 – –	 1.0 200	pA dc
Gate-Source cutoff voltage ($I_D = 0.5$ μA, $V_{DS} = 10$ V) ($I_D = 2.0$ μA, $V_{DS} = 10$ V)	 2N3796 2N3797	$V_{GS(off)}$	 – –	 −3.0 −5.0	 −4.0 −7.0	V dc
Drain-Gate reverse current ($V_{DG} = 10$ V, $I_S = 0$)		I_{DGO}	–	–	1.0	pA dc
ON Characteristics						
Zero-Gate-Voltage drain current ($V_{DS} = 10$ V, $V_{GS} = 0$)	2N3796 2N3797	I_{DSS}	 0.5 2.0	 1.5 2.9	 3.0 6.0	mA dc
On-State drain current ($V_{DS} = 10$ V, $V_{GS} = +3.5$ V)	2N3796 2N3797	$I_{D(on)}$	 7.0 9.0	 8.3 14	 14 18	mA dc
Small-Signal Characteristics						
Forward-transfer admittance ($V_{DS} = 10$ V, $V_{GS} = 0$, $f = 1.0$ kHz) ($V_{DS} = 10$ V, $V_{GS} = 0$, $f = 1.0$ MHz)	 2N3796 2N3797 2N3796 2N3797	$|y_{fs}|$	 900 1500 900 1500	 1200 2300 – –	 1800 3000 – –	μmhos or μS
Output admittance ($V_{DS} = 10$ V, $V_{GS} = 0$, $f = 1.0$ kHz)	2N3796 2N3797	$|y_{os}|$	 – –	 12 27	 25 60	μmhos or μS
Input capacitance ($V_{DS} = 10$ V, $V_{GS} = 0$, $f = 1.0$ MHz)	2N3796 2N3797	C_{iss}	 – –	 5.0 6.0	 7.0 8.0	pF
Reverse transfer capacitance ($V_{DS} = 10$ V, $V_{GS} = 0$, $f = 1.0$ MHz)		C_{rss}	–	0.5	0.8	pF
Functional Characteristics						
Noise figure ($V_{DS} = 10$ V, $V_{GS} = 0$, $f = 1.0$ kHz, $R_S = 3$ megohms)		NF	–	3.8	–	dB

2N3796

2N3797

▲ **FIGURE 7–52**

Partial data sheet for the 2N3797 D-MOSFET.

(a) Equivalent circuit (the voltage
polarity is reversible)

(b) Output voltage versus pH level

▲ **FIGURE 7–53**

Equivalent circuit and output characteristic for the pH sensor.

▲ **FIGURE 7–54**

Test bench setup for the pH sensor circuit board.

▲ **FIGURE 7–55**

Test results for two faulty circuit boards.

SUMMARY OF FIELD-EFFECT TRANSISTORS

JFETs

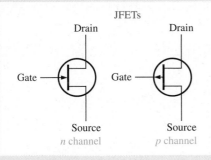

JFETs

Drain   Drain

Gate   Gate

Source   Source

n channel   *p* channel

- Gate-source junction must be reverse-biased.

- V_{GS} controls I_D.

- Value of V_{DS} at which I_D becomes constant is the pinch-off voltage.

- Value of V_{GS} at which I_D becomes zero is the cutoff voltage, $V_{GS(off)}$.

- I_{DSS} is drain current when $V_{GS} = 0$.

- Transfer characteristic:

$$I_D = I_{DSS}\left(1 - \frac{V_{GS}}{V_{GS(off)}}\right)^2$$

- Forward transconductance:

$$g_m = g_{m0}\left(1 - \frac{V_{GS}}{V_{GS(off)}}\right)$$

$$g_{m0} = \frac{2I_{DSS}}{|V_{GS(off)}|}$$

D-MOSFETs

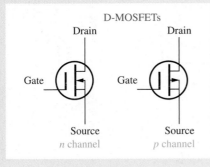

D-MOSFETs

Drain   Drain

Gate   Gate

Source   Source

n channel   *p* channel

Except that it can be operated in enhancement mode, the D-MOSFET characteristics are the same as JFET.

- *Depletion mode:*

  n channel: V_{GS} negative

  p channel: V_{GS} positive

- *Enhancement mode:*

  n channel: V_{GS} positive

  p channel: V_{GS} negative

- V_{GS} controls I_D.

- Value of V_{GS} at which I_D becomes zero is the cutoff voltage, $V_{GS(off)}$.

- I_{DSS} is drain current when $V_{GS} = 0$.

- Transfer characteristic:

$$I_D = I_{DSS}\left(1 - \frac{V_{GS}}{V_{GS(off)}}\right)^2$$

SUMMARY OF FIELD-EFFECT TRANSISTORS, *continued*

E-MOSFETs

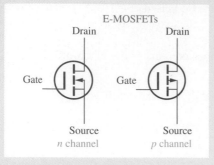

E-MOSFETs

n channel

p channel

There is no depletion mode and characteristics differ from D-MOSFET.

- *Enhancement mode:*

  n channel: V_{GS} positive

  p channel: V_{GS} negative

- V_{GS} controls I_D.

- Value of V_{GS} at which I_D begins is the threshold voltage, $V_{GS(th)}$.

- Transfer characteristic:

$$I_D = K(V_{GS} - V_{GS(th)})^2$$

- *K* in formula can be calculated by substituting data sheet values $I_{D(on)}$ for I_D and V_{GS} at which $I_{D(on)}$ is specified for V_{GS}.

FET BIASING (voltage polarities and current directions reverse for *p* channel)

JFET

Self-bias

Voltage-divider bias

D-MOSFET

Zero bias

E-MOSFET

Drain-feedback bias

Voltage-divider bias

CHAPTER SUMMARY

- Field-effect transistors are unipolar devices (one-charge carrier).
- The three FET terminals are source, drain, and gate.
- The JFET operates with a reverse-biased *pn* junction (gate-to-source).
- The high input resistance of a JFET is due to the reverse-biased gate-source junction.
- Reverse bias of a JFET produces a depletion region within the channel, thus increasing channel resistance.
- For an *n*-channel JFET, V_{GS} can vary from zero negatively to cutoff, $V_{GS(off)}$. For a *p*-channel JFET, V_{GS} can vary from zero positively to $V_{GS(off)}$.
- I_{DSS} is the constant drain current when $V_{GS} = 0$. This is true for both JFETs and D-MOSFETs.
- A FET is called a *square-law device* because of the relationship of I_D to the square of a term containing V_{GS}.
- Unlike JFETs and D-MOSFETs, the E-MOSFET cannot operate with $V_{GS} - 0$ V.
- Midpoint bias for a JFET is $I_D = I_{DSS}/2$, obtained by setting $V_{GS} \cong V_{GS(off)}/3.4$.
- The Q-point in a JFET with voltage-divider bias is more stable than in a self-biased JFET.
- MOSFETs differ from JFETs in that the gate of a MOSFET is insulated from the channel by an SiO$_2$ layer, whereas the gate and channel in a JFET are separated by a *pn* junction.
- A depletion MOSFET (D-MOSFET) can operate with a zero, positive, or negative gate-to-source voltage.
- The D-MOSFET has a physical channel between the drain and source.
- For an *n*-channel D-MOSFET, negative values of V_{GS} produce the depletion mode and positive values produce the enhancement mode.
- The enhancement MOSFET (E-MOSFET) has no physical channel.
- A channel is induced in an E-MOSFET by the application of a V_{GS} greater than the threshold value, $V_{GS(th)}$.
- Midpoint bias for a D-MOSFET is $I_D = I_{DSS}$, obtained by setting $V_{GS} = 0$.
- An E-MOSFET has no I_{DSS} parameter.
- An *n*-channel E-MOSFET has a positive $V_{GS(th)}$. A *p*-channel E-MOSFET has a negative $V_{GS(th)}$.
- LD MOSFET, VMOSFET, and TMOSFET are E-MOSFET technologies developed for higher power dissipation than a conventional E-MOSFET.

KEY TERMS

Key terms and other bold terms in the chapter are defined in the end-of-book glossary.

Depletion   In a MOSFET, the process of removing or depleting the channel of charge carriers and thus decreasing the channel conductivity.

Drain   One of the three terminals of a FET analogous to the collector of a BJT.

Enhancement   In a MOSFET, the process of creating a channel or increasing the conductivity of the channel by the addition of charge carriers.

Gate   One of the three terminals of a FET analogous to the base of a BJT.

JFET   Junction field-effect transistor; one of two major types of field-effect transistors.

MOSFET   Metal oxide semiconductor field-effect transistor; one of two major types of FETs; sometimes called IGFET for insulated-gate FET.

Pinch-off voltage   The value of the drain-to-source voltage of a FET at which the drain current becomes constant when the gate-to-source voltage is zero.

Source   One of the three terminals of a FET analogous to the emitter of a BJT.

Transconductance (g_m)   The ratio of a change in drain current for a change in gate-to-source voltage in a FET.

KEY FORMULAS

7–1 $\quad I_D = I_{DSS}\left(1 - \dfrac{V_{GS}}{V_{GS(off)}}\right)^2 \quad$ JFET/D-MOSFET transfer characteristic

7–2 $\quad g_m = g_{m0}\left(1 - \dfrac{V_{GS}}{V_{GS(off)}}\right) \quad$ Transconductance

7–3 $\quad g_{m0} = \dfrac{2I_{DSS}}{|V_{GS(off)}|} \quad$ Transconductance at $V_{GS} = 0$

7–4 $\quad I_D = K(V_{GS} - V_{GS(th)})^2 \quad$ E-MOSFET transfer characteristic

SELF-TEST

Answers are at the end of the chapter.

1. The JFET is
   (a) a unipolar device   (b) a voltage-controlled device
   (c) a current-controlled device   (d) answers (a) and (c)
   (e) answers (a) and (b)

2. The channel of a JFET is between the
   (a) gate and drain   (b) drain and source   (c) gate and source   (d) input and output

3. A JFET always operates with
   (a) the gate-to-source *pn* junction reverse-biased
   (b) the gate-to-source *pn* junction forward-biased
   (c) the drain connected to ground
   (d) the gate connected to the source

4. For $V_{GS} = 0$ V, the drain current becomes constant when V_{DS} exceeds
   (a) cutoff   (b) V_{DD} (c) V_P (d) 0 V

5. The constant-current area of a FET lies between
   (a) cutoff and saturation   (b) cutoff and pinch-off
   (c) 0 and I_{DSS} (d) pinch-off and breakdown

6. I_{DSS} is
   (a) the drain current with the source shorted   (b) the drain current at cutoff
   (c) the maximum possible drain current   (d) the midpoint drain current

7. Drain current in the constant-current area increases when
   (a) the gate-to-source bias voltage decreases   (b) the gate-to-source bias voltage increases
   (c) the drain-to-source voltage increases   (d) the drain-to-source voltage decreases

8. In a certain FET circuit, $V_{GS} = 0$ V, $V_{DD} = 15$ V, $I_{DSS} = 15$ mA, and $R_D = 470\ \Omega$. If R_D is decreased to $330\ \Omega$, I_{DSS} is
   (a) 19.5 mA   (b) 10.5 mA   (c) 15 mA   (d) 1 mA

9. At cutoff, the JFET channel is
   (a) at its widest point   (b) completely closed by the depletion region
   (c) extremely narrow   (d) reverse-biased

10. A certain JFET data sheet gives $V_{GS(off)} = -4$ V. The pinch-off voltage, V_P,
    (a) cannot be determined   (b) is -4 V   (c) depends on V_{GS} (d) is $+4$ V

11. The JFET in Question 10
    (a) is an *n* channel   (b) is a *p* channel   (c) can be either

12. For a certain JFET, $I_{GSS} = 10$ nA at $V_{GS} = 10$ V. The input resistance is
    (a) 100 MΩ (b) 1 MΩ (c) 1000 MΩ (d) 1000 mΩ

13. For a certain p-channel JFET, $V_{GS(off)} = 8$ V. The value of V_{GS} for an approximate midpoint bias is

    (a) 4 V   **(b)** 0 V   **(c)** 1.25 V   **(d)** 2.34 V

14. A MOSFET differs from a JFET mainly because

    (a) of the power rating   **(b)** the MOSFET has two gates

    (c) the JFET has a *pn* junction   **(d)** MOSFETs do not have a physical channel

15. A certain D-MOSFET is biased at $V_{GS} = 0$ V. Its data sheet specifies $I_{DSS} = 20$ mA and $V_{GS(off)} = -5$ V. The value of the drain current

    (a) is 0 A   **(b)** cannot be determined   **(c)** is 20 mA

16. An n-channel D-MOSFET with a positive V_{GS} is operating in

    (a) the depletion mode   **(b)** the enhancement mode

    (c) cutoff   **(d)** saturation

17. A certain p-channel E-MOSFET has a $V_{GS(th)} = -2$ V. If $V_{GS} = 0$ V, the drain current is

    (a) 0 A   **(b)** $I_{D(on)}$ **(c)** maximum   **(d)** I_{DSS}

18. A TMOSFET is a special type of

    (a) D-MOSFET   **(b)** JFET   **(c)** E-MOSFET   **(d)** answers (a) and (c)

PROBLEMS

Answers to all odd-numbered problems are at the end of the book.

BASIC PROBLEMS

SECTION 7–1 The JFET

    1. The V_{GS} of a p-channel JFET is increased from 1 V to 3 V.

        (a) Does the depletion region narrow or widen?

        (b) Does the resistance of the channel increase or decrease?

    2. Why must the gate-to-source voltage of an n-channel JFET always be either 0 or negative?

    3. Sketch the schematic diagrams for a p-channel and an n-channel JFET. Label the terminals.

    4. Show how to connect bias voltages between the gate and source of the JFETs in Figure 7–56.

    ▶ **FIGURE 7–56**

          (a)   (b)

SECTION 7–2 JFET Characteristics and Parameters

    5. A JFET has a specified pinch-off voltage of 5 V. When $V_{GS} = 0$, what is V_{DS} at the point where the drain current becomes constant?

    6. A certain n-channel JFET is biased such that $V_{GS} = -2$ V. What is the value of $V_{GS(off)}$ if V_P is specified to be 6 V? Is the device on?

    7. A certain JFET data sheet gives $V_{GS(off)} = -8$ V and $I_{DSS} = 10$ mA. When $V_{GS} = 0$, what is I_D for values of V_{DS} above pinch off? $V_{DD} = 15$ V.

    8. A certain p-channel JFET has a $V_{GS(off)} = 6$ V. What is I_D when $V_{GS} = 8$ V?

9. The JFET in Figure 7–57 has a $V_{GS(off)} = -4$ V. Assume that you increase the supply voltage, V_{DD}, beginning at zero until the ammeter reaches a steady value. What does the voltmeter read at this point?

▶ FIGURE 7–57

10. The following parameters are obtained from a certain JFET data sheet: $V_{GS(off)} = -8$ V and $I_{DSS} = 5$ mA. Determine the values of I_D for each value of V_{GS} ranging from 0 V to -8 V in 1 V steps. Plot the transfer characteristic curve from these data.

11. For the JFET in Problem 10, what value of V_{GS} is required to set up a drain current of 2.25 mA?

12. For a particular JFET, $g_{m0} = 3200\ \mu S$. What is g_m when $V_{GS} = -4$ V, given that $V_{GS(off)} = -8$ V?

13. Determine the forward transconductance of a JFET biased at $V_{GS} = -2$ V. From the data sheet, $V_{GS(off)} = -7$ V and $g_m = 2000\ \mu S$ at $V_{GS} = 0$ V. Also determine the forward transfer admittance, y_{fs}.

14. A p-channel JFET data sheet shows that $I_{GSS} = 5$ nA at $V_{GS} = 10$ V. Determine the input resistance.

15. Using Equation 7–1, plot the transfer characteristic curve for a JFET with $I_{DSS} = 8$ mA and $V_{GS(off)} = -5$ V using at least four points.

SECTION 7–3 JFET Biasing

16. An n-channel self-biased JFET has a drain current of 12 mA and a 100 Ω source resistor. What is the value of V_{GS}?

(a) (b) (c)

▲ FIGURE 7–58

File circuits are identified with a CD logo and are in the Problems folder on your CD-ROM. Filenames correspond to figure numbers (e.g., F07-58).

17. Determine the value of R_S required for a self-biased JFET to produce a V_{GS} of -4 V when $I_D = 5$ mA.

18. Determine the value of R_S required for a self-biased JFET to produce $I_D = 2.5$ mA when $V_{GS} = -3$ V.

19. $I_{DSS} = 20$ mA and $V_{GS(off)} = -6$ V for a particular JFET.

   (a) What is I_D when $V_{GS} = 0$ V?

   (b) What is I_D when $V_{GS} = V_{GS(off)}$?

   (c) If V_{GS} is increased from -4 V to -1 V, does I_D increase or decrease?

20. For each circuit in Figure 7–58, determine V_{DS} and V_{GS}.

21. Using the curve in Figure 7–59, determine the value of R_S required for a 9.5 mA drain current.

▶ FIGURE 7–59

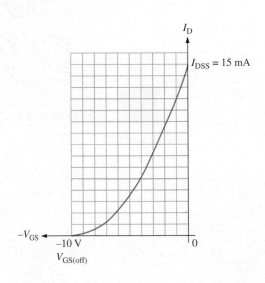

22. Set up a midpoint bias for a JFET with $I_{DSS} = 14$ mA and $V_{GS(off)} = -10$ V. Use a 24 V dc source as the supply voltage. Show the circuit and resistor values. Indicate the values of I_D, V_{GS}, and V_{DS}.

23. Determine the total input resistance in Figure 7–60. $I_{GSS} = 20$ nA at $V_{GS} = -10$ V.

▶ FIGURE 7–60

24. Graphically determine the Q-point for the circuit in Figure 7–61(a) using the transfer characteristic curve in Figure 7–61(b).

(a)

(b)

▲ **FIGURE 7–61**

25. Find the Q-point for the *p*-channel JFET circuit in Figure 7–62.

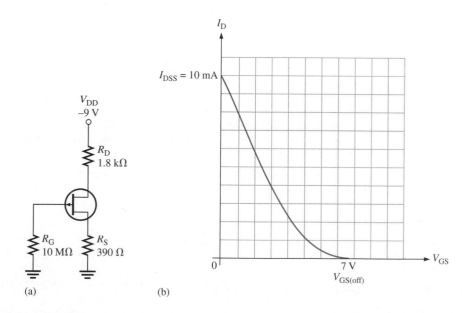

(a)

(b)

▲ **FIGURE 7–62**

26. Given that the drain-to-ground voltage in Figure 7–63 is 5 V, determine the Q-point of the circuit.

27. Find the Q-point values for the JFET with voltage-divider bias in Figure 7–64.

► FIGURE 7–63

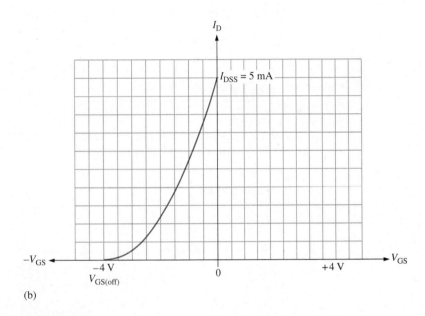

(a)   (b)

▲ FIGURE 7–64

SECTION 7–4   The MOSFET

28. Sketch the schematic symbols for n-channel and p-channel D-MOSFETs and E-MOSFETs. Label the terminals.

29. In what mode is an n-channel D-MOSFET with a positive V_{GS} operating?

30. Describe the basic difference between a D-MOSFET and an E-MOSFET.

31. Explain why both types of MOSFETs have an extremely high input resistance at the gate.

SECTION 7–5   MOSFET Characteristics and Parameters

32. The data sheet for a certain D-MOSFET gives $V_{GS(off)} = -5$ V and $I_{DSS} = 8$ mA.

   (a) Is this device p channel or n channel?

   (b) Determine I_D for values of V_{GS} ranging from -5 V to $+5$ V in increments of 1 V.

   (c) Plot the transfer characteristic curve using the data from part (b).

33. Determine I_{DSS}, given $I_D = 3$ mA, $V_{GS} = -2$ V, and $V_{GS(off)} = -10$ V.

34. The data sheet for an E-MOSFET reveals that $I_{D(on)} = 10$ mA at $V_{GS} = -12$ V and $V_{GS(th)} = -3$ V. Find I_D when $V_{GS} = -6$ V.

SECTION 7–6 MOSFET Biasing

35. Determine in which mode (depletion, enhancement or neither) each D-MOSFET in Figure 7–65 is biased.

▲ **FIGURE 7–65**

36. Each E-MOSFET in Figure 7–66 has a $V_{GS(th)}$ of +5 V or −5 V, depending on whether it is an *n*-channel or a *p*-channel device. Determine whether each MOSFET is on or off.

▶ **FIGURE 7–66**

37. Determine V_{DS} for each circuit in Figure 7–67. I_{DSS} = 8 mA.

▲ **FIGURE 7–67**

38. Find V_{GS} and V_{DS} for the E-MOSFETs in Figure 7–68. Data sheet information is listed with each circuit.

$I_{D(on)} = 3$ mA @ $V_{GS} = 4$ V
$V_{GS(th)} = 2$ V

(a)

$I_{D(on)} = 2$ mA @ $V_{GS} = 3$ V
$V_{GS(th)} = 1.5$ V

(b)

▲ **FIGURE 7–68**

39. Based on the identical V_{GS} measurements, determine the drain current and drain-to-source voltage for each circuit in Figure 7–69.

(a)

(b)

▲ **FIGURE 7–69**

40. Determine the actual gate-to-source voltage in Figure 7–70 by taking into account the gate leakage current, I_{GSS}. Assume that I_{GSS} is 50 pA and I_D is 1 mA under the existing bias conditions.

▶ **FIGURE 7–70**

TROUBLESHOOTING PROBLEMS

SECTION 7–7 Troubleshooting

41. The current reading in Figure 7–58(a) suddenly goes to zero. What are the possible faults?

42. The current reading in Figure 7–58(b) suddenly jumps to approximately 16 mA. What are the possible faults?

43. If the supply voltage in Figure 7–58(c) is accidentally changed to −20 V, what would you see on the ammeter?

44. You measure +10 V at the drain of the MOSFET in Figure 7–66(a). The transistor checks good and the ground connections are okay. What can be the problem?

45. You measure approximately 0 V at the drain of the MOSFET in Figure 7–66(b). You can find no shorts and the transistor checks good. What is the most likely problem?

SYSTEM APPLICATION PROBLEMS

46. The 100 μF capacitor on the pH sensor circuit board in Figure 7–51 has opened. What effect could this have on the circuit operation? Explain.

47. Refer to Figure 7–53. What should be the output voltage of the pH sensor for a pH of 5? For a pH of 9?

48. Refer to the test bench setup in Figure 7–54. When measuring the voltage inputs to the pH sensor circuits from the pH sensor in a test solution, you notice that the voltmeter indicates values that are approximately half of what they should be for each circuit. After trying a new sensor, the voltages are still half the expected value. What do you think is wrong?

49. Determine the voltage that would be measured at test point 6 on the circuit board in Figure 7–55 for a pH of 7 if the rheostat is incorrectly set to a value of 1 kΩ. Assume typical values.

DATA SHEET PROBLEMS

50. What type of FET is the 2N5457?

51. Referring to the data sheet in Figure 7–14, determine the following:

    (a) Minimum $V_{GS(off)}$ for the 2N5457.

    (b) Maximum drain-to-source voltage for the 2N5457.

    (c) Maximum power dissipation for the 2N5458 at an ambient temperature of 25°C.

    (d) Maximum reverse gate-to-source voltage for the 2N5459.

52. Referring to Figure 7–14, determine the maximum power dissipation for a 2N5457 at an ambient temperature of 65°C.

53. Referring to Figure 7–14, determine the minimum g_{m0} for the 2N5459 at a frequency of 1 kHz.

54. Referring to Figure 7–14, what is the typical drain current in a 2N5459 for $V_{GS} = 0$ V?

55. Referring to the data sheet in Figure 7–41, determine the minimum gate-to-source voltage at which the MOSFET begins to conduct current.

56. Referring to Figure 7–41, what is the drain current when $V_{GS} = 10$ V?

57. Referring to the data sheet in Figure 7–52, determine I_D in a 2N3797 when $V_{GS} = +3$ V. Determine I_D when $V_{GS} = -2$ V.

58. Referring to Figure 7–52, how much does the maximum forward transconductance of a 2N3796 change over a range of signal frequencies from 1 kHz to 1 MHz?

59. Referring to Figure 7–52, determine the typical value of gate-to-source voltage at which the 2N3796 will go into cutoff.

ADVANCED PROBLEMS

60. Find V_{DS} and V_{GS} in Figure 7–71 using minimum data sheet values.

▶ **FIGURE 7–71**

V_{DD}
+12 V

R_D
10 kΩ

2N5457

R_S
5.6 kΩ

61. Determine the maximum I_D and V_{GS} for the circuit in Figure 7–72.

▶ **FIGURE 7–72**

V_{DD}
+9 V

R_1
10 kΩ

R_D
4.7 kΩ

2N5459

R_2
3.3 kΩ

R_S
1.8 kΩ

62. Determine the range of possible Q-point values from minimum to maximum for the circuit in Figure 7–71.

63. Find the typical drain-to-source voltage for the pH sensor circuits in Figure 7–54 when a pH of 5 is measured. Assume the rheostats are set for 7 V at the drains when a pH of 7 is measured.

64. Design a MOSFET circuit with zero bias using a 2N3797 that operates from a +9 V dc supply and produces a V_{DS} of 4.5 V. The maximum current drawn from the source is to be 1 mA.

65. Design a circuit using a 2N7008 MOSFET and a +12 V dc supply voltage with voltage-divider bias that will produce +8 V at the drain and draw a maximum current from the supply of 20 mA.

EWB/MULTISIM TROUBLESHOOTING PROBLEMS

These file circuits are in the Troubleshooting Problems folder on your CD-ROM.

66. Open file TSP07-66 and determine the fault.

67. Open file TSP07-67 and determine the fault.

68. Open file TSP07-68 and determine the fault.

69. Open file TSP07-69 and determine the fault.

70. Open file TSP07-70 and determine the fault.

71. Open file TSP07-71 and determine the fault.

72. Open file TSP07-72 and determine the fault.

73. Open file TSP07-73 and determine the fault.

74. Open file TSP07-74 and determine the fault.

SECTION REVIEWS

SECTION 7–1 The JFET

1. Drain, source, and gate
2. An n-channel JFET requires a negative V_{GS}.
3. I_D is controlled by V_{GS}.

SECTION 7–2 JFET Characteristics and Parameters

1. When $V_{DS} = 7$ V at pinch-off and $V_{GS} = 0$ V, $V_P = -7$ V.
2. As V_{GS} increases negatively, I_D decreases.
3. For $V_P = -3$ V, $V_{GS(off)} = +3$ V.

SECTION 7–3 JFET Biasing

1. A p-channel JFET requires a positive V_{GS}.
2. $V_{GS} = V_G - V_S = 0\,\text{V} - (8\text{ mA})(1.0\text{ k}\Omega) = -8$ V
3. $V_{GS} = V_G - V_S = 3\,\text{V} - 5\,\text{V} = -2$ V

SECTION 7–4 The MOSFET

1. Depletion MOSFET (D-MOSFET) and enhancement MOSFET (E-MOSFET)
2. When $V_{GS} = 0$ V for a D-MOSFET, $I_D = I_{DSS}$.
3. When $V_{GS} = 0$ V for an E-MOSFET, $I_D \cong 0$ A.

SECTION 7–5 MOSFET Characteristics and Parameters

1. The D-MOSFET has a structural channel; the E-MOSFET does not.
2. $V_{GS(th)}$ and K are not specified for D-MOSFETs.
3. ESD is ElectroStatic Discharge.

SECTION 7–6 MOSFET Biasing

1. When $V_{GS} = 0$ V, the drain current is equal to I_{DSS}.
2. V_{GS} must exceed $V_{GS(th)} = 2$ V for conduction to occur.

SECTION 7–7 Troubleshooting

1. R_S open, no ground connection
2. Because V_{GS} remains at approximately zero
3. The device is off and $V_D = V_{DD}$.

RELATED PROBLEMS FOR EXAMPLES

7–1 I_D remains at approximately 12 mA.

7–2 $V_P = -4$ V

7–3 $I_D = 3.51$ mA

7–4 $g_m = 1800\ \mu\text{S}$; $I_D = 4.32$ mA

7–5 $R_{IN} = 150{,}000$ MΩ

7–6 $V_{DS} = 2$ V; $V_{GS} = -3.12$ V

7–7 $R_S = 231\ \Omega$

7–8 $R_S = 889\ \Omega$

7–9 $R_S = 586\ \Omega$; $R_D = 1500\ \Omega$

7–10 $V_{GS} \cong -1.8$ V, $I_D \cong 1.8$ mA

7–11 $I_D = 1.81$ mA, $V_{GS} = -1.99$ V

7–12 $I_D \cong 1.25$ mA, $V_{GS} \cong -2.25$ V

7–13 **(a)** p channel   **(b)** 6.48 mA   **(c)** 35.3 mA

7–14 $I_D = 25$ mA

7–15 $V_{DS} = 5.6$ V

7–16 $V_{GS} = 3.13$ V; $V_{DS} = 22.3$ V

7–17 $I_D = 2.13$ mA

SELF-TEST

1. (e)   **2.** (b)   **3.** (a)   **4.** (c)   **5.** (d)

6. (c)   **7.** (a)   **8.** (c)   **9.** (b)   **10.** (d)

11. (a)   **12.** (c)   **13.** (d)   **14.** (c)   **15.** (c)

16. (b)   **17.** (a)   **18.** (c)

8

FET AMPLIFIERS

INTRODUCTION

The things you learned about biasing FETs in the last chapter are carried forward into this chapter where FET circuits are used as small-signal amplifiers. Because of their high input resistance and other characteristics, FETs are often preferred over BJTs (bipolar junction transistors) for certain types of applications.

Many of the concepts that relate to amplifiers using BJTs apply as well to FET amplifiers. The three FET amplifier configurations are common-source, common-drain, and common-gate. These are analogous to the common-emitter, common-collector, and common-base configurations in BJT amplifiers.

- Describe the amplification properties of a FET

- Explain and analyze the operation of common-source FET amplifiers

- Explain and analyze the operation of common-drain FET amplifiers

- Explain and analyze the operation of common-gate FET amplifiers

- Troubleshoot FET amplifiers

KEY TERMS

- Common-source

- Common-drain

- Source-follower

- Common-gate

■■■ SYSTEM APPLICATION PREVIEW

For the system application at the end of the chapter, you will be evaluating the feasibility of replacing the BJT amplifier in a public address system with a FET amplifier.

WWW. VISIT THE COMPANION WEBSITE

Study aids for this chapter are available at
http://www.prenhall.com/floyd

8–1 FET AMPLIFICATION

In this section, you will learn about the amplification properties of FETs and how the gain is affected by certain parameters and circuit components. We will simplify the FET to an equivalent circuit to get to the essence of its operation.

After completing this section, you should be able to

- **Describe the amplification properties of a FET**
  - Discuss FET parameters using an equivalent circuit
  - Discuss the voltage gain of a FET
  - Describe the effect of internal drain-to-source resistance on the voltage gain
  - Describe the effect of external source resistance on the voltage gain

The transconductance is defined as $g_m = \Delta I_D / \Delta V_{GS}$. In ac quantities, $g_m = I_d / V_{gs}$. By rearranging the terms,

Equation 8–1

$$I_d = g_m V_{gs}$$

This equation states that the output current, I_d, equals the input voltage, V_{gs}, multiplied by the transconductance, g_m.

Equivalent Circuit

A FET equivalent circuit representing the relationship in Equation 8–1 is shown in Figure 8–1. In part (a), the internal resistance, r'_{gs}, appears between the gate and source, and a current source equal to $g_m V_{gs}$ appears between the drain and source. Also, the internal drain-to-source resistance, r'_{ds}, is included. In part (b), a simplified ideal model is shown. The resistance, r'_{gs}, is assumed to be infinitely large so that there is an open circuit between the gate and source. Also, r'_{ds} is assumed large enough to neglect.

▶ **FIGURE 8–1**

Internal FET equivalent circuits.

(a) Complete (b) Simplified

Voltage Gain

An ideal FET equivalent circuit with an external ac drain resistance is shown in Figure 8–2. The ac voltage gain of this circuit is V_{out}/V_{in}, where $V_{in} = V_{gs}$ and $V_{out} = V_{ds}$. The voltage gain expression is, therefore,

$$A_v = \frac{V_{ds}}{V_{gs}}$$

► **FIGURE 8-2**

Simplified FET equivalent circuit with an external ac drain resistance.

From the equivalent circuit,

$$V_{ds} = I_d R_d$$

and from the definition of transconductance in Chapter 7,

$$V_{gs} = \frac{I_d}{g_m}$$

Substituting the two preceding expressions into the first equation yields

$$A_v = \frac{I_d R_d}{I_d / g_m} = \frac{g_m I_d R_d}{I_d}$$

$$A_v = g_m R_d$$

Equation 8-2

EXAMPLE 8-1

A certain JFET has a $g_m = 4$ mS. With an external ac drain resistance of 1.5 kΩ, what is the ideal voltage gain?

Solution
$$A_v = g_m R_d = (4 \text{ mS})(1.5 \text{ k}\Omega) = \mathbf{6}$$

Related Problem* What is the ideal voltage gain when $g_m = 6000$ μS and $R_d = 2.2$ kΩ?

*Answers are at the end of the chapter.

Effect of r'_{ds} on Gain

If the internal drain-to-source resistance of a FET is taken into account, it appears in parallel with R_d, as indicated in Figure 8-3. If r'_{ds} is not sufficiently greater than R_d (at least 10 times greater), the gain is reduced from the ideal case of Equation 8-2 to the following:

$$A_v = g_m \left(\frac{R_d r'_{ds}}{R_d + r'_{ds}} \right)$$

► **FIGURE 8-3**

FET equivalent circuit including the internal drain-to-source resistance, r'_{ds}, which appears in parallel with R_d.

EXAMPLE 8–2

The JFET in Example 8–1 has an $r'_{ds} = 10 \text{ k}\Omega$. Determine the voltage gain when r'_{ds} is taken into account.

Solution   The r'_{ds} is effectively in parallel with the external ac drain resistance R_d. Therefore,

$$A_v = g_m\left(\frac{R_d r'_{ds}}{R_d + r'_{ds}}\right) = (4 \text{ mS})\left(\frac{(1.5 \text{ k}\Omega)(10 \text{ k}\Omega)}{1.5 \text{ k}\Omega + 10 \text{ k}\Omega}\right) = (4 \text{ mS})(1.3 \text{ k}\Omega) = \mathbf{5.2}$$

The voltage gain is reduced from a value of 6 (Example 8–1) because r'_{ds} is in parallel with R_d.

Related Problem   A JFET has a $g_m = 6 \text{ mS}$, an $r'_{ds} = 5 \text{ k}\Omega$, and an external ac drain resistance of $1.0 \text{ k}\Omega$. What is the voltage gain?

Effect of External Source Resistance on Gain

Including an external resistance from a FET's source terminal to ground results in the equivalent circuit of Figure 8–4. Examination of this circuit shows that the total input voltage between the gate and ground is

$$V_{in} = V_{gs} + I_d R_s$$

The output voltage taken across R_d is

$$V_{out} = I_d R_d$$

Therefore, the formula for voltage gain is developed as follows:

$$A_v = \frac{V_{out}}{V_{in}} = \frac{I_d R_d}{V_{gs} + I_d R_s} = \frac{g_m V_{gs} R_d}{V_{gs} + g_m V_{gs} R_s} = \frac{g_m V_{gs} R_d}{V_{gs}(1 + g_m R_s)}$$

Equation 8–3

$$A_v = \frac{g_m R_d}{1 + g_m R_s}$$

▶ **FIGURE 8–4**

EXAMPLE 8–3

A FET equivalent circuit is shown in Figure 8–4. Determine the voltage gain when the output is taken across R_d. Neglect r'_{ds}.

Solution   There is an external source resistor, so the voltage gain is

$$A_v = \frac{g_m R_d}{1 + g_m R_s} = \frac{(4 \text{ mS})(1.5 \text{ k}\Omega)}{1 + (4 \text{ mS})(560 \text{ }\Omega)} = \frac{6}{1 + 2.24} = \frac{6}{3.24} = \mathbf{1.85}$$

This is the same circuit as in Example 8–1 except for R_s. As you can see, R_s reduces the voltage gain from 6 (Example 8–1) to 1.85.

Related Problem   For the circuit in Figure 8–4, $g_m = 3.5 \text{ mS}$, $R_s = 330 \text{ }\Omega$, and $R_d = 1.8 \text{ k}\Omega$. Find the voltage gain when the output is taken across R_d. Neglect r'_{ds}.

SECTION 8–1
REVIEW
Answers are at the end of the chapter.

1. One FET has a transconductance of 3000 μS and another has a transconductance of 3.5 mS. Which one can produce the higher voltage gain, with all other circuit components the same?
2. A FET circuit has a $g_m = 2500$ μS and an $R_d = 10$ kΩ. Ideally, what voltage gain can it produce?
3. Two FETs have the same g_m. One has an $r'_{ds} = 50$ kΩ and the other has an $r'_{ds} = 100$ kΩ under the same conditions. Which FET can produce the higher voltage gain when used in a circuit with $R_d = 10$ kΩ?

8–2   COMMON-SOURCE AMPLIFIERS

Now that you have an idea of how a FET functions as a voltage amplifying device, let's look at a complete amplifier circuit. The common-source (CS) amplifier discussed in this section is comparable to the common-emitter BJT amplifier that you studied in Chapter 6.

After completing this section, you should be able to

- **Explain and analyze the operation of common-source FET amplifiers**

- Analyze JFET and MOSFET CS amplifiers

- Determine the dc values of a CS amplifier

- Develop an ac equivalent circuit and determine the voltage gain of a CS amplifier

- Describe the effect of an ac load on the voltage gain

- Discuss phase inversion in a CS amplifier

- Determine the input resistance of a CS amplifier

JFET Amplifier

A self-biased **common-source** n-channel JFET with an ac source capacitively coupled to the gate is shown in Figure 8–5(a). The resistor, R_G, serves two purposes: It keeps the gate at approximately 0 V dc (because I_{GSS} is extremely small), and its large value (usually several megohms) prevents loading of the ac signal source. The bias voltage is produced by the drop across R_S. The bypass capacitor, C_2, keeps the source of the FET effectively at ac ground.

The input signal voltage causes the gate-to-source voltage to swing above and below its Q-point value (V_{GSQ}), causing a corresponding swing in drain current. As the drain current increases, the voltage drop across R_D also increases, causing the drain voltage to decrease. The drain current swings above and below its Q-point value in phase with the gate-to-source voltage. The drain-to-source voltage swings above and below its Q-point value (V_{DSQ}) and is 180° out of phase with the gate-to-source voltage, as illustrated in Figure 8–5(b).

A Graphical Picture   The operation just described for an n-channel JFET is illustrated graphically on both the transfer characteristic curve and the drain characteristic curve in Figure 8–6. Part (a) shows how a sinusoidal variation, V_{gs}, produces a corresponding sinusoidal variation in I_d. As V_{gs} swings from its Q-point value to a more negative value, I_d decreases from its Q-point value. As V_{gs} swings to a less negative value, I_d increases. Figure 8–6(b) shows a view of the same operation using the drain curves. The signal at

▲ **FIGURE 8–5**

JFET common–source amplifier.

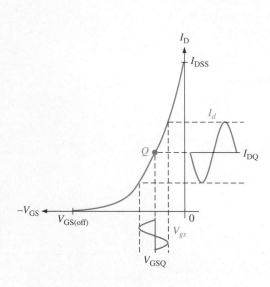

(a) JFET (*n*-channel) transfer characteristic curve
showing signal operation

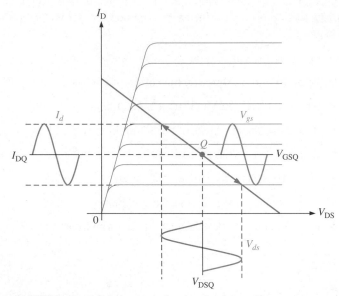

(b) JFET (*n*-channel) drain curves showing signal operation

▲ **FIGURE 8–6**

JFET characteristic curves.

the gate drives the drain current equally above and below the Q-point on the load line, as indicated by the arrows. Lines projected from the peaks of the gate voltage across to the I_D axis and down to the V_{DS} axis indicate the peak-to-peak variations of the drain current and drain-to-source voltage, as shown.

DC Analysis

To analyze the amplifier in Figure 8–7, you must first determine the dc bias values. To do this, develop a dc equivalent circuit by replacing all capacitors with opens, as shown in Figure 8–8. First, you must determine I_D before you can do any analysis. If the circuit is biased at the midpoint of the load line, you can calculate I_D using I_{DSS} from the FET data sheet as follows:

$$I_D = \frac{I_{DSS}}{2}$$

▲ **FIGURE 8–7**

JFET common-source amplifier.

▲ **FIGURE 8–8**

DC equivalent circuit for the amplifier in Figure 8–7.

Otherwise, you must determine I_D from circuit parameter values, which is tedious because Equation 8–5 must be solved for I_D. (This equation is derived by the substitution of $V_{GS} = I_D R_S$ into Equation 7–1.) A solution of the equation for I_D involves expanding it into a quadratic form and then finding the root of the quadratic, as developed in Appendix B.

$$I_D = I_{DSS}\left(1 - \frac{I_D R_S}{V_{GS(off)}}\right)^2$$

Equation 8–5

The equation can be solved for I_D using a calculator. For calculators other than the TI-86, consult the user's manual to determine if it has equation solving capability. The sequence of keystrokes for a TI-86 is as follows:

The screen will appear as follows:

```
ID=IDSS*(1−(-ID*RS/V...
ID=
IDSS=
RS=
VGSOFF=
bound=(-1E99,1E99)

GRAPH  WIND  ZOOM  TRACE  SOLVE
```

Enter Variable Values Use the up/down arrow keys to move the blinking cursor to the IDSS= and enter the value in amps. Next, move the cursor to the RS= and enter the value in ohms. Finally, move the cursor to the VGSOFF= and enter the absolute value in volts (do not include the sign if negative). Move the cursor back up to ID= but do not enter a value. Select SOLVE with the **F5** key. The value of ID will appear.

AC Equivalent Circuit

To analyze the signal operation of the amplifier in Figure 8–7, develop an ac equivalent circuit as follows. Replace the capacitors by effective shorts, based on the simplifying assumption that $X_C \cong 0$ at the signal frequency. Replace the dc source by a ground, based on the assumption that the voltage source has a zero internal resistance. The V_{DD} terminal is at a zero-volt ac potential and therefore acts as an ac ground.

The ac equivalent circuit is shown in Figure 8–9(a). Notice that the $+V_{DD}$ end of R_d and the source terminal are both effectively at ac ground. Recall that in ac analysis, the ac ground and the actual circuit ground are treated as the same point.

(a)   (b)

▲ **FIGURE 8–9**

AC equivalent for the amplifier in Figure 8–7.

Signal Voltage at the Gate An ac voltage source is shown connected to the input in Figure 8–9(b). Since the input resistance to a FET is extremely high, practically all of the input voltage from the signal source appears at the gate with very little voltage dropped across the internal source resistance.

$$V_{gs} = V_{in}$$

Voltage Gain The expression for FET voltage gain that was given in Equation 8–2 applies to the common-source amplifier.

Equation 8–6

$$A_v = g_m R_d$$

The output signal voltage V_{ds} at the drain is

$$V_{out} = V_{ds} = A_v V_{gs}$$

or

$$V_{out} = g_m R_d V_{in}$$

where $R_d = R_D \parallel R_L$ and $V_{in} = V_{gs}$.

EXAMPLE 8–4

What is the total output voltage of the unloaded amplifier in Figure 8–10? For this particular JFET, I_{DSS} is 12 mA and $V_{GS(off)}$ is -3 V.

▲ **FIGURE 8–10**

Solution   First, find the dc output current using a calculator. When Equation 8–5 is solved with the parameter values given, $I_D \cong 1.96$ mA. Using this value, calculate V_D.

$$V_D = V_{DD} - I_D R_D = 12 \text{ V} - (1.96 \text{ mA})(3.3 \text{ k}\Omega) = 5.53 \text{ V}$$

Next, calculate g_m as follows:

$$V_{GS} = -I_D R_S = -(1.96 \text{ mA})(910 \text{ }\Omega) = -1.78 \text{ V}$$

$$g_{m0} = \frac{2I_{DSS}}{|V_{GS(off)}|} = \frac{2(12 \text{ mA})}{3 \text{ V}} = 8 \text{ mS}$$

$$g_m = g_{m0}\left(1 - \frac{V_{GS}}{V_{GS(off)}}\right) = (8 \text{ mS})\left(1 - \frac{-1.78 \text{ V}}{-3 \text{ V}}\right) = 3.25 \text{ mS}$$

Finally, find the ac output voltage.

$$V_{out} = A_v V_{in} = g_m R_D V_{in} = (3.25 \text{ mS})(3.3 \text{ k}\Omega)(100 \text{ mV}) = \textbf{1.07 V rms}$$

The total output voltage is an ac signal with a peak-to-peak value of 1.07 V × 2.828 = 3.03 V, riding on a dc level of 5.53 V.

Related Problem   What will happen in the amplifier of Figure 8–10 if a transistor with $V_{GS(off)} = -2$ V is used? Assume the other parameters are the same.

Open file E08-04 in the Examples folder on your CD-ROM. Using the specified input voltage, measure the output voltage and compare with the calculated value.

Effect of an AC Load on Voltage Gain

When a load is connected to an amplifier's output through a coupling capacitor, as shown in Figure 8–11(a), the ac drain resistance is effectively R_D in parallel with R_L because the upper end of R_D is at ac ground. The ac equivalent circuit is shown in Figure 8–11(b). The total ac drain resistance is

$$R_d = \frac{R_D R_L}{R_D + R_L}$$

The effect of R_L is to reduce the unloaded voltage gain, as Example 8–5 illustrates.

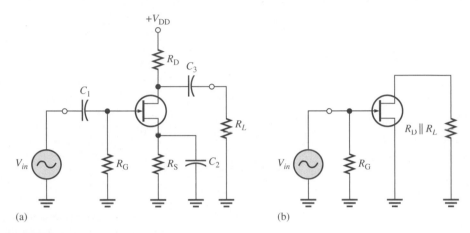

▲ FIGURE 8–11

JFET amplifier and its ac equivalent.

EXAMPLE 8–5

If a 4.7 kΩ load resistor is ac coupled to the output of the amplifier in Example 8–4, what is the resulting rms output voltage?

Solution The ac drain resistance is

$$R_d = \frac{R_D R_L}{R_D + R_L} = \frac{(3.3 \text{ k}\Omega)(4.7 \text{ k}\Omega)}{8 \text{ k}\Omega} = 1.94 \text{ k}\Omega$$

Calculation of V_{out} yields

$$V_{out} = g_m R_d V_{in} = (3.25 \text{ mS})(1.94 \text{ k}\Omega)(100 \text{ mV}) = \textbf{631 mV rms}$$

The unloaded ac output voltage was 1.07 V rms in Example 8–4.

Related Problem If a 10 kΩ load resistor is ac coupled to the output of the amplifier in Example 8–4 and the JFET is replaced with one having a $g_m = 3000 \text{ } \mu\text{S}$, what is the resulting rms output voltage?

Phase Inversion

The output voltage (at the drain) is 180° out of phase with the input voltage (at the gate). The phase inversion can be designated by a negative voltage gain, $-A_v$. Recall that the common-emitter BJT amplifier also exhibited a phase inversion.

Input Resistance

Because the input to a common-source amplifier is at the gate, the input resistance is extremely high. Ideally, it approaches infinity and can be neglected. As you know, the high input resistance is produced by the reverse-biased pn junction in a JFET and by the insulated gate structure in a MOSFET. The actual input resistance seen by the signal source is the gate-to-ground resistor, R_G, in parallel with the FET's input resistance, V_{GS}/I_{GSS}. The reverse leakage current, I_{GSS}, is typically given on the data sheet for a specific value of V_{GS} so that the input resistance of the device can be calculated.

$$R_{in} = R_G \| \left(\frac{V_{GS}}{I_{GSS}} \right)$$

Equation 8–7

EXAMPLE 8–6

What input resistance is seen by the signal source in Figure 8–12? $I_{GSS} = 30$ nA at $V_{GS} = 10$ V.

▲ **FIGURE 8–12**

Solution The input resistance at the gate of the JFET is

$$R_{IN(gate)} = \frac{V_{GS}}{I_{GSS}} = \frac{10\ V}{30\ nA} = 333\ M\Omega$$

The input resistance seen by the signal source is

$$R_{in} = R_G \| R_{IN(gate)} = 10\ M\Omega \| 333\ M\Omega = \textbf{9.7 M}\boldsymbol{\Omega}$$

Related Problem How much is the total input resistance if $I_{GSS} = 1$ nA at $V_{GS} = 10$ V?

D-MOSFET Amplifier Operation

A zero-biased common-source n-channel D-MOSFET with an ac source capacitively coupled to the gate is shown in Figure 8–13. The gate is at approximately 0 V dc and the source terminal is at ground, thus making $V_{GS} = 0$ V.

The signal voltage causes V_{gs} to swing above and below its zero value, producing a swing in I_d, as shown in Figure 8–14. The negative swing in V_{gs} produces the depletion mode, and I_d decreases. The positive swing in V_{gs} produces the enhancement mode, and I_d increases. Note that the enhancement mode is to the right of the vertical axis ($V_{GS} = 0$), and the depletion mode is to the left. The dc analysis of this amplifier is somewhat easier

▶ **FIGURE 8–13**

Zero-biased D-MOSFET common-source amplifier.

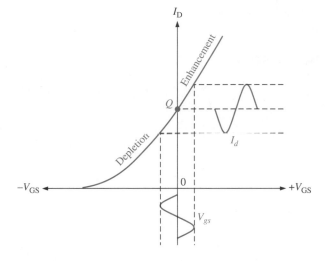

▲ **FIGURE 8–14**

Depletion-enhancement operation of D-MOSFET shown on transfer characteristic curve.

than for a JFET because $I_D = I_{DSS}$ at $V_{GS} = 0$. Once I_D is known, the analysis involves calculating only V_D.

$$V_D = V_{DD} - I_D R_D$$

The ac analysis is the same as for the JFET amplifier.

EXAMPLE 8–7

The particular D-MOSFET used in the amplifier of Figure 8–15 has an I_{DSS} of 200 mA and a g_m of 200 mS. Determine both the dc drain voltage and ac output voltage. $V_{in} = 500$ mV.

▶ **FIGURE 8–15**

Solution Since the amplifier is zero-biased,

$$I_D = I_{DSS} = 200 \text{ mA}$$

and, therefore,

$$V_D = V_{DD} - I_D R_D = 15 \text{ V} - (200 \text{ mA})(33 \ \Omega) = \textbf{8.4 V}$$

$$R_d = R_D \| R_L = 33 \ \Omega \| 8.2 \text{ k}\Omega = 32.9 \ \Omega$$

The ac output voltage is

$$V_{out} = g_m R_d V_{in} = (200 \text{ mS})(32.9 \ \Omega)(500 \text{ mV}) = \textbf{3.29 V}$$

Related Problem If a D-MOSFET with $g_m = 100$ mS and $I_{DSS} = 100$ mA replaces the one in this example, what is the ac output voltage when $V_{in} = 500$ mV?

Open file E08-07 in the Examples folder on your CD-ROM. Measure the dc drain voltage with no input signal. Then measure the ac output voltage with an rms input voltage of 500 mV. Compare with the calculated values.

E-MOSFET Amplifier Operation

A common-source *n*-channel E-MOSFET with voltage-divider bias with an ac source capacitively coupled to the gate is shown in Figure 8–16. The gate is biased with a positive voltage such that $V_{GS} > V_{GS(th)}$.

As with the JFET and D-MOSFET, the signal voltage produces a swing in V_{gs} above and below its Q-point value, V_{GSQ}. This, in turn, causes a swing in I_d above and below its Q-point value, I_{DQ}, as illustrated in Figure 8–17. Operation is entirely in the enhancement mode.

▶ **FIGURE 8–16**

Common-source E-MOSFET amplifier with voltage-divider bias.

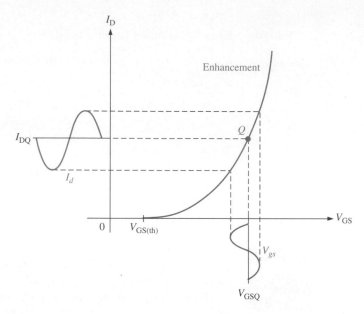

▲ **FIGURE 8–17**

E-MOSFET (*n*-channel) operation shown on transfer characteristic curve.

EXAMPLE 8–8

Transfer characteristic curves for a particular *n*-channel JFET, D-MOSFET, and E-MOSFET are shown in Figure 8–18. Determine the peak-to-peak variation in I_d when V_{gs} is varied ± 1 V about its Q-point value for each curve.

(a) JFET

(b) D-MOSFET

(c) E-MOSFET

▲ **FIGURE 8–18**

Solution (a) The JFET Q-point is at $V_{GS} = -2$ V and $I_D = 2.5$ mA. From the graph in Figure 8–18(a), $I_D = 3.4$ mA when $V_{GS} = -1$ V, and $I_D = 1.8$ mA when $V_{GS} = -3$ V. The peak-to-peak drain current is therefore **1.6 mA**.

(b) The D-MOSFET Q-point is at $V_{GS} = 0$ V and $I_D = I_{DSS} = 4$ mA. From the graph in Figure 8–18(b), $I_D = 2.5$ mA when $V_{GS} = -1$ V, and $I_D = 5.3$ mA when $V_{GS} = +1$ V. The peak-to-peak drain current is therefore **2.8 mA**.

(c) The E-MOSFET Q-point is at $V_{GS} = +8$ V and $I_D = 2.5$ mA. From the graph in Figure 8–18(c), $I_D = 3.9$ mA when $V_{GS} = +9$ V, and $I_D = 1.7$ mA when $V_{GS} = +7$ V. The peak-to-peak drain current is therefore **2.2 mA.**

Related Problem As the Q-point is moved toward the bottom end of the curves in Figure 8–18, does the variation in I_D increase or decrease for the same ± 1 V variation in V_{GS}? In addition to the change in the amount that I_D varies, what else will happen?

The circuit in Figure 8–16 uses voltage-divider bias to achieve a V_{GS} above threshold. The general dc analysis proceeds as follows using the E-MOSFET characteristic equation (Equation 7–4) to solve for I_D.

$$V_{GS} = \left(\frac{R_2}{R_1 + R_2}\right)V_{DD}$$
$$I_D = K(V_{GS} - V_{GS(th)})^2$$
$$V_{DS} = V_{DD} - I_D R_D$$

The voltage gain expression is the same as for the JFET and D-MOSFET circuits. The ac input resistance is

$$R_{in} = R_1 \| R_2 \| R_{IN(gate)}$$

Equation 8–8

where $R_{IN(gate)} = V_{GS}/I_{GSS}$.

EXAMPLE 8–9

A common-source amplifier using an E-MOSFET is shown in Figure 8–19. Find V_{GS}, I_D, V_{DS}, and the ac output voltage. Assume that for this particular device, $I_{D(on)} = 200$ mA at $V_{GS} = 4$ V, $V_{GS(th)} = 2$ V, and $g_m = 23$ mS. $V_{in} = 25$ mV.

▶ **FIGURE 8–19**

Solution

$$V_{GS} = \left(\frac{R_2}{R_1 + R_2}\right)V_{DD} = \left(\frac{8.2\ \text{k}\Omega}{55.2\ \text{k}\Omega}\right)15\ \text{V} = \textbf{2.23 V}$$

For $V_{GS} = 4$ V,

$$K = \frac{I_{D(on)}}{(V_{GS} - V_{GS(th)})^2} = \frac{200\ \text{mA}}{(4\ \text{V} - 2\ \text{V})^2} = 50\ \text{mA/V}^2$$

Therefore,

$$I_D = K(V_{GS} - V_{GS(th)})^2 = (50\ \text{mA}/\text{V}^2)(2.23\ \text{V} - 2\ \text{V})^2 = \textbf{2.65 mA}$$
$$V_{DS} = V_{DD} - I_D R_D = 15\ \text{V} - (2.65\ \text{mA})(3.3\ \text{k}\Omega) = \textbf{6.26 V}$$
$$R_d = R_D \| R_L = 3.3\ \text{k}\Omega \| 33\ \text{k}\Omega = 3\ \text{k}\Omega$$

The ac output voltage is

$$V_{out} = A_v V_{in} = g_m R_d V_{in} = (23 \text{ mS})(3 \text{ k}\Omega)(25 \text{ mV}) = \mathbf{1.73 \text{ V}}$$

Related Problem   For the E-MOSFET in Figure 8–19, $I_{D(on)} = 100$ mA at $V_{GS} = 5$ V, $V_{GS(th)} = 1$ V, and $g_m = 10$ mS. Find V_{GS}, I_D, V_{DS}, and the ac output voltage. $V_{in} = 25$ mV.

Open file E08-09 in the Examples folder on your CD-ROM. Determine I_D, V_{DS}, and V_{out} using the specified value of V_{in}. Compare with the calculated values.

SECTION 8–2 REVIEW

1. When V_{gs} is at its positive peak, at what points are I_d and V_{ds}?
2. What is the difference between V_{gs} and V_{GS}?
3. Which of the three types of FETs can operate with a gate-to-source Q-point value of 0 V?
4. What factors determine the voltage gain of a common-source FET amplifier?
5. A certain amplifier has an $R_D = 1.0$ kΩ. When a load resistance of 1.0 kΩ is capacitively coupled to the drain, how much does the gain change?

8–3   COMMON-DRAIN AMPLIFIERS

The common-drain (CD) amplifier covered in this section is comparable to the common-collector BJT amplifier. Recall that the CC amplifier is called an emitter-follower. Similarly, the common-drain amplifier is called a source-follower because the voltage at the source is approximately the same amplitude as the input (gate) voltage and is in phase with it. In other words, the source voltage follows the gate input voltage.

After completing this section, you should be able to

- **Explain and analyze the operation of common-drain FET amplifiers**
- Analyze a CD amplifier
- Determine the voltage gain of a CD amplifier
- Determine the input resistance of a CD amplifier

    A **common-drain** JFET amplifier is shown in Figure 8–20. Self-biasing is used in this particular circuit. The input signal is applied to the gate through a coupling capacitor, C_1, and the output signal is coupled to the load resistor through C_2. There is no drain resistor.

Voltage Gain

As in all amplifiers, the voltage gain is $A_v = V_{out}/V_{in}$. For the **source-follower**, V_{out} is $I_d R_s$ and V_{in} is $V_{gs} + I_d R_s$, as shown in Figure 8–21. Therefore, the gate-to-source voltage gain is $I_d R_s/(V_{gs} + I_d R_s)$. Substituting $I_d = g_m V_{gs}$ into the expression gives the following result:

$$A_v = \frac{g_m V_{gs} R_s}{V_{gs} + g_m V_{gs} R_s}$$

► **FIGURE 8–20**

JFET common-drain amplifier (source-follower).

► **FIGURE 8–21**

Voltages in a common-drain amplifier with a load resistor shown combined with R_S.

The V_{gs} terms cancel, so

$$A_v = \frac{g_m R_s}{1 + g_m R_s}$$

Equation 8–9

Notice here that the gain is always slightly less than one. If $g_m R_s \gg 1$, then a good approximation is $A_v \cong 1$. Since the output voltage is at the source, it is in phase with the gate (input) voltage.

Input Resistance

Because the input signal is applied to the gate, the input resistance seen by the input signal source is extremely high, just as in the common-source amplifier configuration. The gate resistor, R_G, in parallel with the input resistance looking in at the gate is the total input resistance.

$$R_{in} = R_G \parallel R_{IN(gate)}$$

Equation 8–10

where $R_{IN(gate)} = V_{GS}/I_{GSS}$.

EXAMPLE 8–10

Determine the voltage gain of the amplifier in Figure 8–22 using the data sheet information in Figure 8–23. Also, determine the input resistance. Use minimum data sheet values where available. V_{DD} is negative because it is a p-channel device.

► **FIGURE 8–22**

Electrical Characteristics (T_A = 25°C unless otherwise noted.)

Characteristic		Symbol	Min	Typ	Max	Unit
OFF Characteristics						
Gate-Source breakdown voltage		$V_{(BR)GSS}$		–		V dc
(I_G = 10 μA dc, V_{DS} = 0)	2N5460, 2N5461, 2N5462		40	–	–	
	2N5463, 2N5464, 2N5465		60	–	–	
Gate reverse current		I_{GSS}				
(V_{GS} = 20 V dc, V_{DS} = 0)	2N5460, 2N5461, 2N5462		–	–	5.0	nA dc
(V_{GS} = 30 V dc, V_{DS} = 0)	2N5463, 2N5464, 2N5465		–	–	5.0	
(V_{GS} = 20 V dc, V_{DS} = 0, T_A = 100°C)	2N5460, 2N5461, 2N5462		–	–	1.0	μA dc
(V_{GS} = 30 V dc, V_{DS} = 0, T_A = 100°C)	2N5463, 2N5464, 2N5465		–	–	1.0	
Gate-Source cutoff voltage		$V_{GS(off)}$				V dc
(V_{DS} = 15 V dc, I_D = 1.0 μA dc)	2N5460, 2N5463		0.75	–	6.0	
	2N5461, 2N5464		1.0	–	7.5	
	2N5462, 2N5465		1.8	–	9.0	
Gate-Source voltage		V_{GS}				
(V_{DS} = 15 V dc, I_D = 0.1 mA dc)	2N5460, 2N5463		0.5	–	4.0	V dc
(V_{DS} = 15 V dc, I_D = 0.2 mA dc)	2N5461, 2N5464		0.8	–	4.5	
(V_{DS} = 15 V dc, I_D = 0.4 mA dc)	2N5462, 2N5465		1.5	–	6.0	
ON Characteristics						
Zero-gate-voltage drain current		I_{DSS}				mA dc
(V_{DS} = 15 V dc, V_{GS} = 0,	2N5460, 2N5463		– 1.0	–	– 5.0	
f = 1.0 kHz)	2N5461, 2N5464		– 2.0	–	– 9.0	
	2N5462, 2N5465		– 4.0	–	– 16	
Small-Signal Characteristics						
Forward transfer admittance		$\|Y_{fs}\|$				μmhos
(V_{DS} = 15 V dc, V_{GS} = 0, f = 1.0 kHz)	2N5460, 2N5463		1000	–	4000	or
	2N5461, 2N5464		1500	–	5000	μS
	2N5462, 2N5465		2000	–	6000	
Output admittance		$\|Y_{os}\|$	–	–	75	μmhos or
(V_{DS} = 15 V dc, V_{GS} = 0, f = 1.0 kHz)						μS
Input capacitance		C_{iss}	–	5.0	7.0	pF
(V_{DS} = 15 V dc, V_{GS} = 0, f = 1.0 MHz)						
Reverse transfer capacitance		C_{rss}	–	1.0	2.0	pF
(V_{DS} = 15 V dc, V_{GS} = 0, f = 1.0 MHz)						

▲ **FIGURE 8–23**

Partial data sheet for the 2N5460 *p*-channel JFET.

Solution Since $R_L \gg R_S$, $R_s \cong R_S$. From the partial data sheet in Figure 8–23, $g_m = y_{fs} = 1000$ μS (minimum). The voltage gain is

$$A_v = \frac{g_m R_S}{1 + g_m R_S} = \frac{(1000 \text{ μS})(10 \text{ k}\Omega)}{1 + (1000 \text{ μS})(10 \text{ k}\Omega)} = \mathbf{0.909}$$

From the data sheet, I_{GSS} = 5 nA (maximum) at V_{GS} = 20 V. Therefore,

$$R_{IN(gate)} = \frac{V_{GS}}{I_{GSS}} = \frac{20 \text{ V}}{5 \text{ nA}} = 4000 \text{ M}\Omega$$

$$R_{IN} = R_G \| R_{IN(gate)} = 10 \text{ M}\Omega \| 4000 \text{ M}\Omega \cong \mathbf{10 \text{ M}\Omega}$$

Related Problem If the maximum value of g_m of the 2N5460 JFET in the source-follower of Figure 8–22 is used, what is the voltage gain?

Open file E08-10 in the Examples folder on your CD-ROM. Measure the voltage gain using an input voltage of 10 mV rms to see how it compares with the calculated value.

**SECTION 8–3
REVIEW**

1. What is the ideal maximum voltage gain of a common-drain amplifier?
2. What factors influence the voltage gain of a common-drain amplifier?

8–4   COMMON-GATE AMPLIFIERS

The common-gate FET amplifier configuration introduced in this section is comparable to the common-base BJT amplifier. Like the CB, the common-gate (CG) amplifier has a low input resistance. This is different from the CS and CD configurations, which have very high input resistances.

After completing this section, you should be able to

- **Explain and analyze the operation of common-gate FET amplifiers**
- Analyze a CG amplifier
- Determine the voltage gain of a CG amplifier
- Determine the input resistance of a CG amplifier

A self-biased **common-gate** amplifier is shown in Figure 8–24. The gate is connected directly to ground. The input signal is applied at the source terminal through C_1. The output is coupled through C_2 from the drain terminal.

▶ **FIGURE 8–24**

JFET common-gate amplifier.

Voltage Gain

The voltage gain from source to drain is developed as follows.

$$A_v = \frac{V_{out}}{V_{in}} = \frac{V_d}{V_{gs}} = \frac{I_d R_d}{V_{gs}} = \frac{g_m V_{gs} R_d}{V_{gs}}$$

$$A_v = g_m R_d$$

Equation 8–11

where $R_d = R_D \parallel R_L$. Notice that the gain expression is the same as for the common-source JFET amplifier.

Input Resistance

As you have seen, both the common-source and common-drain configurations have extremely high input resistances because the gate is the input terminal. In contrast, the common-gate configuration where the source is the input terminal has a low input resistance, as shown in the following steps.

First, the input current is equal to the drain current.

$$I_{in} = I_s = I_d = g_m V_{gs}$$

Second, the input voltage equals V_{gs}.

$$V_{in} = V_{gs}$$

Therefore, the input resistance at the source terminal is

$$R_{in(source)} = \frac{V_{in}}{I_{in}} = \frac{V_{gs}}{g_m V_{gs}}$$

Equation 8–12

$$R_{in(source)} = \frac{1}{g_m}$$

If, for example, g_m has a value of 4000 μS, then

$$R_{in(source)} = \frac{1}{4000 \ \mu S} = 250 \ \Omega$$

EXAMPLE 8–11

Determine the minimum voltage gain and input resistance of the amplifier in Figure 8–25.

▶ **FIGURE 8–25**

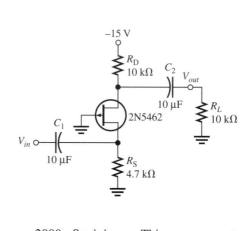

Solution From the data sheet in Figure 8–23, $g_m = 2000 \ \mu$S minimum. This common-gate amplifier has a load resistor, so the effective drain resistance is $R_D \parallel R_L$ and the minimum voltage gain is

$$A_v = g_m(R_D \parallel R_L) = (2000 \ \mu S)(10 \ k\Omega \parallel 10 \ k\Omega) = \mathbf{10}$$

The input resistance at the source terminal is

$$R_{in(source)} = \frac{1}{g_m} = \frac{1}{2000 \ \mu S} = 500 \ \Omega$$

The signal source actually sees R_S in parallel with $R_{in(source)}$, so the total input resistance is

$$R_{in} = R_{in(source)} \parallel R_S = 500 \ \Omega \parallel 4.7 \ k\Omega = \mathbf{452 \ \Omega}$$

Related Problem What is the input resistance in Figure 8–25 if R_S is changed to 10 kΩ?

Open file E08-11 in the Examples folder on your CD-ROM. Measure the voltage using a 10 mV rms input voltage.

SECTION 8-4 REVIEW

1. What is a major difference between a common-gate amplifier and the other two configurations?
2. What common factor determines the voltage gain and the input resistance of a common-gate amplifier?

8-5 TROUBLESHOOTING

A technician who understands the basics of circuit operation and who can, if necessary, perform basic analysis on a given circuit is much more valuable than one who is limited to carrying out routine test procedures. In this section, you will see how to test a circuit board that has only a schematic with no specified test procedure or voltage levels. In this case, basic knowledge of how the circuit operates and the ability to do a quick circuit analysis are useful.

After completing this section, you should be able to

- **Troubleshoot FET amplifiers**
- Troubleshoot a two-stage CS amplifier
- Relate a schematic to a circuit board

Assume that you are given a circuit board pulled from the audio amplifier section of a sound system and told simply that it is not working properly. The first step is to obtain the system schematic and locate this particular circuit on it. The circuit is a two-stage FET amplifier, as shown in Figure 8–26.

▲ **FIGURE 8-26**

A two-stage FET amplifier circuit.

The problem is approached in the following sequence.

Step 1: Determine what the voltage levels in the circuit should be so that you know what to look for. First, pull a data sheet on the particular transistor (assume both Q_1 and Q_2 are found to be the same type of transistor) and determine the g_m so that you can calculate the typical voltage gain. Assume that for this particular device, a typical g_m of 5000 μS is specified. Calculate the expected typical voltage gain of each stage (notice they are identical) based on the typical value of g_m. The g_m of actual devices may be any value between the specified minimum and maximum values. Because the input resistance is very high, the second stage does not significantly load the first stage, as in a BJT amplifier. So, the unloaded voltage gain for each stage is

$$A_v = g_m R_2 = (5000 \ \mu S)(1.5 \ k\Omega) = 7.5$$

Since the stages are identical, the typical overall gain should be

$$A'_v = (7.5)(7.5) = 56.3$$

We will ignore dc levels at this time and concentrate on signal tracing.

Step 2: Arrange a test setup to permit connection of an input test signal, a dc supply voltage, and ground to the circuit. The schematic shows that the dc supply voltage must be +12 V. Choose 10 mV rms as an input test signal. This value is arbitrary (although the capability of your signal source is a factor), but small enough that the expected output signal voltage is well below the absolute peak-to-peak limit of 12 V set by the supply voltage and ground (you know that the output voltage swing cannot go higher than 12 V or lower than 0 V). Set the frequency of the sinusoidal signal source to an arbitrary value in the audio range (say 10 kHz) because you know this is an audio amplifier. The audio frequency range is generally accepted as 20 Hz to 20 kHz.

Step 3: Check the input signal at the gate of Q_1 and the output signal at the drain of Q_2 with an oscilloscope. The results are shown in Figure 8–27. The measured output voltage has a peak value of 226 mV. The expected typical peak output voltage is

$$V_{out} = V_{in} A'_v = (14.14 \ mV)(56.3) = 796 \ mV \text{ peak}$$

The output is much less than it should be.

Step 4: Trace the signal from the output back toward the input to determine the fault. Figure 8–27 shows the oscilloscope displays of the measured signal voltages. The voltage at the gate of Q_2 is 106 mV peak, as expected (14.14 mV × 7.5 = 106 mV). This signal is properly coupled from the drain of Q_1. Therefore, the problem lies in the second stage. From the oscilloscope displays, the gain of Q_2 is much lower than it should be (213 mV/100 mV = 2.13 instead of 7.5).

Step 5: Analyze the possible causes of the observed malfunction. There are three possible reasons the gain is low:

1. Q_2 has a lower transconductance (g_m) than the specified typical value. Check the data sheet to see if the minimum g_m accounts for the lower measured gain.

2. R_5 has a lower value than shown on the schematic.

3. The bypass capacitor C_4 is open.

The only way to check the g_m is by replacing Q_2 with a new transistor of the same type and rechecking the output signal. You can make certain that R_5 is the proper value by removing one end of the resistor from the circuit board and measuring the resistance with an ohmmeter. To avoid having to unsolder a component, the best way to start isolating the fault is by checking the signal voltage at the source of Q_2. If the capacitor is working properly, there will be only a dc voltage at the source. The presence of a signal voltage at the source indicates that C_4 is open. With R_6 unbypassed, the gain expression is $g_m R_d/(1 + g_m R_s)$ rather than simply $g_m R_d$, thus resulting in less gain.

EWB/Multisim Troubleshooting Exercises

These file circuits are in the Troubleshooting Exercises folder on your CD-ROM.

1. Open file TSE08-01. Determine if the circuit is working properly and, if not, determine the fault.

▲ **FIGURE 8–27**

Oscilloscope displays of signals in the two-stage FET amplifier.

2. Open file TSE08-02. Determine if the circuit is working properly and, if not, determine the fault.

3. Open file TSE08-03. Determine if the circuit is working properly and, if not, determine the fault.

4. Open file TSE08-04. Determine if the circuit is working properly and, if not, determine the fault.

**SECTION 8–5
REVIEW**

1. What is the prerequisite to effective troubleshooting?

2. Assume that C_2 in the amplifier of Figure 8–26 opened. What symptoms would indicate this failure?

3. If C_3 opened in the amplifier, would the voltage gain of the first stage be affected?

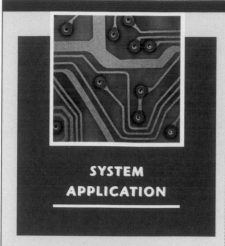

SYSTEM APPLICATION

You are assigned to investigate the possibility of redesigning the audio preamplifier used in the public address system currently being produced to replace the BJTs with FETs. You are asked to evaluate a design and breadboard the circuit to evaluate its performance compared to the existing BJT amplifier. You will apply the knowledge you have gained in this chapter in completing your assignment.

Review of the BJT Amplifier

A schematic of the BJT (bipolar junction transistor) amplifier currently used in the public address/audio paging system is shown in Figure 8–28. (See the system application in Chapter 6.)

Performance of the BJT Amplifier

The overall voltage gain of the two-stage BJT amplifier was previously determined to be adjustable over a range from 145 to 733 as follows:

Stage 1: $V_B = \left(\dfrac{R_2 \| \beta_{DC}R_4}{R_1 + R_2 \| \beta_{DC}R_4}\right)V_{CC} = 1.93\text{ V}$

$V_E = V_B - 0.7\text{ V} = 1.23\text{ V}$

$I_E = \dfrac{V_E}{R_E} = 1.23\text{ mA}$

$r'_e = 20.3\ \Omega$

$A_v = \dfrac{R_c}{r'_e} = \dfrac{R_3 \| R_{IN(base2)}}{r'_e} = \dfrac{1.93\text{ k}\Omega}{20.3\ \Omega} = 95$

Stage 2: $V_B = \left(\dfrac{R_6 \| \beta_{DC}(R_8 + R_9)}{R_5 + R_6 \| \beta_{DC}(R_8 + R_9)}\right)V_{CC} = 1.88\text{ V}$

$V_E = V_B - 0.7\text{ V} = 1.18\text{ V}$

$I_E = \dfrac{V_E}{R_E} = 0.97\text{ mA}$

$r'_e = 25.8\ \Omega$

$A_{v(max)} = \dfrac{R_c}{r'_e + R_8} = \dfrac{4.7\text{ k}\Omega}{245.8\ \Omega} = 19.1$

$A_{v(min)} = \dfrac{R_c}{r'_e + R_8 + R_9} = \dfrac{4.7\text{ k}\Omega}{1245.8\ \Omega} = 3.77$

Attenuation of Input: The microphone input has a 30 Ω resistance.

$\text{Attenuation} = \dfrac{20.3\ \Omega}{20.3\ \Omega + 30\ \Omega} = 0.404$

Total Voltage Gain: $A_{v(tot)} = (0.404)(95)(19.1) = 733$ maximum

$A_{v(tot)} = (0.404)(95)(3.77) = 145$ minimum

$+12\text{ V}$

R_1 68 kΩ	R_3 2.2 kΩ	R_5 100 kΩ	R_7 4.7 kΩ

C_2 1 μF C_1 100 μF Q_1 C_3 1 μF Q_2

R_2 15 kΩ R_4 1.0 kΩ R_6 22 kΩ R_8 220 Ω R_9 1 kΩ C_4 100 μF

▲ **FIGURE 8–28**

BJT audio preamplifier. Both transistors are 2N3904. Assume $\beta_{DC} = \beta_{ac} = 100$.

Basic MOSFET Amplifier Design

The basic design for one stage of a MOS-FET amplifier using a 2N3797 is shown in Figure 8–29. You are to evaluate it to determine if it can be used in a multistage amplifier to provide the same voltage gain as the bipolar version.

- Using the data sheet in Figure 8–30, determine the minimum and maximum dc drain-to-source voltages. Remember, for a zero-biased D-MOS-FET, the drain current equals I_{DSS}.

- Using the data sheet in Figure 8–30, determine the minimum and maximum voltage gains.

- Discuss the problem presented by the variation in I_{DSS} from one device to the next and recommend an approach to minimize the problem.

- Discuss the problem presented by the fact that the voltage gain of the MOS-FET amplifier stage is dependent on g_m, and recommend an approach to minimize the problem.

▲ **FIGURE 8–29**

Schematic of one MOSFET amplifier stage.

2N3796
2N3797

Case 22-03, Style 2
TO-18 (TO-206AA)

MOSFETs
Low Power Audio
N-channel — Depletion

Maximim Ratings

Rating	Symbol	Value	Unit
Drain-Source voltage 2N3796 2N3797	V_{DS}	25 20	V dc
Gate-Source voltage	V_{GS}	±10	V dc
Drain current	I_D	20	mA dc
Total device dissipation @ $T_A = 25°C$ Derate above 25°C	P_D	200 1.14	mW mW/°C
Junction temperature	T_J	+175	°C
Storage channel temperature range	T_{stg}	−65 to +200	°C

Electrical Characteristics ($T_A = 25°C$ unless otherwise noted.)

Characteristic	Symbol	Min	Typ	Max	Unit
OFF Characteristics					
Drain-Source breakdown voltage ($V_{GS} = -4.0$ V, $I_D = 5.0$ µA) 2N3796 ($V_{GS} = -7.0$ V, $I_D = 5.0$ µA) 2N3797	$V_{(BR)DSX}$	25 20	30 25	– –	V dc
Gate reverse current ($V_{GS} = -10$ V, $V_{DS} = 0$) ($V_{GS} = -10$ V, $V_{DS} = 0$, $T_A = 150°C$)	I_{GSS}	– –	– –	1.0 200	pA dc
Gate-Source cutoff voltage ($I_D = 0.5$ µA, $V_{DS} = 10$ V) 2N3796 ($I_D = 2.0$ µA, $V_{DS} = 10$ V) 2N3797	$V_{GS(off)}$	– –	−3.0 −5.0	−4.0 −7.0	V dc
Drain-Gate reverse current ($V_{DG} = 10$ V, $I_S = 0$)	I_{DGO}	–	–	1.0	pA dc
ON Characteristics					
Zero-gate-voltage drain current ($V_{DS} = 10$ V, $V_{GS} = 0$) 2N3796 2N3797	I_{DSS}	0.5 2.0	1.5 2.9	3.0 6.0	mA dc
On-State drain current ($V_{DS} = 10$ V, $V_{GS} = +3.5$ V) 2N3796 2N3797	$I_{D(on)}$	7.0 9.0	8.3 14	14 18	mA dc
Small-Signal Characteristics					
Forward transfer admittance ($V_{DS} = 10$ V, $V_{GS} = 0$, $f = 1.0$ kHz) 2N3796 2N3797	$|Y_{fs}|$	900 1500	1200 2300	1800 3000	µmhos or µS
($V_{DS} = 10$ V, $V_{GS} = 0$, $f = 1.0$ MHz) 2N3796 2N3797		900 1500	– –	– –	
Output admittance ($V_{DS} = 10$ V, $V_{GS} = 0$, $f = 1.0$ kHz) 2N3796 2N3797	$|Y_{os}|$	– –	12 27	25 60	µmhos or µS
Input capacitance ($V_{DS} = 10$ V, $V_{GS} = 0$, $f = 1.0$ MHz) 2N3796 2N3797	C_{iss}	– –	5.0 6.0	7.0 8.0	pF
Reverse transfer capacitance ($V_{DS} = 10$ V, $V_{GS} = 0$, $f = 1.0$ MHz)	C_{rss}	–	0.5	0.8	pF

▲ **FIGURE 8–30**

Partial data sheet for the 2N3797 D-MOSFET.

Amplifier Performance on the Test Bench

The breadboard test bench for the MOS-FET amplifier is shown in Figure 8–31. Two stages have been wired on the connector board for testing. The two 2N3797 transistors in the circuit have been randomly picked from a large selection of the devices.

Verify that the circuit is properly wired on the connector board.

- Measurements taken on the bread-boarded circuit are shown in Figure 8–32 where each set of measurements is made with different transistors. That is, the first set of measurements is made with two randomly selected 2N3797s and then the transistors are replaced with two different randomly selected devices for the second set of measurements. The circled numbers indicate test point connections to the circuit.

- Explain the difference in the two sets of dc voltage readings.

- Explain the difference in the two sets of ac voltage readings.

- Can I_{DSS} and g_m be determined for each transistor from the measurements in Figure 8–32? If so, determine the values for each of the four different MOSFETs.

- Determine the gain of each stage for each set of measurements in Figure 8–32.

- Assuming the maximum theoretical gain can be achieved for V_{DS} centered at 6 V by careful selection of the MOSFETs for maximum g_m and typical I_{DSS}, how many stages will be required to match the maximum gain of the BJT amplifier in Figure 8–28? Remember to take into account the attenuation, if any, of the input circuit.

+12 V from power supply

Sinusoidal voltage from function generator

Power supply ground

▲ FIGURE 8–31

Breadboard test bench for a two-stage MOSFET amplifier.

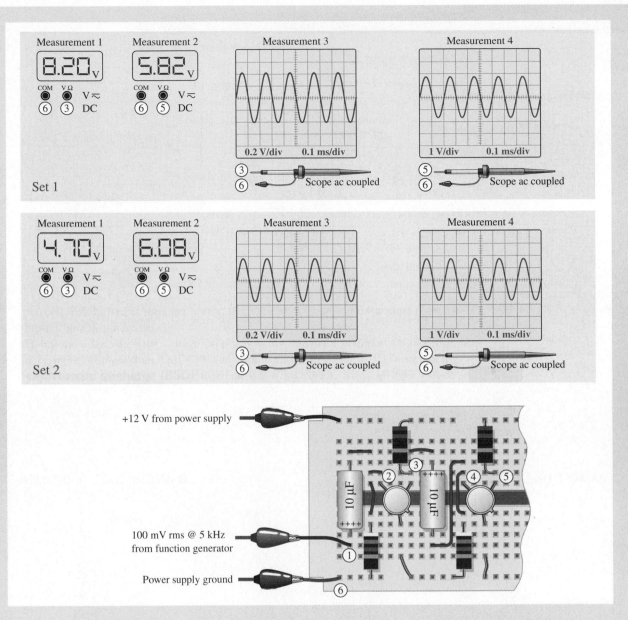

▲ FIGURE 8–32

Measurements of amplifier voltages for two sets of MOSFETs.

Recommendation (Optional)

Submit a recommendation containing the following items:

1. A summary of your analysis and test results for the MOSFET amplifier.

2. A comparison of the performance of the MOSFET amplifier to the BJT amplifier.

3. A statement of recommendation to either replace the BJT amplifier with a MOSFET amplifier or to retain the BJT amplifier in the system.

4. Reasons justifying your recommendation.

SUMMARY OF FET AMPLIFIERS

N channels are shown. V_{DD} is negative for p channel.

COMMON-SOURCE AMPLIFIERS

JFET *Self-bias*

- $I_D = I_{DSS}\left(1 - \dfrac{I_D R_S}{V_{GS(off)}}\right)^2$
- $A_v = g_m R_d$
- $R_{in} = R_G \| \left(\dfrac{V_{GS}}{I_{GSS}}\right)$

D-MOSFET *Zero-bias*

- $I_D = I_{DSS}$
- $A_v = g_m R_d$
- $R_{in} = R_G \| \left(\dfrac{V_{GS}}{I_{GSS}}\right)$

E-MOSFET *Voltage-divider bias*

- $I_D = K(V_{GS} - V_{GS(th)})^2$
- $A_v = g_m R_d$
- $R_{in} = R_1 \| R_2 \| \left(\dfrac{V_{GS}}{I_{GSS}}\right)$

COMMON-DRAIN AMPLIFIER

JFET *Self-bias*

- $I_D = I_{DSS}\left(1 - \dfrac{I_D R_S}{V_{GS(off)}}\right)^2$
- $A_v = \dfrac{g_m R_s}{1 + g_m R_s}$
- $R_{in} = R_G \| \left(\dfrac{V_{GS}}{I_{GSS}}\right)$

COMMON-GATE AMPLIFIER

JFET *Self-bias*

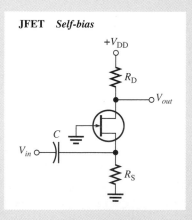

- $I_D = I_{DSS}\left(1 - \dfrac{I_D R_S}{V_{GS(off)}}\right)^2$
- $A_v = g_m R_d$
- $R_{in} = \left(\dfrac{1}{g_m}\right) \| R_S$

CHAPTER SUMMARY

- The drain of a FET is analogous to the collector of a BJT, the source of a FET is analogous to the emitter of a BJT, and the gate of a FET is analogous to the base of a BJT.
- The transconductance, g_m, of a FET relates the output current, I_d, to the input voltage, V_{gs}.
- The voltage gain of a common-source amplifier is determined largely by the transconductance, g_m, and the drain resistance, R_d.
- The internal drain-to-source resistance, r'_{ds}, of a FET influences (reduces) the gain if it is not sufficiently greater than R_d so that it can be neglected.
- An unbypassed resistance between source and ground (R_S) reduces the voltage gain of a FET amplifier.
- A load resistance connected to the drain of a common-source amplifier reduces the voltage gain.
- There is a 180° phase inversion between gate and drain voltages.
- The input resistance at the gate of a FET is extremely high.
- The voltage gain of a common-drain amplifier (source-follower) is always slightly less than 1.
- There is no phase inversion between gate and source in a source-follower.
- The input resistance of a common-gate amplifier is the reciprocal of g_m.
- The total voltage gain of a multistage amplifier is the product of the individual voltage gains (sum of dB gains).
- Generally, higher voltage gains can be achieved with BJT amplifiers than with FET amplifiers.

KEY TERMS

Key terms and other bold terms in the chapter are defined in the end-of-book glossary.

Common-drain   A FET amplifier configuration in which the drain is the grounded terminal.
Common-gate   A FET amplifier configuration in which the gate is the grounded terminal.
Common-source   A FET amplifier configuration in which the source is the grounded terminal.
Source-follower   The common-drain amplifier.

KEY FORMULAS

FET Amplification

8–1	$I_d = g_m V_{gs}$	Drain current
8–2	$A_v = g_m R_d$	Voltage gain with source grounded or R_s bypassed
8–3	$A_v = \dfrac{g_m R_d}{1 + g_m R_s}$	Voltage gain with R_s unbypassed

Common-Source Amplifier

8–4	$I_D = \dfrac{I_{DSS}}{2}$	For centered Q-point
8–5	$I_D = I_{DSS}\left(1 - \dfrac{I_D R_S}{V_{GS(off)}}\right)^2$	Self-biased JFET current
8–6	$A_v = g_m R_d$	Voltage gain
8–7	$R_{in} = R_G \| \left(\dfrac{V_{GS}}{I_{GSS}}\right)$	Input resistance, self-bias and zero-bias
8–8	$R_{in} = R_1 \| R_2 \| R_{IN(gate)}$	Input resistance, voltage-divider bias

Common-Drain Amplifier

8–9	$A_v = \dfrac{g_m R_s}{1 + g_m R_s}$	Voltage gain

8–10 $R_{in} = R_G \parallel R_{IN(gate)}$ Input resistance

Common-Gate Amplifier

8–11 $A_v = g_m R_d$ Voltage gain

8–12 $R_{in(source)} = \dfrac{1}{g_m}$ Input resistance

Answers are at the end of the chapter.

1. In a common-source amplifier, the output voltage is
   (a) 180° out of phase with the input   (b) in phase with the input
   (c) taken at the source   (d) taken at the drain
   (e) answers (a) and (c)   (f) answers (a) and (d)

2. In a certain common-source (CS) amplifier, $V_{ds} = 3.2$ V rms and $V_{gs} = 280$ mV rms. The voltage gain is
   (a) 1   (b) 11.4   (c) 8.75   (d) 3.2

3. In a certain CS amplifier, $R_D = 1.0$ kΩ, $R_S = 560$ Ω, $V_{DD} = 10$ V, and $g_m = 4500$ μS. If the source resistor is completely bypassed, the voltage gain is
   (a) 450   (b) 45   (c) 4.5   (d) 2.52

4. Ideally, the equivalent circuit of a FET contains
   (a) a current source in series with a resistance
   (b) a resistance between drain and source terminals
   (c) a current source between gate and source terminals
   (d) a current source between drain and source terminals

5. The value of the current source in Question 4 is dependent on the
   (a) transconductance and gate-to-source voltage
   (b) dc supply voltage
   (c) external drain resistance
   (d) answers (b) and (c)

6. A certain common-source amplifier has a voltage gain of 10. If the source bypass capacitor is removed,
   (a) the voltage gain will increase   (b) the transconductance will increase
   (c) the voltage gain will decrease   (d) the Q-point will shift

7. A CS amplifier has a load resistance of 10 kΩ and $R_D = 820$ Ω. If $g_m = 5$ mS and $V_{in} = 500$ mV, the output signal voltage is
   (a) 1.89 V   (b) 2.05 V   (c) 25 V   (d) 0.5 V

8. If the load resistance in Question 7 is removed, the output voltage will
   (a) stay the same   (b) decrease   (c) increase   (d) be zero

9. A certain common-drain (CD) amplifier with $R_S = 1.0$ kΩ has a transconductance of 6000 μS. The voltage gain is
   (a) 1   (b) 0.86   (c) 0.98   (d) 6

10. The data sheet for the transistor used in a CD amplifier specifies $I_{GSS} = 5$ nA at $V_{GS} = 10$ V. If the resistor from gate to ground, R_G, is 50 MΩ, the total input resistance is approximately
    (a) 50 MΩ   (b) 200 MΩ   (c) 40 MΩ   (d) 20.5 MΩ

11. The common-gate (CG) amplifier differs from both the CS and CD configurations in that it has a

   (a) much higher voltage gain   (b) much lower voltage gain

   (c) much higher input resistance   (d) much lower input resistance

12. If you are looking for both good voltage gain and high input resistance, you must use a

   (a) CS amplifier   (b) CD amplifier   (c) CG amplifier

13. For small-signal operation, an *n*-channel JFET must be biased at

   (a) $V_{GS} = 0$ V   (b) $V_{GS} = V_{GS(off)}$

   (c) $-V_{GS(off)} < V_{GS} < 0$ V   (d) 0 V $< V_{GS} < +V_{GS(off)}$

14. Two FET amplifiers are cascaded. The first stage has a voltage gain of 5 and the second stage has a voltage gain of 7. The overall voltage gain is

   (a) 35   (b) 12   (c) dependent on the second stage loading

15. If there is an internal open between the drain and source in a CS amplifier, the drain voltage is equal to

   (a) 0 V   (b) V_{DD} (c) a value less than normal   (d) V_{GS}

PROBLEMS

Answers to all odd-numbered problems are at the end of the book.

BASIC PROBLEMS

SECTION 8–1   FET Amplification

1. A FET has a $g_m = 6000$ μS. Determine the rms drain current for each of the following rms values of V_{gs}.

   (a) 10 mV   (b) 150 mV   (c) 0.6 V   (d) 1 V

2. The gain of a certain JFET amplifier with a source resistance of zero is 20. Determine the drain resistance if the g_m is 3500 μS.

3. A certain FET amplifier has a g_m of 4.2 mS, $r'_{ds} = 12$ kΩ, and $R_D = 4.7$ kΩ. What is the voltage gain? Assume the source resistance is 0 Ω.

4. What is the gain for the amplifier in Problem 3 if the source resistance is 1.0 kΩ?

SECTION 8–2   Common-Source Amplifiers

5. Identify the type of FET and its bias arrangement in Figure 8–33. Ideally, what is V_{GS}?

▲ **FIGURE 8–33**

6. Calculate the dc voltages from each terminal to ground for the FETs in Figure 8–33.

7. Identify each characteristic curve in Figure 8–34 by the type of FET that it represents.

8. Refer to the JFET transfer characteristic curve in Figure 8–18(a) and determine the peak-to-peak value of I_d when V_{gs} is varied ±1.5 V about its Q-point value.

9. Repeat Problem 8 for the curves in Figure 8–18(b) and Figure 8–18(c).

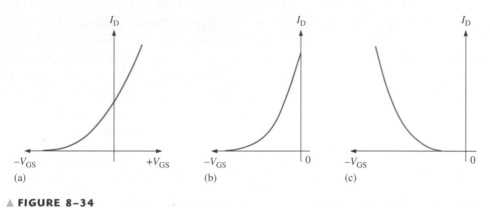

(a)   (b)   (c)

▲ **FIGURE 8–34**

10. Given that $I_D = 2.83$ mA in Figure 8–35, find V_{DS} and V_{GS}. $V_{GS(off)} = -7$ V and $I_{DSS} = 8$ mA.

11. If a 50 mV rms input signal is applied to the amplifier in Figure 8–35, what is the peak-to-peak output voltage? $g_m = 5000$ μS.

12. If a 1500 Ω load is ac coupled to the output in Figure 8–35, what is the resulting output voltage (rms) when a 50 mV rms input is applied? $g_m = 5000$ μS.

▲ **FIGURE 8–35**

File circuits are identified with a CD logo and are in the Problems folder on your CD-ROM. Filenames correspond to figure numbers (e.g., F08-35).

13. Determine the voltage gain of each common-source amplifier in Figure 8–36.

14. Draw the dc and ac equivalent circuits for the amplifier in Figure 8–37.

15. Determine the drain current in Figure 8–37 given that $I_{DSS} = 12.7$ mA and $V_{GS(off)} = -4$ V. The Q-point is centered.

16. What is the gain of the amplifier in Figure 8–37 if C_2 is removed?

17. A 4.7 kΩ resistor is connected in parallel with R_L in Figure 8–37. What is the voltage gain?

▲ FIGURE 8–36

▲ FIGURE 8–37

18. For the common-source amplifier in Figure 8–38, determine I_D, V_{GS}, and V_{DS} for a centered Q-point. $I_{DSS} = 9$ mA, and $V_{GS(off)} = -3$ V.

19. If a 10 mV rms signal is applied to the input of the amplifier in Figure 8–38, what is the rms value of the output signal?

▲ FIGURE 8–38

20. Determine V_{GS}, I_D, and V_{DS} for the amplifier in Figure 8–39. $I_{D(on)} = 18$ mA at $V_{GS} = 10$ V, $V_{GS(th)} = 2.5$ V, and $g_m = 3000$ μS.

▶ **FIGURE 8–39**

21. Determine the input resistance seen by the signal source in Figure 8–40. Assume $I_{GSS} = 25$ nA at $V_{GS} = -15$ V.

▶ **FIGURE 8–40**

22. Determine the total drain voltage waveform (dc and ac) and the V_{out} waveform in Figure 8–41. $g_m = 4.8$ mS and $I_{DSS} = 15$ mA. Observe that $V_{GS} = 0$.

▲ **FIGURE 8–41**

23. For the unloaded amplifier in Figure 8–42, find V_{GS}, I_D, V_{DS}, and the rms output voltage V_{ds}. $I_{D(on)} = 8$ mA at $V_{GS} = 12$ V, $V_{GS(th)} = 4$ V, and $g_m = 4500$ μS.

▶ **FIGURE 8–42**

SECTION 8–3   Common-Drain Amplifiers

24. For the source-follower in Figure 8–43, determine the voltage gain and input resistance. $I_{GSS} = 50$ pA at $V_{GS} = -15$ V and $g_m = 5500$ μS.

25. If the JFET in Figure 8–43 is replaced with one having a g_m of 3000 μS, what are the gain and the input resistance with all other conditions the same?

▶ **FIGURE 8–43**

26. Find the gain of each amplifier in Figure 8–44.

27. Determine the voltage gain of each amplifier in Figure 8–44 when the capacitively coupled load is changed to 10 kΩ.

(a)

(b)

▲ **FIGURE 8–44**

SECTION 8–4   Common-Gate Amplifiers

28. A common-gate amplifier has a $g_m = 4000$ μS and $R_d = 1.5$ kΩ. What is its gain?

29. What is the input resistance of the amplifier in Problem 28?

30. Determine the voltage gain and input resistance of the common-gate amplifier in Figure 8–45.

▲ **FIGURE 8–45**

TROUBLESHOOTING PROBLEMS

SECTION 8–5 **Troubleshooting**

31. What symptom(s) would indicate each of the following failures when a signal voltage is applied to the input in Figure 8–46?

(a) Q_1 open from drain to source   (b) R_3 open

(c) C_2 shorted   (d) C_3 open

(e) Q_2 open from drain to source

32. If $V_{in} = 10$ mV rms in Figure 8–46, what is V_{out} for each of the following faults?

(a) C_1 open   (b) C_4 open

(c) a short from the source of Q_2 to ground   (d) Q_2 has an open gate

▲ **FIGURE 8–46**

SYSTEM APPLICATION PROBLEMS

33. A 100 mV rms voltage at a frequency of 1 kHz is applied to test point 1 in Figure 8–47. A dc voltage of 6.75 V is measured at test point 5, but there is no ac voltage at this point. Determine the one and only fault in the circuit.

+12 V from power supply

Signal voltage

Power supply ground

10 μF

10 μF

++++

++++

▲ **FIGURE 8–47**

34. Assume that the fault in Problem 33 for the circuit of Figure 8–47 has been corrected and the input voltage is increased to 250 mV rms. If the following voltages are measured at the specified test points, determine the possible fault(s) and state what you would do to repair the circuit.

Test point 2: 250 mV rms

Test point 3: 800 mV rms

Test point 4: 530 mV rms

Test point 5: 2.12 V rms

35. Assume the previous faults in the circuit of Figure 8–47 have been corrected. Determine the dc and ac voltages that should be measured at test point 5 if the first stage transistor has an $I_{DSS} = 2.85$ mA and a $g_m = 2200$ μS and the second stage transistor has an $I_{DSS} = 5.10$ mA and a $g_m = 2600$ μS. The input signal is 100 mV rms.

DATA SHEET PROBLEMS

36. What type of FET is the 2N3796?

37. Referring to the data sheet in Figure 8–30, determine the following:

(a) typical $V_{GS(off)}$ for the 2N3796

(b) maximum drain-to-source voltage for the 2N3797

(c) maximum power dissipation for the 2N3797 at an ambient temperature of 25°C

(d) maximum gate-to-source voltage for the 2N3797

38. Referring to Figure 8–30, determine the maximum power dissipation for a 2N3796 at an ambient temperature of 55°C.

39. Referring to Figure 8–30, determine the minimum g_{m0} for the 2N3796 at a frequency of 1 kHz.

40. What is the drain current when $V_{GS} = +3.5$ V for the 2N3797?

41. Typically, what is the drain current for a zero-biased 2N3796?

42. What is the maximum possible voltage gain for a 2N3796 common-source amplifier with $R_d = 2.2$ kΩ?

ADVANCED PROBLEMS

43. The MOSFET in a certain single-stage common-source amplifier has a range of forward transconductance values from 2.5 mS to 7.5 mS. If the amplifier is capacitively coupled to a variable load that ranges from 4 kΩ to 10 kΩ and the dc drain resistance is 1.0 kΩ, determine the minimum and maximum voltage gains.

44. Design an amplifier using a 2N3797 that operates from a 24 V supply voltage. The typical dc drain-to-source voltage should be approximately 12 V and the typical voltage gain should be approximately 9.

45. Modify the amplifier you designed in Problem 44 so that the voltage gain can be set at 9 for any randomly selected 2N3797.

EWB/MULTISIM TROUBLESHOOTING PROBLEMS

These file circuits are in the Troubleshooting Problems folder on your CD-ROM.

46. Open file TSP08-46 and determine the fault.

47. Open file TSP08-47 and determine the fault.

48. Open file TSP08-48 and determine the fault.

49. Open file TSP08-49 and determine the fault.

50. Open file TSP08-50 and determine the fault.

51. Open file TSP08-51 and determine the fault.

52. Open file TSP08-52 and determine the fault.

53. Open file TSP08-53 and determine the fault.

54. Open file TSP08-54 and determine the fault.

ANSWERS

SECTION REVIEWS

SECTION 8–1   FET Amplification

1. The FET with g_m = 3.5 mS can produce the higher gain.
2. $A_v = g_m R_d = (2500\ \mu S)(10\ k\Omega) = 25$
3. The FET with R_{ds} = 100 kΩ can produce the higher gain.

SECTION 8–2   Common-Source Amplifiers

1. I_d is at its positive peak and V_{ds} is at its negative peak when V_{gs} is at its positive peak.
2. V_{gs} is an ac quantity, V_{GS} is a dc quantity.
3. The D-MOSFET can operate with V_{GS} = 0 V at the Q-point.
4. Voltage gain of a CS amplifier is determined by g_m and R_d.
5. The gain is halved because $R_d = R_D/2$.

SECTION 8–3   Common-Drain Amplifiers

1. The ideal maximum voltage gain of a CD amplifier is 1.
2. The voltage gain of a CD amplifier is determined by g_m and R_s.

SECTION 8–4   Common-Gate Amplifiers

1. The CG amplifier has a low input resistance ($1/g_m$).
2. g_m affects both voltage gain and input resistance.

SECTION 8–5   Troubleshooting

1. To be a good troubleshooter, you must understand the circuit.
2. There would be a lower than normal first-stage gain if C_2 opens.
3. No, but there would be a loss of signal to the second stage.

RELATED PROBLEMS FOR EXAMPLES

8–1 13.2

8–2 5

8–3 2.92

8–4 I_D will be 1.44 mA. V_D will increase to 7.25 V. V_{out} will increase to 1.37 V rms.

8–5 744 mV

8–6 $R_{in} = 9.99$ MΩ

8–7 1.65 V

8–8 ΔI_D decreases; distortion and clipping at cutoff

8–9 $V_{GS} = 2.23$ V; $I_D = 0.588$ mA; $V_{DS} = 13.1$ V; $V_{out} = 750$ mV

8–10 0.976

8–11 $R_{in} = 476$ Ω

SELF-TEST

1. (f) **2.** (b) **3.** (c) **4.** (d) **5.** (a)

6. (c) **7.** (a) **8.** (c) **9.** (b) **10.** (a)

11. (d) **12.** (a) **13.** (c) **14.** (a) **15.** (b)

9

POWER AMPLIFIERS

INTRODUCTION

Power amplifiers are large-signal amplifiers. This generally means that a much larger portion of the load line is used during signal operation than in a small-signal amplifier. In this chapter, we will cover four classes of power amplifiers: class A, class B, class AB, and class C. These amplifier classifications are based on the percentage of the input cycle for which the amplifier operates in its linear region. Each class has a unique circuit configuration because of the way it must be operated. The emphasis is on power amplification.

Power amplifiers are normally used as the final stage of a communications receiver or transmitter to provide signal power to speakers or to a transmitting antenna.

CHAPTER OUTLINE

9–1 Class A Power Amplifiers
9–2 Class B and Class AB Push-Pull Amplifiers
9–3 Class C Amplifiers
9–4 Troubleshooting
▪▪▪ System Application

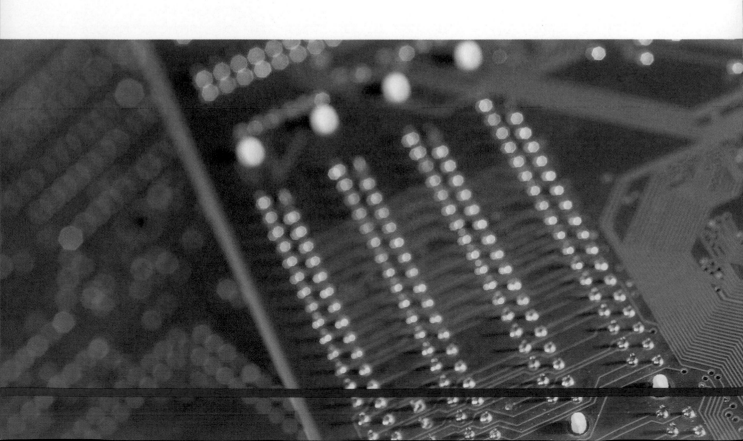

- Explain and analyze the operation of class A power amplifiers

- Explain and analyze the operation of class B and class AB amplifiers

- Discuss and analyze the operation of class C amplifiers

- Troubleshoot power amplifiers

KEY TERMS

- Class A

- Power gain

- Efficiency

- Class B

- Push-pull

- Class AB

- Electrostatic discharge (ESD)

- Class C

The system application in this chapter continues with the public address and paging system started in Chapter 6. Recall that the complete system includes the preamplifier, a power amplifier, and a dc power supply. You will focus on the power amplifier in this chapter and complete the total system. The first step in your assignment is to learn all you can about power amplifier operation. You will then apply your knowledge to the system application at the end of the chapter.

www. VISIT THE COMPANION WEBSITE
Study aids for this chapter are available at
http://www.prenhall.com/floyd

9-1 CLASS A POWER AMPLIFIERS

When an amplifier is biased such that it always operates in the linear region where the output signal is an amplified replica of the input signal, it is a class A amplifier. The discussion of amplifiers in the previous chapters apply to class A operation. Power amplifiers are those amplifiers that have the objective of delivering power to a load. This means that components must be considered in terms of their ability to dissipate heat.

After completing this section, you should be able to

- ■ **Explain and analyze the operation of class A power amplifiers**
- ■ Explain why a centered Q-point is important for a class A power amplifier
- ■ Determine the voltage gain and power gain for a multistage amplifier
- ■ Determine the efficiency of a class A power amplifier

In a small-signal amplifier, the ac signal moves over a small percentage of the total ac load line. When the output signal is larger and approaches the limits of the ac load line, the amplifier is a **large-signal** type. Both large-signal and small signal amplifiers are considered to be **class A** if they operate in the linear region at all times, as illustrated in Figure 9–1. Class A power amplifiers are large-signal amplifiers with the objective of providing power (rather than voltage) to a load. As a rule of thumb, an amplifier may be considered to be a power amplifier if it is necessary to consider the problem of heat dissipation in components.

▲ **FIGURE 9–1**

Basic class A amplifier operation. Output is shown 180° out of phase with the input (inverted).

Heat Dissipation

Power transistors (and other power devices) must dissipate a large amount of internally generated heat. For BJT power transistors, the collector terminal is the critical junction; for this reason, the transistor's case is always connected to the collector terminal. The case of all power transistors is designed to provide a large contact area between it and an external heat sink. Heat from the transistor flows through the case to the heat sink and then dissipates in the surrounding air. Heat sinks vary in size, number of fins, and type of material. Their size depends on the heat dissipation requirement and the maximum ambient temperature in which the transistor is to operate. In high-power applications (a few hundred watts), a cooling fan may be necessary.

Centered Q-Point

Recall that the dc and ac load lines cross at the Q-point. When the Q-point is at the center of the ac load line, a maximum class A signal can be obtained. You can see this concept by examining the graph of the load line for a given amplifier in Figure 9–2(a). This graph shows the ac load line with the Q-point at its center. The collector current can vary from its Q-point value, I_{CQ}, up to its saturation value, $I_{c(sat)}$, and down to its cutoff value of zero. Likewise, the collector-to-emitter voltage can swing from its Q-point value, V_{CEQ},

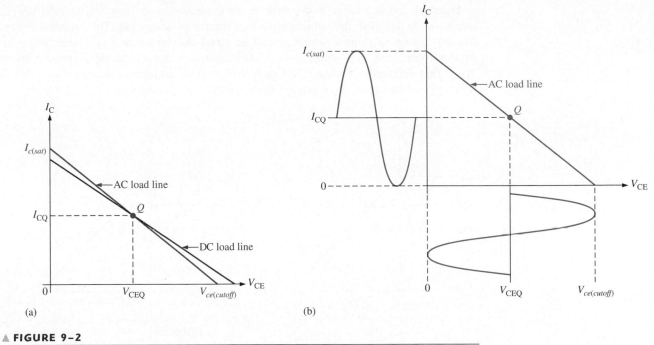

(a) (b)

▲ FIGURE 9–2

Maximum class A output occurs when the Q-point is centered on the ac load line.

up to its cutoff value, $V_{ce(cutoff)}$, and down to its saturation value of near zero. This operation is indicated in Figure 9–2(b). The peak value of the collector current equals I_{CQ}, and the peak value of the collector-to-emitter voltage equals V_{CEQ} in this case. This signal is the maximum that can be obtained from the class A amplifier. Actually, the output cannot quite reach saturation or cutoff, so the practical maximum is slightly less.

If the Q-point is not centered on the ac load line, the output signal is limited. Figure 9–3 shows an ac load line with the Q-point moved away from center toward cutoff. The output variation is limited by cutoff in this case. The collector current can only swing down to near zero and an equal amount above I_{CQ}. The collector-to-emitter voltage can only swing up to its cutoff value and an equal amount below V_{CEQ}. This situation is illustrated in Figure 9–3(a). If the amplifier is driven any further than this, it will "clip" at cutoff, as shown in Figure 9–3(b).

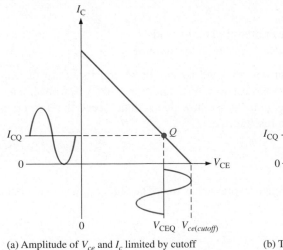

(a) Amplitude of V_{ce} and I_c limited by cutoff

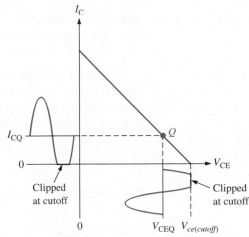

(b) Transistor driven into cutoff by a further increase in input amplitude

▲ FIGURE 9–3

Q-point closer to cutoff.

Figure 9–4 shows an ac load line with the Q-point moved away from center toward saturation. In this case, the output variation is limited by saturation. The collector current can only swing up to near saturation and an equal amount below I_{CQ}. The collector-to-emitter voltage can only swing down to its saturation value and an equal amount above V_{CEQ}. This situation is illustrated in Figure 9–4(a). If the amplifier is driven any further, it will "clip" at saturation, as shown in Figure 9–4(b).

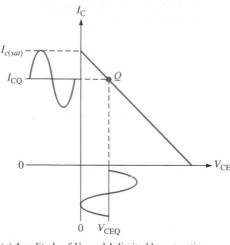

(a) Amplitude of V_{ce} and I_c limited by saturation

(b) Transistor driven into saturation by a further increase in input amplitude

▲ **FIGURE 9–4**

Q-point closer to saturation.

Power Gain

A power amplifier delivers power to a load. The **power gain** of an amplifier is the ratio of the power delivered to the load to the input power. In general, power gain is

Equation 9–1

$$A_p = \frac{P_L}{P_{in}}$$

where A_p = power gain
P_L = signal power delivered to the load
P_{in} = signal power delivered to the amplifier

The power gain can be computed by any of several formulas, depending on what is known. Frequently, the easiest way to obtain power gain is from input resistance, load resistance, and voltage gain. To see how this is done, recall that power can be expressed in terms of voltage and resistance as

$$P = \frac{V^2}{R}$$

For ac power, the voltage is expressed as rms. The output power delivered to the load is

$$P_L = \frac{V_L^2}{R_L}$$

The input power delivered to the amplifier is

$$P_{in} = \frac{V_{in}^2}{R_{in}}$$

By substituting into Equation 9–1, the following useful relationship is produced:

$$A_p = \frac{V_L^2}{V_{in}^2}\left(\frac{R_{in}}{R_L}\right)$$

Since $V_L/V_{in} = A_v$,

$$A_p = A_v^2\left(\frac{R_{in}}{R_L}\right)$$

Equation 9–2

Recall from Chapter 6 that for a voltage-divider biased amplifier

$$R_{in(tot)} = R_1 \,\|\, R_2 \,\|\, R_{in(base)}$$

and that for a CE or CC amplifier

$$R_{in(base)} = \beta_{ac}R_e$$

Equation 9–2 shows that the power gain to an amplifier is the voltage gain squared times the ratio of the input resistance to the output load resistance. It can be applied to any amplifier. For example, assume a common-collector (CC) amplifier has an input resistance of 10 kΩ and a load resistance of 100 Ω. Since a CC amplifier has a voltage gain of approximately 1, the power gain is

$$A_p = A_v^2\left(\frac{R_{in}}{R_L}\right) = 1^2\left(\frac{10\text{ k}\Omega}{100\text{ }\Omega}\right) = 100$$

For a CC amplifier, A_p is just the ratio of the input resistance to the output load resistance.

DC Quiescent Power

The power dissipation of a transistor with no signal input is the product of its Q-point current and voltage.

$$P_{DQ} = I_{CQ}V_{CEQ}$$

Equation 9–3

The only way a class A power amplifier can supply power to a load is to maintain a quiescent current that is at least as large as the peak current requirement for the load current. A signal will not increase the power dissipated by the transistor but actually causes less total power to be dissipated. The quiescent power, given in Equation 9–3, is the maximum power that a class A amplifier must handle. The transistor's power rating must exceed this value.

Output Power

In general, the output signal power is the product of the rms load current and the rms load voltage. The maximum unclipped ac signal occurs when the Q-point is centered on the ac load line. For a CE amplifier with a centered Q-point, the maximum peak voltage swing is

$$V_{c(max)} = I_{CQ}R_c$$

The rms value is $0.707V_{c(max)}$.

The maximum peak current swing is

$$I_{c(max)} = \frac{V_{CEQ}}{R_c}$$

The rms value is $0.707I_{c(max)}$.

To find the maximum signal power output, you use the rms values of maximum current and voltage. The maximum power out from a class A amplifier is

$$P_{out(max)} = (0.707I_c)(0.707V_c)$$

Equation 9–4

$$P_{out(max)} = 0.5I_{CQ}V_{CEQ}$$

EXAMPLE 9–1

Determine the voltage gain and the power gain of the class A power amplifier in Figure 9–5. Assume $\beta_{ac(Q1)} = \beta_{ac(Q2)} = 200$ and $\beta_{ac(Q3)} = 50$.

▲ **FIGURE 9–5**

Solution   Notice that the first stage (Q_1) is a voltage-divider biased common-emitter with a swamping resistor (R_{E1}). The second stage (Q_2 and Q_3) is a darlington voltage-follower configuration. The speaker is the load.

First stage:

The ac collector resistance of the first stage is R_C in parallel with the input resistance to the second stage.

$$R_{c1} = R_C \| [R_3 \| R_4 \| \beta_{ac(Q2)} \beta_{ac(Q3)} (R_{E3} \| R_L)]$$
$$= 1.0\,\text{k}\Omega \| [5.1\,\text{k}\Omega \| 15\,\text{k}\Omega \| (200)(50)(16\,\Omega \| 16\,\Omega)]$$
$$= 1.0\,\text{k}\Omega \| (5.1\,\text{k}\Omega \| 15\,\text{k}\Omega \| 80\,\text{k}\Omega) = 1.0\,\text{k}\Omega \| 3.63\,\text{k}\Omega = 784\,\Omega$$

The voltage gain of the first stage is the ac collector resistance, R_{c1}, divided by the ac emitter resistance, which is the sum of $R_{E1} + r'_{e(Q1)}$. The approximate value of $r'_{e(Q1)}$ is determined by first finding I_E.

$$V_B = \left(\frac{R_2 \| (\beta_{ac(Q1)}(R_{E1} + R_{E2})}{R_1 + R_2 \| (\beta_{ac(Q1)}(R_{E1} + R_{E2}))} \right) V_{CC}$$
$$= \left(\frac{5.1\,\text{k}\Omega \| 200(377\,\Omega)}{20\,\text{k}\Omega + 5.1\,\text{k}\Omega \| 200(377\,\Omega)} \right) 15\,\text{V}$$
$$= \left(\frac{4.78\,\text{k}\Omega}{20\,\text{k}\Omega + 4.78\,\text{k}\Omega} \right) 15\,\text{V} = 2.89\,\text{V}$$
$$I_E = \frac{V_B - 0.7\,\text{V}}{R_{E1} + R_{E2}} = \frac{2.89\,\text{V} - 0.7\,\text{V}}{377\,\Omega} = 5.81\,\text{mA}$$
$$r'_{e(Q1)} = \frac{25\,\text{mV}}{I_E} = \frac{25\,\text{mV}}{5.81\,\text{mA}} = 4.3\,\Omega$$

The voltage gain of the first stage with the loading of the second stage taken into account is

$$A_{v1} = -\frac{R_{c1}}{R_{E1} + r'_{e(Q1)}} = -\frac{784\,\Omega}{47\,\Omega + 4.3\,\Omega} = -15.3$$

The negative sign is for inversion.

The total input resistance of the first stage is equal to the bias resistors in parallel with the ac input resistance at the base of Q_1.

$$R_{in(tot)1} = R_1 \| R_2 \| \beta_{ac(Q1)}(R_{E1} + r'_{e(Q1)})$$
$$= 20 \text{ k}\Omega \| 5.1 \text{ k}\Omega \| 200(47 \text{ }\Omega + 4.3 \text{ }\Omega) = 2.9 \text{ k}\Omega$$

Second stage:

The voltage gain of the darlington emitter-follower is approximately equal to 1.

$$A_{v2} \cong 1$$

Overall amplifier:

The overall voltage gain is the product of the first and second stage voltage gains. Since the second stage has a gain of approximately 1, the overall gain is approximately equal to the gain of the first stage.

$$A_{v(tot)} = A_{v1}A_{v2} = (-15.3)(1) = \mathbf{-15.3}$$

Power gain:

The power gain of the amplifier can be calculated using Equation 9–2.

$$A_p = A_{v(tot)}^2 \left(\frac{R_{in(tot)1}}{R_L}\right) = (-15.3)^2 \left(\frac{2.9 \text{ k}\Omega}{16 \text{ }\Omega}\right) = \mathbf{42,429}$$

Related Problem *** Express the power gain as a decibel power gain.

*Answers are at the end of the chapter.

Efficiency

The **efficiency** of any amplifier is the ratio of the signal power supplied to a load to the power from the dc supply. The maximum signal power that can be obtained is given by Equation 9–4. The average power supply current, I_{CC}, is equal to I_{CQ} and the supply voltage is at least $2V_{CEQ}$. Therefore, the dc power is

$$P_{DC} = I_{CC}V_{CC} = 2I_{CQ}V_{CEQ}$$

The maximum efficiency of a capacitively coupled class A amplifier is

$$eff_{max} = \frac{P_{out}}{P_{DC}} = \frac{0.5I_{CQ}V_{CEQ}}{2I_{CQ}V_{CEQ}} = 0.25$$

The maximum efficiency of a capacitively coupled class A amplifier cannot be higher than 0.25, or 25%, and, in practice, is usually considerably less (about 10%). Although the efficiency can be made higher by transformer coupling the signal to the load, there are drawbacks to transformer coupling. These drawbacks include the size and cost of transformers as well as potential distortion problems when the transformer core begins to saturate. In general, the low efficiency of class A power amplifiers limits their usefulness to small power applications that require only a few watts of load power.

EXAMPLE 9–2

Determine the efficiency of the power amplifier in Figure 9–5.

Solution The efficiency is the ratio of the signal power in the load to the power supplied by the dc source. The input voltage is 500 mV peak-to-peak which is 176 mV rms. The input power is, therefore,

$$P_{in} = \frac{V_{in}^2}{R_{in}} = \frac{(176 \text{ mV})^2}{2.9 \text{ k}\Omega} = 10.7 \ \mu\text{W}$$

The output power is

$$P_{out} = P_{in}A_p = (10.7 \ \mu\text{W})(42,429) = 0.454 \text{ W}$$

Most of the power from the dc source is supplied to the output stage. The current in the output stage can be computed from the emitter voltage of Q_3 which is approximately 9.5 V, taking the loading into account. This results in a Q_3 emitter current of approximately 0.6 A. Neglecting the other transistor and bias currents, which are very small, the total dc supply current is about 0.6 A. The power from the dc source is

$$P_{\text{DC}} = I_{\text{CC}}V_{\text{CC}} = (0.6 \text{ A})(15 \text{ V}) = 9 \text{ W}$$

Therefore, the efficiency of the amplifier for this input is

$$eff = \frac{P_{out}}{P_{\text{DC}}} = \frac{0.454 \text{ W}}{9 \text{ W}} \cong \mathbf{0.05}$$

This represents an efficiency of 5%.

Related Problem   Explain what happens to the efficiency if R_{E3} were replaced with the speaker? What disadvantage does this have?

SECTION 9–1 REVIEW

Answers are at the end of the chapter.

1. What is the purpose of a heat sink?
2. Which lead of a BJT is connected to the case?
3. What are the two types of clipping with a class A power amplifier?
4. What is the maximum theoretical efficiency for a class A amplifier?
5. How can the power gain of a CC amplifier be expressed to terms of a ratio of resistances?

9–2   CLASS B AND CLASS AB PUSH-PULL AMPLIFIERS

When an amplifier is biased at cutoff so that it operates in the linear region for 180° of the input cycle and is in cutoff for 180°, it is a class B amplifier. Class AB amplifiers are biased to conduct for slightly more than 180°. The primary advantage of a class B or class AB amplifier over a class A amplifier is that either one is more efficient than a class A amplifier; you can get more output power for a given amount of input power. A disadvantage of class B or class AB is that it is more difficult to implement the circuit in order to get a linear reproduction of the input waveform. As you will see in this section, the term *push-pull* refers to a common type of class B or class AB amplifier circuit in which the input wave shape is reproduced at the output.

After completing this section, you should be able to

- **Explain and analyze the operation of class B and class AB amplifiers**
- Explain class B operation
- Discuss Q-point location for class B amplifiers

- Describe class B push-pull operation
- Explain crossover distortion and its cause
- Explain class AB operation
- Analyze class AB push-pull amplifiers
- Calculate maximum output power
- Calculate dc input power
- Determine class B maximum efficiency
- Find the input resistance
- Analyze a darlington push-pull amplifier
- Discuss methods of driving a push-pull amplifier

Class B Operation

The **class B** operation is illustrated in Figure 9–6, where the output waveform is shown relative to the input in terms of time (t).

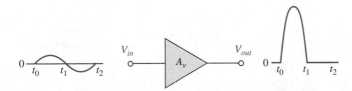

▲ **FIGURE 9–6**

Basic class B amplifier operation (noninverting).

The Q-Point Is at Cutoff   The class B amplifier is biased at the cutoff point so that $I_{CQ} = 0$ and $V_{CEQ} = V_{CE(cutoff)}$. It is brought out of cutoff and operates in its linear region when the input signal drives the transistor into conduction. This is illustrated in Figure 9–7 with an emitter-follower circuit where, as you can see, the output is not a replica of the input.

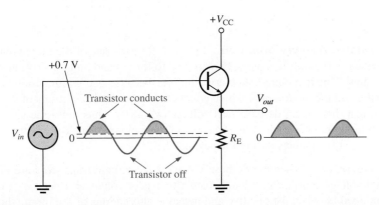

▲ **FIGURE 9–7**

Common–collector class B amplifier.

Class B Push-Pull Operation

As you can see, the circuit in Figure 9–7 only conducts for the positive half of the cycle. To amplify the entire cycle, it is necessary to add a second class B amplifier that operates on the negative half of the cycle. The combination of two class B amplifiers working together is called **push-pull** operation.

There are two common approaches for using push-pull amplifiers to reproduce the entire waveform. The first approach uses transformer coupling. The second uses two **complementary symmetry transistors**; these are a matching pair of *npn/pnp* BJTs or a matching pair of *n*-channel/*p*-channel FETs.

Transformer Coupling   Transformer coupling is illustrated in Figure 9–8. The input transformer has a center-tapped secondary that is connected to ground, producing phase inversion of one side with respect to the other. The input transformer thus converts the input signal to two out-of-phase signals for the transistors. Notice that both transistors are *npn* types. Because of the signal inversion, Q_1 will conduct on the positive part of the cycle and Q_2 will conduct on the negative part. The output transformer combines the signals by permitting current in both directions, even though one transistor is always cut off. The positive power supply signal is connected to the center tap of the output transformer.

▲ **FIGURE 9–8**

Transformer coupled push-pull amplifiers. Q_1 conducts during the positive half-cycle; Q_2 conducts during the negative half-cycle. The two halves are combined by the output transformer.

Complementary Symmetry Transistors   Figure 9–9 shows one of the most popular types of push-pull class B amplifiers using two emitter-followers and both positive and negative power supplies. This is a complementary amplifier because one emitter-follower uses an *npn* transistor and the other a *pnp,* which conduct on opposite alternations of the input cycle. Notice that there is no dc base bias voltage ($V_B = 0$). Thus, only the signal voltage drives the transistors into conduction. Transistor Q_1 conducts during the positive half of the input cycle, and Q_2 conducts during the negative half.

Crossover Distortion   When the dc base voltage is zero, both transistors are off and the input signal voltage must exceed V_{BE} before a transistor conducts. Because of this, there is a time interval between the positive and negative alternations of the input when neither transistor is conducting, as shown in Figure 9–10. The resulting distortion in the output waveform is called **crossover distortion**.

(a) During a positive half-cycle

(b) During a negative half-cycle

▲ **FIGURE 9–9**

Class B push–pull ac operation.

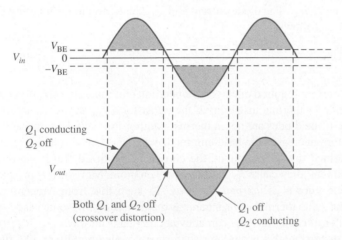

▲ **FIGURE 9–10**

Illustration of crossover distortion in a class B push-pull amplifier. The transistors conduct only during portions of the input indicated by the shaded areas.

Biasing the Push-Pull Amplifier for Class AB Operation

To overcome crossover distortion, the biasing is adjusted to just overcome the V_{BE} of the transistors; this results in a modified form of operation called **class AB**. In class AB operation, the push-pull stages are biased into slight conduction, even when no input signal is present. This can be done with a voltage divider and diode arrangement, as shown in Figure 9–11. When the diode characteristics of D_1 and D_2 are closely matched to the characteristics of the transistor base-emitter junctions, the current in the diodes and the current in the transistors are the same; this is called a **current mirror**. This current mirror produces the desired class AB operation and eliminates crossover distortion.

In the bias path, R_1 and R_2 are of equal value, as are the positive and negative supply voltages. This forces the voltage at point A (between the diodes) to equal 0 V and eliminates

▶ **FIGURE 9–11**

Biasing the push–pull amplifier to eliminate crossover distortion.

the need for an input coupling capacitor. The dc voltage on the output is also 0 V. Assuming that both diodes and both transistors are identical, the drop across D_1 equals the V_{BE} of Q_1, and the drop across D_2 equals the V_{BE} of Q_2. Since they are matched, the diode current will be the same as I_{CQ}. The diode current and I_{CQ} can be found by applying Ohm's law to either R_1 or R_2 as follows:

$$I_{CQ} = \frac{V_{CC} - 0.7\text{ V}}{R_1}$$

This small current required of class AB operation eliminates the crossover distortion but has the potential for thermal instability if the transistor's V_{BE} drops are not matched to the diode drops or if the diodes are not in thermal equilibrium with the transistors. Heat in the power transistors decreases the base-emitter voltage and tends to increase current. If the diodes are warmed the same amount, the current is stabilized; but if the diodes are in a cooler environment, they cause I_{CQ} to increase even more. More heat is produced in an unrestrained cycle known as *thermal runaway.* To keep this from happening, the diodes should have the same thermal environment as the transistors. In some cases, a small resistor in the emitter of each transistor can alleviate thermal runaway.

Crossover distortion also occurs in transformer-coupled amplifiers like the one shown in Figure 9–8. To eliminate it in this case, a 0.7 V is applied to the input transformer's secondary that just biases both transistors into conduction. The bias voltage to produce this drop can be derived from the power supply using a single diode as shown in Figure 9–12.

AC Operation   Consider the ac load line for Q_1 of the class AB amplifier in Figure 9–11. The Q-point is slightly above cutoff. (In a true class B amplifier, the Q-point is at cutoff.) The ac cutoff voltage for a two-supply operation is at V_{CC} with an I_{CQ} as given earlier. The ac saturation current for a two-supply operation with a push-pull amplifier is

Equation 9–5

$$I_{c(sat)} = \frac{V_{CC}}{R_L}$$

The ac load line for the *npn* transistor is as shown in Figure 9–13. The dc load line can be found by drawing a line that passes through V_{CEQ} and the dc saturation current, $I_{C(sat)}$. However, the saturation current for dc is the current if the collector to emitter is shorted on both transistors! This assumed short across the power supplies obviously would cause maximum current from the supplies and implies the dc load line passes almost vertically through the cutoff as shown. Operation along the dc load line, such as caused by thermal runaway, could produce such a high current that the transistors are destroyed.

▲ **FIGURE 9–12**

Eliminating crossover distortion in a transformer-coupled push-pull amplifier. The diode compensates for the base-emitter drop of the transistors and produces class AB operation.

▶ **FIGURE 9–13**

Load lines for a complementary symmetry push-pull amplifier. Only the load lines for the *npn* transistor are shown.

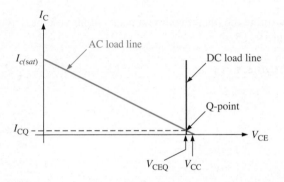

Figure 9–14(a) illustrates the ac load line for Q_1 of the class AB amplifier in Figure 9–14(b). In the case illustrated, a signal is applied that swings over the region of the ac load line shown in bold. At the upper end of the ac load line, the voltage across the transistor (V_{ce}) is a minimum, and the output voltage is maximum.

Under maximum conditions, transistors Q_1 and Q_2 are alternately driven from near cutoff to near saturation. During the positive alternation of the input signal, the Q_1 emitter is

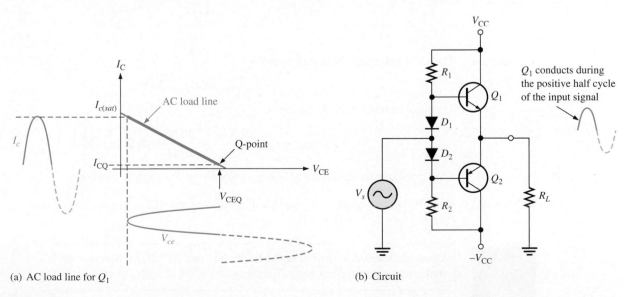

(a) AC load line for Q_1

(b) Circuit

▲ **FIGURE 9–14**

driven from its Q-point value of 0 to nearly V_{CC}, producing a positive peak voltage a little less than V_{CC}. Likewise, during the negative alternation of the input signal, the Q_2 emitter is driven from its Q-point value of 0 V, to near $-V_{CC}$, producing a negative peak voltage almost equal to $-V_{CC}$. Although it is possible to operate close to the saturation current, this type of operation results in increased distortion of the signal.

The ac saturation current (Equation 9–5) is also the peak output current. Each transistor can essentially operate over its entire load line. Recall that in class A operation, the transistor can also operate over the entire load line but with a significant difference. In class A operation, the Q-point is near the middle and there is significant current in the transistors even with no signal. In class B operation, when there is no signal, the transistors have only a very small current and therefore dissipate very little power. Thus, the efficiency of a class B amplifier can be much higher than a class A amplifier. It will be shown later that the maximum theoretical efficiency of a class B amplifier is 79%.

EXAMPLE 9–3

Determine the ideal maximum peak output voltage and current for the circuit shown in Figure 9–15.

▶ **FIGURE 9–15**

Solution   The ideal maximum peak output voltage is

$$V_{out(peak)} \cong V_{CEQ} \cong V_{CC} = \textbf{20 V}$$

The ideal maximum peak current is

$$I_{out(peak)} \cong I_{c(sat)} \cong \frac{V_{CC}}{R_L} = \frac{20\ \text{V}}{16\ \Omega} = \textbf{1.25 A}$$

The actual maximum values of voltage and current are slightly smaller.

Related Problem   What is the maximum peak output voltage and current if the supply voltages are changed to $+30$ V and -30 V?

Open file E09-03 in the Examples folder on your CD-ROM. Measure the maximum peak-to-peak output voltage.

Single-Supply Push-Pull Amplifier

Push-pull amplifiers using complementary symmetry transistors can be operated from a single voltage source as shown in Figure 9–16. The circuit operation is the same as that described previously, except the bias is set to force the output emitter voltage to be $V_{CC}/2$ instead of zero volts used with two supplies. Because the output is not biased at zero volts, capacitive coupling for the input and output is necessary to block the bias voltage from the source and the load resistor. Ideally, the output voltage can swing from zero to V_{CC}, but in practice it does not quite reach these ideal values.

▶ **FIGURE 9–16**

Single-ended push-pull amplifier.

EXAMPLE 9–4

Determine the maximum ideal peak values for the output voltage and current in Figure 9–17.

▲ **FIGURE 9–17**

Solution   The maximum peak output voltage is

$$V_{out(peak)} \cong V_{CEQ} = \frac{V_{CC}}{2} = \frac{20\ V}{2} = \mathbf{10\ V}$$

The maximum peak output current is

$$I_{out(peak)} \cong I_{c(sat)} = \frac{V_{CEQ}}{R_L} = \frac{10 \text{ V}}{16 \text{ }\Omega} = \mathbf{625 \text{ mA}}$$

Related Problem Find the maximum peak values for the output voltage and current in Figure 9–17 if V_{CC} is lowered to 15 V and the load resistance is changed to 8 Ω.

Open file E09-04 in the Examples folder on your CD-ROM. Measure the maximum peak-to-peak output voltage.

Class B/AB Power

Maximum Output Power You have seen that the maximum peak output current for both dual-supply and single-supply push-pull amplifiers is approximately $I_{c(sat)}$, and the maximum peak output voltage is approximately V_{CEQ}. The maximum *average* output power is, therefore,

$$P_{out} = I_{out(rms)}V_{out(rms)}$$

Since

$$I_{out(rms)} = 0.707 I_{out(peak)} \cong 0.707 I_{c(sat)}$$

and

$$V_{out(rms)} = 0.707 V_{out(peak)} \cong 0.707 V_{CEQ}$$

then

$$P_{out} = 0.5 I_{c(sat)} V_{CEQ}$$

Substituting $V_{CC}/2$ for V_{CEQ}, the maximum average output power is

Equation 9–6

$$P_{out} = 0.25 I_{c(sat)} V_{CC}$$

DC Input Power The dc input power comes from the V_{CC} supply and is

$$P_{DC} = I_{CC} V_{CC}$$

Since each transistor draws current for a half-cycle, the current is a half-wave signal with an average value of

$$I_{CC} = \frac{I_{c(sat)}}{\pi}$$

So,

$$P_{DC} = \frac{I_{c(sat)} V_{CC}}{\pi}$$

Efficiency An advantage of push-pull class B and class AB amplifiers over class A is a much higher efficiency. This advantage usually overrides the difficulty of biasing the class AB push-pull amplifier to eliminate crossover distortion. Recall that efficiency is defined as the ratio of ac output power to dc input power.

$$\text{Efficiency} = \frac{P_{out}}{P_{DC}}$$

The maximum efficiency for a class B amplifier (class AB is slightly less) is designated η_{max} and is developed as follows, starting with Equation 9–6.

$$P_{out} = 0.25I_{c(sat)}V_{CC}$$

$$\eta_{max} = \frac{P_{out}}{P_{DC}} = \frac{0.25I_{c(sat)}V_{CC}}{I_{c(sat)}V_{CC}/\pi} = 0.25\pi$$

$$\eta_{max} = 0.79$$

Equation 9–7

or, as a percentage,

$$\eta_{max} = 79\%$$

Recall that the maximum efficiency for class A is 0.25 (25 percent).

Input Resistance The complementary push-pull configuration used in class B/class AB amplifiers is, in effect, two emitter-followers. The input resistance is, therefore, the same as developed in Chapter 6 for the emitter-follower:

$$R_{in} = \beta_{ac}(r'_e + R_E)$$

Since $R_E = R_L$, the formula is

$$R_{in} = \beta_{ac}(r'_e + R_L)$$

Equation 9–8

EXAMPLE 9–5

Find the maximum ac output power and the dc input power of the amplifier in Figure 9–18. Also, determine the input resistance assuming $\beta_{ac} = 50$ and $r'_e = 6\ \Omega$.

▲ **FIGURE 9–18**

Solution The maximum peak output voltage is

$$V_{out(peak)} \cong V_{CEQ} = \frac{V_{CC}}{2} = \frac{20\ V}{2} = 10\ V$$

The maximum peak output current is

$$I_{out(peak)} \cong I_{c(sat)} = \frac{V_{CEQ}}{R_L} = \frac{10\ V}{8\ \Omega} = 1.25\ A$$

The ac output power and the dc input power are

$$P_{out} = 0.25 I_{c(sat)} V_{CC} = 0.25(1.25 \text{ A})(20 \text{ V}) = \textbf{6.25 W}$$

$$P_{DC} = \frac{I_{c(sat)} V_{CC}}{\pi} = \frac{(1.25 \text{ A})(20 \text{ V})}{\pi} = \textbf{7.96 W}$$

The input resistance is

$$R_{in} = \beta_{ac}(r'_e + R_L) = 50(6 \text{ }\Omega + 8 \text{ }\Omega) = \textbf{700 }\boldsymbol{\Omega}$$

Related Problem   Determine the maximum ac output power and the dc input power in Figure 9–18 for $V_{CC} = 15$ V and $R_L = 16 \text{ }\Omega$.

Darlington Class AB Amplifier

In many applications where the push-pull configuration is used, the load resistance is relatively small. For example, an 8 Ω speaker is a common load for a class AB push-pull amplifier.

As a result of low-resistance loading, push-pull amplifiers can present a quite low input resistance to the preceding amplifier that drives it. Depending on the output resistance of the preceding amplifier, the low push-pull input resistance can load it severely and significantly reduce the voltage gain. As an example, if the complementary transistors in a push-pull amplifier exhibit an ac beta of 50 and the load resistance is 8 Ω, the input resistance (assuming $r'_e = 5 \text{ }\Omega$) is

$$R_{in} = \beta_{ac}(r'_e + R_L) = 50(5 \text{ }\Omega + 8 \text{ }\Omega) = 650 \text{ }\Omega$$

If the collector resistance of the driving amplifier is, for example, 1.0 kΩ, the input resistance of the push-pull amplifier reduces the effective collector resistance of the driving amplifier (assuming a common-emitter) to $R_c = R_C \parallel R_{in} = 1.0 \text{ k}\Omega \parallel 650 \text{ }\Omega = 394 \text{ }\Omega$. This drastically reduces the voltage gain of the driving amplifier because its gain is R_c / r'_e.

In certain applications with low-resistance loads, a push-pull amplifier using darlington transistors can be used to increase the input resistance presented to the driving amplifier and avoid severely reducing the voltage gain. The ac beta of a darlington pair is generally in excess of a thousand.

In the previous case, for example, if $\beta_{ac} = 50$ for each transistor in a darlington pair, the overall ac beta is $\beta_{ac} = (50)(50) = 2500$. The input resistance is greatly increased, as the following calculation shows.

$$R_{in} = \beta_{ac}(r'_e + R_L) = 2500(5 \text{ }\Omega + 8 \text{ }\Omega) = 32.5 \text{ k}\Omega$$

A darlington class AB push-pull amplifier is shown in Figure 9–19. Four diodes are required in the bias circuit to match the four base-emitter junctions of the two darlington pairs.

MOSFET Push-Pull Amplifiers

When MOSFETs were first introduced to the commercial market, they were unable to handle the large currents required of power devices. In recent years, the advances in MOSFET technology have made high-power MOSFETs available and offer some important advantages in the design of power amplifiers for both digital and analog designs. MOSFETs are very reliable, providing their specified voltage, current, and temperature ratings are not exceeded.

Comparing MOSFETs to BJTs, there are several important advantages but also some disadvantages to MOSFETs. The principal advantages of MOSFETs over BJTs are that their biasing networks are simpler, their drive requirements are simpler, and they can be connected in parallel for added drive capability. In addition, MOSFETs are not generally

▶ **FIGURE 9–19**

A darlington class AB push-pull amplifier.

prone to thermal instability; as they get hotter, they tend to have less current (just the opposite of bipolar junction transistors).

The BJT has the edge when the voltage drop across the transistor is important and, as a result, may be more efficient than a MOSFET in certain cases. In addition, BJTs are not as prone to **electrostatic discharge (ESD)** that can kill a MOSFET. Most MOSFETs are shipped with the pins shorted together with a ring; they should be soldered into a circuit before the shorting ring is removed.

A simplified class B amplifier using complementary symmetry E-MOSFETs and two-supply operation is shown in Figure 9–20(a). Recall that an E-MOSFET is normally off but can be turned on when the input exceeds the threshold voltage. For logic devices, the *on* voltage is typically between 1 V and 2 V; for standard devices the threshold is higher. When the signal exceeds the positive threshold voltage of Q_1, it conducts; likewise, when

(a) Class B

(b) Class AB

▲ **FIGURE 9–20**

MOSFET push-pull amplifiers.

the signal is below the negative threshold voltage of Q_2, it conducts. Thus, the n-channel device conducts on the positive cycle; the p-channel device conducts on the negative cycle.

As in the case of the BJT push-pull amplifier, the transistor does not conduct just above zero signal voltage, which causes crossover distortion. If each transistor is biased just at the threshold voltage, the MOSFETs will operate in class AB, as shown in the circuit in Figure 9–20(b). This amplifier includes a BJT amplifier as a driver and other components to assure a reasonably linear output from an E-MOSFET push-pull stage. Of course, there are additional features to this basic design in commerical amplifiers.

The basic class AB push-pull amplifier shown in Figure 9–20(b) includes a common-emitter stage that amplifies the input signal and couples the signal to the gates of the push-pull stage, consisting of Q_2 and Q_3. Notice that R_6 is bypassed with capacitor C_3 to allow identical ac signals to be applied to the push-pull stage. Potentiometer R_6 is used to develop the proper dc voltage to set the bias to the threshold voltages of Q_2 and Q_3. It is adjusted to minimize crossover distortion. Potentiometer R_1 is adjusted to zero the output dc output voltage with no input signal.

This type of amplifier can give increased power out by simply adding another pair of MOSFETs in parallel; however, this can sometimes cause unwanted oscillations. The cure for this is to use gate resistors to isolate the MOSFETs from each other. Although not strictly required in this simplified amplifier, they are shown as R_8 and R_9. Power amplifiers with parallel E-MOSFET transistors can deliver over 100 W of power.

EXAMPLE 9–6

The n-channel E-MOSFET shown in Figure 9–21 has a threshold voltage of $+2.0$ V and the p-channel E-MOSFET has a threshold voltage of -2.0 V. What resistance setting for R_6 will bias the transistors to class AB operation? At this setting, what power is delivered to the load if the input signal is 100 mV? Assume that potentiometer R_1 is set to 440 Ω.

▲ **FIGURE 9–21**

Solution Start by computing the dc parameters for the CE amplifier. The base voltage is determined by the voltage divider composed of R_1, R_2, and R_3.

$$I_{R1} \cong \frac{V_{DD} - (-V_{DD})}{R_1 + R_2 + R_3} = \frac{48 \text{ V}}{105.54 \text{ k}\Omega} = 455 \text{ }\mu\text{A}$$

$$V_B = V_{DD} - I_{R1}(R_1 + R_2) = 24 \text{ V} - 455 \text{ }\mu\text{A}(5.54 \text{ k}\Omega) = 21.5 \text{ V}$$

The emitter voltage is one diode drop higher than the base voltage (because the transistor is a *pnp* type).

$$V_E = V_B + 0.7 \text{ V} = 21.5 \text{ V} + 0.7 \text{ V} = 22.2 \text{ V}$$

The emitter current is calculated from Ohm's law.

$$I_E = \frac{V_{DD} - V_E}{R_4 + R_5} = \frac{24 \text{ V} - 22.2 \text{ V}}{1.1 \text{ k}\Omega} = 1.64 \text{ mA}$$

The required drop across R_6 is the difference in the threshold voltages.

$$V_{R6} = V_{TH(Q1)} - V_{TH(Q2)} = 2.0 \text{ V} - (-2.0 \text{ V}) = 4.0 \text{ V}$$

The required setting for R_6 can be determined from Ohm's law.

$$R_6 = \frac{V_{R6}}{I_{R6}} = \frac{4.0 \text{ V}}{1.64 \text{ mA}} = 2.4 \text{ k}\Omega$$

This setting produces class AB operation, so the output voltage replicates the input of the MOSFET (less a small drop across the internal MOSFET resistance). The gain of the CE amplifier is determined using the ratio of unbypassed collector resistance (R_7) to the unbypassed emitter resitance (R_5 and r'_e).

$$r'_e = \frac{25 \text{ mV}}{I_E} = \frac{25 \text{ mV}}{1.64 \text{ mA}} = 15.2 \text{ }\Omega$$

and

$$A_v = \frac{R_7}{R_5 + r'_e} = \frac{15 \text{ k}\Omega}{100 \text{ }\Omega + 15.2 \text{ }\Omega} = 130$$

Assuming no internal drop in the MOSFETs, the output voltage is

$$V_{out} = A_v V_{in} = (130)(100 \text{ mV}) = 13 \text{ V}$$

The power out is

$$P_L = \frac{V_{out}^2}{R_L} = \frac{13 \text{ V}^2}{33 \text{ }\Omega} = \textbf{5.1 W}$$

Related Problem Calculate the setting of R_6 if the threshold voltages for the MOSFETs are $+1.5$ V and -1.5 V.

SECTION 9–2 REVIEW

1. Where is the Q-point for a class B amplifier?
2. What causes crossover distortion?
3. What is the maximum efficiency of a push–pull class B amplifier?
4. Explain the purpose of the push–pull configuration for class B.
5. How does a class AB differ from a class B amplifier?

9–3 CLASS C AMPLIFIERS

Class C amplifiers are biased so that conduction occurs for much less than 180°. Class C amplifiers are more efficient than either class A or push-pull class B and class AB, which means that more output power can be obtained from class C operation. Because the output waveform is severely distorted, class C amplifiers are normally limited to applications as tuned amplifiers at radio frequencies (RF) as you will see in this section.

After completing this section, you should be able to

- **Discuss and analyze the operation of class C amplifiers**
- Explain class C operation
- Discuss class C power dissipation
- Describe tuned operation
- Calculate maximum output power
- Determine efficiency
- Explain clamper bias in a class C amplifier

Basic Class C Operation

The basic concept of **class C** operation is illustrated in Figure 9–22. A common-emitter class C amplifier with a resistive load is shown in Figure 9–23(a). It is biased below cutoff with the negative V_{BB} supply. The ac source voltage has a peak value that is slightly greater than $V_{BB} + V_{BE}$ so that the base voltage exceeds the barrier potential of the base-emitter junction for a short time near the positive peak of each cycle, as illustrated in Figure 9–23(b). During this short interval, the transistor is turned on. When the entire ac load line is used, as shown in Figure 9–23(c), the ideal maximum collector current is $I_{c(sat)}$, and the ideal minimum collector voltage is $V_{ce(sat)}$.

▲ **FIGURE 9–22**

Basic class C amplifier operation (noninverting).

Power Dissipation

The power dissipation of the transistor in a class C amplifier is low because it is on for only a small percentage of the input cycle. Figure 9–24(a) shows the collector current pulses. The time between the pulses is the period (T) of the ac input voltage. To avoid complex mathematics, we will use ideal pulse approximations for the collector current and the collector voltage during the *on* time of the transistor, as shown in Figure 9–24(b). Using this simplification, the maximum current amplitude is $I_{c(sat)}$ and the minimum volt-

(b) Input voltage and output current waveforms

(a) Basic class C amplifier circuit

(c) Load line operation

▲ FIGURE 9–23

Basic class C operation.

age amplitude is $V_{ce(sat)}$ during the time the transistor is on, if the output swings over the entire load line. The power dissipation during the *on* time is, therefore,

$$P_{D(on)} = I_{c(sat)}V_{ce(sat)}$$

The transistor is on for a short time, t_{on}, and off for the rest of the input cycle. Therefore, assuming the entire load line is used, the power dissipation averaged over the entire cycle is

$$P_{D(avg)} = \left(\frac{t_{on}}{T}\right)P_{D(on)} = \left(\frac{t_{on}}{T}\right)I_{c(sat)}V_{ce(sat)}$$

(a) Collector current pulses

(b) Ideal class C waveforms

▲ FIGURE 9–24

Class C waveforms.

EXAMPLE 9–7

A class C amplifier is driven by a 200 kHz signal. The transistor is on for 1 μs, and the amplifier is operating over 100 percent of its load line. If $I_{c(sat)} = 100$ mA and $V_{ce(sat)} = 0.2$ V, what is the average power dissipation?

Solution   The period is

$$T = \frac{1}{200 \text{ kHz}} = 5 \; \mu s$$

Therefore,

$$P_{D(avg)} = \left(\frac{t_{on}}{T}\right) I_{c(sat)} V_{ce(sat)} = (0.2)(100 \text{ mA})(0.2 \text{ V}) = \mathbf{4 \; mW}$$

Related Problem   If the frequency is reduced from 200 kHz to 150 kHz, what is the average power dissipation?

Tuned Operation

Because the collector voltage (output) is not a replica of the input, the resistively loaded class C amplifier alone is of no value in linear applications. It is therefore necessary to use a class C amplifier with a parallel resonant circuit (tank), as shown in Figure 9–25(a). The resonant frequency of the tank circuit is determined by the formula $f_r = 1/(2\pi\sqrt{LC})$. The short pulse of collector current on each cycle of the input initiates and sustains the oscillation of the tank circuit so that an output sinusoidal voltage is produced, as illustrated in Figure 9–25(b).

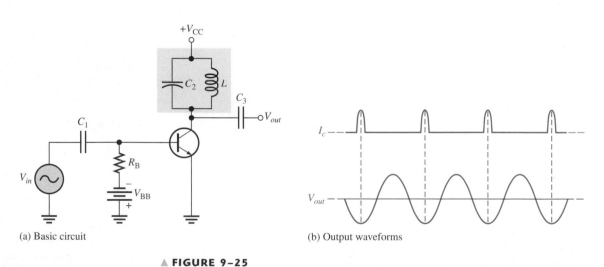

(a) Basic circuit   (b) Output waveforms

▲ **FIGURE 9–25**

Tuned class C amplifier.

The current pulse charges the capacitor to approximately $+V_{CC}$, as shown in Figure 9–26(a). After the pulse, the capacitor quickly discharges, thus charging the inductor. Then, after the capacitor completely discharges, the inductor's magnetic field collapses and then quickly recharges C to near V_{CC} in a direction opposite to the previous charge. This completes one half-cycle of the oscillation, as shown in parts (b) and (c) of Figure 9–26. Next, the capacitor discharges again, increasing the inductor's magnetic field. The inductor then quickly recharges the capacitor back to a positive peak slightly less than the

(a) C charges to +V_{CC} at the input peak
when transistor is conducting.

(b) C discharges to 0 volts.

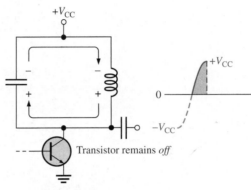

(c) L recharges C in opposite direction.

(d) C discharges to 0 volts.

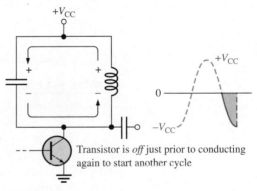

(e) L recharges C.

▲ **FIGURE 9–26**

Resonant circuit action.

previous one, due to energy loss in the winding resistance. This completes one full cycle, as shown in parts (d) and (e) of Figure 9–26. The peak-to-peak output voltage is therefore approximately equal to $2V_{CC}$.

The amplitude of each successive cycle of the oscillation will be less than that of the previous cycle because of energy loss in the resistance of the tank circuit, as shown in Figure 9–27(a), and the oscillation will eventually die out. However, the regular recurrences of the collector current pulse re-energizes the resonant circuit and sustains the oscillations at a constant amplitude.

When the tank circuit is tuned to the frequency of the input signal (fundamental), re-energizing occurs on each cycle of the tank voltage, V_r, as shown in Figure 9–27(b).

(a) An oscillation will gradually die out (decay) due to energy loss. The rate of decay depends on the efficiency of the tank circuit.

(b) Oscillation at the fundamental frequency can be sustained by short pulses of collector current.

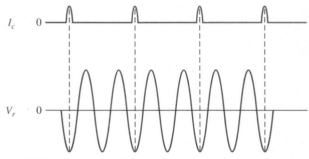

(c) Oscillation at the second harmonic frequency

▲ **FIGURE 9–27**

Tank circuit oscillations. V_r is the voltage across the tank circuit.

When the tank circuit is tuned to the second harmonic of the input signal, re-energizing occurs on alternate cycles as shown in Figure 9–27(c). In this case, a class C amplifier operates as a frequency multiplier ($\times 2$). By tuning the resonant tank circuit to higher harmonics, further frequency multiplication factors are achieved.

Maximum Output Power

Since the voltage developed across the tank circuit has a peak-to-peak value of approximately $2V_{CC}$, the maximum output power can be expressed as

$$P_{out} = \frac{V_{rms}^2}{R_c} = \frac{(0.707V_{CC})^2}{R_c}$$

Equation 9–9

$$P_{out} = \frac{0.5V_{CC}^2}{R_c}$$

R_c is the equivalent parallel resistance of the collector tank circuit and represents the parallel combination of the coil resistance and the load resistance. It usually has a low value. The total power that must be supplied to the amplifier is

$$P_T = P_{out} + P_{D(avg)}$$

Therefore, the efficiency is

$$\eta = \frac{P_{out}}{P_{out} + P_{D(avg)}}$$

Equation 9–10

When $P_{out} \gg P_{D(avg)}$, the class C efficiency closely approaches 1 (100 percent).

EXAMPLE 9–8

Suppose the class C amplifier described in Example 9–7 has a V_{CC} equal to 24 V, and the R_c is 100 Ω. Determine the efficiency.

Solution From Example 9–7, $P_{D(avg)} = 4$ mW.

$$P_{out} = \frac{0.5V_{CC}^2}{R_c} = \frac{0.5(24 \text{ V})^2}{100 \text{ }\Omega} = 2.88 \text{ W}$$

Therefore,

$$\eta = \frac{P_{out}}{P_{out} + P_{D(avg)}} = \frac{2.88 \text{ W}}{2.88 \text{ W} + 4 \text{ mW}} = \mathbf{0.999}$$

or

$$\eta \times 100\% = 99.9\%$$

Related Problem What happens to the efficiency of the amplifier if R_c is increased?

Clamper Bias for a Class C Amplifier

Figure 9–28 shows a class C amplifier with a base bias clamping circuit. The base-emitter junction functions as a diode. When the input signal goes positive, capacitor C_1 is charged to the peak value with the polarity shown in Figure 9–29(a). This action produces an average voltage at the base of approximately $-V_p$. This places the transistor in cutoff except at the positive peaks, when the transistor conducts for a short interval. For good clamping action, the R_1C_1 time constant of the clamping circuit must be much greater than the period of the input signal. Parts (b) through (f) of Figure 9–29 illustrate the bias clamping action in more detail. During the time up to the positive peak of the input (t_0 to t_1), the capacitor charges to $V_p - 0.7$ V through the base-emitter diode, as shown in part (b). During the time from t_1 to t_2, as shown in part (c), the capacitor discharges very little because of the large RC time constant. The capacitor, therefore, maintains an average charge slightly less than $V_p - 0.7$ V.

▶ **FIGURE 9–28**

Tuned class C amplifier with clamper bias.

▲ **FIGURE 9–29**

Clamper bias action.

Since the dc value of the input signal is zero (positive side of C_1), the dc voltage at the base (negative side of C_1) is slightly more positive than $-(V_p - 0.7 \text{ V})$, as indicated in Figure 9–29(d). As shown in Figure 9–29(e), the capacitor couples the ac input signal through to the base so that the voltage at the transistor's base is the ac signal riding on a dc level slightly more positive than $-(V_p - 0.7 \text{ V})$. Near the positive peaks of the input voltage, the base voltage goes slightly above 0.7 V and causes the transistor to conduct for a short time, as shown in Figure 9–29(f).

SECTION 9–3 REVIEW

1. At what point is a class C amplifier normally biased?
2. What is the purpose of the tuned circuit in a class C amplifier?
3. A certain class C amplifier has a power dissipation of 100 mW and an output power of 1 W. What is its percent efficiency?

9–4 TROUBLESHOOTING

In this section, examples of isolating a component failure in a circuit are presented. We will use a class A amplifier and a class AB amplifier with the output voltage monitored by an oscilloscope. Several incorrect output waveforms will be examined and the most likely faults will be discussed.

After completing this section, you should be able to

- **Troubleshoot power amplifiers**

- Detect various faults in a class A power amplifier and a class AB amplifier

Case 1: Class A

As shown in Figure 9–30, the class A power amplifier should have a normal sinusoidal output when a sinusoidal input signal is applied.

▶ **FIGURE 9–30**

Class A power amplifier with correct output voltage swing.

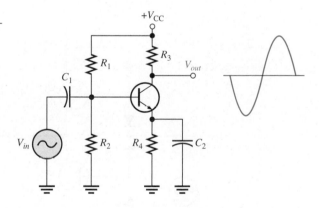

Now let's consider four incorrect output waveforms and the most likely causes in each case. In Figure 9–31(a), the scope displays a dc level equal to the dc supply voltage, indicating that the transistor is in cutoff. The two most likely causes of this condition are (1) the transistor has an open *pn* junction, or (2) R_4 is open, preventing collector and emitter current.

In Figure 9–31(b), the scope displays a dc level at the collector approximately equal to the dc emitter voltage. The two probable causes of this indication are (1) the transistor is shorted from collector to emitter, or (2) R_2 is open, causing the transistor to be biased in saturation. In the second case, a sufficiently large input signal can bring the transistor out of saturation on its negative peaks, resulting in short pulses on the output.

In Figure 9–31(c), the scope displays an output waveform that indicates the transistor is cut off except during a small portion of the input cycle. Possible causes of this indication are (1) the Q-point has shifted down due to a drastic out-of-tolerance change in a resistor value, or (2) R_1 is open, biasing the transistor in cutoff. The display shows that the input signal is sufficient to bring it out of cutoff for a small portion of the cycle.

(a) Transistor in cutoff

(b) CE short or R_2 open

(c) Q-point shift or R_1 open

(d) Transistor in saturation

▲ **FIGURE 9–31**

Oscilloscope displays showing output voltage for the amplifier in Figure 9–30, for several types of failures.

In Figure 9–31(d), the scope displays an output waveform that indicates the transistor is saturated except during a small portion of the input cycle. Again, it is possible that a resistance change has caused a drastic shift in the Q-point up toward saturation, or R_2 is open, causing the transistor to be biased in saturation, and the input signal is bringing it out of saturation for a small portion of the cycle.

Case 2: Class AB

As shown in Figure 9–32, the class AB push-pull amplifier should have a sinusoidal output when a sinusoidal input signal is applied.

▲ **FIGURE 9–32**

A class AB push-pull amplifier with correct output voltage.

Two incorrect output waveforms are shown in Figure 9–33. The waveform in part (a) shows that only the positive half of the input signal is present on the output. One possible cause is that diode D_1 is open. If this is the fault, the positive half of the input signal forward-biases D_2 and causes transistor Q_2 to conduct. Another possible cause is that the base-emitter junction of Q_2 is open so only the positive half of the input signal appears on the output because Q_1 is still working.

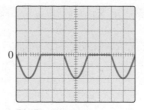

(a) D_1 open or
  Q_2 base-emitter open

(b) D_2 open or
  Q_1 base-emitter open

▲ **FIGURE 9–33**

Incorrect output waveforms for the amplifier in Figure 9–32.

The waveform in Figure 9–33(b) shows that only the negative half of the input signal is present on the output. One possible cause is that diode D_2 is open. If this is the fault, the negative half of the input signal forward-biases D_1 and places the half-wave signal on the base of Q_1. Another possible cause is that the base-emitter junction of Q_1 is open so only the negative half of the input signal appears on the output because Q_2 is still working.

EWB/Multisim Troubleshooting Exercises

These file circuits are in the Troubleshooting Exercises folder on your CD-ROM.

1. Open file TSE09-01. Determine if the circuit is working properly and, if not, determine the fault.

2. Open file TSE09-02. Determine if the circuit is working properly and, if not, determine the fault.

3. Open file TSE09-03. Determine if the circuit is working properly and, if not, determine the fault.

SECTION 9–4 REVIEW

1. What would you check for if you noticed clipping at both peaks of the output waveform?

2. A significant loss of gain in the amplifier of Figure 9–30 would most likely be caused by what type of failure?

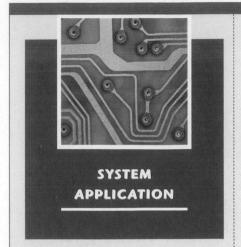

SYSTEM APPLICATION

In the system application in Chapter 6, you were assigned responsibility for a public address system and you completed the audio preamplifier circuit. In this phase of the project, you will complete work on the power amplifier circuit and then put the complete system, including the power supply, together. You will apply the knowledge you have gained in this chapter in completing your assignment.

The Public Address System

Recall from Chapter 6 that this system includes a magnetic microphone, a two-stage preamplifier, a power amplifier, and a horn speaker as shown in the block diagram of Figure 9–34. The focus of this assignment is the power amplifier circuit board.

Basic Operation   The schematic of the push-pull power amplifier is shown in Figure 9–35. The circuit is a class AB amplifier implemented with darlington power transistors. The base-emitter junctions of two additional darlington transistors of the same type are used in the bias circuit to assure matched thermal characteristics.

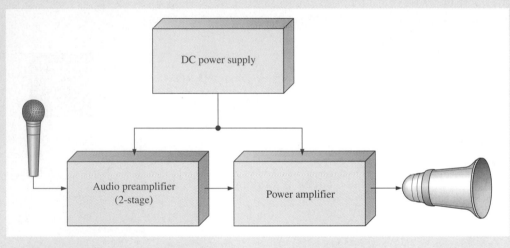

▲ **FIGURE 9–34**

Basic public address/paging system block diagram.

▶ **FIGURE 9–35**

Schematic of the power amplifier circuit.

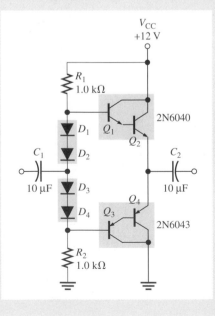

Rather than applying the signal to each base using two input coupling capacitors, this design capacitively couples the output of the preamplifier to the midpoint between the bias diodes. Since the diodes are forward-biased, the signal voltage is equally developed at the bases of Q_1 and Q_3. Darlington transistors are used in this application because of their high betas to prevent excessive loading of the preamplifier and a resulting loss of voltage gain.

Each darlington pair is packaged in a single case which looks like one transistor. The darlington transistors are mounted to a heat sink to prevent overheating and reduction of maximum power dissipation. Heat sinks come in many forms. In this application, a metal bracket with cooling fins is attached to the printed circuit board. The transistors are mounted with their collectors thermally connected but electrically isolated from the heat sink using a mica insulator.

Analysis of the Power Amplifier Circuit

Refer to the partial data sheet in Figure 9–36.

- Determine the minimum input resistance of the push-pull amplifier.
- Determine the overall maximum voltage gain of the preamplifier and the power amplifier when connected together.

- Verify that the power rating of each transistor in the push-pull amplifier is not exceeded under maximum signal conditions.

The Printed Circuit Board

- Check out the printed circuit board in Figure 9–37 to verify that it is correct according to the schematic. The backside interconnection is shown as a darker trace.
- Label a copy of the board with component and input/output designations in agreement with the schematic.

Test Procedure

- Develop a step-by-step set of instructions on how to check the power amplifier circuit board for proper operation at a test frequency of 5 kHz using the test points (circled numbers) indicated in the test bench setup of Figure 9–38. A function generator is the signal source.
- Specify voltage values for all the measurements to be made. Provide a fault analysis for all possible component failures.
- Combine the power amplifier test procedure with the test procedure developed for the preamplifier in Chapter 6 and the test procedure for the power supply board from Chapter 2 to form a complete test procedure for the sys-

tem which is shown in its physical form in Figure 9–39 on page 462.

- Make sure that the interconnecting wiring in Figure 9–40 on page 463 is correct and produce a complete schematic for the system.

Troubleshooting

Problems have developed in two prototype boards. Based on the sequence of test bench measurements for each board indicated in Figure 9–40, determine the most likely fault in each case. The circled numbers indicate test point connections to the circuit board.

Final Report (Optional)

Submit a final written report on the power amplifier circuit board using an organized format that includes the following:

1. A physical description of the circuits.
2. A discussion of the operation of the circuits.
3. A list of the specifications.
4. A list of parts with part numbers if available.
5. A list of the types of problems on the two faulty circuit boards.
6. A complete description of how you determined the problem on each of the faulty circuit boards.

Plastic Medium-Power
Complementary Silicon Transistors

. . . designed for general-purpose amplifier and low-speed switching applications.

- High DC current gain —
  h_{FE} = 2500 (Typ) @ I_C = 4.0 A dc
- Collector-Emitter sustaining voltage — @ 100 mA dc
  $V_{CEO(sus)}$ = 60 V dc (Min) — 2N6040, 2N6043
  = 80 V dc (Min) — 2N6041, 2N6044
  = 100 V dc (Min) — 2N6042, 2N6045
- Low collector-emitter saturation voltage
  $V_{CE(sat)}$ = 2.0 V dc (Max) @ I_C = 4.0 A dc — 2N6040,41,2N6043,44
  = 2.0 V dc (Max) @ I_C = 3.0 A dc — 2N6042, 2N6045
- Monolithic construction with built-in base-emitter shunt resistors

Darlington
8 ampere
Complementary Silicon
Power Transistors
60-80-100 VOLTS
75 WATTS

2N6040 thru 2n6045

CASE 221A-04
TO-220AB

Maximum Ratings

Rating	Symbol	2N6040 2N6043 MJE6040 MJE6043	2N6041 2N6044 MJE6041 MJE6044	2N6042 2N6045 MJE6045	Unit
Collector-Emitter voltage	V_{CEO}	60	80	100	V dc
Collector-Base voltage	V_{CB}	60	80	100	V dc
Emitter-Base voltage	V_{EB}	← 5.0 →			V dc
Collector Current—Continuous Peak	I_C	← 8.0 → 16			A dc
Base current	I_B	← 120 →			mA dc
Total power dissipation @ T_C = 25°C Derate above 25°C	P_D	← 75 → 0.60			Watts W/°C
Total power dissipation @ T_A = 25°C Derate above 25°C	P_D	← 2.2 → 0.0175			Watts W/°C
Operating and storage junction, Temperature range	T_J, T_{stg}	← −65 to +150 →			°C

Electrical Characteristics (T_C = 25°C unless otherwise noted)

Characteristic	Symbol	Min	Max	Unit
DC current gain (I_C = 4.0 A dc, V_{CE} = 4.0 V dc) 2N6040,41,2N6043,44,MJE6040,41, MJE6043,44 (I_C = 3.0 A dc, V_{CE} = 4.0 V dc) 2N6042, 2N6045, MJE6045 (I_C = 8.0 A dc, V_{CE} = 4.0 V dc) All Types	h_{FE}	1000 1000 100	20,000 20,000 —	—
Collector-Emitter saturation voltage (I_C = 4.0 A dc, I_B = 16 mA dc) 2N6040,41,2N6043,44,MJE6040,41,MJE6043,44 (I_C = 3.0 A dc, I_B = 12 mA dc) 2N6042,2N6045,MJE6045 (I_C = 8.0 A dc, I_B = 80 mA dc) All Types	$V_{CE(sat)}$	— — —	2.0 2.0 4.0	V dc
Base-Emitter saturation voltage (I_C = 8.0 A dc, I_B = 80 mA dc)	$V_{CE(sat)}$		4.5	V dc
Base-Emitter on voltage (I_C = 4.0 A dc, V_{CE} = 4.0 V dc)	$V_{BE(on)}$	—	2.8	V dc

Dynamic Characteristics

Small-signal current gain (I_C = 3.0 A dc, V_{CE} = 4.0 V dc, f = 1.0 MHz)	$|h_{fe}|$	4.0	—	
Output capacitance (V_{CB} = 10 V dc, I_E = 0, f = 1.0 MHz) 2N6040/2N6042, MJE6040 2N6043/2N6045, MJE6043/MJE6045	C_{ob}	— —	300 200	pF
Small-signal current gain (I_C = 3.0 A dc, V_{CE} = 4.0 V dc, f = 1.0 kHz)	h_{fe}	300	—	—

▲ **FIGURE 9–36**

Partial data sheet for complementary darlington transistors 2N6040 (*npn*) and 2N6043 (*pnp*).

▲ FIGURE 9–37

Power amplifier circuit board and transistor pin configuration.

+12 V from power supply

Sinusoidal output from
function generator

Power supply ground

8 Ω   10 W

▲ FIGURE 9–38

Test bench setup for the power amplifier board.

▲ **FIGURE 9–39**

The complete public address/paging system before packaging.

▲ FIGURE 9–40

Test results for two faulty circuit boards.

CHAPTER SUMMARY

- A class A power amplifier operates entirely in the linear region of the transistor's characteristic curves. The transistor conducts during the full 360° of the input cycle.
- The Q-point must be centered on the load line for maximum class A output signal swing.
- The maximum efficiency of a class A power amplifier is 25 percent.
- A class B amplifier operates in the linear region for half of the input cycle (180°), and it is in cutoff for the other half.
- The Q-point is at cutoff for class B operation.
- Class B amplifiers are normally operated in a push-pull configuration in order to produce an output that is a replica of the input.
- The maximum efficiency of a class B amplifier is 79 percent.
- A class AB amplifier is biased slightly above cutoff and operates in the linear region for slightly more than 180° of the input cycle.
- Class AB eliminates crossover distortion found in pure class B.
- A class C amplifier operates in the linear region for only a small part of the input cycle.
- The class C amplifier is biased below cutoff.
- Class C amplifiers are normally operated as tuned amplifiers to produce a sinusoidal output.
- The maximum efficiency of a class C amplifier is higher than that of either class A or class B amplifiers. Under conditions of low power dissipation and high output power, the efficiency can approach 100 percent.

KEY TERMS

Key terms and other bold terms in the chapter are defined in the end-of-book glossary.

Class A   A type of amplifier that operates entirely in its linear (active) region.

Class AB   A type of amplifier that is biased into slight conduction.

Class B   A type of amplifier that operates in the linear region for 180° of the input cycle because it is biased at cutoff.

Class C   A type of amplifier that operates only for a small portion of the input cycle.

Efficiency   The ratio of the signal power delivered to a load to the power from the power supply of an amplifier.

Electrostatic discharge (ESD)   The discharge of a high voltage through an insulating path that can destroy an electronic device.

Power gain   The ratio of output power to input power of an amplifier.

Push-Pull   A type of class B amplifier with two transistors in which one transistor conducts for one half-cycle and the other conducts for the other half-cycle.

KEY FORMULAS

Class A Power Amplifiers

9–1 $A_p = \dfrac{P_L}{P_{in}}$ Power gain

9–2 $A_p = A_v^2\left(\dfrac{R_{in}}{R_L}\right)$ Power gain in terms of voltage gain

9–3 $P_{DQ} = I_{CQ}V_{CEQ}$ DC quiescent power

9–4 $P_{out(max)} = 0.5I_{CQ}V_{CEQ}$ Maximum output power

Class B/AB Push-pull Amplifiers

9–5 $I_{C(sat)} = \dfrac{V_{CC}}{R_L}$ AC saturation current

9–6 $P_{out} = 0.25I_{c(sat)}V_{CC}$ Maximum average output power

9–7 $\eta_{max} = 0.79$ Maximum efficiency

9–8 $R_{in} = \beta_{ac}(r'_e + R_L)$ Input resistance

Class C Amplifiers

9–9 $P_{out} = \dfrac{0.5V_{CC}^2}{R_c}$ Output power

9–10 $\eta = \dfrac{P_{out}}{P_{out} + P_{D(avg)}}$ Efficiency

SELF-TEST

Answers are at the end of the chapter.

1. An amplifier that operates in the linear region at all times is

 (a) Class A   (b) Class AB   (c) Class B   (d) Class C

2. A certain class A power amplifier delivers 5 W to a load with an input signal power of 100 mW. The power gain is

 (a) 100   (b) 50   (c) 250   (d) 5

3. The peak current a class A power amplifier can deliver to a load depends on the

 (a) maximum rating of the power supply

 (b) quiescent current

 (c) current in the bias resistors

 (d) size of the heat sink

4. For maximum output, a class A power ampliifer must maintain a value of quiescent current that is

 (a) one-half the peak load current

 (b) twice the peak load current

 (c) at least as large as the peak load current

 (d) just above the cutoff value

5. A certain class A power amplifier has $V_{CEQ} = 12$ V and $I_{CQ} = 1$ A. The maximum signal power output is

 (a) 6 W   (b) 12 W   (c) 1 W   (d) 0.707 W

6. The efficiency of a power amplifier is the ratio of the power delivered to the load to the

 (a) input signal power   (b) power dissipated in the last stage

 (c) power from the dc power supply   (d) none of these answers

7. The maximum efficiency of a class A power amplifier is

 (a) 25%   (b) 50%   (c) 79%   (d) 98%

8. The transistors in a class B amplifier are biased

 (a) into cutoff

 (b) in saturation

 (c) at midpoint of the load line

 (d) right at cutoff

9. Crossover distortion is a problem for

 (a) class A amplifiers   (b) class AB amplifiers

 (c) class B amplifiers   (d) all of these amplifiers

10. A BJT class B push-pull amplifier with no transformer coupling uses

 (a) two *npn* transistors

 (b) two *pnp* transistors

 (c) complementary symmetry transitors

 (d) none of these

11. A current mirror in a push-pull amplifier should give an I_{CQ} that is
    (a) equal to the current in the bias resistors and diodes
    (b) twice the current in the bias resistors and diodes
    (c) half the current in the bias resistors and diodes
    (d) zero

12. The maximum efficiency of a class B push-pull amplifier is
    (a) 25% (b) 50% (c) 79% (d) 98%

13. The output of a certain two-supply class B push-pull amplifier has a V_{CC} of 20 V. If the load resistance is 50 Ω, the value of $I_{c(sat)}$ is
    (a) 5 mA (b) 0.4 A (c) 4 mA (d) 40 mA

14. The maximum efficiency of a class AB amplifier is
    (a) higher than a class B   (b) the same as a class B
    (c) about the same as a class A   (d) slightly less than a class B

15. The power dissipation of a class C amplifier is normally
    (a) very low   (b) very high
    (c) the same as a class B (d) the same as a class A

16. The efficiency of a class C amplifier is
    (a) less than class A   (b) less than class B
    (c) less than class AB (d) greater than classes A, B, or AB

17. The transistor in a class C amplifier conducts for
    (a) more than 180° of the input cycle
    (b) one-half of the input cycle
    (c) a very small percentage of the input cycle
    (d) all of the input cycle

PROBLEMS

Answers to all odd-numbered problems are at the end of the book.

BASIC PROBLEMS

SECTION 9–1   Class A Power Amplifiers

1. Figure 9–41 shows a CE power amplifier in which the collector resistor serves also as the load resistor. Assume $\beta_{DC} = \beta_{ac} = 100$.
   (a) Determine the dc Q-point (I_{CQ} and V_{CEQ}).
   (b) Determine the voltage gain and the power gain.

2. For the circuit in Figure 9–41, determine the following:
   (a) the power dissipated in the transistor with no load
   (b) the total power from the power supply with no load
   (c) the signal power in the load with a 500 mV input

3. Refer to the circuit in Figure 9–41. What changes would be necessary to convert the circuit to a *pnp* transistor with a positive supply? What advantage would this have?

4. Assume a CC amplifier has an input resistance of 2.2 kΩ and drives an output load of 50 Ω. What is the power gain?

5. What is the maximum peak value of collector current that can be realized in each circuit of Figure 9–42? What is the maximum peak value of output voltage?

6. Find the power gain for each circuit in Figure 9–42. Neglect r'_e.

7. Determine the minimum power rating for the transistor in Figure 9–43.

8. Find the maximum output signal power to the load and efficiency for the amplifier in Figure 9–43 with a 500 Ω load resistor.

▲ FIGURE 9–41

File circuits are identified with a CD logo and are in the Problems folder on your CD-ROM. Filenames correspond to figure numbers (e.g., F09-41).

(a) $\beta_{ac} = \beta_{DC} = 125$

(b) $\beta_{ac} = \beta_{DC} = 120$

▲ FIGURE 9–42

▶ FIGURE 9–43

$\beta_{ac} = \beta_{DC} = 90;\ r'_e = 10\ \Omega$

SECTION 9–2   Class B and Class AB Push-Pull Amplifiers

9. Refer to the Class AB amplifier in Figure 9–44.

   (a) Determine the dc parameters $V_{B(Q1)}$, $V_{B(Q2)}$, V_E, I_{CQ}, $V_{CEQ(Q1)}$, $V_{CEQ(Q2)}$.

   (b) For the 5 V rms input, determine the power delivered to the load resistor.

10. Draw the load line for the *npn* transistor in Figure 9–44. Label the saturation current, $I_{c(sat)}$, and show the Q-point.

▶ **FIGURE 9–44**

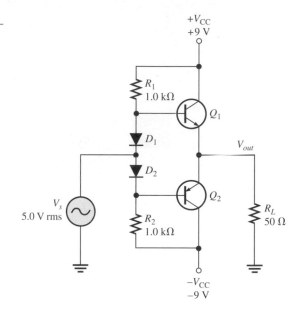

11. Refer to the Class AB amplifier in Figure 9–45 operating with a single power supply.

   (a) Determine the dc parameters $V_{B(Q1)}$, $V_{B(Q2)}$, V_E, I_{CQ}, $V_{CEQ(Q1)}$, $V_{CEQ(Q2)}$.

   (b) Assuming the input voltage is 10 V pp, determine the power delivered to the load resistor.

12. Refer to the Class AB amplifier in Figure 9–45.

   (a) What is the maximum power that could be delivered to the load resistor?

   (b) Assume the power supply voltage is raised to 24 V. What is the new maximum power that could be delivered to the load resistor?

▶ **FIGURE 9–45**

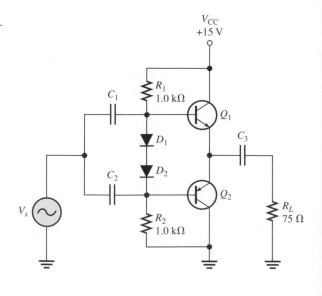

13. Refer to the Class AB amplifier in Figure 9–45. What fault or faults could account for each of the following troubles?

(a) a positive half-wave output signal

(b) zero volts on both bases and the emitters

(c) no output; emitter voltage $= +15$ V

(d) crossover distortion observed on the output waveform

14. Assume the n-channel E-MOSFET shown in Figure 9–46 has a threshold voltage of 2.75 V and the p-channel E-MOSFET has a threshold voltage of -2.75 V. Assume R_1 is set to 600 Ω.

(a) What resistance setting for R_6 will bias the output transistors to class AB operation?

(b) Assuming the input voltage is 50 mV rms, what is the rms voltage delivered to the load?

(c) What is the power delivered to the load with this setting?

▲ **FIGURE 9–46**

SECTION 9–3   Class C Amplifiers

15. A certain class C amplifier transistor is on for 10 percent of the input cycle. If $V_{ce(sat)} = 0.18$ V and $I_{c(sat)} = 25$ mA, what is the average power dissipation for maximum output?

16. What is the resonant frequency of a tank circuit with $L = 10$ mH and $C = 0.001$ μF?

17. What is the maximum peak-to-peak output voltage of a tuned class C amplifier with $V_{CC} = 12$ V?

18. Determine the efficiency of the class C amplifier described in Problem 17 if $V_{CC} = 15$ V and the equivalent parallel resistance in the collector tank circuit is 50 Ω.

TROUBLESHOOTING PROBLEMS

SECTION 9–4 **Troubleshooting**

19. Refer to Figure 9–47. What would you expect to observe across R_L if C_1 opened?

20. Your oscilloscope displays a half-wave output when connected across R_L in Figure 9–47. What is the probable cause?

21. Determine the possible fault or faults, if any, for each circuit in Figure 9–48 based on the indicated dc voltage measurements.

▶ **FIGURE 9–47**

▶ **FIGURE 9–48**

(a)

(b)

(c)

(d)

SYSTEM APPLICATION PROBLEMS

22. Assume that the public address/paging system represented by the block diagram in Figure 9–34 has quit working. You find there is no signal output from the power amplifier or the preamplifier, but you have verified that the microphone is working. Which two blocks are the most likely to be the problem? How would you narrow the choice down to one block?

23. Describe the output that would be observed in the push-pull amplifier of Figure 9–35 with a 3 V rms sinusoidal input voltage if the base-emitter junction of the 2N6043 opened.

24. Describe the output that would be observed in Figure 9–35 if the base-emitter junction of the 2N6040 opened for the same input as in Problem 23.

25. After visually inspecting the power amplifier circuit board in Figure 9–49, describe any problems.

▲ **FIGURE 9–49**

DATA SHEET PROBLEMS

26. Referring to the data sheet in Figure 9–36, determine the following:

    (a) minimum β_{DC} for the 2N6040 and the conditions

    (b) maximum collector-to-emitter voltage for the 2N6041

    (c) maximum power dissipation for the 2N6043 at a case temperature of 25°C

    (d) maximum continuous collector current for the 2N6040

27. Determine the maximum power dissipation for a 2N6040 at a case temperature of 65°C.

28. Determine the maximum power dissipation for a 2N6043 at an ambient temperature of 80°C.

29. Describe what happens to the small-signal current gain as the frequency increases.

30. Determine the approximate h_{fe} for the 2N6040 at a frequency of 2 kHz. At 100 kHz.

ADVANCED PROBLEMS

31. Explain why the specified maximum power dissipation of a power transistor at an ambient temperature of 25°C is much less than maximum power dissipation at a case temperature of 25°C.

32. Draw the dc and the ac load lines for the amplifier in Figure 9–50.

▶ **FIGURE 9–50**

33. Design a swamped class A power amplifier that will operate from a dc supply of +15 V with an approximate voltage gain of 50. The quiescent collector current should be approximately 500 mA, and the total dc current from the supply should not exceed 750 mA. The output power must be at least 1 W.

34. The public address/paging system in Figure 9–39 is to be converted to a portable unit that is independent of 115 V ac power. Determine the modifications necessary for the system to operate for 8 hours on a continuous basis.

EWB/MULTISIM TROUBLESHOOTING PROBLEMS

These file circuits are in the Troubleshooting Problems folder on your CD-ROM.

35. Open file TSP09-35 and determine the fault.

36. Open file TSP09-36 and determine the fault.

37. Open file TSP09-37 and determine the fault.

38. Open file TSP09-38 and determine the fault.

39. Open file TSP09-39 and determine the fault.

ANSWERS

SECTION REVIEWS

SECTION 9–1 **Class A Power Amplifiers**

1. To dissipate excessive heat
2. The collector
3. Cutoff and saturation clipping
4. 25%
5. The ratio of input resistance to output resistance

SECTION 9–2 **Class B and Class AB Push–Pull Amplifiers**

1. The class B Q-point is at cutoff.
2. The barrier potential of the base-emitter junction causes crossover distortion.
3. Maximum efficiency of a class B amplifier is 79%.
4. Push-pull reproduces both positive and negative alternations of the input signal with greater efficiency.
5. Both transistors in class AB are biased slightly above cutoff. In class B they are biased at cutoff.

SECTION 9-3 Class C Amplifiers

1. Class C is biased well into cutoff.
2. The purpose of the tuned circuit is to produce a sinusoidal voltage output.
3. $\eta = [1 \text{ W}/(1 \text{ W} + 0.1 \text{ W})]100 = 90.9\%$

SECTION 9-4 Troubleshooting

1. Excess input signal voltage
2. Open bypass capacitor, C_2

RELATED PROBLEMS FOR EXAMPLES

9-1 $dB = 10 \log A_p = 10 \log 42{,}429 = 46.23 \text{ dB}$

9-2 The efficiency goes up because no power is wasted in R_{E3}. The disadvantage is that the speaker has direct current in the coil.

9-3 30 V, 1.875 A

9-4 7.5 V; 1.875 A

9-5 $P_{out} = 1.76 \text{ W}$; $P_{DC} = 2.24 \text{ W}$

9-6 1.8 kΩ

9-7 3 mW

9-8 The efficiency decreases.

SELF-TEST

1. (a)   2. (b)   3. (b)   4. (a)   5. (b)
6. (c)   7. (a)   8. (e)   9. (c)   10. (c)
11. (a)   12. (c)   13. (b)   14. (d)   15. (a)
16. (d)   17. (c)

10

AMPLIFIER FREQUENCY RESPONSE

INTRODUCTION

In the previous chapters on amplifiers, the effects of the input frequency on an amplifier's operation due to capacitive elements in the circuit were neglected in order to focus on other concepts. The coupling and bypass capacitors were considered to be ideal shorts and the internal transistor capacitances were considered to be ideal opens. This treatment is valid when the frequency is in an amplifier's midrange.

As you know, capacitive reactance decreases with increasing frequency and vice versa. When the frequency is low enough, the coupling and bypass capacitors can no longer be considered as shorts because their reactances are large enough to have a significant effect. Also, when the frequency is high enough, the internal transistor capacitances can no longer be considered as opens because their reactances become small enough to have a significant effect on the amplifier operation. A complete picture of an amplifier's response must take into account the full range of frequencies over which the amplifier may operate.

In this chapter, you will study the frequency effects on amplifier gain and phase shift. The coverage applies to both BJT and FET amplifiers, and a mix of both are included to illustrate the concepts.

CHAPTER OUTLINE

- Discuss the frequency response of an amplifier

- Express the gain of an amplifier in decibels (dB)

- Analyze the low-frequency response of amplifiers

- Analyze the high-frequency response of amplifiers

- Analyze an amplifier for total frequency response

- Analyze multistage amplifiers for frequency response

- Measure the frequency response of an amplifier

KEY TERMS

- Decibel

- Midrange

- Critical frequency

- Roll-off

- Decade

- Bode Plot

- Bandwidth

■■■ SYSTEM APPLICATION PREVIEW

For the system application at the end of the chapter, you will apply knowledge gained in this chapter to determine how the voltage gain of a two-stage amplifier circuit varies with the frequency of the input voltage.

www. VISIT THE COMPANION WEBSITE

Study aids for this chapter are available at
http://www.prenhall.com/floyd

10-1 BASIC CONCEPTS

In the previous coverage of amplifiers, the capacitive reactance of the coupling and bypass capacitors was assumed to be 0 Ω at the signal frequency and, therefore, had no effect on an amplifier's gain or phase shift. Also, the internal transistor capacitances were assumed to be small enough to neglect at the operating frequency. All of these simplifying assumptions are valid and necessary for studying amplifier theory. However, these simplifying assumptions give a limited picture of an amplifier's total operation, so in this section you begin to study the frequency effects of these capacitances. The frequency response of an amplifier is the change in gain or phase shift over a specified range of input signal frequencies.

After completing this section, you should be able to

- **Discuss the frequency response of an amplifier**
- Explain the effect of coupling capacitors
- Explain the effect of bypass capacitors
- Discuss the internal transistor capacitances and explain their effects
- Use Miller's theorem to determine amplifier capacitances

Effect of Coupling Capacitors

Recall from basic circuit theory that $X_C = 1/2\pi fC$. This formula shows that the capacitive reactance varies inversely with frequency. At lower frequencies the reactance is greater, and it decreases as the frequency increases. At lower frequencies—for example, audio frequencies below 10 Hz—capacitively coupled amplifiers such as those in Figure 10–1 have less voltage gain than they have at higher frequencies. The reason is that at lower frequencies more signal voltage is dropped across C_1 and C_3 because their reactances are higher. This higher signal voltage drop at lower frequencies reduces the voltage gain. Also, a phase shift is introduced by the coupling capacitors because C_1 forms a lead circuit with the R_{in} of the amplifier, and C_3 forms a lead circuit with R_L in parallel with R_C or R_D. Recall that a *lead circuit* is an RC circuit in which the output voltage across R leads the input voltage in phase.

(a) BJT (b) JFET

▲ **FIGURE 10–1**

Examples of capacitively coupled BJT and FET amplifiers.

Effect of Bypass Capacitors

At lower frequencies, the reactance of the emitter bypass capacitor, C_2 in Figure 10–1, becomes significant and the emitter (or FET source terminal) is no longer at ac ground. The capacitive reactance X_{C2} in parallel with R_E (or R_S) creates an impedance that reduces the gain. This is illustrated in Figure 10–2.

For example, when the frequency is sufficiently high, $X_C \cong 0\ \Omega$ and the voltage gain of the CE amplifier is $A_v = R_C/r'_e$. At lower frequencies, $X_C \gg 0\ \Omega$ and the voltage gain is $A_v = R_C/(r'_e + Z_e)$.

▶ **FIGURE 10–2**

Nonzero reactance of the bypass capacitor in parallel with R_E creates an emitter impedance, (Z_e), which reduces the voltage gain.

Effect of Internal Transistor Capacitances

At high frequencies, the coupling and bypass capacitors become effective ac shorts and do not affect an amplifier's response. Internal transistor junction capacitances, however, do come into play, reducing an amplifier's gain and introducing phase shift as the signal frequency increases.

Figure 10–3 shows the internal *pn* junction capacitances for both a bipolar junction transistor and a JFET. In the case of the BJT, C_{be} is the base-emitter junction capacitance and C_{bc} is the base-collector junction capacitance. In the case of the JFET, C_{gs} is the capacitance between gate and source and C_{gd} is the capacitance between gate and drain.

▶ **FIGURE 10–3**

Internal transistor capacitances.

(a) BJT   (b) JFET

Data sheets often refer to the BJT capacitance C_{bc} as the output capacitance, often designated C_{ob}. The capacitance C_{be} is often designated as the input capacitance C_{ib}. Data sheets for FETs normally specify input capacitance C_{iss} and reverse transfer capacitance C_{rss}. From these, C_{gs} and C_{gd} can be calculated, as you will see in Section 10–4.

At lower frequencies, the internal capacitances have a very high reactance because of their low capacitance value (usually only a few picofarads) and the low frequency value. Therefore, they look like opens and have no effect on the transistor's performance. As the frequency goes up, the internal capacitive reactances go down, and at some point they begin to have a significant effect on the transistor's gain. When the reactance of C_{be} (or C_{gs}) becomes small enough, a significant amount of the signal voltage is lost due to a voltage-divider effect of the signal source resistance and the reactance of C_{be}, as illustrated in Figure 10–4(a). When the reactance of C_{bc} (or C_{gd}) becomes small enough, a significant amount of output signal voltage is fed back out of phase with the input (negative feedback), thus effectively reducing the voltage gain. This is illustrated in Figure 10–4(b).

(a) Effect of C_{be}, where V_b is reduced by the voltage-divider action of R_s and $X_{C_{be}}$.

(b) Effect of C_{bc}, where part of V_{out} (V_{fb}) goes back through C_{bc} to the base and reduces the input signal because it is approximately 180° out of phase with V_{in}.

▲ **FIGURE 10–4**

AC equivalent circuit for a BJT amplifier showing effects of the internal capacitances C_{be} and C_{bc}.

Miller's Theorem

Miller's theorem will be used later to simplify the analysis of inverting amplifiers at high frequencies where the internal transistor capacitances are important. The capacitance C_{bc} in BJTs (C_{gd} in FETs) between the input (base or gate) and the output (collector or drain) is shown in Figure 10–5(a) in a generalized form. A_v is the absolute voltage gain of the amplifier at midrange frequencies, and C represents either C_{bc} or C_{gd}.

(a)   (b)

▲ **FIGURE 10–5**

General case of Miller input and output capacitances. C represents C_{bc} or C_{gd}.

Miller's theorem states that C effectively appears as a capacitance from input to ground, as shown in Figure 10–5(b), that can be expressed as follows:

Equation 10–1

$$C_{in(Miller)} = C(A_v + 1)$$

This formula shows that C_{bc} (or C_{gd}) has a much greater impact on input capacitance than its actual value. For example, if $C_{bc} = 6$ pF and the amplifier gain is 50, then $C_{in(Miller)} = 306$ pF. Figure 10–6 shows how this effective input capacitance appears in the actual ac equivalent circuit in parallel with C_{be} (or C_{gs}).

(a)   (b)

▲ **FIGURE 10–6**

Amplifier ac equivalent circuits showing internal capacitances and effective Miller capacitances.

Miller's theorem also states that C effectively appears as a capacitance from output to ground, as shown in Figure 10–5(b), that can be expressed as follows:

$$C_{out(Miller)} = C\left(\frac{A_v + 1}{A_v}\right)$$

Equation 10–2

This formula indicates that if the voltage gain is 10 or greater, $C_{out(Miller)}$ is approximately equal to C_{bc} or C_{gd} because $(A_v + 1)/A_v$ is approximately equal to 1. Figure 10–6 also shows how this effective output capacitance appears in the ac equivalent circuit for BJTs and FETs. Equations 10–1 and 10–2 are derived in Appendix B.

SECTION 10–1 REVIEW Answers are at the end of the chapter.	1. In an ac amplifier, which capacitors affect the low-frequency gain? 2. How is the high-frequency gain of an amplifier limited? 3. When can coupling and bypass capacitors be neglected? 4. Determine $C_{in(Miller)}$ if $A_v = 50$ and $C_{bc} = 5$ pF. 5. Determine $C_{out(Miller)}$ if $A_v = 25$ and $C_{bc} = 3$ pF.

10–2   THE DECIBEL

Decibels are a form of gain measurement and are commonly used to express amplifier response.

After completing this section, you should be able to

- **Express the gain of an amplifier in decibels (dB)**
- Express power gain in dB
- Express voltage gain in dB
- State the distinction between a positive and a negative dB gain
- Define *critical frequency*
- Express power in terms of dBm

The use of decibels to express gain was introduced in Chapter 6. The decibel unit is important in amplifier measurements. The basis for the decibel unit stems from the logarithmic response of the human ear to the intensity of sound. The **decibel** is a logarithmic measurement of the ratio of one power to another or one voltage to another. Power gain is expressed in decibels (dB) by the following formula:

Equation 10–3

$$A_{p(dB)} = 10 \log A_p$$

where A_p is the actual power gain, P_{out}/P_{in}. Voltage gain is expressed in decibels by the following formula:

Equation 10–4

$$A_{v(dB)} = 20 \log A_v$$

If A_v is greater than 1, the dB gain is positive. If A_v is less than 1, the dB gain is negative and is usually called *attenuation*. You can use the ▉LOG▉ key on your calculator when working with these formulas.

EXAMPLE 10–1

Express each of the following ratios in dB:

(a) $\dfrac{P_{out}}{P_{in}} = 250$ (b) $\dfrac{P_{out}}{P_{in}} = 100$ (c) $A_v = 10$

(d) $A_p = 0.5$ (e) $\dfrac{V_{out}}{V_{in}} = 0.707$

Solution

(a) $A_{p(dB)} = 10 \log(250) = $ **24 dB** (b) $A_{p(dB)} = 10 \log(100) = $ **20 dB**

(c) $A_{v(dB)} = 20 \log(10) = $ **20 dB** (d) $A_{p(dB)} = 10 \log(0.5) = $ **−3 dB**

(e) $A_{v(dB)} = 20 \log(0.707) = $ **−3 dB**

Related Problem* Express each of the following gains in dB: (a) $A_v = 1200$, (b) $A_p = 50$, (c) $A_v = 125,000$.

---

*Answers are at the end of the chapter.

0 dB Reference

It is often convenient in amplifiers to assign a certain value of gain as the 0 dB reference. This does not mean that the actual voltage gain is 1 (which is 0 dB); it means that the reference gain, no matter what its actual value, is used as a reference with which to compare other values of gain and is therefore assigned a 0 dB value.

Many amplifiers exhibit a maximum gain over a certain range of frequencies and a reduced gain at frequencies below and above this range. The maximum gain is called the *midrange gain* in this case and is assigned a 0 dB value. Any value of gain below **midrange** can be referenced to 0 dB and expressed as a negative dB value. For example, if the midrange voltage gain of a certain amplifier is 100 and the gain at a certain frequency below midrange is 50, then this reduced voltage gain can be expressed as $20 \log(50/100) = 20 \log(0.5) = -6$ dB. This indicates that it is 6 dB *below* the 0 dB reference. Halving the output voltage for a steady input voltage is always a 6 dB *reduction* in the gain. Correspondingly, a doubling of the output voltage is always a 6 dB *increase* in the gain. Figure 10–7 illustrates a normalized gain-versus-frequency curve showing several dB points. The term *normalized* means that the midrange voltage gain is assigned a value of 1 or 0 dB.

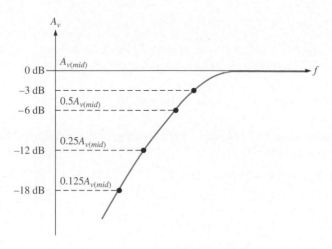

▲ **FIGURE 10–7**

Normalized voltage gain versus frequency curve.

Table 10–1 shows how doubling or halving voltage gains translates into dB values. Notice in the table that every time the voltage gain is doubled, the dB value increases by 6 dB, and every time the gain is halved, the dB value decreases by 6 dB.

◄ **TABLE 10–1**

Decibel values corresponding to doubling and halving of the voltage gain.

VOLTAGE GAIN (A_v)	dB (WITH RESPECT TO ZERO REFERENCE)
32	$20 \log(32) = 30$ dB
16	$20 \log(16) = 24$ dB
8	$20 \log(8) = 18$ dB
4	$20 \log(4) = 12$ dB
2	$20 \log(2) = 6$ dB
1	$20 \log(1) = 0$ dB
0.707	$20 \log(0.707) = -3$ dB
0.5	$20 \log(0.5) = -6$ dB
0.25	$20 \log(0.25) = -12$ dB
0.125	$20 \log(0.125) = -18$ dB
0.0625	$20 \log(0.0625) = -24$ dB
0.03125	$20 \log(0.03125) = -30$ dB

The Critical Frequency

The **critical frequency** (also known as **cutoff frequency** or *corner frequency*) is the frequency at which the output power drops to one-half of its midrange value. This corresponds to a 3 dB reduction in the power gain, as expressed in dB by the following formula:

$$A_{p(\text{dB})} = 10 \log(0.5) = -3 \text{ dB}$$

Also, at the critical frequency the output voltage is 70.7 percent of its midrange value and is expressed in dB as

$$A_{v(\text{dB})} = 20 \log(0.707) = -3 \text{ dB}$$

At the critical frequency, the voltage gain is down 3 dB or is 70.7% of its midrange value. At this same frequency, the power is one-half of its midrange value.

EXAMPLE 10–2

A certain amplifier has a midrange rms output voltage of 10 V. What is the rms output voltage for each of the following dB gain reductions with a constant rms input voltage?

(a) -3 dB   (b) -6 dB   (c) -12 dB   (d) -24 dB

Solution   Multiply the midrange output voltage by the voltage gain corresponding to the specified dB value in Table 10–1.

(a) At -3 dB, $V_{out} = 0.707(10 \text{ V}) = \textbf{7.07 V}$

(b) At -6 dB, $V_{out} = 0.5(10 \text{ V}) = \textbf{5 V}$

(c) At -12 dB, $V_{out} = 0.25(10 \text{ V}) = \textbf{2.5 V}$

(d) At -24 dB, $V_{out} = 0.0625(10 \text{ V}) = \textbf{0.625 V}$

Related Problem   Determine the output voltage at the following dB levels for a midrange value of 50 V:

(a) 0 dB   (b) -18 dB   (c) -30 dB

dBm Power Measurement

A unit that is often used in measuring power is the dBm. The term *dBm* means decibels referenced to 1 mW of power. When dBm is used, all power measurements are relative to a reference level of 1 mW. A 3 dBm increase corresponds to doubling of the power, and a 3 dBm decrease corresponds to a halving of the power. For example, +3 dBm corresponds to 2 mW (twice 1 mW), and –3 dBm corresponds to 0.5 mW (half of 1 mW). Table 10–2 shows several dBm values.

▷ **TABLE 10–2**

Power in terms of dBm.

POWER	dBm
32 mW	15 dBm
16 mW	12 dBm
8 mW	9 dBm
4 mW	6 dBm
2 mW	3 dBm
1 mW	0 dBm
0.5 mW	-3 dBm
0.25 mW	-6 dBm
0.125 mW	-9 dBm
0.0625 mW	-12 dBm
0.03125 mW	-15 dBm

SECTION 10–2 REVIEW

1. How much increase in actual voltage gain corresponds to +12 dB?
2. Convert a power gain of 25 to decibels.
3. What power corresponds to 0 dBm?

10–3 LOW-FREQUENCY AMPLIFIER RESPONSE

In this section, we will examine how the voltage gain and phase shift of a capacitively coupled amplifier are affected by frequencies below which the reactance of the coupling capacitors becomes too large to neglect. At the end of the section, a comparison is made to direct-coupled amplifiers.

After completing this section, you should be able to

- **Analyze the low-frequency response of amplifiers**
- Determine midrange voltage gain
- Generally describe how *RC* circuits affect the gain
- Identify the input *RC* circuit
- Determine the lower critical frequency of the input *RC* circuit
- Discuss gain roll-off in terms of dB/decade and dB/octave
- Determine phase shift of the input *RC* circuit
- Identify the output *RC* circuit
- Determine the lower critical frequency of the output *RC* circuit
- Determine phase shift of the output *RC* circuit
- Identify the bypass *RC* circuit
- Determine the lower critical frequency of the bypass *RC* circuit
- Describe a Bode plot
- Analyze an amplifier for total low-frequency response
- Compare direct-coupled amplifiers to capacitively coupled amplifiers

BJT Amplifiers

A typical capacitively coupled common-emitter amplifier is shown in Figure 10–8. Assuming that the coupling and bypass capacitors are ideal shorts at the midrange signal frequency, you can determine the midrange voltage gain using Equation 10–5, where $R_c = R_C \| R_L$.

$$A_{v(mid)} = \frac{R_c}{r'_e}$$

Equation 10–5

▶ **FIGURE 10–8**

A capacitively coupled amplifier.

The BJT amplifier in Figure 10–8 has three high-pass RC circuits that affect its gain as the frequency is reduced below midrange. These are shown in the low-frequency ac equivalent circuit in Figure 10–9. Unlike the ac equivalent circuit used in previous chapters, which represented midrange response ($X_C \cong 0\ \Omega$), the low-frequency equivalent circuit retains the coupling and bypass capacitors because X_C is not small enough to neglect when the signal frequency is sufficiently low.

▲ **FIGURE 10–9**

The low–frequency ac equivalent circuit of the amplifier in Figure 10–8 consists of three high–pass RC circuits.

One RC circuit is formed by the input coupling capacitor C_1 and the input resistance of the amplifier. The second RC circuit is formed by the output coupling capacitor C_3, the resistance looking in at the collector, and the load resistance. The third RC circuit that affects the low-frequency response is formed by the emitter-bypass capacitor C_2 and the resistance looking in at the emitter.

The Input RC Circuit The input RC circuit for the BJT amplifier in Figure 10–8 is formed by C_1 and the amplifier's input resistance and is shown in Figure 10–10. (Input resistance was discussed in Chapter 6.) As the signal frequency decreases, X_{C1} increases. This causes less voltage across the input resistance of the amplifier at the base because more voltage is dropped across C_1 and because of this, the overall voltage gain of the amplifier is reduced. The base voltage for the input RC circuit in Figure 10–10 (neglecting the internal resistance of the input signal source) can be stated as

$$V_{base} = \left(\frac{R_{in}}{\sqrt{R_{in}^2 + X_{C1}^2}} \right) V_{in}$$

▶ **FIGURE 10–10**

Input RC circuit formed by the input coupling capacitor and the amplifier's input resistance.

As previously mentioned, a critical point in the amplifier's response occurs when the output voltage is 70.7 percent of its midrange value. This condition occurs in the input RC circuit when $X_{C1} = R_{in}$.

$$V_{base} = \left(\frac{R_{in}}{\sqrt{R_{in}^2 + R_{in}^2}}\right)V_{in} = \left(\frac{R_{in}}{\sqrt{2R_{in}^2}}\right)V_{in} = \left(\frac{R_{in}}{\sqrt{2}R_{in}}\right)V_{in} = \left(\frac{1}{\sqrt{2}}\right)V_{in} = 0.707V_{in}$$

In terms of measurement in decibels,

$$20\log\left(\frac{V_{base}}{V_{in}}\right) = 20\log(0.707) = -3 \text{ dB}$$

Lower critical frequency   The condition where the gain is down 3 dB is logically called the *−3 dB point* of the amplifier response; the overall gain is 3 dB less than at midrange frequencies because of the attenuation of the input RC circuit. The frequency, f_c, at which this condition occurs is called the *lower critical frequency* (also known as the *lower cutoff frequency, lower corner frequency,* or *lower break frequency*) and can be calculated as follows:

$$X_{C1} = \frac{1}{2\pi f_c C_1} = R_{in}$$

$$f_c = \frac{1}{2\pi R_{in} C_1}$$

Equation 10–6

If the resistance of the input source is taken into account, Equation 10–6 becomes

$$f_c = \frac{1}{2\pi(R_s + R_{in})C_1}$$

EXAMPLE 10–3

For an input RC circuit in a certain amplifier, $R_{in} = 1.0$ kΩ and $C_1 = 1$ μF. Neglect the source resistance.

(a) Determine the lower critical frequency.

(b) What is the attenuation of the input RC circuit at the lower critical frequency?

(c) If the midrange voltage gain of the amplifier is 100, what is the gain at the lower critical frequency?

Solution   (a) $f_c = \dfrac{1}{2\pi R_{in} C_1} = \dfrac{1}{2\pi(1.0 \text{ k}\Omega)(1 \text{ }\mu\text{F})} = \textbf{159 Hz}$

(b) At f_c, $X_{C1} = R_{in}$. Therefore,

$$\text{Attenuation} = \frac{V_{base}}{V_{in}} = \textbf{0.707}$$

(c) $A_v = 0.707A_{v(mid)} = 0.707(100) = \textbf{70.7}$

Related Problem   For an input RC circuit in a certain amplifier, $R_{in} = 10$ kΩ and $C_1 = 2.2$ μF.

(a) What is f_c?   (b) What is the attenuation at f_c?

(c) If $A_{v(mid)} = 500$, what is A_v at f_c?

Voltage gain roll-off at low frequencies As you have seen, the input RC circuit reduces the overall voltage gain of an amplifier by 3 dB when the frequency is reduced to the critical value f_c. As the frequency continues to decrease below f_c, the overall voltage gain also continues to decrease. The decrease in voltage gain with frequency is called **roll-off.** *For each ten times reduction in frequency below f_c, there is a 20 dB reduction in voltage gain.*

Let's take a frequency that is one-tenth of the critical frequency ($f = 0.1f_c$). Since $X_{C1} = R_{in}$ at f_c, then $X_{C1} = 10R_{in}$ at $0.1f_c$ because of the inverse relationship of X_{C1} and f. The attenuation of the input RC circuit is, therefore,

$$\text{Attenuation} = \frac{V_{base}}{V_{in}} = \frac{R_{in}}{\sqrt{R_{in}^2 + X_{C1}^2}} = \frac{R_{in}}{\sqrt{R_{in}^2 + (10R_{in})^2}} = \frac{R_{in}}{\sqrt{R_{in}^2 + 100R_{in}^2}}$$

$$= \frac{R_{in}}{\sqrt{R_{in}^2(1 + 100)}} = \frac{R_{in}}{R_{in}\sqrt{101}} = \frac{1}{\sqrt{101}} \cong \frac{1}{10} = 0.1$$

The dB attenuation is

$$20 \log\left(\frac{V_{base}}{V_{in}}\right) = 20 \log(0.1) = -20 \text{ dB}$$

dB/decade A ten-times change in frequency is called a **decade.** So, for the input RC circuit, the attenuation is reduced by 20 dB for each decade that the frequency decreases below the critical frequency. This causes the overall voltage gain to drop 20 dB per decade. For example, if the frequency is reduced to one-hundredth of f_c (a two-decade decrease), the amplifier voltage gain drops 20 dB for each decade, giving a total decrease in voltage gain of -20 dB $+ (-20$ dB$) = -40$ dB. This is illustrated in Figure 10–11, which is a graph of dB voltage gain versus frequency. This graph is the low-frequency response curve for an amplifier showing the effect of the input RC circuit on the voltage gain.

▶ **FIGURE 10–11**

dB voltage gain versus frequency for the input RC circuit.

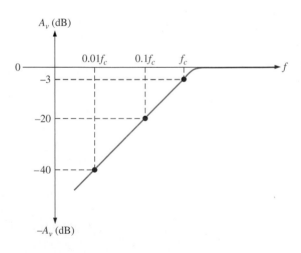

dB/octave Sometimes, the voltage gain roll-off of an amplifier is expressed in dB/octave rather than dB/decade. An **octave** corresponds to a doubling or halving of the frequency. For example, an increase in frequency from 100 Hz to 200 Hz is an octave. Likewise, a decrease in frequency from 100 kHz to 50 kHz is also an octave. A rate of -20 dB/decade is approximately equivalent to -6 dB/octave, a rate of -40 dB/decade is approximately equivalent to -12 dB/octave, and so on.

EXAMPLE 10–4

The midrange voltage gain of a certain amplifier is 100. The input RC circuit has a lower critical frequency of 1 kHz. Determine the actual voltage gain at $f = 1$ kHz, $f = 100$ Hz, and $f = 10$ Hz.

Solution   When $f = 1$ kHz, the voltage gain is 3 dB less than at midrange. At -3 dB, the voltage gain is reduced by a factor of 0.707.

$$A_v = (0.707)(100) = \mathbf{70.7}$$

When $f = 100$ Hz $= 0.1f_c$, the voltage gain is 20 dB less than at f_c. The voltage gain at -20 dB is one-tenth of that at the midrange frequencies.

$$A_v = (0.1)(100) = \mathbf{10}$$

When $f = 10$ Hz $= 0.01f_c$, the voltage gain is 20 dB less than at $f = 0.1f_c$ or -40 dB. The voltage gain at -40 dB is one-tenth of that at -20 dB or one-hundredth that at the midrange frequencies.

$$A_v = (0.01)(100) = \mathbf{1}$$

Related Problem   The midrange voltage gain of an amplifier is 300. The lower critical frequency of the input RC circuit is 400 Hz. Determine the actual voltage gain at 400 Hz, 40 Hz, and 4 Hz.

Phase shift in the input RC circuit   In addition to reducing the voltage gain, the input RC circuit also causes an increasing phase shift through an amplifier as the frequency decreases. At midrange frequencies, the phase shift through the input RC circuit is approximately zero because $X_{C1} \cong 0$ Ω. At lower frequencies, higher values of X_{C1} cause a phase shift to be introduced, and the output voltage of the RC circuit leads the input voltage. As you learned in ac circuit theory, the phase angle in an input RC circuit is expressed as

$$\theta = \tan^{-1}\left(\frac{X_{C1}}{R_{in}}\right)$$

Equation 10–7

For midrange frequencies, $X_{C1} \cong 0$ Ω, so

$$\theta = \tan^{-1}\left(\frac{0\ \Omega}{R_{in}}\right) = \tan^{-1}(0) = 0°$$

At the critical frequency, $X_{C1} = R_{in}$, so

$$\theta = \tan^{-1}\left(\frac{R_{in}}{R_{in}}\right) = \tan^{-1}(1) = 45°$$

A decade below the critical frequency, $X_{C1} = 10R_{in}$, so

$$\theta = \tan^{-1}\left(\frac{10R_{in}}{R_{in}}\right) = \tan^{-1}(10) = 84.3°$$

A continuation of this analysis will show that the phase shift through the input *RC* circuit approaches 90° as the frequency approaches zero. A plot of phase angle versus frequency is shown in Figure 10–12. The result is that the voltage at the base of the transistor *leads* the input signal voltage in phase below midrange, as shown in Figure 10–13.

▶ **FIGURE 10–12**

Phase angle versus frequency for the input *RC* circuit.

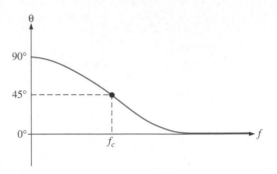

▶ **FIGURE 10–13**

Input *RC* circuit causes the base voltage to lead the input voltage below midrange by an amount equal to the circuit phase angle, θ.

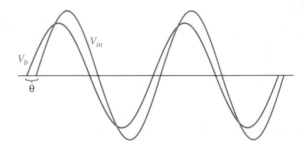

The Output RC Circuit The second high-pass *RC* circuit in the BJT amplifier of Figure 10–8 is formed by the coupling capacitor C_3, the resistance looking in at the collector, and the load resistance R_L, as shown in Figure 10–14(a). In determining the output resistance, looking in at the collector, the transistor is treated as an ideal current source

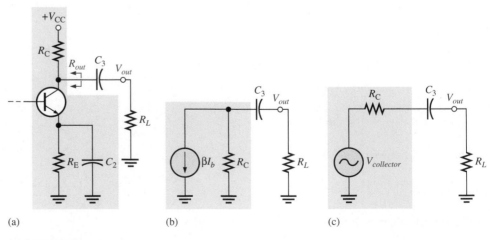

(a) (b) (c)

▲ **FIGURE 10–14**

Development of the equivalent low-frequency output *RC* circuit.

(with infinite internal resistance), and the upper end of R_C is effectively at ac ground, as shown in Figure 10–14(b). Therefore, thevenizing the circuit to the left of capacitor C_3 produces an equivalent voltage source equal to the collector voltage and a series resistance equal to R_C, as shown in Figure 10–14(c). The critical frequency of this output RC circuit is

$$f_c = \frac{1}{2\pi(R_C + R_L)C_3}$$

Equation 10–8

The effect of the output RC circuit on the amplifier voltage gain is similar to that of the input RC circuit. As the signal frequency decreases, X_{C3} increases. This causes less voltage across the load resistance because more voltage is dropped across C_3. The signal voltage is reduced by a factor of 0.707 when frequency is reduced to the lower critical value, f_c, for the circuit. This corresponds to a 3 dB reduction in voltage gain.

EXAMPLE 10–5

For an output RC circuit in a certain amplifier, $R_C = 10$ kΩ, $C_3 = 0.1$ μF, and $R_L = 10$ kΩ.

(a) Determine the critical frequency.

(b) What is the attenuation of the output RC circuit at the critical frequency?

(c) If the midrange voltage gain of the amplifier is 50, what is the gain at the critical frequency?

Solution (a) $f_c = \dfrac{1}{2\pi(R_C + R_L)C_3} = \dfrac{1}{2\pi(20 \text{ k}\Omega)(0.1 \text{ }\mu\text{F})} = $ **79.6 Hz**

(b) For the midrange frequencies, $X_{C3} \cong 0$ Ω; thus, the attenuation of the circuit as determined from Figure 10–14(c) is

$$\frac{V_{out}}{V_{collector}} = \frac{R_L}{R_C + R_L} = \frac{10 \text{ k}\Omega}{20 \text{ k}\Omega} = 0.5$$

or, in dB, $V_{out}/V_{collector} = 20 \log(0.5) = -6$ dB. This shows that, in this case, the midrange voltage gain is reduced by 6 dB because of the load resistor. At the critical frequency, $X_{C3} = R_C + R_L$ and the attenuation is

$$\frac{V_{out}}{V_{in}} = \frac{R_L}{\sqrt{(R_C + R_L)^2 + X_{C3}^2}} = \frac{10 \text{ k}\Omega}{\sqrt{(20 \text{ k}\Omega)^2 + (20 \text{ k}\Omega)^2}} = \textbf{0.354}$$

or, in dB, $V_{out}/V_{collector} = 20 \log(0.354) = -9$ dB. As you can see, the gain at f_c is 3 dB less than the gain at midrange.

(c) $A_v = 0.707A_{v(mid)} = 0.707(50) = $ **35.4**

Related Problem The output RC circuit in a certain amplifier has the following values: $R_C = 3.9$ kΩ, $C_3 = 1$ μF, and $R_L = 8.2$ kΩ.

(a) Find the critical frequency. (b) What is the attenuation at f_c?

(c) If $A_{v(mid)}$ is 100, what is the gain at f_c?

Phase shift in the output RC circuit The phase shift in the output RC circuit is

Equation 10–9

$$\theta = \tan^{-1}\left(\frac{X_{C3}}{R_C + R_L}\right)$$

$\theta \cong 0°$ for the midrange frequencies and approaches $90°$ as the frequency approaches zero (X_{C3} approaches infinity). At the critical frequency f_c, the phase shift is $45°$.

The Bypass RC Circuit The third RC circuit that affects the low-frequency gain of the BJT amplifier in Figure 10–8 includes the bypass capacitor C_2. For midrange frequencies, it is assumed that $X_{C2} \cong 0\ \Omega$, effectively shorting the emitter to ground so that the amplifier gain is R_c/r'_e, as you already know. As the frequency is reduced, X_{C2} increases and no longer provides a sufficiently low reactance to effectively place the emitter at ac ground. This is illustrated in Figure 10–15. Because the impedance from emitter to ground increases, the gain decreases. In this case, R_e in the formula, $A_v = R_c/(r'_e + R_e)$, is replaced by an impedance formed by R_E in parallel with X_{C2}.

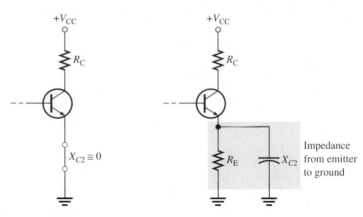

(a) For midrange frequencies, C_2 effectively shorts the emitter to ground.

(b) Below f_c, X_{C2} and R_E form an impedance between the emitter and ground.

▲ **FIGURE 10–15**

At low frequencies, X_{C2} in parallel with R_E creates an impedance that reduces the voltage gain.

The bypass RC circuit is formed by C_2 and the resistance looking in at the emitter, $R_{in(emitter)}$, as shown in Figure 10–16(a). The resistance looking in at the emitter is derived as follows. First, Thevenin's theorem is applied looking from the base of the transistor toward the input source V_{in}, as shown in Figure 10–16(b). This results in an equivalent resistance (R_{th}) and an equivalent voltage source ($V_{th(1)}$) in series with the base, as shown in Figure 10–16(c). The resistance looking in at the emitter is determined with the equivalent input source shorted, as shown in Figure 10–16(d), and is expressed as follows:

$$R_{in(emitter)} = \frac{V_e}{I_e} + r'_e \cong \frac{V_b}{\beta_{ac}I_b} + r'_e = \frac{I_bR_{th}}{\beta_{ac}I_b} + r'_e$$

$$R_{in(emitter)} = \frac{R_{th}}{\beta_{ac}} + r'_e$$

Equation 10–10

Looking from the capacitor C_2, $R_{th}/\beta_{ac} + r'_e$ is in parallel with R_E, as shown in Figure 10–16(e). Thevenizing again, we get the equivalent RC circuit shown in Figure 10–16(f). The critical frequency for this equivalent bypass RC circuit is

$$f_c = \frac{1}{2\pi[(r'_e + R_{th}/\beta_{ac}) \| R_E]C_2}$$

Equation 10–11

(a)

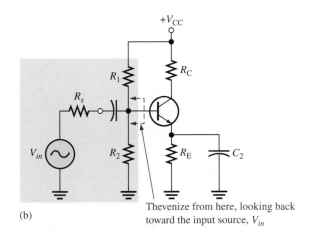

(b)

Thevenize from here, looking back toward the input source, V_{in}

(c)

(d)

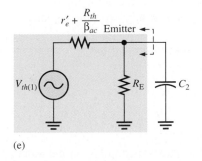

(e)

(f)

▲ **FIGURE 10–16**

Development of the equivalent bypass RC circuit.

EXAMPLE 10–6

Determine the critical frequency of the bypass RC circuit for the amplifier in Figure 10–17. ($r'_e = 12\ \Omega$.)

▲ **FIGURE 10–17**

Solution   Thevenize the base circuit (looking from the base toward the input source).

$$R_{th} = R_1 \parallel R_2 \parallel R_s = 62\ \text{k}\Omega \parallel 22\ \text{k}\Omega \parallel 1.0\ \text{k}\Omega \cong 942\ \Omega$$

The resistance looking in at the emitter is

$$R_{in(emitter)} = r'_e + \frac{R_{th}}{\beta_{ac}} = 12\ \Omega + 9.42\ \Omega = 21.4\ \Omega$$

The resistance of the equivalent bypass RC circuit is $R_{in(emitter)} \parallel R_E$.

$$R_{in(emitter)} \parallel R_E = 21.4\ \Omega \parallel 1000\ \Omega = 21\ \Omega$$

The critical frequency of the bypass RC circuit is

$$f_c = \frac{1}{2\pi(R_{in(emitter)} \parallel R_E)C_2} = \frac{1}{2\pi(21\ \Omega)(100\ \mu\text{F})} = \mathbf{75.8\ Hz}$$

Related Problem   In Figure 10–17, the source resistance, R_s, is 50 Ω, and the transistor ac beta is 150. Determine the critical frequency of the bypass RC circuit.

> Open file E10-06 in the Examples folder on your CD-ROM. Measure the critical frequency of the bypass RC circuit which is the dominant low critical frequency of the amplifier. Compare your results with the calculated value.

FET Amplifiers

A zero-biased D-MOSFET amplifier with capacitive coupling on the input and output is shown in Figure 10–18. As you learned in Chapter 8, the midrange voltage gain of a zero-biased amplifier is

$$A_{v(mid)} = g_m R_d$$

▶ **FIGURE 10–18**

▶ **FIGURE 10–18**

Zero-biased D-MOSFET amplifier.

This is the gain at frequencies high enough so that the capacitive reactances are approximately zero. The amplifier in Figure 10–18 has only two high-pass RC circuits that influence its low-frequency response. One RC circuit is formed by the input coupling capacitor C_1 and the input resistance, as shown in Figure 10–19. The other circuit is formed by the output coupling capacitor C_2 and the output resistance looking in at the drain.

▶ **FIGURE 10–19**

Input RC circuit for the amplifier in Figure 10–18.

The Input RC Circuit Just as in the previous case for the BJT amplifier, the reactance of the input coupling capacitor increases as the frequency decreases. When $X_{C1} = R_{in}$, the gain is down 3 dB below its midrange value. The lowest critical frequency is

$$f_c = \frac{1}{2\pi R_{in} C_1}$$

The input resistance is

$$R_{in} = R_G \| R_{in(gate)}$$

where $R_{in(gate)}$ is determined from data sheet information as

$$R_{in(gate)} = \left| \frac{V_{GS}}{I_{GSS}} \right|$$

So, the critical frequency is

$$f_c = \frac{1}{2\pi (R_G \| R_{in(gate)}) C_1}$$

Equation 10–12

The gain rolls off below f_c at 20 dB/decade, as previously shown. The phase shift in the low-frequency input RC circuit is

$$\theta = \tan^{-1}\left(\frac{X_{C1}}{R_{in}}\right)$$

Equation 10–13

EXAMPLE 10–7

What is the critical frequency of the input RC circuit in the FET amplifier of Figure 10–20?

▲ **FIGURE 10–20**

Solution   First determine R_{in} and then calculate f_c.

$$R_{in(gate)} = \left| \frac{V_{GS}}{I_{GSS}} \right| = \frac{10\,\text{V}}{25\,\text{nA}} = 400\,\text{M}\Omega$$

$$R_{in} = R_G \| R_{in(gate)} = 10\,\text{M}\Omega \| 400\,\text{M}\Omega = 9.8\,\text{M}\Omega$$

$$f_c = \frac{1}{2\pi R_{in} C_1} = \frac{1}{2\pi (9.8\,\text{M}\Omega)(0.001\,\mu\text{F})} = \textbf{16.2 Hz}$$

The critical frequency of the input RC circuit of a FET amplifier is usually very low because of the very high input resistance.

Related Problem   How much does the critical frequency of the input RC circuit change if the FET in Figure 10–20 is replaced by one with $I_{GSS} = 10\,\text{nA}$ @ $V_{GS} = -8\,\text{V}$?

Open file E10-07 in the Examples folder on your CD-ROM and measure the low critical frequency for the input circuit. Compare to the calculated result.

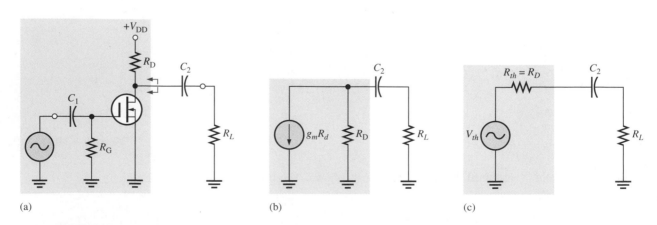

(a)   (b)   (c)

▲ **FIGURE 10–21**

Development of the equivalent low-frequency output RC circuit.

The Output RC Circuit The second *RC* circuit that affects the low-frequency response of the amplifier in Figure 10–18 is formed by a coupling capacitor C_2 and the output resistance looking in at the drain, as shown in Figure 10–21(a). The load resistor, R_L, is also included. As in the case of the BJT, the FET is treated as a current source, and the upper end of R_D is effectively ac ground, as shown in Figure 10–21(b). The Thevenin equivalent of the circuit to the left of C_2 is shown in Figure 10–21(c). The critical frequency for this *RC* circuit is

$$f_c = \frac{1}{2\pi(R_D + R_L)C_2}$$

Equation 10–14

The effect of the output *RC* circuit on the amplifier's voltage gain below the midrange is similar to that of the input *RC* circuit. The circuit with the highest critical frequency dominates because it is the one that first causes the gain to roll off as the frequency drops below its midrange values. The phase shift in the low-frequency output *RC* circuit is

$$\theta = \tan^{-1}\left(\frac{X_{C2}}{R_D + R_L}\right)$$

Equation 10–15

Again, at the critical frequency, the phase angle is 45° and approaches 90° as the frequency approaches zero. However, starting at the critical frequency, the phase angle decreases from 45° and becomes very small as the frequency goes higher.

EXAMPLE 10–8

Determine the total low-frequency response of the FET amplifier in Figure 10–22. Assume that the load is another identical amplifier with the same R_{in}. The data sheet shows $I_{GSS} = 100$ nA at $V_{GS} = -12$ V.

▲ **FIGURE 10–22**

Solution First, find the critical frequency for the input *RC* circuit.

$$R_{in(gate)} = \left|\frac{V_{GS}}{I_{GSS}}\right| = \frac{12\ V}{100\ nA} = 120\ M\Omega$$

$$R_{in} = R_G \| R_{in(gate)} = 10\ M\Omega \| 120\ M\Omega = 9.2\ M\Omega$$

$$f_{c(input)} = \frac{1}{2\pi R_{in}C_1} = \frac{1}{2\pi(9.2\ M\Omega)(0.001\ \mu F)} = \textbf{17.3 Hz}$$

The output RC circuit has a critical frequency of

$$f_{c(output)} = \frac{1}{2\pi(R_D + R_L)C_2} = \frac{1}{2\pi(9.21 \text{ M}\Omega)(0.001 \text{ }\mu\text{F})} \cong \mathbf{17.3 \text{ Hz}}$$

Related Problem   If the circuit in Figure 10–22 were operated with no load, how is the low-frequency response affected?

Open file E10-08 in the Examples folder on your CD-ROM. Determine the total low-frequency response of the amplifier.

The Bode Plot

A plot of dB voltage gain versus frequency on semilog graph paper (logarithmic horizontal axis scale and a linear vertical axis scale) is called a **Bode plot**. A generalized Bode plot for an RC circuit like that shown in Figure 10–23(a) appears in part (b) of the figure. The ideal response curve is shown in blue. Notice that it is flat (0 dB) down to the critical frequency, at which point the gain drops at -20 dB/decade as shown. Above f_c are the midrange frequencies. The actual response curve is shown in red. Notice that it decreases gradually in midrange and is down to -3 dB at the critical frequency. Often, the ideal response is used to simplify amplifier analysis. As previously mentioned, the critical frequency at which the curve "breaks" into a -20 dB/decade drop is sometimes called the *lower break frequency*.

(a) An RC circuit

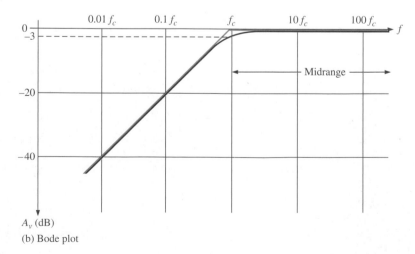

(b) Bode plot

▲ **FIGURE 10–23**

An RC circuit and its low-frequency response. (Blue is ideal; red is actual.)

Total Low-Frequency Response of an Amplifier

Now that we have individually examined the high-pass RC circuits that affect a BJT or FET amplifier's voltage gain at low frequencies, let's look at the combined effect of the three RC circuits in a BJT amplifier. Each circuit has a critical frequency determined by the R and C values. The critical frequencies of the three RC circuits are not necessarily all equal. If one of the RC circuits has a critical (break) frequency higher than the other two, then it is the *dominant RC* circuit. The dominant circuit determines the frequency at which the overall voltage gain of the amplifier begins to drop at -20 dB/decade. The other circuits each cause an additional -20 dB/decade roll-off below their respective critical (break) frequencies.

To get a better picture of what happens at low frequencies, refer to the Bode plot in Figure 10–24, which shows the superimposed ideal responses for the three RC circuits (green lines) of a BJT amplifier. In this example, each RC circuit has a different critical frequency. The input RC circuit is dominant (highest f_c) in this case, and the bypass RC circuit has the lowest f_c. The overall response is shown as the blue line.

▲ **FIGURE 10–24**

Composite Bode plot of a BJT amplifier response for three low-frequency RC circuits with different critical frequencies. Total response is shown by the blue curve.

Here is what happens. As the frequency is reduced from midrange, the first "break point" occurs at the critical frequency of the input RC circuit, $f_{c(input)}$, and the gain begins to drop at -20 dB/decade. This constant roll-off rate continues until the critical frequency of the output RC circuit, $f_{c(output)}$, is reached. At this break point, the output RC circuit adds another -20 dB/decade to make a total roll-off of -40 dB/decade. This constant -40 dB/decade roll-off continues until the critical frequency of the bypass RC circuit, $f_{c(bypass)}$, is reached. At this break point, the bypass RC circuit adds still another -20 dB/decade, making the gain roll-off at -60 dB/decade.

If all RC circuits have the same critical frequency, the response curve has one break point at that value of f_c, and the voltage gain rolls off at -60 dB/decade below that value, as shown by the ideal blue curve in Figure 10–25. Actually, the midrange voltage gain does not extend down to the dominant critical frequency but is really at −9 dB below the midrange voltage gain at that point (−3 dB for each *RC* circuit), as shown by the red curve.

▲ **FIGURE 10–25**

Composite Bode plot of an amplifier response where all *RC* circuits have the same f_c. (Blue is ideal; red is actual.)

EXAMPLE 10–9

Determine the total low-frequency response of the BJT amplifier in Figure 10–26. $\beta_{ac} = 100$ and $r'_e = 16\ \Omega$.

▲ **FIGURE 10–26**

Solution Each *RC* circuit is analyzed to determine its critical frequency. For the input *RC* circuit with the source resistance, R_s, taken into account:

$$R_{in} = R_1 \| R_2 \| \beta_{ac} r'_e = 62\ \text{k}\Omega \| 22\ \text{k}\Omega \| 1.6\ \text{k}\Omega = 1.46\ \text{k}\Omega$$

$$f_{c(input)} = \frac{1}{2\pi(R_s + R_{in})C_1} = \frac{1}{2\pi(600\ \Omega + 1.46\ \text{k}\Omega)(0.1\ \mu\text{F})} = \textbf{773 Hz}$$

For the bypass RC circuit:

$$R_{th} = R_1 \| R_2 \| R_s = 62 \text{ k}\Omega \| 22 \text{ k}\Omega \| 600 \text{ }\Omega = 579 \text{ }\Omega$$

$$R_{in(emitter)} = \frac{R_{th}}{\beta_{ac}} + r_e' = \frac{579 \text{ }\Omega}{100} + 16 \text{ }\Omega = 21.8 \text{ }\Omega$$

$$f_{c(bypass)} = \frac{1}{2\pi(R_{in(emitter)} \| R_E)C_2} = \frac{1}{2\pi(21.8 \text{ }\Omega \| 1.0 \text{ k}\Omega)(10 \text{ }\mu\text{F})}$$

$$= \frac{1}{2\pi(21.3 \text{ }\Omega)(10 \text{ }\mu\text{F})} = \textbf{747 Hz}$$

For the output RC circuit:

$$f_{c(output)} = \frac{1}{2\pi(R_C + R_L)C_3} = \frac{1}{2\pi(2.2 \text{ k}\Omega + 10 \text{ k}\Omega)(0.1 \text{ }\mu\text{F})} = \textbf{130.5 Hz}$$

The above analysis shows that the input circuit produces the dominant (which is the highest f_c) lower critical frequency. The midrange voltage gain of the amplifier is

$$A_{v(mid)} = \frac{R_c}{r_e'} = \frac{2.2 \text{ k}\Omega \| 10 \text{ k}\Omega}{16 \text{ }\Omega} = 113$$

The midrange attenuation of the input circuit is

$$\frac{R_1 \| R_2 \| \beta_{ac} r_e'}{R_s + R_1 \| R_2 \| \beta_{ac} r_e'} = \frac{62 \text{ k}\Omega \| 22 \text{ k}\Omega \| 1600 \text{ }\Omega}{600 \text{ }\Omega + 62 \text{ k}\Omega \| 22 \text{ k}\Omega \| 1600 \text{ }\Omega} = \frac{1456}{2056} = 0.708$$

The overall voltage gain is

$$A_{v(mid)}' = 0.708(113) = 80$$

and is expressed in dB as

$$A_{v(mid)(dB)}' = 20 \log(80) = 38.1 \text{ dB}$$

The ideal Bode plot of the low-frequency response of this amplifier is shown in Figure 10–27. As a practical matter, $f_{c(input)}$ and $f_{c(bypass)}$ are so close in value that the difference would be difficult to measure.

▶ **FIGURE 10–27**

Ideal Bode plot for the total low-frequency response of the amplifier in Figure 10–26.

Related Problem   If the overall voltage gain of the amplifier is reduced, how are the lower critical frequencies affected?

**SECTION 10–3
REVIEW**

1. A certain BJT amplifier exhibits three critical frequencies in its low–frequency response: $f_{c1} = 130$ Hz, $f_{c2} = 167$ Hz, and $f_{c3} = 75$ Hz. Which is the dominant critical frequency?

2. If the midrange voltage gain of the amplifier in Question 1 is 50 dB, what is the gain at the dominant f_c?

3. A certain RC circuit has an $f_c = 235$ Hz, above which the attenuation is 0 dB. What is the dB attenuation at 23.5 Hz?

4. What is the amount of phase shift contributed by an input circuit when $X_C = 0.5R_{in}$ at a certain frequency below f_{c1}?

5. What is the critical fequency when $R_D = 1.5$ kΩ, $R_L = 5$ kΩ, and $C_2 = 0.0022$ μF in a circuit like Figure 10–22?

10–4 HIGH-FREQUENCY AMPLIFIER RESPONSE

You have seen how the coupling and bypass capacitors affect the voltage gain of an amplifier at lower frequencies where the reactances of the coupling and bypass capacitors are significant. In the midrange of an amplifier, the effects of the capacitors are minimal and can be neglected. If the frequency is increased sufficiently, a point is reached where the transistor's internal capacitances begin to have a significant effect on the gain. The basic differences between BJTs and FETs are the specifications of the internal capacitances and the input resistance.

After completing this section, you should be able to

- **Analyze the high-frequency response of amplifiers**

- Apply Miller's theorem

- Generally describe how internal capacitances affect the gain

- Identify the input RC circuit

- Determine the upper critical frequency of the input RC circuit

- Determine phase shift of the input RC circuit

- Identify the output RC circuit

- Determine the upper critical frequency of the output RC circuit

- Determine phase shift of the output RC circuit

- Analyze an amplifier for total high-frequency response

BJT Amplifiers

A high-frequency ac equivalent circuit for the BJT amplifier in Figure 10–28(a) is shown in Figure 10–28(b). Notice that the coupling and bypass capacitors are treated as effective shorts and do not appear in the equivalent circuit. The internal capacitances, C_{be} and C_{bc}, which are significant only at high frequencies, do appear in the diagram. As previously

(a) Capacitively coupled amplifier
(b) High-frequency equivalent circuit

▲ **FIGURE 10–28**

Capacitively coupled amplifier and its high-frequency equivalent circuit.

mentioned, C_{be} is sometimes called the input capacitance C_{ib}, and C_{bc} is sometimes called the output capacitance C_{ob}. C_{be} is specified on data sheets at a certain value of V_{BE}. Often, a data sheet will list C_{ib} as C_{ibo} and C_{ob} as C_{obo}. The o as the last letter in the subscript indicates the capacitance is measured with the base open. For example, a 2N2222A has a C_{be} of 25 pF at $V_{EB} = 0.5$ V dc, $I_C = 0$, and $f = 1$ MHz. Also, C_{bc} is specified at a certain value of V_{CB}. The 2N2222A has a maximum C_{bc} of 8 pF at $V_{CB} = 10$ V dc.

Miller's Theorem in High-Frequency Analysis   By applying Miller's theorem to the circuit in Figure 10–28(b) and using the midrange voltage gain, you have a circuit that can be analyzed for high-frequency response. Looking in from the signal source, the capacitance C_{bc} appears in the Miller input capacitance from base to ground.

$$C_{in(Miller)} = C_{bc}(A_v + 1)$$

C_{be} simply appears as a capacitance to ac ground, as shown in Figure 10–29, in parallel with $C_{in(Miller)}$. Looking in at the collector, C_{bc} appears in the Miller output capacitance from collector to ground. As shown in Figure 10–29, it appears in parallel with R_c.

$$C_{out(Miller)} = C_{bc}\left(\frac{A_v + 1}{A_v}\right)$$

These two Miller capacitances create a high-frequency input RC circuit and a high-frequency output RC circuit. These two circuits differ from the low-frequency input and output circuits, which act as high-pass filters, because the capacitances go to ground and therefore act as low-pass filters. The equivalent circuit in Figure 10–29 is an ideal model because stray capacitances that are due to circuit interconnections are neglected.

▲ **FIGURE 10–29**

High-frequency equivalent circuit after applying Miller's theorem.

The Input RC Circuit   At high frequencies, the input circuit is as shown in Figure 10–30(a), where $\beta_{ac}r'_e$ is the input resistance at the base of the transistor because the bypass capacitor effectively shorts the emitter to ground. By combining C_{be} and $C_{in(Miller)}$ in parallel and repositioning, you get the simplified circuit shown in Figure 10–30(b). Next, by thevenizing the circuit to the left of the capacitor, as indicated, the input RC circuit is reduced to the equivalent form shown in Figure 10–30(c).

(a)

(b)

(c)

▲ **FIGURE 10–30**

Development of the equivalent high-frequency input RC circuit.

As the frequency increases, the capacitive reactance becomes smaller. This causes the signal voltage at the base to decrease, so the amplifier's voltage gain decreases. The reason for this is that the capacitance and resistance act as a voltage divider and, as the frequency increases, more voltage is dropped across the resistance and less across the capacitance. At the critical frequency, the gain is 3 dB less than its midrange value. Just as with the low-frequency response, the critical high frequency, f_c, is the frequency at which the capacitive reactance is equal to the total resistance.

$$X_{C_{tot}} = R_s \| R_1 \| R_2 \| \beta_{ac}r'_e$$

Therefore,

$$\frac{1}{2\pi f_c C_{tot}} = R_s \| R_1 \| R_2 \| \beta_{ac}r'_e$$

and

Equation 10–16

$$f_c = \frac{1}{2\pi(R_s \| R_1 \| R_2 \| \beta_{ac}r'_e)C_{tot}}$$

where R_s is the resistance of the signal source and $C_{tot} = C_{be} + C_{in(Miller)}$. As the frequency goes above f_c, the input RC circuit causes the gain to roll off at a rate of -20 dB/decade just as with the low-frequency response.

EXAMPLE 10–10

Derive the input RC circuit for the BJT amplifier in Figure 10–31. Also determine the critical frequency. The transistor's data sheet provides the following: $\beta_{ac} = 125$, $C_{be} = 20$ pF, and $C_{bc} = 2.4$ pF.

▲ **FIGURE 10–31**

Solution   First, find r'_e as follows:

$$V_B = \left(\frac{R_2}{R_1 + R_2}\right)V_{CC} = \left(\frac{4.7\ \text{k}\Omega}{26.7\ \text{k}\Omega}\right)10\ \text{V} = 1.76\ \text{V}$$

$$V_E = V_B - 0.7\ \text{V} = 1.06\ \text{V}$$

$$I_E = \frac{V_E}{R_E} = \frac{1.06\ \text{V}}{470\ \Omega} = 2.26\ \text{mA}$$

$$r'_e = \frac{25\ \text{mV}}{I_E} = 11.1\ \Omega$$

The total resistance of the input circuit is

$$R_{in(tot)} = R_s \| R_1 \| R_2 \| \beta_{ac}r'_e = 600\ \Omega \| 22\ \text{k}\Omega \| 4.7\ \text{k}\Omega \| 125(11.1\ \Omega) = 378\ \Omega$$

Next, in order to determine the capacitance, you must calculate the midrange gain of the amplifier so that you can apply Miller's theorem.

$$A_{v(mid)} = \frac{R_c}{r'_e} = \frac{R_C \| R_L}{r'_e} = \frac{1.1\ \text{k}\Omega}{11.1\ \Omega} = 99$$

Apply Miller's theorem.

$$C_{in(Miller)} = C_{bc}(A_{v(mid)} + 1) = (2.4\ \text{pF})(100) = 240\ \text{pF}$$

The total input capacitance is $C_{in(Miller)}$ in parallel with C_{be}.

$$C_{in(tot)} = C_{in(Miller)} + C_{be} = 240\ \text{pF} + 20\ \text{pF} = 260\ \text{pF}$$

The resulting high-frequency input RC circuit is shown in Figure 10–32. The critical frequency is

$$f_c = \frac{1}{2\pi(R_{in(tot)})(C_{in(tot)})} = \frac{1}{2\pi(378\ \Omega)(260\ \text{pF})} = \mathbf{1.62\ MHz}$$

$R_s \parallel R_1 \parallel R_2 \parallel \beta_{ac}r'_e = 378 \, \Omega$

Base

V_{th}

$C_{be} + C_{in(Miller)} = 260 \, pF$

▲ **FIGURE 10–32**

High-frequency equivalent input *RC* circuit for the amplifier in Figure 10–31.

Related Problem   Determine the input *RC* circuit for Figure 10–31 and find its critical frequency if a transistor with the following specifications is used: $\beta_{ac} = 75$, $C_{be} = 15 \, pF$, $C_{bc} = 2 \, pF$.

Open file E10-10 in the Examples folder on your CD-ROM. Measure the critical frequency for the amplifier's high-frequency response and compare to the calculated result.

Phase shift of the input RC circuit   Because the output voltage of a high-frequency input *RC* circuit is across the capacitor, the output of the circuit lags the input. The phase angle is expressed as

Equation 10–17

$$\theta = \tan^{-1}\left(\frac{R_s \parallel R_1 \parallel R_2 \parallel \beta_{ac}r'_e}{X_{C_{tot}}}\right)$$

At the critical frequency, the phase angle is 45° with the signal voltage at the base of the transistor lagging the input signal. As the frequency increases above f_c, the phase angle increases above 45° and approaches 90° when the frequency is sufficiently high.

The Output RC Circuit   The high-frequency output *RC* circuit is formed by the Miller output capacitance and the resistance looking in at the collector, as shown in Figure 10–33(a). In determining the output resistance, the transistor is treated as a current source (open) and one end of R_C is effectively ac ground, as shown in Figure 10–33(b). By rearranging the position of the capacitance in the diagram and thevenizing the circuit to the left, as shown in Figure 10–33(c), you get the equivalent circuit in Figure 10–33(d). The equivalent output *RC* circuit consists of a resistance equal to R_C and R_L in parallel in series with a capacitance which is determined by the following Miller formula:

$$C_{out(Miller)} = C_{bc}\left(\frac{A_v + 1}{A_v}\right)$$

If the voltage gain is at least 10, this formula can be approximated as

$$C_{out(Miller)} \cong C_{bc}$$

The critical frequency is determined with the following equation, where $R_c = R_C \parallel R_L$.

Equation 10–18

$$f_c = \frac{1}{2\pi R_c C_{out(Miller)}}$$

(a)

(b)

(c)

(d)

▲ **FIGURE 10–33**

Development of the equivalent high-frequency output *RC* circuit.

Just as in the input *RC* circuit, the output *RC* circuit reduces the gain by 3 dB at the critical frequency. When the frequency goes above the critical value, the gain drops at a −20 dB/decade rate. The phase shift introduced by the output *RC* circuit is

$$\theta = \tan^{-1}\left(\frac{R_c}{X_{C_{out(Miller)}}}\right)$$

Equation 10–19

EXAMPLE 10–11

Determine the critical frequency of the amplifier in Example 10–10 (Figure 10–31) due to its output *RC* circuit.

Solution   The Miller output capacitance is calculated as follows:

$$C_{out(Miller)} = C_{bc}\left(\frac{A_v + 1}{A_v}\right) = (2.4 \text{ pF})\left(\frac{99 + 1}{99}\right) \cong 2.4 \text{ pF}$$

The equivalent resistance is

$$R_c = R_C \| R_L = 2.2 \text{ k}\Omega \| 2.2 \text{ k}\Omega = 1.1 \text{ k}\Omega$$

The equivalent output *RC* circuit is shown in Figure 10–34, and the critical frequency is determined as follows ($C_{out(Miller)} \cong C_{bc}$):

$$f_c = \frac{1}{2\pi R_c C_{bc}} = \frac{1}{2\pi(1.1 \text{ k}\Omega)(2.4 \text{ pF})} = \textbf{60.3 MHz}$$

▶ FIGURE 10–34

Related Problem   If another transistor with $C_{bc} = 5$ pF is used in the amplifier, what is f_c?

FET Amplifiers

The approach to the high-frequency analysis of a FET amplifier is very similar to that of a BJT amplifier. The basic differences are the specifications of the internal FET capacitances and the determination of the input resistance.

Figure 10–35(a) shows a JFET common-source amplifier that will be used to illustrate high-frequency analysis. A high-frequency equivalent circuit for the amplifier is shown in Figure 10–35(b). Notice that the coupling and bypass capacitors are assumed to have negligible reactances and are considered to be shorts. The internal capacitances C_{gs} and C_{gd} appear in the equivalent circuit because their reactances are significant at high frequencies.

(a)   (b)

▲ **FIGURE 10–35**

Example of a JFET amplifier and its high-frequency equivalent circuit.

Values of C_{gs}, C_{gd}, and C_{ds}   FET data sheets do not normally provide values for C_{gs}, C_{gd}, or C_{ds}. Instead, three other values are usually specified because they are easier to measure. These are C_{iss}, the input capacitance; C_{rss}, the reverse transfer capacitance; and C_{oss}, the output capacitance. Because of the manufacturer's method of measurement, the following relationships allow you to determine the capacitor values needed for analysis.

Equation 10–20 $C_{gd} = C_{rss}$

$$C_{gs} = C_{iss} - C_{rss}$$

Equation 10–21

$$C_{ds} = C_{oss} - C_{rss}$$

Equation 10–22

C_{oss} is not specified as often as the other values on data sheets. Sometimes, it is designated as $C_{d(sub)}$, the drain-to-substrate capacitance. In cases where a value is not available, you must either assume a value or neglect C_{ds}.

EXAMPLE 10–12

The data sheet for a 2N3823 JFET gives $C_{iss} = 6$ pF and $C_{rss} = 2$ pF. Determine C_{gd} and C_{gs}.

Solution

$$C_{gd} = C_{rss} = \mathbf{2\ pF}$$
$$C_{gs} = C_{iss} - C_{rss} = 6\ pF - 2\ pF = \mathbf{4\ pF}$$

Related Problem Although C_{oss} is not specified on the data sheet for the 2N3823 JFET, assume a value of 3 pF and determine C_{ds}.

Using Miller's Theorem Miller's theorem is applied the same way in FET amplifier high-frequency analysis as was done in BJT amplifiers. Looking in from the signal source in Figure 10–35(b), C_{gd} effectively appears in the Miller input capacitance, which was given in Equation 10–1, as follows:

$$C_{in(Miller)} = C_{gd}(A_v + 1)$$

C_{gs} simply appears as a capacitance to ac ground in parallel with $C_{in(Miller)}$, as shown in Figure 10–36. Looking in at the drain, C_{gd} effectively appears in the Miller output capacitance (from Equation 10–2) from drain to ground in parallel with R_d, as shown in Figure 10–36.

$$C_{out(Miller)} = C_{gd}\left(\frac{A_v + 1}{A_v}\right)$$

These two Miller capacitances contribute to a high-frequency input RC circuit and a high-frequency output RC circuit. Both are low-pass filters which produce phase lag.

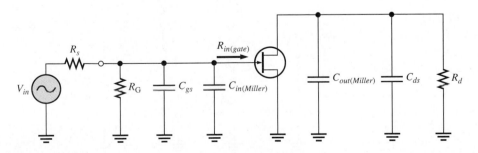

▲ **FIGURE 10–36**

High-frequency equivalent circuit after applying Miller's theorem.

The Input RC Circuit The high-frequency input circuit forms a low-pass type of filter and is shown in Figure 10–37(a). Because both R_G and the input resistance at the gate of FETs are extremely high, the controlling resistance for the input circuit is the resistance of the input source as long as $R_s \ll R_{in}$. This is because R_s appears in parallel with R_{in} when Thevenin's theorem is applied. The simplified input *RC* circuit appears in Figure 10–37(b). The critical frequency is

Equation 10–23

$$f_c = \frac{1}{2\pi R_s C_{tot}}$$

where $C_{tot} = C_{gs} + C_{in(Miller)}$. The input *RC* circuit produces a phase shift of

Equation 10–24

$$\theta = \tan^{-1}\left(\frac{R_s}{X_{C_{tot}}}\right)$$

The effect of the input *RC* circuit is to reduce the midrange gain of the amplifier by 3 dB at the critical frequency and to cause the gain to decrease at -20 dB/decade above f_c.

(a) Thevenizing

(b) Thevenin equivalent input circuit, neglecting R_{in}

▲ **FIGURE 10–37**

Input *RC* circuit.

EXAMPLE 10–13

Find the critical frequency of the input *RC* circuit for the FET amplifier in Figure 10–38. $C_{iss} = 8$ pF, $C_{rss} = 3$ pF, and $g_m = 6500$ μS.

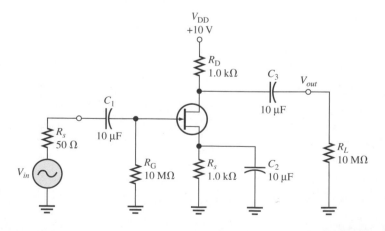

▲ **FIGURE 10–38**

Solution Determine C_{gd} and C_{gs}.

$$C_{gd} = C_{rss} = 3 \text{ pF}$$
$$C_{gs} = C_{iss} - C_{rss} = 8 \text{ pF} - 3 \text{ pF} = 5 \text{ pF}$$

Determine the critical frequency for the input RC circuit as follows:

$$A_v = g_m R_d = g_m(R_D \parallel R_L) \cong (6500 \text{ } \mu\text{S})(1 \text{ k}\Omega) = 6.5$$
$$C_{in(Miller)} = C_{gd}(A_v + 1) = (3 \text{ pF})(7.5) = 22.5 \text{ pF}$$

The total input capacitance is

$$C_{in(tot)} = C_{gs} + C_{in(Miller)} = 5 \text{ pF} + 22.5 \text{ pF} = 27.5 \text{ pF}$$

The critical frequency is

$$f_c = \frac{1}{2\pi R_s C_{tot}} = \frac{1}{2\pi(50 \text{ }\Omega)(27.5 \text{ pF})} = \textbf{116 MHz}$$

Related Problem If the gain of the amplifier in Figure 10–38 is increased to 10, what happens to f_c?

The Output RC Circuit The high-frequency output RC circuit is formed by the Miller output capacitance and the output resistance looking in at the drain, as shown in Figure 10–39(a). As in the case of the BJT, the FET is treated as a current source. When you apply Thevenin's theorem, you get an equivalent output RC circuit consisting of R_D in parallel with R_L and an equivalent output capacitance.

$$C_{out(Miller)} = C_{gd}\left(\frac{A_v + 1}{A_v}\right)$$

This equivalent output circuit is shown in Figure 10–39(b). The critical frequency of the output RC lag circuit is

$$f_c = \frac{1}{2\pi R_d C_{out(Miller)}}$$

Equation 10–25

The output circuit produces a phase shift of

$$\theta = \tan^{-1}\left(\frac{R_d}{X_{C_{out(Miller)}}}\right)$$

Equation 10–26

(a)

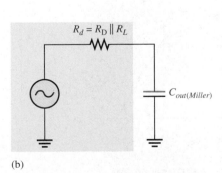

(b)

▲ **FIGURE 10–39**

Output RC circuit.

EXAMPLE 10–14

Determine the critical frequency of the output RC circuit for the FET amplifier in Figure 10–38. What is the phase shift introduced by this circuit at the critical frequency? Which RC circuit is dominant?

Solution Since R_L is very large compared to R_D, it can be neglected, and the equivalent output resistance is

$$R_d \cong R_D = 1.0 \text{ k}\Omega$$

The equivalent output capacitance is

$$C_{out(Miller)} = C_{gd}\left(\frac{A_v + 1}{A_v}\right) = (3 \text{ pF})\left(\frac{7.5}{6.5}\right) = 3.46 \text{ pF}$$

Therefore, the critical frequency is

$$f_c = \frac{1}{2\pi R_d C_{out(Miller)}} = \frac{1}{2\pi(1.0 \text{ k}\Omega)(3.46 \text{ pF})} = \textbf{46 MHz}$$

Although it has been neglected, any stray wiring capacitance could significantly affect the frequency response because $C_{out(Miller)}$ is very small.

The phase angle is always **45°** at f_c for an RC circuit and the output lags.

In Example 10–13, the critical frequency of the input RC circuit was found to be 116 MHz. Therefore, the critical frequency for the output circuit is dominant since it is the lesser of the two.

Related Problem If A_v of the amplifier in Figure 10–38 is increased to 10, what is the f_c of the output circuit?

Total High-Frequency Response of an Amplifier

As you have seen, the two RC circuits created by the internal transistor capacitances influence the high-frequency response of both BJT and FET amplifiers. As the frequency increases and reaches the high end of its midrange values, one of the RC circuits will cause the amplifier's gain to begin dropping off. The frequency at which this occurs is the dominant critical frequency; it is the lower of the two critical high frequencies. An ideal high-frequency Bode plot is shown in Figure 10–40(a). It shows the first break

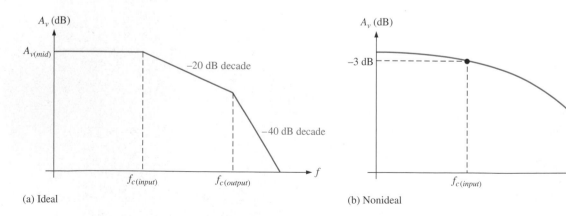

(a) Ideal

(b) Nonideal

▲ **FIGURE 10–40**

High-frequency Bode plots.

point at $f_{c(input)}$ where the voltage gain begins to roll off at -20 dB/decade. At $f_{c(output)}$, the gain begins dropping at -40 dB/decade because each RC circuit is providing a -20 dB/decade roll-off. Figure 10–40(b) shows a nonideal Bode plot where the voltage gain is actually -3 dB below midrange at $f_{c(input)}$. Other possibilities are that the output RC circuit is dominant or that both circuits have the same critical frequency.

SECTION 10–4
REVIEW

1. What determines the high-frequency response of an amplifier?
2. If an amplifier has a midrange voltage gain of 80, the transistor's C_{bc} is 4 pF, and $C_{be} = 8$ pF, what is the total input capacitance?
3. A certain amplifier has $f_{c(input)} = 3.5$ MHz and $f_{c(output)} = 8.2$ MHz. Which circuit dominates the high-frequency response?
4. What are the capacitances that are usually specified on a FET data sheet?
5. If $C_{gs} = 4$ pF and $C_{gd} = 3$ pF, what is the total input capacitance of a FET amplifier whose voltage gain is 25?

10–5   TOTAL AMPLIFIER FREQUENCY RESPONSE

In the previous sections, you learned how each RC circuit in an amplifier affects the frequency response. In this section, we will bring these concepts together and examine the total response of typical amplifiers and the specifications relating to their performance.

After completing this section, you should be able to

- **Analyze an amplifier for total frequency response**
- Determine the bandwidth
- Define *gain-bandwidth product*
- Explain the half-power points

Figure 10–41(b) shows a generalized ideal response curve (Bode plot) for the BJT amplifier shown in Figure 10–41(a). As previously discussed, the three break points at the lower critical frequencies (f_{c1}, f_{c2}, and f_{c3}) are produced by the three low-frequency RC circuits formed by the coupling and bypass capacitors. The break points at the upper critical frequencies, f_{c4} and f_{c5}, are produced by the two high-frequency RC circuits formed by the transistor's internal capacitances.

Of particular interest are the two dominant critical frequencies, f_{c3} and f_{c4}, in Figure 10–41(b). These two frequencies are where the voltage gain of the amplifier is 3 dB below its midrange value. These dominant frequencies are referred to as the *lower critical frequency*, f_{cl}, and the *upper critical frequency*, f_{cu}.

▲ FIGURE 10–41

A BJT amplifier and its generalized ideal response curve (Bode plot).

Bandwidth

An amplifier normally operates with signal frequencies between f_{cl} and f_{cu}. As you know, when the input signal frequency is at f_{cl} or f_{cu}, the output signal voltage level is 70.7 percent of its midrange value or -3 dB. If the signal frequency drops below f_{cl}, the gain and thus the output signal level drops at 20 dB/decade until the next critical frequency is reached. The same occurs when the signal frequency goes above f_{cu}.

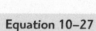

The range (band) of frequencies lying between f_{cl} and f_{cu} is defined as the **bandwidth** of the amplifier, as illustrated in Figure 10–42. Only the dominant critical frequencies appear in the response curve because they determine the bandwidth. Also, sometimes the other critical frequencies are far enough away from the dominant frequencies that they play no significant role in the total amplifier response and can be neglected. The amplifier's bandwidth is expressed in units of hertz as

$$BW = f_{cu} - f_{cl}$$

Ideally, all signal frequencies lying in an amplifier's bandwidth are amplified equally. For example, if a 10 mV rms signal is applied to an amplifier with a voltage gain of 20, it is amplified to 200 mV rms for all frequencies in the bandwidth. Actually, the gain is down 3 dB at f_{cl} and f_{cu}.

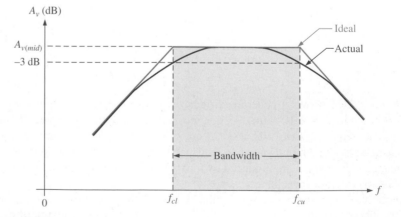

▲ FIGURE 10–42

Response curve illustrating the bandwidth of an amplifier.

EXAMPLE 10–15

What is the bandwidth of an amplifier having an f_{cl} of 200 Hz and an f_{cu} of 2 kHz?

Solution

$$BW = f_{cu} - f_{cl} = 2000 \text{ Hz} - 200 \text{ Hz} = \mathbf{1800 \text{ Hz}}$$

Notice that bandwidth has the unit of hertz.

Related Problem If f_{cl} is increased, does the bandwidth increase or decrease? If f_{cu} is increased, does the bandwidth increase or decrease?

Gain-Bandwidth Product

One characteristic of amplifiers is that the product of the voltage gain and the bandwidth is always constant when the roll-off is -20 dB/decade. This characteristic is called the **gain-bandwidth product**. Let's assume that the lower critical frequency of a particular amplifier is much less than the upper critical frequency.

$$f_{cl} << f_{cu}$$

The bandwidth can then be approximated as

$$BW = f_{cu} - f_{cl} \cong f_{cu}$$

Unity-Gain Frequency The simplified Bode plot for this condition is shown in Figure 10–43. Notice that f_{cl} is neglected because it is so much smaller than f_{cu}, and the bandwidth approximately equals f_{cu}. Beginning at f_{cu}, the gain rolls off until unity gain (0 dB) is reached. The frequency at which the amplifier's gain is 1 is called the *unity-gain frequency, f_T*. The significance of f_T is that it always equals the product of the midrange voltage gain times the bandwidth and is constant for a given transistor.

$$f_T = A_{v(mid)}BW$$

Equation 10–28

For the case shown in Figure 10–43, $f_T = A_{v(mid)}f_{cu}$. For example, if a transistor data sheet specifies $f_T = 100$ MHz, this means that the transistor is capable of producing a voltage gain of 1 up to 100 MHz, or a gain of 100 up to 1 MHz, or any combination of gain and bandwidth that produces a product of 100 MHz.

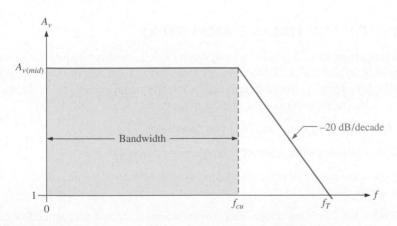

▲ **FIGURE 10–43**

Simplified response curve where f_{cl} is negligible (assumed to be zero) compared to f_{cu}.

EXAMPLE 10–16

A certain transistor has an f_T of 175 MHz. When this transistor is used in an amplifier with a midrange voltage gain of 50, what bandwidth can be achieved ideally?

Solution

$$f_T = A_{v(mid)}BW$$

$$BW = \frac{f_T}{A_{v(mid)}} = \frac{175 \text{ MHz}}{50} = \textbf{3.5 MHz}$$

Related Problem An amplifier has a midrange voltage gain of 20 and a bandwidth of 1 MHz. What is the f_T of the transistor?

Half-Power Points

The upper and lower critical frequencies are sometimes called the *half-power frequencies.* This term is derived from the fact that the output power of an amplifier at its critical frequencies is one-half of its midrange power, as previously mentioned. This can be shown as follows, starting with the fact that the output voltage is 0.707 of its midrange value at the critical frequencies.

$$V_{out(f_c)} = 0.707V_{out(mid)}$$

$$P_{out(f_c)} = \frac{V_{out(f_c)}^2}{R_{out}} = \frac{(0.707V_{out(mid)})^2}{R_{out}} = \frac{0.5V_{out(mid)}^2}{R_{out}} = 0.5P_{out(mid)}$$

**SECTION 10–5
REVIEW**

1. What is the voltage gain of an amplifier at f_T?
2. What is the bandwidth of an amplifier when $f_{cu} = 25$ kHz and $f_{cl} = 100$ Hz?
3. The f_T of a certain transistor is 130 MHz. What voltage gain can be achieved with a bandwidth of 50 MHz?

10–6 FREQUENCY RESPONSE OF MULTISTAGE AMPLIFIERS

To this point, you have seen how the voltage gain of a single-stage amplifier changes over frequency. When two or more stages are cascaded to form a multistage amplifier, the overall frequency response is determined by the frequency response of each stage depending on the relationships of the critical frequencies.

After completing this section, you should be able to

■ **Analyze multistage amplifiers for frequency response**

   ■ Determine the dominant critical frequencies when the critical frequencies of each stage are the same

   ■ Determine the dominant critical frequencies when the critical frequencies of each stage differ

   ■ Determine the bandwidth of a multistage amplifier

When amplifier stages are cascaded to form a multistage amplifier, the dominant frequency response is determined by the responses of the individual stages. There are two cases to consider:

1. Each stage has a different lower critical frequency and a different upper critical frequency.

2. Each stage has the same lower critical frequency and the same upper critical frequency.

Different Critical Frequencies

When the lower critical frequency, f_{cl}, of each amplifier stage is different, the dominant lower critical frequency, f'_{cl}, equals the critical frequency of the stage with the highest f_{cl}.

When the upper critical frequency, f_{cu}, of each amplifier stage is different, the dominant upper critical frequency, f'_{cu}, equals the critical frequency of the stage with the lowest f_{cu}.

Overall Bandwidth The bandwidth of a multistage amplifier is the difference between the dominant lower critical frequency and the dominant upper critical frequency.

$$BW = f'_{cu} - f'_{cl}$$

EXAMPLE 10–17

In a certain 2-stage amplifier, one stage has a lower critical frequency of 850 Hz and an upper critical frequency of 100 kHz. The other has a lower critical frequency of 1 kHz and an upper critical frequency of 230 kHz. Determine the overall bandwidth of the 2-stage amplifier.

Solution

$$f'_{cl} = 1 \text{ kHz}$$
$$f'_{cu} = 100 \text{ kHz}$$
$$BW = f'_{cu} - f'_{cl} = 100 \text{ kHz} - 1 \text{ kHz} = \textbf{99 kHz}$$

Related Problem A certain 3-stage amplifier has the following critical frequencies for each stage: $f_{cl(1)} = 1500$ Hz, $f_{cl(2)} = 980$ Hz, and $f_{cl(3)} = 130$ Hz. What is the dominant lower critical frequency?

Equal Critical Frequencies

When each amplifier stage in a multistage arrangement has equal critical frequencies, you may think that the dominant critical frequency is equal to the critical frequency of each stage. This is not the case, however.

When the lower critical frequencies of each stage in a multistage amplifier are all the same, the dominant lower critical frequency is increased by a factor of $1/\sqrt{2^{1/n} - 1}$ as shown by the following formula (n is the number of stages in the multistage amplifier):

$$f'_{cl} = \frac{f_{cl}}{\sqrt{2^{1/n} - 1}}$$

Equation 10–29

When the upper critical frequencies of each stage are all the same, the dominant upper critical frequency is reduced by a factor of $\sqrt{2^{1/n} - 1}$, as shown by the following formula:

$$f'_{cu} = f_{cu}\sqrt{2^{1/n} - 1}$$

Equation 10–30

The proofs of these formulas are quite involved and are given in Appendix B.

EXAMPLE 10–18

Both stages in a certain 2-stage amplifier have a lower critical frequency of 500 Hz and an upper critical frequency of 80 kHz. Determine the overall bandwidth.

Solution

$$f'_{cl} = \frac{f_{cl}}{\sqrt{2^{1/n} - 1}} = \frac{500 \text{ Hz}}{\sqrt{2^{0.5} - 1}} = \frac{500 \text{ Hz}}{0.644} = 776 \text{ Hz}$$

$$f'_{cu} = f_{cu}\sqrt{2^{1/n} - 1} = (80 \text{ kHz})(0.644) = 51.5 \text{ kHz}$$

$$BW = f'_{cu} - f'_{cl} = 51.5 \text{ kHz} - 776 \text{ Hz} = \mathbf{50.7 \text{ kHz}}$$

Related Problem

If a third identical stage is connected in cascade to the 2-stage amplifier in this example, what is the resulting overall bandwidth?

SECTION 10–6 REVIEW

1. One stage in an amplifier has $f_{cl} = 1$ kHz and the other stage has $f_{cl} = 325$ Hz. What is the dominant lower critical frequency?

2. In a certain 3-stage amplifier $f_{cu(1)} = 50$ kHz, $f_{cu(2)} = 55$ kHz, and $f_{cu(3)} = 49$ kHz. What is the dominant upper critical frequency?

3. When more identical stages are added to a multistage amplifier with each stage having the same critical frequency, does the bandwidth increase or decrease?

10–7 FREQUENCY RESPONSE MEASUREMENT

In this section, two basic methods of measuring the frequency response of an amplifier are covered. The methods apply to both BJT and FET amplifiers although a BJT amplifier is used as an example. You will concentrate on determining the two dominant critical frequencies. From these values, you can get the bandwidth.

After completing this section, you should be able to

- **Measure the frequency response of an amplifier**

- Use frequency and amplitude measurement to determine the critical frequencies of an amplifier

- Relate pulse characteristics to frequency

- Identify the effects of frequency response on pulse shape

- Use step-response measurement to determine the critical frequencies of an amplifier

Frequency and Amplitude Measurement

Figure 10–44(a) shows the test setup for an amplifier circuit board. The schematic for the circuit board is also shown. The amplifier is driven by a sinusoidal voltage source with a dual-channel oscilloscope connected to the input and to the output. The input frequency is set to a midrange value, and its amplitude is adjusted to establish an output signal reference level, as shown in Figure 10–44(b). This output voltage reference level for midrange should be set at a convenient value within the linear operation of the amplifier, for example, 100 mV, 1 V, 10 V, and so on. In this case, set the output signal to a peak value of 1 V.

(a) Circuit and test setup for measuring the frequency response of an amplifier

Input frequency control
on function generator

Ch1 50 mV   Ch2 0.5 V   50 µs

Amplifier input and output voltages

(b) Frequency set to a midrange value (6.67 kHz in this case).
    Input voltage adjusted for an output of 1 V peak.

Input frequency control
on function generator

Ch1 50 mV   Ch2 0.5 mV   2 ms

Amplifier input and output voltages

(c) Frequency is reduced until the output is 0.707 V peak.
    This is the lower critical frequency.

Input frequency control
on function generator

Ch1 50 mV   Ch2 0.5 mV   1 µs

Amplifier input and output voltages

(d) Frequency is increased until the output is again 0.707 V peak.
    This is the upper critical frequency.

▲ FIGURE 10–44

A general procedure for measuring an amplifier's frequency response.

Next, the frequency of the input voltage is decreased until the peak value of the output drops to 0.707 V. The amplitude of the input voltage must be kept constant as the frequency is reduced. Readjustment may be necessary because of changes in loading of the voltage source with frequency. When the output is 0.707 V, the frequency is measured, and you have the value for f_{cl} as indicated in Figure 10–44(c).

Next, the input frequency is increased back up through midrange and beyond until the peak value of the output voltage again drops to 0.707 V. Again, the amplitude of the input must be kept constant as the frequency is increased. When the output is 0.707 V, the frequency is measured and you have the value for f_{cu} as indicated in Figure 10–44(d). From these two frequency measurements, you can find the bandwidth by the formula $BW = f_{cu} - f_{cl}$.

Step-Response Measurement

The lower and upper critical frequencies of an amplifier can be determined using the *step-response method* by applying a voltage step to the input of the amplifier and measuring the rise and fall times of the resulting output voltage. The basic test setup shown in Figure 10–44(a) is used except that the pulse output of the function generator is selected. The input step is created by the rising edge of a pulse that has a long duration compared to the rise and fall times to be measured.

High-Frequency Measurement   When a step input is applied, the amplifier's high-frequency RC circuits (internal capacitances) prevent the output from responding immediately to the step input. As a result, the output voltage has a rise time (t_r) associated with it, as shown in Figure 10–45(a). In fact, the rise time is inversely related to the upper critical frequency (f_{cu}) of the amplifier. As f_{cu} becomes lower, the rise time of the output becomes greater. The oscilloscope display illustrates how the rise time is measured from the 10% amplitude point to the 90% amplitude point. The scope must be set on a short time base so the relatively short interval of the rise time can be accurately observed. Once this measurement is made, f_{cu} can be calculated with the following formula:

Equation 10–31

$$f_{cu} = \frac{0.35}{t_r}$$

Low-Frequency Measurement   To determine the lower critical frequency (f_{cl}) of the amplifier, the step input must be of sufficiently long duration to observe the full charging time of the low-frequency RC circuits (coupling capacitances), which cause the "sloping" of the output and which we will refer to as the fall time (t_f). This is illustrated in Figure 10–45(b). The fall time is inversely related to the low critical frequency of the amplifier.

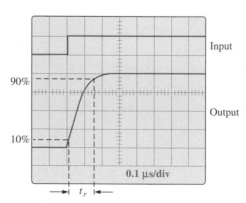

(a) Measurement of output rise time to determine the upper critical frequency

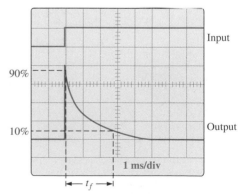

(b) Measurement of output fall time to determine the lower critical frequency

▲ **FIGURE 10–45**

Measurement of the rise and fall times associated with the amplifier's step response. The outputs are inverted.

As f_{cl} becomes higher, the fall time of the output becomes less. The scope display illustrates how the fall time is measured from the 90% point to the 10% point. The scope must be set on a long time base so the complete interval of the fall time can be observed. Once this measurement is made, f_{cl} can be determined with the following formula.

$$f_{cl} = \frac{0.35}{t_f}$$

Equation 10–32

The derivations of Equations 10–31 and 10–32 are in Appendix B.

SECTION 10–7 REVIEW

1. In Figure 10–44, what are the lower and upper critical frequencies?
2. The rise time and the fall time of an amplifier's output voltage are measured between what two points on the voltage transition?
3. In Figure 10–45, what is the rise time?
4. In Figure 10–45, what is the fall time?
5. What is the bandwidth of the amplifier whose step response is measured in Figure 10–45?

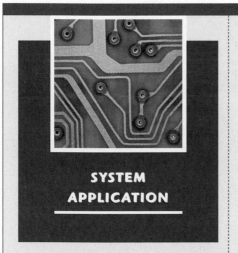

SYSTEM APPLICATION

An audio amplifier circuit board is being finalized for production and will be used in a new intercom system. You are given the board with no documentation and asked to check the frequency response of the amplifier. You will apply the knowledge you have gained in this chapter in completing your assignment.

The Schematic

The amplifier circuit board is shown in Figure 10–46. Draw the schematic of the circuit and label all component values.

▲ **FIGURE 10–46**

Amplifier circuit board. Both transistors are 2N3904.

Electrical Characteristics ($T_A = 25°C$ unless otherwise noted.)

Characteristic		Symbol	Min	Max	Unit
Output capacitance ($V_{CB} = 5.0$ V dc, $I_E = 0$, $f = 1.0$ MHz)		C_{obo}	–	4.0	pF
Input capacitance ($V_{BE} = 0.5$ V dc, $I_C = 0$, $f = 1.0$ MHz)		C_{ibo}	–	8.0	pF
Input impedance ($I_C = 1.0$ mA dc, $V_{CE} = 10$ V dc, $f = 1.0$ kHz)	2N3903 2N3904	h_{ie}	1.0 1.0	8.0 10	k ohms
Voltage feedback ratio ($I_C = 1.0$ mA dc, $V_{CE} = 10$ V dc, $f = 1.0$ kHz)	2N3903 2N3904	h_{re}	0.1 0.5	5.0 8.0	$\times 10^{-4}$
Small-signal current gain ($I_C = 1.0$ mA dc, $V_{CE} = 10$ V dc, $f = 1.0$ kHz)	2N3903 2N3904	h_{fe}	50 100	200 400	–
Output admittance ($I_C = 1.0$ mA dc, $V_{CE} = 10$ V dc, $f = 1.0$ kHz)		h_{oe}	1.0	40	µmhos or µS
Noise figure ($I_C = 100$ µA dc, $V_{CE} = 5.0$ V dc, $R_S = 1.0$ k ohms, $f = 1.0$ kHz)	2N3903 2N3904	NF	– –	6.0 5.0	dB
DC current gain ($I_C = 0.1$ mA dc, $V_{CE} = 1.0$ V dc)	2N3903 2N3904	h_{FE}	20 40	– –	–
($I_C = 1.0$ mA dc, $V_{CE} = 1.0$ V dc)	2N3903 2N3904		35 70	– –	
($I_C = 10$ mA dc, $V_{CE} = 1.0$ V dc)	2N3903 2N3904		50 100	150 300	
($I_C = 50$ mA dc, $V_{CE} = 1.0$ V dc)	2N3903 2N3904		30 60	– –	
($I_C = 100$ mA dc, $V_{CE} = 1.0$ V dc)	2N3903 2N3904		15 30	– –	
Collector-Emitter saturation voltage ($I_C = 10$ mA dc, $I_B = 1.0$ mA dc) ($I_C = 50$ mA dc, $I_B = 5.0$ mA dc)		$V_{CE(sat)}$	– –	0.2 0.3	V dc
Base-Emitter saturation voltage ($I_C = 10$ mA dc, $I_B = 1.0$ mA dc) ($I_C = 50$ mA dc, $I_B = 5.0$ mA dc)		$V_{BE(sat)}$	0.65 –	0.85 0.95	V dc
Small-Signal Characteristics					
Current-gain – Bandwidth product ($I_C = 10$ mA dc, $V_{CE} = 20$ V dc, $f = 100$ MHz)	2N3903 2N3904	f_T	250 300	– –	MHz

▲ **FIGURE 10–47**

Partial data sheet for the 2N3903 and 2N3904 transistors.

Analysis of the Amplifier Circuit

Refer to the partial data sheet in Figure 10–47 when necessary. $V_{CC} = 12$ V. Use a 10 kΩ load resistor on the final output.

■ Determine the lower critical frequency of the first stage.

■ Determine the upper critical frequency of the first stage.

■ Determine the lower critical frequency of the second stage.

■ Determine the upper critical frequency of the second stage.

■ Calculate the dominant lower critical frequency.

■ Calculate the dominant upper critical frequency.

■ Calculate the overall bandwidth.

Test Procedure

■ Specify a procedure for measuring the frequency response of the amplifier board.

■ Indicate the point-to-point hookup with the circled numbers for the circuit board and measuring instruments on the test bench setup of Figure 10–48.

■ Specify voltage and frequency values for all the measurements to be made with the input voltage adjusted to produce a midrange peak output voltage of 1 V.

The Frequency Response on the Test Bench

Frequency response measurements of the amplifier circuit board are shown in Figure 10–49.

■ Assume that the input voltage is held at a constant amplitude over the frequency range.

■ Assume that the frequency response of the oscilloscope is sufficient to accurately measure signals with frequencies well above the upper critical frequency.

■ Determine the correct SEC/DIV settings for each oscilloscope display.

▲ **FIGURE 10–48**

Test bench setup for frequency response measurements of the amplifier board.

Final Report (Optional)

Submit a final written report on the amplifier circuit board using an organized format that includes the following:

1. A description of the circuit.

2. A discussion of the operation of the circuit.

3. A list of the specifications including gain, critical frequencies, and bandwidth.

4. A list of parts with part numbers if available.

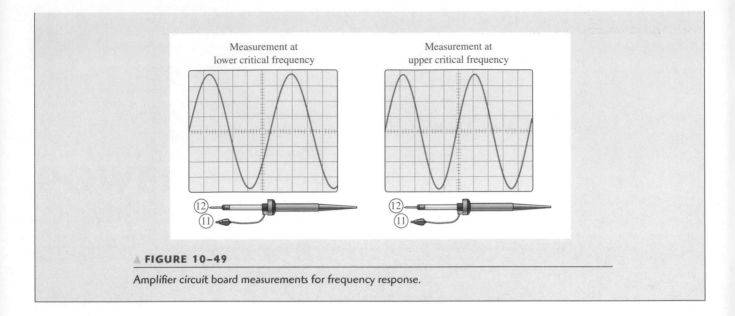

▲ **FIGURE 10–49**

Amplifier circuit board measurements for frequency response.

CHAPTER SUMMARY

- The coupling and bypass capacitors of an amplifier affect the low-frequency response.
- The internal transistor capacitances affect the high-frequency response.
- Critical frequencies are values of frequency at which the RC circuits reduce the voltage gain to 70.7% of its midrange value.
- Each RC circuit causes the gain to drop at a rate of 20 dB/decade.
- For the low-frequency RC circuits, the *highest* critical frequency is the dominant critical frequency.
- For the high-frequency RC circuits, the *lowest* critical frequency is the dominant critical frequency.
- A decade of frequency change is a ten-times change (increase or decrease).
- An octave of frequency change is a two-times change (increase or decrease).
- The bandwidth of an amplifier is the range of frequencies between the lower critical frequency and the upper critical frequency.
- The gain-bandwidth product is a transistor parameter that is constant and equal to the unity-gain frequency.

KEY TERMS

Key terms and other bold terms in the chapter are defined in the end-of-book glossary.

Bandwidth   The characteristic of certain types of electronic circuits that specifies the usable range of frequencies that pass from input to output.

Bode plot   An idealized graph of the gain in dB versus frequency used to graphically illustrate the response of an amplifier or filter.

Critical frequency   The frequency at which the response of an amplifier or filter is 3 dB less than at midrange.

Decade   A ten-times increase or decrease in the value of a quantity such as frequency.

Decibel   A logarithmic measure of the ratio of one power to another or one voltage to another.

Midrange   The portion of a response curve between the lower and upper critical frequencies.

Roll-off   The decrease in the gain of an amplifier above or below the critical frequencies.

Miller's Theorem

10–1 $\qquad C_{in(Miller)} = C(A_v + 1)$ $\qquad\qquad$ Miller input capacitance, where $C = C_{bc}$ or C_{gd}

10–2 $\qquad C_{out(Miller)} = C\left(\dfrac{A_v + 1}{A_v}\right)$ $\qquad$ Miller output capacitance, where $C = C_{bc}$ or C_{gd}

The Decibel

10–3 $\qquad A_{p(dB)} = 10 \log A_p$ $\qquad\qquad$ Power gain in decibels

10–4 $\qquad A_{v(dB)} = 20 \log A_v$ $\qquad\qquad$ Voltage gain in decibels

BJT Amplifier Low-Frequency Response

10–5 $\qquad A_{v(mid)} = \dfrac{R_c}{r_e'}$ $\qquad\qquad$ Midrange voltage gain

10–6 $\qquad f_c = \dfrac{1}{2\pi R_{in}C_1}$ $\qquad\qquad$ Critical frequency, input RC circuit

10–7 $\qquad \theta = \tan^{-1}\left(\dfrac{X_{C1}}{R_{in}}\right)$ $\qquad\qquad$ Phase angle, input RC circuit

10–8 $\qquad f_c = \dfrac{1}{2\pi(R_C + R_L)C_3}$ $\qquad\qquad$ Critical frequency, output RC circuit

10–9 $\qquad \theta = \tan^{-1}\left(\dfrac{X_{C3}}{R_C + R_L}\right)$ $\qquad\qquad$ Phase angle, output RC circuit

10–10 $\qquad R_{in(emitter)} = \dfrac{R_{th}}{\beta_{ac}} + r_e'$ $\qquad\qquad$ Resistance looking in at emitter

10–11 $\qquad f_c = \dfrac{1}{2\pi[(r_e' + R_{th}/\beta_{ac}) \| R_E]C_2}$ $\qquad$ Critical frequency, bypass RC circuit

FET Amplifier Low-Frequency Response

10–12 $\qquad f_c = \dfrac{1}{2\pi(R_G \| R_{in(gate)})C_1}$ $\qquad\qquad$ Critical frequency, input RC circuit

10–13 $\qquad \theta = \tan^{-1}\left(\dfrac{X_{C1}}{R_{in}}\right)$ $\qquad\qquad$ Phase angle, input RC circuit

10–14 $\qquad f_c = \dfrac{1}{2\pi(R_D + R_L)C_2}$ $\qquad\qquad$ Critical frequency, output RC circuit

10–15 $\qquad \theta = \tan^{-1}\left(\dfrac{X_{C2}}{R_D + R_L}\right)$ $\qquad\qquad$ Phase angle, output RC circuit

BJT Amplifier High-Frequency Response

10–16 $\qquad f_c = \dfrac{1}{2\pi(R_s \| R_1 \| R_2 \| \beta_{ac}r_e')C_{tot}}$ $\qquad$ Critical frequency, input RC circuit

10–17 $\qquad \theta = \tan^{-1}\left(\dfrac{R_s \| R_1 \| R_2 \| \beta_{ac}r_e'}{X_{C_{tot}}}\right)$ $\qquad$ Phase angle, input RC circuit

10–18 $\qquad f_c = \dfrac{1}{2\pi R_c C_{out(Miller)}}$ $\qquad\qquad$ Critical frequency, output RC circuit

10–19 $\qquad \theta = \tan^{-1}\left(\dfrac{R_c}{X_{C_{out(Miller)}}}\right)$ $\qquad\qquad$ Phase angle, output RC circuit

FET Amplifier High-Frequency Response

10–20 $\qquad C_{gd} = C_{rss}$ $\qquad\qquad$ Gate-to-drain capacitance

10–21 $\qquad C_{gs} = C_{iss} - C_{rss}$ $\qquad\qquad$ Gate-to-source capacitance

10–22	$C_{ds} = C_{oss} - C_{rss}$	Drain-to-source capacitance
10–23	$f_c = \dfrac{1}{2\pi R_s C_{tot}}$	Critical frequency, input RC circuit
10–24	$\theta = \tan^{-1}\left(\dfrac{R_s}{X_{C_{tot}}}\right)$	Phase angle, input RC circuit
10–25	$f_c = \dfrac{1}{2\pi R_d C_{out(Miller)}}$	Critical frequency, output RC circuit
10–26	$\theta = \tan^{-1}\left(\dfrac{R_d}{X_{C_{out(Miller)}}}\right)$	Phase angle, output RC circuit

Total Response

10–27	$BW = f_{cu} - f_{cl}$	Bandwidth
10–28	$f_T = A_{v(mid)}BW$	Gain-bandwidth product

Multistage Response

10–29	$f'_{cl} = \dfrac{f_{cl}}{\sqrt{2^{1/n} - 1}}$	Lower critical frequency for case of equal critical frequencies
10–30	$f'_{cu} = f_{cu}\sqrt{2^{1/n} - 1}$	Upper critical frequency for case of equal critical frequencies

Measurement Techniques

10–31	$f_{cu} = \dfrac{0.35}{t_r}$	Upper critical frequency
10–32	$f_{cl} = \dfrac{0.35}{t_f}$	Lower critical frequency

SELF-TEST

Answers are at the end of the chapter.

1. The low-frequency response of an amplifier is determined in part by
   (a) the voltage gain   (b) the type of transistor
   (c) the supply voltage   (d) the coupling capacitors

2. The high-frequency response of an amplifier is determined in part by
   (a) the gain-bandwidth product
   (b) the bypass capacitor
   (c) the internal transistor capacitances
   (d) the roll-off

3. The bandwidth of an amplifier is determined by
   (a) the midrange gain   (b) the critical frequencies
   (c) the roll-off rate   (d) the input capacitance

4. The gain of a certain amplifier decreases by 6 dB when the frequency is reduced from 1 kHz to 10 Hz. The roll-off is
   (a) −3 dB/decade   (b) −6 dB/decade
   (c) −3 dB/octave   (d) −6 dB/octave

5. The gain of a particular amplifier at a given frequency decreases by 6 dB when the frequency is doubled. The roll-off is
   (a) −12 dB/decade   (b) −20 dB/decade
   (c) −6 dB/octave   (d) answers (b) and (c)

6. The Miller input capacitance of an amplifier is dependent, in part, on

   (a) the input coupling capacitor

   (b) the voltage gain

   (c) the bypass capacitor

   (d) none of these

7. An amplifier has the following critical frequencies: 1.2 kHz, 950 Hz, 8 kHz, and 8.5 kHz. The bandwidth is

   (a) 7550 Hz   (b) 7300 Hz   (c) 6800 Hz   (d) 7050 Hz

8. Ideally, the midrange gain of an amplifier

   (a) increases with frequency

   (b) decreases with frequency

   (c) remains constant with frequency

   (d) depends on the coupling capacitors

9. The frequency at which an amplifier's gain is 1 is called the

   (a) unity-gain frequency   (b) midrange frequency

   (c) corner frequency   (d) break frequency

10. When the voltage gain of an amplifier is increased, the bandwidth

    (a) is not affected   (b) increases

    (c) decreases   (d) becomes distorted

11. If the f_T of the transistor used in a certain amplifier is 75 MHz and the bandwidth is 10 MHz, the voltage gain must be

    (a) 750   (b) 7.5   (c) 10   (d) 1

12. In the midrange of an amplifier's bandwidth, the peak output voltage is 6 V. At the lower critical frequency, the peak output voltage is

    (a) 3 V   (b) 3.82 V   (c) 8.48 V   (d) 4.24 V

13. At the upper critical frequency, the peak output voltage of a certain amplifier is 10 V. The peak voltage in the midrange of the amplifier is

    (a) 7.07 V   (b) 6.37 V   (c) 14.14 V   (d) 10 V

14. In the step response of a noninverting amplifier, a longer rise time means

    (a) a narrower bandwidth   (b) a lower f_{cl}

    (c) a higher f_{cu} (d) answers (a) and (b)

15. The lower critical frequency of a direct-coupled amplifier with no bypass capacitor is

    (a) variable   (b) 0 Hz

    (c) dependent on the bias   (d) none of these

PROBLEMS

Answers to all odd-numbered problems are at the end of the book.

BASIC PROBLEMS

SECTION 10–1   **Basic Concepts**

1. In a capacitively coupled amplifier, the input coupling capacitor and the output coupling capacitor form two of the circuits (along with the respective resistances) that determine the low-frequency response. Assuming that the input and output impedances are the same and neglecting the bypass circuit, which circuit will first cause the gain to drop from its midrange value as the frequency is lowered?

2. Explain why the coupling capacitors do not have a significant effect on gain at sufficiently high-signal frequencies.

3. List the capacitances that affect high-frequency gain in both BJT and FET amplifiers.

4. In the amplifier of Figure 10–50, list the capacitances that affect the low-frequency response of the amplifier and those that affect the high-frequency response.

5. Determine the Miller input capacitance in Figure 10–50.

6. Determine the Miller output capacitance in Figure 10–50.

7. Determine the Miller input and output capacitances for the amplifier in Figure 10–51.

▲ **FIGURE 10–50**

File circuits are identified with a CD logo and are in the Problems folder on your CD-ROM. File-names correspond to figure numbers (e.g., F10-50).

▲ **FIGURE 10–51**

SECTION 10–2   The Decibel

8. A certain amplifier exhibits an output power of 5 W with an input power of 0.5 W. What is the power gain in dB?

9. If the output voltage of an amplifier is 1.2 V rms and its voltage gain is 50, what is the rms input voltage? What is the gain in dB?

10. The midrange voltage gain of a certain amplifier is 65. At a certain frequency beyond mid-range, the gain drops to 25. What is the gain reduction in dB?

11. What are the dBm values corresponding to the following power values?

    (a) 2 mW   (b) 1 mW   (c) 4 mW   (d) 0.25 mW

12. Express the midrange voltage gain of the amplifier in Figure 10–50 in decibels. Also express the voltage gain in dB for the critical frequencies.

SECTION 10–3 Low-Frequency Amplifier Response

13. Determine the critical frequencies of each *RC* circuit in Figure 10–52.

(a)   **(b)**

▲ **FIGURE 10–52**

14. Determine the critical frequencies associated with the low-frequency response of the BJT amplifier in Figure 10–53. Which is the dominant critical frequency? Sketch the Bode plot.

15. Determine the voltage gain of the amplifier in Figure 10–53 at one-tenth of the dominant critical frequency, at the dominant critical frequency, and at ten times the dominant critical frequency for the low-frequency response.

16. Determine the phase shift at each of the frequencies used in Problem 15.

▶ **FIGURE 10–53**

17. Determine the critical frequencies associated with the low-frequency response of the FET amplifier in Figure 10–54. Indicate the dominant critical frequency and draw the Bode plot.

▲ **FIGURE 10–54**

18. **(a)** Find the voltage gain of the amplifier in Figure 10–54 at the following frequencies: f_c, $0.1f_c$, and $10f_c$, where f_c is the dominant critical frequency.

    (b) Find the phase shift at each frequency in part (a).

SECTION 10–4 High-Frequency Amplifier Response

19. Determine the critical frequencies associated with the high-frequency response of the amplifier in Figure 10–53. Identify the dominant critical frequency and sketch the Bode plot.

20. Determine the voltage gain of the amplifier in Figure 10–53 at the following frequencies: $0.1f_c$, f_c, $10f_c$, and $100f_c$, where f_c is the dominant critical frequency in the high-frequency response.

21. The data sheet for the FET in Figure 10–54 gives $C_{rss} = 4$ pF and $C_{iss} = 10$ pF. Determine the critical frequencies associated with the high-frequency response of the amplifier, and indicate the dominant frequency.

22. Determine the voltage gain in dB and the phase shift at each of the following multiples of the dominant critical frequency in Figure 10–54 for the high-frequency response: $0.1f_c$, f_c, $10f_c$, and $100f_c$.

SECTION 10–5 Total Amplifier Frequency Response

23. A particular amplifier has the following low critical frequencies: 25 Hz, 42 Hz, and 136 Hz. It also has high critical frequencies of 8 kHz and 20 kHz. Determine the upper and lower critical frequencies.

24. Determine the bandwidth of the amplifier in Figure 10–53.

25. $f_T = 200$ MHz is taken from the data sheet of a transistor used in a certain amplifier. If the midrange gain is determined to be 38 and if f_{cl} is low enough to be neglected compared to f_{cu}, what bandwidth would you expect? What value of f_{cu} would you expect?

26. If the midrange gain of a given amplifier is 50 dB and therefore 47 dB at f_{cu}, how much gain is there at $2f_{cu}$? At $4f_{cu}$? At $10f_{cu}$?

SECTION 10–6 Frequency Response of Multistage Amplifiers

27. In a certain two-stage amplifier, the first stage has critical frequencies of 230 Hz and 1.2 MHz. The second stage has critical frequencies of 195 Hz and 2 MHz. What are the dominant critical frequencies?

28. What is the bandwidth of the two-stage amplifier in Problem 27?

29. Determine the bandwidth of a two-stage amplifier in which each stage has a lower critical frequency of 400 Hz and an upper critical frequency of 800 kHz.

30. What is the dominant lower critical frequency of a three-stage amplifier in which $f_{cl} = 50$ Hz for each stage.

31. In a certain two-stage amplifier, the lower critical frequencies are $f_{cl(1)} = 125$ Hz and $f_{cl(2)} = 125$ Hz, and the upper critical frequencies are $f_{cu(1)} = 3$ MHz and $f_{cu(2)} = 2.5$ MHz. Determine the bandwidth.

SECTION 10–7 Frequency Response Measurement

32. In a step response test of a certain amplifier, $t_r = 20$ ns and $t_f = 1$ ms. Determine f_{cl} and f_{cu}.

33. Suppose you are measuring the frequency response of an amplifier with a signal source and an oscilloscope. Assume that the signal level and frequency are set such that the oscilloscope indicates an output voltage level of 5 V rms in the midrange of the amplifier's response. If you wish to determine the upper critical frequency, indicate what you would do and what scope indication you would look for.

34. Determine the approximate bandwidth of an amplifier from the indicated results of the step-response test in Figure 10–55.

5 μs/div   0.1 ms/div

▲ **FIGURE 10–55**

SYSTEM APPLICATION PROBLEMS

35. Determine the dominant lower critical frequency for the amplifier in Figure 10–46 if the coupling capacitors are changed to 1 μF.

36. Does the change in Problem 35 significantly affect the overall bandwidth?

37. How does a change from 10 kΩ to 100 kΩ in load resistance on the final output of the amplifier in Figure 10–46 affect the dominant lower critical frequency?

38. Determine the step response of the amplifier in Figure 10–46 by specifying the rise and fall times on the output.

DATA SHEET PROBLEMS

39. Referring to the partial data sheet in Figure 10–47, determine the total input capacitance for a 2N3903 amplifier if the voltage gain is 25.

40. A certain amplifier uses a 2N3904 and has a midrange voltage gain of 50. Referring to the partial data sheet in Figure 10–47, determine its minimum bandwidth.

41. The data sheet for a 2N4351 MOSFET specifies the maximum values of internal capacitances as follows: $C_{iss} = 5$ pF, $C_{rss} = 1.3$ pF, and $C_{d(sub)} - 5$ pF. Determine C_{gd}, C_{gs}, and C_{ds}.

ADVANCED PROBLEMS

42. Two single-stage capacitively coupled amplifiers like the one in Figure 10–50 are connected as a two-stage amplifier (R_L is removed from the first stage). Determine whether or not this configuration will operate as a linear amplifier with an input voltage of 10 mV rms. If not, modify the design to achieve maximum gain without distortion.

43. Two stages of the amplifier in Figure 10–54 are connected in cascade. Determine the overall bandwidth.

44. Redesign the amplifier in Figure 10–46 for an adjustable voltage gain of 50 to 500 and a lower critical frequency of 1 kHz.

EWB/MULTISIM TROUBLESHOOTING PROBLEMS

These file circuits are in the Troubleshooting Problems folder on your CD-ROM.

45. Open file TSP10-45 and determine the fault.

46. Open file TSP10-46 and determine the fault.

47. Open file TSP10-47 and determine the fault.

48. Open file TSP10-48 and determine the fault.

ANSWERS

SECTION REVIEWS

SECTION 10–1 Basic Concepts

1. The coupling and bypass capacitors affect the low-frequency gain.
2. The high-frequency gain is limited by internal capacitances.
3. Coupling and bypass capacitors can be neglected at frequencies for which their reactances are negligible.
4. $C_{in(Miller)} = (5 \text{ pF})(51) = 255 \text{ pF}$
5. $C_{out(Miller)} = (3 \text{ pF})(1.04) = 3.12 \text{ pF}$

SECTION 10–2 The Decibel

1. $+12 \text{ dB}$ corresponds to a voltage gain of approximately 4.
2. $A_p = 10 \log(25) = 13.98 \text{ dB}$
3. 0 dBm corresponds to 1 mW.

SECTION 10–3 Low-Frequency Amplifier Response

1. $f_{c2} = 167 \text{ Hz}$ is dominant.
2. $A_{v(dB)} = 50 \text{ dB} - 3 \text{ dB} = 47 \text{ dB}$
3. -20 dB attenuation at one decade below f_c.
4. $\theta = \tan^{-1}(0.5) = 26.6°$
5. $f_c = 1/(2\pi(6500 \ \Omega)(0.0022 \ \mu\text{F})) = 11.1 \text{ kHz}$

SECTION 10–4 High-Frequency Amplifier Response

1. The internal transistor capacitances determine the high-frequency response.
2. $C_{in(tot)} = C_{in(Miller)} + C_{ce} = (4 \text{ pF})(81) + 8 \text{ pF} = 342 \text{ pF}$
3. The input RC circuit dominates.
4. C_{iss} and C_{rss} are usually specified on a FET data sheet.
5. $C_{in(tot)} = (3 \text{ pF})(26) + 4 \text{ pF} = 82 \text{ pF}$

SECTION 10–5 Total Amplifier Frequency Response

1. The gain is 1 at f_T.
2. $BW = 25 \text{ kHz} - 100 \text{ Hz} = 24.9 \text{ kHz}$
3. $A_v = 130 \text{ MHz}/50 \text{ MHz} = 2.6$

SECTION 10–6 Frequency Response of Multistage Amplifiers

1. $f'_{cl} = 1 \text{ kHz}$
2. $f'_{cu} = 49 \text{ kHz}$
3. BW decreases.

SECTION 10–7 Frequency Response Measurement

1. $f_{cl} = 125 \text{ Hz}; f_{cu} = 500 \text{ kHz}$
2. Rise time is between the 10% and 90% points and fall time is between the 90% and 10% points.
3. $t_r = 150 \text{ ns}$
4. $t_f = 2.8 \text{ ms}$
5. Since $f_{cu} >> f_{cl}$, $BW \cong f_{cu} = 2.5 \text{ MHz}$.

RELATED PROBLEMS FOR EXAMPLES

10–1   **(a)** 61.6 dB   **(b)** 17 dB   **(c)** 102 dB

10–2   **(a)** 50 V   **(b)** 6.25 V   **(c)** 1.56 V

10–3   **(a)** 7.23 Hz   **(b)** 0.707   **(c)** 354

10–4   212 @ 400 Hz; 30 @ 40 Hz; 3 @ 4 Hz

10–5   **(a)** 13.2 Hz   **(b)** 0.479   **(c)** 70.7

10–6   13.1 Hz

10–7 f_{cl} changes from 16.2 Hz to 16.1 Hz.

10–8   Ideally, the low-frequency response is not affected because the input RC circuit is unchanged.

10–9   No effect

10–10   320 Ω in series with 215 pF, $f_c = 2.31$ MHz

10–11   28.9 MHz

10–12   1 pF

10–13 f_c decreases to 83.8 MHz.

10–14   48.2 MHz

10–15 BW decreases; BW increases

10–16   20 MHz

10–17   1500 Hz

10–18   39.8 kHz

SELF-TEST

1. (d)   **2.** (c)   **3.** (b)   **4.** (a)   **5.** (d)

6. (b)   **7.** (c)   **8.** (c)   **9.** (a)   **10.** (c)

11. (b)   **12.** (d)   **13.** (c)   **14.** (a)   **15.** (b)

11

THYRISTORS AND OTHER DEVICES

INTRODUCTION

In this chapter, several types of semiconductor devices are introduced. Some or all of the topics in this chapter may be considered optional in those electronics programs where these devices will be studied in more depth in a later industrial electronics course.

First, we discuss a family of devices that are constructed of four semiconductor layers (*pnpn*) known as thyristors. Thyristors include the 4-layer diode, the silicon–controlled rectifier (SCR), the diac, the triac, and the silicon–controlled switch (SCS). These various types of thyristors share certain common characteristics in addition to their four-layer construction. They act as open circuits capable of withstanding a certain rated voltage until they are triggered. When triggered, they turn on and become low-resistance current paths and remain so, even after the trigger is removed, until the current is reduced to a certain level or until they are triggered off, depending on the type of device. Other devices described in this chapter include the unijunction transistor (UJT), the programmable unijunction transistor (PUT), and several optoelectronic devices.

Thyristors can be used to control the amount of ac power to a load and are used in lamp dimmers, motor speed controls, ignition systems, and charging circuits, to name a few. UJTs and PUTs are also used as trigger devices for thyristors and also in oscillators and timing circuits.

- Describe the basic structure and operation of a 4-layer diode

- Describe the basic structure and operation of an SCR

- Discuss several SCR applications

- Describe the basic structure and operation of diacs and triacs

- Describe the basic operation of an SCS

- Describe the basic structure and operation of a UJT

- Describe the structure and operation of a PUT

- Describe a phototransistor and its operation

- Describe the LASCR and its operation

- Discuss various types of optical couplers

KEY TERMS

- 4-layer diode

- Thyristor

- Forward-breakover voltage ($V_{BR(F)}$)

- Holding current (I_H)

- SCR (silicon-controlled rectifier)

- Diac

- Triac

- SCS (silicon-controlled switch)

- UJT (unijunction transistor)

- Standoff ratio

- PUT (programmable unijunction transistor)

- Phototransistor

- LASCR (light-activated silicon-controlled rectifier)

The system application in this chapter is a speed-control system for a production line conveyor. The system senses the number of parts passing a point in a specified period of time and adjusts the rate of movement of the conveyor belt to achieve a desired rate of flow of the parts. The focus is on the conveyor motor speed-control circuit. You will apply the knowledge gained in this chapter to the system application at the end of the chapter.

WWW. VISIT THE COMPANION WEBSITE
Study aids for this chapter are available at
http://www.prenhall.com/floyd

11–1 THE BASIC 4-LAYER DEVICE

The basic thyristor is a 4-layer device with two terminals, the anode and the cathode. It is constructed of four semiconductor layers that form a *pnpn* structure. The device acts as a switch and remains off until the forward voltage reaches a certain value; then it turns on and conducts. Conduction continues until the current is reduced below a specified value. This basic thyristor is also known as a silicon unilateral switch (SUS), Shockley diode, or 4-layer diode.

After completing this section, you should be able to

- **Describe the basic structure and operation of a 4-layer diode**
- Identify the 4-layer diode symbol
- Define *forward-breakover voltage*
- Define *holding current*
- Define *switching current*
- Discuss an application

The **4-layer diode** (also known as Shockley diode and SUS) is a type of **thyristor**, which is a class of devices constructed of four semiconductor layers. The basic construction of a 4-layer diode and its schematic symbol are shown in Figure 11–1.

The *pnpn* structure can be represented by an equivalent circuit consisting of a *pnp* transistor and an *npn* transistor, as shown in Figure 11–2(a). The upper *pnp* layers form Q_1 and the lower *npn* layers form Q_2, with the two middle layers shared by both equivalent transistors. Notice that the base-emitter junction of Q_1 corresponds to *pn* junction 1 in Figure 11–1, the base-emitter junction of Q_2 corresponds to *pn* junction 3, and the base-collector junctions of both Q_1 and Q_2 correspond to *pn* junction 2.

When a positive bias voltage is applied to the anode with respect to the cathode, as shown in Figure 11–2(b), the base-emitter junctions of Q_1 and Q_2 (*pn* junctions 1 and 3 in Figure 11–1(a)) are forward-biased, and the common base-collector junction (*pn* junction 2 in Figure 11–1(a)) is reverse-biased. Therefore, both equivalent transistors are in the linear region.

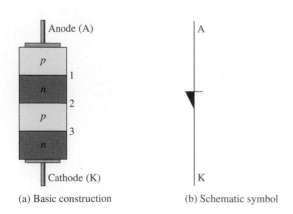

(a) Basic construction (b) Schematic symbol

▲ FIGURE 11–1

The 4-layer diode.

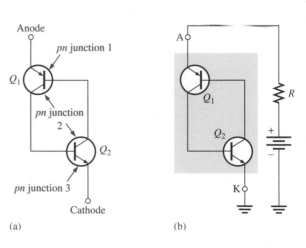

(a) (b)

▲ FIGURE 11–2

A 4-layer diode equivalent circuit.

The currents in a 4-layer diode are shown in the equivalent circuit in Figure 11–3. At low-bias levels, there is very little anode current, and thus it is in the *off* state or forward-blocking region.

▶ **FIGURE 11–3**

Currents in a 4-layer diode equivalent circuit.

EXAMPLE 11–1

A certain 4-layer diode is biased in the forward-blocking region with an anode-to-cathode voltage of 20 V. Under this bias condition, the anode current is 1 μs. Determine the resistance of the diode in the forward-blocking region.

Solution   The resistance is

$$R_{AK} = \frac{V_{AK}}{I_A} = \frac{20 \text{ V}}{1 \ \mu\text{A}} = \textbf{20 M}\boldsymbol{\Omega}$$

Related Problem*   If the anode current is 2 μA and $V_{AK} = 20$ V, what is the 4-layer diode's resistance in the forward-blocking region?

---

*Answers are at the end of the chapter.

Forward–Breakover Voltage   The operation of the 4-layer diode may seem strange because when it is forward-biased, it can act essentially as an open switch. As previously mentioned, there is a region of forward bias, called the *forward-blocking region,* in which the device has a very high forward resistance (ideally an open) and is in the *off* state. The forward-blocking region exists from $V_{AK} = 0$ V up to a value of V_{AK} called the **forward-breakover voltage,** $V_{BR(F)}$. This is indicated on the 4-layer diode characteristic curve in Figure 11–4.

▶ **FIGURE 11–4**

A 4-layer diode characteristic curve. The beige area is the forward-blocking region and the blue area is the forward-conduction region.

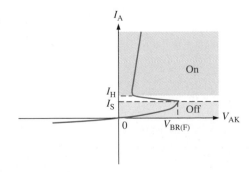

As V_{AK} is increased from 0, the anode current, I_A, gradually increases, as shown on the graph. As I_A increases, a point is reached where $I_A = I_S$, the switching current. At this point, $V_{AK} = V_{BR(F)}$, and the internal transistor structures become saturated. When this happens, the forward voltage drop, V_{AK}, suddenly decreases to a low value, and the 4-layer diode enters the forward-conduction region as indicated in Figure 11–4. Now, the device is in the *on* state and acts as a closed switch. When the anode current drops back below the holding value, I_H, the device turns off.

Holding Current   Once the 4-layer diode is conducting (in the *on* state), it will continue to conduct until the anode current is reduced below a specified level, called the **holding current**, I_H. This parameter is also indicated on the characteristic curve in Figure 11–4. When I_A falls below I_H, the device rapidly switches back to the *off* state and enters the forward-blocking region.

Switching Current   The value of the anode current at the point where the device switches from the forward-blocking region (off) to the forward-conduction region (on) is called the **switching current**, I_S. This value of current is always less than the holding current, I_H.

EXAMPLE 11–2

Determine the value of anode current in Figure 11–5 when the device is on. $V_{BR(F)} = 100$ V. Assume $V_{BE} = 0.7$ V and $V_{CE(sat)} = 0.1$ V for the internal transistor structure.

▶ **FIGURE 11–5**

Solution   The voltage at the anode is

$$V_A = V_{BE} + V_{CE(sat)} = 0.7\,\text{V} + 0.1\,\text{V} = 0.8\,\text{V}$$

The voltage across R_S is

$$V_{R_S} = V_{BIAS} - V_A = 110\,\text{V} - 0.8\,\text{V} = 109.2\,\text{V}$$

The anode current is

$$I_A = \frac{V_{R_S}}{R_S} = \frac{109.2\,\text{V}}{1.0\,\text{k}\Omega} = \textbf{109.2 mA}$$

Related Problem   What is the resistance in the forward-conduction region of the 4-layer diode in Figure 11–5?

Open file E11-02 in the EWB Examples folder on your CD-ROM. Compare the measured current with the calculated value. Change V_{BIAS} to 90 V and observe the current. Explain.

An Application

Although the 4-layer diode is rarely, if ever, used in new designs, the principles apply to other thyristors that you will study. The circuit in Figure 11–6(a) is a **relaxation oscillator**. The operation is as follows. When the switch is closed, the capacitor charges through R until its voltage reaches the forward-breakover voltage of the 4-layer diode. At this point the diode switches into conduction, and the capacitor rapidly discharges through the diode. Discharging continues until the current through the diode falls below the holding value. At this point, the diode switches back to the *off* state, and the capacitor begins to charge again. The result of this action is a voltage waveform across C like that shown in Figure 11–6(b).

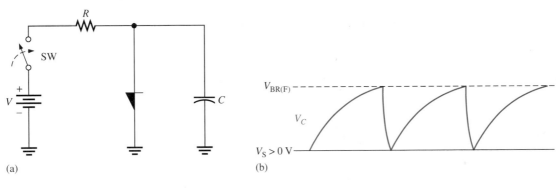

(a)   (b)

▲ **FIGURE 11–6**

A 4-layer diode relaxation oscillator.

**SECTION 11–1
REVIEW**

Answers are at the end of the chapter.

1. Why is the 4-layer diode classified as a thyristor?
2. What is the forward-blocking region?
3. What happens when the anode-to-cathode voltage exceeds the forward-breakover voltage?
4. Once it is on, how can the 4-layer diode be turned off?

11–2 THE SILICON-CONTROLLED RECTIFIER (SCR)

The silicon-controlled rectifer (SCR) is another 4-layer *pnpn* device similar to the 4-layer diode except with three terminals: anode, cathode, and gate. Like the 4-layer diode, the SCR has two possible states of operation. In the *off* state, it acts ideally as an open circuit between the anode and the cathode; actually, rather than an open, there is a very high resistance. In the *on* state, the SCR acts ideally as a short from the anode to the cathode; actually, there is a small *on* (forward) resistance. The SCR is used in many applications, including motor controls, time-delay circuits, heater controls, phase controls, and relay controls, to name a few.

After completing this section, you should be able to

- **Describe the basic structure and operation of an SCR**
- Identify an SCR by the schematic symbol
- Draw the equivalent circuit of an SCR
- Explain how to turn an SCR on and off
- Explain the characteristic curves of an SCR
- Define *forced commutation*
- Define various SCR parameters

The basic structure of an **SCR (silicon-controlled rectifier)** is shown in Figure 11–7(a), and the schematic symbol is shown in Figure 11–7(b). Typical SCR packages are shown in Figure 11–7(c). Other types of thyristors are found in the same or similar packages.

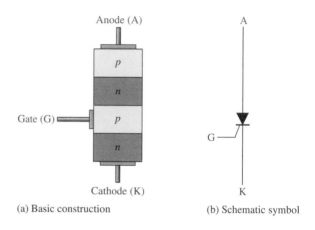

(a) Basic construction   (b) Schematic symbol

(c) Typical packages

▲ **FIGURE 11–7**

The silicon-controlled rectifier (SCR).

SCR Equivalent Circuit

Like the 4-layer diode operation, the SCR operation can best be understood by thinking of its internal *pnpn* structure as a two-transistor arrangement, as shown in Figure 11–8. This structure is like that of the 4-layer diode except for the gate connection. The upper *pnp* layers act as a transistor, Q_1, and the lower *npn* layers act as a transistor, Q_2. Again, notice that the two middle layers are "shared."

▶ FIGURE 11–8

SCR equivalent circuit.

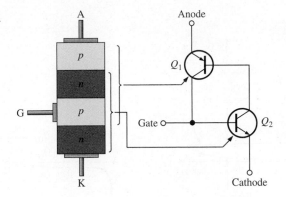

▶ FIGURE 11–8

SCR equivalent circuit.

Turning the SCR On

When the gate current, I_G, is zero, as shown in Figure 11–9(a), the device acts as a 4-layer diode in the *off* state. In this state, the very high resistance between the anode and cathode can be approximated by an open switch, as indicated. When a positive pulse of current (**trigger**) is applied to the gate, both transistors turn on (the anode must be more positive than the cathode). This action is shown in Figure 11–9(b). I_{B2} turns on Q_2, providing a path for I_{B1} into the Q_2 collector, thus turning on Q_1. The collector current of Q_1 provides additional base current for Q_2 so that Q_2 stays in conduction after the trigger pulse is removed from the gate. By this regenerative action, Q_2 sustains the saturated conduction of Q_1 by providing a path for I_{B1}; in turn, Q_1 sustains the saturated conduction of Q_2 by providing I_{B2}. Thus, the device stays on (latches) once it is triggered on, as shown in Figure 11–9(c). In this state, the very low resistance between the anode and cathode can be approximated by a closed switch, as indicated.

(a) SCR off

(b) SCR triggered on

(c) SCR stays on after trigger pulse

▲ FIGURE 11–9

The SCR turn-on process with the switch equivalents shown.

Like the 4-layer diode, an SCR can also be turned on without gate triggering by increasing anode-to-cathode voltage to a value exceeding the forward-breakover voltage $V_{BR(F)}$, as shown on the characteristic curve in Figure 11–10(a). The forward-breakover voltage decreases as I_G is increased above 0 V, as shown by the set of curves in Figure 11–10(b). Eventually, a value of I_G is reached at which the SCR turns on at a very low anode-to-cathode voltage. So, as you can see, the gate current controls the value of forward breakover voltage, $V_{BR(F)}$, required for turn-on.

(a) For $I_G = 0$

(b) For various I_G values

▲ **FIGURE 11–10**

SCR characteristic curves.

Although anode-to-cathode voltages in excess of $V_{BR(F)}$ will not damage the device if current is limited, this situation should be avoided because the normal control of the SCR is lost. It should normally be triggered on only with a pulse at the gate.

Turning the SCR Off

When the gate returns to 0 V after the trigger pulse is removed, the SCR cannot turn off; it stays in the forward-conduction region. The anode current must drop below the value of the holding current, I_H, in order for turn-off to occur. The holding current is indicated in Figure 11–10.

There are two basic methods for turning off an SCR: *anode current interruption* and *forced commutation*. The anode current can be interrupted by either a momentary series or parallel switching arrangement, as shown in Figure 11–11. The series switch in part (a) simply reduces the anode current to zero and causes the SCR to turn off. The parallel

▶ **FIGURE 11–11**

SCR turn-off by anode current interruption.

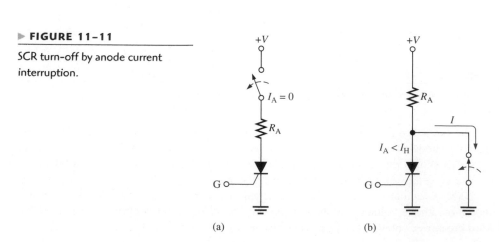

(a)

(b)

SCR turn–off by forced
commutation.

(a) On   (b) Off

switch in part (b) routes part of the total current away from the SCR, thereby reducing the
anode current to a value less than I_H.

The **forced commutation** method basically requires momentarily forcing current
through the SCR in the direction opposite to the forward conduction so that the net forward current is reduced below the holding value. The basic circuit, as shown in Figure
11–12, consists of a switch (normally a transistor switch) and a battery in parallel with
the SCR. While the SCR is conducting, the switch is open, as shown in part (a). To turn
off the SCR, the switch is closed, placing the battery across the SCR and forcing current
through it opposite to the forward current, as shown in part (b). Typically, turn-off times
for SCRs range from a few microseconds up to about 30 μs.

SCR Characteristics and Ratings

Several of the most important SCR characteristics and ratings are defined as follows. Use
the curve in Figure 11–10(a) for reference where appropriate.

Forward-breakover voltage, $V_{BR(F)}$ This is the voltage at which the SCR enters the
forward-conduction region. The value of $V_{BR(F)}$ is maximum when $I_G = 0$ and is designated $V_{BR(F0)}$. When the gate current is increased, $V_{BR(F)}$ decreases and is designated
$V_{BR(F1)}$, $V_{BR(F2)}$, and so on, for increasing steps in gate current (I_{G1}, I_{G2}, and so on).

Holding current, I_H   This is the value of anode current below which the SCR switches
from the forward-conduction region to the forward-blocking region. The value increases with decreasing values of I_G and is maximum for $I_G = 0$.

Gate trigger current, I_{GT}   This is the value of gate current necessary to switch the
SCR from the forward-blocking region to the forward-conduction region under specified conditions.

Average forward current, $I_{F(avg)}$ This is the maximum continuous anode current (dc)
that the device can withstand in the conduction state under specified conditions.

Forward-conduction region   This region corresponds to the *on* condition of the SCR
where there is forward current from anode to cathode through the very low resistance
(approximate short) of the SCR.

Forward-blocking and reverse-blocking regions   These regions correspond to the *off*
condition of the SCR where the forward current from anode to cathode is blocked by
the effective open circuit of the SCR.

Reverse-breakdown voltage, $V_{BR(R)}$ This parameter specifies the value of reverse
voltage from cathode to anode at which the device breaks into the avalanche region
and begins to conduct heavily (the same as in a *pn* junction diode).

SECTION 11–2 REVIEW

1. What is an SCR?
2. Name the SCR terminals.
3. How can an SCR be turned on (made to conduct)?
4. How can an SCR be turned off?

11–3 SCR APPLICATIONS

The SCR has many uses in the areas of power control and switching applications. A few basic applications are described in this section.

After completing this section, you should be able to

- **Discuss several SCR applications**

- Explain how an SCR is used to control current

- Describe half-wave power control

- Explain a basic phase-control circuit

- Discuss the function of an SCR in a lighting system for power interruptions

- Explain an over-voltage protection or "crowbar" circuit

On-Off Control of Current

Figure 11–13 shows an SCR circuit that permits current to be switched to a load by the momentary closure of switch SW1 and removed from the load by the momentary closure of switch SW2.

▶ **FIGURE 11–13**

On-off SCR control circuit.

Assuming the SCR is initially off, momentary closure of SW1 provides a pulse of current into the gate, thus triggering the SCR on so that it conducts current through R_L. The SCR remains in conduction even after the momentary contact of SW1 is removed if the anode current is equal to or greater than the holding current, I_H. When SW2 is momentarily closed, current is shunted around the SCR, thus reducing its anode current below the holding value, I_H. This turns the SCR off and reduces the load current to zero.

EXAMPLE 11–3

Determine the gate current and the anode current when the switch, SW1, is momentarily closed in Figure 11–14. Assume $V_{AK} = 0.8$ V, $V_{GK} = 0.7$ V, and $I_H = 20$ mA.

▲ **FIGURE 11–14**

Solution

$$I_G = \frac{V_{TRIG} - V_{GK}}{R_G} = \frac{3 \text{ V} - 0.7 \text{ V}}{560 \text{ } \Omega} = \textbf{4.1 mA}$$

$$I_A = \frac{V_A - V_{AK}}{R_A} = \frac{24 \text{ V} - 0.8 \text{ V}}{1.0 \text{ k}\Omega} = \textbf{23.2 mA}$$

Related Problem Will the SCR turn on if V_A is reduced to 12 V? Explain.

Open file E11-03 in the Examples folder on your CD-ROM. With SW2 open, momentarily close SW1. Compare the measured anode current with the calculated value. Notice that the anode current continues even after SW1 is opened. Close SW2 and observe the anode current. Explain your observation.

Half-Wave Power Control

A common application of SCRs is in the control of ac power for lamp dimmers, electric heaters, and electric motors. A half-wave, variable-resistance, phase-control circuit is shown in Figure 11–15; 120 V ac are applied across terminals *A* and *B; R_L* represents the resistance of the load (for example, a heating element or lamp filament). Resistor R_1 limits the current, and potentiometer R_2 sets the trigger level for the SCR.

▶ **FIGURE 11–15**

Half-wave, variable–resistance, phase–control circuit.

(a) 180° conduction

(b) 90° conduction

(c) 135° conduction

▲ **FIGURE 11–16**

Operation of the phase-control circuit.

By adjusting R_2, the SCR can be made to trigger at any point on the positive half-cycle of the ac waveform between 0° and 90°, as shown in Figure 11–16.

When the SCR triggers near the beginning of the cycle (approximately 0°), as in Figure 11–16(a), it conducts for approximately 180° and maximum power is delivered to the load. When it triggers near the peak of the positive half-cycle (90°), as in Figure 11–16(b), the SCR conducts for approximately 90° and less power is delivered to the load. By adjusting R_2, triggering can be made to occur anywhere between these two extremes, and therefore, a variable amount of power can be delivered to the load. Figure 11–16(c) shows triggering at the 45° point as an example. When the ac input goes negative, the SCR turns off and does not conduct again until the trigger point on the next positive half-cycle. The diode prevents the negative ac voltage from being applied to the gate of the SCR.

EXAMPLE 11–4

Show the voltage waveform across the SCR in Figure 11–17 from anode to cathode (ground) in relation to the load current for 180°, 45°, and 90° conduction. Assume an ideal SCR.

▶ **FIGURE 11–17**

Solution When there is load current, the SCR is conducting and the voltage across it is ideally zero. When there is no load current, the voltage across the SCR is the same as the applied voltage. The waveforms are shown in Figure 11–18.

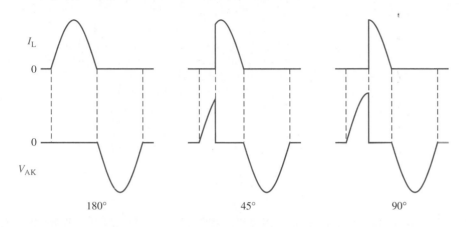

180°   45°   90°

▲ **FIGURE 11–18**

Related Problem What is the voltage across the SCR if it is never triggered?

Open file E11-04 in the EWB Examples folder on your CD-ROM. View the voltage across the SCR with the oscilloscope. Vary the potentiometer setting and observe how V_{AK} changes.

Lighting System for Power Interruptions

As another example of SCR applications, let's examine a circuit that will maintain lighting by using a backup battery when there is an ac power failure. Figure 11–19 shows a center-tapped full-wave rectifier used for providing ac power to a low-voltage lamp. As long as the ac power is available, the battery charges through diode D_3 and R_1.

(a) ac power on

(b) Back-up battery power (ac power off)

▲ **FIGURE 11–19**

Automatic back–up lighting circuit.

The SCR's cathode voltage is established when the capacitor charges to the peak value of the full-wave rectified ac (6.3 V rms less the drops across R_2 and D_1). The anode is at the 6 V battery voltage, making it less positive than the cathode, thus preventing conduction. The SCR's gate is at a voltage established by the voltage divider made up of R_2 and R_3. Under these conditions the lamp is illuminated by the ac input power and the SCR is off, as shown in Figure 11–19(a).

When there is an interruption of ac power, the capacitor discharges through the closed path D_3, R_1, and R_3, making the cathode less positive than the anode or the gate. This action establishes a triggering condition, and the SCR begins to conduct. Current from the battery is through the SCR and the lamp, thus maintaining illumination, as shown in Figure 11–19(b). When ac power is restored, the capacitor recharges and the SCR turns off. The battery begins recharging.

An Over-Voltage Protection Circuit

Figure 11–20 shows a simple over-voltage protection circuit, sometimes called a "crowbar" circuit, in a dc power supply. The dc output voltage from the regulator is monitored by the zener diode (D_1) and the resistive voltage divider (R_1 and R_2). The upper limit of the output voltage is set by the zener voltage. If this voltage is exceeded, the zener conducts and the voltage divider produces an SCR trigger voltage. The trigger voltage turns on the SCR, which is connected across the line voltage. The SCR current causes the fuse to blow, thus disconnecting the line voltage from the power supply.

▲ **FIGURE 11–20**

A basic SCR over-voltage protection circuit (shown in blue).

SECTION 11–3 REVIEW

1. If the potentiometer in Figure 11–16 is set at its midpoint, during what part of the input cycle will the SCR conduct?

2. In Figure 11–19, what is the purpose of diode D_3?

11–4   THE DIAC AND TRIAC

Both the diac and the triac are types of thyristors that can conduct current in both directions (bilateral). The difference between the two devices is that a diac has two terminals, while a triac has a third terminal, which is the gate for triggering. The diac functions basically like two parallel 4-layer diodes turned in opposite directions. The triac functions basically like two parallel SCRs turned in opposite directions with a common gate terminal.

After completing this section, you should be able to

- **Describe the basic structure and operation of diacs and triacs**
- Identify a diac or triac by the schematic symbol
- Discuss the equivalent circuit and bias conditions
- Explain the characteristic curve
- Discuss an application

The Diac

The basic construction and schematic symbol for a **diac** are shown in Figure 11–21. Notice that there are two terminals, labelled A_1 and A_2. Conduction occurs in a diac when the breakover voltage is reached with either polarity across the two terminals. The curve in Figure 11–22 illustrates this characteristic. Once breakover occurs, current is in a direction depending on the polarity of the voltage across the terminals. The device turns off when the current drops below the holding value.

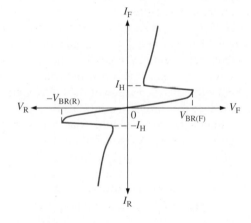

▲ **FIGURE 11–21**

The diac.

▲ **FIGURE 11–22**

Diac characteristic curve.

The equivalent circuit of a diac consists of four transistors arranged as shown in Figure 11–23(a). When the diac is biased as in Figure 11–23(b), the *pnpn* structure from A_1 to A_2 provides the same operation as was described for the 4-layer diode. In the equivalent circuit, Q_1 and Q_2 are forward-biased, and Q_3 and Q_4 are reverse-biased. The device operates on the upper right portion of the characteristic curve in Figure 11–22 under this bias condition. When the diac is biased as shown in Figure 11–23(c), the *pnpn* structure from A_2 and A_1 is used. In the equivalent circuit, Q_3 and Q_4 are forward-biased, and Q_1 and Q_2 are reverse-biased. Under this bias condition, the device operates on the lower left portion of the characteristic curve, as shown in Figure 11–22.

(a)   (b)   (c)

▲ FIGURE 11–23

Diac equivalent circuit and bias conditions.

The Triac

A **triac** is like a diac with a gate terminal. A triac can be turned on by a pulse of gate current and does not require the breakover voltage to initiate conduction, as does the diac. Basically, a triac can be thought of simply as two SCRs connected in parallel and in opposite directions with a common gate terminal. Unlike the SCR, the triac can conduct current in either direction when it is triggered on, depending on the polarity of the voltage across its A_1 and A_2 terminals. Figure 11–24 shows the basic construction and schematic symbol for a triac. The characteristic curve is shown in Figure 11–25. Notice that the breakover potential decreases as the gate current increases, just as with the SCR.

As with other thyristors, the triac ceases to conduct when the anode current drops below the specified value of the holding current, I_H. The only way to turn off the triac is to reduce the current to a sufficiently low level.

Figure 11–26 shows the triac being triggered into both directions of conduction. In part (a), terminal A_1 is biased positive with respect to A_2, so the triac conducts as shown when triggered by a positive pulse at the gate terminal. The transistor equivalent circuit in part (b) shows that Q_1 and Q_2 conduct when a positive trigger pulse is applied. In part (c), terminal A_2 is biased positive with respect to A_1, so the triac conducts as shown. In this case, Q_3 and Q_4 conduct as indicated in part (d) upon application of a positive trigger pulse.

▶ FIGURE 11–24

The triac.

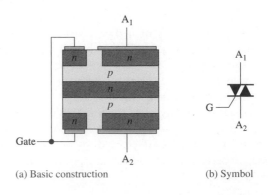

(a) Basic construction   (b) Symbol

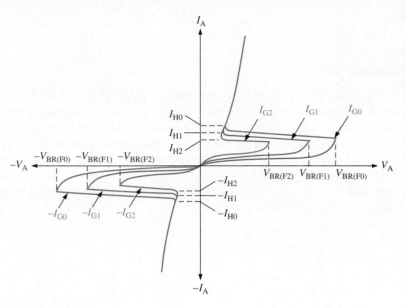

▲ **FIGURE 11–25**

Triac characteristic curves.

(a)

(b) Q_1 and Q_2 on

(c)

(d) Q_3 and Q_4 on

▲ **FIGURE 11–26**

Bilateral operation of a triac.

Applications

Like the SCR, triacs are also used to control average power to a load by the method of phase control. The triac can be triggered such that the ac power is supplied to the load for a controlled portion of each half-cycle. During each positive half-cycle of the ac, the triac is off for a certain interval, called the *delay angle* (measured in degrees), and then it is triggered on and conducts current through the load for the remaining portion of the positive half-cycle, called the *conduction angle.* Similar action occurs on the negative half-cycle except that, of course, current is conducted in the opposite direction through the load. Figure 11–27 illustrates this action.

▲ **FIGURE 11–27**

Basic triac phase control.

One example of phase control using a triac is illustrated in Figure 11–28(a). Diodes are used to provide trigger pulses to the gate of the triac. Diode D_1 conducts during the positive half-cycle. The value of R_1 sets the point on the positive half-cycle at which the triac triggers. Notice that during this portion of the ac cycle, A_1 and G are positive with respect to A_2.

Diode D_2 conducts during the negative half-cycle, and R_1 sets the trigger point. Notice that during this portion of the ac cycle, A_2 and G are positive with respect to A_1. The resulting waveform across R_L is shown in Figure 11–28(b).

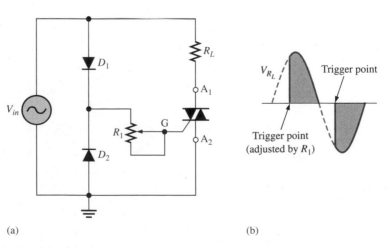

(a)   (b)

▲ **FIGURE 11–28**

Triac phase-control circuit.

In the phase-control circuit, it is necessary that the triac turn off at the end of each positive and each negative alternation of the ac. Figure 11–29 illustrates that there is an interval near each 0 crossing where the triac current drops below the holding value, thus turning the device off.

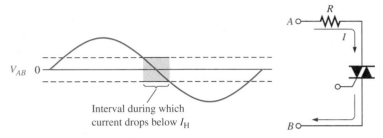

▲ FIGURE 11–29

Triac turn-off interval.

**SECTION 11–4
REVIEW**

1. Compare the diac to the 4-layer diode in terms of basic operation.
2. Compare the triac with the SCR in terms of basic operation.
3. How does a triac differ from a diac?

11–5 THE SILICON-CONTROLLED SWITCH (SCS)

The silicon-controlled switch (SCS) is similar in construction to the SCR. The SCS, however, has two gate terminals, the cathode gate and the anode gate. The SCS can be turned on and off using either gate terminal. Remember that the SCR can be only turned on using its gate terminal. Normally, the SCS is available in power ratings lower than those of the SCR.

After completing this section, you should be able to

- **Describe the basic operation of an SCS**

- Identify an SCS by its schematic symbol

- Use an equivalent circuit to describe SCS operation

- Compare the SCS to the SCR

The symbol and terminal identification for an **SCS (silicon-controlled switch)** are shown in Figure 11–30.

As with the previous thyristors, the basic operation of the SCS can be understood by referring to the transistor equivalent, shown in Figure 11–31. To start, assume that both Q_1 and Q_2 are off, and therefore that the SCS is not conducting. A positive pulse on the cathode gate drives Q_2 into conduction and thus provides a path for Q_1 base current. When Q_1 turns on, its collector current provides base current to Q_2, thus sustaining the *on* state of the device. This regenerative action is the same as in the turn-on process of the SCR and the 4-layer diode and is illustrated in Figure 11–31(a).

▲ **FIGURE 11–30**

The silicon-controlled switch (SCS).

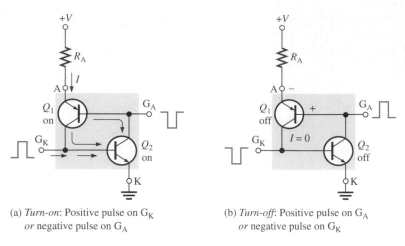

(a) *Turn-on*: Positive pulse on G_K
or negative pulse on G_A

(b) *Turn-off*: Positive pulse on G_A
or negative pulse on G_K

▲ **FIGURE 11–31**

SCS operation.

The SCS can also be turned on with a negative pulse on the anode gate, as indicated in Figure 11–31(a). This drives Q_1 into conduction which, in turn, provides base current for Q_2. Once Q_2 is on, it provides a path for Q_1 base current, thus sustaining the *on* state.

To turn the SCS off, a positive pulse is applied to the anode gate. This reverse-biases the base-emitter junction of Q_1 and turns it off. Q_2, in turn, cuts off and the SCS ceases conduction, as shown in Figure 11–31(b). The device can also be turned off with a negative pulse on the cathode gate, as indicated in part (b). The SCS typically has a faster turn-off time than the SCR.

In addition to the positive pulse on the anode gate or the negative pulse on the cathode gate, there are other methods for turning off an SCS. Figure 11–32(a) and (b) shows two switching methods to reduce the anode current below the holding value. In each case, the bipolar junction transistor (BJT) acts as a switch.

(a) Series switch turns off SCS

(b) Shunt switch turns off SCS

▲ **FIGURE 11–32**

The transistor switch in both series and shunt configurations reduces I_A below I_H and turns off the SCS.

Applications

The SCS and SCR are used in similar applications. The SCS has the advantage of faster turn-off with pulses on either gate terminal; however, it is more limited in terms of maximum current and voltage ratings. Also, the SCS is sometimes used in digital applications such as counters, registers, and timing circuits.

**SECTION 11–5
REVIEW**

1. Explain the difference between an SCS and an SCR.
2. How can an SCS be turned on?
3. Describe four ways an SCS can be turned off.

11–6 THE UNIJUNCTION TRANSISTOR (UJT)

The unijunction transistor does not belong to the thyristor family because it does not have a four-layer type of construction. The term *unijunction* refers to the fact that the UJT has a single *pn* junction. As you will see in this section, the UJT is useful in certain oscillator applications and as a triggering device in thyristor circuits.

After completing this section, you should be able to

- **Describe the basic structure and operation of a UJT**
- Identify a UJT by its symbol
- Draw the equivalent circuit
- Explain why a UJT is not a thyristor
- Define *standoff ratio*
- Analyze the operation of a UJT relaxation oscillator

The **UJT (unijunction transistor)** is a three-terminal device whose basic construction is shown in Figure 11–33(a); the schematic symbol appears in Figure 11–33(b). Notice the terminals are labelled Emitter (E), Base 1 (B_1), and Base 2 (B_2). Do not confuse this symbol with that of a JFET; *the difference is that the arrow is at an angle for the UJT.* The UJT has only one *pn* junction, and therefore, the characteristics of this device are very different from those of either the BJT or the FET, as you will see.

▶ **FIGURE 11–33**

The unijunction transistor (UJT).

(a) Basic construction

(b) Symbol

Equivalent Circuit

The equivalent circuit for the UJT, shown in Figure 11–34(a), will aid in understanding the basic operation. The diode shown in the figure represents the *pn* junction, r'_{B1} represents the internal dynamic resistance of the silicon bar between the emitter and base 1,

▲ **FIGURE 11–34**

UJT equivalent circuit.

and r'_{B2} represents the dynamic resistance between the emitter and base 2. The sum $r'_{B1} + r'_{B2}$ is the total resistance between the base terminals and is called the *interbase resistance*, r'_{BB}.

$$r'_{BB} = r'_{B1} + r'_{B2}$$

The value of r'_{B1} varies inversely with emitter current I_E, and therefore, it is shown as a variable resistor. Depending on I_E, the value of r'_{B1} can vary from several thousand ohms down to tens of ohms. The internal resistances r'_{B1} and r'_{B2} form a voltage divider when the device is biased, as shown in Figure 11–34(b). The voltage across the resistance r'_{B1} can be expressed as

$$V_{r'_{B1}} = \left(\frac{r'_{B1}}{r'_{BB}} \right) V_{BB}$$

Standoff Ratio

The ratio r'_{B1}/r'_{BB} is a UJT characteristic called the intrinsic **standoff ratio** and is designated by η (Greek *eta*).

Equation 11–1

$$\eta = \frac{r'_{B1}}{r'_{BB}}$$

As long as the applied emitter voltage V_{EB1} is less than $V_{r'_{B1}} + V_{pn}$, there is no emitter current because the *pn* junction is not forward-biased (V_{pn} is the barrier potential of the *pn* junction). The value of emitter voltage that causes the *pn* junction to become forward-biased is called V_P (peak-point voltage) and is expressed as

Equation 11–2

$$V_P = \eta V_{BB} + V_{pn}$$

When V_{EB1} reaches V_P, the *pn* junction becomes forward-biased and I_E begins. Holes are injected into the *n*-type bar from the *p*-type emitter. This increase in holes causes an increase in free electrons, thus increasing the conductivity between emitter and B_1 (decreasing r'_{B1}).

After turn-on, the UJT operates in a negative resistance region up to a certain value of I_E, as shown by the characteristic curve in Figure 11–35. As you can see, after the peak point ($V_E = V_P$ and $I_E = I_P$), V_E decreases as I_E continues to increase, thus producing the negative resistance characteristic. Beyond the valley point ($V_E = V_V$ and $I_E = I_V$), the device is in saturation, and V_E increases very little with an increasing I_E.

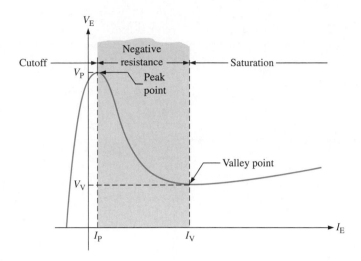

▲ **FIGURE 11–35**

UJT characteristic curve for a fixed value of V_{BB}.

EXAMPLE 11–5

The data sheet of a certain UJT gives $\eta = 0.6$. Determine the peak-point emitter voltage V_P if $V_{BB} = 20$ V.

Solution

$$V_P = \eta V_{BB} + V_{pn} = 0.6(20\text{ V}) + 0.7\text{ V} = \mathbf{12.7\ V}$$

Related Problem How can the peak-point emitter voltage of a UJT be increased?

A UJT Application

The UJT can be used as a trigger device for SCRs and triacs. Other applications include nonsinusoidal oscillators, sawtooth generators, phase control, and timing circuits. Figure 11–36 shows a UJT relaxation oscillator as an example of one application.

The operation is as follows. When dc power is applied, the capacitor C charges exponentially through R_1 until it reaches the peak-point voltage V_P. At this point, the pn junction becomes forward-biased, and the emitter characteristic goes into the negative resistance region (V_E decreases and I_E increases). The capacitor then quickly discharges through the forward-biased junction, r'_B, and R_2. When the capacitor voltage decreases to the valley-point voltage V_V, the UJT turns off, the capacitor begins to charge again, and the cycle is repeated, as shown in the emitter voltage waveform in Figure 11–37 (top). During the discharge time of the capacitor, the UJT is conducting. Therefore, a voltage is developed across R_2, as shown in the waveform diagram in Figure 11–37 (bottom).

▲ **FIGURE 11–36**

Relaxation oscillator.

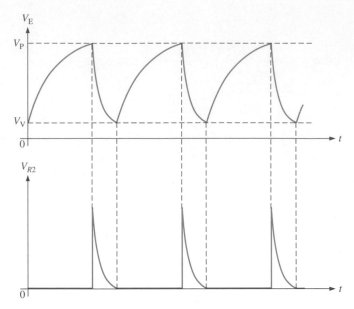

▲ FIGURE 11–37

Waveforms for UJT relaxation oscillator.

Conditions for Turn-On and Turn-Off

In the relaxation oscillator of Figure 11–36, certain conditions must be met for the UJT to reliably turn on and turn off. First, to ensure turn-on, R_1 must not limit I_E at the peak point to less than I_P. To ensure this, the voltage drop across R_1 at the peak point should be greater than $I_P R_1$. Thus, the condition for turn-on is

$$V_{BB} - V_P > I_P R_1$$

or

$$R_1 < \frac{V_{BB} - V_P}{I_P}$$

To ensure turn-off of the UJT at the valley point, R_1 must be large enough that I_E (at the valley point) can decrease below the specified value of I_V. This means that the voltage across R_1 at the valley point must be less than $I_V R_1$. Thus, the condition for turn-off is

$$V_{BB} - V_V < I_V R_1$$

or

$$R_1 > \frac{V_{BB} - V_V}{I_V}$$

So, for a proper turn-on and turn-off, R_1 must be in the range

$$\frac{V_{BB} - V_P}{I_P} > R_1 > \frac{V_{BB} - V_V}{I_V}$$

EXAMPLE 11–6

Determine a value of R_1 in Figure 11–38 that will ensure proper turn-on and turn-off of the UJT. The characteristic of the UJT exhibits the following values: $\eta = 0.5$, $V_V = 1$ V, $I_V = 10$ mA, $I_P = 20$ μA, and $V_P = 14$ V.

▶ **FIGURE 11–38**

Solution

$$\frac{V_{BB} - V_P}{I_P} > R_1 > \frac{V_{BB} - V_V}{I_V}$$

$$\frac{30 \text{ V} - 14 \text{ V}}{20 \text{ } \mu\text{A}} > R_1 > \frac{30 \text{ V} - 1 \text{ V}}{10 \text{ mA}}$$

$$\mathbf{800 \text{ k}\Omega > R_1 > 2.9 \text{ k}\Omega}$$

As you can see, R_1 has quite a wide range of possible values that will work.

Related Problem Determine a value of R_1 in Figure 11–38 that will ensure proper turn-on and turn-off for the following values:

$\eta = 0.33$, $V_V = 0.8$ V, $I_V = 15$ mA, $I_P = 35$ μA, and $V_P = 18$ V.

SECTION 11–6
REVIEW

1. Name the UJT terminals.

2. What is the intrinsic standoff ratio?

3. In a basic UJT relaxation oscillator such as in Figure 11–36, what three factors determine the period of oscillation?

11–7 THE PROGRAMMABLE UNIJUNCTION TRANSISTOR (PUT)

The programmable unijunction transistor (PUT) is actually a type of thyristor and not like the UJT at all in terms of structure. The only similarity to a UJT is that the PUT can be used in some oscillator applications to replace the UJT. The PUT is similar to an SCR except that its anode-to-gate voltage can be used to both turn on and turn off the device.

After completing this section, you should be able to

- **Describe the structure and operation of a PUT**

- Compare PUT structure to that of the SCR

- Explain the difference between a PUT and a UJT

- State how to set the PUT trigger voltage

- Discuss an application

The structure of the **PUT (programmable unijunction transistor)** is similar to that of an SCR (four-layer) except that the gate is brought out as shown in Figure 11–39. Notice that the gate is connected to the n region adjacent to the anode. This pn junction controls the *on* and *off* states of the device. The gate is always biased positive with respect to the cathode. When the anode voltage exceeds the gate voltage by approximately 0.7 V, the pn junction is forward-biased and the PUT turns on. The PUT stays on until the anode voltage falls back below this level, then the PUT turns off.

▶ **FIGURE 11–39**

The programmable unijunction transistor (PUT).

(a) Basic construction   (b) Symbol

Setting the Trigger Voltage

The gate can be biased to a desired voltage with an external voltage divider, as shown in Figure 11–40(a), so that when the anode voltage exceeds this "programmed" level, the PUT turns on.

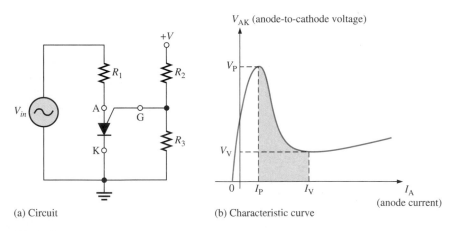

(a) Circuit   (b) Characteristic curve

▲ **FIGURE 11–40**

PUT biasing.

An Application

A plot of the anode-to-cathode voltage, V_{AK}, versus anode current, I_A, in Figure 11–40(b) reveals a characteristic curve similar to that of the UJT. Therefore, the PUT replaces the UJT in many applications. One such application is the relaxation oscillator in Figure 11–41(a). Its basic operation is as follows.

The gate is biased at +9 V by the voltage divider consisting of resistors R_2 and R_3. When dc power is applied, the PUT is off and the capacitor charges toward +18 V through R_1. When the capacitor reaches $V_G + 0.7$ V, the PUT turns on and the capacitor rapidly discharges through the low *on* resistance of the PUT and R_4. A voltage spike is developed across R_4 during the discharge. As soon as the capacitor discharges, the PUT turns off and the charging cycle starts over, as shown by the waveforms in Figure 11–41(b).

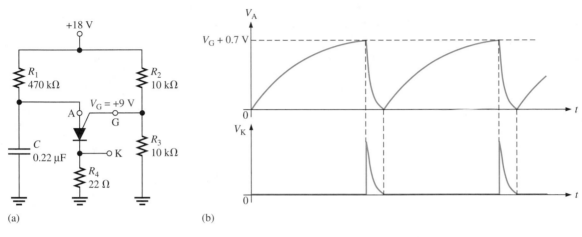

(a)   (b)

▲ **FIGURE 11–41**

PUT relaxation oscillator.

SECTION 11–7 REVIEW

1. What does the term *programmable* mean as used in programmable unijunction transistor (PUT)?

2. Compare the structure and the operation of a PUT to those of other devices such as the UJT and SCR.

11–8   THE PHOTOTRANSISTOR

The phototransistor has a light-sensitive, collector-base *pn* junction. It is exposed to incident light through a lens opening in the transistor package. When there is no incident light, there is only a small thermally generated collector-to-emitter leakage current, I_{CEO}; this is called the dark current and is typically in the nA range. When light strikes the collector-base *pn* junction, a base current, I_λ, is produced that is directly proportional to the light intensity. This action produces a collector current that increases with I_λ. Except for the way base current is generated, the phototransistor behaves as a conventional BJT. In many cases, there is no electrical connection to the base.

After completing this section, you should be able to

- **Describe a phototransistor and its operation**

- Explain how the base current is produced

- Discuss how phototransistors are used

The relationship between the collector current and the light-generated base current in a phototransistor is

Equation 11–3

$$I_C = \beta_{DC}I_\lambda$$

The schematic symbol and some typical phototransistors are shown in Figure 11–42. Since the actual photogeneration of base current occurs in the collector-base region, the larger the physical area of this region, the more base current is generated. Thus, a typical phototransistor is designed to offer a large area to the incident light, as the simplified structure diagram in Figure 11–43 illustrates.

▶ **FIGURE 11–42**

Phototransistor.

(a) Schematic symbol   (b) Typical packages

▶ **FIGURE 11–43**

Typical phototransistor chip structure.

A **phototransistor** can be either a two-lead or a three-lead device. In the three-lead configuration, the base lead is brought out so that the device can be used as a conventional BJT with or without the additional light-sensitivity feature. In the two-lead configuration, the base is not electrically available, and the device can be used only with light as the input. In many applications, the phototransistor is used in the two-lead version. Figure 11–44 shows a phototransistor with a biasing circuit and typical collector

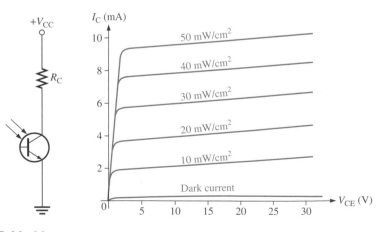

▲ **FIGURE 11–44**

Phototransistor bias circuit and typical collector characteristic curves.

characteristic curves. Notice that each individual curve on the graph corresponds to a certain value of light intensity (in this case, the units are mW/cm^2) and that the collector current increases with light intensity.

Phototransistors are not sensitive to all light but only to light within a certain range of wavelengths. They are most sensitive to particular wavelengths, as shown by the peak of the spectral response curve in Figure 11–45.

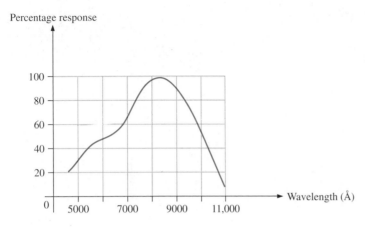

▲ FIGURE 11–45

Typical phototransistor spectral response.

Photodarlington

The photodarlington consists of a phototransistor connected in a darlington arrangement with a conventional BJT, as shown in Figure 11–46. Because of the higher current gain, this device has a much higher collector current and exhibits a greater light sensitivity than does a regular phototransistor.

Applications

Phototransistors are used in a wide variety of applications. A light-operated relay circuit is shown in Figure 11–47. The phototransistor Q_1 drives the BJT Q_2. When there is sufficient incident light on Q_1, transistor Q_2 is driven into saturation, and collector current through the relay coil energizes the relay. The diode across the relay coil prevents, by its limiting action, a large voltage transient from occurring at the collector of Q_2 when the transistor turns off.

▲ FIGURE 11–46

Photodarlington.

▲ FIGURE 11–47

Light-operated relay circuit.

Figure 11–48 shows a circuit in which a relay is de-energized by incident light on the phototransistor. When there is insufficient light, transistor Q_2 is biased on, keeping the relay energized. When there is sufficient light, phototransistor Q_1 turns on; this pulls the base of Q_2 low, thus turning Q_2 off and de-energizing the relay.

▶ **FIGURE 11–48**

Darkness–operated relay circuit.

These relay circuits can be used in a variety of applications such as automatic door activators, process counters, and various alarm systems. Another simple application is illustrated in Figure 11–49. The phototransistor is normally on, holding the gate of the SCR low. When the light is interrupted, the phototransistor turns off. The high-going transition on the collector triggers the SCR and sets off the alarm mechanism. The momentary contact switch SW1 provides for resetting the alarm. Smoke detection and intrusion detection are possible uses for this circuit.

▶ **FIGURE 11–49**

Light–interruption alarm.

SECTION 11–8 REVIEW

1. How does a phototransistor differ from a conventional BJT?
2. Some, but not all, phototransistors have external _____ leads.
3. The collector current in a phototransistor circuit depends on what two factors?

11-9 THE LIGHT-ACTIVATED SCR (LASCR)

The light-activated silicon-controlled rectifier (LASCR) operates essentially as does the conventional SCR except that it can also be light-triggered. Most LASCRs have an available gate terminal so that the device can also be triggered by an electrical pulse just as a conventional SCR.

After completing this section, you should be able to

- **Describe the LASCR and its operation**
- Compare the LASCR to the conventional SCR
- Discuss an application

Figure 11–50 shows a **LASCR (light-activated silicon-controlled rectifier)** schematic symbol and typical packages. The LASCR is most sensitive to light when the gate terminal is open. If necessary, a resistor from the gate to the cathode can be used to reduce the sensitivity. Figure 11–51 shows a LASCR used to energize a latching relay. The input source turns on the lamp; and the resulting incident light triggers the LASCR. The anode current energizes the relay and closes the contact. Notice that the input source is electrically isolated from the rest of the circuit.

▶ **FIGURE 11–50**

LASCRs.

(a) Symbol   (b) Typical packages

▲ **FIGURE 11–51**

A LASCR circuit.

SECTION 11–9 REVIEW

1. Can most LASCRs be operated as conventional SCRs?
2. What is required in Figure 11–51 to turn off the LASCR and de-energize the relay?

11–10 OPTICAL COUPLERS

Optical couplers are designed to provide complete electrical isolation between an input circuit and an output circuit. The usual purpose of isolation is to provide protection from high-voltage transients, surge voltage, or low-level noise that could possibly result in an erroneous output or damage to the device. Optical couplers also allow interfacing circuits with different voltage levels, different grounds, and so on.

After completing this section, you should be able to

- ■ **Discuss various types of optical couplers**
- ■ Explain these parameters: isolation voltage, dc current transfer ratio, LED trigger current, and transfer gain
- ■ Discuss fiber optics
- ■ Explain refraction and reflection of light

The input circuit of an optical coupler is typically an LED, but the output circuit can take several forms, such as the phototransistor shown in Figure 11–52(a). When the input voltage forward-biases the LED, light transmitted to the phototransistor turns it on, producing current through the external load, as shown in Figure 11–52(b). Some typical devices are shown in Figure 11–52(c).

Several other types of couplers are shown in Figure 11–53. The darlington transistor coupler in Figure 11–53(a) can be used when increased output current capability is needed beyond that provided by the phototransistor output. The disadvantage is that the photodarlington has a switching speed less than that of the phototransistor.

(a) Basic device

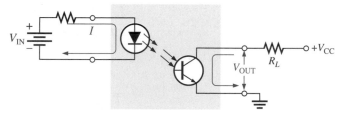

(b) Device with external connections

(c) Typical packages

▲ **FIGURE 11–52**

Phototransistor couplers.

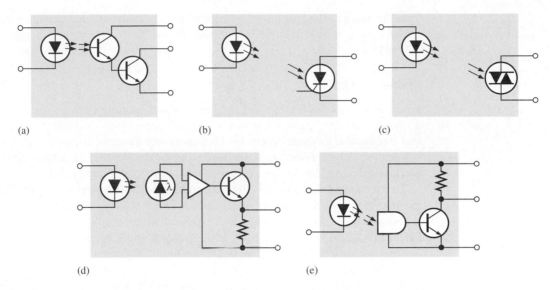

(a) (b) (c)

(d) (e)

▲ **FIGURE 11–53**

Common types of optical-coupling devices.

An LASCR output coupler is shown in Figure 11–53(b). This device can be used in applications where, for example, a low-level input voltage is required to latch a high-voltage relay for the purpose of activating some type of electromechanical device.

A phototriac output coupler is illustrated in Figure 11–53(c). This device is designed for applications that require isolated triac triggering, such as switching a 110 V ac line from a low-level input.

Figure 11–53(d) shows an optically isolated ac linear coupler. This device converts an input current variation to an output voltage variation. The output circuit consists of an amplifier with a photodiode across its input terminals. Light variations emitted from the LED are picked up by the photodiode, providing an input signal to the amplifier. The output of the amplifier is buffered with the emitter-follower stage. The optically isolated ac linear coupler can be used for telephone line coupling, peripheral equipment isolation, and audio applications.

Part (e) of Figure 11–53 shows a digital output coupler. This device consists of a high-speed detector circuit followed by a transistor buffer stage. When there is current through the input LED, the detector is light-activated and turns on the output transistor, so that the collector switches to a low-voltage level. When there is no current through the LED, the output is at the high-voltage level. The digital output coupler can be used in applications that require compatibility with digital circuits, such as interfacing computer terminals with peripheral devices.

Isolation Voltage   The *isolation voltage* of an optical coupler is the maximum voltage that can exist between the input and output terminals without dielectric breakdown occurring. Typical values are about 7500 V ac peak.

DC Current Transfer Ratio   This parameter is the ratio of the output current to the input current through the LED. It is usually expressed as a percentage. For a phototransistor output, typical values range from 2 percent to 100 percent. For a photodarlington output, typical values range from 50 percent to 500 percent.

LED Trigger Current   This parameter applies to the LASCR output coupler and the phototriac output device. The *trigger current* is the value of current required to trigger the thyristor output device. Typically, the trigger current is in the mA range.

Transfer Gain This parameter applies to the optically isolated ac linear coupler. The *transfer gain* is the ratio of output voltage to input current, and a typical value is 200 mV/mA.

Fiber Optics

Fiber optics provides a means for coupling a photo-emitting device to a photodetector via a light-transmitting cable consisting of optical fibers. Typical applications include medical electronics, industrial controls, microprocessor systems, security systems, and communications. Glass fiber cables are used to maximize the optical coupling between opto-electronic devices. Fiber optics is based on the principle of internal reflection.

Every material that conducts light has an **index of refraction**, *n*. A ray of light striking the interface between two materials with different indices of refraction will either be reflected or refracted, depending on the angle at which the light strikes the interface surface, as illustrated in Figure 11–54. If the incident angle, θ_i, is equal to or greater than a certain value called the *critical angle, θ_c*, the light is reflected. If θ_i is less than the critical angle, the light is refracted.

A glass fiber is clad with a layer of glass with a lower index of refraction than the fiber. A ray of light entering the end of the cable will be refracted as shown in Figure 11–55. If, after refraction, it approaches and hits the glass interface at an angle greater than the critical angle, it is reflected within the fiber as shown. Since, according to a law of physics, the angle of reflection must equal the angle of incidence, the light ray will bounce down the fiber and emerge, refracted, at the other end, as shown in the figure.

▶ **FIGURE 11–54**

Refraction and reflection of light rays.

(a) Refraction ($\theta_i < \theta_c$)   (b) Reflection ($\theta_i > \theta_c$)

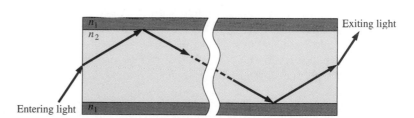

▲ **FIGURE 11–55**

A light ray in optical fiber.

SECTION 11–10 REVIEW

1. What type of input device is normally used in an optical coupler?

2. List five types of output devices used in optical couplers.

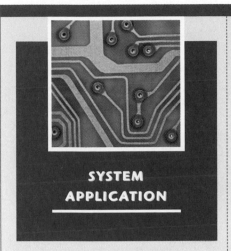

SYSTEM APPLICATION

This system is a variation of the system in Chapter 3 for counting baseballs for packing control. In this system, objects passing on an assembly line conveyor belt are counted, and the speed of the conveyor is adjusted according to a predetermined rate. The focus of this application is the conveyor motor speed-control circuit. The knowledge you have gained in this chapter will be applied in completing your assignment.

Basic Operation of the System

The system controls the speed of the conveyor so that a preset average number of randomly spaced parts flow past a point on the production line in a specified period of time. This is to allow a proper amount of time for the production line workers to perform certain tasks on each part. A basic diagram of the conveyor speed-control system is shown in Figure 11–56.

Each time a part on the moving conveyor belt passes the optical detector and interrupts the infrared light beam, a digital counter in the processing circuits is advanced by one. The count of the passing parts is accumulated over a specified period of time and converted to a proportional voltage by the processing circuits based on the desired number of parts in a specified time interval. The more parts that pass the sensor, the higher the voltage. The proportional voltage is applied to the motor speed-control circuit which, in turn, adjusts the speed of the electric motor that drives the conveyor belt in order to maintain the desired number of parts per unit time.

The Motor Speed-Control Circuit The proportional voltage from the processing circuits is applied to the gate of a PUT on the motor speed-control circuit board. This voltage determines the point in the ac cycle that the SCR is triggered on. For a higher PUT gate voltage, the SCR turns on later in the half-cycle and therefore delivers less average power to the motor causing its speed to decrease. For a lower PUT gate voltage, the SCR turns on earlier in the half-cycle delivering more average power to the motor and increasing its speed. This process continually adjusts the motor speed to maintain the required number of objects per unit time. The potentiometer is used for calibration of the SCR trigger point.

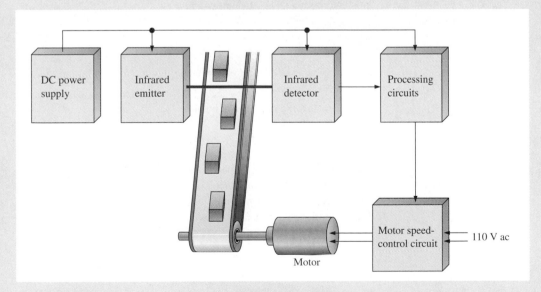

▲ **FIGURE 11–56**

Diagram of the conveyor speed-control system.

(a)

(b)

▲ **FIGURE 11–57**

Motor speed-control circuit board.

The Motor Speed-Control Circuit Board

■ Make sure that the circuit board shown in Figure 11–57(a) is correctly assembled by comparison with the schematic in part (b). Device pin diagrams are shown.

■ Label a copy of the board with component and input/output designations in agreement with the schematic.

Analysis of the Motor Speed-Control Circuit

Refer to the schematic in Figure 11–57(b). A 1.0 kΩ resistor is connected in place of the motor and 110 V, 60 Hz voltage is applied across the input terminals.

■ Determine the voltage waveforms at the anode, cathode, and gate of the SCR and the PUT with respect to ground for a PUT gate voltage of 0 V and with the potentiometer set at 25 kΩ.

■ Determine voltage across the 1.0 kΩ resistor, for each of the following PUT gate voltages: 0 V, 2 V, 4 V, 6 V, 8 V, and 10 V. The potentiometer is still at 25 kΩ.

Test Procedure

■ Develop a step-by-step set of instructions on how to check the motor speed-control circuit board for proper operation using the numbered test points indicated in the test bench setup of Figure 11–58.

■ Specify voltage values for all the measurements to be made.

■ Provide a fault analysis for all possible component failures.

Troubleshooting

Problems have developed in four boards. Based on the test bench measurements for each board indicated in Figure 11–59, determine the most likely fault in each

case. The circled numbers indicate test point connections to the circuit board. Assume that each board has 110 V ac applied.

Final Report (Optional)

Submit a final written report on the motor speed-control circuit board using an organized format that includes the following:

1. A physical description of the circuits.

2. A discussion of the operation of the circuits.

3. A list of the specifications.

4. A list of parts with part numbers if available.

5. A list of the types of problems on the four faulty circuit boards.

6. A description of how you determined the problem on each of the faulty circuit boards.

▲ **FIGURE 11–58**

Test bench setup for the motor speed-control circuit board.

▲ **FIGURE 11–59**

Test results for four faulty circuit boards.

CHAPTER SUMMARY

- Thyristors are devices constructed with four semiconductor layers (*pnpn*).
- Thyristors include 4-layer diodes, SCRs, diacs, triacs, SCSs, and PUTs.
- The 4-layer diode is a thyristor that conducts when the voltage across its terminals exceeds the breakover potential.
- The silicon-controlled rectifier (SCR) can be triggered on by a pulse at the gate and turned off by reducing the anode current below the specified holding value.
- The diac can conduct current in either direction and is turned on when a breakover voltage is exceeded. It turns off when the current drops below the holding value.
- The triac, like the diac, is a bidirectional device. It can be turned on by a pulse at the gate and conducts in a direction depending on the voltage polarity across the two anode terminals.
- The silicon-controlled switch (SCS) has two gate terminals and can be turned on by a pulse at the cathode gate and turned off by a pulse at the anode gate.
- The intrinsic standoff ratio of a unijunction transistor (UJT) determines the voltage at which the device will trigger on.
- The programmable unijunction transistor (PUT) can be externally programmed to turn on at a desired anode-to-gate voltage level.
- In a phototransistor, base current is generated by light input.
- Light acts as the trigger source in light-activated SCRs (LASCRs).
- Optical coupling devices provide electrical isolation between a source and an output circuit.
- Fiber optics provides a light path from a light-emitting device to a light-activated device.
- Device symbols are shown in Figure 11–60.

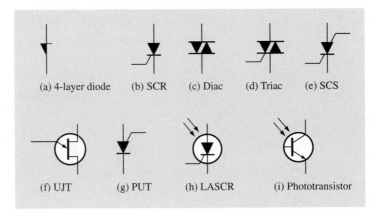

(a) 4-layer diode (b) SCR (c) Diac (d) Triac (e) SCS
(f) UJT (g) PUT (h) LASCR (i) Phototransistor

▲ FIGURE 11–60

KEY TERMS

Key terms and other bold terms in the chapter are defined in the end-of-book glossary.

Diac A two-terminal four-layer semiconductor device (thyristor) that can conduct current in either direction when properly activated.

Forward-breakover voltage ($V_{BR(F)}$) The voltage at which a device enters the forward-blocking region.

4-layer diode The type of two-terminal thyristor that conducts current when the anode-to-cathode voltage reaches a specified "breakover" value.

Holding current (I_H) The value of the anode current below which a device switches from the forward-conduction region to the forward-blocking region.

LASCR Light-activated silicon-controlled rectifier; a four-layer semiconductor device (thyristor) that conducts current in one direction when activated by a sufficient amount of light and continues to conduct until the current falls below a specified value.

Phototransistor A transistor in which base current is produced when light strikes the photosensitive semiconductor base region.

PUT Programmable unijunction transistor; a type of three-terminal thyristor (more like an SCR than a UJT) that is triggered into conduction when the voltage at the anode exceeds the voltage at the gate.

SCR Silicon-controlled rectifier; a type of three-terminal thyristor that conducts current when triggered on by a voltage at the single gate terminal and remains on until the anode current falls below a specified value.

SCS Silicon-controlled switch; a type of four-terminal thyristor that has two gate terminals that are used to trigger the device on and off.

Standoff ratio The characteristic of a UJT that determines its turn-on point.

Thyristor A class of four-layer (*pnpn*) semiconductor devices.

Triac A three-terminal thyristor that can conduct current in either direction when properly activated.

UJT Unijunction transistor; a three-terminal single *pn* junction device that exhibits a negative resistance characteristic.

KEY FORMULAS

11–1	$\eta = \dfrac{r'_{B1}}{r'_{BB}}$	UJT intrinsic standoff ratio
11–2	$V_P = \eta V_{BB} + V_{pn}$	UJT peak-point voltage
11–3	$I_C = \beta_{DC} I_\lambda$	Phototransistor collector current

SELF-TEST

Answers are at the end of the chapter.

1. A thyristor has
   - (a) two *pn* junctions
   - (b) three *pn* junctions
   - (c) four *pn* junctions
   - (d) only two terminals

2. Common types of thyristors include
   - (a) BJTs and SCRs
   - (b) UJTs and PUTs
   - (c) FETs and triacs
   - (d) diacs and triacs

3. A 4-layer diode turns on when the anode to cathode voltage exceeds
   - (a) 0.7 V
   - (b) the gate voltage
   - (c) the forward-breakover voltage
   - (d) the forward-blocking voltage

4. Once it is conducting, a 4-layer diode can be turned off by
   - (a) reducing the current below a certain value
   - (b) disconnecting the anode voltage
   - (c) answers (a) and (b)
   - (d) neither answer (a) nor (b)

5. An SCR differs from the 4-layer diode because
   - (a) it has a gate terminal
   - (b) it is not a thyristor
   - (c) it does not have four layers
   - (d) it cannot be turned on and off

6. An SCR can be turned off by
   - (a) forced commutation
   - (b) a negative pulse on the gate
   - (c) anode current interruption
   - (d) answers (a), (b), and (c)
   - (e) answers (a) and (c)

7. In the forward-blocking region, the SCR is
   - (a) reverse-biased
   - (b) in the *off* state
   - (c) in the *on* state
   - (d) at the point of breakdown

8. The specified value of holding current for an SCR means that
   - (a) the device will turn on when the anode current exceeds this value
   - (b) the device will turn off when the anode current falls below this value
   - (c) the device may be damaged if the anode current exceeds this value
   - (d) the gate current must equal or exceed this value to turn the device on

9. The diac is

    (a) a thyristor

    (b) a bilateral, two-terminal device

    (c) like two parallel 4-layer diodes in reverse directions

    (d) answers (a), (b), and (c)

10. The triac is

    (a) like a bidirectional SCR   (b) a four-terminal device

    (c) not a thyristor   (d) answers (a) and (b)

11. The SCS differs from the SCR because

    (a) it does not have a gate terminal   (b) its holding current is less

    (c) it can handle much higher currents   (d) it has two gate terminals

12. The SCS can be turned on by

    (a) an anode voltage that exceeds forward-breakover voltage

    (b) a positive pulse on the cathode gate

    (c) a negative pulse on the anode gate

    (d) either (b) or (c)

13. The SCS can be turned off by

    (a) a negative pulse on the cathode gate and a positive pulse on the anode gate

    (b) reducing the anode current to below the holding value

    (c) answers (a) and (b)

    (d) a positive pulse on the cathode gate and a negative pulse on the anode gate

14. Which of the following is *not* a characteristic of the UJT?

    (a) intrinsic standoff ratio   (b) negative resistance

    (c) peak-point voltage   (d) bilateral conduction

15. The PUT is

    (a) much like the UJT   (b) not a thyristor

    (c) triggered on and off by the gate-to-anode voltage   (d) not a four-layer device

16. In a phototransistor, base current is

    (a) set by a bias voltage   (b) directly proportional to light

    (c) inversely proportional to light   (d) not a factor

PROBLEMS

Answers to all odd-numbered problems are at the end of the book.

BASIC PROBLEMS

SECTION 11–1 The Basic 4-Layer Device

1. The 4-layer diode in Figure 11–61 is biased such that it is in the forward-conduction region. Determine the anode current for $V_{BR(F)} = 20$ V, $V_{BE} = 0.7$ V, and $V_{CE(sat)} = 0.2$ V.

▶ **FIGURE 11–61**

File circuits are identified with a CD logo and are in the Problems folder on your CD-ROM. Filenames correspond to figure numbers (e.g., F11-61). This circuit is EWB only.

2. (a) Determine the resistance of a certain 4-layer diode in the forward-blocking region if $V_{AK} = 15$ V and $I_A = 1$ μA.

(b) If the forward-breakover voltage is 50 V, how much must V_{AK} be increased to switch the diode into the forward-conduction region?

SECTION 11–2 The Silicon-Controlled Rectifier (SCR)

3. Explain the operation of an SCR in terms of its transistor equivalent.

4. To what value must the variable resistor be adjusted in Figure 11–62 in order to turn the SCR off? Assume $I_H = 10$ mA and $V_{AK} = 0.7$ V.

 ▶ **FIGURE 11–62**

SECTION 11–3 SCR Applications

5. Describe how you would modify the circuit in Figure 11–15 so that the SCR triggers and conducts on the negative half-cycle of the input.

6. What is the purpose of diodes D_1 and D_2 in Figure 11–19?

7. Sketch the V_R waveform for the circuit in Figure 11–63, given the indicated relationship of the input waveforms.

▶ **FIGURE 11–63**

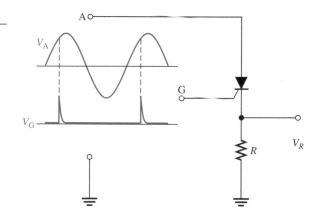

SECTION 11–4 The Diac and Triac

8. Sketch the current waveform for the circuit in Figure 11–64. The diac has a breakover potential of 20 V. $I_H = 20$ mA.

9. Repeat Problem 8 for the triac circuit in Figure 11–65. The breakover potential is 25 V and $I_H = 1$ mA.

▶ **FIGURE 11–64**

15 V peak

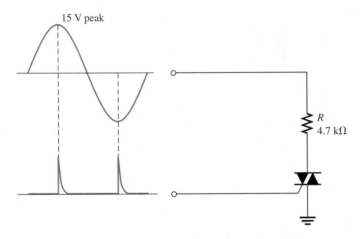

▲ **FIGURE 11–65**

SECTION 11–5 The Silicon-Controlled Switch (SCS)

10. Explain the turn-on and turn-off operation of an SCS in terms of its transistor equivalent.

11. Name the terminals of an SCS.

SECTION 11–6 The Unijunction Transistor (UJT)

12. In a certain UJT, $r'_{B1} = 2.5\ \text{k}\Omega$ and $r'_{B2} = 4\ \text{k}\Omega$. What is the intrinsic standoff ratio?

13. Determine the peak-point voltage for the UJT in Problem 12 if $V_{BB} = 15\ \text{V}$.

14. Find the range of values of R_1 in Figure 11–66 that will ensure proper turn-on and turn-off of the UJT. $\eta = 0.68$, $V_V = 0.8\ \text{V}$, $I_V = 15\ \text{mA}$, $I_P = 10\ \mu\text{A}$, and $V_P = 10\ \text{V}$.

▶ **FIGURE 11–66**

SECTION 11–7   **The Programmable Unijunction Transistor (PUT)**

15. At what anode voltage (V_A) will each PUT in Figure 11–67 begin to conduct?

16. Draw the current waveform for each circuit in Figure 11–67 when there is a 10 V peak sinusoidal voltage at the anode. Neglect the forward voltage of the PUT.

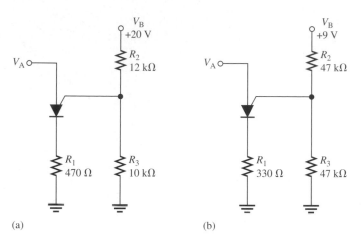

(a)   (b)

▲ **FIGURE 11–67**

17. Sketch the voltage waveform across R_1 in Figure 11–68 in relation to the input voltage waveform.

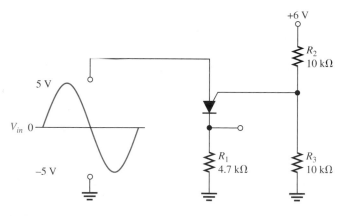

▲ **FIGURE 11–68**

SECTION 11–8   **The Phototransistor**

18. A certain phototransistor in a circuit has a $\beta_{DC} = 200$. If $I_\lambda = 100\ \mu A$, what is the collector current?

19. Determine the output voltage in Figure 11–69, **(a)** when the light source is off, and **(b)** when it is on (assuming the transistor saturates).

20. Determine the emitter current in the photodarlington circuit in Figure 11–70 if, for each $1m/m^2$ of light intensity, $1\ \mu A$ of base current is produced in Q_1.

▲ FIGURE 11–69   ▲ FIGURE 11–70

SECTION 11–9   The Light-Activated SCR (LASCR)

21. By examination of the circuit in Figure 11–71, explain its purpose and basic operation.

▲ FIGURE 11–71

22. Determine the voltage waveform across R_K in Figure 11–72.

▶ FIGURE 11–72

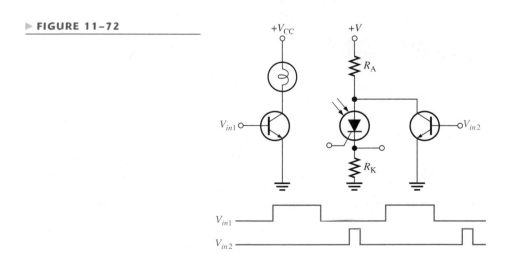

SECTION 11–10   Optical Couplers

23. A particular optical coupler has a current transfer ratio of 30 percent. If the input current is 100 mA, what is the output current?

24. The optical coupler shown in Figure 11–73 is required to deliver at least 10 mA to the external load. If the current transfer ratio is 60 percent, how much current must be supplied to the input?

▲ **FIGURE 11–73**

SYSTEM APPLICATION PROBLEMS

25. In the motor speed-control circuit of Figure 11–57, at which PUT gate voltage does the electric motor run at the fastest speed: 0 V, 2 V, or 5 V?

26. Does the SCR in the motor speed-control circuit turn on earlier or later in the ac cycle if the resistance of the rheostat is reduced?

27. Describe the SCR action as the PUT gate voltage is increased in the motor speed-control circuit.

ADVANCED PROBLEMS

28. Refer to the SCR over-voltage protection circuit in Figure 11–20. For a +12 V output dc power supply, specify the component values that will provide protection for the circuit if the output voltage exceeds +15 V. Assume the fuse is rated at 1 A.

29. Design an SCR crowbar circuit to protect electronic circuits against a voltage from the power supply in excess of 6.2 V.

30. Design a relaxation oscillator to produce a frequency of 2.5 kHz using a UJT with $\eta = 0.75$ and a valley voltage of 1 V. The circuit must operate from a +12 V dc source. Design values of $I_V = 10$ mA and $I_P = 20$ μA are to be used.

EWB/MULTISIM TROUBLESHOOTING PROBLEMS

These file circuits are in the Troubleshooting Problems folder on your CD-ROM.

31. Open file TSP11-31 (EWB only) and determine the fault.

32. Open file TSP11-32 and determine the fault.

33. Open file TSP11-33 and determine the fault.

ANSWERS

SECTION REVIEWS

SECTION 11–1   The Basic 4-Layer Device

1. The 4-layer diode is a thyristor because it has four semiconductor layers in a *pnpn* configuration.

2. In the forward-blocking region, the diode is off (nonconducting).

3. The device turns on and conducts when V_{AK} exceeds the forward-breakover voltage.

4. When the anode current is reduced below the holding current value, the device turns off.

SECTION 11–2   The Silicon-Controlled Rectifier (SCR)

1. An SCR (silicon-controlled rectifier) is a three-terminal thyristor.

2. The SCR terminals are anode, cathode, and gate.

3. A positive gate pulse turns the SCR on.

4. Reduce the anode current below I_H (holding current) to turn a conducting SCR off.

SECTION 11–3   SCR Applications

1. The SCR will conduct for more than 90° but less than 180°.

2. To block discharge of the battery through that path

SECTION 11–4 The Diac and Triac

1. The diac is like two parallel 4-layer diodes connected in opposite directions.

2. A triac is like two parallel SCRs having a common gate and connected in opposite directions.

3. A triac has a gate terminal, but a diac does not.

SECTION 11–5 The Silicon-Controlled Switch (SCS)

1. An SCS can be turned off with the application of a gate pulse, but an SCR cannot.

2. A positive pulse on the cathode gate or a negative pulse on the anode gate turns the SCS on.

3. An SCS can be turned off by any of the following:

    (a) positive pulse on anode gate   **(b)** negative pulse on cathode gate

    (c) reduce anode current below holding value   **(d)** completely interrupt anode current

SECTION 11–6 The Unijunction Transistor (UJT)

1. The UJT terminals are base 1, base 2, and emitter.

2. $\eta = r'_{B1}/r'_{BB}$

3. R, C, and η determine the period.

SECTION11–7 The Programmable Unijunction Transistor (PUT)

1. *Programmable* means that the turn-on voltage can be adjusted to a desired value.

2. The PUT is a thyristor, similar in structure to an SCR, but it is turned on by the anode-to-gate voltage. It has a negative resistance characteristic like the UJT.

SECTION 11–8 The Phototransistor

1. The base current of a phototransistor is light induced.

2. Base

3. The collector current depends on β_{DC} and I_λ.

SECTION 11–9 The Light-Activated SCR (LASCR)

1. Most LASCRs can be operated as conventional SCRs.

2. A series switch to shut off the anode current is required.

SECTION 11–10 Optical Couplers

1. An LED

2. Phototransistor, photodarlington, LASCR, phototriac, linear amplifier

RELATED PROBLEMS FOR EXAMPLES

11–1 10 MΩ

11–2 7.29 Ω

11–3 No. The current is less than I_H.

11–4 $V_{AK} = V_s$

11–5 By increasing V_{BB}

11–6 343 kΩ $> R_1 >$ 1.95 kΩ

SELF-TEST

  1. (b)   **2.** (d)   **3.** (c)   **4.** (c)   **5.** (a)

  6. (e)   **7.** (b)   **8.** (b)   **9.** (d)   **10.** (a)

11. (d)   **12.** (d)   **13.** (c)   **14.** (d)   **15.** (c)

16. (b)

12

OPERATIONAL AMPLIFIERS

INTRODUCTION

So far, you have studied a number of important electronic devices. These devices, such as the diode and the transistor, are separate devices that are individually packaged and interconnected in a circuit with other devices to form a complete, functional unit. Such devices are referred to as *discrete components*.

Now you will begin the study of linear integrated circuits (ICs), where many transistors, diodes, resistors, and capacitors are fabricated on a single tiny chip of semiconductive material and packaged in a single case to form a functional circuit. An integrated circuit, such as an operational amplifier (op-amp), is treated as a single device. This means that you will be concerned with what the circuit does more from an external viewpoint than from an internal, component-level viewpoint.

In this chapter, you will learn the basics of op-amps, which are the most versatile and widely used of all linear integrated circuits. You will also learn about open-loop and closed-loop frequency response, bandwidth, phase shift, and other frequency-related parameters. The effects of negative feedback will be examined.

- Describe the basic op-amp and its characteristics

- Discuss op-amp modes and several parameters

- Explain negative feedback in op-amp circuits

- Analyze the noninverting, voltage-follower, and inverting op-amp configurations

- Describe impedances of the three op-amp configurations

- Discuss op-amp compensation

- Analyze the open-loop response of an op-amp

- Analyze the closed-loop response of an op-amp

- Troubleshoot op-amp circuits

KEY TERMS

- Operational amplifier (op-amp)

- Single-ended

- Differential

- Common mode

- CMRR

- Open-loop voltage gain

- Slew rate

- Negative feedback

- Closed-loop voltage gain

- Noninverting amplifier

- Voltage-follower

- Inverting amplifier

- Phase shift

- Gain-bandwidth product

■ ■ ■ SYSTEM APPLICATION PREVIEW

For the system application at the end of the chapter, the op-amp is used as an audio preamplifier in an AM receiver. The AM receiver receives amplitude-modulated frequencies from 535 kHz to 1605 kHz, extracts the audio signal from the modulated carrier frequency, and amplifies the audio signal to drive a speaker. AM is a process by which the amplitude of a higher frequency signal (carrier) is varied (modulated) by a lower frequency signal (audio in this case). The focus of the system application is on the audio amplifier circuit, which includes an op-amp as well as a push-pull power amplifier.

WWW. VISIT THE COMPANION WEBSITE

Study aids for this chapter are available at

http://www.prenhall.com/floyd

12–1 INTRODUCTION TO OPERATIONAL AMPLIFIERS

Early operational amplifiers (op-amps) were used primarily to perform mathematical operations such as addition, subtraction, integration, and differentiation—thus the term *operational*. These early devices were constructed with vacuum tubes and worked with high voltages. Today's op-amps are linear integrated circuits (ICs) that use relatively low dc supply voltages and are reliable and inexpensive.

After completing this section, you should be able to

- **Describe the basic op-amp and its characteristics**
- Recognize the op-amp symbol
- Identify the terminals on op-amp packages
- Describe the ideal op-amp
- Describe the practical op-amp

Symbol and Terminals

The standard **operational amplifier (op-amp)** symbol is shown in Figure 12–1(a). It has two input terminals, the inverting (−) input and the noninverting (+) input, and one output terminal. The typical op-amp operates with two dc supply voltages, one positive and the other negative, as shown in Figure 12–1(b). Usually these dc voltage terminals are left off the schematic symbol for simplicity but are understood to be there. Some typical op-amp IC packages are shown in Figure 12–1(c).

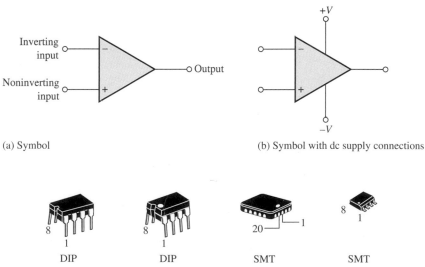

(a) Symbol

(b) Symbol with dc supply connections

(c) Typical packages. Pin 1 is indicated by a notch or dot on dual-in-line (DIP) and surface-mount technology (SMT) packages, as shown.

▲ FIGURE 12–1

Op-amp symbols and packages.

The Ideal Op-Amp

To illustrate what an op-amp is, let's consider its *ideal* characteristics. A practical op-amp, of course, falls short of these ideal standards, but it is much easier to understand and analyze the device from an ideal point of view.

First, the ideal op-amp has *infinite voltage gain* and *infinite bandwidth*. Also, it has an *infinite input impedance* (open) so that it does not load the driving source. Finally, it has a *zero output impedance*. These characteristics are illustrated in Figure 12–2(a). The input voltage, V_{in}, appears between the two input terminals, and the output voltage is $A_v V_{in}$, as indicated by the internal voltage source symbol. The concept of infinite input impedance is a particularly valuable analysis tool for the various op-amp configurations, which will be discussed in Section 12–4.

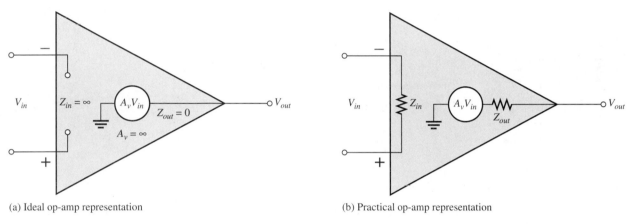

(a) Ideal op-amp representation   (b) Practical op-amp representation

▲ **FIGURE 12–2**

Basic op-amp representations.

The Practical Op-Amp

Although modern **integrated circuit (IC)** op-amps approach parameter values that can be treated as ideal in many cases, the ideal device can never be made.

Any device has limitations, and the IC op-amp is no exception. Op-amps have both voltage and current limitations. Peak-to-peak output voltage, for example, is usually limited to slightly less than the two supply voltages. Output current is also limited by internal restrictions such as power dissipation and component ratings.

Characteristics of a practical op-amp are *very high voltage gain, very high input impedance, very low output impedance,* and *wide bandwidth*. Three of these are labelled in Figure 12–2(b).

Internal Block Diagram of an Op-Amp

A typical op-amp is made up of three types of amplifier circuit: a *differential amplifier*, a *voltage amplifier*, and a *push-pull amplifier*, as shown in Figure 12 3.

A differential amplifier is the input stage for the op-amp. It has two inputs and provides amplification of the difference voltage between the two inputs. The voltage amplifier is usually a class A amplifier that provides additional op-amp gain. Some op-amps may have more than one voltage amplifier stage. A push-pull class B amplifier is used for the output stage.

▲ FIGURE 12–3

Basic internal arrangement of an op-amp.

SECTION 12–1 REVIEW

Answers are at the end of the chapter.

1. What are the connections to a basic op-amp?
2. Describe some of the characteristics of a practical op-amp.
3. List the amplifier stages in a typical op-amp.

12–2 OP-AMP INPUT MODES AND PARAMETERS

In this section, important op-amp input modes and several parameters are defined. Also several common IC op-amps are compared in terms of these parameters.

After completing this section, you should be able to

- **Discuss op-amp modes and several parameters**
- Explain single-ended input operation
- Explain differential-input operation
- Explain common-mode operation
- Define *common-mode rejection ratio*
- Discuss open-loop voltage gain
- Discuss common-mode input voltage range
- Define *input offset voltage*
- Discuss input offset voltage drift with temperature
- Define *input bias current*
- Define *input impedance*
- Define *input offset current*
- Define *output impedance*
- Define *slew rate*
- Discuss frequency response
- Compare the parameters of several types of IC op-amps

Input Signal Modes

Single-Ended Input   When an op-amp is operated in the **single-ended** mode, one input is grounded and the signal voltage is applied only to the other input, as shown in Figure 12–4. In the case where the signal voltage is applied to the inverting input as in part (a), an inverted, amplified signal voltage appears at the output. In the case where the signal is applied to the noninverting input with the inverting input grounded, as in Figure 12–4(b), a noninverted, amplified signal voltage appears at the output.

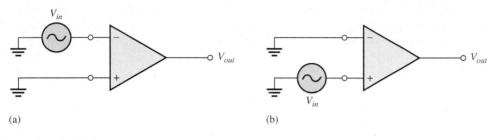

(a)   (b)

▲ **FIGURE 12–4**

Single-ended input mode.

Differential Input   In the **differential** mode, two opposite-polarity (out-of-phase) signals are applied to the inputs, as shown in Figure 12–5. This type of operation is also referred to as *double-ended.* The amplified difference between the two inputs appears on the output.

▶ **FIGURE 12–5**

Differential input mode.

Common–Mode Input   In the **common mode**, two signal voltages of the same phase, frequency, and amplitude are applied to the two inputs, as shown in Figure 12–6. When equal input signals are applied to both inputs, they cancel, resulting in a zero output voltage.

This action is called *common-mode rejection*. Its importance lies in the situation where an unwanted signal appears commonly on both op-amp inputs. Common-mode rejection means that this unwanted signal will not appear on the output and distort the desired signal. Common-mode signals (noise) generally are the result of the pick-up of radiated energy on the input lines, from adjacent lines, the 60 Hz power line, or other sources.

▶ **FIGURE 12–6**

Common-mode operation.

Common-Mode Rejection Ratio

Desired signals can appear on only one input or with opposite polarities on both input lines. These desired signals are amplified and appear on the output as previously discussed. Unwanted signals (noise) appearing with the same polarity on both input lines are essentially cancelled by the op-amp and do not appear on the output. The measure of an amplifier's ability to reject common-mode signals is a parameter called the **CMRR (common-mode rejection ratio)**.

Ideally, an op-amp provides a very high gain for desired signals (single-ended or differential) and zero gain for common-mode signals. Practical op-amps, however, do exhibit a very small common-mode gain (usually much less than 1), while providing a high open-loop voltage gain (usually several thousand). The higher the open-loop gain with respect to the common-mode gain, the better the performance of the op-amp in terms of rejection of common-mode signals. This suggests that a good measure of the op-amp's performance in rejecting unwanted common-mode signals is the ratio of the open-loop voltage gain, A_{ol}, to the common-mode gain, A_{cm}. This ratio is the common-mode rejection ratio, CMRR.

Equation 12–1

$$CMRR = \frac{A_{ol}}{A_{cm}}$$

The higher the CMRR, the better. A very high value of CMRR means that the open-loop gain, A_{ol}, is high and the common-mode gain, A_{cm}, is low.

The CMRR is often expressed in decibels (dB) as

Equation 12–2

$$CMRR = 20 \log\left(\frac{A_{ol}}{A_{cm}}\right)$$

The **open-loop voltage gain, A_{ol},** of an op-amp is the internal voltage gain of the device and represents the ratio of output voltage to input voltage when there are no external components. The open-loop voltage gain is set entirely by the internal design. Open-loop voltage gain can range up to 200,000 and is *not a well-controlled parameter*. Data sheets often refer to the open-loop voltage gain as the *large-signal voltage gain*.

EXAMPLE 12–1

A certain op-amp has an open-loop voltage gain of 100,000 and a common-mode gain of 0.2. Determine the CMRR and express it in decibels.

Solution A_{ol} = 100,000, and A_{cm} = 0.2. Therefore,

$$CMRR = \frac{A_{ol}}{A_{cm}} = \frac{100,000}{0.2} = \mathbf{500,000}$$

Expressed in decibels,

$$CMRR = 20 \log(500,000) = \mathbf{114\ dB}$$

Related Problem*   Determine the CMRR and express it in dB for an op-amp with an open-loop voltage gain of 85,000 and a common-mode gain of 0.25.

---

*Answers are at the end of the chapter.

A CMRR of 100,000, for example, means that the desired input signal (differential) is amplified 100,000 times more than the unwanted noise (common-mode). So, as an example, if the amplitudes of the differential input signal and the common-mode noise are equal, the desired signal will appear on the output 100,000 times greater in amplitude than the noise. Thus, the noise or interference has been essentially eliminated.

Common-Mode Input Voltage Range

All op-amps have limitations on the range of voltages over which they will operate. The *common-mode input voltage range* is the range of input voltages which, when applied to both inputs, will not cause clipping or other output distortion. Many op-amps have common-mode input voltage ranges of ±10 V with dc supply voltages of ±15 V.

Input Offset Voltage

The ideal op-amp produces zero volts out for zero volts in. In a practical op-amp, however, a small dc voltage, $V_{OUT(error)}$, appears at the output when no differential input voltage is applied. Its primary cause is a slight mismatch of the base-emitter voltages of the differential amplifier input stage of an op-amp.

As specified on an op-amp data sheet, the *input offset voltage, V_{OS}*, is the differential dc voltage required between the inputs to force the output to zero volts. Typical values of input offset voltage are in the range of 2 mV or less. In the ideal case, it is 0 V.

Input Offset Voltage Drift with Temperature

The *input offset voltage drift* is a parameter related to V_{OS} that specifies how much change occurs in the input offset voltage for each degree change in temperature. Typical values range anywhere from about 5 μV per degree Celsius to about 50 μV per degree Celsius. Usually, an op-amp with a higher nominal value of input offset voltage exhibits a higher drift.

Input Bias Current

You have seen that the input terminals of a bipolar differential amplifier are the transistor bases and, therefore, the input currents are the base currents.

The *input bias current* is the dc current required by the inputs of the amplifier to properly operate the first stage. By definition, the input bias current is the *average* of both input currents and is calculated as follows:

$$I_{BIAS} = \frac{I_1 + I_2}{2}$$

Equation 12–3

The concept of input bias current is illustrated in Figure 12–7.

▶ **FIGURE 12–7**

Input bias current is the average of the two op-amp input currents.

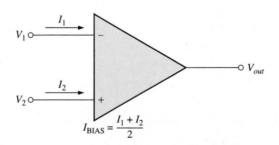

Input Impedance

Two basic ways of specifying the input impedance of an op-amp are the differential and the common mode. The *differential input impedance* is the total resistance between the inverting and the noninverting inputs, as illustrated in Figure 12–8(a). Differential impedance is measured by determining the change in bias current for a given change in differential input voltage. The *common-mode input impedance* is the resistance between each input and ground and is measured by determining the change in bias current for a given change in common-mode input voltage. It is depicted in Figure 12–8(b).

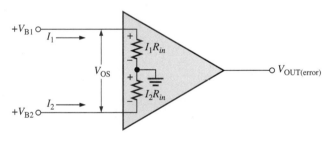

▲ FIGURE 12–8

Op-amp input impedance.

Input Offset Current

Ideally, the two input bias currents are equal, and thus their difference is zero. In a practical op-amp, however, the bias currents are not exactly equal.

The *input offset current, I_{OS},* is the difference of the input bias currents, expressed as an absolute value.

Equation 12–4

$$I_{OS} = |I_1 - I_2|$$

Actual magnitudes of offset current are usually at least an order of magnitude (ten times) less than the bias current. In many applications, the offset current can be neglected. However, high-gain, high-input impedance amplifiers should have as little I_{OS} as possible, because the difference in currents through large input resistances develops a substantial offset voltage, as shown in Figure 12–9.

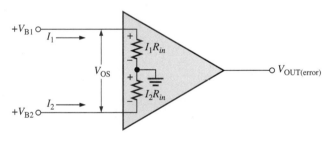

▲ FIGURE 12–9

Effect of input offset current.

The offset voltage developed by the input offset current is

$$V_{OS} = I_1 R_{in} - I_2 R_{in} = (I_1 - I_2)R_{in}$$

Equation 12–5

$$V_{OS} = I_{OS}R_{in}$$

The error created by I_{OS} is amplified by the gain A_v of the op-amp and appears in the output as

Equation 12–6

$$V_{OUT(error)} = A_v I_{OS}R_{in}$$

A change in offset current with temperature affects the error voltage. Values of temperature coefficient for the offset current in the range of 0.5 nA per degree Celsius are common.

Output Impedance

The *output impedance* is the resistance viewed from the output terminal of the op-amp, as indicated in Figure 12–10.

▶ **FIGURE 12–10**

Op-amp output impedance.

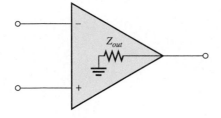

Slew Rate

The maximum rate of change of the output voltage in response to a step input voltage is the **slew rate** of an op-amp. The slew rate is dependent upon the high-frequency response of the amplifier stages within the op-amp.

Slew rate is measured with an op-amp connected as shown in Figure 12–11(a). This particular op-amp connection is a unity-gain, noninverting configuration that will be discussed in Section 12–4. It gives a worst-case (slowest) slew rate. Recall that the high-frequency components of a voltage step are contained in the rising edge and that the upper critical frequency of an amplifier limits its response to a step input. For a step input, the slope on the output is inversely proportional to the upper critical frequency. Slope increases as upper critical frequency decreases.

(a) Test circuit

(b) Step input voltage and the resulting output voltage

▲ **FIGURE 12–11**

Slew-rate measurement.

A pulse is applied to the input as shown, and the ideal output voltage is measured as indicated in Figure 12–11(b). The width of the input pulse must be sufficient to allow the output to "slew" from its lower limit to its upper limit, as shown. As you can see, a certain time interval, Δt, is required for the output voltage to go from its lower limit $-V_{max}$ to its upper limit $+V_{max}$, once the input step is applied. The slew rate is expressed as

$$\text{Slew rate} = \frac{\Delta V_{out}}{\Delta t}$$

Equation 12–7

where $\Delta V_{out} = +V_{max} - (-V_{max})$. The unit of slew rate is volts per microsecond (V/μs).

EXAMPLE 12–2

The output voltage of a certain op-amp appears as shown in Figure 12–12 in response to a step input. Determine the slew rate.

▶ **FIGURE 12–12**

Solution The output goes from the lower to the upper limit in 1 μs. Since this response is not ideal, the limits are taken at the 90% points, as indicated. So, the upper limit is +9 V and the lower limit is −9 V. The slew rate is

$$\text{Slew rate} = \frac{\Delta V_{out}}{\Delta t} = \frac{+9\ \text{V} - (-9\ \text{V})}{1\ \mu\text{s}} = \textbf{18 V/}\boldsymbol{\mu}\textbf{s}$$

Related Problem When a pulse is applied to an op-amp, the output voltage goes from −8 V to +7 V in 0.75 μs. What is the slew rate?

Frequency Response

The internal amplifier stages that make up an op-amp have voltage gains limited by junction capacitances, as discussed in Chapter 10. Although the differential amplifiers used in op-amps are somewhat different from the basic amplifiers discussed earlier, the same principles apply. An op-amp has no internal coupling capacitors, however; therefore, the low-frequency response extends down to dc (0 Hz).

Comparison of Op-Amp Parameters

Table 12–1 provides a comparison of values of some of the parameters just described for several common IC op-amps. Any values not listed were not given on the manufacturer's data sheet.

Other Features

Most available op-amps have three important features: short-circuit protection, no latch-up, and input offset nulling. Short-circuit protection keeps the circuit from being damaged if the output becomes shorted, and the no latch-up feature prevents the op-amp from hanging up in one output state (high or low voltage level) under certain input conditions. Input offset nulling is achieved by an external potentiometer that sets the output voltage at precisely zero with zero input.

▼ TABLE 12-1

OP-AMP	CMRR (dB) (MIN)	OPEN-LOOP GAIN (TYP)	INPUT OFFSET VOLTAGE (mV) (MAX)	INPUT BIAS CURRENT (nA) (MAX)	INPUT IMPEDANCE (MΩ) (MIN)	SLEW RATE (V/μs) (TYP)	COMMENT
LM741C	70	200,000	6	500	0.3	0.5	Industry standard
LM101A	80	160,000	7.5	250	1.5	–	General-purpose
OP113	100	2,400,000	0.075	600	–	1.2	Low noise, low drift
OP177	130	12,000,000	0.01	1.5	26	0.3	Ultra precision
OP184	60	240,000	0.065	350	–	2.4	Precision, rail-to-rail*
AD8009	50	–	5	150	–	5500	BW = 700 MHz, ultra fast, low distortion, current feedback
AD8041	74	56,000	7	2000	.16	160	BW = 160 MHz, rail-to-rail
AD8055	82	3500	5	1200	10	1400	Very fast voltage feedback

*Rail-to-rail means that the output voltage can go as high as the supply voltages.

SECTION 12-2 REVIEW

1. Distinguish between differential and single-ended inputs.
2. Define *common-mode rejection?*
3. For a given value of open-loop gain, does a higher CMRR result in a higher or lower common-mode gain?
4. List at least ten op-amp parameters.
5. Which two parameters, not including the frequency response, are frequency dependent?

12-3 NEGATIVE FEEDBACK

Negative feedback is one of the most useful concepts in electronics, particularly in op-amp applications. Negative feedback is the process whereby a portion of the output voltage of an amplifier is returned to the input with a phase angle that opposes (or subtracts from) the input signal.

After completing this section, you should be able to

- **Explain negative feedback in op-amp circuits**
- Discuss why negative feedback is used
- Describe the effects of negative feedback

Negative feedback is illustrated in Figure 12–13. The inverting (−) input effectively makes the feedback signal 180° out of phase with the input signal.

▶ FIGURE 12–13

Illustration of negative feedback.

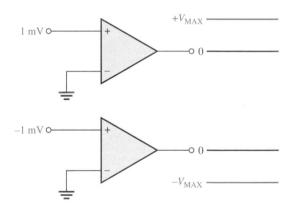

Why Use Negative Feedback?

As you have seen, the inherent open-loop voltage gain of a typical op-amp is very high (usually greater than 100,000). Therefore, an extremely small input voltage drives the op-amp into its saturated output states. In fact, even the input offset voltage of the op-amp can drive it into saturation. For example, assume $V_{IN} = 1$ mV and $A_{ol} = 100,000$. Then,

$$V_{IN}A_{ol} = (1 \text{ mV})(100,000) = 100 \text{ V}$$

Since the output level of an op-amp can never reach 100 V, it is driven deep into saturation and the output is limited to its maximum output levels, as illustrated in Figure 12–14 for both a positive and a negative input voltage of 1 mV.

▶ FIGURE 12–14

Without negative feedback, a small input voltage drives the op-amp to its output limits and it becomes nonlinear.

▼ TABLE 12–2

	VOLTAGE GAIN	INPUT Z	OUTPUT Z	BANDWIDTH
Without negative feedback	A_{ol} is too high for linear amplifier applications	Relatively high (see Table 12–1)	Relatively low	Relatively narrow (because the gain is so high)
With negative feedback	A_{cl} is set to desired value by the feedback circuit	Can be increased or reduced to a desired value depending on type of circuit	Can be reduced to a desired value	Significantly wider

The usefulness of an op-amp operated without negative feedback is severely restricted and is generally limited to comparator and other applications (to be studied in Chapter 13). With negative feedback, the closed-loop voltage gain (A_{cl}) can be reduced and controlled so that the op-amp can function as a linear amplifier. In addition to providing a controlled, stable voltage gain, negative feedback also provides for control of the input and output impedances and amplifier bandwidth. Table 12–2 summarizes the general effects of negative feedback on op-amp performance.

**SECTION 12–3
REVIEW**

1. What are the benefits of negative feedback in an op-amp circuit?
2. Why is it necessary to reduce the gain of an op-amp from its open-loop value?

12–4 OP-AMPS WITH NEGATIVE FEEDBACK

In this section, you will learn several basic ways in which an op-amp can be connected using negative feedback to stabilize the gain and increase frequency response. The extremely high open-loop gain of an op-amp creates an unstable situation because a small noise voltage on the input can be amplified to a point where the amplifier is driven out of its linear region. Also, unwanted oscillations can occur. In addition, the open-loop gain parameter of an op-amp can vary greatly from one device to the next. Negative feedback takes a portion of the output and applies it back out of phase with the input, creating an effective reduction in gain. This closed-loop gain is usually much less than the open-loop gain and independent of it.

After completing this section, you should be able to

- **Analyze the noninverting, voltage-follower, and inverting op-amp configurations**

- Define *closed-loop voltage gain*

- Identify the noninverting amplifier configuration

- Determine the closed-loop voltage gain of a noninverting amplifier

- Identify the voltage-follower configuration

- Identify the inverting amplifier configuration

- Determine the closed-loop voltage gain of an inverting amplifier

Closed-Loop Voltage Gain, A_{cl}

The **closed-loop voltage gain** is the voltage gain of an op-amp with external feedback. The amplifier configuration consists of the op-amp and an external negative feedback circuit that connects the output to the inverting input. The closed-loop voltage gain is determined by the external component values and can be precisely controlled by them.

Noninverting Amplifier

An op-amp connected in a **closed-loop** configuration as a **noninverting amplifier** with a controlled amount of voltage gain is shown in Figure 12–15. The input signal is applied

Noninverting amplifier.

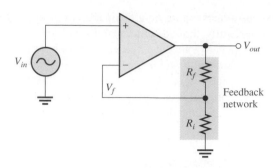

to the noninverting (+) input. The output is applied back to the inverting (−) input through the feedback circuit (closed loop) formed by the input resistor R_i and the feedback resistor R_f. This creates negative feedback as follows. Resistors R_i and R_f form a voltage-divider circuit, which reduces V_{out} and connects the reduced voltage V_f to the inverting input. The feedback voltage is expressed as

$$V_f = \left(\frac{R_i}{R_i + R_f}\right)V_{out}$$

The difference of the input voltage, V_{in}, and the feedback voltage, V_f, is the differential input to the op-amp, as shown in Figure 12–16. This differential voltage is amplified by the open-loop voltage gain of the op-amp (A_{ol}) and produces an output voltage expressed as

$$V_{out} = A_{ol}(V_{in} - V_f)$$

Letting $R_i/(R_i + R_f) = B$, which is the attenuation of the feedback circuit, and then substituting BV_{out} for V_f in the previous equation,

$$V_{out} = A_{ol}(V_{in} - BV_{out})$$

Then applying basic algebra,

$$V_{out} = A_{ol}V_{in} - A_{ol}BV_{out}$$
$$V_{out} + A_{ol}BV_{out} = A_{ol}V_{in}$$
$$V_{out}(1 + A_{ol}B) = A_{ol}V_{in}$$

Differential input, $V_{in} - V_f$.

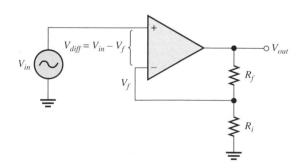

Since the overall voltage gain of the amplifier in Figure 12–15 is V_{out}/V_{in}, it can be expressed as

$$\frac{V_{out}}{V_{in}} = \frac{A_{ol}}{1 + A_{ol}B}$$

The product $A_{ol}B$ is typically much greater than 1, so the equation simplifies to

$$\frac{V_{out}}{V_{in}} = \frac{A_{ol}}{A_{ol}B} = \frac{1}{B}$$

The closed-loop gain of the noninverting (NI) amplifier is the reciprocal of the attenuation (B) of the feedback circuit (voltage-divider).

$$A_{cl(NI)} = \frac{V_{out}}{V_{in}} = \frac{1}{B} = \frac{R_i + R_f}{R_i}$$

Therefore,

$$A_{cl(NI)} = 1 + \frac{R_f}{R_i}$$

Equation 12–8

Notice that the closed-loop voltage gain is not at all dependent on the op-amp's open-loop voltage gain under the condition $A_{ol}B \gg 1$. The closed-loop gain can be set by selecting values of R_i and R_f.

EXAMPLE 12–3

Determine the gain of the amplifier in Figure 12–17. The open-loop voltage gain of the op-amp is 100,000.

▶ **FIGURE 12–17**

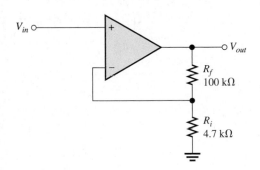

Solution   This is a noninverting op-amp configuration. Therefore, the closed-loop voltage gain is

$$A_{cl(NI)} = 1 + \frac{R_f}{R_i} = 1 + \frac{100\ k\Omega}{4.7\ k\Omega} = \mathbf{22.3}$$

Related Problem   If the open-loop gain of the amplifier in Figure 12–17 is 150,000 and R_f is increased to 150 kΩ, determine the closed-loop gain.

Open file E12-03 in the Examples folder on your CD-ROM. Measure the closed-loop voltage gain of the amplifier and compare with the calculated value.

Voltage-Follower

The **voltage-follower** configuration is a special case of the noninverting amplifier where all of the output voltage is fed back to the inverting (−) input by a straight connection, as shown in Figure 12–18. As you can see, the straight feedback connection has a voltage gain of 1 (which means there is no gain). The closed-loop voltage gain of a noninverting amplifier is $1/B$ as previously derived. Since $B = 1$ for a voltage-follower, the closed-loop voltage gain of the voltage-follower is

Equation 12–9

$$A_{cl(VF)} = 1$$

▶ **FIGURE 12–18**

Op-amp voltage-follower.

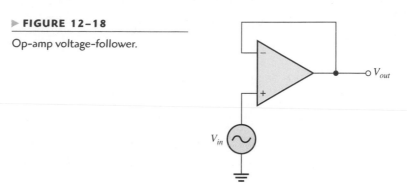

The most important features of the voltage-follower configuration are its very high input impedance and its very low output impedance. These features make it a nearly ideal buffer amplifier for interfacing high-impedance sources and low-impedance loads. This is discussed further in Section 12–5.

Inverting Amplifier

An op-amp connected as an **inverting amplifier** with a controlled amount of voltage gain is shown in Figure 12–19. The input signal is applied through a series input resistor R_i to the inverting (−) input. Also, the output is fed back through R_f to the same input. The noninverting (+) input is grounded.

▶ **FIGURE 12–19**

Inverting amplifier.

At this point, the ideal op-amp parameters mentioned earlier are useful in simplifying the analysis of this circuit. In particular, the concept of infinite input impedance is of great value. An infinite input impedance implies zero current at the inverting input. If there is zero current through the input impedance, then there must be *no* voltage drop between the inverting and noninverting inputs. This means that the voltage at the inverting (−) input is zero because the noninverting (+) input is grounded. This zero voltage at the inverting input terminal is referred to as *virtual ground*. This condition is illustrated in Figure 12–20(a).

(a) Virtual ground

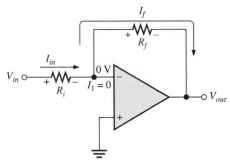

(b) $I_{in} = I_f$ and current into the inverting input (I_1) is 0.

▲ **FIGURE 12–20**

Virtual ground concept and closed-loop voltage gain development for the inverting amplifier.

Since there is no current at the inverting input, the current through R_i and the current through R_f are equal, as shown in Figure 12–20(b).

$$I_{in} = I_f$$

The voltage across R_i equals V_{in} because of virtual ground on the other side of the resistor. Therefore,

$$I_{in} = \frac{V_{in}}{R_i}$$

Also, the voltage across R_f equals $-V_{out}$ because of virtual ground, and therefore,

$$I_f = \frac{-V_{out}}{R_f}$$

Since $I_f = I_{in}$,

$$\frac{-V_{out}}{R_f} = \frac{V_{in}}{R_i}$$

Rearranging the terms,

$$\frac{V_{out}}{V_{in}} = -\frac{R_f}{R_i}$$

Of course, V_{out}/V_{in} is the overall gain of the inverting (I) amplifier.

$$A_{cl(I)} = -\frac{R_f}{R_i}$$

Equation 12–10

Equation 12–10 shows that the closed-loop voltage gain of the inverting amplifier ($A_{cl(I)}$) is the ratio of the feedback resistance (R_f) to the input resistance (R_i). *The closed-loop gain is independent of the op-amp's internal open-loop gain.* Thus, the negative feedback stabilizes the voltage gain. The negative sign indicates inversion.

EXAMPLE 12–4

Given the op-amp configuration in Figure 12–21, determine the value of R_f required to produce a closed-loop voltage gain of -100.

▶ **FIGURE 12–21**

Solution

Knowing that $R_i = 2.2$ kΩ and the absolute value of the closed-loop gain is $|A_{cl(I)}| = 100$, calculate R_f as follows:

$$|A_{cl(I)}| = \frac{R_f}{R_i}$$

$$R_f = |A_{cl(I)}|R_i = (100)(2.2 \text{ k}\Omega) = \textbf{220 k}\boldsymbol{\Omega}$$

Related Problem

If R_i is changed to 2.7 kΩ in Figure 12–21, what value of R_f is required to produce a closed-loop gain with an absolute value of 25?

Open file E12-04 in the Examples folder on your CD-ROM. The circuit has a value of R_f which was calculated to be 220 kΩ. Measure the closed-loop voltage gain and see if it agrees with the specified value.

SECTION 12–4 REVIEW

1. What is the main purpose of negative feedback?

2. The closed-loop voltage gain of each of the op-amp configurations discussed is dependent on the internal open-loop voltage gain of the op-amp. (True or False)

3. The attenuation of the negative feedback circuit of a noninverting op-amp configuration is 0.02. What is the closed-loop gain of the amplifier?

12–5 EFFECTS OF NEGATIVE FEEDBACK ON OP-AMP IMPEDANCES

In this section, you will see how negative feedback affects the input and output impedances of an op-amp. The effects on both inverting and noninverting amplifiers are examined.

After completing this section, you should be able to

- **Describe impedances of the three op-amp configurations**

- Determine input and output impedances of a noninverting amplifier

- Determine input and output impedances of a voltage-follower

- Determine input and output impedances of an inverting amplifier

Impedances of a Noninverting Amplifier

Input Impedance The input impedance of a noninverting amplifier can be developed with the aid of Figure 12–22. For this analysis, assume a small differential voltage, V_d, exists between the two inputs, as indicated. This means that you cannot assume the op-amp's input impedance to be infinite or the input current to be zero. Express the input voltage as

$$V_{in} = V_d + V_f$$

Substituting BV_{out} for V_f yields

$$V_{in} = V_d + BV_{out}$$

Since $V_{out} \cong A_{ol}V_d$ (A_{ol} is the open-loop gain of the op-amp),

$$V_{in} = V_d + A_{ol}BV_d = (1 + A_{ol}B)V_d$$

Now substituting $I_{in}Z_{in}$ for V_d,

$$V_{in} = (1 + A_{ol}B)I_{in}Z_{in}$$

where Z_{in} is the open-loop input impedance of the op-amp (without feedback connections).

$$\frac{V_{in}}{I_{in}} = (1 + A_{ol}B)Z_{in}$$

V_{in}/I_{in} is the overall input impedance of a closed-loop noninverting amplifier configuration.

$$Z_{in(NI)} = (1 + A_{ol}B)Z_{in}$$

Equation 12–11

This equation shows that the input impedance of the noninverting amplifier configuration with negative feedback is much greater than the internal input impedance of the op-amp itself (without feedback).

▶ **FIGURE 12–22**

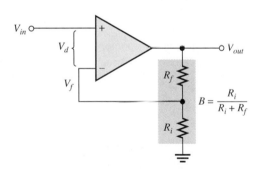

Output Impedance An expression for output impedance of a noninverting amplifier can be developed with the aid of Figure 12–23. By applying Kirchhoff's law to the output circuit,

$$V_{out} = A_{ol}V_d - Z_{out}I_{out}$$

The differential input voltage is $V_d = V_{in} - V_f$; therefore, by assuming that $A_{ol}V_d \gg Z_{out}I_{out}$, you can express the output voltage as

$$V_{out} \cong A_{ol}(V_{in} - V_f)$$

Substituting BV_{out} for V_f,

$$V_{out} \cong A_{ol}(V_{in} - BV_{out})$$

▶ FIGURE 12–23

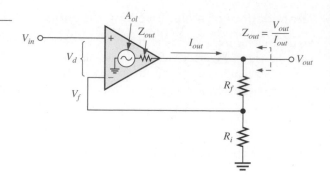

Remember, B is the attenuation of the negative feedback circuit. Expanding and factoring yields

$$V_{out} \cong A_{ol}V_{in} - A_{ol}BV_{out}$$
$$A_{ol}V_{in} \cong V_{out} + A_{ol}BV_{out} \cong (1 + A_{ol}B)V_{out}$$

Since the output impedance of the noninverting amplifier configuration is $Z_{out(NI)} = V_{out}/I_{out}$, you can substitute $I_{out}Z_{out(NI)}$ for V_{out}; therefore,

$$A_{ol}V_{in} = (1 + A_{ol}B)I_{out}Z_{out(NI)}$$

Dividing both sides of the above expression by I_{out},

$$\frac{A_{ol}V_{in}}{I_{out}} = (1 + A_{ol}B)Z_{out(NI)}$$

The term on the left is the internal output impedance of the op-amp (Z_{out}) because, without feedback, $A_{ol}V_{in} = V_{out}$. Therefore,

$$Z_{out} = (1 + A_{ol}B)Z_{out(NI)}$$

Thus,

Equation 12–12

$$Z_{out(NI)} = \frac{Z_{out}}{1 + A_{ol}B}$$

This equation shows that the output impedance of the noninverting amplifier configuration with negative feedback is much less than the internal output impedance, Z_{out}, of the op-amp itself (without feedback) because Z_{out} is divided by the factor $1 + A_{ol}B$.

EXAMPLE 12–5

(a) Determine the input and output impedances of the amplifier in Figure 12–24. The op-amp data sheet gives $Z_{in} = 2 \text{ M}\Omega$, $Z_{out} = 75 \ \Omega$, and $A_{ol} = 200,000$.

(b) Find the closed-loop voltage gain.

▶ FIGURE 12–24

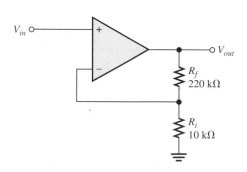

Solution   **(a)** The attenuation, B, of the feedback circuit is

$$B = \frac{R_i}{R_i + R_f} = \frac{10 \text{ k}\Omega}{230 \text{ k}\Omega} = 0.0435$$

$$Z_{in(\text{NI})} = (1 + A_{ol}B)Z_{in} = [1 + (200,000)(0.0435)](2 \text{ M}\Omega)$$

$$= (1 + 8700)(2 \text{ M}\Omega) = \textbf{17.4 G}\Omega$$

$$Z_{out(\text{NI})} = \frac{Z_{out}}{1 + A_{ol}B} = \frac{75 \ \Omega}{1 + 8700} = \textbf{8.6 m}\Omega$$

(b) $A_{cl(\text{NI})} = \dfrac{1}{B} = \dfrac{1}{0.0435} \cong \textbf{23.0}$

Related Problem   **(a)** Determine the input and output impedances in Figure 12–24 for op-amp data sheet values of $Z_{in} = 3.5 \text{ M}\Omega$, $Z_{out} = 82 \ \Omega$, and $A_{ol} = 135,000$.

(b) Find A_{cl}.

Open file E12-05 in the Examples folder on your CD-ROM. Measure the closed-loop voltage gain and compare with the calculated value.

Voltage-Follower Impedances

Since a voltage-follower is a special case of the noninverting amplifier configuration, the same impedance formulas are used but with $B = 1$.

$$Z_{in(\text{VF})} = (1 + A_{ol})Z_{in}$$

Equation 12–13

$$Z_{out(\text{VF})} = \frac{Z_{out}}{1 + A_{ol}}$$

Equation 12–14

As you can see, the voltage-follower input impedance is greater for a given A_{ol} and Z_{in} than for the noninverting amplifier configuration with the voltage-divider feedback circuit. Also, its output impedance is much smaller.

EXAMPLE 12–6

The same op-amp in Example 12–5 is used in a voltage-follower configuration. Determine the input and output impedances.

Solution   Since $B = 1$,

$$Z_{in(\text{VF})} = (1 + A_{ol})Z_{in} = (1 + 200,000)(2 \text{ M}\Omega) \cong \textbf{400 G}\Omega$$

$$Z_{out(\text{VF})} = \frac{Z_{out}}{1 + A_{ol}} = \frac{75 \ \Omega}{1 + 200,000} = \textbf{375 } \mu\Omega$$

Notice that $Z_{in(\text{VF})}$ is much greater than $Z_{in(\text{NI})}$, and $Z_{out(\text{VF})}$ is much less than $Z_{out(\text{NI})}$ from Example 12–5.

Related Problem   If the op-amp in this example is replaced with one having a higher open-loop gain, how are the input and output impedances affected?

Impedances of an Inverting Amplifier

The input and output impedances of an inverting op-amp configuration are developed with the aid of Figure 12–25. Both the input signal and the negative feedback are applied, through resistors, to the inverting (−) terminal as shown.

▶ **FIGURE 12–25**

Inverting amplifier.

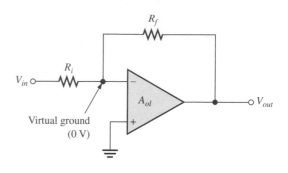

Input Impedance   The input impedance for an inverting amplifier is

Equation 12–15

$$Z_{in(I)} \cong R_i$$

This is because the inverting input of the op-amp is at virtual ground (0 V), and the input source simply sees R_i to ground, as shown in Figure 12–26.

▶ **FIGURE 12–26**

Output Impedance   As with a noninverting amplifier, the output impedance of an inverting amplifier is decreased by the negative feedback. In fact, the expression is the same as for the noninverting case.

Equation 12–16

$$Z_{out(I)} = \frac{Z_{out}}{1 + A_{ol}B}$$

The output impedance of both the noninverting and the inverting amplifier configurations is very low; in fact, it is almost zero in practical cases. Because of this near zero output impedance, any load impedance connected to the op-amp output can vary greatly and not change the output voltage at all.

EXAMPLE 12–7

Find the values of the input and output impedances in Figure 12–27. Also, determine the closed-loop voltage gain. The op-amp has the following parameters: $A_{ol} = 50{,}000$; $Z_{in} = 4 \text{ M}\Omega$; and $Z_{out} = 50 \text{ }\Omega$.

▶ **FIGURE 12–27**

Solution

$$Z_{in(I)} \cong R_i = \mathbf{1.0\ k\Omega}$$

The feedback attenuation, B, is

$$B = \frac{R_i}{R_i + R_f} = \frac{1.0\ k\Omega}{101\ k\Omega} = 0.001$$

Then

$$Z_{out(I)} = \frac{Z_{out}}{1 + A_{ol}B} = \frac{50\ \Omega}{1 + (50,000)(0.001)}$$

$$= \mathbf{980\ m\Omega} \quad \text{(zero for all practical purposes)}$$

The closed-loop voltage gain is

$$A_{cl(I)} = -\frac{R_f}{R_i} = -\frac{100\ k\Omega}{1.0\ k\Omega} = \mathbf{-100}$$

Related Problem Determine the input and output impedances and the closed-loop voltage gain in Figure 12–27. The op-amp parameters and circuit values are as follows: $A_{ol} = 100,000$; $Z_{in} = 5\ M\Omega$; $Z_{out} = 75\ \Omega$; $R_i = 560\ \Omega$; and $R_f = 82\ k\Omega$.

Open file E12-07 in the Examples folder on your CD-ROM and measure the closed-loop voltage gain. Compare to the calculated result.

SECTION 12–5 REVIEW

1. How does the input impedance of a noninverting amplifier configuration compare to the input impedance of the op-amp itself?

2. When an op-amp is connected in a voltage-follower configuration, does the input impedance increase or decrease?

3. Given that $R_f = 100\ k\Omega$; $R_i = 2\ k\Omega$; $A_{ol} = 120,000$; $Z_{in} = 2\ M\Omega$; and $Z_{out} = 60\ \Omega$, what are $Z_{in(I)}$ and $Z_{out(I)}$ for an inverting amplifier configuration?

12–6 BIAS CURRENT AND OFFSET VOLTAGE COMPENSATION

Until now, we have treated the op-amp as an ideal device in many of our discussions. However, certain deviations from the ideal op-amp must be recognized because of their effects on its operation. Transistors within the op-amp must be biased so that they have the correct values of base and collector currents and collector-to-emitter voltages. The ideal op-amp has no input current at its terminals, but in fact, the practical op-amp has small input bias currents typically in the nA range. Also, small internal imbalances in the transistors effectively produce a small offset voltage between the inputs. These nonideal parameters were described in Section 12–2.

After completing this section, you should be able to

- **Discuss op-amp compensation**
- Describe the effect of input bias current
- Explain bias current compensation
- Describe the effect of input offset voltage
- Explain input offset voltage compensation

Effect of an Input Bias Current

Figure 12–28 is an inverting amplifier with zero input voltage. Ideally, the current through R_i is zero because the input voltage is zero and the voltage at the inverting ($-$) terminal is zero. The small input bias current, I_1, is through R_f from the output terminal. I_1 creates a voltage drop across R_f, as indicated. The positive side of R_f is the output terminal, and therefore, the output error voltage is I_1R_f when it should be zero.

Figure 12–29 is a voltage-follower with zero input voltage and a source resistance, R_s. In this case, an input bias current, I_1, produces a drop across R_s and creates an output voltage error as shown. The voltage at the inverting input terminal decreases to $-I_1R_s$ because the negative feedback tends to maintain a differential voltage of zero, as indicated. Since the inverting terminal is connected directly to the output terminal, the output error voltage is $-I_1R_s$.

▲ FIGURE 12–28

Input bias current creates output error voltage (I_1R_f) in inverting amplifier.

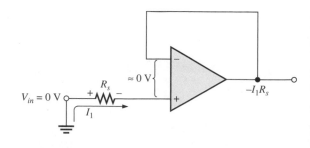

▲ FIGURE 12–29

Input bias current creates output error voltage in a voltage-follower.

Figure 12–30 is a noninverting amplifier with zero input voltage. Ideally, the voltage at the inverting terminal is also zero, as indicated. The input bias current, I_1, produces a voltage drop across R_f and thus creates an output error voltage of $I_1 R_f$, just as with the inverting amplifier.

▶ **FIGURE 12–30**

Input bias current creates output error voltage in noninverting amplifier.

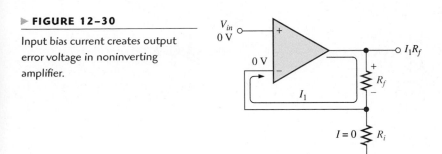

Bias Current Compensation in a Voltage-Follower

The output error voltage due to bias currents in a voltage-follower can be sufficiently reduced by adding a resistor, R_f, equal to the source resistance, R_s, in the feedback path, as shown in Figure 12–31. The voltage drop created by I_1 across the added resistor subtracts from the $-I_2 R_s$ output error voltage. If $I_1 = I_2$, then the output voltage is zero. Usually I_1 does not quite equal I_2; but even in this case, the output error voltage is reduced as follows because I_{OS} is less than I_2.

$$V_{OUT(error)} = |I_1 - I_2| R_s = I_{OS} R_s$$

where I_{OS} is the input offset current.

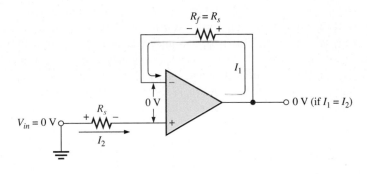

▲ **FIGURE 12–31**

Bias current compensation in a voltage-follower.

Bias Current Compensation in Other Op-Amp Configurations

To compensate for the effect of bias current in the noninverting amplifier, a resistor R_c is added, as shown in Figure 12–32(a). The compensating resistor value equals the parallel combination of R_i and R_f. The input current creates a voltage drop across R_c that offsets the voltage across the combination of R_i and R_f, thus sufficiently reducing the output error voltage. The inverting amplifier is similarly compensated, as shown in Figure 12–32(b).

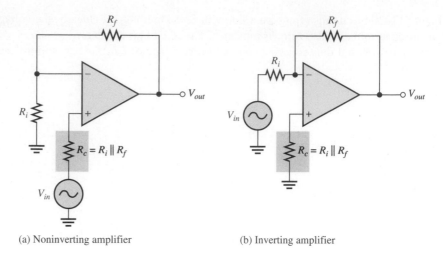

(a) Noninverting amplifier   (b) Inverting amplifier

▲ **FIGURE 12–32**

Bias current compensation in the noninverting and inverting amplifier configurations.

Use of a BIFET Op-Amp to Eliminate the Need for Bias Current Compensation

The BIFET op-amp uses both bipolar junction transistors and JFETs in its internal circuitry. The JFETs are used as the input devices to achieve a higher input impedance than is possible with standard BJT amplifiers. Because of their very high input impedance, BIFETs typically have input bias currents that are much smaller than in BJT op-amps, thus reducing or eliminating the need for bias current compensation.

Effect of Input Offset Voltage

The output voltage of an op-amp should be zero when the differential input is zero. However, there is always a small output error voltage present whose value typically ranges from microvolts to millivolts. This is due to unavoidable imbalances within the internal op-amp transistors aside from the bias currents previously discussed. In a negative feedback configuration, the input offset voltage V_{IO} can be visualized as an equivalent small dc voltage source, as illustrated in Figure 12–33 for a voltage-follower. Generally, the output error voltage due to the input offset voltage is

$$V_{OUT(error)} = A_{cl}V_{IO}$$

For the case of the voltage-follower, $A_{cl} = 1$, so

$$V_{OUT(error)} = V_{IO}$$

▶ **FIGURE 12–33**

Input offset voltage equivalent.

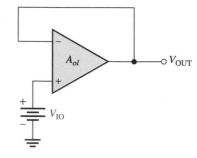

Input Offset Voltage Compensation

Most integrated circuit op-amps provide a means of compensating for offset voltage. This is usually done by connecting an external potentiometer to designated pins on the IC package, as illustrated in Figure 12–34(a) and (b) for a 741 op-amp. The two terminals are labelled *offset null*. With no input, the potentiometer is simply adjusted until the output voltage reads 0, as shown in Figure 12–34(c).

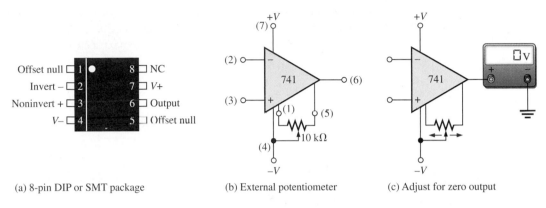

(a) 8-pin DIP or SMT package   (b) External potentiometer   (c) Adjust for zero output

▲ **FIGURE 12–34**

Input offset voltage compensation for a 741 op-amp.

SECTION 12–6 REVIEW

1. What are two sources of dc output error voltages?
2. How do you compensate for bias current in a voltage-follower?

12–7 OPEN-LOOP RESPONSE

You have learned how closed-loop voltage gains of the basic op-amp configurations are determined, and the distinction between open-loop gain and closed-loop gain was established. In this section, you will learn about the open-loop frequency response and the open-loop phase response of an op-amp. Open-loop responses relate to an op-amp with no external feedback. The frequency response indicates how the voltage gain changes with frequency, and the phase response indicates how the phase shift between the input and output signal changes with frequency. The open-loop gain, like the β of a transistor, varies greatly from one device to the next of the same type and cannot be depended upon to have a constant value.

After completing this section, you should be able to

- **Analyze the open-loop response of an op-amp**
- Discuss the frequency dependency of gain
- Explain the open-loop bandwidth
- Explain the unity-gain bandwidth
- Determine phase shift
- Discuss how internal stages affect the overall response
- Discuss critical frequencies and roll-off rates
- Determine overall phase response

Review of Op-Amp Gains

Figure 12–35 illustrates the open-loop and closed-loop amplifier configurations. As shown in part (a), the open-loop voltage gain, A_{ol}, of an op-amp is the internal voltage gain of the device and represents the ratio of output voltage to input voltage. Notice that there are no external components, so the open-loop voltage gain is set entirely by the internal design. In the closed-loop op-amp configuration shown in part (b), the closed-loop voltage gain, A_{cl}, is the voltage gain of an op-amp with external feedback. The closed-loop voltage gain is determined by the external component values for an inverting amplifier configuration. The closed-loop voltage gain can be precisely controlled by external component values. The closed-loop response of op-amps is covered in the next section.

(a) Open-loop   (b) Closed-loop (inverting configuration)

▲ **FIGURE 12–35**

Open-loop and closed-loop op-amp configurations.

Voltage Gain Is Frequency Dependent

In the previous sections, all of the voltage gain expressions were based on the midrange gain and were considered independent of the frequency. The midrange open-loop gain of an op-amp extends from zero frequency (dc) up to a critical frequency at which the gain is 3 dB less than the midrange value. This concept should be familiar from your study of Chapter 10. Op-amps are dc amplifiers (no capacitive coupling between stages), and therefore, there is no lower critical frequency. This means that the midrange gain extends down to zero frequency (dc), and dc voltages are amplified the same as midrange signal frequencies.

An open-loop response curve (Bode plot) for a certain op-amp is shown in Figure 12–36. Most op-amp data sheets show this type of curve or specify the midrange open-loop gain. Notice that the curve rolls off (decreases) at -20 dB per decade (-6 dB per octave). The midrange gain is 200,000, which is 106 dB, and the critical (cutoff) frequency is approximately 10 Hz.

3 dB Open-Loop Bandwidth

Recall from Chapter 10 that the bandwidth of an ac amplifier is the frequency range between the points where the gain is 3 dB less than the midrange gain. In general, the bandwidth equals the upper critical frequency (f_{cu}) minus the lower critical frequency (f_{cl}).

$$BW = f_{cu} - f_{cl}$$

Since f_{cl} for an op-amp is zero, the bandwidth is simply equal to the upper critical frequency.

Equation 12–17

$$BW = f_{cu}$$

From now on, we will refer to f_{cu} as simply f_c; and we will use open-loop (*ol*) or closed-loop (*cl*) subscript designators, for example, $f_{c(ol)}$.

▲ FIGURE 12–36

Ideal plot of open-loop voltage gain versus frequency for a typical op-amp. The frequency scale is logarithmic.

Unity-Gain Bandwidth

Notice in Figure 12–36 that the gain steadily decreases to a point where it is equal to one (0 dB). The value of the frequency at which this unity gain occurs is the *unity-gain bandwidth*.

Gain-Versus-Frequency Analysis

The *RC* lag (low-pass) circuits within an op-amp are responsible for the roll-off in gain as the frequency increases, just as was discussed for the discrete amplifiers in Chapter 10. From basic ac circuit theory, the attenuation of an *RC* lag circuit, such as in Figure 12–37, is expressed as

$$\frac{V_{out}}{V_{in}} = \frac{X_C}{\sqrt{R^2 + X_C^2}}$$

Dividing both the numerator and denominator to the right of the equals sign by X_C,

$$\frac{V_{out}}{V_{in}} = \frac{1}{\sqrt{1 + R^2/X_C^2}}$$

▶ FIGURE 12–37

RC lag circuit.

The critical frequency of an *RC* circuit is

$$f_c = \frac{1}{2\pi RC}$$

Dividing both sides by *f* gives

$$\frac{f_c}{f} = \frac{1}{2\pi RCf} = \frac{1}{(2\pi fC)R}$$

Since $X_C = 1/(2\pi fC)$, the above expression can be written as

$$\frac{f_c}{f} = \frac{X_C}{R}$$

Substituting this result in the previous equation for V_{out}/V_{in} produces the following expression for the attenuation of an RC lag circuit in terms of frequency:

Equation 12–18

$$\frac{V_{out}}{V_{in}} = \frac{1}{\sqrt{1 + f^2/f_c^2}}$$

If an op-amp is represented by a voltage gain element with a gain of $A_{ol(mid)}$ plus a single RC lag circuit, as shown in Figure 12–38, then the total open-loop gain of the op-amp is the product of the midrange open-loop gain, $A_{ol(mid)}$, and the attenuation of the RC circuit.

Equation 12–19

$$A_{ol} = \frac{A_{ol(mid)}}{\sqrt{1 + f^2/f_c^2}}$$

▶ **FIGURE 12–38**

Op-amp represented by a gain element and an internal RC circuit.

As you can see from Equation 12–19, the open-loop gain equals the midrange gain when the signal frequency f is much less than the critical frequency f_c and drops off as the frequency increases. Since f_c is part of the open-loop response of an op-amp, we will refer to it as $f_{c(ol)}$.

The following example demonstrates how the open-loop gain decreases as the frequency increases above $f_{c(ol)}$.

EXAMPLE 12–8

Determine A_{ol} for the following values of f. Assume $f_{c(ol)} = 100$ Hz and $A_{ol(mid)} = 100,000$.

(a) $f = 0$ Hz   (b) $f = 10$ Hz   (c) $f = 100$ Hz   (d) $f = 1000$ Hz

Solution   (a) $A_{ol} = \dfrac{A_{ol(mid)}}{\sqrt{1 + f^2/f_{c(ol)}^2}} = \dfrac{100,000}{\sqrt{1 + 0}} =$ **100,000**

(b) $A_{ol} = \dfrac{100,000}{\sqrt{1 + (0.1)^2}} =$ **99,503**

(c) $A_{ol} = \dfrac{100,000}{\sqrt{1 + (1)^2}} = \dfrac{100,000}{\sqrt{2}} =$ **70,710**

(d) $A_{ol} = \dfrac{100,000}{\sqrt{1 + (10)^2}} =$ **9950**

Related Problem   Find A_{ol} for the following frequencies. Assume $f_{c(ol)} = 200$ Hz and $A_{ol(mid)} = 80,000$.

(a) $f = 2$ Hz   (b) $f = 10$ Hz   (c) $f = 2500$ Hz

Phase Shift

As you know from Chapter 10, an *RC* circuit causes a propagation delay from input to output, thus creating a **phase shift** between the input signal and the output signal. An *RC* lag circuit such as found in an op-amp stage causes the output signal voltage to lag the input, as shown in Figure 12–39. From basic ac circuit theory, the phase shift, θ, is

$$\theta = -\tan^{-1}\left(\frac{R}{X_C}\right)$$

Since $R/X_C = f/f_c$,

$$\theta = -\tan^{-1}\left(\frac{f}{f_c}\right)$$

Equation 12–20

The negative sign indicates that the output lags the input. This equation shows that the phase shift increases with frequency and approaches $-90°$ as f becomes much greater than f_c.

▲ **FIGURE 12–39**

Output voltage lags input voltage.

EXAMPLE 12–9

Calculate the phase shift for an *RC* lag circuit for each of the following frequencies, and then plot the curve of phase shift versus frequency. Assume $f_c = 100$ Hz.

(a) $f = 1$ Hz   (b) $f = 10$ Hz   (c) $f = 100$ Hz

(d) $f = 1000$ Hz   (e) $f = 10,000$ Hz

Solution   (a) $\theta = -\tan^{-1}\left(\frac{f}{f_c}\right) = -\tan^{-1}\left(\frac{1\ \text{Hz}}{100\ \text{Hz}}\right) = -0.573°$

(b) $\theta = -\tan^{-1}\left(\frac{10\ \text{Hz}}{100\ \text{Hz}}\right) = -5.71°$

(c) $\theta = -\tan^{-1}\left(\frac{100\ \text{Hz}}{100\ \text{Hz}}\right) = -45°$

(d) $\theta = -\tan^{-1}\left(\frac{1000\ \text{Hz}}{100\ \text{Hz}}\right) = -84.3°$

(e) $\theta = -\tan^{-1}\left(\frac{10,000\ \text{Hz}}{100\ \text{Hz}}\right) = -89.4°$

The phase shift-versus-frequency curve is plotted in Figure 12–40. Note that the frequency axis is logarithmic.

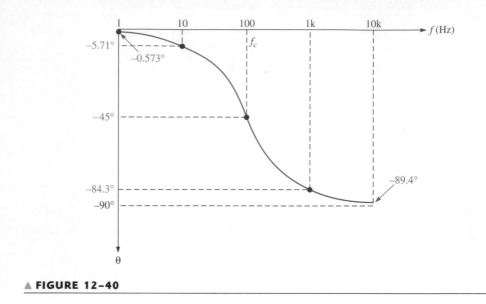

▲ **FIGURE 12–40**

Related Problem At what frequency, in this example, is the phase shift 60°?

Complete Frequency Response

Previously, an op-amp was assumed to have a constant roll-off of −20 dB/decade above its critical frequency. For most op-amps this is the case; for some, however, the situation is more complex. The more complex IC operational amplifier may consist of two or more cascaded amplifier stages. The gain of each stage is frequency dependent and rolls off at −20 dB/decade above its critical frequency. Therefore, the total response of an op-amp is a composite of the individual responses of the internal stages. As an example, a three-stage op-amp is represented in Figure 12–41(a), and the frequency response of each stage is shown in Figure 12–41(b). As you know, dB gains are added so that the total op-amp frequency response is as shown in Figure 12–41(c). Since the roll-off rates are additive, the total roll-off rate increases by −20 dB/decade (−6 dB/octave) as each critical frequency is reached.

Complete Phase Response

In a multistage amplifier, each stage contributes to the total phase lag. As you have seen, each *RC* lag circuit can produce up to a −90° phase shift. Since each stage in an op-amp includes an *RC* lag circuit, a three-stage op-amp, for example, can have a maximum phase lag of −270°. Also, the phase lag of each stage is less than −45° when the frequency is below the critical frequency, equal to −45° at the critical frequency, and greater than −45° when the frequency is above the critical frequency. The phase lags of the stages of an op-amp are added to produce a total phase lag, according to the following formula for three stages:

$$\theta_{tot} = -\tan^{-1}\left(\frac{f}{f_{c1}}\right) - \tan^{-1}\left(\frac{f}{f_{c2}}\right) - \tan^{-1}\left(\frac{f}{f_{c3}}\right)$$

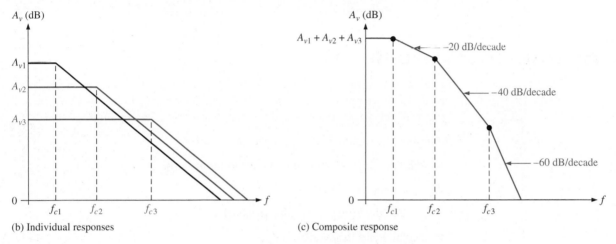

(a) Representation of an op-amp with three internal stages

(b) Individual responses

(c) Composite response

▲ **FIGURE 12–41**

Op-amp open-loop frequency response.

EXAMPLE 12–10

A certain op-amp has three internal amplifier stages with the following gains and critical frequencies:

Stage 1: $A_{v1} = 40$ dB, $f_{c1} = 2$ kHz

Stage 2: $A_{v2} = 32$ dB, $f_{c2} = 40$ kHz

Stage 3: $A_{v3} = 20$ dB, $f_{c3} = 150$ kHz

Determine the open-loop midrange gain in decibels and the total phase lag when $f = f_{c1}$.

Solution $A_{ol(mid)} = A_{v1} + A_{v2} + A_{v3} = 40$ dB $+ 32$ dB $+ 20$ dB $= 92$ dB

$$\theta_{tot} = -\tan^{-1}\left(\frac{f}{f_{c1}}\right) - \tan^{-1}\left(\frac{f}{f_{c2}}\right) - \tan^{-1}\left(\frac{f}{f_{c3}}\right)$$

$$= -\tan^{-1}(1) - \tan^{-1}\left(\frac{2}{40}\right) - \tan^{-1}\left(\frac{2}{150}\right) = -45° - 2.86° - 0.76° = \mathbf{-48.6°}$$

Related Problem The internal stages of a two-stage amplifier have the following characteristics: $A_{v1} = 50$ dB, $A_{v2} = 25$ dB, $f_{c1} = 1500$ Hz, and $f_{c2} = 3000$ Hz. Determine the open-loop midrange gain in decibels and the total phase lag when $f = f_{c1}$.

**SECTION 12–7
REVIEW**

1. How do the open-loop voltage gain and the closed-loop voltage gain of an op-amp differ?

2. The upper critical frequency of a particular op-amp is 100 Hz. What is its open-loop 3 dB bandwidth?

3. Does the open-loop gain increase or decrease with frequency above the critical frequency?

4. If the individual stage gains of an op-amp are 20 dB and 30 dB, what is the total gain in decibels?

5. If the individual phase lags are $-49°$ and $-5.2°$, what is the total phase lag?

12–8 CLOSED-LOOP RESPONSE

Op-amps are normally used in a closed-loop configuration with negative feedback in order to achieve precise control of the gain and bandwidth. In this section, you will see how feedback affects the gain and frequency response of an op-amp.

After completing this section, you should be able to

- **Analyze the closed-loop response of an op-amp**
- Determine the closed-loop gain
- Explain the effect of negative feedback on bandwidth
- Explain gain-bandwidth product

Recall that midrange gain of an op-amp is reduced by negative feedback, as indicated by the following closed-loop gain expressions for the three amplifier configurations previously covered, where B is the feedback attenuation. For a noninverting amplifier,

$$A_{cl(\text{NI})} = \frac{A_{ol}}{1 + A_{ol}B} \cong \frac{1}{B} = 1 + \frac{R_f}{R_i}$$

For an inverting amplifier,

$$A_{cl(\text{I})} \cong -\frac{R_f}{R_i}$$

For a voltage-follower,

$$A_{cl(\text{VF})} = 1$$

Effect of Negative Feedback on Bandwidth

You know how negative feedback affects the gain; now you will learn how it affects the amplifier's bandwidth. The closed-loop critical frequency of an op-amp is

Equation 12–21
$$f_{c(cl)} = f_{c(ol)}(1 + BA_{ol(mid)})$$

This expression shows that the closed-loop critical frequency, $f_{c(cl)}$, is higher than the open-loop critical frequency $f_{c(ol)}$ by the factor $1 + BA_{ol(mid)}$. You will find a derivation of Equation 12–21 in Appendix B.

Since $f_{c(cl)}$ equals the bandwidth for the closed-loop amplifier, the closed-loop bandwidth (BW_{cl}) is also increased by the same factor.

Equation 12–22
$$BW_{cl} = BW_{ol}(1 + BA_{ol(mid)})$$

EXAMPLE 12–11

A certain amplifier has an open-loop midrange gain of 150,000 and an open-loop 3 dB bandwidth of 200 Hz. The attenuation of the feedback loop is 0.002. What is the closed-loop bandwidth?

Solution $BW_{cl} = BW_{ol}(1 + BA_{ol(mid)}) = 200 \text{ Hz}[1 + (0.002)(150,000)] = \mathbf{60.2 \text{ kHz}}$

Related Problem If $A_{ol(mid)} = 200,000$ and $B = 0.05$, what is the closed-loop bandwidth?

Figure 12–42 graphically illustrates the concept of closed-loop response. When the open-loop gain of an op-amp is reduced by negative feedback, the bandwidth is increased. The closed-loop gain is independent of the open-loop gain up to the point of intersection of the two gain curves. This point of intersection is the critical frequency, $f_{c(cl)}$, for the closed-loop response. Notice that the closed-loop gain has the same roll-off rate as the open-loop gain, beyond the closed-loop critical frequency.

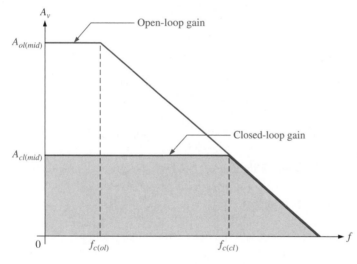

▲ **FIGURE 12–42**

Closed–loop gain compared to open–loop gain.

Gain-Bandwidth Product

An increase in closed-loop gain causes a decrease in the bandwidth and vice versa, such that *the product of gain and bandwidth is a constant*. This is true as long as the roll-off rate is fixed. If you let A_{cl} represent the gain of any of the closed-loop configurations and $f_{c(cl)}$ represent the closed-loop critical frequency (same as the bandwidth), then

$$A_{cl}f_{c(cl)} = A_{ol}f_{c(ol)}$$

The **gain-bandwidth product** is always equal to the frequency at which the op-amp's open-loop gain is unity or 0 dB (unity-gain bandwidth).

Unity-gain bandwidth $= A_{cl}f_{c(cl)}$

Equation 12–23

EXAMPLE 12–12

Determine the bandwidth of each of the amplifiers in Figure 12–43. Both op-amps have an open-loop gain of 100 dB and a unity-gain bandwidth of 3 MHz.

▶ FIGURE 12–43

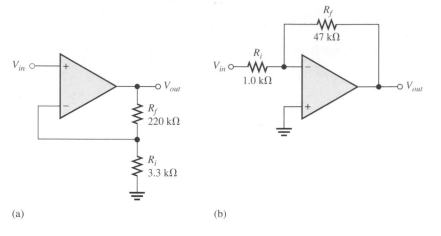

(a)   (b)

Solution   **(a)** For the noninverting amplifier in Figure 12–43(a), the closed-loop gain is

$$A_{cl} \cong \frac{1}{B} = 1 + \frac{R_f}{R_i} = 1 + \frac{220 \text{ k}\Omega}{3.3 \text{ k}\Omega} = 67.7$$

Use Equation 12–23 and solve for $f_{c(cl)}$ (where $f_{c(cl)} = BW_{cl}$).

$$f_{c(cl)} = BW_{cl} = \frac{\text{unity-gain } BW}{A_{cl}}$$

$$BW_{cl} = \frac{3 \text{ MHz}}{67.7} = \mathbf{44.3 \text{ kHz}}$$

(b) For the inverting amplifier in Figure 12–43(b), the closed-loop gain is

$$A_{cl} = -\frac{R_f}{R_i} = -\frac{47 \text{ k}\Omega}{1.0 \text{ k}\Omega} = -47$$

Using the absolute value of A_{cl}, the closed-loop bandwidth is

$$BW_{cl} = \frac{3 \text{ MHz}}{47} = \mathbf{63.8 \text{ kHz}}$$

Related Problem   Determine the bandwidth of each of the amplifiers in Figure 12–43. Both op-amps have an A_{ol} of 90 dB and a unity-gain bandwidth of 2 MHz.

Open file E12-12 in the EWB Examples folder on your CD-ROM. Measure the bandwidth of each amplifier and compare with the calculated values.

SECTION 12–8 REVIEW

1. Is the closed-loop gain always less than the open-loop gain?
2. A certain op-amp is used in a feedback configuration having a gain of 30 and a bandwidth of 100 kHz. If the external resistor values are changed to increase the gain to 60, what is the new bandwidth?
3. What is the unity-gain bandwidth of the op-amp in Question 2?

12–9 TROUBLESHOOTING

As a technician, you may encounter situations in which an op-amp or its associated circuitry has malfunctioned. The op-amp is a complex integrated circuit with many types of internal failures possible. However, since you cannot troubleshoot the op-amp internally, you treat it as a single device with only a few connections to it. If it fails, you replace it just as you would a resistor, capacitor, or transistor.

After completing this section, you should be able to

- **Troubleshoot op-amp circuits**
- Analyze faults in a noninverting amplifier
- Analyze faults in a voltage-follower
- Analyze faults in an inverting amplifier

In the basic op-amp configurations, there are only a few external components that can fail. These are the feedback resistor, the input resistor, and the potentiometer used for off-set voltage compensation. Also, of course, the op-amp itself can fail or there can be faulty contacts in the circuit. Let's examine the three basic configurations for possible faults and the associated symptoms.

Faults in the Noninverting Amplifier

The first thing to do when you suspect a faulty circuit is to check for the proper supply voltage and ground. Having done that, several other possible faults are as follows.

Open Feedback Resistor If the feedback resistor, R_f, in Figure 12–44 opens, the op-amp is operating with its very high open-loop gain, which causes the input signal to drive

(a) R_f open

(b) R_i open

(c) Input to op-amp open internally

▲ **FIGURE 12–44**

Faults in the noninverting amplifier.

the device into nonlinear operation and results in a severely clipped output signal as shown in part (a).

Open Input Resistor In this case, you still have a closed-loop configuration. But, since R_i is open and effectively equal to infinity (∞), the closed-loop gain from Equation 12–8 is

$$A_{cl(NI)} = 1 + \frac{R_f}{R_i} = 1 + \frac{R_f}{\infty} = 1 + 0 = 1$$

This shows that the amplifier acts like a voltage-follower. You would observe an output signal that is the same as the input, as indicated in Figure 12–44(b).

Internally Open Noninverting Op–Amp Input In this situation, because the input voltage is not applied to the op-amp, the output is zero. This is indicated in Figure 12–44(c).

Other Op–Amp Faults In general, an internal failure will result in a loss or distortion of the output signal. The best approach is to first make sure that there are no external failures or faulty conditions. If everything else is good, then the op-amp must be bad.

Faults in the Voltage-Follower

The voltage-follower is a special case of the noninverting amplifier. Except for a bad op-amp, a bad external connection, or a problem with the offset null potentiometer, about the only thing that can happen in a voltage-follower circuit is an open feedback loop. This would have the same effect as an open feedback resistor as previously discussed.

Faults in the Inverting Amplifier

Open Feedback Resistor If R_f opens, as indicated in Figure 12–45(a), the input signal still feeds through the input resistor and is amplified by the high open-loop gain of the op-amp. This forces the device to be driven into nonlinear operation, and you will see an output something like that shown. This is the same result as in the noninverting amplifier configuration.

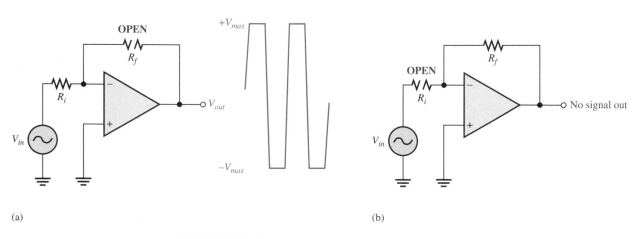

(a)

(b)

▲ **FIGURE 12–45**

Faults in the inverting amplifier.

Open Input Resistor   This prevents the input signal from getting to the op-amp input, so there will be no output signal, as indicated in Figure 12–45(b).

Failures in the op-amp itself or the offset null potentiometer have the same effects as previously discussed for the noninverting amplifier configuration.

EWB/Multisim Troubleshooting Exercises

These file circuits are in the Troubleshooting Exercises folder on your CD-ROM.

1. Open file TSE12-01. Determine if the circuit is working properly, and, if not, determine the fault.

2. Open file TSE12-02. Determine if the circuit is working properly, and, if not, determine the fault.

3. Open file TSE12-03. Determine if the circuit is working properly, and, if not, determine the fault.

4. Open file TSE12-04. Determine if the circuit is working properly, and, if not, determine the fault.

SECTION 12–9 REVIEW

1. If you notice that the op-amp output signal is beginning to clip on one peak as you increase the input signal, what should you check?

2. For a noninverting amplifier, if there is no op-amp output signal when there is a verified input signal at the input pin, what would you suspect as being faulty?

SYSTEM APPLICATION

In this system application you will learn something about an AM receiver. In particular, you will concentrate on the audio amplifier board. The circuit is designed with both discrete components and an integrated circuit (LM101A op-amp). You will apply the knowledge of op-amps acquired in this chapter and knowledge from previous chapters to this assignment.

Basic Operation of the System

Figure 12–46 shows the block diagram of a superheterodyne AM (amplitude modulation) receiver. The antenna picks up all signals across the AM broadcast band from 535 kHz to 1605 kHz and feeds them to the **RF** (radio frequency) amplifier. A signal in the AM band is called a carrier and it is modulated with the audio signal at the transmitting station. Amplitude modulation is accomplished by varying the amplitude of the carrier proportional to the audio signal.

The RF amplifier selects a desired frequency from all of those received and amplifies the extremely small signal coming from the antenna. Since it is a tuned amplifier, it is highly frequency selective and eliminates essentially all signals but the one to which it is tuned. The amplified AM signal from the RF amplifier goes to the mixer where it is combined with the output of the local oscillator which has a frequency of 455 kHz above the frequency of the selected carrier frequency. The mixer, through a process called *heterodyning*, produces output frequencies equal to both the sum and difference of the selected carrier frequency and the local oscillator frequency. The sum frequency is filtered out and only the difference frequency of 455 kHz is used. The 455 kHz is called the intermediate frequency (IF) and still carries the same audio modulation that was on the carrier frequency.

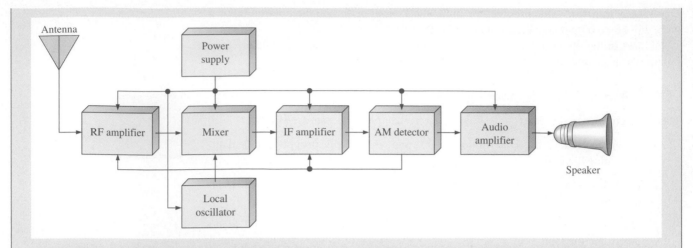

▲ **FIGURE 12–46**

Block diagram of basic superheterodyne AM receiver.

▲ **FIGURE 12–47**

Audio amplifier board.

The IF amplifier is tuned to 455 kHz and amplifies the signal. The detector takes the AM signal from the IF amplifier and recovers the audio signal while eliminating the intermediate frequency. The small audio signal from the output of the detector is amplified by the audio amplifier circuits, which includes both a preamplifier and a power amplifier. The power amplifier drives the speaker which converts the audio signal to sound.

The Photocell/Amplifier Circuit Board

■ Make sure that the circuit board shown in Figure 12–47 is correctly assembled by checking it against the schematic in Figure 12–48. A pin diagram for the power transistors is shown for reference. Backside interconnections are shown as darker traces.

■ Label a copy of the board with component and input/output designations in agreement with the schematic.

▲ **FIGURE 12–48**

Audio amplifier schematic.

Analysis of the Audio Amplifier Circuit

- Determine the midrange voltage gain of the amplifier.

- Determine the lower critical frequency. Given that the upper critical frequency is 15 kHz, what is the bandwidth?

- Determine the maximum peak-to-peak input voltage that can be applied to the audio amplifier without producing a distorted output signal. Assume that the maximum output peaks are 1 V less than the dc supply voltages.

- Calculate the power to the speaker for the maximum voltage output.

Test Procedure

- Develop a step-by-step set of instructions on how to check the audio amplifier circuit board for proper operation.

- Specify voltage values for all the measurements to be made.

- Provide a fault analysis for all possible component failures.

Troubleshooting

Problems have developed in three boards. Based on the test bench measurements for each board indicated in Figure 12–49, determine the most likely fault in each case. The circled numbers indicate test point connections to the circuit board. Assume that each board has the proper dc supply voltage.

Final Report (Optional)

Submit a final written report on the photocell/amplifier circuit board using an organized format that includes the following:

1. A physical description of the circuit.

2. A discussion of the operation of the circuit.

3. A list of the specifications.

4. A list of parts with part numbers if available.

5. A list of the types of problems on the three faulty circuit boards.

6. A description of how you determined the problem on each of the faulty circuit boards.

▲ FIGURE 12–49

Test results for three faulty circuit boards.

SUMMARY OF OP-AMP CONFIGURATIONS

BASIC OP-AMP

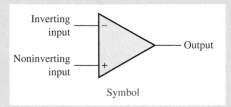

- Very high open-loop voltage gain
- Very high input impedance
- Very low output impedance

NONINVERTING AMPLIFIER

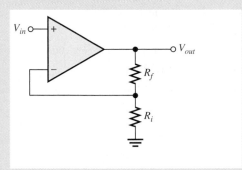

- Voltage gain:

$$A_{cl(\text{NI})} = 1 + \frac{R_f}{R_i}$$

- Input impedance:

$$Z_{in(\text{NI})} = (1 + A_{ol}B)Z_{in}$$

- Output impedance:

$$Z_{out(\text{NI})} = \frac{Z_{out}}{1 + A_{ol}B}$$

VOLTAGE-FOLLOWER

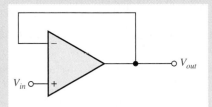

- Voltage gain:

$$A_{cl(\text{VF})} = 1$$

- Input impedance:

$$Z_{in(\text{VF})} = (1 + A_{ol})Z_{in}$$

- Output impedance:

$$Z_{out(\text{VF})} = \frac{Z_{out}}{1 + A_{ol}}$$

INVERTING AMPLIFIER

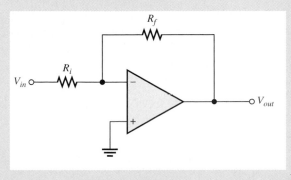

- Voltage gain:

$$A_{cl(\text{I})} = -\frac{R_f}{R_i}$$

- Input impedance:

$$Z_{in(\text{I})} \cong R_i$$

- Output impedance:

$$Z_{out(1)} = \frac{Z_{out}}{1 + A_{ol}B}$$

CHAPTER SUMMARY

- The basic op-amp has three terminals not including power and ground: inverting $(-)$ input, non-inverting $(+)$ input, and output.
- Most op-amps require both a positive and a negative dc supply voltage.
- The ideal op-amp has infinite input impedance, zero output impedance, infinite open-loop voltage gain, infinite bandwidth, and infinite CMRR.
- A practical op-amp has very high input impedance, very low output impedance, very high open-loop voltage gain, and a wide bandwidth.
- Two types of op-amp input operation are the single-ended mode and the differential mode.
- Common-mode occurs when equal in-phase voltages are applied to both input terminals.
- The common-mode rejection ratio (CMRR) is a measure of an op-amp's ability to reject common-mode inputs.
- Input offset voltage produces an output error voltage (with no input voltage).
- Input bias current also produces an output error voltage (with no input voltage).
- Input offset current is the difference between the two bias currents.
- Open-loop voltage gain is the gain of an op-amp with no external feedback connections.
- Closed-loop voltage gain is the gain of an op-amp with external feedback.
- Slew rate is the rate in volts per microsecond at which the output voltage of an op-amp can change in response to a step input.
- There are three basic op-amp configurations: inverting, noninverting, and voltage-follower.
- The three basic op-amp configurations employ negative feedback. Negative feedback occurs when a portion of the output voltage is connected back to the inverting input such that it subtracts from the input voltage, thus reducing the voltage gain but increasing the stability and bandwidth.
- A noninverting amplifier configuration has a higher input impedance and a lower output impedance than the op-amp itself (without feedback).
- An inverting amplifier configuration has an input impedance approximately equal to the input resistor R_i and an output impedance approximately equal to the output impedance of the op-amp itself.
- The voltage-follower has the highest input impedance and the lowest output impedance of the three amplifier configurations.
- All practical op-amps have small input bias currents and input offset voltages that produce small output error voltages.
- The input bias current effect can be compensated for with external resistors.
- The input offset voltage can be compensated for with an external potentiometer between the two offset null pins provided on the IC op-amp package and as recommended by the manufacturer.
- The closed-loop voltage gain is always less than the open-loop voltage gain.
- The midrange gain of an op-amp extends down to dc.
- The gain of an op-amp decreases as frequency increases above the critical frequency.
- The bandwidth of an op-amp equals the upper critical frequency.
- The internal RC lag circuits that are inherently part of the amplifier stages cause the gain to roll off as frequency goes up.
- The internal RC lag circuits also cause a phase shift between input and output signals.
- Negative feedback lowers the gain and increases the bandwidth.
- The product of gain and bandwidth is constant for a given op-amp.
- The gain-bandwidth product equals the frequency at which unity voltage gain occurs.

Key terms and other bold terms in the chapter are defined in the end–of–book glossary.

Closed-loop voltage gain (A_{cl}) The voltage gain of an op-amp with external feedback.

CMRR   Common-mode rejection ratio; the ratio of open-loop gain to common-mode gain; a measure of an op-amp's ability to reject common-mode signals.

Common mode   A condition characterized by the presence of the same signal on both op-amp inputs.

Differential   A mode of op-amp operation in which two opposite-polarity signal voltages are applied to the two inputs.

Gain-bandwidth product   A constant parameter which is always equal to the frequency at which the op-amp's open-loop gain is unity (1).

Inverting amplifier   An op-amp closed-loop configuration in which the input signal is applied to the inverting input.

Negative feedback   The process of returning a portion of the output signal to the input of an amplifier such that it is out of phase with the input signal.

Noninverting amplifier   An op-amp closed-loop configuration in which the input signal is applied to the noninverting input.

Open-loop voltage gain (A_{ol}) The voltage gain of an op-amp without external feedback.

Operational amplifier (op-amp)   A type of amplifier that has very high voltage gain, very high input impedance, very low output impedance, and good rejection of common-mode signals.

Phase shift   The relative angular displacement of a time-varying function relative to a reference.

Single-ended   A mode of op-amp operation in which a signal voltage is applied to only one input.

Slew rate   The rate of change of the output voltage of an op-amp in response to a step input.

Voltage-follower   A closed-loop, noninverting op-amp with a voltage gain of 1.

Op-Amp Input Modes and Parameters

12–1 $\text{CMRR} = \dfrac{A_{ol}}{A_{cm}}$ Common-mode rejection ratio

12–2 $\text{CMRR} = 20 \log\left(\dfrac{A_{ol}}{A_{cm}}\right)$ Common-mode rejection ratio (dB)

12–3 $I_{\text{BIAS}} = \dfrac{I_1 + I_2}{2}$ Input bias current

12–4 $I_{\text{OS}} = |I_1 - I_2|$ Input offset current

12–5 $V_{\text{OS}} = I_{\text{OS}}R_{in}$ Offset voltage

12–6 $V_{\text{OUT(error)}} = A_v I_{\text{OS}} R_{in}$ Output error voltage

12–7 $\text{Slew rate} = \dfrac{\Delta V_{out}}{\Delta t}$ Slew rate

Op-Amp Configurations

12–8 $A_{cl(\text{NI})} = 1 + \dfrac{R_f}{R_i}$ Voltage gain (noninverting)

12–9 $A_{cl(\text{VF})} = 1$ Voltage gain (voltage-follower)

12–10 $A_{cl(\text{I})} = -\dfrac{R_f}{R_i}$ Voltage gain (inverting)

Op-Amp Impedances

12–11 $Z_{in(\text{NI})} = (1 + A_{ol}B)Z_{in}$ Input impedance (noninverting)

$$12\text{–}12 \qquad Z_{out(NI)} = \frac{Z_{out}}{1 + A_{ol}B}$$ Output impedance (noninverting)

$$12\text{–}13 \qquad Z_{in(VF)} = (1 + A_{ol})Z_{in}$$ Input impedance (voltage-follower)

$$12\text{–}14 \qquad Z_{out(VF)} = \frac{Z_{out}}{1 + A_{ol}}$$ Output impedance (voltage-follower)

$$12\text{–}15 \qquad Z_{in(I)} \cong R_i$$ Input impedance (inverting)

$$12\text{–}16 \qquad Z_{out(I)} = \frac{Z_{out}}{1 + A_{ol}B}$$ Output impedance (inverting)

Op-Amp Frequency Responses

$$12\text{–}17 \qquad BW = f_{cu}$$ Op-amp bandwidth

$$12\text{–}18 \qquad \frac{V_{out}}{V_{in}} = \frac{1}{\sqrt{1 + f^2/f_c^2}}$$ RC attenuation

$$12\text{–}19 \qquad A_{ol} = \frac{A_{ol(mid)}}{\sqrt{1 + f^2/f_c^2}}$$ Open-loop voltage gain

$$12\text{–}20 \qquad \theta = -\tan^{-1}\left(\frac{f}{f_c}\right)$$ RC phase shift

$$12\text{–}21 \qquad f_{c(cl)} = f_{c(ol)}(1 + BA_{ol(mid)})$$ Closed-loop critical frequency

$$12\text{–}22 \qquad BW_{cl} = BW_{ol}(1 + BA_{ol(mid)})$$ Closed-loop bandwidth

$$12\text{–}23 \qquad \textbf{Unity-gain bandwidth} = A_{cl}f_{c(cl)}$$ Unity-gain bandwidth

SELF-TEST

Answers are at the end of the chapter.

1. An integrated circuit (IC) op-amp has
   (a) two inputs and two outputs   (b) one input and one output
   (c) two inputs and one output

2. Which of the following characteristics does not necessarily apply to an op-amp?
   (a) High gain   (b) Low power
   (c) High input impedance   (d) Low output impedance

3. A differential amplifier
   (a) is part of an op-amp   (b) has one input and one output
   (c) has two outputs   (d) answers (a) and (c)

4. When an op-amp is operated in the single-ended mode,
   (a) the output is grounded
   (b) one input is grounded and a signal is applied to the other
   (c) both inputs are connected together
   (d) the output is not inverted

5. In the differential mode,
   (a) opposite polarity signals are applied to the inputs
   (b) the gain is 1
   (c) the outputs are different amplitudes
   (d) only one supply voltage is used

6. In the common mode,
   (a) both inputs are grounded   (b) the outputs are connected together
   (c) an identical signal appears on both inputs   (d) the output signals are in-phase

7. Common-mode gain is
   (a) very high   (b) very low   (c) always unity   (d) unpredictable

8. If $A_{v(d)} = 3500$ and $A_{cm} = 0.35$, the CMRR is

    (a) 1225   (b) 10,000   (c) 80 dB   (d) answers (b) and (c)

9. With zero volts on both inputs, an op-amp ideally should have an output equal to

    (a) the positive supply voltage   (b) the negative supply voltage

    (c) zero   (d) the CMRR

10. Of the values listed, the most realistic value for open-loop gain of an op-amp is

    (a) 1   (b) 2000   (c) 80 dB   (d) 100,000

11. A certain op-amp has bias currents of 50 μA and 49.3 μA. The input offset current is

    (a) 700 nA   (b) 99.3 μA   (c) 49.7 μA   (d) none of these

12. The output of a particular op-amp increases 8 V in 12 μs. The slew rate is

    (a) 96 V/μs   (b) 0.67 V/μs   (c) 1.5 V/μs   (d) none of these

13. The purpose of offset nulling is to

    (a) reduce the gain   (b) equalize the input signals

    (c) zero the output error voltage   (d) answers (b) and (c)

14. For an op-amp with negative feedback, the output is

    (a) equal to the input   (b) increased

    (c) fed back to the inverting input   (d) fed back to the noninverting input

15. The use of negative feedback

    (a) reduces the voltage gain of an op-amp

    (b) makes the op-amp oscillate

    (c) makes linear operation possible

    (d) answers (a) and (c)

16. Negative feedback

    (a) increases the input and output impedances

    (b) increases the input impedance and the bandwidth

    (c) decreases the output impedance and the bandwidth

    (d) does not affect impedances or bandwidth

17. A certain noninverting amplifier has an R_i of 1.0 kΩ and an R_f of 100 kΩ. The closed-loop gain is

    (a) 100,000   (b) 1000   (c) 101   (d) 100

18. If the feedback resistor in Question 17 is open, the voltage gain

    (a) increases   (b) decreases   (c) is not affected   (d) depends on R_i

19. A certain inverting amplifier has a closed-loop gain of 25. The op-amp has an open-loop gain of 100,000. If another op-amp with an open-loop gain of 200,000 is substituted in the configuration, the closed-loop gain

    (a) doubles   (b) drops to 12.5   (c) remains at 25   (d) increases slightly

20. A voltage-follower

    (a) has a gain of 1   (b) is noninverting

    (c) has no feedback resistor   (d) has all of these

21. The bandwidth of an ac amplifier having a lower critical frequency of 1 kHz and an upper critical frequency of 10 kHz is

    (a) 1 kHz   (b) 9 kHz   (c) 10 kHz   (d) 11 kHz

22. The bandwidth of a dc amplifier having an upper critical frequency of 100 kHz is

    (a) 100 kHz   (b) unknown   (c) infinity   (d) 0 kHz

23. The midrange open-loop gain of an op-amp

    (a) extends from the lower critical frequency to the upper critical frequency

    (b) extends from 0 Hz to the upper critical frequency

    (c) rolls off at 20 dB/decade beginning at 0 Hz

    (d) answers (b) and (c)

24. The frequency at which the open-loop gain is equal to 1 is called

    (a) the upper critical frequency   **(b)** the cutoff frequency

    (c) the notch frequency   **(d)** the unity-gain frequency

25. Phase shift through an op-amp is caused by

    (a) the internal *RC* circuits   **(b)** the external *RC* circuits

    (c) the gain roll-off   **(d)** negative feedback

26. Each *RC* circuit in an op-amp

    (a) causes the gain to roll off at −6 dB/octave

    (b) causes the gain to roll off at −20 dB/decade

    (c) reduces the midrange gain by 3 dB

    (d) answers (a) and (b)

27. When negative feedback is used, the gain-bandwidth product of an op-amp

    (a) increases   **(b)** decreases   **(c)** stays the same   **(d)** fluctuates

28. If a certain op-amp has a midrange open-loop gain of 200,000 and a unity-gain frequency of 5 MHz, the gain-bandwidth product is

    (a) 200,000 Hz   **(b)** 5,000,000 Hz

    (c) 1×10^{12} Hz   **(d)** not determinable from the information

29. If a certain op-amp has a closed-loop gain of 20 and an upper critical frequency of 10 MHz, the gain-bandwidth product is

    (a) 200 MHz   **(b)** 10 MHz   **(c)** the unity-gain frequency   **(d)** answers (a) and (c)

PROBLEMS

Answers to all odd-numbered problems are at the end of the book.

BASIC PROBLEMS

SECTION 12–1   **Introduction to Operational Amplifiers**

    1. Compare a practical op-amp to an ideal op-amp.

    2. Two IC op-amps are available to you. Their characteristics are listed below. Choose the one you think is more desirable.

        Op-amp 1: $Z_{in} = 5$ MΩ, $Z_{out} = 100$ Ω, $A_{ol} = 50,000$

        Op-amp 2: $Z_{in} = 10$ MΩ, $Z_{out} = 75$ Ω, $A_{ol} = 150,000$

SECTION 12–2   **Op-Amp Input Modes and Parameters**

    3. Identify the type of input mode for each op-amp in Figure 12–50.

    4. A certain op-amp has a CMRR of 250,000. Convert this to decibels.

    5. The open-loop gain of a certain op-amp is 175,000. Its common-mode gain is 0.18. Determine the CMRR in decibels.

(a)

(b)

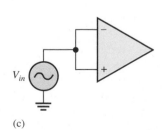

(c)

▲ **FIGURE 12–50**

6. An op-amp data sheet specifies a CMRR of 300,000 and an A_{ol} of 90,000. What is the common-mode gain?

7. Determine the bias current, I_{BIAS}, given that the input currents to an op-amp are 8.3 μA and 7.9 μA.

8. Distinguish between input bias current and input offset current, and then calculate the input offset current in Problem 7.

9. Figure 12–51 shows the output voltage of an op-amp in response to a step input. What is the slew rate?

10. How long does it take the output voltage of an op-amp to go from -10 V to $+10$ V, if the slew rate is 0.5 V/μs?

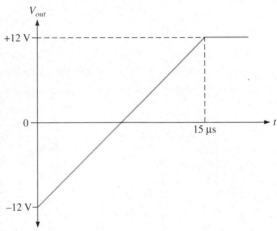

▶ **FIGURE 12–51**

SECTION 12–4   Op-Amps with Negative Feedback

11. Identify each of the op-amp configurations in Figure 12–52.

(a)   (b)   (c)

▲ **FIGURE 12–52**

12. A noninverting amplifier has an R_i of 1.0 kΩ and an R_f of 100 kΩ. Determine V_f and B, if $V_{out} = $ 5 V.

13. For the amplifier in Figure 12–53, determine the following:

    (a) $A_{cl(NI)}$ **(b)** V_{out} **(c)** V_f

▶ **FIGURE 12–53**

File circuits are identified with a CD logo and are in the Problems folder on your CD-ROM. Filenames correspond to figure numbers (e.g., F12–53).

14. Determine the closed-loop gain of each amplifier in Figure 12–54.

(a)

(b)

(c)

(d)

▲ **FIGURE 12–54**

15. Find the value of R_f that will produce the indicated closed-loop gain in each amplifier in Figure 12–55.

(a)   (b)   (c)   (d)

▲ **FIGURE 12–55**

16. Find the gain of each amplifier in Figure 12–56.

17. If a signal voltage of 10 mV rms is applied to each amplifier in Figure 12–56, what are the output voltages and what is their phase relationship with inputs?

(a)   (b)   (c)   (d)

▲ **FIGURE 12–56**

18. Determine the approximate values for each of the following quantities in Figure 12–57.

    (a) I_{in} **(b)** I_f **(c)** V_{out} **(d)** closed-loop gain

▶ **FIGURE 12–57**

SECTION 12–5   Effects of Negative Feedback on Op-Amp Impedances

19. Determine the input and output impedances for each amplifier configuration in Figure 12–58.

(a)   (b)   (c)

▲ **FIGURE 12–58**

20. Repeat Problem 19 for each circuit in Figure 12–59.

21. Repeat Problem 19 for each circuit in Figure 12–60.

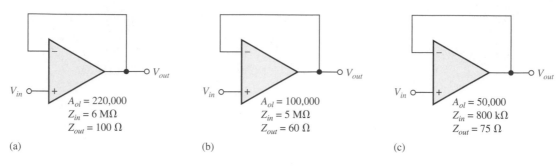

(a)

$A_{ol} = 220,000$
$Z_{in} = 6\ \text{M}\Omega$
$Z_{out} = 100\ \Omega$

(b)

$A_{ol} = 100,000$
$Z_{in} = 5\ \text{M}\Omega$
$Z_{out} = 60\ \Omega$

(c)

$A_{ol} = 50,000$
$Z_{in} = 800\ \text{k}\Omega$
$Z_{out} = 75\ \Omega$

▲ **FIGURE 12–59**

(a)

R_f 150 kΩ
R_i 10 kΩ
$A_{ol} = 125,000$
$Z_{in} = 1.5\ \text{M}\Omega$
$Z_{out} = 40\ \Omega$

(b)

R_f 10 MΩ
R_i 100 kΩ
$A_{ol} = 75,000$
$Z_{in} = 1\ \text{M}\Omega$
$Z_{out} = 50\ \Omega$

(c)

R_f 10 kΩ
R_i 470 Ω
$A_{ol} = 250,000$
$Z_{in} = 3\ \text{M}\Omega$
$Z_{out} = 70\ \Omega$

▲ **FIGURE 12–60**

SECTION 12–6 Bias Current and Offset Voltage Compensation

22. A voltage-follower is driven by a voltage source with a source resistance of 75 Ω.

    (a) What value of compensating resistor is required for bias current, and where should the resistor be placed?

    (b) If the two input currents after compensation are 42 μA and 40 μA, what is the output error voltage?

23. Determine the compensating resistor value for each amplifier configuration in Figure 12–58, and indicate the placement of the resistor.

24. A particular op-amp voltage-follower has an input offset voltage of 2 nV. What is the output error voltage?

25. What is the input offset voltage of an op-amp if a dc output voltage of 35 mV is measured when the input voltage is zero? The op-amp's open-loop gain is specified to be 200,000.

SECTION 12–7 Open-Loop Response

26. The midrange open-loop gain of a certain op-amp is 120 dB. Negative feedback reduces this gain by 50 dB. What is the closed-loop gain?

27. The upper critical frequency of an op-amp's open-loop response is 200 Hz. If the midrange gain is 175,000, what is the ideal gain at 200 Hz? What is the actual gain? What is the op-amp's open-loop bandwidth?

28. An RC lag circuit has a critical frequency of 5 kHz. If the resistance value is 1.0 kΩ, what is X_C when $f = 3$ kHz?

29. Determine the attenuation of an RC lag circuit with $f_c = 12$ kHz for each of the following frequencies.

    (a) 1 kHz   **(b)** 5 kHz   **(c)** 12 kHz   **(d)** 20 kHz   **(e)** 100 kHz

30. The midrange open-loop gain of a certain op-amp is 80,000. If the open-loop critical frequency is 1 kHz, what is the open-loop gain at each of the following frequencies?

    (a) 100 Hz   **(b)** 1 kHz   **(c)** 10 kHz   **(d)** 1 MHz

31. Determine the phase shift through each circuit in Figure 12–61 at a frequency of 2 kHz.

▲ **FIGURE 12–61**

32. An RC lag circuit has a critical frequency of 8.5 kHz. Determine the phase shift for each frequency and plot a graph of its phase angle versus frequency.

    (a) 100 Hz   **(b)** 400 Hz   **(c)** 850 Hz

    (d) 8.5 kHz   **(e)** 25 kHz   **(f)** 85 kHz

33. A certain op-amp has three internal amplifier stages with midrange gains of 30 dB, 40 dB, and 20 dB. Each stage also has a critical frequency associated with it as follows: $f_{c1} = 600$ Hz, $f_{c2} = 50$ kHz, and $f_{c3} = 200$ kHz.

    (a) What is the midrange open-loop gain of the op-amp, expressed in dB?

    (b) What is the total phase shift through the amplifier, including inversion, when the signal frequency is 10 kHz?

34. What is the gain roll-off rate in Problem 33 between the following frequencies?

    (a) 0 Hz and 600 Hz   **(b)** 600 Hz and 50 kHz

    (c) 50 kHz and 200 kHz   **(d)** 200 kHz and 1 MHz

SECTION 12–8   **Closed-Loop Response**

35. Determine the midrange gain in dB of each amplifier in Figure 12–62. Are these open-loop or closed-loop gains?

▲ **FIGURE 12–62**

36. A certain amplifier has an open-loop gain in midrange of 180,000 and an open-loop critical frequency of 1500 Hz. If the attenuation of the feedback path is 0.015, what is the closed-loop bandwidth?

37. Given that $f_{c(ol)}$ = 750 Hz, A_{ol} = 89 dB, and $f_{c(cl)}$ = 5.5 kHz, determine the closed-loop gain in decibels.

38. What is the unity-gain bandwidth in Problem 37?

39. For each amplifier in Figure 12–63, determine the closed-loop gain and bandwidth. The op-amps in each circuit exhibit an open-loop gain of 125 dB and a unity-gain bandwidth of 2.8 MHz.

40. Which of the amplifiers in Figure 12–64 has the smaller bandwidth?

▲ **FIGURE 12–63**

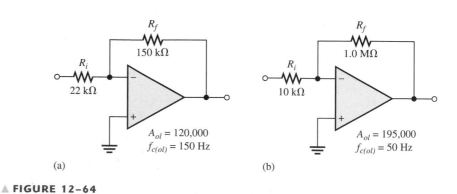

▲ **FIGURE 12–64**

TROUBLESHOOTING PROBLEMS

SECTION 12–9 **Troubleshooting**

41. Determine the most likely fault(s) for each of the following symptoms in Figure 12–65 with a 100 mV signal applied.

    (a) No output signal.

    (b) Output severely clipped on both positive and negative swings.

42. Determine the effect on the output if the circuit in Figure 12–65 has the following fault (one fault at a time).

    (a) Output pin is shorted to the inverting input.

    (b) R_3 is open.

    (c) R_3 is 10 kΩ instead of 910 Ω.

    (d) R_1 and R_2 are swapped.

43. On the circuit board in Figure 12–66, what happens if the middle lead (wiper) of the 100 kΩ potentiometer is broken?

► **FIGURE 12–65**

▲ **FIGURE 12–66**

Electrical Characteristics

Parameter	Conditions	LM741A			LM741			LM741C			Units
		Min	Typ	Max	Min	Typ	Max	Min	Typ	Max	
Input offset voltage	$T_A = 25°C$	—	—	—	—	—	—	—	—	—	
	$R_S \leq 10 \text{ k}\Omega$	—	—	—	—	1.0	5.0	—	2.0	6.0	mV
	$R_S \leq 50 \ \Omega$	—	0.8	3.0	—	—	—	—	—	—	mV
	$T_{AMIN} \leq T_A \leq T_{AMAX}$	—	—	—	—	—	—	—	—	—	
	$R_S \leq 50 \ \Omega$	—	—	4.0	—	—	—	—	—	—	mV
	$R_S \leq 10 \text{ k}\Omega$	—	—	—	—	—	6.0	—	—	7.5	mV
Average input offset voltage drift		—	—	15	—	—	—	—	—	—	µV/°C
Input offset voltage adjustment range	$T_A = 25°C, V_S = \pm 20$ V	±10	—	—	—	±15	—	—	±15	—	mV
Input offset current	$T_A = 25°C$	—	3.0	30	—	20	200	—	20	200	nA
	$T_{AMIN} \leq T_A \leq T_{AMAX}$	—	—	70	—	85	500	—	—	300	nA
Average input offset current drift		—	—	0.5	—	—	—	—	—	—	nA/°C
Input bias current	$T_A = 25°C$	—	30	80	—	80	500	—	80	500	nA
	$T_{AMIN} \leq T_A \leq T_{AMAX}$	—	—	0.210	—	—	1.5	—	—	0.8	µA
Input resistance	$T_A = 25°C, V_S = \pm 20$ V	1.0	6.0	—	0.3	2.0	—	0.3	2.0	—	MΩ
	$T_{AMIN} \leq T_A \leq T_{AMAX}$	0.5	—	—	—	—	—	—	—	—	MΩ
	$V_S = \pm 20$ V	—									
Input voltage range	$T_A = 25°C$	—	—	—	—	—	—	—	±12	±13	V
	$T_{AMIN} \leq T_A \leq T_{AMAX}$	—	—	—	±12	±13	—	—	—	—	V
Large-signal voltage gain	$T_A = 25°C, R_L \geq 2 \text{ k}\Omega$	—	—	—	—	—	—	—	—	—	
	$V_S = \pm 20$ V, $V_O = \pm 15$ V	50	—	—	—	—	—	—	—	—	V/mV
	$V_S = \pm 15$ V, $V_O = \pm 10$ V	—	—	—	50	200	—	20	200	—	V/mV
	$T_{AMIN} \leq T_A \leq T_{AMAX}$	—	—	—	—	—	—	—	—	—	
	$R_L \geq 2 \text{ k}\Omega$	—	—	—	—	—	—	—	—	—	
	$V_S = \pm 20$ V, $V_O = \pm 15$ V	32	—	—	—	—	—	—	—	—	V/mV
	$V_S = \pm 15$ V, $V_O = \pm 10$ V	—	—	—	25	—	—	15	—	—	V/mV
	$V_S = \pm 5$ V, $V_O = \pm 2$ V	10	—	—	—	—	—	—	—	—	V/mV
Output voltage swing	$V_S = \pm 20$ V	—	—	—	—	—	—	—	—	—	
	$R_L \geq 10 \text{ k}\Omega$	±16	—	—	—	—	—	—	—	—	V
	$R_L \geq 2 \text{ k}\Omega$	±15	—	—	—	—	—	—	—	—	V
	$V_S = \pm 15$ V	—	—	—	—	—	—	—	—	—	
	$R_L \geq 10 \text{ k}\Omega$	—	—	—	±12	±14	—	±12	±14	—	V
	$R_L \geq 2 \text{ k}\Omega$	—	—	—	±10	±13	—	±10	±13	—	V
Output short circuit Current	$T_A = 25°C$	10	25	35	—	25	—	—	25	—	mA
	$T_{AMIN} \leq T_A \leq T_{AMAX}$	10	—	40	—	—	—	—	—	—	mA
Common-mode rejection ratio	$T_{AMIN} \leq T_A \leq T_{AMAX}$	—	—	—	—	—	—	—	—	—	
	$R_S \leq 10 \text{ k}\Omega$, $V_{CM} = \pm 12$ V	—	—	—	70	90	—	70	90	—	dB
	$R_S \leq 50 \ \Omega$, $V_{CM} = \pm 12$ V	80	95	—	—	—	—	—	—	—	dB
Supply voltage rejection ratio	$T_{AMIN} \leq T_A \leq T_{AMAX}$,	—	—	—	—	—	—	—	—	—	
	$V_S = \pm 20$ V to $V_S = \pm 5$ V	—	—	—	—	—	—	—	—	—	
	$R_S \leq 50 \ \Omega$	86	96	—	—	—	—	—	—	—	dB
	$R_S \leq 10 \text{ k}\Omega$	—	—	—	77	96	—	77	96	—	dB
Transient response Rise time Overshoot	$T_A = 25°C$, Unity gain	—	—	—	—	—	—	—	—	—	
		—	0.25	0.8	—	0.3	—	—	0.3	—	µs
		—	6.0	20	—	5	—	—	5	—	%
Bandwidth	$T_A = 25°C$	0.437	1.5	—	—	—	—	—	—	—	MHz
Slew rate	$T_A = 25°C$, Unity gain	0.3	0.7	—	—	0.5	—	—	0.5	—	V/µs
Supply current	$T_A = 25°C$			—	—	1.7	2.8	—	1.7	2.8	mA
Power consumption	$T_A = 25°C$	—	—	—	—	—	—	—	—	—	
	$V_S = \pm 20$ V	—	80	150	—	—	—	—	—	—	mW
	$V_S = \pm 15$ V	—	—	—	—	50	85	—	50	85	mW
LM741A	$V_S = \pm 20$ V	—	—	—	—	—	—	—	—	—	
	$T_A = T_{AMIN}$	—	—	165	—	—	—	—	—	—	mW
	$T_A = T_{AMAX}$	—	—	135	—	—	—	—	—	—	mW
LM741	$V_S = \pm 15$ V	—	—	—	—	—	—	—	—	—	
	$T_A = T_{AMIN}$	—	—	—	—	60	100	—	—	—	mW
	$T_A = T_{AMAX}$	—	—	—	—	45	75	—	—	—	mW

▲ **FIGURE 12–67**

SYSTEM APPLICATION PROBLEMS

44. In the amplifier circuit of Figure 12–48, list the possible faults that will cause the push-pull stage to operate nonlinearly.

45. What indication would you observe if a 2.2 MΩ resistor is incorrectly installed for R_3 in Figure 12–48?

46. What voltage will you measure on the output of the amplifier in Figure 12–48 if diode D_1 opens?

DATA SHEET PROBLEMS

47. Refer to the partial 741 data sheet (LM741) in Figure 12–67. Determine the input resistance (impedance) of a noninverting amplifier which uses a 741 op-amp with $R_f = 47$ kΩ and $R_i = 470$ Ω. Use typical values.

48. Refer to the partial data sheet in Figure 12–67. Determine the input impedances of a LM741 op-amp connected as an inverting amplifier with a closed-loop voltage gain of 100 and $R_f = 100$ kΩ.

49. Refer to Figure 12–67 and determine the minimum open-loop voltage gain for an LM741 expressed as a ratio of output volts to input volts.

50. Refer to Figure 12–67. How long does it typically take the output voltage of an LM741 to make a transition from -8 V to $+8$ V in response to a step input?

ADVANCED PROBLEMS

51. Design a noninverting amplifier with an appropriate closed-loop voltage gain of 150 and a minimum input impedance of 100 MΩ using a 741 op-amp.

52. Design an inverting amplifier using a 741 op-amp. The voltage gain must be $68 \pm 5\%$ and the input impedance must be approximately 10 kΩ.

53. Design a noninverting amplifier with an upper critical frequency, f_{cu}, of 10 kHz using a 741 op-amp. The dc supply voltages are ± 15 V. Refer to Figure 12–68.

▲ **FIGURE 12–68**

54. For the circuit you designed in Problem 53, determine the minimum load resistance if the minimum output voltage swing is to be ±10 V. Refer to the data sheet graphs in Figure 12–69.

55. Design an inverting amplifier using a 741 op-amp if a midrange voltage gain of 50 and a bandwidth of 20 kHz is required.

56. What is the maximum closed-loop voltage gain that can be achieved with a 741 op-amp if the bandwidth must be no less than 5 kHz?

(a) Positive output voltage swing versus load resistance

(b) Negative output voltage swing versus load resistance

▲ **FIGURE 12–69**

EWB/MULTISIM TROUBLESHOOTING PROBLEMS

These file circuits are in the Troubleshooting Problems folder on your CD-ROM.

57. Open file TSP12-57 and determine the fault.

58. Open file TSP12-58 and determine the fault.

59. Open file TSP12-59 and determine the fault.

60. Open file TSP12-60 and determine the fault.

61. Open file TSP12-61 and determine the fault.

62. Open file TSP12-62 and determine the fault.

63. Open file TSP12-63 and determine the fault.

64. Open file TSP12-64 and determine the fault.

65. Open file TSP12-65 and determine the fault.

66. Open file TSP12-66 and determine the fault.

67. Open file TSP12-67 and determine the fault.

68. Open file TSP12-68 and determine the fault.

69. Open file TSP12-69 and determine the fault.

70. Open file TSP12-70 and determine the fault.

71. Open file TSP12-71 and determine the fault.

72. Open file TSP12-72 and determine the fault.

ANSWERS

SECTION REVIEWS

SECTION 12–1   Introduction to Operational Amplifiers

1. Inverting input, noninverting input, output, positive and negative supply voltages
2. A practical op-amp has very high input impedance, very low output impedance, and very high voltage gain.
3. Differential amplifier, voltage amplifier, and push-pull amplifier.

SECTION 12–2   Op-Amp Input Modes and Parameters

1. Differential input is between two input terminals. Single-ended input is from one input terminal to ground (with other input grounded).
2. Common–mode rejection is the ability of an op-amp to produce very little output when the same signal is applied to both inputs.
3. A higher CMRR results in a lower common-mode gain.
4. Input bias current, input offset voltage, drift, input offset current, input impedance, output impedance, common-mode input voltage range, CMRR, open-loop voltage gain, slew rate, frequency response.
5. Slew rate and voltage gain are both frequency dependent.

SECTION 12–3   Negative Feedback

1. Negative feedback provides a stable controlled voltage gain, control of impedances, and wider bandwidth.
2. The open-loop gain is so high that a very small signal on the input will drive the op-amp into saturation.

SECTION 12–4   Op-Amps with Negative Feedback

1. The main purpose of negative feedback is to stabilize the gain.
2. False
3. $A_{cl} = 1/0.02 = 50$

SECTION 12–5   Effects of Negative Feedback on Op-Amp Inpedances

1. The noninverting configuration has a higher Z_{in} than the op-amp alone.
2. Z_{in} increases in a voltage-follower.
3. $Z_{in(I)} \cong R_i = 2 \text{ k}\Omega$, $Z_{out(I)} = Z_{out}/(1 + A_{ol}B) = 25 \text{ m}\Omega$

SECTION 12–6   Bias Current and Offset Voltage Compensation

1. Input bias current and input offset voltage are sources of output error.
2. Add a resistor in the feedback path equal to the input source resistance.

SECTION 12–7   Open-Loop Response

1. Open-loop voltage gain is without feedback, and closed-loop voltage gain is with negative feedback. Open-loop voltage gain is larger.
2. $BW = 100 \text{ Hz}$
3. A_{ol} decreases.
4. $A_{v(tot)} = 20 \text{ dB} + 30 \text{ dB} = 50 \text{ dB}$
5. $\theta_{tot} = -49° + (-5.2°) = -54.2°$

SECTION 12–8 Closed-Loop Response

1. Yes, A_{cl} is always less than A_{ol}.
2. $BW = 3{,}000 \text{ kHz}/60 = 50 \text{ kHz}$
3. unity-gain $BW = 3{,}000 \text{ kHz}/1 = 3 \text{ MHz}$

SECTION 12–9 Troubleshooting

1. Check the output null adjustment.
2. After a verification that there is power supply voltage to the op-amp, then the absence of an output signal probably indicates a bad op-amp.

RELATED PROBLEMS FOR EXAMPLES

12–1 340,000; 111 dB

12–2 20 V/μs

12–3 32.9

12–4 67.5 kΩ

12–5 (a) 20.6 GΩ, 14 mΩ (b) 23

12–6 Z_{in} increases, Z_{out} decreases.

12–7 $Z_{in(I)} = 560 \ \Omega$; $Z_{out(I)} = 110 \text{ m}\Omega$; $A_{cl} = -146$

12–8 (a) 79,996   (b) 79,900   (c) 6380

12–9 173 Hz

12–10 75 dB; $-71.6°$

12–11 2 MHz

12–12 (a) 29.6 kHz   (b) 42.6 kHz

SELF-TEST

1. (c)	**2.** (b)	**3.** (d)	**4.** (b)	**5.** (a)
6. (c)	**7.** (b)	**8.** (d)	**9.** (c)	**10.** (d)
11. (a)	**12.** (b)	**13.** (c)	**14.** (c)	**15.** (d)
16. (b)	**17.** (c)	**18.** (a)	**19.** (c)	**20.** (d)
21. (b)	**22.** (a)	**23.** (b)	**24.** (d)	**25.** (a)
26. (d)	**27.** (c)	**28.** (b)	**29.** (d)	

13

BASIC OP-AMP APPLICATIONS

INTRODUCTION

In the last chapter, you learned about the principles, operation, and characteristics of the operational amplifier. Op-amps are used in such a wide variety of applications that it is impossible to cover all of them in one chapter, or even in one book. Therefore, in this chapter, three fundamentally important applications are covered to give you a foundation in op-amp circuits.

CHAPTER OBJECTIVES

■ Analyze the operation of several basic comparator circuits

■ Analyze the operation of several types of summing amplifiers

■ Analyze the operation of integrators and differentiators

■ Troubleshoot basic op-amp circuits

KEY TERMS

■ Comparator

■ Hysteresis

■ Schmitt trigger

■ Bounding

■ Summing amplifier

■ Integrator

■ Differentiator

■ ■ ■ SYSTEM APPLICATION PREVIEW

The system application at the end of the chapter is an example of the use of three types of op-amp circuits that you will learn about in this chapter: the summing amplifier, the integrator, and the comparator. The system includes both analog and digital circuits; however, you will focus on the analog-to-digital converter board that incorporates the op-amps. The analog-to-digital converter board converts an audio signal to a digital code in order to record the sound in a digital format. After studying this chapter, you should be able to complete the system application assignment.

WWW. VISIT THE COMPANION WEBSITE

Study aids for this chapter are available at
http://www.prenhall.com/floyd

13–1 COMPARATORS

Operational amplifiers are often used as comparators to compare the amplitude of one voltage with another. In this application, the op-amp is used in the open-loop configuration, with the input voltage on one input and a reference voltage on the other.

After completing this section, you should be able to

- **Analyze the operation of several basic comparator circuits**
- Describe the operation of a zero-level detector
- Describe the operation of a nonzero-level detector
- Discuss how input noise affects comparator operation
- Define *hysteresis*
- Explain how hysteresis reduces noise effects
- Describe a Schmitt trigger circuit
- Describe the operation of bounded comparators
- Discuss two comparator applications including analog-to-digital conversion

Zero-Level Detection

One application of an op-amp used as a **comparator** is to determine when an input voltage exceeds a certain level. Figure 13–1(a) shows a zero-level detector. Notice that the inverting (−) input is grounded to produce a zero level and that the input signal voltage is applied to the noninverting (+) input. Because of the high open-loop voltage gain, a very small difference voltage between the two inputs drives the amplifier into saturation, causing the output voltage to go to its limit. For example, consider an op-amp having $A_{ol} = 100,000$. A voltage difference of only 0.25 mV between the inputs could produce an output voltage of $(0.25 \text{ mV})(100,000) = 25$ V *if* the op-amp were capable. However, since most op-amps have maximum output voltage limitations of ± 15 V because of their dc supply voltages, the device would be driven into saturation.

Figure 13–1(b) shows the result of a sinusoidal input voltage applied to the noninverting (+) input of the zero-level detector. When the sine wave is positive, the output is at its maximum positive level. When the sine wave crosses 0, the amplifier is driven to its

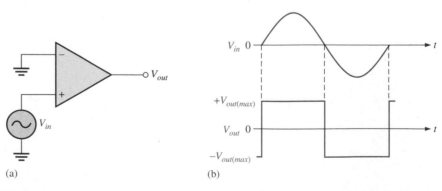

(a)   (b)

▲ **FIGURE 13–1**

The op-amp as a zero-level detector.

opposite state and the output goes to its maximum negative level, as shown. As you can see, the zero-level detector can be used as a squaring circuit to produce a square wave from a sine wave.

Nonzero-Level Detection

The zero-level detector in Figure 13–1 can be modified to detect positive and negative voltages by connecting a fixed reference voltage source to the inverting (−) input, as shown in Figure 13–2(a). A more practical arrangement is shown in Figure 13–2(b) using a voltage divider to set the reference voltage, V_{REF}, as follows:

$$V_{REF} = \frac{R_2}{R_1 + R_2}(+V)$$

where $+V$ is the positive op-amp dc supply voltage. The circuit in Figure 13–2(c) uses a zener diode to set the reference voltage ($V_{REF} = V_Z$). As long as V_{in} is less than V_{REF}, the output remains at the maximum negative level. When the input voltage exceeds the reference voltage, the output goes to its maximum positive voltage, as shown in Figure 13–2(d) with a sinusoidal input voltage.

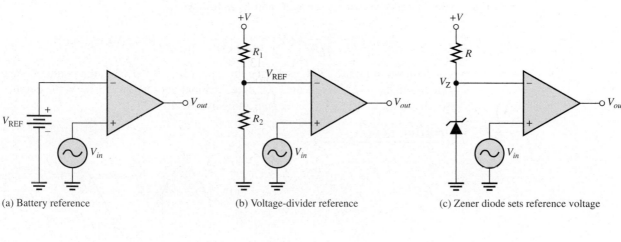

(a) Battery reference (b) Voltage-divider reference (c) Zener diode sets reference voltage

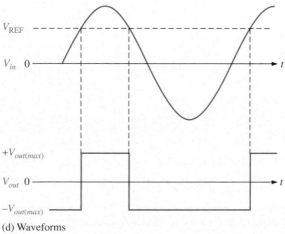

(d) Waveforms

▲ **FIGURE 13–2**

Nonzero-level detectors.

EXAMPLE 13–1

The input signal in Figure 13–3(a) is applied to the comparator circuit in Figure 13–3(b). Make a sketch of the output showing its proper relationship to the input signal. Assume the maximum output levels of the op-amp are ±12 V.

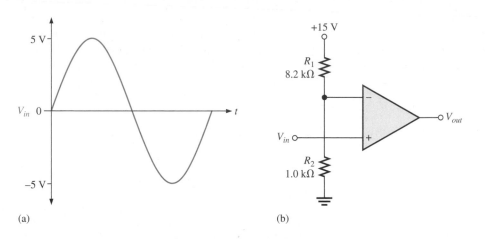

(a)　　　　　　　　　　　　　　　　　　　　　　(b)

▲ **FIGURE 13–3**

Solution　　The reference voltage is set by R_1 and R_2 as follows:

$$V_{REF} = \frac{R_2}{R_1 + R_2}(+V) = \frac{1.0 \text{ k}\Omega}{8.2 \text{ k}\Omega + 1.0 \text{ k}\Omega}(+15 \text{ V}) = 1.63 \text{ V}$$

As shown in Figure 13–4, each time the input exceeds +1.63 V, the output voltage switches to its +12 V level, and each time the input goes below +1.63 V, the output switches back to its −12 V level, neglecting hysteresis.

▶ **FIGURE 13–4**

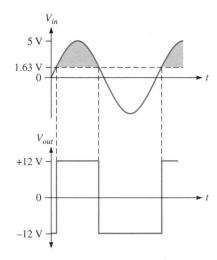

Related Problem *　　Determine the reference voltage in Figure 13–3 if $R_1 = 22$ kΩ and $R_2 = 3.3$ kΩ.

*Answers are at the end of the chapter.

Open file E13-01 in the Examples folder on your CD-ROM. Compare the output waveform to the specified input at any arbitrary frequency and verify that the reference voltage agrees with the calculated value.

Effects of Input Noise on Comparator Operation

In many practical situations, **noise** (unwanted voltage fluctuations) appears on the input line. This noise voltage becomes superimposed on the sinusoidal input voltage, as shown in Figure 13–5, and can cause a comparator to erratically switch output states.

▶ **FIGURE 13–5**

Sine wave with superimposed noise.

In order to understand the potential effects of noise voltage, consider a low-frequency sinusoidal voltage applied to the noninverting (+) input of an op-amp comparator used as a zero-level detector, as shown in Figure 13–6(a). Part (b) of the figure shows the input sine wave plus noise and the resulting output. As you can see, when the sine wave approaches 0, the fluctuations due to noise cause the total input to vary above and below 0 several times, thus producing an erratic output voltage.

(a)

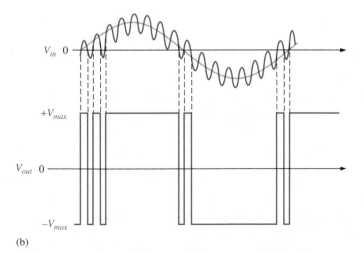

(b)

▲ **FIGURE 13–6**

Effects of noise on comparator circuit.

Reducing Noise Effects with Hysteresis

An erratic output voltage caused by noise on the input occurs because the op-amp comparator switches from its negative output state to its positive output state at the same input voltage level that causes it to switch in the opposite direction, from positive to negative. This unstable condition occurs when the input voltage hovers around the reference voltage, and any small noise fluctuations cause the comparator to switch first one way and then the other.

In order to make the comparator less sensitive to noise, a technique incorporating positive feedback, called **hysteresis**, can be used. Basically, hysteresis means that there is a higher reference level when the input voltage goes from a lower to higher value than when it goes from a higher to a lower value. A good example of hysteresis is a common household thermostat that turns the furnace on at one temperature and off at another.

The two reference levels are referred to as the upper trigger point (UTP) and the lower trigger point (LTP). This two-level hysteresis is established with a positive feedback arrangement, as shown in Figure 13–7. Notice that the noninverting (+) input is connected to a resistive voltage divider such that a portion of the output voltage is fed back to the input. The input signal is applied to the inverting (−) input in this case.

▶ **FIGURE 13–7**

Comparator with positive feedback for hysteresis.

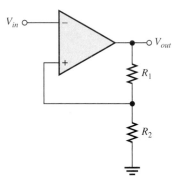

The basic operation of the comparator with hysteresis is as follows and is illustrated in Figure 13–8. Assume that the output voltage is at its positive maximum, $+V_{out(max)}$. The voltage fed back to the noninverting input is V_{UTP} and is expressed as

Equation 13–1

$$V_{UTP} = \frac{R_2}{R_1 + R_2}(+V_{out(max)})$$

When V_{in} exceeds V_{UTP}, the output voltage drops to its negative maximum, $-V_{out(max)}$. Now the voltage fed back to the noninverting input is V_{LTP} and is expressed as

Equation 13–2

$$V_{LTP} = \frac{R_2}{R_1 + R_2}(-V_{out(max)})$$

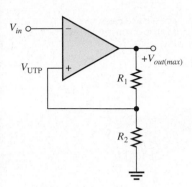

(a) Output at the maximum positive voltage

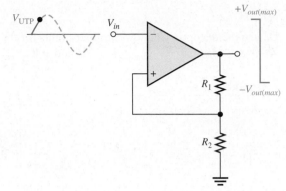

(b) Input exceeds UTP; output switches from the maximum positive voltage to the maximum negative voltage.

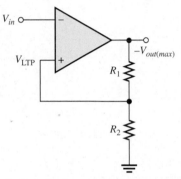

(c) Output at the maximum negative voltage

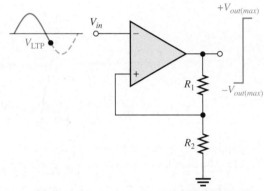

(d) Input goes below LTP; output switches from maximum negative voltage back to maximum positive voltage.

(e) Device triggers only once when UTP or LTP is reached; thus, there is immunity to noise that is riding on the input signal.

▲ **FIGURE 13-8**

Operation of a comparator with hysteresis.

The input voltage must now fall below V_{LTP} before the device will switch from the maximum negative voltage back to the maximum positive voltage. This means that a small amount of noise voltage has no effect on the output, as illustrated by Figure 13–8(e).

A comparator with hysteresis is sometimes known as a **Schmitt trigger**. The amount of hysteresis is defined by the difference of the two trigger levels.

$$V_{HYS} = V_{UTP} - V_{LTP}$$

Equation 13–3

EXAMPLE 13–2

Determine the upper and lower trigger points for the comparator circuit in Figure 13–9. Assume that $+V_{out(max)} = +5$ V and $-V_{out(max)} = -5$ V.

▶ **FIGURE 13–9**

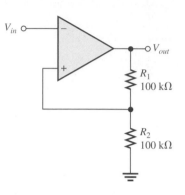

Solution

$$V_{\text{UTP}} = \frac{R_2}{R_1 + R_2}(+V_{out(max)}) = 0.5(5 \text{ V}) = \mathbf{+2.5 \text{ V}}$$

$$V_{\text{LTP}} = \frac{R_2}{R_1 + R_2}(-V_{out(max)}) = 0.5(-5 \text{ V}) = \mathbf{-2.5 \text{ V}}$$

Related Problem

Determine the upper and lower trigger points in Figure 13–9 for $R_1 = 68$ kΩ and $R_2 = 82$ kΩ. The maximum output voltage levels are ±7 V.

Open file E13-02 in the Examples folder on your CD-ROM. Determine the upper and lower trigger points and compare with the calculated values using a 5 V rms, 60 Hz sine wave for the input.

Output Bounding

In some applications, it is necessary to limit the output voltage levels of a comparator to a value less than that provided by the saturated op-amp. A single zener diode can be used as shown in Figure 13–10 to limit the output voltage to the zener voltage in one direction and to the forward diode drop in the other. This process of limiting the output range is called **bounding**.

The operation is as follows. Since the anode of the zener is connected to the inverting (−) input, it is at virtual ground (≅ 0 V). Therefore, when the output voltage reaches a positive value equal to the zener voltage, it limits at that value, as illustrated in Figure 13–11. When the output switches negative, the zener acts as a regular diode and becomes forward-biased at 0.7 V, limiting the negative output voltage to this value, as shown. Turning the zener around limits the output voltage in the opposite direction.

▶ **FIGURE 13–10**

Comparator with output bounding.

(a) Bounded at a positive value

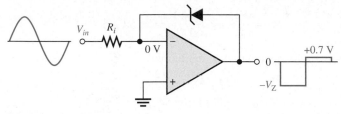

(b) Bounded at a negative value

▲ **FIGURE 13–11**

Operation of a bounded comparator.

Two zener diodes arranged as in Figure 13–12 limit the output voltage to the zener voltage plus the forward voltage drop (0.7 V) of the forward-biased zener, both positively and negatively, as shown in Figure 13–12.

▲ **FIGURE 13–12**

Double-bounded comparator.

EXAMPLE 13–3

Determine the output voltage waveform for Figure 13–13.

▲ **FIGURE 13–13**

Solution This comparator has both hysteresis and zener bounding.

The voltage across D_1 and D_2 in either direction is 4.7 V + 0.7 V = 5.4 V. This is because one zener is always forward-biased with a drop of 0.7 V when the other one is in breakdown.

The voltage at the inverting (−) op-amp input is $V_{out} \pm 5.4$ V. Since the differential voltage is negligible, the voltage at the noninverting (+) op-amp input is also approximately $V_{out} \pm 5.4$ V. Thus,

$$V_{R1} = V_{out} - (V_{out} \pm 5.4 \text{ V}) = \pm 5.4 \text{ V}$$

$$I_{R1} = \frac{V_{R1}}{R_1} = \frac{\pm 5.4 \text{ V}}{100 \text{ k}\Omega} = \pm 54 \text{ }\mu\text{A}$$

Since the noninverting input current is negligible,

$$I_{R2} = I_{R1} = \pm 54 \text{ }\mu\text{A}$$

$$V_{R2} = R_2 I_{R2} = (47 \text{ k}\Omega)(\pm 54 \text{ }\mu\text{A}) = \pm 2.54 \text{ V}$$

$$V_{out} = V_{R1} + V_{R2} = \pm 5.4 \text{ V} \pm 2.54 \text{ V} = \pm 7.94 \text{ V}$$

The upper trigger point (UTP) and the lower trigger point (LTP) are as follows:

$$V_{\text{UTP}} = \left(\frac{R_2}{R_1 + R_2}\right)(+V_{out}) = \left(\frac{47 \text{ k}\Omega}{147 \text{ k}\Omega}\right)(+7.94 \text{ V}) = +2.54 \text{ V}$$

$$V_{\text{LTP}} = \left(\frac{R_2}{R_1 + R_2}\right)(-V_{out}) = \left(\frac{47 \text{ k}\Omega}{147 \text{ k}\Omega}\right)(-7.94 \text{ V}) = -2.54 \text{ V}$$

The output waveform for the given input voltage is shown in Figure 13–14.

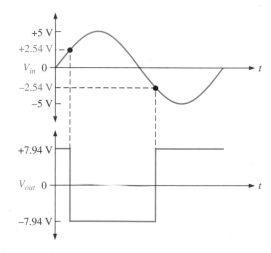

▲ **FIGURE 13–14**

Related Problem Determine the upper and lower trigger points for Figure 13–13 if $R_1 = 150$ kΩ, $R_2 = 68$ kΩ, and the zener diodes are 3.3 V devices.

Open file E13-03 in the Examples folder on your CD-ROM. Compare the output waveform to the specified input at any arbitrary frequency and see if the upper and lower trigger points agree with the calculated values.

A Comparator Application: Over-Temperature Sensing Circuit

Figure 13–15 shows an op-amp comparator used in a precision over-temperature sensing circuit to determine when the temperature reaches a certain critical value. The circuit consists of a Wheatstone bridge with the op-amp used to detect when the bridge is balanced. One leg of the bridge contains a thermistor (R_1), which is a temperature-sensing resistor with a negative temperature coefficient (its resistance decreases as temperature increases). The potentiometer (R_2) is set at a value equal to the resistance of the thermistor at the critical temperature. At normal temperatures (below critical), R_1 is greater than R_2, thus creating an unbalanced condition that drives the op-amp to its low saturated output level and keeps transistor Q_1 off.

▲ **FIGURE 13–15**

An over-temperature sensing circuit.

As the temperature increases, the resistance of the thermistor decreases. When the temperature reaches the critical value, R_1 becomes equal to R_2, and the bridge becomes balanced (since $R_3 = R_4$). At this point the op-amp switches to its high saturated output level, turning Q_1 on. This energizes the relay, which can be used to activate an alarm or initiate an appropriate response to the over-temperature condition.

A Comparator Application: Analog-to-Digital (A/D) Conversion

A/D conversion is a common interfacing process often used when a linear **analog** system must provide inputs to a **digital** system. Many methods for A/D conversion are available. However, in this discussion, only one type is used to demonstrate the concept.

The *simultaneous*, or *flash*, method of A/D conversion uses parallel comparators to compare the linear input signal with various reference voltages developed by a voltage divider. When the input voltage exceeds the reference voltage for a given comparator, a high level is produced on that comparator's output. Figure 13–16 shows an analog-to-digital converter (ADC) that produces three-digit binary numbers on its output, which represent the values of the analog input voltage as it changes. This converter requires seven comparators. In general, $2^n - 1$ comparators are required for conversion to an n-digit binary number. The large number of comparators necessary for a reasonably sized

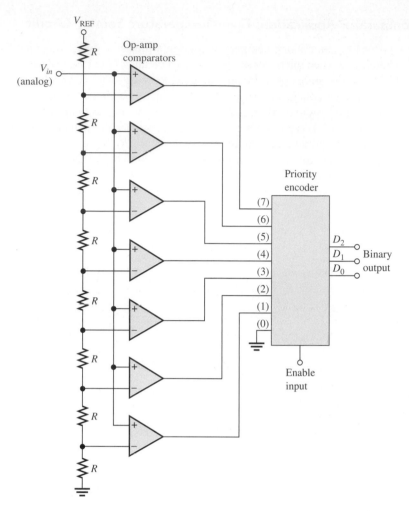

▲ **FIGURE 13–16**

A simultaneous (flash) analog-to-digital converter (ADC) using op-amps as comparators.

binary number is one of the drawbacks of the simultaneous ADC. Its chief advantage is that it provides a fast conversion time.

The reference voltage for each comparator is set by the resistive voltage-divider circuit and V_{REF}. The output of each comparator is connected to an input of the priority encoder. The *priority encoder* is a digital device that produces a binary number on its output representing the highest value input.

The encoder samples its input when a pulse occurs on the enable line (sampling pulse), and a three-digit binary number proportional to the value of the analog input signal appears on the encoder's outputs.

The sampling rate determines the accuracy with which the sequence of binary numbers represents the changing input signal. The more samples taken in a given unit of time, the more accurately the analog signal is represented in digital form.

The following example illustrates the basic operation of the simultaneous ADC in Figure 13–16.

EXAMPLE 13-4

Determine the binary number sequence of the three-digit simultaneous ADC in Figure 13–16 for the input signal in Figure 13–17 and the sampling pulses (encoder enable) shown. Sketch the resulting digital output waveforms.

▲ **FIGURE 13-17**

Sampling of values on analog waveform for conversion to digital.

Solution   The resulting binary output sequence is listed as follows and is shown in the waveform diagram of Figure 13–18 in relation to the sampling pulses.

011, 101, 110, 110, 100, 001, 000, 001, 010, 101, 110, 111

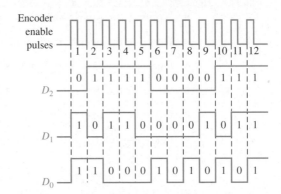

▲ **FIGURE 13-18**

Resulting digital outputs for sampled values in Figure 13–17. D_0 is the least significant digit.

Related Problem   If the frequency of the enable pulses in Figure 13–17 is doubled, does the resulting binary output sequence represent the analog waveform more or less accurately?

SECTION 13-1
REVIEW
Answers are at the end of the chapter.

1. What is the reference voltage for each comparator in Figure 13–19?
2. What is the purpose of hysteresis in a comparator?
3. Define the term *bounding* in relation to a comparator's output.

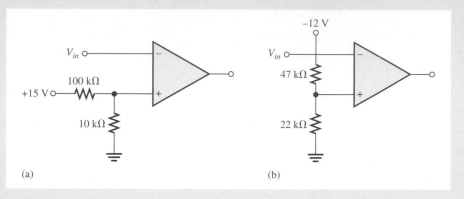

(a)   (b)

▲ **FIGURE 13–19**

13-2   SUMMING AMPLIFIERS

The summing amplifier is an application of the inverting op-amp configuration covered in Chapter 12. The summing amplifier has two or more inputs, and its output voltage is proportional to the negative of the algebraic sum of its input voltages. In this section, you will see how a summing amplifier works, and you will learn about the averaging amplifier and the scaling amplifier, which are variations of the basic summing amplifier.

After completing this section, you should be able to

- **Analyze the operation of several types of summing amplifiers**
- Describe the operation of a unity-gain summing amplifier
- Discuss how to achieve any specified gain greater than unity
- Describe the operation of an averaging amplifier
- Describe the operation of a scaling adder
- Discuss a scaling adder used as a digital-to-analog converter

Summing Amplifier with Unity Gain

A two-input **summing amplifier** is shown in Figure 13–20, but any number of inputs can be used. The operation of the circuit and derivation of the output expression are as follows. Two voltages, V_{IN1} and V_{IN2}, are applied to the inputs and produce currents I_1 and I_2, as shown.

Using the concepts of infinite input impedance and virtual ground, you can see that the inverting ($-$) input of the op-amp is approximately 0 V, and there is no current at the input. This means that both input currents I_1 and I_2 combine at this summing point and form the total current (I_T), which goes through R_f, as indicated in Figure 13–20.

$$I_T = I_1 + I_2$$

▶ **FIGURE 13-20**

Two-input inverting summing amplifier.

Since $V_{OUT} = -I_T R_f$, the following steps apply:

$$V_{OUT} = -(I_1 + I_2)R_f = -\left(\frac{V_{IN1}}{R_1} + \frac{V_{IN2}}{R_2}\right)R_f$$

If all three of the resistors are equal ($R_1 = R_2 = R_f = R$), then

$$V_{OUT} = -\left(\frac{V_{IN1}}{R} + \frac{V_{IN2}}{R}\right)R = -(V_{IN1} + V_{IN2})$$

The previous equation shows that the output voltage has the same magnitude as the sum of the two input voltages but with a negative sign, indicating inversion.

A general expression is given in Equation 13–4 for a unity-gain summing amplifier with n inputs, as shown in Figure 13–21 where all resistors are equal in value.

$$V_{OUT} = -(V_{IN1} + V_{IN2} + V_{IN3} + \cdots + V_{INn})$$

Equation 13–4

▶ **FIGURE 13-21**

Summing amplifier with n inputs.

EXAMPLE 13-5

Determine the output voltage in Figure 13–22.

▲ **FIGURE 13-22**

Solution $V_{OUT} = -(V_{IN1} + V_{IN2} + V_{IN3}) = -(3\text{ V} + 1\text{ V} + 8\text{ V}) = -12\text{ V}$

Related Problem   If a fourth input of +0.5 V is added to Figure 13–22 with a 10 kΩ resistor, what is the output voltage?

> Open file E13-05 in the Examples folder on your CD-ROM. Apply the indicated dc voltages to the inputs of the summing amplifier and verify that the output is the inverted sum of the inputs.

Summing Amplifier with Gain Greater Than Unity

When R_f is larger than the input resistors, the amplifier has a gain of R_f/R, where R is the value of each equal-value input resistor. The general expression for the output is

Equation 13–5

$$V_{OUT} = -\frac{R_f}{R}(V_{IN1} + V_{IN2} + \cdot \cdot \cdot + V_{INn})$$

As you can see, the output has the same magnitude as the sum of all the input voltages multiplied by a constant determined by the ratio $-R_f/R$.

EXAMPLE 13–6

Determine the output voltage for the summing amplifier in Figure 13–23.

▲ **FIGURE 13–23**

Solution $R_f = 10$ kΩ and $R = R_1 = R_2 = 1.0$ kΩ. Therefore,

$$V_{OUT} = -\frac{R_f}{R}(V_{IN1} + V_{IN2}) = -\frac{10\text{ k}\Omega}{1.0\text{ k}\Omega}(0.2\text{ V} + 0.5\text{ V}) = -10(0.7\text{ V}) = -7\text{ V}$$

Related Problem   Determine the output voltage in Figure 13–23 if the two input resistors are 2.2 kΩ and the feedback resistor is 18 kΩ.

> Open file E13-06 in the Examples folder on your CD-ROM. Apply the indicated dc voltages to the inputs of the summing amplifier and verify that the output is the inverted sum of the inputs times a gain of 10.

Averaging Amplifier

A summing amplifier can be made to produce the mathematical average of the input voltages. This is done by setting the ratio R_f/R equal to the reciprocal of the number of inputs (n).

$$\frac{R_f}{R} = \frac{1}{n}$$

You obtain the average of several numbers by first adding the numbers and then dividing by the quantity of numbers you have. Examination of Equation 13–5 and a little thought will convince you that a summing amplifier can be designed to do this. The next example will illustrate.

EXAMPLE 13–7

Show that the amplifier in Figure 13–24 produces an output whose magnitude is the mathematical average of the input voltages.

▲ **FIGURE 13–24**

Solution   Since the input resistors are equal, $R = 100 \text{ k}\Omega$. The output voltage is

$$V_{OUT} = -\frac{R_f}{R}(V_{IN1} + V_{IN2} + V_{IN3} + V_{IN4})$$

$$= -\frac{25 \text{ k}\Omega}{100 \text{ k}\Omega}(1 \text{ V} + 2 \text{ V} + 3 \text{ V} + 4 \text{ V}) = -\frac{1}{4}(10 \text{ V}) = \mathbf{-2.5 \text{ V}}$$

A simple calculation shows that the average of the input values is the same magnitude as V_{OUT} but of opposite sign.

$$V_{IN(avg)} = \frac{1 \text{ V} + 2 \text{ V} + 3 \text{ V} + 4 \text{ V}}{4} = \frac{10 \text{ V}}{4} = 2.5 \text{ V}$$

Related Problem   Specify the changes required in the averaging amplifier in Figure 13–24 in order to handle five inputs.

Open file E13-07 in the Examples folder on your CD-ROM. Apply the indicated dc voltages to the inputs of the summing amplifier and verify that the output is the inverted average of the inputs.

Scaling Adder

A different weight can be assigned to each input of a summing amplifier by simply adjusting the values of the input resistors. As you have seen, the output voltage can be expressed as

Equation 13–6

$$V_{OUT} = -\left(\frac{R_f}{R_1}V_{IN1} + \frac{R_f}{R_2}V_{IN2} + \cdots + \frac{R_f}{R_n}V_{INn}\right)$$

The weight of a particular input is set by the ratio of R_f to the resistance, R_x, for that input ($R_x = R_1, R_2, \cdots R_n$). For example, if an input voltage is to have a weight of 1, then $R_x = R_f$. Or, if a weight of 0.5 is required, $R_x = 2R_f$. The smaller the value of input resistance R_x, the greater the weight, and vice versa.

EXAMPLE 13–8

Determine the weight of each input voltage for the scaling adder in Figure 13–25 and find the output voltage.

▲ **FIGURE 13–25**

Solution   Weight of input 1: $\dfrac{R_f}{R_1} = \dfrac{10 \text{ k}\Omega}{47 \text{ k}\Omega} = \mathbf{0.213}$

Weight of input 2: $\dfrac{R_f}{R_2} = \dfrac{10 \text{ k}\Omega}{100 \text{ k}\Omega} = \mathbf{0.100}$

Weight of input 3: $\dfrac{R_f}{R_3} = \dfrac{10 \text{ k}\Omega}{10 \text{ k}\Omega} = \mathbf{1.00}$

The output voltage is

$$V_{OUT} = -\left(\frac{R_f}{R_1}V_{IN1} + \frac{R_f}{R_2}V_{IN2} + \frac{R_f}{R_3}V_{IN3}\right) = -[0.213(3 \text{ V}) + 0.100(2 \text{ V}) + 1.00(8 \text{ V})]$$

$$= -(0.639 \text{ V} + 0.2 \text{ V} + 8 \text{ V}) = \mathbf{-8.84 \text{ V}}$$

Related Problem   Determine the weight of each input voltage in Figure 13–25 if $R_1 = 22 \text{ k}\Omega$, $R_2 = 82 \text{ k}\Omega$, $R_3 = 56 \text{ k}\Omega$, and $R_f = 10 \text{ k}\Omega$. Also find V_{OUT}.

Open file E13-08 in the Examples folder on your CD-ROM. Apply the indicated dc voltages to the inputs of the summing amplifier and verify that the output agrees with the calculated value.

A Scaling Adder Application: Digital-to-Analog (D/A) Conversion

D/A conversion is an important interface process for converting digital signals to analog (linear) signals. An example is a voice signal that is digitized for storage, processing, or transmission and must be changed back into an approximation of the original audio signal in order to drive a speaker.

One method of D/A conversion uses a scaling adder with input resistor values that represent the binary weights of the digital input code. Figure 13–26 shows a four-digit digital-to-analog converter (DAC) of this type (called a *binary-weighted resistor DAC*). The switch symbols represent transistor switches for applying each of the four binary digits to the inputs.

The inverting (−) input is at virtual ground, and so the output voltage is proportional to the current through the feedback resistor R_f (sum of input currents).

The lowest-value resistor R corresponds to the highest weighted binary input (2^3). All of the other resistors are multiples of R and correspond to the binary weights 2^2, 2^1, and 2^0.

▲ **FIGURE 13–26**

A scaling adder as a four-digit digital-to-analog converter (DAC).

EXAMPLE 13–9

Determine the output voltage of the DAC in Figure 13–27(a). The sequence of four-digit binary codes represented by the waveforms in Figure 13–27(b) are applied to the inputs. A high level is a binary 1, and low level is a binary 0. The least significant binary digit is D_0.

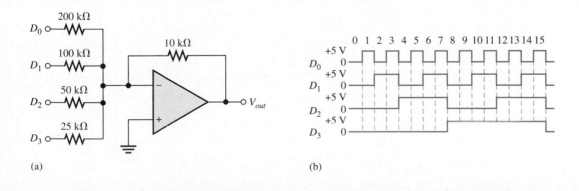

(a)

(b)

▲ **FIGURE 13–27**

Solution   First, determine the current for each of the weighted inputs. Since the inverting input of the op-amp is at 0 V (virtual ground), and a binary 1 corresponds to a high level (+5 V), the current through any of the input resistors equals 5 V divided by the resistance value.

$$I_0 = \frac{5\ V}{200\ k\Omega} = 0.025\ mA$$

$$I_1 = \frac{5\ V}{100\ k\Omega} = 0.05\ mA$$

$$I_2 = \frac{5\ V}{50\ k\Omega} = 0.1\ mA$$

$$I_3 = \frac{5\ V}{25\ k\Omega} = 0.2\ mA$$

There is almost no current at the inverting op-amp input because of its extremely high impedance. Therefore, assume that all of the input current goes through R_f. Since one end of R_f is at 0 V (virtual ground), the drop across R_f equals the output voltage, which is negative with respect to virtual ground.

$$V_{OUT(D0)} = (10\ k\Omega)(-0.025\ mA) = \textbf{-0.25 V}$$

$$V_{OUT(D1)} = (10\ k\Omega)(-0.05\ mA) = \textbf{-0.5 V}$$

$$V_{OUT(D2)} = (10\ k\Omega)(-0.1\ mA) = \textbf{-1 V}$$

$$V_{OUT(D3)} = (10\ k\Omega)(-0.2\ mA) = \textbf{-2 V}$$

From Figure 13–27(b), the first binary input code is 0000, which produces an output voltage of 0 V. The next input code is 0001 (it stands for decimal 1). For this, the output voltage is −0.25 V. The next code is 0010, which produces an output voltage of −0.5 V. The next code is 0011, which produces an output voltage of −0.25 V + (−0.5 V) = −0.75 V. Each successive binary code increases the output voltage by −0.25 V. So, for this particular straight binary sequence on the inputs, the output is a stairstep waveform going from 0 V to −3.75 V in −0.25 V steps, as shown in Figure 13–28. If the steps are very small, the output approximates a straight line (linear).

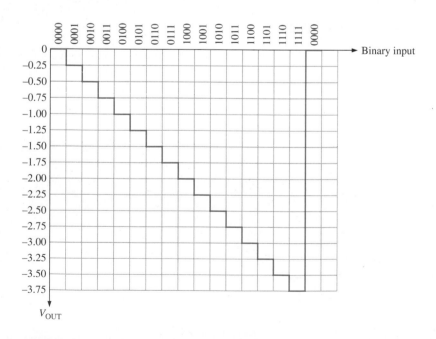

▲ **FIGURE 13-28**

Related Problem If the 200 kΩ resistor in Figure 13–27(a) is changed to 400 kΩ, would the other resistor values have to be changed? If so, specify the values.

SECTION 13–2 REVIEW

1. Define *summing point*.
2. What is the value of R_f/R for a five-input averaging amplifier?
3. A certain scaling adder has two inputs, one having twice the weight of the other. If the resistor value for the lower-weighted input is 10 kΩ, what is the value of the other input resistor?

13–3 INTEGRATORS AND DIFFERENTIATORS

An op-amp integrator simulates mathematical integration, which is basically a summing process that determines the total area under the curve of a function. An op-amp differentiator simulates mathematical differentiation, which is a process of determining the instantaneous rate of change of a function. It is not necessary for you to understand mathematical integration or differentiation, at this point, in order to learn how an integrator and differentiator work. The integrators and differentiators shown in this section are idealized to show basic principles. Practical integrators often have an additional resistor or other circuitry in parallel with the feedback capacitor to prevent saturation. Practical differentiators may include a series resistor to reduce high frequency noise.

After completing this section, you should be able to

- **Analyze the operation of integrators and differentiators**

- Identify an integrator

- Discuss how a capacitor charges

- Determine the rate of change of an integrator's output

- Identify a differentiator

- Determine the output voltage of a differentiator

The Op-Amp Integrator

An ideal **integrator** is shown in Figure 13–29. Notice that the feedback element is a capacitor that forms an *RC* circuit with the input resistor. Although a large-value resistor is normally used in parallel with the capacitor to limit the gain, it does not affect the basic operation and is not shown for purposes of this analysis.

▶ **FIGURE 13–29**

An op-amp integrator.

How a Capacitor Charges To understand how the integrator works, it is important to review how a capacitor charges. Recall that the charge Q on a capacitor is proportional to the charging current (I_C) and the time (t).

$$Q = I_C t$$

Also, in terms of the voltage, the charge on a capacitor is

$$Q = CV_C$$

From these two relationships, the capacitor voltage can be expressed as

$$V_C = \left(\frac{I_C}{C}\right)t$$

This expression has the form of an equation for a straight line that begins at zero with a constant slope of I_C/C. Remember from algebra that the general formula for a straight line is $y = mx + b$. In this case, $y = V_C$, $m = I_C/C$, $x = t$, and $b = 0$.

Recall that the capacitor voltage in a simple *RC* circuit is not linear but is exponential. This is because the charging current continuously decreases as the capacitor charges and causes the rate of change of the voltage to continuously decrease. The key thing about using an op-amp with an *RC* circuit to form an integrator is that the capacitor's charging current is made constant, thus producing a straight-line (linear) voltage rather than an exponential voltage. Now let's see why this is true.

In Figure 13–30, the inverting input of the op-amp is at virtual ground (0 V), so the voltage across R_i equals V_{in}. Therefore, the input current is

$$I_{in} = \frac{V_{in}}{R_i}$$

▶ **FIGURE 13–30**

Currents in an integrator.

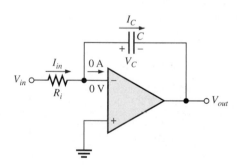

If V_{in} is a constant voltage, then I_{in} is also a constant because the inverting input always remains at 0 V, keeping a constant voltage across R_i. Because of the very high input impedance of the op-amp, there is negligible current at the inverting input. This makes all of the input current go through the capacitor, as indicated in Figure 13–30, so

$$I_C = I_{in}$$

The Capacitor Voltage Since I_{in} is constant, so is I_C. The constant I_C charges the capacitor linearly and produces a linear voltage across C. The positive side of the capacitor is held at 0 V by the virtual ground of the op-amp. The voltage on the negative side of the capacitor, which is the op-amp output voltage, decreases linearly from zero as the capacitor charges, as shown in Figure 13–31. This voltage is called a *negative ramp* and is the consequence of a constant positive input.

The Output Voltage V_{out} is the same as the voltage on the negative side of the capacitor. When a constant positive input voltage in the form of a step or pulse (a pulse has a constant amplitude when high) is applied, the output ramp decreases negatively until the op-amp saturates at its maximum negative level. This is indicated in Figure 13–32.

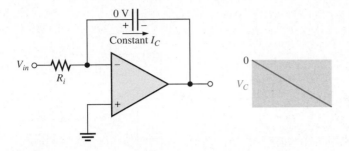

▲ FIGURE 13–31

A linear ramp voltage is produced across C by the constant charging current.

▲ FIGURE 13–32

A constant input voltage produces a ramp on the output of the integrator.

Rate of Change of the Output   The rate at which the capacitor charges, and therefore the slope of the output ramp, is set by the ratio I_C/C, as you have seen. Since $I_C = V_{in}/R_i$, the rate of change or slope of the integrator's output voltage is $\Delta V_{out}/\Delta t$.

$$\frac{\Delta V_{out}}{\Delta t} = -\frac{V_{in}}{R_i C}$$

Equation 13–7

Integrators are especially useful in triangular-wave generators as you will see in Chapter 16.

EXAMPLE 13–10

(a) Determine the rate of change of the output voltage in response to the first input pulse in a pulse waveform, as shown for the integrator in Figure 13–33(a). The output voltage is initially zero.

(b) Describe the output after the first pulse. Draw the output waveform.

Solution   (a) The rate of change of the output voltage during the time that the input pulse is HIGH is

$$\frac{\Delta V_{out}}{\Delta t} = -\frac{V_{in}}{R_i C} = -\frac{5 \text{ V}}{(10 \text{ k}\Omega)(0.01 \text{ }\mu\text{F})} = -50 \text{ kV/s} = \textbf{−50 mV/}\mu\textbf{s}$$

(a)

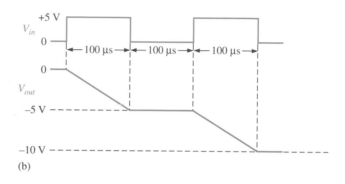

(b)

▲ **FIGURE 13–33**

(b) The rate of change was found to be −50 mV/μs in part (a). When the input is at +5 V, the output is a negative-going ramp. In 100 μs, the voltage decreases.

$$\Delta V_{out} = (-50 \text{ mV/}\mu s)(100 \text{ }\mu s) = -5 \text{ V}$$

Therefore, the negative-going ramp reaches −5 V at the end of the pulse. When the input is at 0 V, the output is a positive-going ramp. The output voltage then goes from −5 V to 0 V for the time that the input is zero. When the next pulse occurs, the output again becomes a negative-going ramp which reaches −5 V at the end of the second pulse. Then the output will return again to 0 V while the input is at 0 V, and so on. The waveforms are shown in Figure 13–33(b).

Related Problem Modify the integrator in Figure 13–33 to make the output change from 0 to −5 V in 50 μs with the same input.

Open file E13-10 in the Examples folder on your CD-ROM. Using the function generator, apply the indicated pulse waveform to the input and verify that the output voltage is correct.

The Op-Amp Differentiator

An ideal **differentiator** is shown in Figure 13–34. Notice how the placement of the capacitor and resistor differ from the integrator. The capacitor is now the input element. A differentiator produces an output that is proportional to the rate of change of the input voltage.

To see how the differentiator works, apply a positive-going ramp voltage to the input as indicated in Figure 13–35. In this case, $I_C = I_{in}$ and the voltage across the capacitor is equal to V_{in} at all times ($V_C = V_{in}$) because of virtual ground on the inverting input.

▲ FIGURE 13–34

An op-amp differentiator.

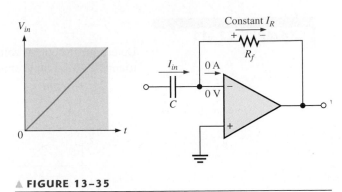

▲ FIGURE 13–35

A differentiator with a ramp input.

From the basic formula, $V_C = (I_C/C)t$, the capacitor current is

$$I_C = \left(\frac{V_C}{t}\right)C$$

Since the current at the inverting input is negligible, $I_R = I_C$. Both currents are constant because the slope of the capacitor voltage (V_C/t) is constant. The output voltage is also constant and equal to the voltage across R_f because one side of the feedback resistor is always 0 V (virtual ground).

$$V_{out} = I_R R_f = I_C R_f$$

$$V_{out} = -\left(\frac{V_C}{t}\right)R_f C$$

Equation 13–8

The output is negative when the input is a positive-going ramp and positive when the input is a negative-going ramp, as illustrated in Figure 13–36. During the positive slope of the input, the capacitor is charging from the input source and the constant current through the feedback resistor is in the direction shown. During the negative slope of the input, the current is in the opposite direction because the capacitor is discharging.

▲ FIGURE 13–36

Output of a differentiator with a series of positive and negative ramps (triangle wave) on the input.

Notice in Equation 13–8 that the term V_C/t is the slope of the input. If the slope increases, V_{out} increases. If the slope decreases, V_{out} decreases. So, the output voltage is proportional to the slope (rate of change) of the input. The constant of proportionality is the time constant, $R_f C$.

EXAMPLE 13–11

Determine the output voltage of the op-amp differentiator in Figure 13–37 for the triangular-wave input shown.

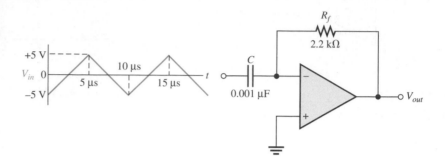

▲ **FIGURE 13–37**

Solution Starting at $t = 0$, the input voltage is a positive-going ramp ranging from -5 V to $+5$ V (a $+10$ V change) in 5 μs. Then it changes to a negative-going ramp ranging from $+5$ V to -5 V (a -10 V change) in 5 μs.

The time constant is

$$R_f C = (2.2 \text{ k}\Omega)(0.001 \text{ }\mu\text{F}) = 2.2 \text{ }\mu\text{s}$$

The slope or rate of change (V_C/t) of the positive-going ramp is determined, and the output voltage is calculated as follows:

$$\frac{V_C}{t} = \frac{10 \text{ V}}{5 \text{ }\mu\text{s}} = 2 \text{ V}/\mu\text{s}$$

$$V_{out} = -\left(\frac{V_C}{t}\right)R_f C = -(2 \text{ V}/\mu\text{s})2.2 \text{ }\mu\text{s} = \textbf{-4.4 V}$$

Likewise, the slope of the negative-going ramp is -2 V/μs, and the output voltage is

$$V_{out} = -(-2 \text{ V}/\mu\text{s})2.2 \text{ }\mu\text{s} = \textbf{+4.4 V}$$

Finally, the output voltage waveform is graphed relative to the input as shown in Figure 13–38.

▶ **FIGURE 13–38**

Related Problem What would the output voltage be if the feedback resistor in Figure 13–37 is changed to 3.3 kΩ?

SECTION 13–3
REVIEW

1. What is the feedback element in an op-amp integrator?
2. For a constant input voltage to an integrator, why is the voltage across the capacitor linear?
3. What is the feedback element in an op-amp differentiator?
4. How is the output of a differentiator related to the input?

13–4 TROUBLESHOOTING

Although integrated circuit op-amps are extremely reliable and trouble-free, failures do occur from time to time. One type of internal failure mode is a condition where the op-amp output is in a saturated state resulting in a constant high or constant low level, regardless of the input. Also, external component failures will produce various types of failure modes in op-amp circuits. Some examples are presented in this section.

After completing this section, you should be able to

■ **Troubleshoot basic op-amp circuits**

■ Identify failures in comparator circuits

■ Identify failures in summing amplifiers

Figure 13–39 illustrates an internal failure of a comparator circuit that results in a "stuck" output.

(a) Output failed in the HIGH state   (b) Output failed in the LOW state

▲ **FIGURE 13–39**

Internal comparator failures typically result in the output being "stuck" in the HIGH or LOW state.

Symptoms of External Component Failures in Comparator Circuits

Comparators with zener-bounding and hysteresis are shown in Figure 13–40. In addition to a failure of the op-amp itself, a zener diode or one of the resistors could be faulty. For example, suppose one of the zener diodes opens. This effectively eliminates both zeners, and the circuit operates as an unbounded comparator, as indicated in Figure 13–41(a). With a shorted diode, the output is limited to the zener voltage (bounded) only in one direction, depending on which diode remains operational, as illustrated in Figure 13–41(b). In the other direction, the output is held at the forward diode voltage.

Recall that R_1 and R_2 set the UTP and LTP for the hysteresis comparator. Now, suppose that R_2 opens. Essentially all of the output voltage is fed back to the noninverting

▶ **FIGURE 13–40**

A bounded comparator.

▶ **FIGURE 13–41**

Examples of comparator circuit failures and their effects.

(a) The effect of an open zener

(b) The effect of a shorted zener

(c) Open R_2 causes output to "stick" in one state

(d) Open R_1 forces the circuit to operate as a zero-level detector

($+$) input, and, since the input voltage will never exceed the output, the device will remain in one of its saturated states. This symptom can also indicate a faulty op-amp, as mentioned before. Now, assume that R_1 opens. This leaves the noninverting input near ground potential and causes the circuit to operate as a zero-level detector. These conditions are shown in parts (c) and (d) of Figure 13–41.

EXAMPLE 13–12

One channel of a dual-trace oscilloscope is connected to the comparator output and the other channel is connected to the input, as shown in Figure 13–42. From the observed waveforms, determine if the circuit is operating properly, and if not, what the most likely failure is.

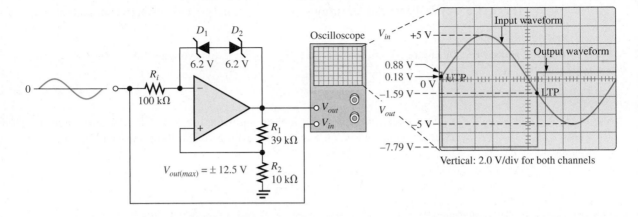

▲ **FIGURE 13–42**

Solution The output should be limited to ±8.67 V. However, the positive maximum is +0.88 V and the negative maximum is −7.79 V. This indicates that D_2 is shorted. Refer to Example 13–3 for analysis of the bounded comparator.

Related Problem What would the output voltage look like if D_1 shorted rather than D_2?

Symptoms of Component Failures in Summing Amplifiers

If one of the input resistors in a unity-gain summing amplifier opens, the output will be less than the normal value by the amount of the voltage applied to the open input. Stated another way, the output will be the sum of the remaining input voltages.

If the summing amplifier has a nonunity gain, an open input resistor causes the output to be less than normal by an amount equal to the gain times the voltage at the open input.

EXAMPLE 13–13

(a) What is the normal output voltage in Figure 13–43?

(b) What is the output voltage if R_2 opens?

(c) What happens if R_5 opens?

▶ **FIGURE 13–43**

Solution   **(a)** $V_{OUT} = -(V_{IN1} + V_{IN2} + \cdots + V_{INn}) = -(1\,V + 0.5\,V + 0.2\,V + 0.1\,V) = \mathbf{-1.8\,V}$

    (b) $V_{OUT} = -(1\,V + 0.2\,V + 0.1\,V) = \mathbf{-1.3\,V}$

    (c) If the feedback resistor opens, the circuit becomes a comparator and the output goes to $-V_{max}$.

Related Problem   In Figure 13–43, $R_5 = 47\,k\Omega$. What is the output voltage if R_1 opens?

As another example, let's look at an averaging amplifier. An open input resistor will result in an output voltage that is the average of all the inputs with the open input averaged in as a zero.

EXAMPLE 13–14

(a) What is the normal output voltage for the averaging amplifier in Figure 13–44?

(b) If R_4 opens, what is the output voltage? What does the output voltage represent?

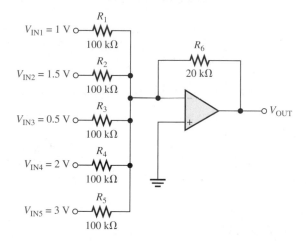

▲ **FIGURE 13–44**

Solution   Since the input resistors are equal, $R = 100\,k\Omega$. $R_f = R_6$.

(a) $V_{OUT} = -\dfrac{R_f}{R}(V_{IN1} + V_{IN2} + \cdots + V_{INn})$

$$= -\frac{20\,k\Omega}{100\,k\Omega}(1\,V + 1.5\,V + 0.5\,V + 2\,V + 3\,V) = -\frac{1}{5}(8\,V) = \mathbf{-1.6\,V}$$

(b) $V_{OUT} = -\dfrac{20\,k\Omega}{100\,k\Omega}(1\,V + 1.5\,V + 0.5\,V + 3\,V) = -\dfrac{1}{5}(6\,V) = \mathbf{-1.2\,V}$

1.2 V is the average of five voltages with the 2 V input replaced by 0 V. Notice that the output is not the average of the four remaining input voltages.

Related Problem   If R_4 is open, as was the case in this example, what would you have to do to make the output equal to the average of the remaining four input voltages?

EWB/Multisim Troubleshooting Exercises

These file circuits are in the Troubleshooting Exercises folder on your CD-ROM.

1. Open file TSE13-01. Determine if the circuit is working properly and, if not, determine the fault.

2. Open file TSE13-02. Determine if the circuit is working properly and, if not, determine the fault.

3. Open file TSE13-03. Determine if the circuit is working properly and, if not, determine the fault.

4. Open file TSE13-04. Determine if the circuit is working properly and, if not, determine the fault.

5. Open file TSE13-05. Determine if the circuit is working properly and, if not, determine the fault.

6. Open file TSE13-06 (Multisim). Determine if the circuit is working properly and, if not, determine the fault.

7. Open file TSE13-07. Determine if the circuit is working properly and, if not, determine the fault.

8. Open file TSE13-08. Determine if the circuit is working properly and, if not, determine the fault.

9. Open file TSE13-09. Determine if the circuit is working properly and, if not, determine the fault.

SECTION 13–4 REVIEW

1. Describe one type of internal op-amp failure.
2. If a certain malfunction is attributable to more than one possible component failure, what would you do to isolate the problem?

SYSTEM APPLICATION

This application involves an analog-to-digital converter (ADC) that is used in a recording system for changing the audio signal into digital form for recording. You were introduced to one method of A/D conversion in this chapter, called the simultaneous, or flash, method. There are several other methods of A/D conversion, and this system application uses a method called dual-slope. Although many parts of this system are digital, you will focus on the ADC circuit board which incorporates types of op-amp circuits with which you are familiar.

Basic Operation of the System

The dual-slope ADC in Figure 13–45 accepts an audio signal voltage and converts it to a series of digital codes for the purpose of recording. The audio signal voltage is applied to the sample-and-hold circuit. At fixed intervals, sample pulses cause the instantaneous amplitude of the audio waveform to be converted to proportional dc levels that are processed by the rest of the circuits and represented by a series of digital codes.

The sample pulses occur at a much higher frequency than the audio signal so that a sufficient number of points on the audio waveform are sampled and converted to obtain an accurate digital representation of the audio signal. A rough approximation of the sampling process is illustrated in Figure 13–46. As the

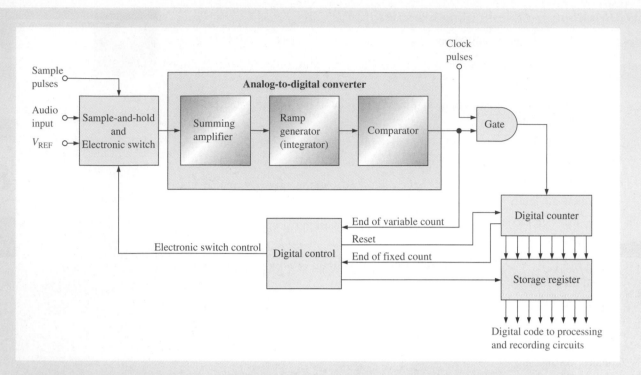

▲ FIGURE 13–45

Block diagram of dual-slope analog-to-digital converter.

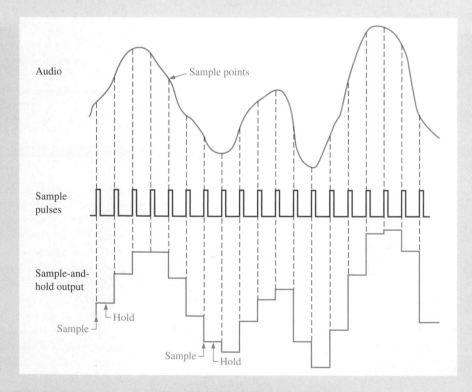

▲ FIGURE 13–46

Illustration of the sample-and-hold process. The output of the sample-and-hold circuit is shown as a rough approximation of the audio voltage for simplification.

frequency of the sample pulses increases relative to the audio frequency, an increasingly accurate representation is achieved.

The summing amplifier has only one input active at a time. For example, when the audio input is switched in, the reference voltage input is zero and vice versa.

During the time between each sample pulse, the dc level from the sample-and-hold circuit is switched electronically into the summing amplifier on the ADC board. The output of the summing amplifier goes to the ramp generator which is an integrator. At the same time, the digital counter starts counting up from zero. During the fixed time in terval of the counting sequence, the integrator (ramp generator) produces a positive-going ramp voltage with a slope that depends on the level of the sampled audio voltage. At the end of the fixed time interval, the ramp voltage at the output of the integrator has reached a value that is proportional to the sampled audio voltage. At this time, the digital control logic switches from the sample-and-hold input to the negative dc reference voltage input (V_{REF}) and resets the digital counter to zero.

The summing amplifier inverts the negative dc reference voltage and applies it to the integrator input, which starts a negative-going ramp on the integrator output. This ramp voltage has a slope that is fixed by the value of V_{REF}. When the negative-going ramp starts, the digital counter begins to count up again from zero and will continue to count up until the integrator output reaches zero volts.

At the time when the negative-going ramp reaches zero, the comparator switches to its negative saturated output voltage and disables the gate so that there are no additional clock pulses to the counter. At this time, the digital code in the counter is proportional to the time that was required for the negative-going ramp at the integrator output to reach zero from its maximum value. The code in the counter will vary for each different sampled value of the audio.

Recall that the negative-going ramp started at a positive voltage that was dependent on the sampled value of the audio signal. Therefore, the digital code in the counter is also proportional to and represents the amplitude of the sampled audio voltage. The code is shifted out of the counter for temporary storage in the register from which it is processed and recorded.

The conversion process for each sampled value is repeated many times during the period of the highest audio harmonic frequency that is to be recorded accurately. The result is a sequence of digital codes that represents the audio voltage as it varies with time.

Figure 13–47 illustrates this process for several sampled values. In this application, you will focus on the ADC circuit which contains the summing amplifier, integrator, and comparator.

The Analog-to-Digital Converter Board

■ Make sure that the circuit board shown in Figure 13–48 is correctly assembled by checking it against the schematic in Figure 13–49. A pin diagram for the power transistors is shown for reference. Backside interconnections are shown as darker traces.

■ Label a copy of the board with component and input/output designations in agreement with the schematic.

Analysis of the ADC Circuit

■ Determine the gain of the summing amplifier.

■ Determine the slope of the integrator ramp in volts per microsecond when a sampled audio voltage of $+2$ V is applied.

■ Determine the slope of the integrator ramp in volts per microsecond when a dc reference voltage of -8 V is applied.

■ Given that the reference voltage is -8 V and the fixed time interval of the positive-going slope is 1 μs, sketch the dual-slope output of the integrator when an instantaneous audio voltage of $+3$ V is applied.

■ Assuming that the maximum audio voltage to be sampled is $+6$ V, determine the maximum audio frequency that can be sampled by this particular system if there are to be 100 samples per cycle. What is the sample pulse rate in this case?

Test Procedure

■ Develop a step-by-step set of instructions on how to check the ADC circuit board for proper operation independent of the rest of the system.

■ Specify voltage values for all the measurements to be made.

■ Provide a fault analysis for all possible component failures.

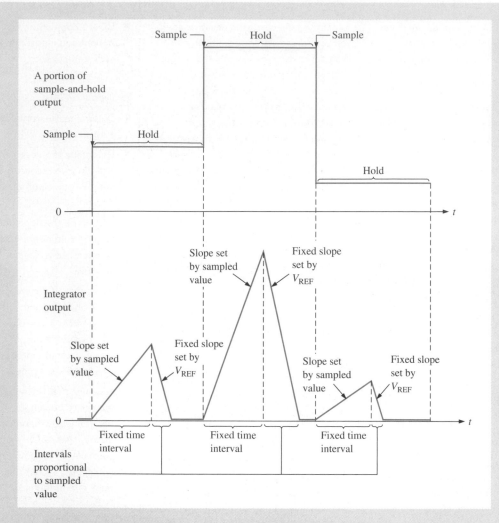

▲ **FIGURE 13–47**

During the fixed time interval of the positive-going ramp, the sampled audio input is applied to the integrator. During the variable time interval of the fixed-slope negative-going ramp, the reference voltage is applied to the integrator. The counter controls the fixed time interval and is reset. Another count begins during the variable time interval, and the digital code in the counter at the end of this interval represents the sampled audio value.

Troubleshooting

Faults have developed in three boards. Based on the test bench measurements for each board with specified test signals applied as indicated in Figure 13–50, determine the most likely fault in each case. The circled numbers indicate test point connections to the circuit board. Assume that each board has the proper dc supply voltage.

Final Report (Optional)

Submit a final written report on the ADC board using an organized format that includes the following:

1. A physical description of the circuit.
2. A discussion of the operation of the circuit.
3. A list of the specifications.
4. A list of parts with part numbers if available.
5. A list of the types of problems on the three faulty circuit boards.
6. A description of how you determined the problem on each of the faulty circuit boards.

▲ **FIGURE 13–48**

Analog-to-digital converter (ADC) board.

▲ **FIGURE 13–49**

Schematic of the ADC.

▲ FIGURE 13–50

Test results for three faulty ADC boards.

SUMMARY OF OP-AMP CIRCUITS

COMPARATORS

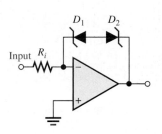

Zero-level detector

Nonzero-level detector

$$V_{\text{UTP}} = \frac{R_2}{R_1 + R_2}(+V_{out(max)})$$

$$V_{\text{LTP}} = \frac{R_2}{R_1 + R_2}(-V_{out(max)})$$

Comparator with hysteresis

$$+V_{out(max)} = V_{Z1} + 0.7 \text{ V}$$
$$-V_{out(max)} = -(V_{Z2} + 0.7 \text{ V})$$

Bounded comparator

SUMMING AMPLIFIERS

Summing amplifier

- Unity-gain amplifier:

$$R_f = R_1 = R_2 = R_3 = \cdots = R_n$$
$$V_{\text{OUT}} = -(V_{\text{IN1}} + V_{\text{IN2}} + V_{\text{IN3}} + \cdots + V_{\text{IN}n})$$

- Greater than unity-gain amplifier:

$$R_f > R$$
$$R = R_1 = R_2 = R_3 = \cdots = R_n$$
$$V_{\text{OUT}} = -\frac{R_f}{R}(V_{\text{IN1}} + V_{\text{IN2}} + V_{\text{IN3}} + \cdots + V_{\text{IN}n})$$

- Averaging amplifier:

$$\frac{R_f}{R} = \frac{1}{n}$$
$$R = R_1 = R_2 = R_3 = \cdots = R_n$$
$$V_{\text{OUT}} = -\frac{R_f}{R}(V_{\text{IN1}} + V_{\text{IN2}} + V_{\text{IN3}} + \cdots + V_{\text{IN}n})$$

- Scaling adder:

$$V_{\text{OUT}} = -\left(\frac{R_f}{R_1}V_{\text{IN1}} + \frac{R_f}{R_2}V_{\text{IN2}} + \frac{R_f}{R_3}V_{\text{IN3}} + \cdots + \frac{R_f}{R_n}V_{\text{IN}n}\right)$$

SUMMARY OF OP-AMP CIRCUITS, *continued*

INTEGRATOR AND DIFFERENTIATOR

■ Slope of output voltage:

$$\frac{\Delta V_{out}}{\Delta t} = -\frac{V_{in}}{R_i C}$$

Integrator

■ Output voltage:

$$V_{out} = -\left(\frac{V_C}{t}\right)R_f C$$

$$V_C = V_{in}$$

Differentiator

CHAPTER SUMMARY

- In an op-amp comparator, when the input voltage exceeds a specified reference voltage, the output changes state.
- Hysteresis gives an op-amp noise immunity.
- A comparator switches to one state when the input reaches the upper trigger point (UTP) and back to the other state when the input drops below the lower trigger point (LTP).
- The difference between the UTP and the LTP is the hysteresis voltage.
- Bounding limits the output amplitude of a comparator.
- The output voltage of a summing amplifier is proportional to the sum of the input voltages.
- An averaging amplifier is a summing amplifier with a closed-loop gain equal to the reciprocal of the number of inputs.
- In a scaling adder, a different weight can be assigned to each input, thus making the input contribute more or contribute less to the output.
- Integration is a mathematical process for determining the area under a curve.
- Integration of a step produces a ramp with a slope proportional to the amplitude.
- Differentiation is a mathematical process for determining the rate of change of a function.
- Differentiation of a ramp produces a step with an amplitude proportional to the slope.

KEY TERMS

Key terms and other bold terms in the chapter are defined in the end-of-book glossary.

Bounding   The process of limiting the output range of an amplifier or other circuit.

Comparator   A circuit that compares two input voltages and produces an output in either of two states indicating the greater than or less than relationship of the inputs.

Differentiator   A circuit that produces an output which approximates the instantaneous rate of change of the input function.

Hysteresis   Characteristic of a circuit in which two different trigger levels create an offset or lag in the switching action.

Integrator   A circuit that produces an output which approximates the area under the curve of the input function.

Schmitt trigger   A comparator with hysteresis.

Summing amplifier   An op-amp configuration with two or more inputs that produces an output equal to the sum of the inputs.

KEY FORMULAS

Comparator

13–1 $\qquad V_{\text{UTP}} = \dfrac{R_2}{R_1 + R_2}(+V_{out(max)})$ $\qquad$ Upper trigger point

13–2 $\qquad V_{\text{LTP}} = \dfrac{R_2}{R_1 + R_2}(-V_{out(max)})$ $\qquad$ Lower trigger point

13–3 $\qquad V_{\text{HYS}} = V_{\text{UTP}} - V_{\text{LTP}}$ $\qquad$ Hysteresis voltage

Summing Amplifier

13–4 $\qquad V_{\text{OUT}} = -(V_{\text{IN1}} + V_{\text{IN2}} + \cdots + V_{\text{IN}n})$ $\qquad$ *n*-input adder

13–5 $\qquad V_{\text{OUT}} = \dfrac{R_f}{R}(V_{\text{IN1}} + V_{\text{IN2}} + \cdots + V_{\text{IN}n})$ $\qquad$ Adder with gain

13–6 $\qquad V_{\text{OUT}} = -\left(\dfrac{R_f}{R_1}V_{\text{IN1}} + \dfrac{R_f}{R_2}V_{\text{IN2}} + \cdots + \dfrac{R_f}{R_n}V_{\text{IN}n}\right)$ $\qquad$ Scaling adder with gain

Integrator and Differentiator

13–7 $\qquad \dfrac{\Delta V_{out}}{\Delta t} = -\dfrac{V_{in}}{R_i C}$ $\qquad$ Integrator output rate of change

13–8 $\qquad V_{out} = -\left(\dfrac{V_C}{t}\right)R_f C$ $\qquad$ Differentiator output voltage with ramp input

SELF-TEST

Answers are at the end of the chapter.

1. In a zero-level detector, the output changes state when the input
   - (a) is positive
   - (b) is negative
   - (c) crosses zero
   - (d) has a zero rate of change

2. The zero-level detector is one application of a
   - (a) comparator
   - (b) differentiator
   - (c) summing amplifier
   - (d) diode

3. Noise on the input of a comparator can cause the output to
   - (a) hang up in one state
   - (b) go to zero
   - (c) change back and forth erratically between two states
   - (d) produce the amplified noise signal

4. The effects of noise can be reduced by
   - (a) lowering the supply voltage
   - (b) using positive feedback
   - (c) using negative feedback
   - (d) using hysteresis
   - (e) answers (b) and (d)

5. A comparator with hysteresis
   - (a) has one trigger point
   - (b) has two trigger points
   - (c) has a variable trigger point
   - (d) is like a magnetic circuit

6. In a comparator with hysteresis,
   - (a) a bias voltage is applied between the two inputs
   - (b) only one supply voltage is used
   - (c) a portion of the output is fed back to the inverting input
   - (d) a portion of the output is fed back to the noninverting input

7. Using output bounding in a comparator

(a) makes it faster   (b) keeps the output positive

(c) limits the output levels   (d) stabilizes the output

8. A summing amplifier can have

(a) only one input   (b) only two inputs   (c) any number of inputs

9. If the voltage gain for each input of a summing amplifier with a 4.7 kΩ feedback resistor is unity, the input resistors must have a value of

(a) 4.7 kΩ

(b) 4.7 kΩ divided by the number of inputs

(c) 4.7 kΩ times the number of inputs

10. An averaging amplifier has five inputs. The ratio R_f/R_i must be

(a) 5   (b) 0.2   (c) 1

11. In a scaling adder, the input resistors are

(a) all the same value

(b) all of different values

(c) each proportional to the weight of its input

(d) related by a factor of two

12. In an integrator, the feedback element is a

(a) resistor   (b) capacitor

(c) zener diode   (d) voltage divider

13. For a step input, the output of an integrator is

(a) a pulse   (b) a triangular waveform

(c) a spike   (d) a ramp

14. The rate of change of an integrator's output voltage in response to a step input is set by

(a) the RC time constant

(b) the amplitude of the step input

(c) the current through the capacitor

(d) all of these

15. In a differentiator, the feedback element is a

(a) resistor   (b) capacitor

(c) zener diode   (d) voltage divider

16. The output of a differentiator is proportional to

(a) the RC time constant   (b) the rate at which the input is changing

(c) the amplitude of the input   (d) answers (a) and (b)

17. When you apply a triangular waveform to the input of a differentiator, the output is

(a) a dc level   (b) an inverted triangular waveform

(c) a square waveform   (d) the first harmonic of the triangular waveform

PROBLEMS

Answers to all odd-numbered problems are at the end of the book.

BASIC PROBLEMS

SECTION 13–1   **Comparators**

1. A certain op-amp has an open-loop gain of 80,000. The maximum saturated output levels of this particular device are ±12 V when the dc supply voltages are ±15 V. If a differential voltage of 0.15 mV rms is applied between the inputs, what is the peak-to-peak value of the output?

2. Determine the output level (maximum positive or maximum negative) for each comparator in Figure 13–51.

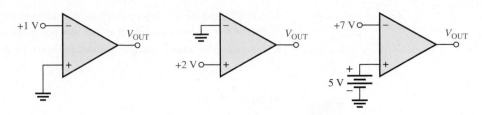

▲ **FIGURE 13–51**

---

3. Calculate the V_{UTP} and V_{LTP} in Figure 13–52. $V_{out(max)} = \pm 10$ V.

4. What is the hysteresis voltage in Figure 13–52?

▶ **FIGURE 13–52**

File circuits are identified with a CD logo and are in the Problems folder on your CD-ROM. Filenames correspond to figure numbers (e.g., F13–52).

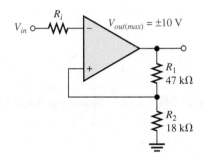

5. Sketch the output voltage waveform for each circuit in Figure 13–53 with respect to the input. Show voltage levels.

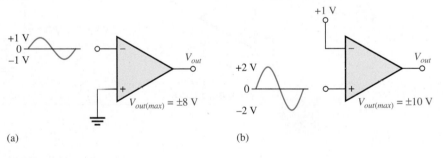

(a)   (b)

▲ **FIGURE 13–53**

---

6. Determine the hysteresis voltage for each comparator in Figure 13–54. The maximum output levels are ± 11 V.

(a)   (b)

▲ **FIGURE 13–54**

7. A 6.2 V zener diode is connected from the output to the inverting input in Figure 13–52 with the cathode at the output. What are the positive and negative output levels?

8. Determine the output voltage waveform in Figure 13–55.

▶ FIGURE 13–55

SECTION 13–2 Summing Amplifiers

9. Determine the output voltage for each circuit in Figure 13–56.

▲ FIGURE 13–56

10. Refer to Figure 13–57. Determine the following:

    (a) V_{R1} and V_{R2} (b) Current through R_f (c) V_{OUT}

11. Find the value of R_f necessary to produce an output that is five times the sum of the inputs in Figure 13–57.

12. Show a summing amplifier that will average eight input voltages. Use input resistances of 10 kΩ each.

▶ FIGURE 13–57

13. Find the output voltage when the input voltages shown in Figure 13–58 are applied to the scaling adder. What is the current through R_f?

▶ **FIGURE 13–58**

14. Determine the values of the input resistors required in a six-input scaling adder so that the lowest weighted input is 1 and each successive input has a weight twice the previous one. Use $R_f = 100$ kΩ.

SECTION 13–3 Integrators and Differentiators

15. Determine the rate of change of the output voltage in response to the step input to the integrator in Figure 13–59.

▲ **FIGURE 13–59**

16. A triangular waveform is applied to the input of the circuit in Figure 13–60 as shown. Determine what the output should be and sketch its waveform in relation to the input.

17. What is the magnitude of the capacitor current in Problem 16?

▲ **FIGURE 13–60**

18. A triangular waveform with a peak-to-peak voltage of 2 V and a period of 1 ms is applied to the differentiator in Figure 13–61(a). What is the output voltage?

19. Beginning in position 1 in Figure 13–61(b), the switch is thrown into position 2 and held there for 10 ms, then back to position 1 for 10 ms, and so forth. Sketch the resulting output waveform if its initial value is 0 V. The saturated output levels of the op-amp are ±12 V.

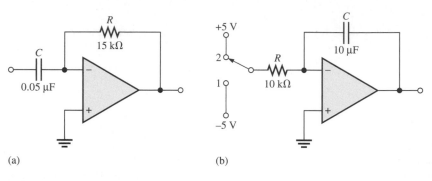

(a)　　　　　　　　　　　　　　(b)

▲ FIGURE 13–61

TROUBLESHOOTING PROBLEMS

SECTION 13–4 Troubleshooting

20. The waveforms given in Figure 13–62(a) are observed at the indicated points in Figure 13–62(b). Is the circuit operating properly? If not, what is a likely fault?

(a)　　　　　　　　　　　　　　　　　　　(b)

▲ FIGURE 13–62

21. The sequences of voltage levels shown in Figure 13–63 are applied to the summing amplifier and the indicated output is observed. First, determine if this output is correct. If it is not correct, determine the fault.

▲ **FIGURE 13–63**

22. The given ramp voltages are applied to the op-amp circuit in Figure 13–64. Is the given output correct? If it isn't, what is the problem?

  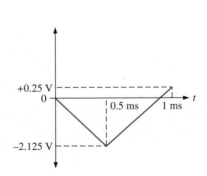

▲ **FIGURE 13–64**

23. The DAC with inputs as shown in Figure 13–27 produces the output shown in Figure 13–65. Determine the fault in the circuit.

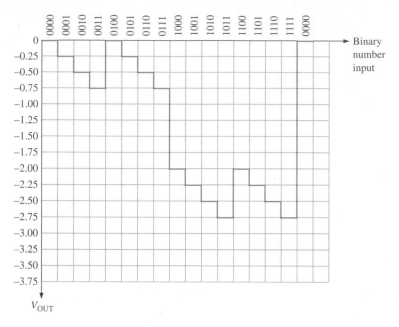

▲ **FIGURE 13–65**

SYSTEM APPLICATION PROBLEMS

24. The ADC board, shown in Figure 13–66, for the system application has just come off the assembly line and a pass/fail test indicates that it doesn't work. The board now comes to you for troubleshooting. What is the very first thing you should do? Can you isolate the problem(s) by this first step in this case?

25. Describe the effect of an open integrator capacitor on the ADC in Figure 13–49.

26. Assume that a 1.0 kΩ resistor is inadvertently used for R_1 in Figure 13–49. What effect does this have on the circuit operation?

▲ **FIGURE 13–66**

ADVANCED PROBLEMS

27. The schematic in Figure 13–49 for the analog-to-digital converter shows 741S op-amps used for the integrator and comparator functions. Data sheets indicate that a 741 has a typical slew rate of 0.5 V/μs and a 741S has a typical slew rate of 12 V/μs. If the sample pulse rate is 500 kHz and the reference voltage is −8 V, determine the maximum input voltage from the sample-and-hold circuit that the ADC can handle.

28. Repeat Problem 27 if the 741S op-amps are replaced by 741 op-amps.

29. Design an integrator that will produce an output voltage with a slope of 100 mV/μs when the input voltage is a constant 5 V. Specify the input frequency of a square wave with an amplitude of 5 V that will result in a 5 V peak-to-peak triangular wave output.

EWB/MULTISIM TROUBLESHOOTING PROBLEMS

These file circuits are in the Troubleshooting Problems folder on your CD-ROM.

30. Open file TSP13-30 and determine the fault.

31. Open file TSP13-31 and determine the fault.

32. Open file TSP13-32 and determine the fault.

33. Open file TSP13-33 and determine the fault.

34. Open file TSP13-34 and determine the fault.

35. Open file TSP13-35 and determine the fault.

36. Open file TSP13-36 and determine the fault.

37. Open file TSP13-37 and determine the fault.

38. Open file TSP13-38 and determine the fault.

39. Open file TSP13-39 and determine the fault.

ANSWERS

SECTION REVIEWS

SECTION 13–1 Comparators

1. **(a)** $V = (10\ \text{k}\Omega/110\ \text{k}\Omega)15\ \text{V} = 1.36\ \text{V}$
   (b) $V = (22\ \text{k}\Omega/69\ \text{k}\Omega)(-12\ \text{V}) = -3.83\ \text{V}$

2. Hysteresis makes the comparator noise free.

3. Bounding limits the output amplitude to a specified level.

SECTION 13–2 Summing Amplifiers

1. The summing point is the point where the input resistors are commonly connected.

2. $R_f/R = 1/5 = 0.2$

3. 5 kΩ

SECTION 13–3 Integrators and Differentiators

1. The feedback element in an integrator is a capacitor.

2. The capacitor voltage is linear because the capacitor current is constant.

3. The feedback element in a differentiator is a resistor.

4. The output of a differentiator is proportional to the rate of change of the input.

SECTION 13–4 Troubleshooting

1. An op-amp can fail with a shorted output.

2. Replace suspected components one by one.

RELATED PROBLEMS FOR EXAMPLES

13–1 1.96 V

13–2 +3.83 V; −3.83 V

13–3 +1.81 V; −1.81 V

13–4 More accurately

13–5 −12.5 V

13–6 −5.73 V

13–7 Changes require an additional 100 kΩ input resistor and a change of R_f to 20 kΩ.

13–8 0.45, 0.12, 0.18; $V_{OUT} = -3.03$ V

13–9 Yes. All should be doubled.

13–10 Change C to 5000 pF.

13–11 Same waveform but with an amplitude of 6.6 V

13–12 A pulse from −0.88 V to +7.79 V

13–13 −3.76 V

13–14 Change R_6 to 25 kΩ.

SELF-TEST

1. (c) **2.** (a) **3.** (c) **4.** (e) **5.** (b) **6.** (d) **7.** (c) **8.** (c)

9. (a) **10.** (b) **11.** (c) **12.** (b) **13.** (d) **14.** (d) **15.** (a) **16.** (d)

17. (c)

14

SPECIAL-PURPOSE OP-AMP CIRCUITS

INTRODUCTION

A general-purpose op-amp, such as the 741, is a versatile and widely used device. However, some specialized IC amplifiers are available that have certain features or characteristics oriented to special applications. Most of these devices are actually derived from the basic op-amp. These special circuits include the instrumentation amplifier that is used in high-noise environments, the isolation amplifier that is used in high-voltage and medical applications, the operational transconductance amplifier (OTA) that is used as a voltage-to-current amplifier, and the logarithmic amplifiers that are used for linearizing certain types of inputs and for mathematical operations.

- Analyze and explain the operation of an instrumentation amplifier

- Analyze and explain the operation of an isolation amplifier

- Analyze and explain the operation of an OTA

- Analyze and explain the operation of log and antilog amplifiers

- Analyze and explain several special types of op-amp circuits

KEY TERMS

- Instrumentation amplifier

- Isolation amplifier

- Operational transconductance amplifier (OTA)

- Transconductance

- Natural logarithm

■■■ SYSTEM APPLICATION PREVIEW

Medical electronics is an important area of application for electronic devices and one of the most beneficial. The electrocardiograph (ECG) is a common instrument used to monitor the heart function of patients to detect irregularities or abnormalities in the heartbeat. Electrode sensors are placed at points on the body to pick up the small electrical signal produced by the heart. This signal is amplified and fed to a video monitor or chart recorder for analysis. Due to the safety hazards related to electrical equipment, it is important that the patient be protected from the possibility of severe electrical shock. For this reason, the isolation amplifier is used in medical equipment that comes in contact with the human body. The block diagram shows a basic ECG system that is the topic of the system application at the end of the chapter.

WWW. VISIT THE COMPANION WEBSITE
Study aids for this chapter are available at
http://www.prenhall.com/floyd

14–1 INSTRUMENTATION AMPLIFIERS

An instrumentation amplifier is a differential voltage-gain device that amplifies the difference between the voltages existing at its two input terminals. The main purpose of an instrumentation amplifier is to amplify small signals that are riding on large common-mode voltages. The key characteristics are high input impedance, high common-mode rejection, low output offset, and low output impedance. A basic instrumentation amplifier is made up of three operational amplifiers and several resistors. The voltage gain is set with an external resistor. Instrumentation amplifiers are commonly used in environments with high common-mode noise such as in data acquisition systems where remote sensing of input variables is required.

After completing this section, you should be able to

- **Analyze and explain the operation of an instrumentation amplifier**
- Explain how op-amps are connected to form an instrumentation amplifier
- Describe how the voltage gain is set
- Discuss an application
- Describe the features of the AD622 instrumentation amplifier

The Basic Instrumentation Amplifier

A basic **instrumentation amplifier** is shown in Figure 14–1. Op-amps A1 and A2 are noninverting configurations that provide high input impedance and voltage gain. Op-amp A3 is used as a unity-gain differential amplifier.

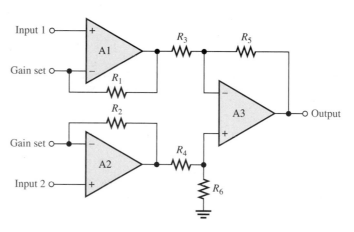

▲ **FIGURE 14–1**

The basic instrumentation amplifier using three op–amps.

The gain-setting resistor, R_G, is connected externally as shown in Figure 14–2. Op-amp A1 receives the differential input signal V_{in1} on its noninverting (+) input and amplifies this signal with a voltage gain of

$$A_v = 1 + \frac{R_1}{R_G}$$

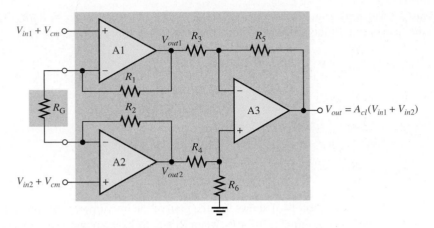

▲ FIGURE 14–2

The instrumentation amplifier with the external gain-setting resistor R_G. Differential and common-mode signals are indicated.

Op-amp A1 also has V_{in2} as an input signal to its inverting (−) input through op-amp A2 and the path formed by R_2 and R_G. The input signal V_{in2} is amplified by op-amp A1 with a voltage gain of

$$A_v = \frac{R_1}{R_G}$$

Also, the common-mode voltage, V_{cm}, on the noninverting input is amplified by the small common-mode gain of op-amp A1. (A_{cm} is typically less than 1.) The total output voltage of op-amp A1 is

$$V_{out1} = \left(1 + \frac{R_1}{R_G}\right)V_{in1} - \left(\frac{R_1}{R_G}\right)V_{in2} + V_{cm}$$

A similar analysis can be applied to op-amp A2 and results in the following output expression:

$$V_{out2} = \left(1 + \frac{R_2}{R_G}\right)V_{in2} - \left(\frac{R_2}{R_G}\right)V_{in1} + V_{cm}$$

Op-amp A3 has V_{out1} on one of its inputs and V_{out2} on the other. Therefore, the differential input voltage to op-amp A3 is $V_{out2} - V_{out1}$.

$$V_{out2} - V_{out1} = \left(1 + \frac{R_2}{R_G} + \frac{R_1}{R_G}\right)V_{in2} - \left(1 + \frac{R_2}{R_G} + \frac{R_1}{R_G}\right)V_{in1} + V_{cm} - V_{cm}$$

For $R_1 = R_2 = R$,

$$V_{out2} - V_{out1} = \left(1 + \frac{2R}{R_G}\right)V_{in2} - \left(1 + \frac{2R}{R_G}\right)V_{in1} + V_{cm} - V_{cm}$$

Notice that, since the common-mode voltages (V_{cm}) are equal, they cancel each other. Factoring out the differential gain gives the following expression for the differential input to op-amp A3:

$$V_{out2} - V_{out1} = \left(1 + \frac{2R}{R_G}\right)(V_{in2} - V_{in1})$$

Op-amp A3 has unity gain because $R_3 = R_5 = R_4 = R_6$ and $A_v = R_5/R_3 = R_6/R_4$. Therefore, the final output of the instrumentation amplifier (the output of op-amp A3) is

$$V_{out} = 1(V_{out2} - V_{out1}) = \left(1 + \frac{2R}{R_G}\right)(V_{in2} - V_{in1})$$

The closed-loop gain is

$$A_{cl} = \frac{V_{out}}{V_{in2} - V_{in1}}$$

Equation 14–1

$$A_{cl} = 1 + \frac{2R}{R_G}$$

where $R_1 = R_2 = R$. Equation 14–1 shows that the gain of the instrumentation amplifier can be set by the value of the external resistor R_G when R_1 and R_2 have known fixed values.

The external gain-setting resistor R_G can be calculated for a desired voltage gain by using Equation 14–1.

Equation 14–2

$$R_G = \frac{2R}{A_{cl} - 1}$$

Instrumentation amplifiers in which the gain is set to specific values using a binary input instead of a resistor are also available.

EXAMPLE 14–1

Determine the value of the external gain-setting resistor R_G for a certain IC instrumentation amplifier with $R_1 = R_2 = 25$ kΩ. The closed-loop voltage gain is to be 500.

Solution

$$R_G = \frac{2R}{A_{cl} - 1} = \frac{50 \text{ k}\Omega}{500 - 1} \cong \mathbf{100 \ \Omega}$$

Related Problem* What value of external gain-setting resistor is required for an instrumentation amplifier with $R_1 = R_2 = 39$ kΩ to produce a gain of 325?

*Answers are at the end of the chapter.

Applications

As mentioned in the introduction to this section, the instrumentation amplifier is normally used to measure small differential signal voltages that are superimposed on a common-mode voltage often much larger than the signal voltage. Applications include situations where a quantity is sensed by a remote device, such as a temperature- or pressure-sensitive transducer, and the resulting small electrical signal is sent over a long line subject to electrical noise that produces common-mode voltages in the line. The instrumentation amplifier at the end of the line must amplify the small signal from the remote sensor and reject the large common-mode voltage. Figure 14–3 illustrates this.

Small differential high-frequency signal riding on a larger low-frequency common-mode signal

Instrumentation amplifier

Amplified differential signal. No common-mode signal.

▲ **FIGURE 14–3**

Illustration of the rejection of large common-mode voltages and the amplification of smaller signal voltages by an instrumentation amplifier.

A Specific Instrumentation Amplifier

Now that you have the basic idea of how an instrumentation amplifier works, let's look at a specific device. A representative device, the AD622, is shown in Figure 14–4 where IC pin numbers are given for reference. This instrumentation amplifier is based on the classic design using three op-amps, as previously discussed.

▶ **FIGURE 14–4**

The AD622 instrumentation amplifier.

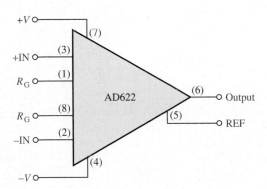

Some of the features of the AD622 are as follows. The voltage gain can be adjusted from 2 to 1000 with an external resistor R_G. There is unity gain with no external resistor. The input impedance is 10 GΩ. The common-mode rejection ratio (CMRR) has a minimum value of 66 dB. Recall that a higher CMRR means better rejection of common-mode voltages. The AD622 has a bandwidth of 800 kHz at a gain of 10 and a slew rate of 1.2 V/μs.

Setting the Voltage Gain   For the AD622, an external resistor must be used to achieve a voltage gain greater than unity, as indicated in Figure 14–5. Resistor R_G is connected between the R_G terminals. (pins 1 and 8). No resistor is required for unity. R_G is selected for the desired gain based on the following formula:

$$R_G = \frac{50.5 \text{ k}\Omega}{A_v - 1}$$

Equation 14–3

Notice that this formula is the same as Equation 14–2 for the classic three-op-amp configuration where the internal resistors R_1 and R_2 are each 25.25 kΩ.

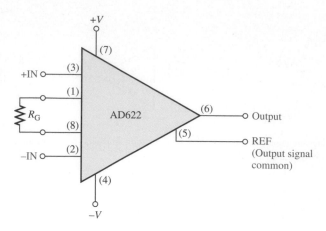

▲ **FIGURE 14–5**

The AD622 with a gain-setting resistor.

Gain versus Frequency   The graph in Figure 14–6 shows how the gain varies with frequency for gains of 1, 10, 100, and 1000. As you can see, the bandwidth decreases as the gain increases.

▶ **FIGURE 14–6**

Gain versus frequency for the AD622 instrumentation amplifier.

EXAMPLE 14–2

Calculate the voltage gain and determine the bandwidth using the graph in Figure 14–6 for the instrumentation amplifier in Figure 14–7.

▶ **FIGURE 14–7**

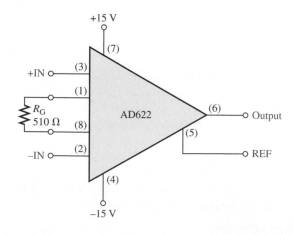

Solution   The voltage gain is determined as follows:

$$R_G = \frac{50.5 \text{ k}\Omega}{A_v - 1}$$

$$A_v - 1 = \frac{50.5 \text{ k}\Omega}{R_G}$$

$$A_v = \frac{50.5 \text{ k}\Omega}{510 \ \Omega} + 1 = \mathbf{100}$$

The approximate bandwidth is determined from the graph in Figure 14–6.

$$BW \cong \mathbf{60 \ kHz}$$

Related Problem   Modify the circuit in Figure 14–7 for a gain of approximately 45.

**SECTION 14–1
REVIEW**
Answers are at the end of the chapter.

1. What is the main purpose of an instrumentation amplifier and what are three of its key characteristics?
2. What components do you need to construct a basic instrumentation amplifier?
3. How is the gain determined in a basic instrumentation amplifier?
4. In a certain AD622 configuration, $R_G = 10 \text{ k}\Omega$. What is the voltage gain?

14–2 ISOLATION AMPLIFIERS

An isolation amplifier provides dc isolation between input and output for the protection of human life or sensitive equipment in those applications where hazardous power-line leakage or high-voltage transients are possible. The principal areas of application for isolation amplifiers are in medical instrumentation, power plant instrumentation, industrial processing, and automated testing.

After completing this section, you should be able to

- **Analyze and explain the operation of an isolation amplifier**
- Explain the basic configuration of an isolation amplifier
- Discuss an application in medical electronics
- Describe the features of the 3656KG isolation amplifier

The Basic Isolation Amplifier

In some ways, the **isolation amplifier** can be viewed as an elaborate op-amp or instrumentation amplifier. The difference is that the isolation amplifier has an input stage, an output stage, and a power supply section that are all electrically isolated from each other. Although there are isolation amplifiers that use optical coupling or capacitive coupling to achieve isolation, the transformer-coupled device is the most commonly used. The circuit is in IC form, but the miniature, multiple-winding, toroid transformer is not fully integrated. This results in a package configuration that deviates somewhat from standard IC packages but is still designed for printed circuit board assembly.

A typical isolation amplifier is capable of operating with three independent grounds. As shown in Figure 14–8, the input and output stages and the power supplies are transformer coupled to achieve isolation of the input and output signals and the power.

The input stage contains an op-amp, a demodulator, a modulator, and dual-polarity power supply. The output stage contains an op-amp, an oscillator, a demodulator, and a dual-polarity power supply.

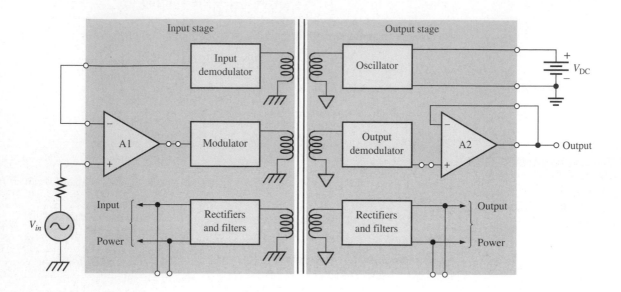

▲ **FIGURE 14–8**

Diagram of a transformer–coupled isolation amplifier.

General Operation As mentioned, there are three isolated independent grounds (indicated by three different symbols) for the input signal, the output signal, and the power supply, as shown in Figure 14–8.

The power to the input and output stages is produced as follows. An external dc supply voltage, V_{DC}, is applied to the oscillator. The oscillator converts the dc power to ac at a relatively high frequency. The ac output of the oscillator is coupled by the transformer to the input power supply (rectifiers and filters) and to the output power supply (rectifiers and filters) where it is rectified and filtered to produced dual-polarity (positive and negative) dc voltages for the input and output stages.

The oscillator output is also coupled to the modulator where it is combined with the input signal from the input op-amp, A1 (modulators are covered in Chapter 17). The purpose of the modulator is to vary the amplitude of the relatively high oscillator frequency with the lower-frequency input signal. The higher modulated frequency allows a very small transformer. To couple the lower-frequency input signal without modulation would require a prohibitively large transformer.

The modulated signal is coupled to the demodulator in the output stage. The demodulator recovers the original input signal from the higher oscillator frequency. The demodulated input signal is then applied to op-amp A2. The demodulator in the input stage is part of a feedback loop that forces the signal at the inverting input of A1 to equal the original input signal at the noninverting input.

Although the isolation amplifier is a fairly complex circuit, in terms of its overall function, it is still simply an amplifier. You apply a dc voltage, put a signal in, and you get an amplified signal out. The isolation function is an unseen process.

Applications

As previously mentioned, the isolation amplifier is used in applications that require no common grounds between a transducer and the processing circuits where interfacing to sensitive equipment is required. In chemical, nuclear, and metal-processing industries, for example, millivolt signals typically exist in the presence of large common-mode voltages that can be in the kilovolt range. In this type of environment, the isolation amplifier can amplify small signals from very noisy equipment and provide a safe output to sensitive equipment such as computers.

Another very important application is in various types of medical equipment. In medical applications where body functions such as heart rate and blood pressure are monitored, the very small monitored signals are combined with large common-mode signals, such as 60 Hz power-line pickup from the skin. In these situations, without isolation, dc leakage or equipment failure could be fatal. Figure 14–9 shows a simplified diagram of an isolation amplifier in a cardiac-monitoring application. In this situation, heart signals, which arc very small, are combined with much larger common-mode signals caused by muscle noise, electrochemical noise, residual electrode voltage, and 60 Hz line pickup form the skin.

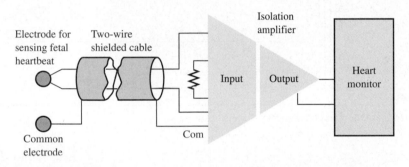

▲ **FIGURE 14–9**

Fetal heartbeat monitoring using an isolation amplifier. The triangle with a "split" down the middle is one way of representing an isolation amplifier by indicating that the input and output stages are separated by transformer coupling.

The monitoring of fetal heartbeat, as illustrated, is the most demanding type of cardiac monitoring because in addition to the fetal heartbeat that typically generates 50 μV, there is also the mother's heartbeat that typically generates 1 mV. The common-mode voltages can run from about 1 mV to about 100 mV. The CMR (common-mode rejection) of the isolation amplifier separates the signal of the fetal heartbeat from that of the mother's heartbeat and from those common-mode signals. Therefore, the signal from the fetal heartbeat is essentially all that the amplifier sends to the monitoring equipment.

A Specific Isolation Amplifier

Now that you know basically what an isolation amplifier is and what it does, let's look at a representative device, the Burr-Brown 3656KG. As you can see in Figure 14–10, the 3656KG is similar to the basic isolation amplifier in Figure 14–8 except that it has a few more inputs and outputs and the oscillator is called a pulse generator. These additional pins provide for gain adjustments, offset adjustments, isolated dc voltage outputs, and other functions.

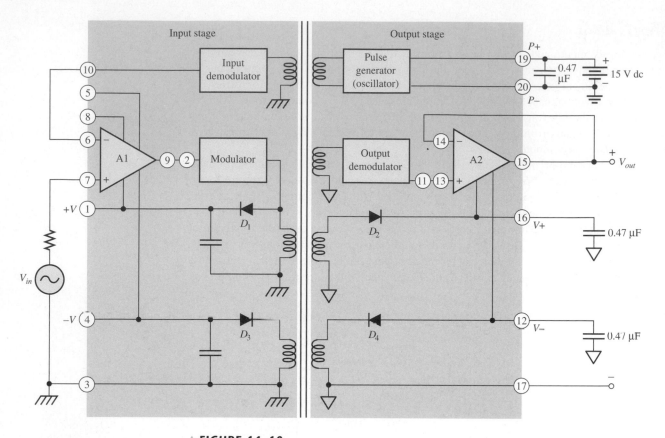

▲ **FIGURE 14–10**

The 3656KG isolation amplifier.

Isolatated Power   The 3656KG provides ±9 V from the filtered rectifiers (D_1 through D_4) at the +V and −V terminals for both the input and output stages. These voltages are used to power the internal circuits and are also available for powering external associated circuits such as preamplifiers and transducers. The power supply connections shown in Figure 14–10 show the three-port isolation configuration with the three separate grounds indicated by the different symbols. Filtering in the output stage is provided by two external capacitors. In the input stage, the filter capacitors are internal.

Instead of using the internally generated dc voltages, external supply voltages can be connected as long as they are greater than the internal voltages. This is necessary in order to reverse-bias the rectifier diodes and, essentially take them out of the picture.

Voltage Gain   The voltage gains of both the input stage and the output stage can be set with external resistors, as shown in Figure 14–11. The gain of the input stage is

Equation 14–4

$$A_{v1} = \frac{R_{f1}}{R_{i1}} + 1$$

The gain of the output stage is

Equation 14–5

$$A_{v2} = \frac{R_{f2}}{R_{i2}} + 1$$

The total amplifier gain is the product of the gains of the input and output stages.

$$A_{v(tot)} = A_{v1}A_{v2}$$

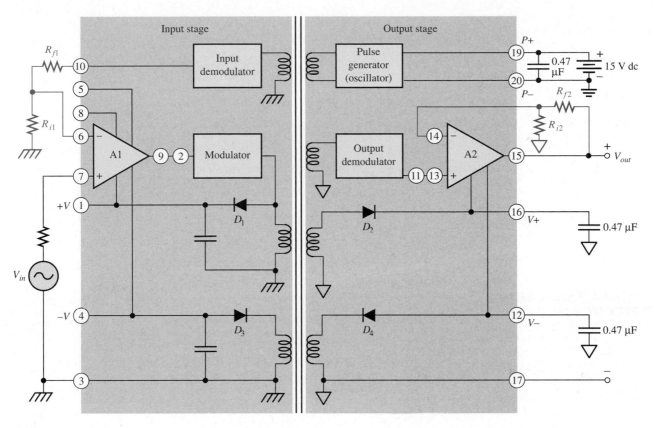

▲ **FIGURE 14–11**

The 3656KG connected as a noninverting amplifier with external gain resistors (shown in blue).

EXAMPLE 14–3

Determine the total voltage gain of the 3656KG isolation amplifier in Figure 14–12.

▲ **FIGURE 14–12**

Solution   The voltage gain of the input stage is

$$A_{v1} = \frac{R_{f1}}{R_{i1}} + 1 = \frac{22 \text{ k}\Omega}{2.2 \text{ k}\Omega} + 1 = 10 + 1 = 11$$

The voltage gain of the output stage is

$$A_{v2} = \frac{R_{f2}}{R_{i2}} + 1 = \frac{47 \text{ k}\Omega}{10 \text{ k}\Omega} + 1 = 4.7 + 1 = 5.7$$

The total voltage gain of the isolation amplifier is

$$A_{v(tot)} = A_{v1}A_{v2} = (11)(5.7) = \mathbf{62.7}$$

Related Problem   Select resistor values in Figure 14–12 that will produce a total voltage gain of approximately 100.

**SECTION 14–2
REVIEW**

1. In what types of applications are isolation amplifiers used?
2. What are the two stages in a typical isolation amplifier?
3. How are the stages in an isolation amplifier connected?
4. What is the purpose of the oscillator in an isolation amplifier?

14–3   OPERATIONAL TRANSCONDUCTANCE AMPLIFIERS (OTAs)

Conventional op-amps are, as you know, primarily voltage amplifiers in which the output voltage equals the gain times the input voltage. The OTA is primarily a voltage-to-current amplifier in which the output current equals the gain times the input voltage.

After completing this section, you should be able to

- **Analyze and explain the operation of an OTA**
- Identify the OTA symbol
- Define *transconductance*
- Discuss the relationship between transconductance and bias current
- Describe the features of the LM13700 OTA
- Discuss OTA applications

Figure 14–13 shows the symbol for an **operational transconductance amplifier (OTA)**. The double circle symbol at the output represents an output current source that is dependent on a bias current. Like the conventional op-amp, the OTA has two differential input terminals, a high input impedance, and a high CMRR. Unlike the conventional op-amp, the OTA has a bias-current input terminal, a high output impedance, and no fixed open-loop voltage gain.

▶ **FIGURE 14–13**

Symbol for an operational transconductance amplifier (OTA).

The Transconductance Is the Gain of an OTA

By definition, the **transconductance** of an electronic device is the ratio of the output current to the input voltage. For an OTA, voltage is the input variable and current is the output variable; therefore, the ratio of output current to input voltage is also its gain. Consequently, the voltage-to-current gain of an OTA is the transconductance, g_m.

$$g_m = \frac{I_{out}}{V_{in}}$$

Equation 14–6

In an OTA, the transconductance is dependent on a constant (K) times the bias current (I_{BIAS}), as indicated in Equation 14–7. The value of the constant is dependent on the internal circuit design.

$$g_m = KI_{BIAS}$$

Equation 14–7

The output current is controlled by the input voltage and the bias current as shown by the following formula:

$$I_{out} = g_m V_{in} - KI_{BIAS}V_{in}$$

The Transconductance Is a Function of Bias Current

The relationship of the transconductance and the bias current in an OTA is an important characteristic. The graph in Figure 14–14 illustrates a typical relationship. Notice that the transconductance increases linearly with the bias current. The constant of proportionality, K, is the slope of the line. In this case, K is approximately 16 $\mu S/\mu A$.

▶ **FIGURE 14–14**

Example of a transconductance versus bias current graph for a typical OTA.

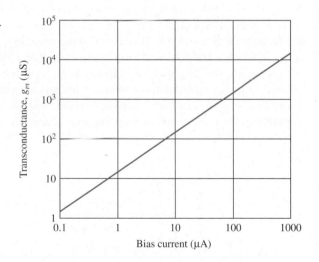

Bias current (μA)

EXAMPLE 14–4

If an OTA has a $g_m = 1000$ μS, what is the output current when the input voltage is 25 mV?

Solution

$$I_{out} = g_m V_{in} = (1000 \ \mu S)(25 \ mV) = \textbf{25 } \mu\textbf{A}$$

Related Problem Based on $K \cong 16$ $\mu S/\mu A$, calculate the approximate bias current required to produce $g_m = 1000$ μS.

Basic OTA Circuits

Figure 14–15 shows the OTA used as an inverting amplifier with a fixed voltage gain. The voltage gain is set by the transconductance and the load resistance as follows.

$$V_{out} = I_{out}R_L$$

Dividing both sides by V_{in},

$$\frac{V_{out}}{V_{in}} = \left(\frac{I_{out}}{V_{in}}\right)R_L$$

Since V_{out}/V_{in} is the voltage gain and $I_{out}/V_{in} = g_m$,

$$A_v = g_mR_L$$

▶ **FIGURE 14–15**

An OTA as an inverting amplifier with a fixed voltage gain.

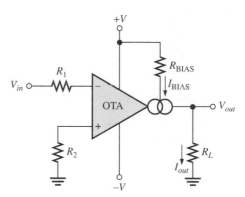

The transconductance of the amplifier in Figure 14–15 is determined by the amount of bias current, which is set by the dc supply voltages and the bias resistor R_{BIAS}.

One of the most useful features of an OTA is that the voltage gain can be controlled by the amount of bias current. This can be done manually, as shown in Figure 14–16(a), by using a variable resistor in series with R_{BIAS} in the circuit of Figure 14–15. By changing the resistance, you can produce a change in I_{BIAS}, which changes the transconductance. A change in the transconductance changes the voltage gain. The voltage gain can also be controlled with an externally applied variable voltage, as shown in Figure 14–16(b). A variation in the applied bias voltage causes a change in the bias current.

(a) Amplifier with resistance-controlled gain   (b) Amplifier with voltage-controlled gain

▲ **FIGURE 14–16**

An OTA as an inverting amplifier with a variable–voltage gain.

A Specific OTA

The LM13700 is a typical OTA and serves as a representative device. The LM13700 is a dual-device package containing two OTAs and buffer circuits. Figure 14–17 shows the pin configuration using a single OTA in the package. The maximum dc supply voltages are ±18 V, and its transconductance characteristic happens to be the same as indicated by the graph in Figure 14–14. For an LM13700, the bias current is determined by the following formula:

$$I_{BIAS} = \frac{+V_{BIAS} - (-V) - 1.4 \text{ V}}{R_{BIAS}}$$

The 1.4 V is due to the internal circuit where a base-emitter junction and a diode connect the external R_{BIAS} with the negative supply voltage $(-V)$. The positive bias voltage, $+V_{BIAS}$, may be obtained from the positive supply voltage, $+V$.

Not only does the transconductance of an OTA vary with bias current, but so do the input and output resistances. Both the input and output resistances decrease as the bias current increases, as shown in Figure 14–18.

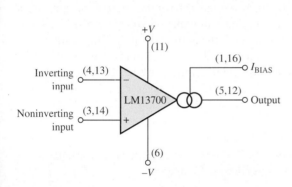

▲ **FIGURE 14–17**

An LM13700 OTA. There are two in an IC package. The buffer transistors are not shown. Pin numbers for both OTAs are given in parentheses.

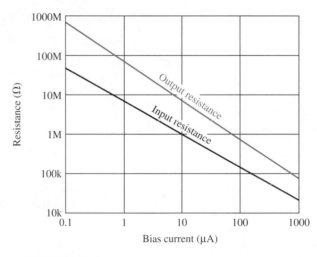

▲ **FIGURE 14–18**

Example of input and output resistances versus bias current.

EXAMPLE 14–5

The OTA in Figure 14–19 is connected as an inverting fixed-gain amplifier where $+V_{BIAS} = +V$. Determine the approximate voltage gain.

▶ **FIGURE 14–19**

Solution Calculate the bias current as follows:

$$I_{BIAS} = \frac{+V_{BIAS} - (-V) - 1.4\ V}{R_{BIAS}} = \frac{9\ V - (-9\ V) - 1.4\ V}{33\ k\Omega} = 503\ \mu A$$

Using $K \cong 16\ \mu S/\mu A$ from the graph in Figure 14–14, the value of transconductance corresponding to $I_{BIAS} = 503\ \mu A$ is approximately

$$g_m = KI_{BIAS} \cong (16\ \mu S/\mu A)(503\ \mu A) = 8.05 \times 10^3\ \mu S$$

Using this value of g_m, calculate the voltage gain.

$$A_v = g_m R_L \cong (8.05 \times 10^3\ \mu S)(10\ k\Omega) = \mathbf{80.5}$$

Related Problem If the OTA in Figure 14–19 is operated with dc supply voltages of ±12 V, will this change the voltage gain and, if so, to what value?

Two OTA Applications

Amplitude Modulator Figure 14–20 illustrates an OTA connected as an amplitude modulator. The voltage gain is varied by applying a modulation voltage to the bias input. When a constant-amplitude input signal is applied, the amplitude of the output signal will vary according to the modulation voltage on the bias input. The gain is dependent on bias current, and bias current is related to the modulation voltage by the following relationship:

$$I_{BIAS} = \frac{V_{MOD} - (-V) - 1.4\ V}{R_{BIAS}}$$

This modulating action is shown in Figure 14–20 for a higher-frequency sinusoidal input voltage and a lower-frequency sinusoidal modulating voltage.

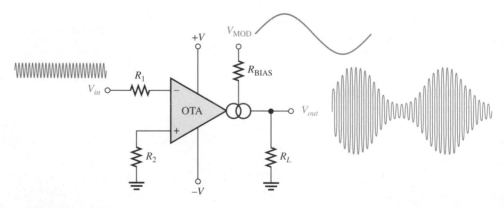

▲ **FIGURE 14–20**

The OTA as an amplitude modulator.

EXAMPLE 14–6

The input to the OTA amplitude modulator in Figure 14–21 is a 50 mV peak-to-peak, 1 MHz sine wave. Determine the output signal, given the modulation voltage shown is applied to the bias input.

▲ FIGURE 14–21

Solution The maximum voltage gain is when I_{BIAS}, and thus g_m, is maximum. This occurs at the maximum peak of the modulating voltage, V_{MOD}.

$$I_{BIAS(max)} = \frac{V_{MOD(max)} - (-V) - 1.4\ V}{R_{BIAS}} = \frac{10\ V - (-9\ V) - 1.4\ V}{56\ k\Omega} = 314\ \mu A$$

From the graph in Figure 14–14, the constant K is approximately 16 $\mu S/\mu A$.

$$g_m = KI_{BIAS(max)} \cong (16\ \mu S/\mu A)(314\ \mu A) = 5.02\ mS$$
$$A_{v(max)} = g_m R_L \cong (5.02\ mS)(10\ k\Omega) = 50.2$$
$$V_{out(max)} = A_{v(max)} V_{in} \cong (50.2)(50\ mV) = 2.51\ V$$

The minimum bias current is

$$I_{BIAS(min)} = \frac{V_{MOD(min)} - (-V) - 1.4\ V}{R_{BIAS}} = \frac{1\ V - (-9\ V) - 1.4\ V}{56\ k\Omega} = 154\ \mu A$$

$$g_m = KI_{BIAS(min)} \cong (16\ \mu S/\mu A)(154\ \mu A) = 2.46\ mS$$
$$A_{v(min)} = g_m R_L \cong (2.46\ mS)(10\ k\Omega) = 24.6$$
$$V_{out(min)} = A_{v(min)} V_{in} \cong (24.6)(50\ mV) = 1.23\ V$$

The resulting output voltage is shown in Figure 14–22.

▲ FIGURE 14–22

Related Problem Repeat this example with the sinusoidal modulating signal replaced by a square wave with the same maximum and minimum levels and a bias resistor of 39 kΩ.

Schmitt Trigger Figure 14–23 shows an OTA used in a Schmitt-trigger configuration. Basically, a Schmitt trigger is a comparator with hysteresis where the input voltage is large enough to drive the device into its saturated states. When the input voltage exceeds a certain threshold value or trigger point, the device switches to one of its saturated output states. When the input falls back below another threshold value, the device switches back to its other saturated output state.

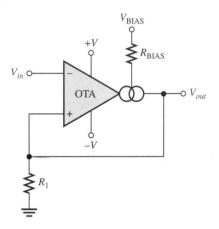

▶ **FIGURE 14–23**

The OTA as a Schmitt trigger.

In the case of the OTA Schmitt trigger, the threshold levels are set by the current through resistor R_1. The maximum output current in an OTA equals the bias current. Therefore, in the saturated output states, $I_{out} = I_{BIAS}$. The maximum positive output voltage is $I_{out}R_1$, and this voltage is the positive threshold value or upper trigger point. When the input voltage exceeds this value, the output switches to its maximum negative voltage, which is $-I_{out}R_1$. Since $I_{out} = I_{BIAS}$, the trigger points can be controlled by the bias current. Figure 14–24 illustrates this operation.

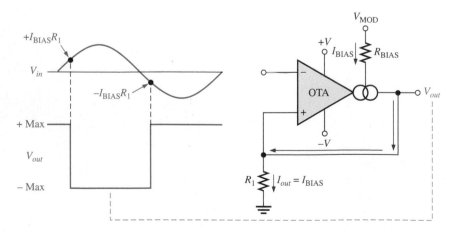

▲ **FIGURE 14–24**

Basic operation of the OTA Schmitt trigger.

SECTION 14–3 REVIEW

1. What does OTA stand for?

2. If the bias current in an OTA is increased, does the transconductance increase or decrease?

3. What happens to the voltage gain if the OTA is connected as a fixed-voltage amplifier and the supply voltages are increased?

4. What happens to the voltage gain if the OTA is connected as a variable-gain voltage amplifier and the voltage at the bias terminal is decreased?

14–4 LOG AND ANTILOG AMPLIFIERS

The **logarithm** of a number is the power to which the base must be raised to get that number. A logarithmic (log) amplifier produces an output that is proportional to the logarithm of the input, and antilogarithmic (antilog) amplifiers take the antilog or inverse log of the input. Log amplifiers are used in applications that require compression of analog input data, linearization of transducers that have exponential outputs, and analog multiplication and division. In this section, we will discuss the principles of logarithmic amplifiers.

After completing this section, you should be able to

- **Analyze and explain the operation of log and antilog amplifiers**

- Describe the feedback configurations

- Define *logarithm, antilogarithm,* and *natural logarithm*

- Discuss signal compression with logarithmic amplifiers

The Basic Logarithmic Amplifier

The key element in a log amplifier is a device that exhibits a logarithmic characteristic that, when placed in the feedback loop of an op-amp, produces a logarithmic response. This means that the output voltage is a function of the logarithm of the input voltage, as expressed by the following general equation. K is a constant, and ln is the **natural logarithm** to the base e.

$$V_{out} = -K \ln(V_{in})$$

Equation 14–8

Although we will use natural logarithms in the formulas in this section, each expression can be converted to a logarithm to the base 10 ($\log_{10}$) using the relationship $\ln x = 2.3 \log_{10} x$.

The semiconductor *pn* junction in the form of either a diode or the base-emitter junction of a BJT provides a logarithmic characteristic. You may recall that a diode has a nonlinear characteristic up to a forward voltage of approximately 0.7 V. Figure 14–25 shows the characteristic curve, where V_F is the forward diode voltage and I_F is the forward diode current.

As you can see on the graph, the diode curve is nonlinear. Not only is the characteristic curve nonlinear, it is logarithmic and is specifically defined by the following formula:

$$I_F \cong I_R e^{q V_F / kT}$$

where I_R is the reverse leakage current, q is the charge on an electron, k is Boltzmann's constant, and T is the absolute temperature in Kelvin. From the previous equation, the

$$I_F \cong I_R e^{qV_F/kT}$$

▲ FIGURE 14–25

A portion of a diode (*pn* junction) characteristic curve (V_F versus I_F).

diode forward voltage, V_F, can be determined as follows. Take the natural logarithm (ln is the logarithm to the base e) of both sides.

$$\ln I_F = \ln I_R e^{qV_F/kT}$$

The ln of a product of two terms equals the sum of the ln of each term.

$$\ln I_F = \ln I_R + \ln e^{qV_F/kT} = \ln I_R + \frac{qV_F}{kT}$$

$$\ln I_F - \ln I_R = \frac{qV_F}{kT}$$

The difference of two ln terms equals the ln of the quotient of the terms.

$$\ln\left(\frac{I_F}{I_R}\right) = \frac{qV_F}{kT}$$

Solving for V_F,

$$V_F = \left(\frac{kT}{q}\right)\ln\left(\frac{I_F}{I_R}\right)$$

Log Amplifier with a Diode When you place a diode in the feedback loop of an op-amp circuit, as shown in Figure 14–26, you have a basic log amplifier. Since the inverting input is at virtual ground (0 V), the output is at $-V_F$ when the input is positive. Since V_F is logarithmic, so is V_{out}. The output is limited to a maximum value of approximately -0.7 V because the diode's logarithmic characteristic is restricted to voltages below 0.7 V. Also, the input must be positive when the diode is connected in the direction shown in the figure. To handle negative inputs, you must turn the diode around.

An analysis of the circuit in Figure 14–26 is as follows, beginning with the facts that $V_{out} = -V_F$ and $I_F = I_{in}$ because there is no current at the inverting input.

$$V_{out} = -V_F$$

$$I_F = I_{in} = \frac{V_{in}}{R_1}$$

► FIGURE 14–26

A basic log amplifier using a diode as the feedback element.

► FIGURE 14–26

A basic log amplifier using a diode as the feedback element.

Substituting into the formula for V_F,

$$V_{out} = -\left(\frac{kT}{q}\right)\ln\left(\frac{V_{in}}{I_R R_1}\right)$$

The term kT/q is a constant equal to approximately 25 mV at 25°C. Therefore, the output voltage can be expressed as

$$V_{out} \cong -(0.025\text{ V})\ln\left(\frac{V_{in}}{I_R R_1}\right)$$

Equation 14–9

From Equation 14–9, you can see that the output voltage is the negative of a logarithmic function of the input voltage. The value of the output is controlled by the value of the input voltage and the value of the resistor R_1. The other factor, I_R, is a constant for a given diode.

EXAMPLE 14–7

Determine the output voltage for the log amplifier in Figure 14–27. Assume $I_R = 50$ nA.

► FIGURE 14–27

Solution   The input voltage and the resistor value are given in Figure 14–27.

$$V_{OUT} = -(0.025\text{ V})\ln\left(\frac{V_{in}}{I_R R_1}\right) = -(0.025\text{ V})\ln\left(\frac{2\text{ V}}{50\text{ nA} \times 100\text{ k}\Omega}\right)$$

$$= -(0.025\text{ V})\ln(400) = -(0.025\text{ V})(5.99) = \mathbf{-0.150\text{ V}}$$

Related Problem   Calculate the output voltage of the log amplifier with a +4 V input.

Open file E14-07 in the Examples folder on your CD-ROM. Apply the specified input voltage and measure the output voltage.

Log Amplifier with a BJT The base-emitter junction of a bipolar junction transistor exhibits the same type of logarithmic characteristic as a diode because it is also a *pn* junction. A log amplifier with a BJT connected in a common-base form in the feedback loop is shown in Figure 14–28. Notice that V_{out} with respect to ground is equal to $-V_{BE}$.

▶ FIGURE 14–28

A basic log amplifier using a transistor as the feedback element.

The analysis for this circuit is the same as for the diode log amplifier except that $-V_{BE}$ replaces V_F, I_C replaces I_F, and I_{EBO} replaces I_R. The expression for the V_{BE} versus I_C characteristic curve is

$$I_C = I_{EBO}e^{qV_{BE}/kT}$$

where I_{EBO} is the emitter-to-base leakage current. The expression for the output voltage is

Equation 14–10

$$V_{out} = -(0.025 \text{ V})\ln\left(\frac{V_{in}}{I_{EBO}R_1}\right)$$

EXAMPLE 14–8

What is V_{out} for a transistor log amplifier with $V_{in} = 3$ V and $R_1 = 68$ kΩ? Assume $I_{EBO} = 40$ nA.

Solution

$$V_{out} = -(0.025 \text{ V})\ln\left(\frac{V_{in}}{I_{EBO}R_1}\right) = -(0.025 \text{ V})\ln\left(\frac{3 \text{ V}}{40 \text{ nA} \times 68 \text{ kΩ}}\right)$$

$$= -(0.025 \text{ V})\ln(1103) = \mathbf{-175.1 \text{ mV}}$$

Related Problem   Calculate V_{out} if R_1 is changed to 33 kΩ.

The Basic Antilog Amplifier

The **antilogarithm** of a number is the result obtained when the base is raised to a power equal to the logarithm of that number. To get the antilogarithm, you must take the exponential of the logarithm (antilogarithm of $x = e^{\ln x}$).

An antilog amplifier is formed by connecting a transistor (or diode) as the input element as shown in Figure 14–29. The exponential formula still applies to the base-emitter *pn* junction. The output voltage is determined by the current (equal to the collector current) through the feedback resistor.

$$V_{out} = -R_f I_C$$

The characteristic equation of the *pn* junction is

$$I_C = I_{EBO}e^{qV_{BE}/kT}$$

► **FIGURE 14–29**

A basic antilog amplifier.

Substituting into the equation for V_{out},

$$V_{out} = -R_f I_{EBO} e^{qV_{BE}/kT}$$

As you can see in Figure 14–29, $V_{in} = V_{BE}$.

$$V_{out} = -R_f I_{EBO} e^{qV_{in}/kT}$$

The exponential term can be expressed as an antilogarithm as follows:

$$V_{out} = -R_f I_{EBO} \text{antilog}\left(\frac{V_{in}q}{kT}\right)$$

Since kT/q is approximately 25 mV,

$$V_{out} = -R_f I_{EBO} \text{antilog}\left(\frac{V_{in}}{25 \text{ mV}}\right)$$

Equation 14–11

EXAMPLE 14–9

For the antilog amplifier in Figure 14–30, find the output voltage. Assume $I_{EBO} = 40$ nA.

► **FIGURE 14–30**

Solution First of all, notice that the input voltage in Figure 14–30 is the inverted output voltage of the log amplifier in Example 14–8, where the output voltage is proportional to the logarithm of the input voltage. In this case, the antilog amplifier reverses the process and produces an output that is proportional to the antilog of the input. Stated another way, the input of an antilog amplifier is proportional to the logarithm of the output. So, the output voltage of the antilog amplifier in Figure 14–30 should have the same magnitude as the input voltage of the log amplifier in Example 14–8 because all the constants are the same. Let's see if it does.

$$V_{OUT} = -I_{EBO}R_f \text{antilog}\left(\frac{V_{in}}{25 \text{ mV}}\right) = -(40 \text{ nA})(68 \text{ k}\Omega)\text{antilog}\left(\frac{175.1 \text{ mV}}{25 \text{ mV}}\right)$$

$$= -(40 \text{ nA})(68 \text{ k}\Omega)(1101) = -3 \text{ V}$$

Related Problem Determine V_{OUT} for the amplifier in Figure 14–30 if the feedback resistor is changed to 100 kΩ.

Open file E14-09 in the Examples folder on your CD-ROM. Apply the specified input voltage and measure the output voltage.

Signal Compression with Logarithmic Amplifiers

In certain applications, a signal may be too large in magnitude for a particular system to handle. The term *dynamic range* is often used to describe the range of voltages contained in a signal. In these cases, the signal voltage must be scaled down by a process called **signal compression** so that it can be properly handled by the system. If a linear circuit is used to scale a signal down in amplitude, the lower voltages are reduced by the same percentage as the higher voltages. Linear signal compression often results in the lower voltages becoming obscured by noise and difficult to accurately distinguish, as illustrated in Figure 14–31(a). To overcome this problem, a signal with a large dynamic range can be compressed using a logarithmic response, as shown in Figure 14–31(b). In logarithmic signal compression, the higher voltages are reduced by a greater percentage than the lower voltages, thus keeping the lower voltage signals from being lost in noise.

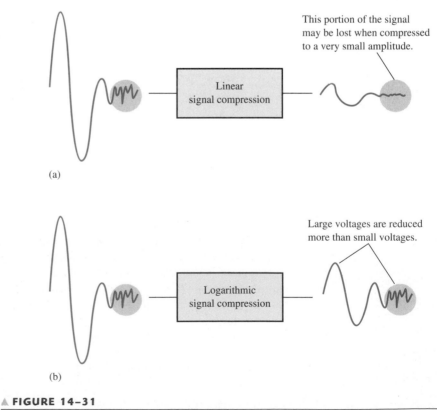

▲ **FIGURE 14–31**

The basic concept of signal compression with a logarithmic amplifier.

**SECTION 14–4
REVIEW**

1. What purpose does the diode or transistor perform in the feedback loop of a log amplifier?
2. Why is the output of a log amplifier limited to about 0.7 V?
3. What are the factors that determine the output voltage of a basic log amplifier?
4. In terms of implementation, how does a basic antilog amplifier differ from a basic log amplifier?

14–5 CONVERTERS AND OTHER OP-AMP CIRCUITS

This section introduces a few more op-amp circuits that represent basic applications of the op-amp. You will learn about the constant-current source, the current-to-voltage converter, the voltage-to-current converter, and the peak detector. This is, of course, not a comprehensive coverage of all possible op-amp circuits but is intended only to introduce you to some common basic uses.

After completing this section, you should be able to

- **Analyze and explain several special types of op-amp circuits**
- Describe how the op-amp is used as a constant-current source
- Explain the operation of a current-to-voltage converter
- Explain the operation of a voltage-to-current converter
- Describe the operation of a peak detector

Constant-Current Source

A constant-current source delivers a load current that remains constant when the load resistance changes. Figure 14–32 shows a basic circuit in which a stable voltage source (V_{IN}) provides a constant current (I_i) through the input resistor (R_i). Since the inverting (−) input of the op-amp is at virtual ground (0 V), the value of I_i is determined by V_{IN} and R_i as

$$I_i = \frac{V_{IN}}{R_i}$$

▶ **FIGURE 14–32**

A basic constant-current source.

Now, since the internal input impedance of the op-amp is extremely high (ideally infinite), practically all of I_i is through R_L, which is connected in the feedback path. Since $I_i = I_L$,

Equation 14–12

$$I_L = \frac{V_{IN}}{R_i}$$

If R_L changes, I_L remains constant as long as V_{IN} and R_i are held constant.

Current-to-Voltage Converter

A current-to-voltage converter converts a variable input current to a proportional output voltage. A basic circuit that accomplishes this is shown in Figure 14–33(a). Since practically all of I_i is through the feedback path, the voltage dropped across R_f is I_iR_f. Because the left side of R_f is at virtual ground (0 V), the output voltage equals the voltage across R_f, which is proportional to I_i.

Equation 14–13

$$V_{out} = I_iR_f$$

(a) Basic circuit

(b) Circuit for sensing light level and converting it to a proportional output voltage

▲ **FIGURE 14–33**

Current-to-voltage converter.

A specific application of this circuit is illustrated in Figure 14–33(b), where a photoconductive cell is used to sense changes in light level. As the amount of light changes, the current through the photoconductive cell varies because of the cell's change in resistance. This change in resistance produces a proportional change in the output voltage ($\Delta V_{out} = \Delta I_iR_f$).

Voltage-to-Current Converter

A basic voltage-to-current converter is shown in Figure 14–34. This circuit is used in applications where it is necessary to have an output (load) current that is controlled by an input voltage.

▶ **FIGURE 14–34**

Voltage-to-current converter.

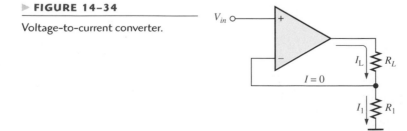

Neglecting the input offset voltage, both inverting and noninverting input terminals of the op-amp are at the same voltage, V_{in}. Therefore, the voltage across R_1 equals V_{in}. Since there is negligible current at the inverting input, the current through R_1 is the same as the current through R_L; thus

$$I_L = \frac{V_{in}}{R_1}$$

Equation 14–14

Peak Detector

An interesting application of the op-amp is in a peak detector circuit such as the one shown in Figure 14–35. In this case the op-amp is used as a comparator. This circuit is used to detect the peak of the input voltage and store that peak voltage on a capacitor. For example, this circuit can be used to detect and store the maximum value of a voltage surge; this value can then be measured at the output with a voltmeter or recording device. The basic operation is as follows. When a positive voltage is applied to the noninverting input of the op-amp through R_i, the high-level output voltage of the op-amp forward-biases the diode and charges the capacitor. The capacitor continues to charge until its voltage reaches a value equal to the input voltage and thus both op-amp inputs are at the same voltage. At this point, the op-amp comparator switches, and its output goes to the low level. The diode is now reverse-biased, and the capacitor stops charging. It has reached a voltage equal to the peak of V_{in} and will hold this voltage until the charge eventually leaks off. If a greater input peak occurs, the capacitor charges to the new peak.

▶ **FIGURE 14–35**

A basic peak detector.

SECTION 14–5 REVIEW

1. For the constant-current source in Figure 14–32, the input reference voltage is 6.8 V and R_i is 10 kΩ. What value of constant current does the circuit supply to a 1.0 kΩ load? To a 5 kΩ load?

2. What element determines the constant of proportionality that relates input current to output voltage in the current-to-voltage converter?

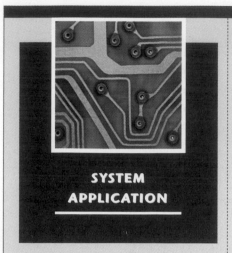

SYSTEM APPLICATION

This system application focuses on the amplifier circuits in an ECG system which include an isolation amplifier, a filter, and a postamplifier with summing capability. The ECG, or electrocardiograph, is used for monitoring heart signals. From the output waveform, a doctor can detect abnormalities in the heartbeat.

Basic Operation of the System

The human heart produces an electrical signal that can be picked up by electrodes in contact with the skin. When the heart signal is displayed on a chart recorder or on a video monitor, it is called an electrocardiograph (ECG). Typically, the heart signal picked up by the electrode is about 1 mV and has significant frequency components from less than 1 Hz to about 100 Hz.

As indicated in the block diagram in Figure 14–36, an ECG system has at least three electrodes. There is a right-arm (RA) electrode, a left-arm (LA) electrode, and a right-leg (RL) electrode that is the common terminal. The isolation amplifier provides for differential inputs from the electrodes, provides a high CMR (common-mode rejection) to eliminate the relatively high common-mode noise voltages associated with heart signals, and provides electrical isolation for protection of the patient. The low-pass filter rejects frequencies above those contained in the heart signal. The postamplifier provides most of the amplification in the system and drives a video monitor and/or a chart recorder. The three op-amp circuits are contained on the amplifier circuit board shown in Figure 14–37.

The inputs from the electrode sensors are connected to the amplifier board with shielded cable to prevent noise pickup. The schematic for the amplifier board is shown in Figure 14–38. The shielded cable is basically a twisted pair of wires surrounded by a braided metal sheathing that is covered by an insulated sleeve. The

braided metal shield serves as the conduit for the common connection. The incoming differential signal is amplified by the fixed gain of the 3656KG isolation amplifier. The 3656KG package used in this system has 20 pins. As shown in Figure 14–37, pin 1 is at the square corner and pin 20 is directly opposite pin 1.

The low-pass filter is an active filter of a type to be studied in Chapter 15. All you need to know for this assignment is how to calculate its critical frequency and gain, and those formulas will be provided.

The postamplifier is an inverting amplifier with an adjustable voltage gain. The inverting input also serves as a summing point for the signal voltage, and a dc voltage is used for adding an adjustable dc level to the output for adjusting the vertical position of the display.

The Amplifier Board

■ Make sure that the circuit board shown in Figure 14–37 is correctly assembled by checking it against the schematic in Figure 14–38. Backside interconnections are shown as darker traces.

▲ FIGURE 14–36

Block diagram of the ECG system.

■ Label a copy of the board with component and input/output designations in agreement with the schematic.

The Circuits

■ Determine the voltage gain of the isolation amplifier.

■ Determine the bandwidth and voltage gain of the filter using the following formulas, given that the response is from dc to the critical frequency.

$$f = \frac{1}{2\pi\sqrt{R_6 R_T C_4 C_5}}$$

$$A_v = \frac{R_8}{R_9} + 1$$

■ Determine the minimum and maximum voltage gains of the postamplifier.

■ Determine the overall gain range of the amplifier board.

■ Determine the dc voltage range at the wiper of the position adjustment potentiometer.

▲ FIGURE 14–37

Amplifier board.

Test Procedure

■ Develop a step-by-step set of instructions on how to check the amplifier board for proper operation independent of the rest of the system.

■ Specify voltage values for all the measurements to be made.

Troubleshooting

Faults have developed in four boards. Based on the indication for each board

listed below, determine the most likely fault in each case. The circled numbers indicate the referenced test points on the circuit board in Figure 14–39. Assume that there is a verified sinusoidal input signal in each case and that each board has the proper dc supply voltages.

Board 1:   There is no voltage at test point 1.

Board 2:   With an input signal of 1 mV, there is an 10.6 mV signal at

test point 2 but no signal at test point 3.

Board 3:   With an input signal of 1 mV, there is a 16.9 mV signal at test point 4, but no signal at test point 5.

Board 4:   There is an approximate square wave voltage with peaks at +14 V and −14 V at test point 1.

▲ FIGURE 14–38

Schematic of the amplifier board.

Final Report (Optional)

Submit a final written report on the amplifier board using an organized format that includes the following:

1. A physical description of the circuit.

2. A discussion of the operation of the circuit.

3. A list of the specifications.

4. A list of parts with part numbers if available.

5. A list of the types of problems on the four faulty circuit boards.

6. A description of how you determined the problem on each of the faulty circuit boards.

▲ **FIGURE 14–39**

Amplifier board with designated test points.

CHAPTER SUMMARY

- A basic instrumentation amplifier is formed by three op-amps and seven resistors, including the gain-setting resistor R_G.
- An instrumentation amplifier has high input impedance, high CMRR, low output offset, and low output impedance.
- The voltage gain of a basic instrumentation amplifier is set by a single external resistor.
- An instrumentation amplifier is useful in applications where small signals are embedded in large common-mode noise.
- A basic isolation amplifier has three electrically isolated parts: input, output, and power.
- Most isolation amplifiers use transformer coupling for isolation.
- Isolation amplifiers are used to interface sensitive equipment with high-voltage environments and to provide protection from electrical shock in certain medical applications.
- The operational transconductance amplifier (OTA) is a voltage-to-current amplifier.
- The output current of an OTA is the input voltage times the transconductance.
- In an OTA, transconductance varies with bias current; therefore, the gain of an OTA can be varied with a bias voltage or a variable resistor.
- The operation of log and antilog amplifiers is based on the nonlinear (logarithmic) characteristics of a pn junction.
- A log amplifier has a BJT in the feedback loop.

- An antilog amplifier has a BJT in series with the input.
- Logarithmic amplifiers are used for analog multiplication and division.

KEY TERMS

Key terms and other bold terms in the chapter are defined in the end-of-book glossary.

Instrumentation amplifier   An amplifier used for amplifying small signals riding on large common-mode voltages.

Isolation amplifier   An amplifier with electrically isolated internal stages.

Natural logarithm   The exponent to which the base e ($e = 2.71828$) must be raised in order to equal a given quantity.

Operational transconductance amplifier (OTA)   A voltage-to-current amplifier.

Transconductance   In an electronic device, the ratio of the output current to the input voltage.

KEY FORMULAS

Instrumentation Amplifier

$$14\text{--}1 \qquad A_{cl} = 1 + \frac{2R}{R_G}$$

$$14\text{--}2 \qquad R_G = \frac{2R}{A_{cl} - 1}$$

$$14\text{--}3 \qquad R_G = \frac{50.5\ \text{k}\Omega}{A_v - 1}$$

Isolation Amplifier

$$14\text{--}4 \qquad A_{v1} = \frac{R_{f1}}{R_{i1}} + 1$$

$$14\text{--}5 \qquad A_{v2} = \frac{R_{f2}}{R_{i2}} + 1$$

Operational Transconductance Ampifier (OTA)

$$14\text{--}6 \qquad g_m = \frac{I_{out}}{V_{in}}$$

$$14\text{--}7 \qquad g_m = KI_{\text{BIAS}}$$

Log and Antilog Amplifiers

$$14\text{--}8 \qquad V_{out} = -K \ln(V_{in})$$

$$14\text{--}9 \qquad V_{out} \cong -(0.025\ \text{V})\ln\left(\frac{V_{in}}{I_R R_1}\right)$$

$$14\text{--}10 \qquad V_{out} = -(0.025\ \text{V})\ln\left(\frac{V_{in}}{I_{\text{EBO}}R_1}\right)$$

$$14\text{--}11 \qquad V_{out} = -R_f I_{\text{EBO}}\text{antilog}\left(\frac{V_{in}}{25\ \text{mV}}\right)$$

Converters and Other Op-Amp Circuits

$$14\text{--}12 \qquad I_L = \frac{V_{IN}}{R_i} \qquad\qquad \text{Constant-current source}$$

$$14\text{--}13 \qquad V_{out} \doteq I_i R_f \qquad\qquad \text{Current-to-voltage converter}$$

$$14\text{--}14 \qquad I_L = \frac{V_{in}}{R_1} \qquad\qquad \text{Voltage-to-current converter}$$

Answers are at the end of the chapter.

1. To make a basic instrumentation amplifier, it takes
   (a) one op-amp with a certain feedback arrangement
   (b) two op-amps and seven resistors
   (c) three op-amps and seven capacitors
   (d) three op-amps and seven resistors

2. Typically, an instrumentation amplifier has an external resistor used for
   (a) establishing the input impedance   (b) setting the voltage gain
   (c) setting the current gain   (d) interfacing with an instrument

3. Instrumentation amplifiers are used primarily in
   (a) high-noise environments   (b) medical equipment
   (c) test instruments   (d) filter circuits

4. Isolation amplifiers are used primarily in
   (a) remote, isolated locations
   (b) systems that isolate a single signal from many different signals
   (c) applications where there are high voltages and sensitive equipment
   (d) applications where human safety is a concern
   (e) answers (c) and (d)

5. The three parts of a basic isolation amplifier are
   (a) amplifier, filter, and power   (b) input, output, and coupling
   (c) input, output, and power   (d) gain, attenuation, and offset

6. The stages of most isolation amplifiers are connected by
   (a) copper strips   (b) transformers
   (c) microwave links   (d) current loops

7. The characteristic that allows an isolation amplifier to amplify small signal voltages in the presence of much greater noise voltages is its
   (a) CMRR   (b) high gain
   (c) high input impedance   (d) magnetic coupling between input and output

8. The term *OTA* means
   (a) operational transistor amplifier
   (b) operational transformer amplifier
   (c) operational transconductance amplifier
   (d) output transducer amplifier

9. In an OTA, the transconductance is controlled by
   (a) the dc supply voltage   (b) the input signal voltage
   (c) the manufacturing process   (d) a bias current

10. The voltage gain of an OTA circuit is set by
    (a) a feedback resistor
    (b) the transconductance only
    (c) the transconductance and the load resistor
    (d) the bias current and supply voltage

11. An OTA is basically a
    (a) voltage-to-current amplifier   (b) current-to-voltage amplifier
    (c) current-to-current amplifier   (d) voltage-to-voltage amplifier

12. The operation of a logarithmic amplifier is based on
   (a) the nonlinear operation of an op-amp
   (b) the logarithmic characteristic of a *pn* junction
   (c) the reverse breakdown characteristic of a *pn* junction
   (d) the logarithmic charge and discharge of an *RC* circuit

13. If the input to a log amplifier is x, the output is proportional to
   (a) e^x (b) $\ln x$ (c) $\log_{10} x$
   (d) $2.3 \log_{10} x$ (e) answers (a) and (c)   (f) answers (b) and (d)

14. If the input to an antilog amplifier is x, the output is proportional to
   (a) $e^{\ln x}$ (b) e^x (c) $\ln x$ (d) e^{-x}

15. The logarithm of the product of two numbers is equal to the
   (a) sum of the two numbers
   (b) sum of the logarithms of each of the numbers
   (c) difference of the logarithms of each of the numbers
   (d) ratio of the logarithms of the numbers

16. If you subtract $\ln y$ from $\ln x$, you get
   (a) $\ln x / \ln y$ (b) $(\ln x)(\ln y)$ (c) $\ln(x/y)$ (d) $\ln(y/x)$

PROBLEMS

Answers to all odd-numbered problems are at the end of the book.

BASIC PROBLEMS

SECTION 14–1   **Instrumentation Amplifiers**

1. Determine the voltage gains of op-amps A1 and A2 for the instrumentation amplifier configuration in Figure 14–40.

2. Find the overall voltage gain of the instrumentation amplifier in Figure 14–40.

3. The following voltages are applied to the instrumentation amplifier in Figure 14–40: $V_{in1} = 5$ mV, $V_{in2} = 10$ mV, and $V_{cm} = 225$ mV. Determine the final output voltage.

4. What value of R_G must be used to change the gain of the instrumentation amplifier in Figure 14–40 to 1000?

▲ **FIGURE 14–40**

File circuits are identified with a CD logo and are in the Problems folder on your CD-ROM. Filenames correspond to figure numbers (e.g., F14-40).

5. What is the voltage gain of the AD622 instrumentation amplifier in Figure 14–41?

6. Determine the approximate bandwidth of the amplifier in Figure 14–41 if the voltage gain is set to 10. Use the graph in Figure 14–6.

7. Specify what you must do to change the gain of the amplifier in Figure 14–41 to approximately 24.

8. Determine the value of R_G in Figure 14–41 for a voltage gain of 20.

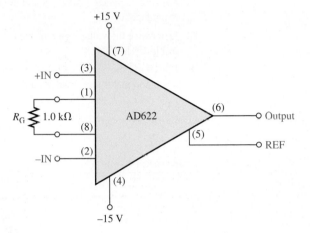

▲ FIGURE 14–41

SECTION 14–2   Isolation Amplifiers

9. The op-amp in the input stage of a certain isolation amplifier has a voltage gain of 30. The output stage is set for a gain of 10. What is the total voltage gain of this device?

10. Determine the total voltage gain of each 3656KG in Figure 14–42.

11. Specify how you would change the total gain of the amplifier in Figure 14–42(a) to approximately 100 by changing only the gain of the input stage.

12. Specify how you would change the total gain in Figure 14–42(b) to approximately 440 by changing only the gain of the output stage.

13. Specify how you would connect each amplifier in Figure 14–42 for unity gain.

(a)   (b)

▲ FIGURE 14–42

SECTION 14–3   Operational Transconductance Amplifiers (OTAs)

14. A certain OTA has an input voltage of 10 mV and an output current of 10 μA. What is the transconductance?

15. A certain OTA with a transconductance of 5000 μS has a load resistance of 10 kΩ. If the input voltage is 100 mV, what is the output current? What is the output voltage?

16. The output voltage of a certain OTA with a load resistance is determined to be 3.5 V. If its transconductance is 4000 μS and the input voltage is 100 mV, what is the value of the load resistance?

17. Determine the voltage gain of the OTA in Figure 14–43. Assume $K = 16$ μS/μA for the graph in Figure 14–44.

18. If a 10 kΩ rheostat is added in series with the bias resistor in Figure 14–43, what are the minimum and maximum voltage gains?

▶ **FIGURE 14–43**

▶ **FIGURE 14–44**

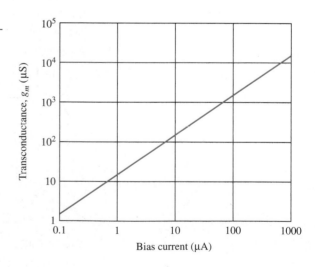

19. The OTA in Figure 14–45 functions as an amplitude modulation circuit. Determine the output voltage waveform for the given input waveforms assuming $K = 16$ μS/μA.

20. Determine the trigger points for the Schmitt trigger in Figure 14–46.

21. Determine the output voltage waveform for the Schmitt trigger in Figure 14–46 in relation to a 1 kHz sine wave with peak values of ± 10 V.

▲ **FIGURE 14–45**

▶ **FIGURE 14–46**

SECTION 14–4 Log and Antilog Amplifiers

22. Using your calculator, find the natural logarithm (ln) of each of the following numbers:

   (a) 0.5   (b) 2   (c) 50   (d) 130

23. Repeat Problem 22 for $\log_{10}$.

24. What is the antilog of 1.6?

25. Explain why the output of a log amplifier is limited to approximately 0.7 V.

26. What is the output voltage of a certain log amplifier with a diode in the feedback path when the input voltage is 3 V? The input resistor is 82 kΩ and the reverse leakage current is 100 nA.

27. Determine the output voltage for the amplifier in Figure 14–47. Assume $I_{EBO} = 60$ nA.

▶ **FIGURE 14–47**

28. Determine the output voltage for the amplifier in Figure 14–48. Assume $I_{EBO} = 60$ nA.

29. Signal compression is one application of logarithmic amplifiers. Suppose an audio signal with a maximum voltage of 1 V and a minimum voltage of 100 mV is applied to the log amplifier in Figure 14–47. What will be the maximum and minimum output voltages? What conclusion can you draw from this result?

▶ **FIGURE 14–48**

SECTION 14–5 Converters and Other Op-Amp Circuits

30. Determine the load current in each circuit of Figure 14–49.

31. Devise a circuit for remotely sensing temperature and producing a proportional voltage that can then be converted to digital form for display. A thermistor can be used as the temperature-sensing element.

(a)   (b)

▲ **FIGURE 14–49**

TROUBLESHOOTING PROBLEMS

32. With a 1 mV, 50 Hz signal applied to the ECG amplifier board in Figure 14–50, what voltage would you expect to see at each of the test points? Assume that the offset voltage is nulled out and the position control is adjusted for zero deflection.

33. Repeat Problem 32 for a 2 mV, 1 kHz input signal.

▲ **FIGURE 14–50**

EWB/MULTISIM TROUBLESHOOTING PROBLEMS

These file circuits are in the Troubleshooting Problems folder on your CD-ROM.

34. Open file TSP14-34 and determine the fault.

35. Open file TSP14-35 and determine the fault.

36. Open file TSP14-36 and determine the fault.

37. Open file TSP14-37 and determine the fault.

38. Open file TSP14-38 and determine the fault.

ANSWERS

SECTION REVIEWS

SECTION 14–1 **Instrumentation Amplifiers**

1. The main purpose of an instrumentation amplifier is to amplify small signals that occur on large common-mode voltages. The key characteristics are high input impedance, high CMRR, low output impedance, and low output offset.

2. Three op-amps and seven resistors including the gain resistor are required to construct a basic instrumentation amplifier (see Figure 14–2).

3. The gain is set by the external resistor R_G.

4. The gain is approximately 6.

SECTION 14–2 **Isolation Amplifiers**

1. Isolation amplifiers are used in medical equipment, power plant instrumentation, industrial processing, and automated testing.

2. The two stages of an isolation amplifier are input and output.

3. The stages are connected by transformer coupling.

4. The oscillator acts as a dc to ac converter so that the dc power can be ac coupled to the input and output stages.

SECTION 14–3 **Operational Transconductance Amplifiers (OTAs)**

1. OTA stands for Operational Transconductance Amplifier.

2. Transconductance increases with bias current.

3. Assuming that the bias input is connected to the supply voltage, the voltage gain increases when the supply voltage is increased because this increases the bias current.

4. The voltage gain decreases as the bias voltage decreases.

SECTION 14–4 **Log and Antilog Amplifiers**

1. A diode or transistor in the feedback loop provides the exponential (nonlinear) characteristic.

2. The output of a log amplifier is limited to the barrier potential of the *pn* junction (about 0.7 V).

3. The output voltage is determined by the input voltage, the input resistor, and the emitter-to-base leakage current.

4. The transistor in an antilog amplifier is in series with the input rather than in the feedback loop.

SECTION 14–5 **Converters and Other Op-Amp Circuits**

1. $I_L = 6.8$ V/10 kΩ = 0.68 mA; same value to 5 kΩ load.

2. The feedback resistor is the constant of proportionality.

RELATED PROBLEMS FOR EXAMPLES

14–1 240 Ω

14–2 Make $R_G = 1.1$ kΩ.

14–3 Many combinations are possible. Here is one: $R_{f1} = 10$ kΩ, $R_{i1} = 1.0$ kΩ, $R_{f2} = 10$ kΩ, and $R_{i2} = 1.0$ kΩ

14–4 $I_{BIAS} \cong 62.5$ μA

14–5 Yes. The gain will change to approximately 110.

14–6 The output is a square-wave modulated signal with a maximum amplitude of approximately 3.6 V and a minimum amplitude of approximately 1.76 V.

14–7 -0.167 V

14–8 -0.193 V

14–9 -4.39 V

SELF-TEST

1. (d)   2. (b)   3. (a)   4. (e)   5. (c)   6. (b)   7. (a)   8. (c)
9. (d)   10. (c)   11. (a)   12. (b)   13. (f)   14. (b)   15. (b)   16. (c)

15

ACTIVE FILTERS

INTRODUCTION

Power supply filters were introduced in Chapter 2. In this chapter, active filters that are used for signal processing are introduced. Filters are circuits that are capable of passing signals with certain selected frequencies while rejecting signals with other frequencies. This property is called *selectivity*.

Active filters use transistors or op-amps combined with passive *RC, RL,* or *RLC* circuits. The active devices provide voltage gain, and the passive circuits provide frequency selectivity. In terms of general response, the four basic categories of active filters are low-pass, high-pass, band-pass, and band-stop. In this chapter, you will study active filters using op-amps and *RC* circuits.

CHAPTER OBJECTIVES

- Describe the gain-versus-frequency responses of the basic filters

- Describe the three basic filter response characteristics and other filter parameters

- Analyze active low-pass filters

- Analyze active high-pass filters

- Analyze active band-pass filters

- Analyze active band-stop filters

- Discuss two methods for measuring frequency response

KEY TERMS

- Low-pass filter

- Pole

- Roll-off

- High-pass filter

- Band-pass filter

- Band-stop filter

- Damping factor

■■■ SYSTEM APPLICATION PREVIEW

The FM stereo receiver accepts carrier signals in the frequency range from 88 MHz to 108 MHz. In frequency modulation, the frequency of the carrier is varied in proportion to the amplitude and frequency of the modulating audio signal. The actual FM stereo multiplex signal that is received is quite complex and beyond the scope of our coverage; therefore, the focus of the system application at the end of the chapter is limited to the filter circuits which are part of the channel separation circuits. You will learn how the active filters covered in the chapter can be applied in the FM system to separate out the audio signals that go to the left and right channel speakers.

WWW. VISIT THE COMPANION WEBSITE

Study aids for this chapter are available at
http://www.prenhall.com/floyd

15–1 BASIC FILTER RESPONSES

Filters are usually categorized by the manner in which the output voltage varies with the frequency of the input voltage. The categories of active filters are low-pass, high-pass, band-pass, and band-stop. We will examine each of these general responses in this section.

After completing this section, you should be able to

- ■ **Describe the gain-versus-frequency responses of the basic filters**
- ■ Explain the low-pass response
- ■ Determine the critical frequency and bandwidth of a low-pass filter
- ■ Explain the high-pass response
- ■ Determine the critical frequency of a high-pass filter
- ■ Explain the band-pass response
- ■ Explain the significance of the quality factor
- ■ Determine the critical frequency, bandwidth, quality factor, and damping factor of a band-pass filter
- ■ Explain the band-stop response

Low-Pass Filter Response

A filter is a circuit that passes certain frequencies and attenuates or rejects all other frequencies. The **passband** of a filter is the range of frequencies that are allowed to pass through the filter with minimum attenuation (usually defined as less than −3 dB of attenuation). The **critical frequency,** f_c, (also called the *cutoff frequency*) defines the end of the passband and is normally specified at the point where the response drops −3 dB (70.7%) from the passband response. Following the passband is a region called *the transition region* that leads into a region called the *stopband*. There is no precise point between the transition region and the stopband.

 A **low-pass filter** is one that passes frequencies from dc to f_c and significantly attenuates all other frequencies. The passband of the ideal low-pass filter is shown in the blue-shaded area of Figure 15–1(a); the response drops to zero at frequencies beyond the passband. This ideal response is sometimes referred to as a "brick-wall" because nothing gets through beyond the wall. The bandwidth of an ideal low-pass filter is equal to f_c.

Equation 15–1
$$BW = f_c$$

The ideal response shown in Figure 15–1(a) is not attainable by any practical filter. Actual filter responses depend on the number of **poles**, a term used with filters to describe the number of *RC* circuits contained in the filter. The most basic low-pass filter is a simple *RC* circuit consisting of just one resistor and one capacitor; the output is taken across the capacitor as shown in Figure 15–1(b). This basic *RC* filter has a single pole and it rolls off at −20 dB/decade beyond the critical frequency. The actual response is indicated by the blue line in Figure 15–1(a). The response is plotted on a standard log plot that is used for filters to show details of the curve as the gain drops. Notice that the gain drops off slowly until the frequency is at the critical frequency; after this, the gain drops rapidly.

 The −20 dB/decade **roll-off** rate for the gain of a basic *RC* filter means that at a frequency of $10f_c$, the output will be −20 dB (10%) of the input. This rather gentle roll-off is not a particularly good filter characteristic because too much of the unwanted frequencies (beyond the passband) are allowed through the filter.

(a) Comparison of an ideal low-pass filter response with actual response

(b) Basic low-pass circuit

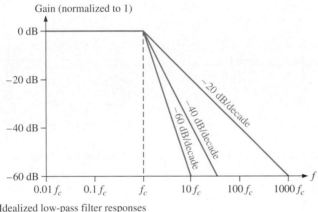

(c) Idealized low-pass filter responses

▲ **FIGURE 15–1**

Low-pass filter responses.

The critical frequency of the simple low-pass RC filter occurs when $X_C = R$, where

$$f_c = \frac{1}{2\pi RC}$$

Recall from your basic dc/ac course that the output at the critical frequency is 70.7% of the input. This response is equivalent to an attenuation of -3 dB.

Figure 15–1(c) illustrates several idealized low-pass response curves including the basic one pole response (-20 dB/decade). The approximations show a *flat* response to the cutoff frequency and a roll-off at a constant rate after the cutoff frequency. Actual filters do not have a perfectly flat response to the cutoff frequency but have dropped to -3 dB at this point as described previously.

In order to produce a filter that has a steeper transition region, (and hence form a more effective filter), it is necessary to add additional circuitry to the basic filter. Responses that are steeper than -20 dB/decade in the transition region cannot be obtained by simply cascading identical RC stages (due to loading effects). However, by combining an op-amp with frequency-selective feedback circuits, filters can be designed with roll-off rates of -40, -60, or more dB/decade. Filters that include one or more op-amps in the design are called **active filters**. These filters can optimize the roll-off rate or other attribute (such as phase response) with a particular filter design. In general, the more poles the filter uses, the steeper its transition region will be. The exact response depends on the type of filter and the number of poles.

High-Pass Filter Response

A **high-pass filter** is one that significantly attenuates or rejects all frequencies below f_c and passes all frequencies above f_c. The critical frequency is, again, the frequency at which the output is 70.7% of the input (or -3 dB) as shown in Figure 15–2(a). The ideal response, indicated by the blue-shaded area, has an instantaneous drop at f_c, which, of course, is not achievable. Ideally, the passband of a high-pass filter is all frequencies above the critical frequency. The high-frequency response of practical circuits is limited by the op-amp or other components that make up the filter.

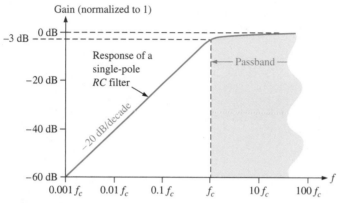

(a) Comparison of an ideal high-pass filter response with actual response

(b) Basic high-pass circuit

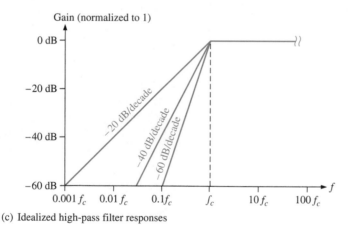

(c) Idealized high-pass filter responses

▲ **FIGURE 15–2**

High-pass filter responses.

A simple RC circuit consisting of a single resistor and capacitor can be configured as a high-pass filter by taking the output across the resistor as shown in Figure 15–2(b). As in the case of the low-pass filter, the basic RC circuit has a roll-off rate of -20 dB/decade as indicated by the blue line in Figure 15–2(a). Also, the critical frequency for the basic high-pass filter occurs when $X_C = R$, where

$$f_c = \frac{1}{2\pi RC}$$

Figure 15–2(c) illustrates several idealized high-pass response curves including the basic one pole response (-20 dB/decade) for a basic RC circuit. As in the case of the low-pass filter, the approximations show a *flat* response to the cutoff frequency and a roll-off at a constant rate after the cutoff frequency. Actual high-pass filters do not have the perfectly flat response indicated or the precise roll-off rate shown. Responses that are steeper than −20 dB/decade in the transition region are also possible with active high-pass filters; the particular response depends on the type of filter and the number of poles.

Band-Pass Filter Response

A **band-pass filter** passes all signals lying within a band between a lower-frequency limit and an upper-frequency limit and essentially rejects all other frequencies that are outside this specified band. A generalized band-pass response curve is shown in Figure 15–3. The bandwidth (*BW*) is defined as the difference between the upper critical frequency (f_{c2}) and the lower critical frequency (f_{c1}).

$$BW = f_{c2} - f_{c1}$$

Equation 15–2

The critical frequencies are, of course, the points at which the response curve is 70.7 percent of its maximum. Recall from Chapter 12 that these critical frequencies are also called *3 dB frequencies*. The frequency about which the passband is centered is called the *center frequency*, f_0, defined as the geometric mean of the critical frequencies.

$$f_0 = \sqrt{f_{c1}f_{c2}}$$

Equation 15–3

► **FIGURE 15–3**

General band-pass response curve.

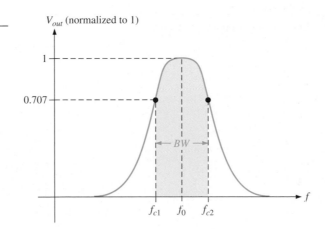

Quality Factor The **quality factor (*Q*)** of a band-pass filter is the ratio of the center frequency to the bandwidth.

$$Q = \frac{f_0}{BW}$$

Equation 15–4

The value of *Q* is an indication of the selectivity of a band-pass filter. The higher the value of *Q*, the narrower the bandwidth and the better the selectivity for a given value of f_0. Band-pass filters are sometimes classified as narrow-band ($Q > 10$) or wide-band ($Q < 10$). The quality factor (*Q*) can also be expressed in terms of the damping factor (*DF*) of the filter as

$$Q = \frac{1}{DF}$$

You will study the damping factor in Section 15–2.

EXAMPLE 15–1

A certain band-pass filter has a center frequency of 15 kHz and a bandwidth of 1 kHz. Determine Q and classify the filter as narrow-band or wide-band.

Solution

$$Q = \frac{f_0}{BW} = \frac{15 \text{ kHz}}{1 \text{ kHz}} = 15$$

Because $Q > 10$, this is a narrow-band filter.

Related Problem*

If the quality factor of the filter is doubled, what will the bandwidth be?

*Answers are at the end of the chapter.

Band-Stop Filter Response

Another category of active filter is the **band-stop filter**, also known as *notch, band-reject,* or *band-elimination* filter. You can think of the operation as opposite to that of the band-pass filter because frequencies within a certain bandwidth are rejected, and frequencies outside the bandwidth are passed. A general response curve for a band-stop filter is shown in Figure 15–4. Notice that the bandwidth is the band of frequencies between the 3 dB points, just as in the case of the band-pass filter response.

▶ **FIGURE 15–4**

General band-stop filter response.

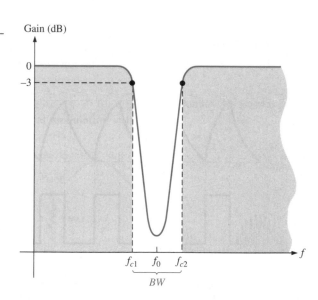

SECTION 15–1 REVIEW

Answers are at the end of the chapter.

1. What determines the bandwidth of a low-pass filter?
2. What limits the bandwidth of an active high-pass filter?
3. How are the Q and the bandwidth of a band-pass filter related? Explain how the selectivity is affected by the Q of a filter.

15–2 FILTER RESPONSE CHARACTERISTICS

Each type of filter response (low-pass, high-pass, band-pass, or band-stop) can be tailored by circuit component values to have either a Butterworth, Chebyshev, or Bessel characteristic. Each of these characteristics is identified by the shape of the response curve, and each has an advantage in certain applications.

After completing this section, you should be able to

- **Describe the three basic filter response characteristics and other filter parameters**
- Describe the Butterworth characteristic
- Describe the Chebyshev characteristic
- Describe the Bessel characteristic
- Define *damping factor* and discuss its significance
- Calculate the damping factor of a filter
- Discuss the order of a filter and its affect on the roll-off rate

Butterworth, Chebyshev, or Bessel response characteristics can be realized with most active filter circuit configurations by proper selection of certain component values. A general comparison of the three response characteristics for a low-pass filter response curve is shown in Figure 15–5. High-pass and band-pass filters can also be designed to have any one of the characteristics.

▶ **FIGURE 15–5**

Comparative plots of three types of filter response characteristics.

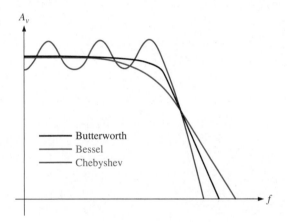

The Butterworth Characteristic   The **Butterworth** characteristic provides a very flat amplitude response in the passband and a roll-off rate of −20 dB/decade/pole. The phase response is not linear, however, and the phase shift (thus, time delay) of signals passing through the filter varies nonlinearly with frequency. Therefore, a pulse applied to a filter with a Butterworth response will cause overshoots on the output because each frequency component of the pulse's rising and falling edges experiences a different time delay. Filters with the Butterworth response are normally used when all frequencies in the passband must have the same gain. The Butterworth response is often referred to as a maximally flat response.

The Chebyshev Characteristic   Filters with the **Chebyshev** response characteristic are useful when a rapid roll-off is required because it provides a roll-off rate greater than −20 dB/decade/pole. This is a greater rate than that of the Butterworth, so filters can be implemented with the Chebyshev response with fewer poles and less complex circuitry for a given roll-off rate. This type of filter response is characterized by overshoot or ripples in the passband (depending on the number of poles) and an even less linear phase response than the Butterworth.

The Bessel Characteristic   The **Bessel** response exhibits a linear phase characteristic, meaning that the phase shift increases linearly with frequency. The result is almost no overshoot on the output with a pulse input. For this reason, filters with the Bessel response are used for filtering pulse waveforms without distorting the shape of the waveform.

The Damping Factor

As mentioned, an active filter can be designed to have either a Butterworth, Chebyshev, or Bessel response characteristic regardless of whether it is a low-pass, high-pass, band-pass, or band-stop type. The **damping factor** (**DF**) of an active filter circuit determines which response characteristic the filter exhibits. To explain the basic concept, a generalized active filter is shown in Figure 15–6. It includes an amplifier, a negative feedback circuit, and a filter section. The amplifier and feedback are connected in a noninverting configuration. The damping factor is determined by the negative feedback circuit and is defined by the following equation:

Equation 15–5

$$DF = 2 - \frac{R_1}{R_2}$$

▶ **FIGURE 15–6**

General diagram of an active filter.

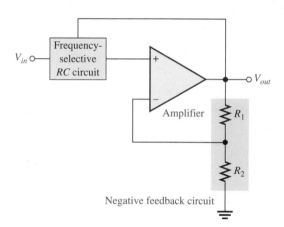

Basically, the damping factor affects the filter response by negative feedback action. Any attempted increase or decrease in the output voltage is offset by the opposing effect of the negative feedback. This tends to make the response curve flat in the passband of the filter if the value for the damping factor is precisely set. By advanced mathematics, which we will not cover, values for the damping factor have been derived for various orders of filters to achieve the maximally flat response of the Butterworth characteristic.

The value of the damping factor required to produce a desired response characteristic depends on the **order** (number of poles) of the filter. A *pole*, for our purposes, is simply a circuit with one resistor and one capacitor. The more poles a filter has, the faster its roll-off rate is. To achieve a second-order Butterworth response, for example, the damping factor must be 1.414. To implement this damping factor, the feedback resistor ratio must be

$$\frac{R_1}{R_2} = 2 - DF = 2 - 1.414 = 0.586$$

This ratio gives the closed-loop gain of the noninverting filter amplifier, $A_{cl(NI)}$, a value of 1.586, derived as follows:

$$A_{cl(NI)} = \frac{1}{B} = \frac{1}{R_2/(R_1 + R_2)} = \frac{R_1 + R_2}{R_2} = \frac{R_1}{R_2} + 1 = 0.586 + 1 = 1.586$$

EXAMPLE 15–2

If resistor R_2 in the feedback circuit of an active two-pole filter of the type in Figure 15–6 is 10 kΩ, what value must R_1 be to obtain a maximally flat Butterworth response?

Solution

$$\frac{R_1}{R_2} = 0.586$$

$$R_1 = 0.586R_2 = 0.586(10 \text{ k}\Omega) = \textbf{5.86 k}\boldsymbol{\Omega}$$

Using the nearest standard 5 percent value of 5.6 kΩ will get very close to the ideal Butterworth response.

Related Problem What is the damping factor for $R_2 = 10$ kΩ and $R_1 = 5.6$ kΩ?

Critical Frequency and Roll-Off Rate

The critical frequency is determined by the values of the resistor and capacitors in the frequency-selective RC circuit, as shown in Figure 15–6. For a single-pole (first-order) filter, as shown in Figure 15–7, the critical frequency is

$$f_c = \frac{1}{2\pi RC}$$

Although we show a low-pass configuration, the same formula is used for the f_c of a single-pole high-pass filter. The number of poles determines the roll-off rate of the filter. A Butterworth response produces -20 dB/decade/pole. So, a first-order (one-pole) filter has a roll-off of -20 dB/decade; a second-order (two-pole) filter has a roll-off rate of -40 dB/decade; a third-order (three-pole) filter has a roll-off rate of -60 dB/ decade; and so on.

▶ **FIGURE 15–7**

First-order (one-pole) low-pass filter.

Single-pole low-pass circuit

Generally, to obtain a filter with three poles or more, one-pole or two-pole filters are cascaded, as shown in Figure 15–8. To obtain a third-order filter, for example, cascade a second-order and a first-order filter; to obtain a fourth-order filter, cascade two second-order filters; and so on. Each filter in a cascaded arrangement is called a *stage* or *section*.

Because of its maximally flat response, the Butterworth characteristic is the most widely used. Therefore, we will limit our coverage to the Butterworth response to illustrate basic filter concepts. Table 15–1 lists the roll-off rates, damping factors, and feedback resistor ratios for up to sixth-order Butterworth filters.

▲ FIGURE 15–8

The number of filter poles can be increased by cascading.

▼ TABLE 15–1

Values for the Butterworth response.

| | | 1ST STAGE | | | 2ND STAGE | | | 3RD STAGE | | |
ORDER	ROLL-OFF dB/DECADE	POLES	DF	R_1/R_2	POLES	DF	R_3/R_4	POLES	DF	R_5/R_6
1	−20	1	Optional							
2	−40	2	1.414	0.586						
3	−60	2	1.00	1	1	1.00	1			
4	−80	2	1.848	0.152	2	0.765	1.235			
5	−100	2	1.00	1	2	1.618	0.382	1	0.618	1.382
6	−120	2	1.932	0.068	2	1.414	0.586	2	0.518	1.482

SECTION 15–2 REVIEW

1. Explain how Butterworth, Chebyshev, and Bessel responses differ.
2. What determines the response characteristic of a filter?
3. Name the basic parts of an active filter.

15–3   ACTIVE LOW-PASS FILTERS

Filters that use op-amps as the active element provide several advantages over passive filters (R, L, and C elements only). The op-amp provides gain, so the signal is not attenuated as it passes through the filter. The high input impedance of the op-amp prevents excessive loading of the driving source, and the low output impedance of the op-amp prevents the filter from being affected by the load that it is driving. Active filters are also easy to adjust over a wide frequency range without altering the desired response.

After completing this section, you should be able to

■ **Analyze active low-pass filters**

■ Identify a single-pole filter and determine its gain and critical frequency

■ Identify a two-pole Sallen-Key filter and determine its gain and critical frequency

■ Explain how a higher roll-off rate is achieved by cascading low-pass filters

A Single-Pole Filter

Figure 15–9(a) shows an active filter with a single low-pass RC frequency-selective circuit that provides a roll-off of -20 dB/decade above the critical frequency, as indicated by the response curve in Figure 15–9(b). The critical frequency of the single-pole filter is $f_c = 1/2\pi RC$. The op-amp in this filter is connected as a noninverting amplifier with the closed-loop voltage gain in the passband set by the values of R_1 and R_2.

$$A_{cl(NI)} = \frac{R_1}{R_2} + 1$$

Equation 15–6

(a)   (b)

▲ **FIGURE 15–9**

Single-pole active low-pass filter and response curve.

The Sallen-Key Low-Pass Filter

The Sallen-Key is one of the most common configurations for a second-order (two-pole) filter. It is also known as a VCVS (voltage-controlled voltage source) filter. A low-pass version of the Sallen-Key filter is shown in Figure 15–10. Notice that there are two low-pass RC circuits that provide a roll-off of -40 dB/decade above the critical frequency (assuming a Butterworth characteristic). One RC circuit consists of R_A and C_A, and the second circuit consists of R_B and C_B. A unique feature is the capacitor C_A that provides feedback for shaping the response near the edge of the passband. The critical frequency for the second-order Sallen-Key filter is

$$f_c = \frac{1}{2\pi\sqrt{R_A R_B C_A C_B}}$$

Equation 15–7

▶ **FIGURE 15–10**

Basic Sallen-Key second-order low-pass filter.

For simplicity, the component values can be made equal so that $R_A = R_B = R$ and $C_A = C_B = C$. In this case, the expression for the critical frequency simplifies to

$$f_c = \frac{1}{2\pi RC}$$

As in the single-pole filter, the op-amp in the second-order Sallen-Key filter acts as a noninverting amplifier with the negative feedback provided by resistors R_1 and R_2. As you have learned, the damping factor is set by the values of R_1 and R_2, thus making the filter response either Butterworth, Chebyshev, or Bessel. For example, from Table 15–1, the R_1/R_2 ratio must be 0.586 to produce the damping factor of 1.414 required for a second-order Butterworth response.

EXAMPLE 15–3

Determine the critical frequency of the low-pass filter in Figure 15–11, and set the value of R_1 for an approximate Butterworth response.

▲ **FIGURE 15–11**

Solution   Since $R_A = R_B = 1.0$ kΩ and $C_A = C_B = 0.02$ μF,

$$f_c = \frac{1}{2\pi RC} = \frac{1}{2\pi(1.0 \text{ k}\Omega)(0.02 \text{ μF})} = \mathbf{7.96 \text{ kHz}}$$

For a Butterworth response, $R_1/R_2 = 0.586$.

$$R_1 = 0.586 R_2 = 0.586(1.0 \text{ k}\Omega) = \mathbf{586 \text{ } \Omega}$$

Select a standard value as near as possible to this calculated value.

Related Problem   Determine f_c for Figure 15–11 if $R_A = R_B = R_2 = 2.2$ kΩ and $C_A = C_B = 0.01$ μF. Also determine the value of R_1 for a Butterworth response.

> Open file E15-03 in the Examples folder on your CD-ROM. Determine the critical frequency and compare with the calculated value.

Cascaded Low-Pass Filters Achieve a Higher Roll-Off Rate

A three-pole filter is required to get a third-order low-pass response (−60 dB/decade). This is done by cascading a two-pole low-pass filter and a single-pole low-pass filter, as shown in Figure 15–12(a). Figure 15–12(b) shows a four-pole configuration obtained by cascading two two-pole filters.

(a) Third-order configuration

(b) Fourth-order configuration

▲ **FIGURE 15–12**

Cascaded low-pass filters.

EXAMPLE 15–4

For the four-pole filter in Figure 15–12(b), determine the capacitance values required to produce a critical frequency of 2680 Hz if all the resistors in the *RC* low-pass circuits are 1.8 kΩ. Also select values for the feedback resistors to get a Butterworth response.

Solution Both stages must have the same f_c. Assuming equal-value capacitors,

$$f_c = \frac{1}{2\pi RC}$$

$$C = \frac{1}{2\pi R f_c} = \frac{1}{2\pi(1.8 \text{ k}\Omega)(2680 \text{ Hz})} = 0.033 \ \mu\text{F}$$

$$C_{A1} = C_{B1} = C_{A2} = C_{B2} = \mathbf{0.033 \ \mu F}$$

Also select $R_2 = R_4 = 1.8$ kΩ for simplicity. Refer to Table 15–1. For a Butterworth response in the first stage, $DF = 1.848$ and $R_1/R_2 = 0.152$. Therefore,

$$R_1 = 0.152R_2 = 0.152(1800\ \Omega) = \mathbf{274\ \Omega}$$

Choose $R_1 = 270\ \Omega$.

In the second stage, $DF = 0.765$ and $R_3/R_4 = 1.235$. Therefore,

$$R_3 = 1.235R_4 = 1.235(1800\ \Omega) = \mathbf{2.22\ k\Omega}$$

Choose $R_3 = 2.2$ kΩ.

Related Problem   For the filter in Figure 15–12(b), determine the capacitance values for $f_c = 1$ kHz if all the filter resistors are 680 Ω. Also specify the values for the feedback resistors to produce a Butterworth response.

**SECTION 15–3
REVIEW**

1. How many poles does a second-order low-pass filter have? How many resistors and how many capacitors are used in the frequency–selective circuit?
2. Why is the damping factor of a filter important?
3. What is the primary purpose of cascading low-pass filters?

15–4   ACTIVE HIGH-PASS FILTERS

In high-pass filters, the roles of the capacitor and resistor are reversed in the RC circuits. Otherwise, the basic parameters are the same as for the low-pass filters.

After completing this section, you should be able to

- **Analyze active high-pass filters**
- Identify a single-pole filter and determine its gain and critical frequency
- Identify a two-pole Sallen-Key filter and determine its gain and critical frequency
- Explain how a higher roll-off rate is achieved by cascading high-pass filters

A Single-Pole Filter

A high-pass active filter with a -20 dB/decade roll-off is shown in Figure 15–13(a). Notice that the input circuit is a single high-pass RC circuit. The negative feedback circuit is the same as for the low-pass filters previously discussed. The high-pass response curve is shown in Figure 15–13(b).

Ideally, a high-pass filter passes all frequencies above f_c without limit, as indicated in Figure 15–14(a), although in practice, this is not the case. As you have learned, all op-amps inherently have internal RC circuits that limit the amplifier's response at high frequencies. Therefore, there is an upper-frequency limit on the high-pass filter's response which, in effect, makes it a band-pass filter with a very wide bandwidth. In the majority of applications, the internal high-frequency limitation is so much greater than that of the filter's critical frequency that the limitation can be neglected. In some applications, discrete transistors are used for the gain element to increase the high-frequency limitation beyond that realizable with available op-amps.

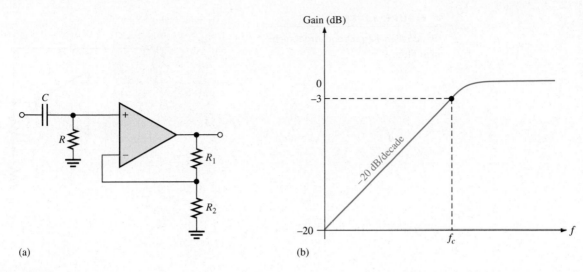

▲ **FIGURE 15–13**

Single-pole active high-pass filter and response curve.

▲ **FIGURE 15–14**

High-pass filter response.

The Sallen-Key High-Pass Filter

A high-pass second-order Sallen-Key configuration is shown in Figure 15–15. The components R_A, C_A, R_B, and C_B form the two-pole frequency-selective circuit. Notice that the positions of the resistors and capacitors in the frequency-selective circuit are opposite to those in the low-pass configuration. As with the other filters, the response characteristic can be optimized by proper selection of the feedback resistors, R_1 and R_2.

► **FIGURE 15–15**

Basic Sallen–Key second-order high-pass filter.

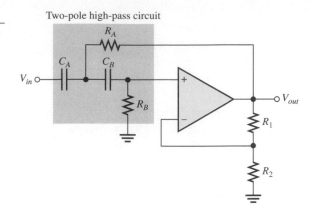

Two-pole high-pass circuit

EXAMPLE 15–5

Choose values for the Sallen-Key high-pass filter in Figure 15–15 to implement an equal-value second-order Butterworth response with a critical frequency of approximately 10 kHz.

Solution   Start by selecting a value for R_A and R_B (R_1 or R_2 can also be the same value as R_A and R_B for simplicity).

$$R = R_A = R_B = R_2 = \textbf{3.3 k}\boldsymbol{\Omega} \qquad \text{(an arbitrary selection)}$$

Next, calculate the capacitance value from $f_c = 1/2\pi RC$.

$$C = C_A = C_B = \frac{1}{2\pi R f_c} = \frac{1}{2\pi(3.3 \text{ k}\Omega)(10 \text{ kHz})} = \textbf{0.0048 } \boldsymbol{\mu}\textbf{F}$$

For a Butterworth response, the damping factor must be 1.414 and $R_1/R_2 = 0.586$.

$$R_1 = 0.586 R_2 = 0.586(3.3 \text{ k}\Omega) = \textbf{1.93 k}\boldsymbol{\Omega}$$

If you had chosen $R_1 = 3.3$ kΩ, then

$$R_2 = \frac{R_1}{0.586} = \frac{3.3 \text{ k}\Omega}{0.586} = 5.63 \text{ k}\Omega$$

Either way, an approximate Butterworth response is realized by choosing the nearest standard values.

Related Problem   Select values for all the components in the high-pass filter of Figure 15–15 to obtain an $f_c = 300$ Hz. Use equal-value components with $R = 10$ kΩ and optimize for a Butterworth response.

Cascading High-Pass Filters

As with the low-pass configuration, first- and second-order high-pass filters can be cascaded to provide three or more poles and thereby create faster roll-off rates. Figure 15–16 shows a six-pole high-pass filter consisting of three two-pole stages. With this configuration optimized for a Butterworth response, a roll-off of -120 dB/decade is achieved.

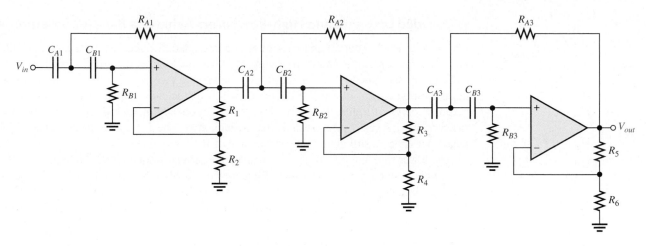

▲ FIGURE 15–16

▲ FIGURE 15–16

Sixth-order high–pass filter.

SECTION 15–4 REVIEW

1. How does a high-pass Sallen-Key filter differ from the low-pass configuration?

2. To increase the critical frequency of a high-pass filter, would you increase or decrease the resistor values?

3. If three two-pole high-pass filters and one single-pole high-pass filter are cascaded, what is the resulting roll-off?

15–5 ACTIVE BAND-PASS FILTERS

As mentioned, band-pass filters pass all frequencies bounded by a lower-frequency limit and an upper-frequency limit and reject all others lying outside this specified band. A band-pass response can be thought of as the overlapping of a low-frequency response curve and a high-frequency response curve.

After completing this section, you should be able to

■ **Analyze active band-pass filters**

■ Describe a band-pass filter composed of a low-pass and a high-pass filter

■ Determine the critical frequencies and center frequency of a cascaded band-pass filter

■ Analyze a multiple-feedback band-pass filter to determine center frequency, bandwidth, and gain

■ Analyze a state-variable band-pass filter

Cascaded Low-Pass and High-Pass Filters Achieve a Band-Pass Response

One way to implement a band-pass filter is a cascaded arrangement of a high-pass filter and a low-pass filter, as shown in Figure 15–17(a), as long as the critical frequencies are sufficiently separated. Each of the filters shown is a two-pole Sallen-Key Butterworth configuration so that the roll-off rates are −40 dB/decade, indicated in the composite response curve of Figure 15–17(b). The critical frequency of each filter is chosen so that the response curves overlap sufficiently, as indicated. The critical frequency of the high-pass filter must be sufficiently lower than that of the low-pass stage.

The lower frequency f_{c1} of the passband is the critical frequency of the high-pass filter. The upper frequency f_{c2} is the critical frequency of the low-pass filter. Ideally, as discussed earlier, the center frequency f_0 of the passband is the geometric mean of f_{c1} and f_{c2}. The following formulas express the three frequencies of the band-pass filter in Figure 15–17.

$$f_{c1} = \frac{1}{2\pi\sqrt{R_{A1}R_{B1}C_{A1}C_{B1}}}$$

$$f_{c2} = \frac{1}{2\pi\sqrt{R_{A2}R_{B2}C_{A2}C_{B2}}}$$

$$f_0 = \sqrt{f_{c1}f_{c2}}$$

(a)

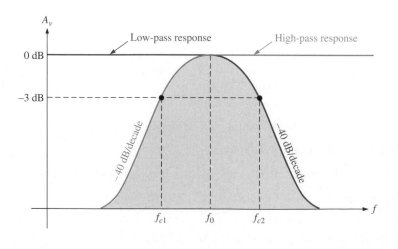

(b)

▲ **FIGURE 15–17**

Band-pass filter formed by cascading a two-pole high-pass and a two-pole low-pass filter (it does not matter in which order the filters are cascaded).

Of course, if equal-value components are used in implementing each filter, the critical frequency equations simplify to the form $f_c = 1/2\pi RC$.

Multiple-Feedback Band-Pass Filter

Another type of filter configuration, shown in Figure 15–18, is a multiple-feedback band-pass filter. The two feedback paths are through R_2 and C_1. Components R_1 and C_1 provide the low-pass response, and R_2 and C_2 provide the high-pass response. The maximum gain, A_0, occurs at the center frequency. Q values of less than 10 are typical in this type of filter. An expression for the center frequency is developed as follows, recognizing that R_1 and R_3 appear in parallel as viewed from the C_1 feedback path (with the V_{in} source replaced by a short).

$$f_0 = \frac{1}{2\pi\sqrt{(R_1\|R_3)R_2C_1C_2}}$$

Making $C_1 = C_2 = C$ yields

$$f_0 = \frac{1}{2\pi\sqrt{(R_1\|R_3)R_2C^2}} = \frac{1}{2\pi C\sqrt{(R_1\|R_3)R_2}}$$

$$= \frac{1}{2\pi C}\sqrt{\frac{1}{R_2(R_1\|R_3)}} = \frac{1}{2\pi C}\sqrt{\left(\frac{1}{R_2}\right)\left(\frac{1}{R_1R_3/(R_1+R_3)}\right)}$$

$$f_0 = \frac{1}{2\pi C}\sqrt{\frac{R_1+R_3}{R_1R_2R_3}}$$

Equation 15–8

▶ **FIGURE 15–18**

Multiple-feedback band-pass filter.

A convenient value for the capacitors is chosen; then the three resistor values are calculated based on the desired values for f_0, BW, and A_0. As you know, the Q can be determined from the relation $Q = f_0/BW$. The resistor values can be found using the following formulas (stated without derivation):

$$R_1 = \frac{Q}{2\pi f_0 C A_0}$$

$$R_2 = \frac{Q}{\pi f_0 C}$$

$$R_3 = \frac{Q}{2\pi f_0 C(2Q^2 - A_0)}$$

To develop a gain expression, solve for Q in the R_1 and R_2 formulas as follows:

$$Q = 2\pi f_0 A_0 C R_1$$
$$Q = \pi f_0 C R_2$$

Then,

$$2\pi f_0 A_0 C R_1 = \pi f_0 C R_2$$

Cancelling yields

$$2A_0 R_1 = R_2$$

Equation 15–9

$$A_0 = \frac{R_2}{2R_1}$$

In order for the denominator of the equation $R_3 = Q/[2\pi f_0 C(2Q^2 - A_0)]$ to be positive, $A_0 < 2Q^2$, which imposes a limitation on the gain.

EXAMPLE 15–6

Determine the center frequency, maximum gain, and bandwidth for the filter in Figure 15–19.

▶ **FIGURE 15–19**

Solution

$$f_0 = \frac{1}{2\pi C}\sqrt{\frac{R_1 + R_3}{R_1 R_2 R_3}} = \frac{1}{2\pi(0.01\ \mu F)}\sqrt{\frac{68\ k\Omega + 2.7\ k\Omega}{(68\ k\Omega)(180\ k\Omega)(2.7\ k\Omega)}} = \textbf{736 Hz}$$

$$A_0 = \frac{R_2}{2R_1} = \frac{180\ k\Omega}{2(68\ k\Omega)} = \textbf{1.32}$$

$$Q = \pi f_0 C R_2 = \pi(736\ Hz)(0.01\ \mu F)(180\ k\Omega) = 4.16$$

$$BW = \frac{f_0}{Q} = \frac{736\ Hz}{4.16} = \textbf{177 Hz}$$

Related Problem If R_2 in Figure 15–19 is increased to 330 kΩ, determine the gain, center frequency, and bandwidth of the filter?

> Open file E15-06 in the Examples folder on your CD-ROM. Measure the center frequency and the bandwidth and compare to the calculated values.

State-Variable Band-Pass Filter

The state-variable or universal active filter is widely used for band-pass applications. As shown in Figure 15–20, it consists of a summing amplifier and two op-amp integrators (which act as single-pole low-pass filters) that are combined in a cascaded arrangement to form a second-order filter. Although used primarily as a band-pass (BP) filter, the state-variable configuration also provides low-pass (LP) and high-pass (HP) outputs. The center frequency is set by the RC circuits in both integrators. When used as a band-pass filter, the critical frequencies of the integrators are usually made equal, thus setting the center frequency of the passband.

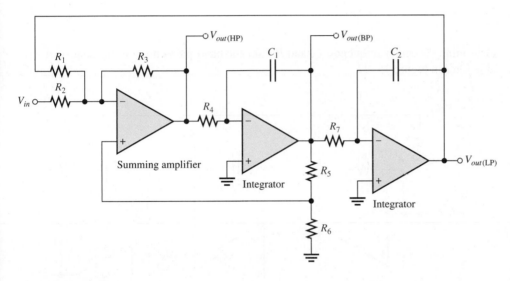

▲ **FIGURE 15–20**

State–variable band–pass filter.

Basic Operation At input frequencies below f_c, the input signal passes through the summing amplifier and integrators and is fed back 180° out of phase. Thus, the feedback signal and input signal cancel for all frequencies below approximately f_c. As the low-pass response of the integrators rolls off, the feedback signal diminishes, thus allowing the input to pass through to the band-pass output. Above f_c, the low-pass response disappears, thus preventing the input signal from passing through the integrators. As a result, the band-pass output peaks sharply at f_c, as indicated in Figure 15–21. Stable Qs up to 100 can be obtained with this type of filter. The Q is set by the feedback resistors R_5 and R_6 according to the following equation:

$$Q = \frac{1}{3}\left(\frac{R_5}{R_6} + 1\right)$$

The state-variable filter cannot be optimized for low-pass, high-pass, and band-pass performance simultaneously for this reason: To optimize for a low-pass or a high-pass Butterworth response, DF must equal 1.414. Since $Q = 1/DF$, a Q of 0.707 will result. Such a low Q provides a very poor band-pass response (large BW and poor selectivity). For optimization as a band-pass filter, the Q must be set high.

▶ **FIGURE 15–21**

General state-variable response curves. BP is for band-pass.

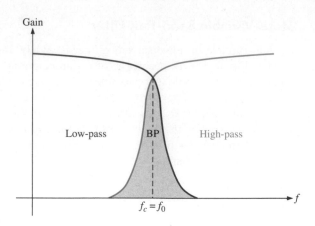

EXAMPLE 15–7

Determine the center frequency, Q, and BW for the band-pass output of the state-variable filter in Figure 15–22.

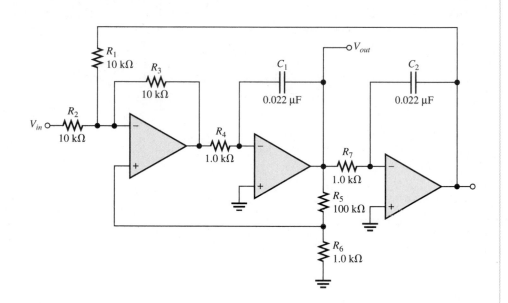

▲ **FIGURE 15–22**

Solution For each integrator,

$$f_c = \frac{1}{2\pi R_4 C_1} = \frac{1}{2\pi R_7 C_2} = \frac{1}{2\pi(1.0 \text{ k}\Omega)(0.022 \ \mu\text{F})} = 7.23 \text{ kHz}$$

The center frequency is approximately equal to the critical frequencies of the integrators.

$$f_0 = f_c = \textbf{7.23 kHz}$$

$$Q = \frac{1}{3}\left(\frac{R_5}{R_6} + 1\right) = \frac{1}{3}\left(\frac{100 \text{ k}\Omega}{1.0 \text{ k}\Omega} + 1\right) = \textbf{33.7}$$

$$BW = \frac{f_0}{Q} = \frac{7.23 \text{ kHz}}{33.7} = \textbf{215 Hz}$$

Related Problem Determine f_0, Q, and BW for the filter in Figure 15–22 if $R_4 = R_6 = R_7 = 330 \ \Omega$ with all other component values the same as shown on the schematic.

Open file E15-07 in the Examples folder on your CD-ROM. Measure the center frequency and the bandwidth and compare to the calculated values.

**SECTION 15–5
REVIEW**

1. What determines selectivity in a band-pass filter?
2. One filter has a $Q = 5$ and another has a $Q = 25$. Which has the narrower band-width?
3. List the elements that make up a state-variable filter.

15–6 ACTIVE BAND-STOP FILTERS

Band-stop filters reject a specified band of frequencies and pass all others. The response is opposite to that of a band-pass filter.

After completing this section, you should be able to

■ **Analyze active band-stop filters**

■ Identify a multiple-feedback band-stop filter

■ Analyze a state-variable band-stop filter

Multiple-Feedback Band-Stop Filter

Figure 15–23 shows a multiple-feedback band-stop filter. Notice that this configuration is similar to the band-pass version in Figure 15–18 except that R_3 has been moved and R_4 has been added.

▶ **FIGURE 15–23**

Multiple-feedback band-stop filter.

State-Variable Band-Stop Filter

Summing the low-pass and the high-pass responses of the state-variable filter covered in Section 15–5 with a summing amplifier creates a band-stop filter, as shown in Figure 15–24. One important application of this filter is minimizing the 60 Hz "hum" in audio systems by setting the center frequency to 60 Hz.

▲ FIGURE 15–24

State-variable band-stop filter.

EXAMPLE 15–8

Verify that the band-stop filter in Figure 15–25 has a center frequency of 60 Hz, and optimize the filter for a Q of 10.

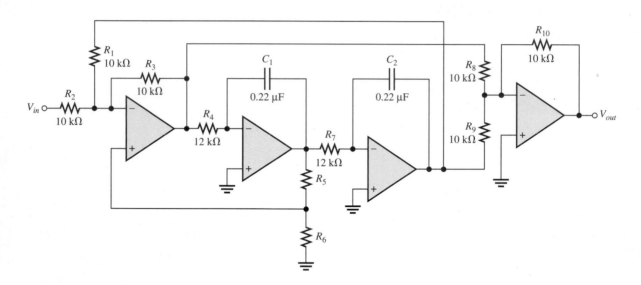

▲ FIGURE 15–25

Solution f_0 equals the f_c of the integrator stages. (In practice, component values are critical.)

$$f_0 = \frac{1}{2\pi R_4 C_1} = \frac{1}{2\pi R_7 C_2} = \frac{1}{2\pi(12 \text{ k}\Omega)(0.22 \text{ }\mu\text{F})} = \mathbf{60\ Hz}$$

You can obtain a $Q = 10$ by choosing R_6 and then calculating R_5.

$$Q = \frac{1}{3}\left(\frac{R_5}{R_6} + 1\right)$$
$$R_5 = (3Q - 1)R_6$$

Choose $R_6 = \mathbf{3.3\ k\Omega}$. Then

$$R_5 = [3(10) - 1]3.3 \text{ k}\Omega = \mathbf{95.7\ k\Omega}$$

Use the nearest standard value of 100 kΩ.

Related Problem   How would you change the center frequency to 120 Hz in Figure 15–25?

Open file E15-08 in the Examples folder on your CD-ROM and verify that the center frequency is approximately 60 Hz.

**SECTION 15–6
REVIEW**

1. How does a band-stop response differ from a band-pass response?
2. How is a state-variable band-pass filter converted to a band-stop filter?

15–7 FILTER RESPONSE MEASUREMENTS

In this section, we discuss two methods of determining a filter's response by measurement—discrete point measurement and swept frequency measurement.

After completing this section, you should be able to

- **Discuss two methods for measuring frequency response**
- Explain the discrete point measurement method
- Explain the swept frequency measurement method

Discrete Point Measurement

Figure 15–26 shows an arrangement for taking filter output voltage measurements at discrete values of input frequency using common laboratory instruments. The general procedure is as follows:

1. Set the amplitude of the sine wave generator to a desired voltage level.

2. Set the frequency of the sine wave generator to a value well below the expected critical frequency of the filter under test. For a low-pass filter, set the frequency as near as possible to 0 Hz. For a band-pass filter, set the frequency well below the expected lower critical frequency.

3. Increase the frequency in predetermined steps sufficient to allow enough data points for an accurate response curve.

4. Maintain a constant input voltage amplitude while varying the frequency.

5. Record the output voltage at each value of frequency.

6. After recording a sufficient number of points, plot a graph of output voltage versus frequency.

If the frequencies to be measured exceed the response of the DMM, an oscilloscope may have to be used instead.

▶ **FIGURE 15–26**

Test setup for discrete point measurement of the filter response. (Readings are arbitrary and for display only.)

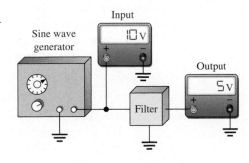

Swept Frequency Measurement

The swept frequency method requires more elaborate test equipment than does the discrete point·method, but it is much more efficient and can result in a more accurate response curve. A general test setup is shown in Figure 15–27(a) using a swept frequency generator and a spectrum analyzer. Figure 15–27(b) shows how the test can be made with an oscilloscope.

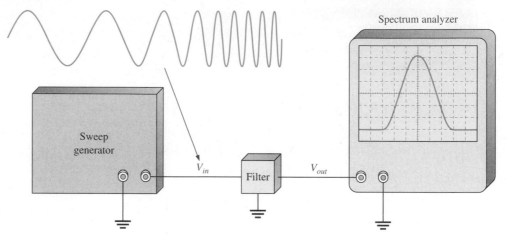

(a) Test setup for a filter response using a spectrum analyzer

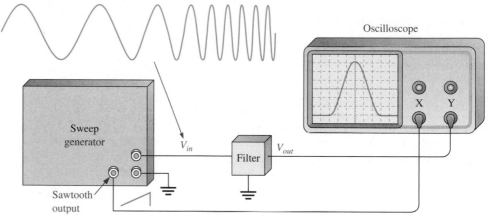

(b) Test setup for a filter response using an oscilloscope. The scope is put in X-Y mode. The sawtooth waveform from the sweep generator drives the X-channel of the oscilloscope.

▲ **FIGURE 15–27**

Test setup for swept frequency measurement of the filter response.

The swept frequency generator produces a constant amplitude output signal whose frequency increases linearly between two preset limits, as indicated in Figure 15–27. The spectrum analyzer is essentially an elaborate oscilloscope that can be calibrated for a desired *frequency span/division* rather than for the usual *time/division* setting. Therefore, as the input frequency to the filter sweeps through a preselected range, the response curve is traced out on the screen of the spectrum analyzer or an oscilloscope.

**SECTION 15–7
REVIEW**

1. What is the purpose of the two tests discussed in this section?

2. Name one disadvantage and one advantage of each test method.

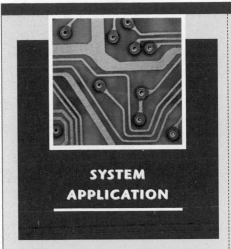

SYSTEM APPLICATION

In this system application, the focus is on the filter board which is part of the channel separation circuit in an FM stereo receiver. In addition to the active filters, the left and right channel separation circuit includes a demodulator, a frequency doubler, and a stereo matrix. Except for a brief statement of their purpose, these circuits will not be emphasized. The matrix, however, is an interesting application of summing amplifiers, with which you are familiar, and will be shown in detail on the schematic.

Basic Operation of the System

Stereo FM **(frequency modulation)** signals are transmitted on a **carrier** frequency of 88 MHz to 108 MHz. The standard transmitted stereo signal consists of three modulating signals. These are the sum of the left and right channel audio $(L + R)$, the difference of the left and right channel audio $(L - R)$, and a 19 kHz pilot subcarrier.

The $L + R$ audio extends from 30 Hz to 15 kHz and the $L - R$ audio is contained in two sidebands extending from 23 kHz to 53 kHz, as indicated in Figure 15–28. These frequencies come from the FM detector and go into the filter circuits where they are separated.

The frequency doubler and demodulator are used to extract the audio signal from the 23 kHz to 53 kHz sidebands after which the 30 Hz to 15 kHz $L + R$ signal is passed through a filter.

The $L + R$ and $L - R$ audio signals are then sent to the matrix where they are applied to the summing circuits to produce the left and right channel audio $(-2L$ and $-2R)$.

The Channel Filter Circuit Board

■ Make sure that the circuit board shown in Figure 15–29 is correct by checking it against the schematic in Figure 15–30.

■ Label a copy of the board with component and input/output designations in agreement with the schematic.

The Filter Circuit

■ Calculate the critical frequencies of each of the Sallen-Key filters.

■ Calculate the center frequency of the multiple-feedback filter.

■ Calculate the bandwidth of each filter.

■ Determine the voltage gain of each filter.

■ Verify that the Sallen-Key filters have an approximate Butterworth response characteristic.

Test Procedure

Develop a basic procedure for independently testing the filter board using only the inputs and outputs available on the

▲ **FIGURE 15–28**

Basic block diagram of an FM receiver system.

terminal strip. Use a sweep generator, a spectrum analyzer, and a dual-polarity power supply.

For the sweep generator, a minimum and a maximum frequency are selected, and the instrument produces an output signal that repetitively sweeps through all frequencies between the preset limits. The spectrum analyzer is essentially a type of oscilloscope that displays a frequency response curve.

Final Report (Optional)

Submit a final written report on the channel filter board using an organized format that includes the following:

1. A physical description of the circuit.

2. A discussion of the operation of the circuit.

3. A list of the specifications.

4. A list of parts with part numbers if available.

(a) Component side of board

(b) Backside of board

▲ **FIGURE 15–29**

Channel filter board.

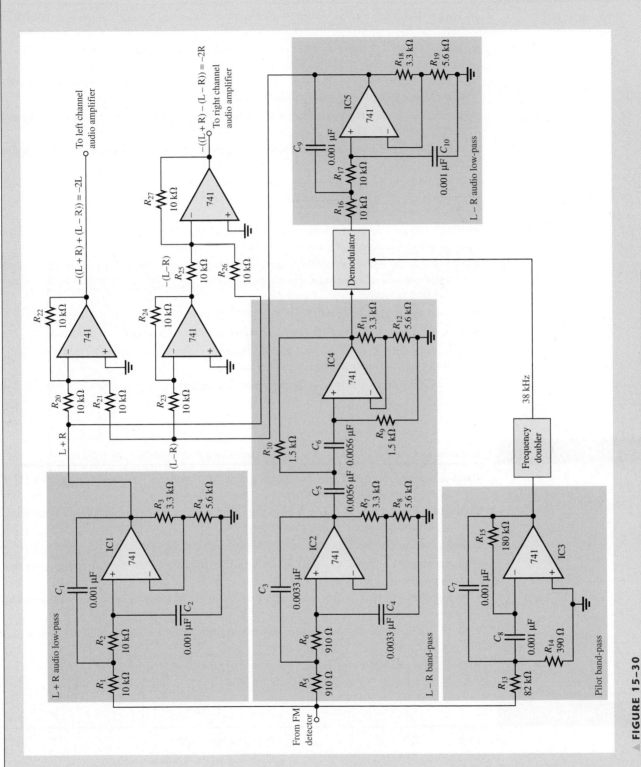

▲ FIGURE 15–30

Channel separation circuits. The circuits on the filter board are shown in the green blocks.

CHAPTER SUMMARY

- The bandwidth in a low-pass filter equals the critical frequency because the response extends to 0 Hz.
- The bandwidth in a high-pass filter extends above the critical frequency and is limited only by the inherent frequency limitation of the active circuit.
- A band-pass filter passes all frequencies within a band between a lower and an upper critical frequency and rejects all others outside this band.
- The bandwidth of a band-pass filter is the difference between the upper critical frequency and the lower critical frequency.
- A band-stop filter rejects all frequencies within a specified band and passes all those outside this band.
- Filters with the Butterworth response characteristic have a very flat response in the passband, exhibit a roll-off of −20 dB/decade/pole, and are used when all the frequencies in the passband must have the same gain.
- Filters with the Chebyshev characteristic have ripples or overshoot in the passband and exhibit a faster roll-off per pole than filters with the Butterworth characteristic.
- Filters with the Bessel characteristic are used for filtering pulse waveforms. Their linear phase characteristic results in minimal waveshape distortion. The roll-off rate per pole is slower than for the Butterworth.
- In filter terminology, a single RC circuit is called a *pole*.
- Each pole in a Butterworth filter causes the output to roll off at a rate of −20 dB/decade.
- The quality factor Q of a band-pass filter determines the filter's selectivity. The higher the Q, the narrower the bandwidth and the better the selectivity.
- The damping factor determines the filter response characteristic (Butterworth, Chebyshev, or Bessel).

KEY TERMS

Key terms and other bold terms in the chapter are defined in the end-of-book glossary.

Band-pass filter   A type of filter that passes a range of frequencies lying between a certain lower frequency and a certain higher frequency.

Band-stop filter   A type of filter that blocks or rejects a range of frequencies lying between a certain lower frequency and a certain higher frequency.

Damping factor   A filter characteristic that determines the type of response.

High-pass filter   A type of filter that passes frequencies above a certain frequency while rejecting lower frequencies.

Low-pass filter   A type of filter that passes frequencies below a certain frequency while rejecting higher frequencies.

Pole   A circuit containing one resistor and one capacitor that contributes −20 dB/decade to a filter's roll-off rate.

Roll-off   The rate of decrease in gain, below or above the critical frequencies of a filter.

KEY FORMULAS

15–1	$BW = f_c$	Low-pass bandwidth
15–2	$BW = f_{c2} - f_{c1}$	Filter bandwidth of a band-pass filter
15–3	$f_0 = \sqrt{f_{c1}f_{c2}}$	Center frequency of a band-pass filter
15–4	$Q = \dfrac{f_0}{BW}$	Quality factor of a band-pass filter
15–5	$DF = 2 - \dfrac{R_1}{R_2}$	Damping factor

15–6 $\quad A_{cl(NI)} = \dfrac{R_1}{R_2} + 1$ $\qquad$ Closed-loop voltage gain

15–7 $\quad f_c = \dfrac{1}{2\pi\sqrt{R_A R_B C_A C_B}}$ $\qquad$ Critical frequency for a second-order Sallen-Key filter

15–8 $\quad f_0 = \dfrac{1}{2\pi C}\sqrt{\dfrac{R_1 + R_3}{R_1 R_2 R_3}}$ $\qquad$ Center frequency of a multiple-feedback filter

15–9 $\quad A_0 = \dfrac{R_2}{2R_1}$ $\qquad$ Gain of a multiple-feedback filter

SELF-TEST

Answers are at the end of the chapter.

1. The term *pole* in filter terminology refers to
   (a) a high-gain op-amp $\qquad$ (b) one complete active filter
   (c) a single *RC* circuit $\qquad$ (d) the feedback circuit

2. A single resistor and a single capacitor can be connected to form a filter with a roll-off rate of
   (a) −20 dB/decade $\qquad$ (b) −40 dB/decade
   (c) −6 dB/octave $\qquad$ (d) answers (a) and (c)

3. A band-pass response has
   (a) two critical frequencies $\qquad$ (b) one critical frequency
   (c) a flat curve in the passband $\qquad$ (d) a wide bandwidth

4. The lowest frequency passed by a low-pass filter is
   (a) 1 Hz $\qquad$ (b) 0 Hz
   (c) 10 Hz $\qquad$ (d) dependent on the critical frequency

5. The quality factor (Q) of a band-pass filter depends on
   (a) the critical frequencies
   (b) only the bandwidth
   (c) the center frequency and the bandwidth
   (d) only the center frequency

6. The damping factor of an active filter determines
   (a) the voltage gain $\qquad$ (b) the critical frequency
   (c) the response characteristic $\qquad$ (d) the roll-off rate

7. A maximally flat frequency response is known as
   (a) Chebyshev $\qquad$ (b) Butterworth
   (c) Bessel $\qquad$ (d) Colpitts

8. The damping factor of a filter is set by
   (a) the negative feedback circuit $\qquad$ (b) the positive feedback circuit
   (c) the frequency-selective circuit $\qquad$ (d) the gain of the op-amp

9. The number of poles in a filter affect the
   (a) voltage gain $\qquad$ (b) bandwidth
   (c) center frequency $\qquad$ (d) roll-off rate

10. Sallen-Key filters are
    (a) single-pole filters $\qquad$ (b) second-order filters
    (c) Butterworth filters $\qquad$ (d) band-pass filters

11. When filters are cascaded, the roll-off rate
    (a) increases $\qquad$ (b) decreases $\qquad$ (c) does not change

12. When a low-pass and a high-pass filter are cascaded to get a band-pass filter, the critical frequency of the low-pass filter must be

 (a) equal to the critical frequency of the high-pass filter

 (b) less than the critical frequency of the high-pass filter

 (c) greater than the critical frequency of the high-pass filter

13. A state-variable filter consists of

 (a) one op-amp with multiple-feedback paths

 (b) a summing amplifier and two integrators

 (c) a summing amplifier and two differentiators

 (d) three Butterworth stages

14. When the gain of a filter is minimum at its center frequency, it is

 (a) a band-pass filter   **(b)** a band-stop filter

 (c) a notch filter   **(d)** answers (b) and (c)

PROBLEMS

Answers to all odd-numbered problems are at the end of the book.

BASIC PROBLEMS

SECTION 15–1   Basic Filter Responses

 1. Identify each type of filter response (low-pass, high-pass, band-pass, or band-stop) in Figure 15–31.

(a)

(b)

(c)

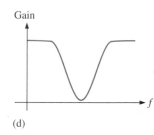
(d)

▲ **FIGURE 15–31**

 2. A certain low-pass filter has a critical frequency of 800 Hz. What is its bandwidth?

 3. A single-pole high-pass filter has a frequency-selective circuit with $R = 2.2$ kΩ and $C = 0.0015$ μF. What is the critical frequency? Can you determine the bandwidth from the available information?

 4. What is the roll-off rate of the filter described in Problem 3?

 5. What is the bandwidth of a band-pass filter whose critical frequencies are 3.2 kHz and 3.9 kHz? What is the Q of this filter?

 6. What is the center frequency of a filter with a Q of 15 and a bandwidth of 1 kHz?

SECTION 15–2   Filter Response Characteristics

 7. What is the damping factor in each active filter shown in Figure 15–32? Which filters are approximately optimized for a Butterworth response characteristic?

 8. For the filters in Figure 15–32 that do not have a Butterworth response, specify the changes necessary to convert them to Butterworth responses. (Use nearest standard values.)

 9. Response curves for second-order filters are shown in Figure 15–33. Identify each as Butterworth, Chebyshev, or Bessel.

(a)

(b)

(c)

▲ **FIGURE 15–32**

File circuits are identified with a CD logo and are in the Problems folder on your CD-ROM. Filenames correspond to figure numbers (e.g., F15-32).

(a)

(b)

(c)

(d)

▲ **FIGURE 15–33**

SECTION 15–3 Active Low-Pass Filters

10. Is the four-pole filter in Figure 15–34 approximately optimized for a Butterworth response? What is the roll-off rate?

11. Determine the critical frequency in Figure 15–34.

▲ FIGURE 15–34

12. Without changing the response curve, adjust the component values in the filter of Figure 15–34 to make it an equal-value filter. Select $C = 0.22\ \mu F$ for both stages.

13. Modify the filter in Figure 15–34 to increase the roll-off rate to −120 dB/decade while maintaining an approximate Butterworth response.

14. Using a block diagram format, show how to implement the following roll-off rates using single-pole and two-pole low-pass filters with Butterworth responses.

    (a) −40 dB/decade   **(b)** −20 dB/decade

    (c) −60 dB/decade   **(d)** −100 dB/decade

    (e) −120 dB/decade

SECTION 15–4 Active High-Pass Filters

15. Convert the filter in Problem 12 to a high-pass with the same critical frequency and response characteristic.

16. Make the necessary circuit modification to reduce by half the critical frequency in Problem 15.

17. For the filter in Figure 15–35, **(a)** how would you increase the critical frequency? **(b)** How would you increase the gain?

▶ FIGURE 15–35

SECTION 15–5 Active Band-Pass Filters

18. Identify each band-pass filter configuration in Figure 15–36.

19. Determine the center frequency and bandwidth for each filter in Figure 15–36.

(a)

(b)

(c)

▲ **FIGURE 15–36**

20. Optimize the state-variable filter in Figure 15–37 for $Q = 50$. What bandwidth is achieved?

▲ **FIGURE 15–37**

SECTION 15–6 Active Band-Stop Filters

21. Show how to make a notch (band-stop) filter using the basic circuit in Figure 15–37.

22. Modify the band-stop filter in Problem 21 for a center frequency of 120 Hz.

EWB/MULTISIM TROUBLESHOOTING PROBLEMS

These file circuits are in the Troubleshooting Problems folder on your CD-ROM.

23. Open file TSP15-23 and determine the fault.

24. Open file TSP15-24 and determine the fault.

25. Open file TSP15-25 and determine the fault.

26. Open file TSP15-26 and determine the fault.

27. Open file TSP15-27 and determine the fault.

28. Open file TSP15-28 and determine the fault.

29. Open file TSP15-29 and determine the fault.

30. Open file TSP15-30 and determine the fault.

31. Open file TSP15-31 and determine the fault.

ANSWERS

SECTION REVIEWS

SECTION 15–1 Basic Filter Responses

1. The critical frequency determines the bandwidth.

2. The inherent frequency limitation of the op-amp limits the bandwidth.

3. Q and BW are inversely related. The higher the Q, the better the selectivity, and vice versa.

SECTION 15–2 Filter Response Characteristics

1. Butterworth is very flat in the passband and has a −20 dB/decade/pole roll-off.
Chebyshev has ripples in the passband and has greater than −20 dB/decade/pole roll-off.
Bessel has a linear phase characteristic and less than −20 dB/decade/pole roll-off.

2. The damping factor

3. Frequency-selection circuit, gain element, and negative feedback circuit are the parts of an active filter.

SECTION 15–3 **Active Low-Pass Filters**

1. A second-order filter has two poles. Two resistors and two capacitors make up the frequency-selective circuit.

2. The damping factor sets the response characteristic.

3. Cascading increases the roll-off rate.

SECTION 15–4 **Active High-Pass Filters**

1. The positions of the Rs and Cs in the frequency-selection circuit are opposite for low-pass and high-pass configurations.

2. Decrease the R values to increase f_c.

3. −140 dB/decade

SECTION 15–5 **Active Band-Pass Filters**

1. Q determines selectivity.

2. $Q = 25$. Higher Q gives narrower BW.

3. A summing amplifier and two integrators make up a state-variable filter.

SECTION 15–6 **Active Band-Stop Filters**

1. A band-stop rejects frequencies within the stopband. A band-pass passes frequencies within the passband.

2. The low-pass and high-pass outputs are summed.

SECTION 15–7 **Filter Response Measurements**

1. To check the frequency response of a filter

2. Discrete point measurement: tedious and less complete; simpler equipment.

Swept frequency measurement: uses more expensive equipment; more efficient, can be more accurate and complete.

RELATED PROBLEMS FOR EXAMPLES

15–1 500 Hz

15–2 1.44

15–3 7.23 kHz; 1.29 kΩ

15–4 $C_{A1} = C_{A2} = C_{B1} = C_{B2} = 0.234$ μF; $R_2 = R_4 = 680$ Ω; $R_1 = 103$ Ω; $R_3 = 840$ Ω

15–5 $R_A = R_B = R_2 = 10$ kΩ; $C_A = C_B = 0.053$ μF; $R_1 = 5.86$ kΩ

15–6 Gain increases to 2.43, frequency decreases to 544 Hz, and bandwidth decreases to 96.5 Hz.

15–7 $f_0 = 21.9$ kHz; $Q = 101$; $BW = 217$ Hz

15–8 Decrease the input resistors or the feedback capacitors of the two integrator stages by half.

SELF-TEST

1. (c) **2.** (d) **3.** (a) **4.** (b) **5.** (c) **6.** (c) **7.** (b) **8.** (a)

9. (d) **10.** (b) **11.** (a) **12.** (c) **13.** (b) **14.** (d)

16

OSCILLATORS

INTRODUCTION

Oscillators are circuits that generate an output signal without the necessity of an input signal. They are used as signal sources in all sorts of applications. Different types of oscillators produce various types of outputs including sine waves, square waves, triangular waves, and sawtooth waves. In this chapter, several types of basic oscillator circuits using both discrete transistors and op-amps as the gain element are introduced. Also, a very popular integrated circuit, called the 555 timer, is discussed in relation to its oscillator applications.

Sinusoidal oscillator operation is based on the principle of positive feedback, where a portion of the output signal is fed back to the input in a way that causes it to reinforce itself and thus sustain a continuous output signal. Oscillators are widely used in most communications systems as well as in digital systems, including computers, to generate required frequencies and timing signals. Also, oscillators are found in many types of test instruments like those used in the laboratory.

CHAPTER OUTLINE

- Describe the basic operating principles of an oscillator

- Discuss the main principle on which feedback oscillators is based

- Describe and analyze the basic operation of three *RC* feedback oscillators

- Describe and analyze the basic operation of *LC* feedback oscillators

- Describe and analyze the basic operation of relaxation oscillators

- Use a 555 timer in an oscillator application

KEY TERMS

- Feedback oscillator

- Relaxation oscillator

- Positive feedback

- Voltage-controlled oscillator (VCO)

- Astable

■ ■ ■ SYSTEM APPLICATION PREVIEW

The system application at the end of the chapter is a function generator that uses a sinusoidal oscillator as its signal source. Other circuits in the system are a zero-level detector and an integrator, with which you are already familiar. The function generator is a laboratory test instrument which is, effectively, a multiple-signal source because it can produce not only sinusoidal output voltages but also square waves and triangular waveforms.

WWW. VISIT THE COMPANION WEBSITE

Study aids for this chapter are available at
http://www.prenhall.com/floyd

16–1 THE OSCILLATOR

An oscillator is a circuit that produces a periodic waveform on its output with only the dc supply voltage as an input. A repetitive input signal is not required except to synchronize oscillations in some applications. The output voltage can be either sinusoidal or nonsinusoidal, depending on the type of oscillator. Two major classifications for oscillators are feedback oscillators and relaxation oscillators.

After completing this section, you should be able to

- **Describe the basic operating principles of an oscillator**
- Explain the purpose of an oscillator
- List the basic elements of an oscillator
- Discuss two important oscillator classifications

Essentially, an **oscillator** converts electrical energy from the dc power supply to periodic waveforms. A basic oscillator is shown in Figure 16–1.

▲ **FIGURE 16–1**

The basic oscillator concept showing three common types of output waveforms: sine wave, square wave, and sawtooth.

  Feedback Oscillators   One type of oscillator is the **feedback oscillator**, which returns a fraction of the output signal to the input with no net phase shift, resulting in a reinforcement of the output signal. After oscillations are started, the loop gain is maintained at 1.0 to maintain oscillations. A feedback oscillator consists of an amplifier for gain (either a discrete transistor or an op-amp) and a positive feedback circuit that produces phase shift and provides attenuation, as shown in Figure 16–2.

▶ **FIGURE 16–2**

Basic elements of a feedback oscillator.

Relaxation Oscillators   A second type of oscillator is the **relaxation oscillator**. A relaxation oscillator uses an *RC* timing circuit to generate a waveform that is generally a square wave or other nonsinusoidal waveform. Typically, a relaxation oscillator uses a Schmitt trigger or other device that changes states to alternately charge and discharge a capacitor through a resistor. Relaxation oscillators are discussed in Section 16–5.

SECTION 16–1 **REVIEW** Answers are at the end of the chapter.	1. What is an oscillator? 2. What type of feedback does a feedback oscillator require? 3. What is the purpose of the feedback circuit? 4. Name the two types of oscillators.

16–2   FEEDBACK OSCILLATOR PRINCIPLES

Feedback oscillator operation is based on the principle of positive feedback. In this section, we will examine this concept and look at the general conditions required for oscillation to occur. Feedback oscillators are widely used to generate sinusoidal waveforms.

After completing this section, you should be able to

- **Discuss the main principle on which feedback oscillators is based**

- Explain positive feedback

- Describe the conditions for oscillation

- Discuss the start-up conditions

Positive Feedback

Positive feedback is characterized by the condition wherein an in-phase portion of the output voltage of an amplifier is fed back to the input with no net phase shift, resulting in a reinforcement of the output signal. This basic idea is illustrated in Figure 16–3. As you can see, the in-phase feedback voltage, V_f, is amplified to produce the output voltage, which in turn produces the feedback voltage. That is, a loop is created in which the signal

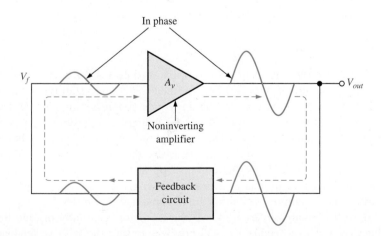

▲ **FIGURE 16–3**

Positive feedback produces oscillation.

sustains itself and a continuous sinusoidal output is produced. This phenomenon is called *oscillation.*

Conditions for Oscillation

Two conditions, illustrated in Figure 16–4, are required for a sustained state of oscillation:

1. The phase shift around the feedback loop must be effectively 0°.

2. The voltage gain, A_{cl}, around the closed feedback loop (loop gain) must equal 1 (unity).

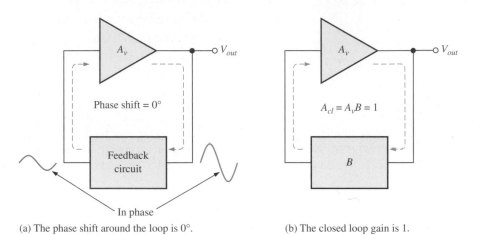

(a) The phase shift around the loop is 0°.   (b) The closed loop gain is 1.

▲ FIGURE 16–4

General conditions to sustain oscillation.

The voltage gain around the closed feedback loop, A_{cl}, is the product of the amplifier gain, A_v, and the attenuation, B, of the feedback circuit.

$$A_{cl} = A_v B$$

If a sinusoidal wave is the desired output, a loop gain greater than 1 will rapidly cause the output to saturate at both peaks of the waveform, producing unacceptable distortion. To avoid this, some form of gain control must be used to keep the loop gain at exactly 1, once oscillations have started. For example, if the attenuation of the feedback circuit is 0.01, the amplifier must have a gain of exactly 100 to overcome this attenuation and not create unacceptable distortion (0.01 × 100 = 1.0). An amplifier gain of greater than 100 will cause the oscillator to limit both peaks of the waveform.

Start-Up Conditions

So far, you have seen what it takes for an oscillator to produce a continuous sinusoidal output. Now let's examine the requirements for the oscillation to start when the dc supply voltage is first turned on. As you know, the unity-gain condition must be met for oscillation to be sustained. For oscillation to *begin,* the voltage gain around the positive feedback loop must be greater than 1 so that the amplitude of the output can build up to a desired level. The gain must then decrease to 1 so that the output stays at the desired level and oscillation is sustained. Ways that certain amplifiers achieve this reduction in gain after start-up are discussed in later sections of this chapter. The voltage gain conditions for both starting and sustaining oscillation are illustrated in Figure 16–5.

A question that normally arises is this: If the oscillator is initially off and there is no output voltage, how does a feedback signal originate to start the positive feedback build-up process? Initially, a small positive feedback voltage develops from thermally produced broad-band noise in the resistors or other components or from power supply turn-on tran-

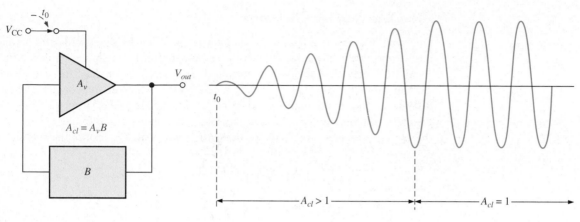

▲ FIGURE 16–5

When oscillation starts at t_0, the condition $A_{cl} > 1$ causes the sinusoidal output voltage amplitude to build up to a desired level. Then A_{cl} decreases to 1 and maintains the desired amplitude.

sients. The feedback circuit permits only a voltage with a frequency equal to the selected oscillation frequency to appear in phase on the amplifier's input. This initial feedback voltage is amplified and continually reinforced, resulting in a buildup of the output voltage as previously discussed.

SECTION 16–2 **REVIEW**	**1.** What are the conditions required for a circuit to oscillate? **2.** Define *positive feedback*. **3.** What is the voltage gain condition for oscillator start-up?

16–3 OSCILLATORS WITH *RC* FEEDBACK CIRCUITS

In this section, you will learn about three types of feedback oscillators that use *RC* circuits to produce sinusoidal outputs: the Wien-bridge oscillator, the phase-shift oscillator, and the twin-T oscillator. Generally, *RC* feedback oscillators are used for frequencies up to about 1 MHz. The Wien-bridge is by far the most widely used type of *RC* feedback oscillator for this range of frequencies.

After completing this section, you should be able to

- **Describe and analyze the basic operation of three *RC* feedback oscillators**
- Identify a Wien-bridge oscillator
- Calculate the resonant frequency of a Wien-bridge oscillator
- Analyze oscillator feedback conditions
- Analyze oscillator start-up conditions
- Describe a self-starting Wien-bridge oscillator
- Identify a phase-shift oscillator
- Calculate the resonant frequency and analyze the feedback conditions for a phase-shift oscillator
- Identify a twin-T oscillator and describe its operation

The Wien-Bridge Oscillator

One type of sinusoidal feedback oscillator is the **Wien-bridge oscillator**. A fundamental part of the Wien-bridge oscillator is a lead-lag circuit like that shown in Figure 16–6(a). R_1 and C_1 together form the lag portion of the circuit; R_2 and C_2 form the lead portion. The operation of this lead-lag circuit is as follows. At lower frequencies, the lead circuit dominates due to the high reactance of C_2. As the frequency increases, X_{C2} decreases, thus allowing the output voltage to increase. At some specified frequency, the response of the lag circuit takes over, and the decreasing value of X_{C1} causes the output voltage to decrease.

(a) Circuit   (b) Response curve

▲ **FIGURE 16–6**

A lead–lag circuit and its response curve.

The response curve for the lead-lag circuit shown in Figure 16–6(b) indicates that the output voltage peaks at a frequency called the resonant frequency, f_r. At this point, the attenuation (V_{out}/V_{in}) of the circuit is 1/3 if $R_1 = R_2$ and $X_{C1} = X_{C2}$ as stated by the following equation (derived in Appendix B):

Equation 16–1

$$\frac{V_{out}}{V_{in}} = \frac{1}{3}$$

The formula for the resonant frequency (also derived in Appendix B) is

Equation 16–2

$$f_r = \frac{1}{2\pi RC}$$

To summarize, the lead-lag circuit in the Wien-bridge oscillator has a resonant frequency, f_r, at which the phase shift through the circuit is 0° and the attenuation is 1/3. Below f_r, the lead circuit dominates and the output leads the input. Above f_r, the lag circuit dominates and the output lags the input.

The Basic Circuit   The lead-lag circuit is used in the positive feedback loop of an op-amp, as shown in Figure 16–7(a). A voltage divider is used in the negative feedback loop. The Wien-bridge oscillator circuit can be viewed as a noninverting amplifier configuration with the input signal fed back from the output through the lead-lag circuit. Recall that the closed-loop gain of the amplifier is determined by the voltage divider.

$$A_{cl} = \frac{1}{B} = \frac{1}{R_2/(R_1 + R_2)} = \frac{R_1 + R_2}{R_2}$$

The circuit is redrawn in Figure 16–7(b) to show that the op-amp is connected across the bridge circuit. One leg of the bridge is the lead-lag circuit, and the other is the voltage divider.

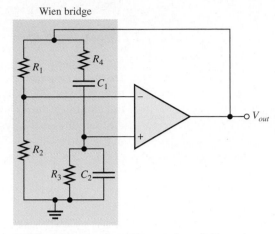

(a)

(b) Wien bridge circuit combines a voltage divider and a lead-lag circuit.

▲ **FIGURE 16–7**

The Wien-bridge oscillator schematic drawn in two different but equivalent ways.

Positive Feedback Conditions for Oscillation As you know, for the circuit to produce a sustained sinusoidal output (oscillate), the phase shift around the positive feedback loop must be 0° and the gain around the loop must equal unity (1). The 0° phase-shift condition is met when the frequency is f_r because the phase shift through the lead-lag circuit is 0° and there is no inversion from the noninverting (+) input of the op-amp to the output. This is shown in Figure 16–8(a).

(a) The phase shift around the loop is 0°.

(b) The voltage gain around the loop is 1.

▲ **FIGURE 16–8**

Conditions for sustained oscillation.

The unity-gain condition in the feedback loop is met when

$$A_{cl} = 3$$

This offsets the 1/3 attenuation of the lead-lag circuit, thus making the total gain around the positive feedback loop equal to 1, as depicted in Figure 16–8(b). To achieve a closed-loop gain of 3,

$$R_1 = 2R_2$$

Then

$$A_{cl} = \frac{R_1 + R_2}{R_2} = \frac{2R_2 + R_2}{R_2} = \frac{3R_2}{R_2} = 3$$

Start-Up Conditions Initially, the closed-loop gain of the amplifier itself must be more than three ($A_{cl} > 3$) until the output signal builds up to a desired level. Ideally, the gain of the amplifier must then decrease to 3 so that the total gain around the loop is 1 and the output signal stays at the desired level, thus sustaining oscillation. This is illustrated in Figure 16–9.

(a) Loop gain greater than 1 causes output to build up.

(b) Loop gain of 1 causes a sustained constant output.

▲ **FIGURE 16–9**

Conditions for start-up and sustained oscillations.

The circuit in Figure 16–10 illustrates a method for achieving sustained oscillations. Notice that the voltage-divider circuit has been modified to include an additional resistor R_3 in parallel with a back-to-back zener diode arrangement. When dc power is first applied, both zener diodes appear as opens. This places R_3 in series with R_1, thus increasing the closed-loop gain of the amplifier as follows ($R_1 = 2R_2$):

$$A_{cl} = \frac{R_1 + R_2 + R_3}{R_2} = \frac{3R_2 + R_3}{R_2} = 3 + \frac{R_3}{R_2}$$

▶ **FIGURE 16–10**

Self-starting Wien-bridge oscillator using back-to-back zener diodes.

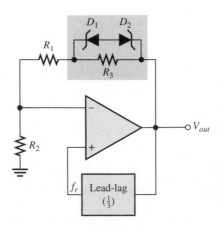

Initially, a small positive feedback signal develops from noise or turn-on transients. The lead-lag circuit permits only a signal with a frequency equal to f_r to appear in phase on the noninverting input. This feedback signal is amplified and continually reinforced, resulting in a buildup of the output voltage. When the output signal reaches the zener breakdown voltage, the zeners conduct and effectively short out R_3. This lowers the amplifier's closed-loop gain to 3. At this point, the total loop gain is 1 and the output signal levels off and the oscillation is sustained.

All practical methods to achieve stability for feedback oscillators require the gain to be self-adjusting. This requirement is a form of automatic gain control (AGC). The zener diodes in Figure 16–10 limit the gain at the onset of nonlinearity, in this case, zener conduction. Although the zener feedback is simple, it suffers from the nonlinearity of the zener diodes that occurs in order to control gain. It is difficult to achieve an undistorted sinusoidal output waveform.

Another method to control the gain uses a JFET as a voltage-controlled resistor in a negative feedback path. This method can produce an excellent sinusoidal waveform that is stable. A JFET operating with a small or zero V_{DS} is operating in the ohmic region. As the gate voltage increases, the drain-source resistance increases. If the JFET is placed in the negative feedback path, automatic gain control can be achieved because of this voltage-controlled resistance.

A JFET stabilized Wien bridge is shown in Figure 16–11. The gain of the op-amp is controlled by the components shown in the shaded box, which include the JFET. The JFET's drain-source resistance depends on the gate voltage. With no output signal, the gate is at zero volts, causing the drain-source resistance to be at the minimum. With this condition, the loop gain is greater than 1. Oscillations begin and rapidly build to a large output signal. Negative excursions of the output signal forward-bias D_1, causing capacitor C_3 to charge to a negative voltage. This voltage increases the drain-source resistance of the JFET and reduces the gain (and hence the output). This is classic negative feedback at work. With the proper selection of components, the gain can be stabilized at the required level. The following example illustrates a JFET stabilized Wien-bridge oscillator.

▲ **FIGURE 16–11**

Self-starting Wien–bridge oscillator using a JFET in the negative feedback loop.

EXAMPLE 16–1

Determine the frequency for the Wien-bridge oscillator in Figure 16–12. Also, calculate the setting for R_f assuming the internal drain-source resistance, r'_{ds}, of the JFET is 500 Ω when oscillations are stable.

▲ **FIGURE 16–12**

Solution   For the lead-lag circuit, $R_1 = R_2 = R = 10$ kΩ and $C_1 = C_2 = C = 0.01$ μF. The frequency is

$$f_r = \frac{1}{2\pi RC} = \frac{1}{2\pi(10 \text{ k}\Omega)(0.01 \text{ } \mu\text{F})} = \textbf{1.59 kHz}$$

The closed-loop gain must be 3.0 for oscillations to be sustained. For an inverting amplifier, the gain is that of a noninverting amplifier.

$$A_v = \frac{R_f}{R_i} + 1$$

R_i is composed of R_3 (the source resistor) and r'_{ds}. Substituting,

$$A_v = \frac{R_f}{R_3 + r'_{ds}} + 1$$

Rearranging and solving for R_f,

$$R_f = (A_v - 1)(R_3 + r'_{ds}) = (3 - 1)(1.0 \text{ k}\Omega + 500 \text{ }\Omega) = \textbf{3.0 k}\Omega$$

Related Problem*   What happens to the oscillations if the setting of R_f is too high? What happens if the setting is too low?

*Answers are at the end of the chapter.

Open file E16-01 in the Examples folder on your CD-ROM. Determine the frequency of oscillation and compare with the calculated value.

The Phase-Shift Oscillator

Figure 16–13 shows a type of sinusoidal feedback oscillator called the **phase-shift oscilla-tor**. Each of the three *RC* circuits in the feedback loop can provide a *maximum* phase shift approaching 90°. Oscillation occurs at the frequency where the total phase shift through the three *RC* circuits is 180°. The inversion of the op-amp itself provides the additional 180° to meet the requirement for oscillation of a 360° (or 0°) phase shift around the feedback loop.

▲ **FIGURE 16–13**

Phase–shift oscillator.

The attenuation, *B*, of the three-section *RC* feedback circuit is

$$B = \frac{1}{29}$$

Equation 16–3

where $B = R_3/R_f$. The derivation of this unusual result is given in Appendix B. To meet the greater-than-unity loop gain requirement, the closed-loop voltage gain of the op-amp must be greater than 29 (set by R_f and R_3). The frequency of oscillation (f_r) is also derived in Appendix B and stated in the following equation, where $R_1 = R_2 = R_3 = R$ and $C_1 = C_2 = C_3 = C$.

$$f_r = \frac{1}{2\pi\sqrt{6}RC}$$

Equation 16–4

EXAMPLE 16–2

(a) Determine the value of R_f necessary for the circuit in Figure 16–14 to operate as an oscillator.

(b) Determine the frequency of oscillation.

▲ **FIGURE 16–14**

Solution   (a) $A_{cl} = 29$, and $B = 1/29 = R_3/R_f$. Therefore,

$$\frac{R_f}{R_3} = 29$$

$$R_f = 29R_3 = 29(10 \text{ k}\Omega) = \mathbf{290 \text{ k}\Omega}$$

(b) $R_1 = R_2 = R_3 = R$ and $C_1 = C_2 = C_3 = C$. Therefore,

$$f_r = \frac{1}{2\pi\sqrt{6}RC} = \frac{1}{2\pi\sqrt{6}(10 \text{ k}\Omega)(0.001 \text{ }\mu\text{F})} \cong \mathbf{6.5 \text{ kHz}}$$

Related Problem   (a) If R_1, R_2, and R_3 in Figure 16–14 are changed to 8.2 kΩ, what value must R_f be for oscillation?

(b) What is the value of f_r?

> Open file E16-02 in the Examples folder on your CD-ROM. Measure the frequency of oscillation and compare to the calculated value.

Twin-T Oscillator

Another type of *RC* feedback oscillator is called the *twin-T* because of the two T-type *RC* filters used in the feedback loop, as shown in Figure 16–15(a). One of the twin-T filters has a low-pass response, and the other has a high-pass response. The combined parallel filters produce a band-stop or notch response with a center frequency equal to the desired frequency of oscillation, f_r, as shown in Figure 16–15(b).

Oscillation cannot occur at frequencies above or below f_r because of the negative feedback through the filters. At f_r, however, there is negligible negative feedback; thus, the positive feedback through the voltage divider (R_1 and R_2) allows the circuit to oscillate.

(a) Oscillator circuit

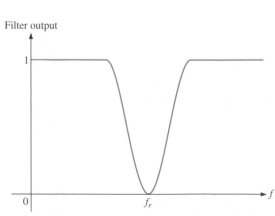

(b) Twin-T filter's frequency response curve

▲ FIGURE 16–15

Twin-T oscillator and twin-T filter response.

**SECTION 16–3
REVIEW**

1. There are two feedback loops in the Wien-bridge oscillator. What is the purpose of each?

2. A certain lead-lag circuit has $R_1 = R_2$ and $C_1 = C_2$. An input voltage of 5 V rms is applied. The input frequency equals the resonant frequency of the circuit. What is the rms output voltage?

3. Why is the phase shift through the *RC* feedback circuit in a phase-shift oscillator 180°?

16–4   OSCILLATORS WITH *LC* FEEDBACK CIRCUITS

Although the *RC* feedback oscillators, particularly the Wien bridge, are generally suitable for frequencies up to about 1 MHz, *LC* feedback elements are normally used in oscillators that require higher frequencies of oscillation. Also, because of the frequency limitation (lower unity-gain frequency) of most op-amps, discrete transistors (BJT or FET) are often used as the gain element in *LC* oscillators. This section introduces several types of resonant *LC* feedback oscillators—the Colpitts, Clapp, Hartley, Armstrong, and crystal-controlled oscillators.

After completing this section, you should be able to

■ **Describe and analyze the basic operation of *LC* feedback oscillators**

■ Identify and analyze a Colpitts oscillator

■ Identify and analyze a Clapp oscillator

■ Identify and analyze a Hartley oscillator

■ Identify and analyze an Armstrong oscillator

■ Discuss the operation of crystal-controlled oscillators

The Colpitts Oscillator

One basic type of resonant circuit feedback oscillator is the Colpitts, named after its inventor—as are most of the others we cover here. As shown in Figure 16–16, this type of oscillator uses an *LC* circuit in the feedback loop to provide the necessary phase shift and to act as a resonant filter that passes only the desired frequency of oscillation.

The approximate frequency of oscillation is the resonant frequency of the *LC* circuit and is established by the values of C_1, C_2, and L according to this familiar formula, which also applies to the Clapp and Hartley oscillators that are covered next.

$$f_r \cong \frac{1}{2\pi\sqrt{LC_T}}$$

Equation 16–5

Because the capacitors effectively appear in series around the tank circuit, the total capacitance (C_T) is

$$C_T = \frac{C_1 C_2}{C_1 + C_2}$$

▲ **FIGURE 16–16**

A basic Colpitts oscillator with a BJT as the gain element.

Conditions for Oscillation and Start-Up The attenuation, *B*, of the resonant feedback circuit in the Colpitts oscillator is basically determined by the values of C_1 and C_2.

Figure 16–17 shows that the circulating tank current is through both C_1 and C_2 (they are effectively in series). The voltage developed across C_2 is the oscillator's output voltage (V_{out}) and the voltage developed across C_1 is the feedback voltage (V_f), as indicated. The expression for the attenuation (*B*) is

$$B = \frac{V_f}{V_{out}} \cong \frac{IX_{C1}}{IX_{C2}} = \frac{X_{C1}}{X_{C2}} = \frac{1/(2\pi f_r C_1)}{1/(2\pi f_r C_2)}$$

▶ **FIGURE 16–17**

The attenuation of the tank circuit is the output of the tank (V_f) divided by the input to the tank (V_{out}). $B = V_f/V_{out} = C_2/C_1$. For $A_v B > 1$, A_v must be greater than C_1/C_2.

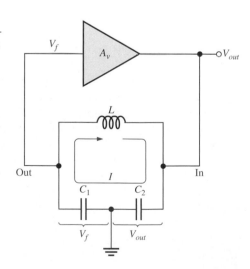

Cancelling the $2\pi f_r$ terms gives

$$B = \frac{C_2}{C_1}$$

As you know, a condition for oscillation is $A_v B = 1$. Since $B = C_2/C_1$,

$$A_v = \frac{C_1}{C_2}$$

Equation 16–6

where A_v is the voltage gain of the amplifier which is represented by the triangle in Figure 16–17. With this condition met, $A_v B = (C_1/C_2)(C_2/C_1) = 1$. Actually, for the oscillator to be self-starting, $A_v B$ must be greater than 1 (that is, $A_v B > 1$). Therefore, the voltage gain must be made slightly greater than C_1/C_2.

$$A_v > \frac{C_1}{C_2}$$

Loading of the Feedback Circuit Affects the Frequency of Oscillation As indicated in Figure 16–18, the input impedance of the amplifier acts as a load on the resonant feedback circuit and reduces the Q of the circuit. Recall from your study of resonance that the resonant frequency of a parallel resonant circuit depends on the Q, according to the following formula:

$$f_r = \frac{1}{2\pi\sqrt{LC_T}}\sqrt{\frac{Q^2}{Q^2+1}}$$

Equation 16–7

As a rule of thumb, for a Q greater than 10, the frequency is approximately $1/(2\pi\sqrt{LC_T})$, as stated in Equation 16–5. When Q is less than 10, however, f_r is reduced significantly.

▶ **FIGURE 16–18**

Z_{in} of the amplifier loads the feedback circuit and lowers its Q, thus lowering the resonant frequency.

A FET can be used in place of a BJT, as shown in Figure 16–19, to minimize the loading effect of the transistor's input impedance. Recall that FETs have much higher input impedances than do bipolar junction transistors. Also, when an external load is connected to the oscillator output, as shown in Figure 16–20(a), f_r may decrease, again because of a reduction in Q. This happens if the load resistance is too small. In some cases, one way to eliminate the effects of a load resistance is by transformer coupling, as indicated in Figure 16–20(b).

▶ **FIGURE 16–19**

A basic FET Colpitts oscillator.

(a) A load capacitively coupled to oscillator output can reduce circuit Q and f_r.

(b) Transformer coupling of load can reduce loading effect by impedance transformation.

▲ **FIGURE 16–20**

Oscillator loading.

EXAMPLE 16–3

(a) Determine the frequency for the oscillator in Figure 16–21. Assume there is negligible loading on the feedback circuit and that its Q is greater than 10.

(b) Find the frequency if the oscillator is loaded to a point where the Q drops to 8.

▶ **FIGURE 16–21**

Solution (a) $C_T = \dfrac{C_1 C_2}{C_1 + C_2} = \dfrac{(0.1\ \mu F)(0.01\ \mu F)}{0.11\ \mu F} = 0.0091\ \mu F$

$$f_r \cong \dfrac{1}{2\pi\sqrt{LC_T}} = \dfrac{1}{2\pi\sqrt{(50\ \text{mH})(0.0091\ \mu F)}} = \textbf{7.46 kHz}$$

(b) $f_r = \dfrac{1}{2\pi\sqrt{LC_T}} \sqrt{\dfrac{Q^2}{Q^2 + 1}} = (7.46\ \text{kHz})(0.9923) = \textbf{7.40 kHz}$

Related Problem What frequency does the oscillator in Figure 16–21 produce if it is loaded to a point where $Q = 4$?

The Clapp Oscillator

The Clapp oscillator is a variation of the Colpitts. The basic difference is an additional capacitor, C_3, in series with the inductor in the resonant feedback circuit, as shown in Figure 16–22. Since C_3 is in series with C_1 and C_2 around the tank circuit, the total capacitance is

$$C_T = \dfrac{1}{\dfrac{1}{C_1} + \dfrac{1}{C_2} + \dfrac{1}{C_3}}$$

and the approximate frequency of oscillation ($Q > 10$) is

$$f_r \cong \dfrac{1}{2\pi\sqrt{LC_T}}$$

If C_3 is much smaller than C_1 and C_2, then C_3 almost entirely controls the resonant frequency ($f_r \cong 1/(2\pi\sqrt{LC_3})$). Since C_1 and C_2 are both connected to ground at one end, the junction capacitance of the transistor and other stray capacitances appear in parallel with C_1 and C_2 to ground, altering their effective values. C_3 is not affected, however, and thus provides a more accurate and stable frequency of oscillation.

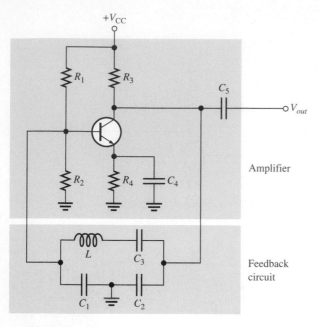

▲ FIGURE 16–22

A basic Clapp oscillator.

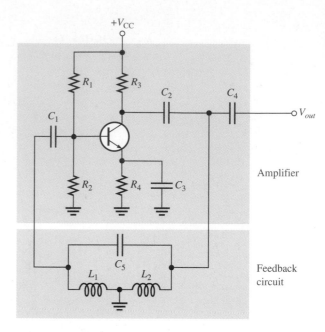

▲ FIGURE 16–23

A basic Hartley oscillator.

The Hartley Oscillator

The Hartley oscillator is similar to the Colpitts except that the feedback circuit consists of two series inductors and a parallel capacitor as shown in Figure 16–23.

In this circuit, the frequency of oscillation for $Q > 10$ is

$$f_r \cong \frac{1}{2\pi\sqrt{L_T C}}$$

where $L_T = L_1 + L_2$. The inductors act in a role similar to C_1 and C_2 in the Colpitts to determine the attenuation, B, of the feedback circuit.

$$B \cong \frac{L_1}{L_2}$$

To assure start-up of oscillation, A_v must be greater than $1/B$.

Equation 16–8

$$A_v > \frac{L_2}{L_1}$$

Loading of the tank circuit has the same effect in the Hartley as in the Colpitts; that is, the Q is decreased and thus f_r decreases.

The Armstrong Oscillator

This type of LC feedback oscillator uses transformer coupling to feed back a portion of the signal voltage, as shown in Figure 16–24. It is sometimes called a "tickler" oscillator in reference to the transformer secondary or "tickler coil" that provides the feedback to keep the oscillation going. The Armstrong is less common than the Colpitts, Clapp, and Hartley, mainly because of the disadvantage of transformer size and cost. The frequency of oscillation is set by the inductance of the primary winding (L_{pri}) in parallel with C_1.

Equation 16–9

$$f_r = \frac{1}{2\pi\sqrt{L_{pri} C_1}}$$

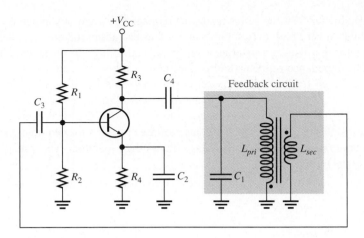

▲ **FIGURE 16–24**

A basic Armstrong oscillator.

Crystal-Controlled Oscillators

The most stable and accurate type of feedback oscillator uses a piezoelectric **crystal** in the feedback loop to control the frequency.

The Piezoelectric Effect Quartz is one type of crystalline substance found in nature that exhibits a property called the **piezoelectric effect**. When a changing mechanical stress is applied across the crystal to cause it to vibrate, a voltage develops at the frequency of mechanical vibration. Conversely, when an ac voltage is applied across the crystal, it vibrates at the frequency of the applied voltage. The greatest vibration occurs at the crystal's natural resonant frequency, which is determined by the physical dimensions and by the way the crystal is cut.

Crystals used in electronic applications typically consist of a quartz wafer mounted between two electrodes and enclosed in a protective "can" as shown in Figure 16–25(a) and (b). A schematic symbol for a crystal is shown in Figure 16–25(c), and an equivalent *RLC* circuit for the crystal appears in Figure 16–25(d). As you can see, the crystal's equivalent circuit is a series-parallel *RLC* circuit and can operate in either series resonance or parallel resonance. At the series resonant frequency, the inductive reactance is cancelled by the reactance of C_s. The remaining series resistor, R_s, determines the impedance of the crystal. Parallel resonance occurs when the inductive reactance and the reactance of the parallel capacitance, C_m, are equal. The parallel resonant frequency is usually

| (a) Typical packaged crystal | (b) Basic construction (without case) | (c) Symbol | (d) Electrical equivalent |

▲ **FIGURE 16–25**

A quartz crystal.

at least 1 kHz higher than the series resonant frequency. A great advantage of the crystal is that it exhibits a very high Q (Qs with values of several thousand are typical).

An oscillator that uses a crystal as a series resonant tank circuit is shown in Figure 16–26(a). The impedance of the crystal is minimum at the series resonant frequency, thus providing maximum feedback. The crystal tuning capacitor, C_C, is used to "fine tune" the oscillator frequency by "pulling" the resonant frequency of the crystal slightly up or down.

A modified Colpitts configuration is shown in Figure 16–26(b) with a crystal acting as a parallel resonant tank circuit. The impedance of the crystal is maximum at parallel resonance, thus developing the maximum voltage across the capacitors. The voltage across C_1 is fed back to the input.

(a)   (b)

▲ **FIGURE 16–26**

Basic crystal oscillators.

Modes of Oscillation in the Crystal Piezoelectric crystals can oscillate in either of two modes—fundamental or overtone. The fundamental frequency of a crystal is the lowest frequency at which it is naturally resonant. The fundamental frequency depends on the crystal's mechanical dimensions, type of cut, and other factors, and is inversely proportional to the thickness of the crystal slab. Because a slab of crystal cannot be cut too thin without fracturing, there is an upper limit on the fundamental frequency. For most crystals, this upper limit is less than 20 MHz. For higher frequencies, the crystal must be operated in the overtone mode. Overtones are approximate integer multiples of the fundamental frequency. The overtone frequencies are usually, but not always, odd multiples (3, 5, 7, . . .) of the fundamental.

SECTION 16–4 REVIEW	1. What is the basic difference between the Colpitts and the Hartley oscillators?
	2. What is the advantage of a FET amplifier in a Colpitts or Hartley oscillator?
	3. How can you distinguish a Colpitts oscillator from a Clapp oscillator?

16–5   RELAXATION OSCILLATORS

The second major category of oscillators is the relaxation oscillator. Relaxation oscillators use an *RC* timing circuit and a device that changes states to generate a periodic waveform. In this section, you will learn about several circuits that are used to produce nonsinusoidal waveforms.

After completing this section, you should be able to

- **Describe and analyze the basic operation of relaxation oscillators**

- Discuss the operation of basic triangular-wave oscillators

- Discuss the operation of a voltage-controlled oscillator (VCO)

- Discuss the operation of a square-wave relaxation oscillator

A Triangular-Wave Oscillator

The op-amp integrator covered in Chapter 13 can be used as the basis for a triangular-wave oscillator. The basic idea is illustrated in Figure 16–27(a) where a dual-polarity, switched input is used. We use the switch only to introduce the concept; it is not a practical way to implement this circuit. When the switch is in position 1, the negative voltage is applied, and the output is a positive-going ramp. When the switch is thrown into position 2, a negative-going ramp is produced. If the switch is thrown back and forth at fixed intervals, the output is a triangular wave consisting of alternating positive-going and negative-going ramps, as shown in Figure 16–27(b).

(a)

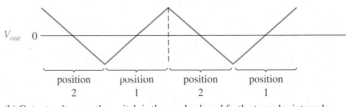

(b) Output voltage as the switch is thrown back and forth at regular intervals

▲ **FIGURE 16–27**

Basic triangular-wave oscillator.

A Practical Triangular-Wave Oscillator   One practical implementation of a triangular-wave oscillator utilizes an op-amp comparator to perform the switching function, as shown in Figure 16–28. The operation is as follows. To begin, assume that the output voltage of the comparator is at its maximum negative level. This output is connected to the inverting input of the integrator through R_1, producing a positive-going ramp on the output of the integrator. When the ramp voltage reaches the upper trigger point (UTP), the comparator switches to its maximum positive level. This positive level causes the integrator ramp to change to a negative-going direction. The ramp continues in this direction until the lower trigger point (LTP) of the comparator is reached. At this point, the comparator output switches back to the maximum negative level and the cycle repeats. This action is illustrated in Figure 16–29.

▲ **FIGURE 16–28**

A triangular-wave oscillator using two op-amps.

▶ **FIGURE 16–29**

Waveforms for the circuit in Figure 16–28.

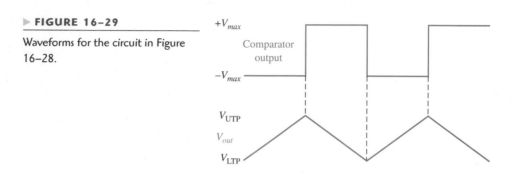

Since the comparator produces a square-wave output, the circuit in Figure 16–28 can be used as both a triangular-wave oscillator and a square-wave oscillator. Devices of this type are commonly known as *function generators* because they produce more than one output function. The output amplitude of the square wave is set by the output swing of the comparator, and the resistors R_2 and R_3 set the amplitude of the triangular output by establishing the UTP and LTP voltages according to the following formulas:

$$V_{\text{UTP}} = +V_{max}\left(\frac{R_3}{R_2}\right)$$

$$V_{\text{LTP}} = -V_{max}\left(\frac{R_3}{R_2}\right)$$

where the comparator output levels, $+V_{max}$ and $-V_{max}$, are equal. The frequency of both waveforms depends on the R_1C time constant as well as the amplitude-setting resistors, R_2 and R_3. By varying R_1, the frequency of oscillation can be adjusted without changing the output amplitude.

Equation 16–10

$$f_r = \frac{1}{4R_1C}\left(\frac{R_2}{R_3}\right)$$

EXAMPLE 16–4

Determine the frequency of oscillation of the circuit in Figure 16–30. To what value must R_1 be changed to make the frequency 20 kHz?

▲ **FIGURE 16–30**

Solution

$$f_r = \frac{1}{4R_1C}\left(\frac{R_2}{R_3}\right) = \left(\frac{1}{4(10\ \text{k}\Omega)(0.01\ \mu\text{F})}\right)\left(\frac{33\ \text{k}\Omega}{10\ \text{k}\Omega}\right) = \textbf{8.25 kHz}$$

To make $f = 20$ kHz,

$$R_1 = \frac{1}{4fC}\left(\frac{R_2}{R_3}\right) = \left(\frac{1}{4(20\ \text{kHz})(0.01\ \mu\text{F})}\right)\left(\frac{33\ \text{k}\Omega}{10\ \text{k}\Omega}\right) = \textbf{4.13 k}\Omega$$

Related Problem

What is the amplitude of the triangular wave in Figure 16–30 if the comparator output is ± 10 V?

> Open file E16-04 in the Examples folder on your CD-ROM. Measure the frequency of oscillation and compare to the calculated value. Change R_1 to the value calculated for 20 kHz and check the frequency.

A Voltage-Controlled Sawtooth Oscillator (VCO)

The **voltage-controlled oscillator (VCO)** is a relaxation oscillator whose frequency can be changed by a variable dc control voltage. VCOs can be either sinusoidal or nonsinusoidal. One way to build a voltage-controlled sawtooth oscillator is with an op-amp integrator that uses a switching device (PUT) in parallel with the feedback capacitor to terminate each ramp at a prescribed level and effectively "reset" the circuit. Figure 16–31(a) shows the implementation.

As you learned in Chapter 11, the PUT is a programmable unijunction transistor with an anode, a cathode, and a gate terminal. The gate is always biased positively with respect to the cathode. When the anode voltage exceeds the gate voltage by approximately 0.7 V, the PUT turns on and acts as a forward-biased diode. When the anode voltage falls below this level, the PUT turns off. Also, the current must be above the holding value to maintain conduction.

The operation of the sawtooth oscillator begins when the negative dc input voltage, $-V_{IN}$, produces a positive-going ramp on the output. During the time that the ramp is increasing, the circuit acts as a regular integrator. The PUT triggers on when the output ramp (at the anode) exceeds the gate voltage by 0.7 V. The gate is set to the approximate

(a) Initially, the capacitor charges, the output ramp begins, and the PUT is off.

(b) The capacitor rapidly discharges when the PUT momentarily turns on.

▲ **FIGURE 16–31**

Voltage-controlled sawtooth oscillator operation.

desired sawtooth peak voltage. When the PUT turns on, the capacitor rapidly discharges, as shown in Figure 16–31(b). The capacitor does not discharge completely to zero because of the PUT's forward voltage, V_F. Discharge continues until the PUT current falls below the holding value. At this point, the PUT turns off and the capacitor begins to charge again, thus generating a new output ramp. The cycle continually repeats, and the resulting output is a repetitive sawtooth waveform, as shown. The sawtooth amplitude and period can be adjusted by varying the PUT gate voltage.

The frequency of oscillation is determined by the R_iC time constant of the integrator and the peak voltage set by the PUT. Recall that the charging rate of the capacitor is V_{IN}/R_iC. The time it takes the capacitor to charge from V_F to V_p is the period, T, of the sawtooth (neglecting the rapid discharge time).

$$T = \frac{V_p - V_F}{|V_{IN}|/R_iC}$$

From $f = 1/T,$

Equation 16–11

$$f = \frac{|V_{IN}|}{R_iC}\left(\frac{1}{V_p - V_F}\right)$$

EXAMPLE 16–5

(a) Find the amplitude and frequency of the sawtooth output in Figure 16–32. Assume that the forward PUT voltage, V_F, is approximately 1 V.

(b) Sketch the output waveform.

▲ **FIGURE 16–32**

Solution   (a) First, find the gate voltage in order to establish the approximate voltage at which the PUT turns on.

$$V_G = \frac{R_4}{R_3 + R_4}(+V) = \frac{10\ k\Omega}{20\ k\Omega}(15\ V) = 7.5\ V$$

This voltage sets the approximate maximum peak value of the sawtooth output (neglecting the 0.7 V).

$$V_p \cong 7.5\ V$$

The minimum peak value (low point) is

$$V_F \cong 1\ V$$

So the peak-to-peak amplitude is

$$V_{pp} = V_p - V_F = 7.5\ V - 1\ V = \mathbf{6.5\ V}$$

The frequency is determined as follows:

$$V_{IN} = \frac{R_2}{R_1 + R_2}(-V) = \frac{10\ k\Omega}{78\ k\Omega}(-15\ V) = -1.92\ V$$

$$f = \frac{|V_{IN}|}{R_iC}\left(\frac{1}{V_p - V_F}\right) = \left(\frac{1.92\ V}{(100\ k\Omega)(0.0047\ \mu F)}\right)\left(\frac{1}{7.5\ V - 1\ V}\right) = \mathbf{628\ Hz}$$

(b) The output waveform is shown in Figure 16–33, where the period is determined as follows:

$$T = \frac{1}{f} = \frac{1}{628\ Hz} = 2\ ms$$

▲ FIGURE 16–33

Output of the circuit in Figure 16–32.

Related Problem   If R_i is changed to 56 kΩ in Figure 16–32, what is the frequency?

A Square-Wave Oscillator

The basic square-wave oscillator shown in Figure 16–34 is a type of relaxation oscillator because its operation is based on the charging and discharging of a capacitor. Notice that the op-amp's inverting input is the capacitor voltage and the noninverting input is a portion of the output fed back through resistors R_2 and R_3. When the circuit is first turned on, the capacitor is uncharged, and thus the inverting input is at 0 V. This makes the output a positive maximum, and the capacitor begins to charge toward V_{out} through R_1. When the capacitor voltage (V_C) reaches a value equal to the feedback voltage (V_f) on the noninverting input, the op-amp switches to the maximum negative state. At this point, the capacitor begins to discharge from $+V_f$ toward $-V_f$. When the capacitor voltage reaches $-V_f$, the op-amp switches back to the maximum positive state. This action continues to repeat, as shown in Figure 16–35, and a square-wave output voltage is obtained.

▲ FIGURE 16–34

A square-wave relaxation oscillator.

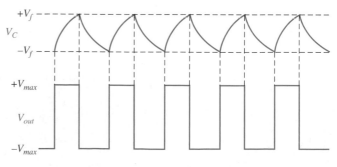

▲ FIGURE 16–35

Waveforms for the square-wave relaxation oscillator.

SECTION 16–5 REVIEW

1. What is a VCO, and basically, what does it do?
2. Upon what principle does a relaxation oscillator operate?

16–6 THE 555 TIMER AS AN OSCILLATOR

The 555 timer is a versatile integrated circuit with many applications. In this section, you will see how the 555 is configured as an astable or free-running multivibrator, which is essentially a square-wave oscillator. The use of the 555 timer as a voltage-controlled oscillator (VCO) is also discussed.

After completing this section, you should be able to

- **Use a 555 timer in an oscillator application**

- Explain what the 555 timer is

- Discuss astable operation of the 555 timer

- Explain how to use the 555 timer as a VCO

The 555 timer consists basically of two comparators, a flip-flop, a discharge transistor, and a resistive voltage divider, as shown in Figure 16–36. The flip-flop (bistable multivibrator) is a digital device that may be unfamiliar to you at this point unless you already have taken a digital fundamentals course. Briefly, it is a two-state device whose output can be at either a high voltage level (set, S) or a low voltage level (reset, R). The state of the output can be changed with proper input signals.

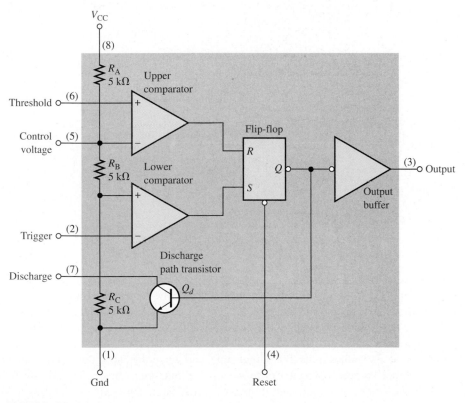

▲ **FIGURE 16–36**

Internal diagram of a 555 integrated circuit timer. (IC pin numbers are in parentheses.)

The resistive voltage divider is used to set the voltage comparator levels. All three resistors are of equal value; therefore, the upper comparator has a reference of $\frac{2}{3}V_{CC}$, and the lower comparator has a reference of $\frac{1}{3}V_{CC}$. The comparators' outputs control the state of the flip-flop. When the trigger voltage goes below $\frac{1}{3}V_{CC}$, the flip-flop sets and the output jumps to its high level. The threshold input is normally connected to an external RC timing circuit. When the external capacitor voltage exceeds $\frac{2}{3}V_{CC}$, the upper comparator resets the flip-flop, which in turn switches the output back to its low level. When the device output is low, the discharge transistor (Q_d) is turned on and provides a path for rapid discharge of the external timing capacitor. This basic operation allows the timer to be configured with external components as an oscillator, a one-shot, or a time-delay element.

Astable Operation

A 555 timer connected to operate in the **astable** mode as a free-running relaxation oscillator (astable multivibrator) is shown in Figure 16–37. Notice that the threshold input (THRESH) is now connected to the trigger input (TRIG). The external components R_1, R_2, and C_{ext} form the timing circuit that sets the frequency of oscillation. The 0.01 μF capacitor connected to the control (CONT) input is strictly for decoupling and has no effect on the operation.

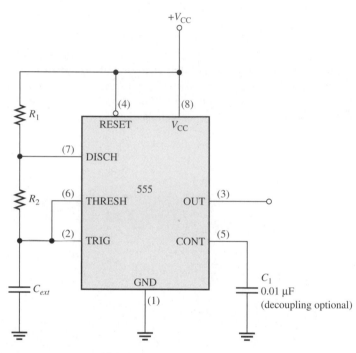

▲ **FIGURE 16–37**

The 555 timer connected as an astable multivibrator.

Initially, when the power is turned on, the capacitor C_{ext} is uncharged and thus the trigger voltage (pin 2) is at 0 V. This causes the output of the lower comparator to be high and the output of the upper comparator to be low, forcing the output of the flip-flop, and thus the base of Q_d, low and keeping the transistor off. Now, C_{ext} begins charging through R_1 and R_2 as indicated in Figure 16–38. When the capacitor voltage reaches $\frac{1}{3}V_{CC}$, the lower comparator switches to its low output state, and when the capacitor voltage reaches $\frac{2}{3}V_{CC}$, the upper comparator switches to its high output state. This resets the flip-flop, causes the base of Q_d to go high, and turns on the transistor. This sequence creates a discharge path for the capacitor through R_2 and the transistor, as indicated. The capacitor now begins to discharge, causing the upper comparator to go low. At the point where the capacitor discharges down to $\frac{1}{3}V_{CC}$, the lower comparator switches high, setting the flip-flop, which makes the base of Q_d low and turns off the transistor. Another charging cycle begins, and the entire process repeats. The result is a rectangular wave output whose duty cycle depends on the values of R_1 and R_2. The frequency of oscillation is given by Equation 16–12, or it can be found using the graph in Figure 16–39.

Equation 16–12

$$f_r = \frac{1.44}{(R_1 + 2R_2)C_{ext}}$$

▲ **FIGURE 16–38**

Operation of the 555 timer in the astable mode.

▲ **FIGURE 16–39**

Frequency of oscillation (free-running frequency) of a 555 timer in the astable mode as a function of C_{ext} and $R_1 + 2R_2$. The sloped lines are values of $R_1 + 2R_2$.

By selecting R_1 and R_2, the duty cycle of the output can be adjusted. Since C_{ext} charges through $R_1 + R_2$ and discharges only through R_2, duty cycles approaching a minimum of 50 percent can be achieved if $R_2 >> R_1$ so that the charging and discharging times are approximately equal.

A formula to calculate the duty cycle is developed as follows. The time that the output is high (t_H) is how long it takes C_{ext} to charge from $\frac{1}{3}V_{CC}$ to $\frac{2}{3}V_{CC}$. It is expressed as

$$t_H = 0.694(R_1 + R_2)C_{ext}$$

The time that the output is low (t_L) is how long it takes C_{ext} to discharge from $\frac{2}{3}V_{CC}$ to $\frac{1}{3}V_{CC}$. It is expressed as

$$t_L = 0.694R_2C_{ext}$$

The period, T, of the output waveform is the sum of t_H and t_L. The following formula for T is the reciprocal of f in Equation 16–12.

$$T = t_H + t_L = 0.694(R_1 + 2R_2)C_{ext}$$

Finally, the percent duty cycle is

$$\text{Duty cycle} = \left(\frac{t_H}{T}\right)100\% = \left(\frac{t_H}{t_H + t_L}\right)100\%$$

Equation 16–13

$$\text{Duty cycle} = \left(\frac{R_1 + R_2}{R_1 + 2R_2}\right)100\%$$

To achieve duty cycles of less than 50 percent, the circuit in Figure 16–37 can be modified so that C_{ext} charges through only R_1 and discharges through R_2. This is achieved with a diode, D_1, placed as shown in Figure 16–40. The duty cycle can be made less than 50 percent by making R_1 less than R_2. Under this condition, the formula for the percent duty cycle is

Equation 16–14

$$\text{Duty cycle} = \left(\frac{R_1}{R_1 + R_2}\right)100\%$$

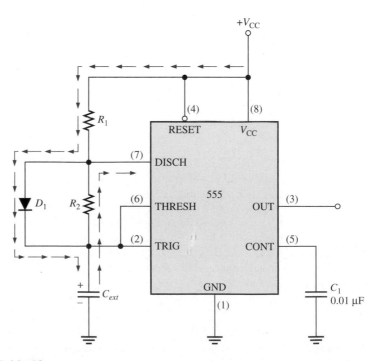

▲ FIGURE 16–40

The addition of diode D_1 allows the duty cycle of the output to be adjusted to less than 50 percent by making $R_1 < R_2$.

EXAMPLE 16–6

A 555 timer configured to run in the astable mode (oscillator) is shown in Figure 16–41. Determine the frequency of the output and the duty cycle.

▲ **FIGURE 16–41**

Solution

$$f_r = \frac{1.44}{(R_1 + 2R_2)C_{ext}} = \frac{1.44}{(2.2\ k\Omega + 9.4\ k\Omega)0.022\ \mu F} = \textbf{5.64 kHz}$$

$$\text{Duty cycle} = \left(\frac{R_1 + R_2}{R_1 + 2R_2}\right)100\% = \left(\frac{2.2\ k\Omega + 4.7\ k\Omega}{2.2\ k\Omega + 9.4\ k\Omega}\right)100\% = \textbf{59.5\%}$$

Related Problem

Determine the duty cycle in Figure 16–41 if a diode is connected across R_2 as indicated in Figure 16–40.

Operation as a Voltage-Controlled Oscillator (VCO)

A 555 timer can be set up to operate as a VCO by using the same external connections as for astable operation, with the exception that a variable control voltage is applied to the CONT input (pin 5), as indicated in Figure 16–42.

As shown in Figure 16–43, the control voltage (V_{CONT}) changes the threshold values of $\frac{1}{3}V_{CC}$ and $\frac{2}{3}V_{CC}$ for the internal comparators. With the control voltage, the upper value is V_{CONT} and the lower value is $\frac{1}{2}V_{CONT}$, as you can see by examining the internal diagram of the 555 timer. When the control voltage is varied, the output frequency also varies. An increase in V_{CONT} increases the charging and discharging time of the external capacitor and causes the frequency to decrease. A decrease in V_{CONT} decreases the charging and discharging time of the capacitor and causes the frequency to increase.

An interesting application of the VCO is in phase-locked loops, which are used in various types of communication receivers to track variations in the frequency of incoming signals. You will learn about the basic operation of a phase-locked loop in Section 17–8.

▲ **FIGURE 16–42**

The 555 timer connected as a voltage-controlled oscillator (VCO). Note the variable control voltage input on pin 5.

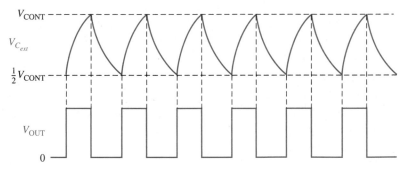

▲ **FIGURE 16–43**

The VCO output frequency varies inversely with V_{CONT} because the charging and discharging time of C_{ext} is directly dependent on the control voltage.

SECTION 16–6 REVIEW	1. Name the five basic elements in a 555 timer IC.
	2. When the 555 timer is configured as an astable multivibrator, how is the duty cycle determined?

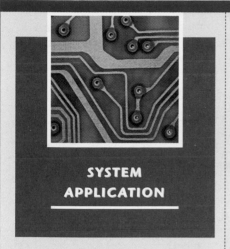

SYSTEM APPLICATION

This system application focuses on a function generator, an instrument that produces sinusoidal waveforms, square waveforms, and triangular waveforms for use in testing circuits in the laboratory.

Basic Operation of the System

The function generator is an electronic system that produces either a sinusoidal waveform, a square waveform, or a triangular waveform depending on the function selected by the front panel switches, as shown in Figure 16–44. The frequency of the selected output waveform can be varied from less than 1 Hz to greater than 80 kHz by using the frequency range switches and the frequency dial. The amplitude of the output waveform can be adjusted up to approximately +10 V with the front panel amplitude control. Also, any dc offset can be nulled out with the front panel dc offset control.

The block diagram of the function generator is shown in Figure 16–45. The concept of this particular design is very simple. The oscillator produces a sinusoidal output voltage that drives a zero-level detector (comparator), which produces a square wave of the same frequency as the oscillator output. The output of the level detector goes to an integrator, which produces a triangular output voltage also with a frequency equal to the oscillator output. The type of output waveform is selected by the function switches on the front panel. The frequency is set by the front panel range switches and the frequency dial, and amplitude is set by the front panel control knob.

The schematic of the function generator is shown in Figure 16–46, where the

▲ **FIGURE 16–44**

Front panel of the function generator.

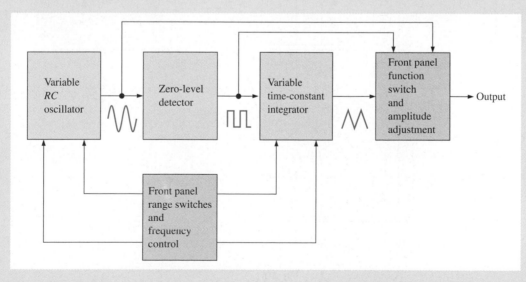

▲ **FIGURE 16–45**

Block diagram of the function generator.

portions in blue are the front panel components. The frequency of the sinusoidal oscillator is controlled by the selection of any two of ten capacitors (C_1 through C_{10}) in the oscillator feedback circuit. These capacitors produce the five frequency ranges indicated on the front panel switches, which are multiplication factors for the setting of the frequency

dial. The adjustment of the frequency within each range is accomplished by varying resistors R_8 and R_9 in the feedback circuit of the oscillator.

The integrator time constant is adjusted in step with the frequency by selection of the appropriate capacitor (C_{11} through C_{15}) and adjustment of resistor R_{10}. Resistors R_8, R_9, and R_{10} are

potentiometers that are ganged together so that they change resistance together as the frequency dial is turned. For example, if the ×1k switch is selected and the frequency dial is set at 5, then the resulting output frequency for any of the three types of waveforms is 1 kHz × 5 = 5 kHz.

▲ FIGURE 16–46

Schematic of the function generator.

1   20   1   17
Board A   Board B

▲ **FIGURE 16–47**

The function generator circuit boards.

The Function Generator Circuit Boards

■ Make sure that the circuit boards shown in Figure 16–47 are correct by checking them against the schematic in Figure 16–46. Backside interconnections are shown as darker traces.

■ Label a copy of the board with component and input/output designations in agreement with the schematic.

■ Develop a board-to-board wire list specifying which terminals on the two boards connect together and which terminals connect to front panel components.

The Function Generator Circuit

■ Determine the maximum frequency of the oscillator for each range switch ($\times 1$, $\times 10$, $\times 100$, $\times 1k$, and $\times 10k$). Only one set of three switches corresponding to a given range setting can be closed at a time. There is a set of ganged switches for $\times 1$, a set of ganged switches for $\times 10$, and so on.

■ Determine the minimum frequency of the oscillator for each range switch.

■ Determine the approximate maximum peak-to-peak output voltages for each function. The dc supply voltages are $+15$ V and -15 V.

Test Procedure

Develop a basic procedure for thoroughly testing the function generator.

Troubleshooting

Based on the results indicated in Figure 16–48 for four faulty function generator units, determine the most likely fault or faults in each case.

Final Report (Optional)

Submit a final written report on the channel filter boards using an organized format that includes the following:

1. A physical description of the circuit.

2. A discussion of the operation of the circuit.

3. A list of the specifications.

4. A list of parts with part numbers if available.

5. A list of the types of problems in the four faulty units.

▲ FIGURE 16–48

Results of tests on four faulty units. The scope screen shows the output voltage in each case.

CHAPTER SUMMARY

- Sinusoidal oscillators operate with positive feedback.
- The two conditions for positive feedback are the phase shift around the feedback loop must be 0° and the voltage gain around the feedback loop must equal 1.
- For initial start-up, the voltage gain around the feedback loop must be greater than 1.
- Sinusoidal *RC* oscillators include the Wien-bridge, phase-shift, and twin-T.
- Sinusoidal *LC* oscillators include the Colpitts, Clapp, Hartley, Armstrong, and crystal-controlled.
- The feedback signal in a Colpitts oscillator is derived from a capacitive voltage divider in the *LC* circuit.
- The Clapp oscillator is a variation of the Colpitts with a capacitor added in series with the inductor.
- The feedback signal in a Hartley oscillator is derived from an inductive voltage divider in the *LC* circuit.
- The feedback signal in an Armstrong oscillator is derived by transformer coupling.
- Crystal oscillators are the most stable type.
- A relaxation oscillator uses an *RC* timing circuit and a device that changes states to generate a periodic waveform.
- The frequency in a voltage-controlled oscillator (VCO) can be varied with a dc control voltage.
- The 555 timer is an integrated circuit that can be used as an oscillator, in addition to many other applications.

KEY TERMS

Key terms and other bold terms in the chapter are defined in the end-of-book glossary.

Astable   Characterized by having no stable states.

Feedback oscillator   An electronic circuit that operates with positive feedback and produces a time-varying output signal without an external input signal.

Positive feedback   The return of a portion of the output signal to the input such that it reinforces and sustains the output.

Relaxation oscillator   An electronic circuit that uses an *RC* timing circuit to generate a nonsinusoidal waveform without an external input signal.

Voltage-controlled oscillator (VCO)   A type of relaxation oscillator whose frequency can be varied by a dc control voltage.

KEY FORMULAS

16–1 $\dfrac{V_{out}}{V_{in}} = \dfrac{1}{3}$ Wien-bridge positive feedback attenuation

16–2 $f_r = \dfrac{1}{2\pi RC}$ Wien-bridge resonant frequency

16–3 $B = \dfrac{1}{29}$ Phase-shift feedback attenuation

16–4 $f_r = \dfrac{1}{2\pi \sqrt{6}RC}$ Phase-shift oscillator frequency

16–5 $f_r = \dfrac{1}{2\pi \sqrt{LC_T}}$ Colpitts, Clapp, and Hartley approximate resonant frequency

16–6 $A_r = \dfrac{C_1}{C_2}$ Colpitts amplifier gain

16–7 $f_r = \dfrac{1}{2\pi \sqrt{LC_T}} \sqrt{\dfrac{Q^2}{Q^2 + 1}}$ Colpitts resonant frequency

16–8 $A_v > \dfrac{L_2}{L_1}$ Hartley self-starting gain

16–9 $f_r = \dfrac{1}{2\pi \sqrt{L_{pri}C_1}}$ Armstrong resonant frequency

16–10 $f_r = \dfrac{1}{4R_1C}\left(\dfrac{R_2}{R_3}\right)$ Triangular-wave oscillator frequency

16–11 $f = \dfrac{|V_{IN}|}{R_iC}\left(\dfrac{1}{V_p - V_F}\right)$ Sawtooth VCO frequency

16–12 $f_r = \dfrac{1.44}{(R_1 + 2R_2)C_{ext}}$ 555 astable frequency

16–13 $\textbf{Duty cycle} = \left(\dfrac{R_1 + R_2}{R_1 + 2R_2}\right)100\%$ 555 astable

16–14 $\textbf{Duty cycle} = \left(\dfrac{R_1}{R_1 + R_2}\right)100\%$ 555 astable (duty cycle < 50%)

SELF-TEST

Answers are at the end of the chapter.

1. An oscillator differs from an amplifier because
   (a) it has more gain   (b) it requires no input signal
   (c) it requires no dc supply   (d) it always has the same output

2. Wien-bridge oscillators are based on
   (a) positive feedback   (b) negative feedback
   (c) the piezoelectric effect   (d) high gain

3. One condition for oscillation is
   (a) a phase shift around the feedback loop of 180°
   (b) a gain around the feedback loop of one-third
   (c) a phase shift around the feedback loop of 0°
   (d) a gain around the feedback loop of less than 1

4. A second condition for oscillation is
   (a) no gain around the feedback loop
   (b) a gain of 1 around the feedback loop
   (c) the attenuation of the feedback circuit must be one-third
   (d) the feedback circuit must be capacitive

5. In a certain oscillator, $A_v = 50$. The attenuation of the feedback circuit must be
   (a) 1   (b) 0.01   (c) 10   (d) 0.02

6. For an oscillator to properly start, the gain around the feedback loop must initially be
   (a) 1   (b) less than 1   (c) greater than 1   (d) equal to B

7. In a Wien-bridge oscillator, if the resistances in the positive feedback circuit are decreased, the frequency
   (a) decreases   (b) increases   (c) remains the same

8. The Wien-bridge oscillator's positive feedback circuit is
   (a) an RL circuit   (b) an LC circuit
   (c) a voltage divider   (d) a lead-lag circuit

9. A phase-shift oscillator has
   (a) three RC circuits   (b) three LC circuits
   (c) a T-type circuit   (d) a π-type circuit

10. Colpitts, Clapp, and Hartley are names that refer to

    (a) types of RC oscillators   (b) inventors of the transistor

    (c) types of LC oscillators   (d) types of filters

11. An oscillator whose frequency is changed by a variable dc voltage is known as

    (a) a crystal oscillator   (b) a VCO

    (c) an Armstrong oscillator   (d) a piezoelectric device

12. The main feature of a crystal oscillator is

    (a) economy   (b) reliability   (c) stability   (d) high frequency

13. The operation of a relaxation oscillator is based on

    (a) the charging and discharging of a capacitor

    (b) a highly selective resonant circuit

    (c) a very stable supply voltage

    (d) low power consumption

14. Which one of the following is *not* an input or output of the 555 timer?

    (a) Threshold   (b) Control voltage   (c) Clock   (d) Trigger

    (e) Discharge   (f) Reset

PROBLEMS

Answers to odd-numbered problems are at the end of the book.

BASIC PROBLEMS

SECTION 16–1   The Oscillator

1. What type of input is required for an oscillator?

2. What are the basic components of an oscillator circuit?

SECTION 16–2   Feedback Oscillator Principles

3. If the voltage gain of the amplifier portion of an oscillator is 75, what must be the attenuation of the feedback circuit to sustain the oscillation?

4. Generally describe the change required in the oscillator of Problem 3 in order for oscillation to begin when the power is initially turned on.

SECTION 16–3   Oscillators with RC Feedback Circuits

5. A certain lead-lag circuit has a resonant frequency of 3.5 kHz. What is the rms output voltage if an input signal with a frequency equal to f_r and with an rms value of 2.2 V is applied to the input?

6. Calculate the resonant frequency of a lead-lag circuit with the following values:
   $R_1 = R_2 = 6.2$ kΩ, and $C_1 = C_2 = 0.02$ μF.

7. Determine the necessary value of R_2 in Figure 16–49 so that the circuit will oscillate. Neglect the forward resistance of the zener diodes. (*Hint:* The total gain of the circuit must be 3 when the zener diodes are conducting.)

8. Explain the purpose of R_3 in Figure 16–49.

► FIGURE 16–49

9. For the Wien-bridge oscillator in Figure 16–50, calculate the setting for R_f, assuming the internal drain-source resistance, r'_{ds}, of the JFET is 350 Ω when oscillations are stable.

10. Find the frequency of oscillation for the Wien-bridge oscillator in Figure 16–50.

▲ FIGURE 16–50

File circuits are identified with a CD logo and are in the Problems folder on your CD-ROM. Filenames correspond to figure numbers (e.g., F16-50).

11. What value of R_f is required in Figure 16–51? What is f_r?

▲ **FIGURE 16–51**

SECTION 16–4 *Oscillators with LC Feedback Circuits*

12. Calculate the frequency of oscillation for each circuit in Figure 16–52 and identify the type of oscillator. Assume $Q > 10$ in each case.

(a)   (b)

▲ **FIGURE 16–52**

13. Determine what the gain of the amplifier stage must be in Figure 16–53 in order to have sustained oscillation.

▶ FIGURE 16–53

SECTION 16–5 **Relaxation Oscillators**

14. What type of signal does the circuit in Figure 16–54 produce? Determine the frequency of the output.

▲ FIGURE 16–54

15. Show how to change the frequency of oscillation in Figure 16–54 to 10 kHz.

16. Determine the amplitude and frequency of the output voltage in Figure 16–55. Use 1 V as the forward PUT voltage.

▶ FIGURE 16–55

17. Modify the sawtooth generator in Figure 16–55 so that its peak-to-peak output is 4 V.

18. A certain sawtooth generator has the following parameter values: $V_{IN} = 3$ V, $R = 4.7$ kΩ, $C = 0.001$ μF. Determine its peak-to-peak output voltage if the period is 10 μs.

SECTION 16–6   The 555 Timer as an Oscillator

19. What are the two comparator reference voltages in a 555 timer when $V_{CC} = 10$ V?

20. Determine the frequency of oscillation for the 555 astable oscillator in Figure 16–56.

▲ **FIGURE 16–56**

21. To what value must C_{ext} be changed in Figure 16–56 to achieve a frequency of 25 kHz?

22. In an astable 555 configuration, the external resistor $R_1 = 3.3$ kΩ. What must R_2 equal to produce a duty cycle of 75 percent?

EWB/MULTISIM TROUBLESHOOTING PROBLEMS

These file circuits are in the Troubleshooting Problems folder on your CD-ROM.

23. Open file TSP16-23 and determine the fault.

24. Open file TSP16-24 and determine the fault.

25. Open file TSP16-25 and determine the fault.

26. Open file TSP16-26 and determine the fault.

27. Open file TSP16-27 and determine the fault.

28. Open file TSP16-28 and determine the fault.

ANSWERS

SECTION REVIEWS

SECTION 16–1   The Oscillator

1. An oscillator is a circuit that produces a repetitive output waveform with only the dc supply voltage as an input.

2. Positive feedback

3. The feedback circuit provides attenuation and phase shift.

4. Feedback and relaxation

SECTION 16–2 Feedback Oscillator Principles

1. Zero phase shift and unity voltage gain around the closed feedback

2. Positive feedback is when a portion of the output signal is fed back to the input of the amplifier such that it reinforces itself.

3. Loop gain greater than 1

SECTION 16–3 Oscillators with *RC* Feedback Circuits

1. The negative feedback loop sets the closed-loop gain; the positive feedback loop sets the frequency of oscillation.

2. 1.67 V

3. The three *RC* circuits contribute a total of 180°, and the inverting amplifier contributes 180° for a total of 360° around the loop.

SECTION 16–4 Oscillators with *LC* Feedback Circuits

1. Colpitts uses a capacitive voltage divider in the feedback circuit; Hartley uses an inductive voltage divider.

2. The higher FET input impedance has less loading effect on the resonant feedback circuit.

3. A Clapp has an additional capacitor in series with the inductor in the feedback circuit.

SECTION 16–5 Relaxation Oscillators

1. A voltage-controlled oscillator exhibits a frequency that can be varied with a dc control voltage.

2. The basis of a relaxation oscillator is the charging and discharging of a capacitor.

SECTION 16–6 The 555 Timer as an Oscillator

1. Two comparators, a flip-flop, a discharge transistor, and a resistive voltage divider

2. The duty cycle is set by the external resistors.

RELATED PROBLEMS FOR EXAMPLES

16–1 Instability may occur in either case.

16–2 **(a)** 238 kΩ   **(b)** 7.92 kHz

16–3 7.24 kHz

16–4 6.06 V peak-to-peak

16–5 1122 Hz

16–6 31.9%

SELF-TEST

1. (b)   **2.** (a)   **3.** (c)   **4.** (b)   **5.** (d)   **6.** (c)   **7.** (b)   **8.** (d)

9. (a)   **10.** (c)   **11.** (b)   **12.** (c)   **13.** (a)   **14.** (c)

17

COMMUNICATIONS CIRCUITS

INTRODUCTION

Communications electronics encompasses a wide range of systems, including both analog (linear) and digital. Any system that sends information from one point to another over relatively long distances can be classified as a communications system. Some of the categories of communications systems are radio (broadcast, ham, CB, marine), television, telephony, radar, navigation, satellite, data (digital), and telemetry.

Many communications systems use either amplitude modulation (AM) or frequency modulation (FM) to send information. Other modulation methods include pulse modulation, phase modulation, and frequency shift keying (FSK) as well as more specialized techniques. By necessity, the scope of this chapter is limited and is intended to introduce you to basic AM and FM communications systems and circuits.

- Describe basic superheterodyne receivers

- Discuss the function of a linear multiplier

- Discuss the fundamentals of amplitude modulation

- Discuss the basic function of a mixer

- Describe AM demodulation

- Describe IF and audio amplifiers

- Describe frequency modulation

- Describe a phase-locked loop (PLL)

KEY TERMS

- Amplitude modulation (AM)

- Frequency modulation (FM)

- Four-quadrant multiplier

- Balanced modulation

- Mixer

- Phase-locked loop (PLL)

- Lock range

- Capture range

■■■ SYSTEM APPLICATION PREVIEW

Digital data consisting of a series of binary digits (1s and 0s) are commonly sent from one computer to another over the telephone lines. Two voltage levels are used to represent the two types of bits, a high-voltage level and a low-voltage level. The data stream is made up of time intervals when the voltage has a constant high value or a constant low value with very fast transitions from one level to the other. In other words, the data stream contains very low frequencies (constant-voltage intervals) and very high frequencies (transitions). Since the standard telephone system has a bandwidth of approximately 300 Hz to 3000 Hz, it cannot handle the very low and the very high frequencies that make up a typical data stream without losing most of the information. Because of the bandwidth limitation of the telephone system, it is necessary to modify digital data before they are sent out; and one method of doing this is with frequency shift keying (FSK), which is a form of frequency modulation.

A simplified block diagram of a digital communications equipment (DCE) system for interfacing digital terminal equipment (DTE), such as a computer, to the telephone network is shown in the system application at the end of the chapter. The system FSK-modulates digital data before they are transmitted over the phone line and demodulates FSK signals received from another computer. Because the DTE's basic function is to *modulate* and *demodulate*, it is called a **modem**. Although the modem performs many associated functions, as indicated by the different blocks, in this system application our focus will be on the modulation and demodulation circuits.

WWW. **VISIT THE COMPANION WEBSITE**

Study aids for this chapter are available at

http://www.prenhall.com/floyd

17–1 BASIC RECEIVERS

Receivers based on the superheterodyne principle are standard in one form or another in most types of communications systems and are found in familiar systems such as standard broadcast radio, stereo, and television. In several of the system applications in previous chapters, we presented the superheterodyne receiver in order to focus on a given circuit; now we cover it from a system viewpoint. This section provides a basic introduction to amplitude modulation and frequency modulation and an overview of the complete AM and FM receiver.

After completing this section, you should be able to

■ **Describe basic superheterodyne receivers**

■ Define *AM* and *FM*

■ Discuss the major functional blocks of an AM receiver

■ Discuss the major functional blocks of an FM receiver

Amplitude Modulation (AM)

Amplitude modulation (AM) is a method for sending audible information, such as voice and music, by electromagnetic waves that are broadcast through the atmosphere. In AM, the amplitude of a signal with a specific frequency (f_c), called the *carrier*, is varied according to a modulating signal which can be an audio signal (such as voice or music), as shown in Figure 17–1. The carrier frequency permits the receiver to be tuned to a specific known frequency. The resulting AM waveform contains the carrier frequency, an upper-side frequency equal to the carrier frequency plus the modulation frequency ($f_c + f_m$), and a lower-side frequency equal to the carrier frequency minus the modulation frequency ($f_c - f_m$). Harmonics of these frequencies are also present. For example, if a 1 MHz carrier is amplitude modulated with a 5 kHz audio signal, the frequency components in the AM waveform are 1 MHz (carrier), 1 MHz + 5 kHz = 1,005,000 Hz (upper side), and 1 MHz − 5 kHz = 995,000 Hz (lower side).

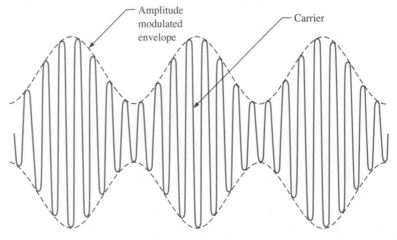

▲ **FIGURE 17–1**

An example of an amplitude modulated signal. In this case, the higher-frequency carrier is modulated by a lower-frequency sinusoidal signal.

The frequency band for AM broadcast receivers is 540 kHz to 1640 kHz. This means that an AM receiver can be tuned to pick up a specific carrier frequency that lies in the broadcast band. Each AM radio station transmits at a specific carrier frequency that is different from any other station in the area, so you can tune the receiver to pick up any desired station.

The Superheterodyne AM Receiver

A block diagram of a superheterodyne AM receiver is shown in Figure 17–2. The receiver shown consists of an antenna, an RF (radio frequency) amplifier, a mixer, a local oscillator (LO), an IF (intermediate frequency) amplifier, a detector, an audio amplifier, a power amplifier, and a speaker.

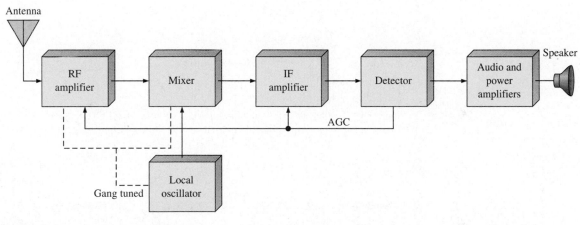

▲ FIGURE 17–2

Superheterodyne AM receiver block diagram.

Antenna The antenna picks up all radiated signals and feeds them into the RF amplifier. These signals are very small (usually only a few microvolts).

RF Amplifier This circuit can be adjusted (tuned) to select and amplify any carrier frequency within the AM broadcast band. Only the selected frequency and its two side bands pass through the amplifier. (Some AM receivers do not have a separate RF amplifier stage.)

Local Oscillator This circuit generates a steady sine wave at a frequency 455 kHz above the selected RF frequency.

Mixer This circuit accepts two inputs, the amplitude modulated RF signal from the output of the RF amplifier (or the antenna when there is no RF amplifier) and the sinusoidal output of the local oscillator (LO). These two signals are then "mixed" by a nonlinear process called *heterodyning* to produce sum and difference frequencies. For example, if the RF carrier has a frequency of 1000 kHz, the LO frequency is 1455 kHz and the sum and difference frequencies out of the mixer are 2455 kHz and 455 kHz, respectively. The difference frequency is always 455 kHz no matter what the RF carrier frequency.

IF Amplifier The input to the IF amplifier is the 455 kHz AM signal, a replica of the original AM carrier signal except that the frequency has been lowered to 455 kHz. The IF amplifier significantly increases the level of this signal.

Detector   This circuit recovers the modulating signal (audio signal) from the 455 kHz IF. At this point the IF is no longer needed, so the output of the detector consists of only the audio signal.

Audio and Power Amplifiers   This circuit amplifies the detected audio signal and drives the speaker to produce sound.

AGC   The automatic gain control (AGC) provides a dc level out of the detector that is proportional to the strength of the received signal. This level is fed back to the IF amplifier, and sometimes to the mixer and RF amplifier, to adjust the gains so as to maintain constant signal levels throughout the system over a wide range of incoming carrier signal strengths.

Figure 17–3 shows the signal flow through an AM superheterodyne receiver. The receiver can be tuned to accept any frequency in the AM band. The RF amplifier, mixer, and local oscillator are tuned simultaneously so that the LO frequency is always 455 kHz above the incoming RF signal frequency. This is called *gang tuning*.

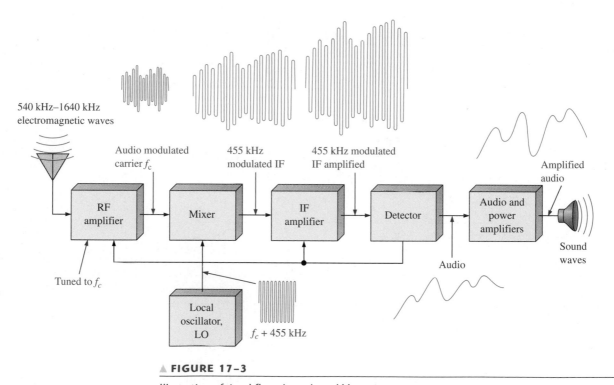

▲ **FIGURE 17–3**

Illustration of signal flow through an AM receiver.

Frequency Modulation (FM)

In this method of **modulation**, the modulating signal (audio) varies the frequency of a carrier as opposed to the amplitude, as in the case of AM. Figure 17–4 illustrates basic **frequency modulation (FM)**. The standard FM broadcast band consists of carrier frequencies from 88 MHz to 108 MHz, which is significantly higher than AM. The FM receiver is similar to the AM receiver in many ways, but there are several differences.

The Superheterodyne FM Receiver

A block diagram of a superheterodyne FM receiver is shown in Figure 17–5. Notice that it includes an RF amplifier, mixer, local oscillator, and IF amplifier just as in the AM

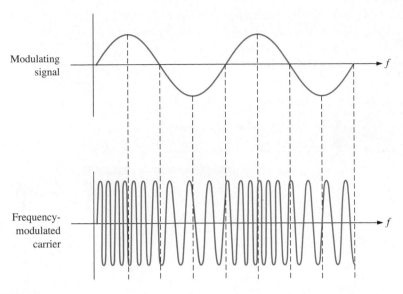

FIGURE 17–4

An example of frequency modulation.

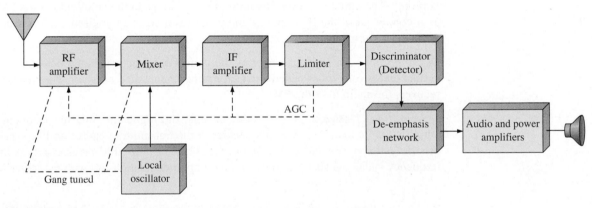

FIGURE 17–5

Superheterodyne FM receiver block diagram.

receiver. These circuits must, however, operate at higher frequencies than in the AM system. A significant difference in FM is the way the audio signal must be recovered from the modulated IF. This is accomplished by the limiter, discriminator, and de-emphasis network. Figure 17–6 depicts the signal flow through an FM receiver.

RF Amplifier This circuit must be capable of amplifying any frequency between 88 MHz and 108 MHz. It is highly selective so that it passes only the selected carrier frequency and significant side-band frequencies that contain the **audio**.

Local Oscillator This circuit produces a sine wave at a frequency 10.7 MHz above the selected RF frequency.

Mixer This circuit performs the same function as in the AM receiver, except that its output is a 10.7 MHz FM signal regardless of the RF carrier frequency.

IF Amplifier This circuit amplifies the 10.7 MHz FM signal.

▲ FIGURE 17–6

Example of signal flow through an FM receiver.

Limiter   The limiter removes any unwanted variations in the amplitude of the FM signal as it comes out of the IF amplifier and produces a constant amplitude FM output at the 10.7 MHz intermediate frequency.

Discriminator   This circuit performs the equivalent function of the detector in an AM system and is sometimes called a detector rather than a discriminator. The **discriminator** recovers the audio from the FM signal.

De-emphasis Network   For certain reasons, the higher modulating frequencies are amplified more than the lower frequencies at the transmitting end of an FM system by a process called *preemphasis*. The de-emphasis circuit in the FM receiver brings the high-frequency audio signals back to the proper amplitude relationship with the lower frequencies.

Audio and Power Amplifiers   This circuit is the same as in the AM system and can be shared when there is a dual AM/FM configuration.

SECTION 17–1 REVIEW

Answers are at the end of the chapter.

1. What do *AM* and *FM* mean?
2. How do AM and FM differ?
3. What are the standard broadcast frequency bands for AM and FM?

17–2 THE LINEAR MULTIPLIER

The linear multiplier is a key circuit in many types of communications systems. In this section, you will examine the basic principles of IC linear multipliers and look at a few applications that are found in communications as well as in other areas. In the following sections, we will concentrate on multiplier applications in AM and FM systems.

After completing this section, you should be able to

- **Discuss the function of a linear multiplier**
- Describe multiplier quadrants and transfer characteristic
- Discuss scale factor
- Show how to use a multiplier circuit as a multiplier, squaring circuit, divide circuit, square root circuit, and mean square circuit

Multiplier Quadrants

There are one-quadrant, two-quadrant, and **four-quadrant multipliers**. The quadrant classification indicates the number of input polarity combinations that the multiplier can handle. A graphical representation of the quadrants is shown in Figure 17–7. A four-quadrant multiplier can accept any of the four possible input polarity combinations and produce an output with the corresponding polarity.

▶ **FIGURE 17–7**

Four-quadrant polarities and their products.

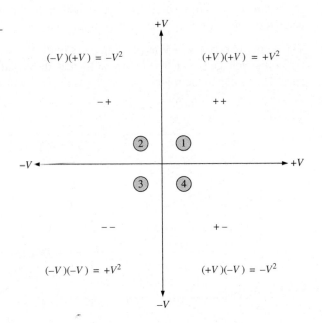

The Multiplier Transfer Characteristic

Figure 17–8 shows the transfer characteristic for a typical IC linear multiplier. To find the output voltage from the transfer characteristic graph, you find the intersection of the two input voltages V_X and V_Y. Values of V_X run along the horizontal axis and values of V_Y are the sloped lines. The output voltage is found by projecting the point of intersection over to the vertical axis. An example will illustrate this.

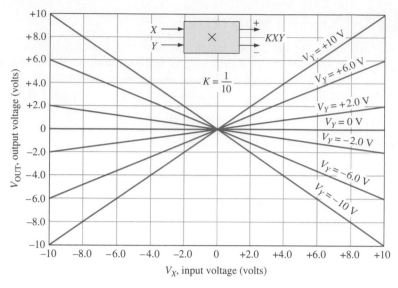

▲ **FIGURE 17–8**

A four-quadrant multiplier transfer characteristic.

EXAMPLE 17–1

Determine the output voltage for a four-quadrant linear multiplier whose transfer characteristic is given in Figure 17–8. The input voltages are $V_X = -4$ V and $V_Y = +10$ V.

Solution   The output voltage is **–4 V** as illustrated in Figure 17–9. For this transfer characteristic, the output voltage is a factor of ten smaller than the actual product of the two input voltages. This is due to the *scale factor* of the multiplier, which is discussed next.

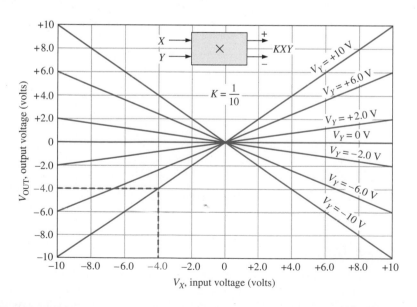

▲ **FIGURE 17–9**

Related Problem*   Find V_{OUT} if $V_X = -6$ V and $V_Y = +6$ V.

*Answers are at the end of the chapter.

The Scale Factor, K

The scale factor, K, is basically an internal attenuation that reduces the output by a fixed amount. The scale factor on most IC multipliers is adjustable and has a typical value of 0.1. Figure 17–10 shows an MC1495 configured as a basic multiplier. The scale factor is determined by external resistors, which include two equal load resistors, according to the following formula:

$$K = \frac{2R_L}{R_X R_Y I_{R2}}$$

The current I_{R2} is set by internal and external parameters according to this formula:

$$I_{R2} = \frac{|-V| - 0.7 \text{ V}}{R_2 + 500 \ \Omega}$$

where R_2 is the combination of the fixed resistor and the potentiometer. The potentiometer provides for fine adjustment by controlling I_{R2}.

▲ **FIGURE 17–10**

Basic MC1495 linear multiplier with external circuitry for setting the scale factor.

The expression for the output voltage of the IC linear multiplier includes the scale factor, K, as indicated in Equation 17–1.

$$V_{OUT} = K V_X V_Y$$

Equation 17–1

EXAMPLE 17–2

Determine the scale factor for the basic MC1495 multiplier in Figure 17–11. Assume the 5 kΩ potentiometer portion of R_2 is set to 2.5 kΩ. Also, determine the output voltage for the given inputs.

▲ **FIGURE 17–11**

Solution   Calculate I_{R2}.

$$I_{R2} = \frac{|-V| - 0.7\text{ V}}{R_2 + 500\ \Omega} = \frac{15\text{ V} - 0.7\text{ V}}{12.5\text{ k}\Omega + 500\ \Omega} = \frac{14.3\text{ V}}{13\text{ k}\Omega} = 1.1\text{ mA}$$

The scale factor is

$$K = \frac{2R_L}{R_X R_Y I_{R2}} = \frac{2(5.6\text{ k}\Omega)}{(10\text{ k}\Omega)(10\text{ k}\Omega)(1.1\text{ mA})} = \mathbf{0.102}$$

The output voltage is

$$V_{OUT} = K V_X V_Y = 0.102(+2\text{ V})(+3\text{ V}) = \mathbf{0.611\text{ V}}$$

Related Problem   What is the output voltage in Figure 17–11 if the 5 kΩ potentiometer is set to its maximum resistance?

Offset Adjustment

Due to internal mismatches, small offset voltages are usually at the inputs and the output of an IC linear multiplier. External circuits to null out the offset voltages are shown in Figure 17–12. The resistive voltage dividers on the inputs allow the actual input voltages to be greater than the recommended maximum for the device. For example, the MC1495 has a maximum input voltage of 5 V. The voltage dividers allow a maximum of 10 V to be applied if the resistors are of equal value. The zener diodes in the input offset adjust circuit keep the inputs on pins 8 and 12 from exceeding the maximum of 5 V.

Basic MC1495 multiplier with both scale factor and offset circuitry.

Basic Applications of the Multiplier

Applications of linear multipliers are numerous. Some basic applications are now presented.

Multiplier The most obvious application of a linear multiplier is, of course, to multiply two voltages as indicated in Figure 17–13.

FIGURE 17–13

Multiplier.

Squaring Circuit A special case of the multiplier is a squaring circuit that is realized by simply applying the same voltage to both inputs by connecting the inputs together as shown in Figure 17–14.

FIGURE 17–14

Squaring circuit.

Divide Circuit The circuit in Figure 17–15 shows the multiplier placed in the feedback loop of an op-amp. The basic operation is as follows. There is a virtual ground at the inverting ($-$) input of the op-amp and therefore the current at the inverting input is negligible. Therefore, I_1 and I_2 are equal. Since the inverting input voltage is 0 V, the voltage across R_1 is KV_YV_{OUT} and the current through R_1 is

$$I_1 = \frac{KV_YV_{OUT}}{R_1}$$

The voltage across R_2 is V_X, so the current through R_2 is

$$I_2 = \frac{V_X}{R_2}$$

Since $I_1 = -I_2$,

$$\frac{KV_YV_{OUT}}{R_1} = -\frac{V_X}{R_2}$$

Solving for V_{OUT},

$$V_{OUT} = -\frac{V_XR_1}{KV_YR_2}$$

If $R_1 = KR_2$,

$$V_{OUT} = -\frac{V_X}{V_Y}$$

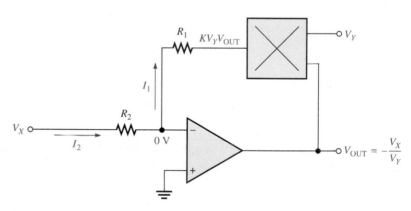

▲ FIGURE 17–15

Divide circuit.

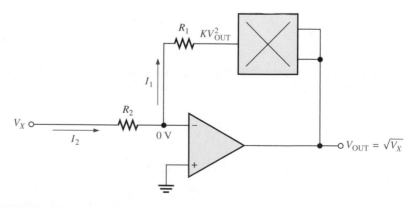

▲ FIGURE 17–16

Square root circuit.

Square Root Circuit   The square root circuit is a special case of the divide circuit where V_{OUT} is applied to both inputs of the multiplier as shown in Figure 17–16.

Mean Square Circuit   In this application, the multiplier is used as a squaring circuit with its output connected to an op-amp integrator as shown in Figure 17–17. The integrator produces the average or mean value of the squared input over time, as indicated by the integration sign ($\int$).

▲ FIGURE 17–17

Mean square circuit.

SECTION 17–2 REVIEW

1. Compare a four-quadrant multiplier to a one-quadrant multiplier in terms of the inputs that can be handled.

2. If 5 V and 1 V are applied to the inputs of a multiplier and its output is 0.5 V, what is the scale factor? What must the scale factor be for an output of 5 V?

3. How do you convert a basic multiplier to a squaring circuit?

17–3   AMPLITUDE MODULATION

Amplitude modulation (AM) is an important method for transmitting information. Of course, the AM superheterodyne receiver is designed to receive transmitted AM signals. In this section, we take a further look at amplitude modulation and show how the linear multiplier can be used as an amplitude-modulated device.

After completing this section, you should be able to

- **Discuss the fundamentals of amplitude modulation**

- Explain how AM is basically a multiplication process

- Describe sum and difference frequencies

- Discuss balanced modulation

- Describe the frequency spectra

- Explain standard AM

As you learned in Section 17–1, amplitude modulation is the process of varying the amplitude of a signal of a given frequency (carrier) with another signal of much lower frequency (modulating signal). One reason that the higher-frequency carrier signal is necessary is because audio or other signals with relatively low frequencies cannot be transmitted with antennas of a practical size. The basic concept of standard amplitude modulation is illustrated in Figure 17–18.

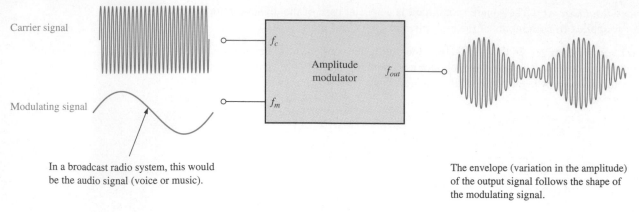

▲ **FIGURE 17–18**

Basic concept of amplitude modulation.

A Multiplication Process

If a signal is applied to the input of a variable-gain device, the resulting output is an amplitude-modulated signal because $V_{out} = A_v V_{in}$. The output voltage is the input voltage multiplied by the voltage gain. For example, if the gain of an amplifier is made to vary sinusoidally at a certain frequency and an input signal is applied at a higher frequency, the output signal will have the higher frequency. However, its amplitude will vary according to the variation in gain as illustrated in Figure 17–19. Amplitude modulation is basically a multiplication process (input voltage multiplied by a variable gain).

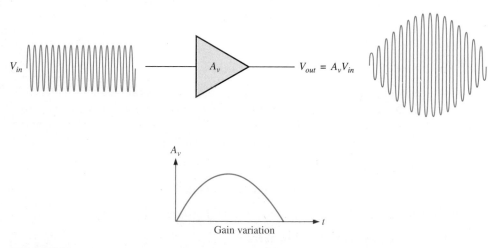

▲ **FIGURE 17–19**

The amplitude of the output voltage varies according to the gain and is the product of voltage gain and input voltage.

Sum and Difference Frequencies

If the expressions for two sinusoidal signals of different frequencies are multiplied mathematically, a term containing both the difference and the sum of the two frequencies is produced. Recall from ac circuit theory that a sinusoidal voltage can be expressed as

$$v = V_p \sin 2\pi ft$$

where V_p is the peak voltage and f is the frequency. Two different sinusoidal signals can be expressed as follows:

$$v_1 = V_{1(p)}\sin 2\pi f_1 t$$
$$v_2 = V_{2(p)}\sin 2\pi f_2 t$$

Multiplying these two sinusoidal wave terms,

$$v_1 v_2 = (V_{1(p)}\sin 2\pi f_1 t)(V_{2(p)}\sin 2\pi f_2 t)$$
$$= V_{1(p)}V_{2(p)}(\sin 2\pi f_1 t)(\sin 2\pi f_2 t)$$

The general trigonometric identity for the product of two sinusoidal functions is

$$(\sin A)(\sin B) = \frac{1}{2}[\cos(A - B) - \cos(A + B)]$$

Applying this identity to the previous formula for $v_1 v_2$,

$$v_1 v_2 = \frac{V_{1(p)}V_{2(p)}}{2}[\cos(2\pi f_1 t - 2\pi f_2 t) - \cos(2\pi f_1 t + 2\pi f_2 t)]$$

$$= \frac{V_{1(p)}V_{2(p)}}{2}[\cos 2\pi(f_1 - f_2)t - \cos 2\pi(f_1 + f_2)t]$$

$$v_1 v_2 = \frac{V_{1(p)}V_{2(p)}}{2}\cos 2\pi(f_1 - f_2)t - \frac{V_{1(p)}V_{2(p)}}{2}\cos 2\pi(f_1 + f_2)t$$

Equation 17–2

You can see in Equation 17–2 that the product of the two sinusoidal voltages V_1 and V_2 contains a difference frequency $(f_1 - f_2)$ and a sum frequency $(f_1 + f_2)$. The fact that the product terms are cosine simply indicates a 90° phase shift in the multiplication process.

Analysis of Balanced Modulation

Since amplitude modulation is simply a multiplication process, the preceding analysis is now applied to carrier and modulating signals. The expression for the sinusoidal carrier signal can be written as

$$v_c = V_{c(p)}\sin 2\pi f_c t$$

Assuming a sinusoidal modulating signal, it can be expressed as

$$v_m = V_{m(p)}\sin 2\pi f_m t$$

Substituting these two signals in Equation 17–2,

$$v_c v_m = \frac{V_{c(p)}V_{m(p)}}{2}\cos 2\pi(f_c - f_m)t - \frac{V_{c(p)}V_{m(p)}}{2}\cos 2\pi(f_c + f_m)t$$

An output signal described by this expression for the product of two sinusoidal signals is produced by a linear multiplier. Notice that there is a difference frequency term $(f_c - f_m)$ and a sum frequency term $(f_c + f_m)$, but the original frequencies, f_c and f_m, do not appear alone in the expression. Thus, the product of two sinusoidal signals contains no signal with the carrier frequency, f_c, or with the modulating frequency, f_m. This form of amplitude modulation is called **balanced modulation** because there is no carrier frequency in the output. The carrier frequency is "balanced out."

The Frequency Spectra of a Balanced Modulator

A graphical picture of the frequency content of a signal is called its frequency spectrum. A frequency spectrum shows voltage on a frequency base rather than on a time base as a waveform diagram does. The frequency spectra of the product of two sinusoidal signals are shown in Figure 17–20. Part (a) shows the two input frequencies and part (b) shows

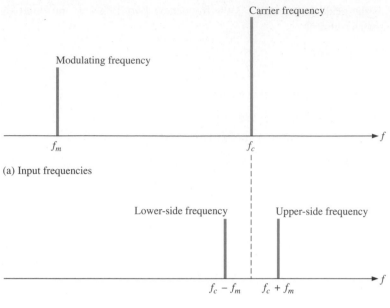

Carrier frequency

Modulating frequency

f_m f_c f

(a) Input frequencies

Lower-side frequency   Upper-side frequency

$f_c - f_m$ $f_c + f_m$ f

(b) Output frequencies

▲ **FIGURE 17–20**

Illustration of the input and output frequency spectra for a linear multiplier.

the output frequencies. In communications terminology, the sum frequency is called the **upper-side frequency** and the difference frequency is called the **lower-side frequency** because the frequencies appear on each side of the missing carrier frequency.

The Linear Multiplier as a Balanced Modulator

As mentioned, the linear multiplier acts as a balanced modulator when a carrier signal and a modulating signal are applied to its inputs, as illustrated in Figure 17–21. A balanced modulator produces an upper-side frequency and a lower-side frequency, but it does not produce a carrier frequency. Since there is no carrier signal, balanced modulation is sometimes known as **suppressed-carrier modulation**. Balanced modulation is used in certain types of communications such as single side-band systems, but it is not used in standard AM broadcast systems.

f_m

f_c

$f_c - f_m$ f_c $f_c + f_m$ f

▲ **FIGURE 17–21**

The linear multiplier as a balanced modulator.

EXAMPLE 17–3

Determine the frequencies contained in the output signal of the balanced modulator in Figure 17–22.

$f_c = 5$ MHz

$f_m = 10$ kHz

f_{out}

▲ **FIGURE 17–22**

Solution   The upper-side frequency is

$$f_c + f_m = 5 \text{ MHz} + 10 \text{ kHz} = \textbf{5.01 MHz}$$

The lower-side frequency is

$$f_c - f_m = 5 \text{ MHz} - 10 \text{ kHz} = \textbf{4.99 MHz}$$

Related Problem   Explain how the separation between the side frequencies can be increased using the same carrier frequency.

Standard Amplitude Modulation (AM)

In standard AM systems, the output signal contains the carrier frequency as well as the sum and difference side frequencies. Standard amplitude modulation is illustrated by the frequency spectrum in Figure 17–23.

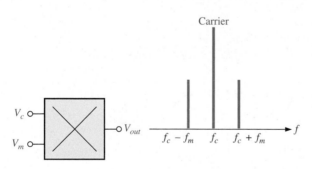

V_c

V_m

V_{out}

Carrier

$f_c - f_m \quad f_c \quad f_c + f_m$

f

▲ **FIGURE 17–23**

The output frequency spectrum of a standard amplitude modulator.

The expression for a standard amplitude-modulated signal is

$$V_{out} = V_{c(p)}^2 \sin 2\pi f_c t + \frac{V_{c(p)}V_{m(p)}}{2}\cos 2\pi(f_c - f_m)t - \frac{V_{c(p)}V_{m(p)}}{2}\cos 2\pi(f_c + f_m)t$$

Equation 17–3

Notice in Equation 17–3 that the first term is for the carrier frequency and the other two terms are for the side frequencies. Let's see how the carrier-frequency term gets into the equation.

If a dc voltage equal to the peak of the carrier voltage is added to the modulating signal before the modulating signal is multiplied by the carrier signal, a carrier-signal term appears in the final result as shown in the following steps. Add the peak carrier voltage to the modulating signal, and you get the following expression:

$$V_{c(p)} + V_{m(p)}\sin 2\pi f_m t$$

Multiply by the carrier signal.

$$V_{out} = (V_{c(p)}\sin 2\pi f_c t)(V_{c(p)} + V_{m(p)}\sin 2\pi f_m t)$$
$$= \underbrace{V_{c(p)}^2 \sin 2\pi f_c t}_{\text{carrier term}} + \underbrace{V_{c(p)}V_{m(p)}(\sin 2\pi f_c t)(\sin 2\pi f_m t)}_{\text{product term}}$$

Apply the basic trigonometric identity to the product term.

$$V_{out} = V_{c(p)}^2 \sin 2\pi f_c t + \frac{V_{c(p)}V_{m(p)}}{2}\cos 2\pi (f_c - f_m)t - \frac{V_{c(p)}V_{m(p)}}{2}\cos 2\pi (f_c + f_m)t$$

This result shows that the output of the multiplier contains a carrier term and two side-frequency terms. Figure 17–24 illustrates how a standard amplitude modulator can be implemented by a summing circuit followed by a linear multiplier. Figure 17–25 shows a possible implementation of the summing circuit.

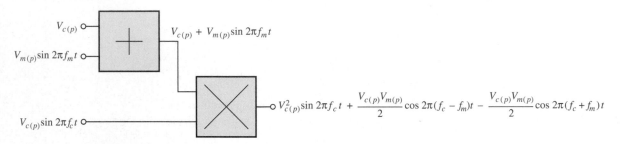

▲ FIGURE 17–24

Basic block diagram of an amplitude modulator.

▲ FIGURE 17–25

Implementation of the summing circuit in the amplitude modulator.

EXAMPLE 17–4

A carrier frequency of 1200 kHz is modulated by a sinusoidal wave with a frequency of 25 kHz by a standard amplitude modulator. Determine the output frequency spectrum.

Solution   The lower-side frequency is

$$f_c - f_m = 1200 \text{ kHz} - 25 \text{ kHz} = \mathbf{1175 \text{ kHz}}$$

The upper-side frequency is

$$f_c + f_m = 1200 \text{ kHz} + 25 \text{ kHz} = \mathbf{1225 \text{ kHz}}$$

The output contains the carrier frequency and the two side frequencies as shown in Figure 17–26.

▲ **FIGURE 17–26**

Related Problem   Compare the output frequency spectrum in this example to that of a balanced modulator having the same inputs.

Amplitude Modulation with Voice or Music

To this point in our discussion, we have considered the modulating signal to be a pure sinusoidal signal just to keep things fairly simple. If you receive an AM signal modulated by a pure sinusoidal signal in the audio frequency range, you will hear a single tone from the receiver's speaker.

A voice or music signal consists of many sinusoidal components within a range of frequencies from about 20 Hz to 20 kHz. For example, if a carrier frequency is amplitude modulated with voice or music with frequencies from 100 Hz to 10 kHz, the frequency spectrum is as shown in Figure 17–27. Instead of one lower-side and one upper-side frequency as in the case of a single-frequency modulating signal, a band of lower-side frequencies and a band of upper-side frequencies correspond to the sum and difference frequencies of each sinusoidal component of the voice or music signal.

Carrier
1000 kHz

Lower-side band   Upper-side band

                            f
990 kHz   999.9 kHz   1000.1 kHz   1010 kHz

▲ FIGURE 17–27

Example of a frequency spectrum for a voice or music signal.

SECTION 17–3 REVIEW	1. What is amplitude modulation?

1. What is amplitude modulation?
2. What is the difference between balanced modulation and standard AM?
3. What two input signals are used in amplitude modulation? Explain the purpose of each signal.
4. What are the upper-side frequency and the lower-side frequency?
5. How can a balanced modulator be changed to a standard amplitude modulator?

17–4 THE MIXER

The mixer in the receiver system discussed in Section 17–1 can be implemented with a linear multiplier as you will see in this section. The basic principles of linear multiplication of sinusoidal signals are covered, and you will see how sum and difference frequencies are produced. The difference frequency is a critical part of the operation of many types of receiver systems.

After completing this section, you should be able to

■ **Discuss the basic function of a mixer**

■ Explain why a mixer is a linear multiplier

■ Describe the frequencies in the mixer and IF portion of a receiver

The **mixer** is basically a frequency converter because it changes the frequency of a signal to another value. The mixer in a receiver system takes the incoming modulated RF signal (which is sometimes amplified by an RF amplifier and sometimes not) along with the signal from the local oscillator and produces a modulated signal with a frequency equal to the difference of its two input frequencies (RF and LO). The mixer also produces a frequency equal to the sum of the input frequencies. The mixer function is illustrated in Figure 17–28.

The Mixer Is a Linear Multiplier

In the case of receiver applications, the mixer must produce an output that has a frequency component equal to the difference of its input frequencies. From the mathematical

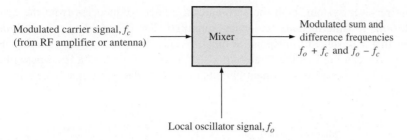

▲ FIGURE 17–28

The mixer function.

analysis in Section 17–3, you can see that if two sinusoidal signals are multiplied, the product contains the difference frequency and the sum frequency. Thus, the mixer is actually a linear multiplier as indicated in Figure 17–29.

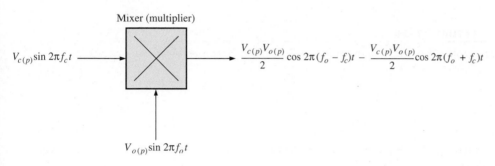

▲ FIGURE 17–29

The mixer as a linear multiplier.

EXAMPLE 17–5

Determine the output expression for a multiplier with one sinusoidal input having a peak voltage of 5 mV and a frequency of 1200 kHz and the other input having a peak voltage of 10 mV and a frequency of 1655 kHz.

Solution The two input expressions are

$$v_1 = (5 \text{ mV})\sin 2\pi(1200 \text{ kHz})t$$
$$v_2 = (10 \text{ mV})\sin 2\pi(1655 \text{ kHz})t$$

Multiplying,

$$v_1 v_2 = (5 \text{ mV})(10 \text{ mV})[\sin 2\pi(1200 \text{ kHz})t][\sin 2\pi(1655 \text{ kHz})t]$$

Applying the trigonometric identity, $(\sin A)(\sin B) = \frac{1}{2}[\cos(A - B) - \cos(A + B)]$,

$$V_{out} = \frac{(5 \text{ mV})(10 \text{ mV})}{2}\cos 2\pi(1655 \text{ kHz} - 1200 \text{ kHz})t$$

$$- \frac{(5 \text{ mV})(10 \text{ mV})}{2}\cos 2\pi(1655 \text{ kHz} + 1200 \text{ kHz})t$$

$$\mathbf{V_{out} = (25 \ \mu V)\cos 2\pi(455 \text{ kHz})t - (25 \ \mu V)\cos 2\pi(2855 \text{ kHz})t}$$

Related Problem What is the value of the peak amplitude and frequency of the difference frequency component in this example?

In the receiver system, both the sum and difference frequencies from the mixer are applied to the IF (intermediate frequency) amplifier. The IF amplifier is actually a tuned amplifier that is designed to respond to the difference frequency while rejecting the sum frequency. You can think of the IF amplifier section of a receiver as a band-pass filter plus an amplifier because it uses resonant circuits to provide the frequency selectivity. This is illustrated in Figure 17–30.

▲ **FIGURE 17–30**

Example of frequencies in the mixer and IF portion of a receiver.

EXAMPLE 17–6

Determine the output frequency of the IF amplifier for the conditions shown in Figure 17–31.

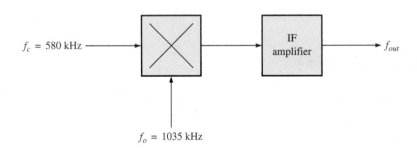

▲ **FIGURE 17–31**

Solution   The IF amplifier produces only the difference frequency signal on its output.

$$f_{out} = f_{diff} = f_o - f_c = 1035 \text{ kHz} - 580 \text{ kHz} = \textbf{455 kHz}$$

Related Problem   Based on your basic knowledge of the superheterodyne receiver from Section 17–1, determine the IF output frequency when the incoming RF signal changes to 1550 kHz.

**SECTION 17–4
REVIEW**

1. What is the purpose of the mixer in a superheterodyne receiver?
2. How does the mixer produce its output?
3. If a mixer has 1000 kHz on one input and 350 kHz on the other, what frequencies appear on the output?

17–5 AM DEMODULATION

The linear multiplier can be used to demodulate or detect an AM signal as well as to perform the modulation process that was discussed in Section 17–3. Demodulation can be thought of as reverse modulation. The purpose is to get back the original modulating signal (sound in the case of standard AM receivers). The detector in the AM receiver can be implemented using a multiplier, although another method using peak envelope detection is common.

After completing this section, you should be able to

- **Describe AM demodulation**
- Discuss a basic AM demodulator
- Discuss the frequency spectra

The Basic AM Demodulator

An AM demodulator can be implemented with a linear multiplier followed by a low-pass filter, as shown in Figure 17–32. The critical frequency of the filter is the highest audio frequency that is required for a given application (15 kHz, for example).

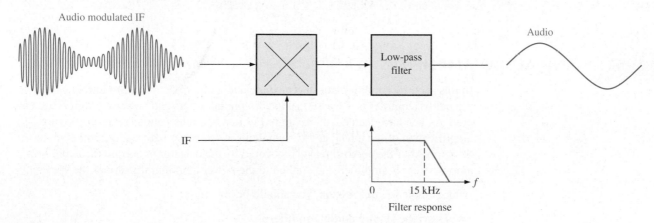

▲ **FIGURE 17–32**

Basic AM demodulator.

Operation in Terms of the Frequency Spectra

Let's assume a carrier modulated by a single tone with a frequency of 10 kHz is received and converted to a modulated intermediate frequency of 455 kHz, as indicated by the frequency spectra in Figure 17–33. Notice that the upper-side and lower-side frequencies are separated from both the carrier and the IF by 10 kHz.

▲ **FIGURE 17–33**

An AM signal converted to IF.

When the modulated output of the IF amplifier is applied to the demodulator along with the IF, sum and difference frequencies for each input frequency are produced as shown in Figure 17–34. Only the 10 kHz audio frequency is passed by the filter. A drawback to this type of AM detection is that a pure IF must be produced to mix with the modulated IF.

▲ FIGURE 17–34

Example of demodulation.

SECTION 17–5 REVIEW

1. What is the purpose of the filter in the linear multiplier demodulator?
2. If a 455 kHz IF modulated by a 1 kHz audio frequency is demodulated, what frequency or frequencies appear on the output of the demodulator?

17–6 IF AND AUDIO AMPLIFIERS

In this section, amplifiers for intermediate and audio frequencies are introduced. A typical IF amplifier is discussed and audio preamplifiers and power amplifiers are covered. As you have learned, the IF amplifier in a communications receiver provides amplification of the modulated IF signal out of the mixer before it is applied to the detector. After the audio signal is recovered by the detector, it goes to the audio preamp where it is amplified and applied to the power amplifier that drives the speaker.

After completing this section, you should be able to

- **Describe IF and audio amplifiers**
- Discuss the function of an IF amplifier
- Explain how the local oscillator and mixer operate with the IF amplifier
- Discuss the MC1350 IF amplifier
- State the purpose of the audio amplifier
- Discuss the LM386 audio power amplifier

The Basic Function of the IF Amplifier

The IF amplifier in a receiver is a tuned amplifier with a specified bandwidth operating at a center frequency of 455 kHz for AM and 10.7 MHz for FM. The IF amplifier is one of the key features of a superheterodyne receiver because it is set to operate at a single resonant frequency that remains the same over the entire band of carrier frequencies that can be received. Figure 17–35 illustrates the basic function of an IF amplifier in terms of the frequency spectra.

Assume, for example, that the received carrier frequency of $f_c = 1$ MHz is modulated by an audio signal with a maximum frequency of $f_m = 5$ kHz, indicated in Figure 17–35

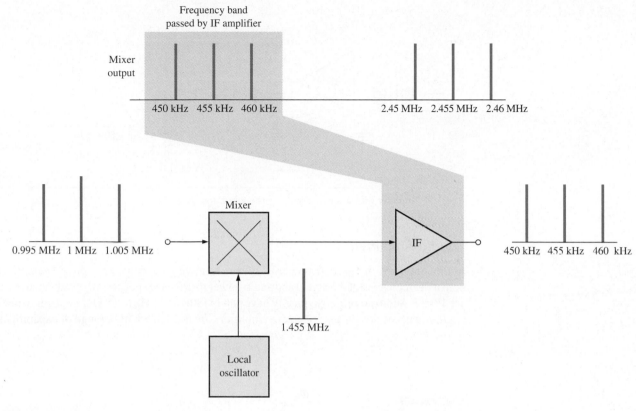

▲ **FIGURE 17–35**

An illustration of the basic function of the IF amplifier in an AM receiver.

by the frequency spectrum on the input to the mixer. For this frequency, the local oscillator is at a frequency of

$$f_o = 1 \text{ MHz} + 455 \text{ kHz} = 1.455 \text{ MHz}$$

The mixer produces the following sum and difference frequencies as indicated in Figure 17–35.

$$f_o + f_c = 1.455 \text{ MHz} + 1 \text{ MHz} = 2.455 \text{ MHz}$$
$$f_o - f_c = 1.455 \text{ MHz} - 1 \text{ MHz} = 455 \text{ kHz}$$
$$f_o + (f_c + f_m) = 1.455 \text{ MHz} + 1.005 \text{ MHz} = 2.46 \text{ MHz}$$
$$f_o + (f_c - f_m) = 1.455 \text{ MHz} + 0.995 \text{ MHz} = 2.45 \text{ MHz}$$
$$f_o - (f_c + f_m) = 1.455 \text{ MHz} - 1.005 \text{ MHz} = 450 \text{ kHz}$$
$$f_o - (f_c - f_m) = 1.455 \text{ MHz} - 0.995 \text{ MHz} = 460 \text{ kHz}$$

Since the IF amplifier is a frequency-selective circuit, it responds only to 455 kHz and any side frequencies lying in the 10 kHz band centered at 455 kHz. So, all of the frequencies out of the mixer are rejected except the 455 kHz IF, all lower-side frequencies down to 450 kHz, and all upper-side frequencies up to 460 kHz. This frequency spectrum is the audio modulated IF.

A Basic IF Amplifier

Although the detailed circuitry of the IF amplifier may differ from one system to another, it always has a tuned (resonant) circuit on the input or on the output or on both. Figure 17–36(a) shows a basic IF amplifier with tuned transformer coupling at the input and output. The general frequency response curve is shown in Figure 17–36(b).

(a)   (b)

▲ **FIGURE 17–36**

A basic IF amplifier with a tuned circuit on the input and output.

Audio Amplifiers

Audio amplifiers are used in a receiver system following the detector to provide amplification of the recovered audio signal and audio power to drive the speaker(s), as indicated in Figure 17–37. Audio amplifiers typically have bandwidths of 3 kHz to 15 kHz depending on the requirements of the system. IC audio amplifiers are available with a range of capabilities.

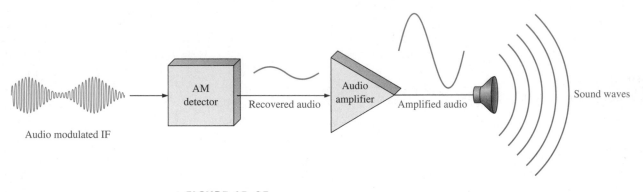

▲ **FIGURE 17–37**

The audio amplifier in a receiver system.

The LM386 Audio Power Amplifier This device is an example of a low-power audio amplifier that is capable of providing several hundred milliwatts to an 8 Ω speaker. It operates from any dc supply voltage in the 4 V to 12 V range, making it a good choice for battery operation. The pin configuration of the LM386 is shown in Figure 17–38(a). The voltage gain of the LM386 is 20 without external connections to the gain terminals, as shown in

(a)

(b) $A_v = 20$ (c) $A_v = 200$ (d) $20 < A_v < 200$

▲ **FIGURE 17–38**

Pin configuration and gain connections for the LM386 audio amplifier.

Figure 17–38(b). A voltage gain of 200 is achieved by connecting a 10 μF capacitor from pin 1 to pin 8, as shown in Figure 17–38(c). Voltage gains between 20 and 200 can be realized by a resistor (R_G) and capacitor (C_G) connected in series from pin 1 to pin 8 as shown in Figure 17–38(d). These external components are effectively placed in parallel with an internal gain-setting resistor.

A typical application of the LM386 as a power amplifier in a radio receiver is shown in Figure 17–39. Here the detected AM signal is fed to the inverting input through the volume control potentiometer, R_1, and resistor R_2. C_1 is the input coupling capacitor and C_2 is the power supply decoupling capacitor. R_2 and C_3 filter out any residual RF or IF signal that may be on the output of the detector. R_3 and C_6 provide additional filtering before the audio signal is applied to the speaker through the coupling capacitor C_7.

▲ **FIGURE 17–39**

The LM386 as an AM audio power amplifier.

SECTION 17–6 REVIEW

1. What is the purpose of the IF amplifier in an AM receiver?
2. What is the center frequency of an AM IF amplifier?
3. Why is the bandwidth of an AM receiver IF amplifier 10 kHz?
4. Why must the audio amplifier follow the detector in a receiver system?
5. Compare the frequency response of the IF amplifier to that of the audio amplifier.

17–7 FREQUENCY MODULATION

As you have seen, modulation is the process of varying a parameter of a carrier signal with an information signal. Recall that in amplitude modulation the parameter of amplitude is varied. In frequency modulation (FM), the frequency of a carrier is varied above and below its normal or at-rest value by a modulating signal. This section provides a further look into FM and discusses the differences between an AM and an FM receiver.

After completing this section, you should be able to

- **Describe frequency modulation**
- Discuss the voltage-controlled oscillator
- Describe the MC2833D FM modulator
- Explain frequency demodulation

In a frequency-modulated (FM) signal, the carrier frequency is increased or decreased according to the modulating signal. The amount of deviation above or below the carrier frequency depends on the amplitude of the modulating signal. The rate at which the frequency deviation occurs depends on the frequency of the modulating signal.

Figure 17–40 illustrates both a square wave and a sine wave modulating the frequency of a carrier. The carrier frequency is highest when the modulating signal is at its maximum positive amplitude and is lowest when the modulating signal is at its maximum negative amplitude.

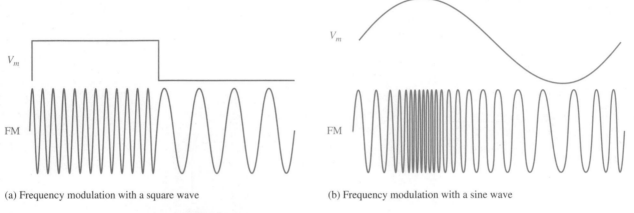

(a) Frequency modulation with a square wave

(b) Frequency modulation with a sine wave

▲ FIGURE 17–40

Examples of frequency modulation.

A Basic Frequency Modulator

Frequency modulation is achieved by varying the frequency of an oscillator with the modulating signal. A **voltage-controlled oscillator (VCO)** is typically used for this purpose, as illustrated in Figure 17–41.

▲ FIGURE 17–41

Frequency modulation with a voltage–controlled oscillator.

Generally, a variable-reactance type of voltage-controlled oscillator is used in FM applications. The variable-reactance VCO uses the varactor diode as a voltage-variable capacitance, as illustrated in Figure 17–42, where the capacitance is varied with the modulating voltage, V_m.

► FIGURE 17–42

Basic variable–reactance VCO.

An Integrated Circuit FM Transmitter

An example of a single-chip FM transmitter is the MC2833D, which is designed for cordless telephone and other FM communications equipment. The 16-pin package configuration showing the basic functional blocks is illustrated in Figure 17–43. The microphone amplifier (mic amp) amplifies the low-level input from a microphone and feeds its output to the variable-reactance circuit which uses a varactor diode tuning circuit to control the RF oscillator. The reference voltage circuit (V_{REF}) provides stable bias to the reactance circuit. The two individual transistors can be connected as tuned amplifiers to boost the power output.

► FIGURE 17–43

The MC2833D FM modulator.

A typical VHF narrow-band FM transmitter using the MC2833D appropriate external circuitry is shown in Figure 17–44. This particular implementation has an output frequency of 49.7 MHz. The frequency of the oscillator is set by the external 16.5667 MHz

▲ FIGURE 17–44

The MC2833D connected as a 49.7 MHz FM transmitter.

crystal. The reactance circuit deviates this frequency with the amplified audio input to produce an FM signal. The 16.5667 MHz output of the oscillator goes to the buffer and is then applied to the input of Q_2, which is operated as a frequency tripler (16.5667 MHz × 3 = 49.7 MHz). A frequency tripler is basically a class C amplifier with the output tuned to a resonant frequency equal to three times the input. The signal from the Q_2 frequency tripler goes to the input of Q_1, which functions as a linear amplifier that drives the transmitting antenna from its resonant output circuit.

FM Demodulation

Except for the higher frequencies, the standard broadcast FM receiver is basically the same as the AM receiver up through the IF amplifier. The main difference between an FM receiver and an AM receiver, other than the frequency band, is the method used to recover the audio signal from the modulated IF.

There are several methods for demodulating an FM signal. These include slope detection, phase-shift discrimination, ratio detection, quadrature detection, and phase-locked loop demodulation. Most of these methods are covered in detail in communications courses. However, because of its importance in many types of applications, we will cover the phase-locked loop (PLL) demodulation in the next section.

SECTION 17–7 REVIEW

1. How does an FM signal carry information?
2. On what principle are most VCOs used in FM based?

17–8 THE PHASE-LOCKED LOOP (PLL)

In the last section, the PLL was mentioned as a way to demodulate an FM signal. In addition to FM demodulation, PLLs are used in a wide variety of communications applications, which include TV receivers, tone decoders, telemetry receivers, modems, and data synchronizers, to name a few. Many of these applications are covered in electronic communications courses. In fact, entire books have been written on the finer points of PLL operation, analysis, and applications. The approach in this section is intended only to present the basic concept and give you an intuitive idea of how PLLs work and how they are used in FM demodulation. A specific PLL integrated circuit is also introduced.

After completing this section, you should be able to

■ **Describe a phase-locked loop (PLL)**

■ Draw a basic block diagram for the PLL

■ Discuss the phase detector and state its purpose

■ State the purpose of the VCO

■ State the purpose of the low-pass filter

■ Explain lock range and capture range

■ Discuss the LM565 PLL and explain how it can be used as an FM demodulator

The Basic PLL Concept

The **phase-locked loop (PLL)** is a feedback circuit consisting of a phase detector, a low-pass filter, and a voltage-controlled oscillator (VCO). Some PLLs also include an amplifier in the loop, and in some applications the filter is not used.

The PLL is capable of locking onto or synchronizing with an incoming signal. When the phase of the incoming signal changes, indicating a change in frequency, the phase detector's output increases or decreases just enough to keep the VCO frequency the same as the frequency of the incoming signal. A basic PLL block diagram is shown in Figure 17–45.

The general operation of a PLL is as follows. The phase detector compares the phase difference between the incoming signal, V_i, and the VCO signal, V_o. When the frequency of the incoming signal, f_i, is different from that of the VCO frequency, f_o, the phase angle between the two signals is also different. The output of the phase detector and the filter is proportional to the phase difference of the two signals. This proportional voltage is fed to the VCO, forcing its frequency to move toward the frequency of the incoming signal until the two frequencies are equal. At this point, the PLL is locked onto the incoming frequency. If f_i changes, the phase difference also changes, forcing the VCO to track the incoming frequency.

▲ **FIGURE 17–45**

Basic PLL block diagram.

The Phase Detector

The phase-detector circuit in a PLL is basically a linear multiplier. The following analysis illustrates how it works in a PLL application. The incoming signal, V_i, and the VCO signal, V_o, applied to the phase detector can be expressed as

$$v_i = V_i\sin(2\pi f_i t + \theta_i)$$
$$v_o = V_o\sin(2\pi f_o t + \theta_o)$$

where θ_i and θ_o are the relative phase angles of the two signals. The phase detector multiplies these two signals and produces a sum and difference frequency output, V_d, as follows:

$$V_d = V_i\sin(2\pi f_i t + \theta_i) \times V_o\sin(2\pi f_o t + \theta_o)$$
$$= \frac{V_i V_o}{2}\cos[(2\pi f_i t + \theta_i) - (2\pi f_o t + \theta_o)] - \frac{V_i V_o}{2}\cos[(2\pi f_i t + \theta_i) + (2\pi f_o t + \theta_o)]$$

When the PLL is locked,

$$f_i = f_o$$

and

$$2\pi f_i t = 2\pi f_o t$$

Therefore, the detector output voltage is

$$V_d = \frac{V_i V_o}{2}[\cos(\theta_i - \theta_o) - \cos(4\pi f_i t + \theta_i + \theta_o)]$$

The second cosine term in the above equation is a second harmonic term ($2 \times 2\pi f_i t$) and is filtered out by the low-pass filter. The control voltage on the output of the filter is expressed as

Equation 17–4

$$V_c = \frac{V_i V_o}{2}\cos\theta_e$$

where $\theta_e = \theta_i - \theta_o$, where θ_e is the *phase error*. The filter output voltage is proportional to the phase difference between the incoming signal and the VCO signal and is used as the control voltage for the VCO. This operation is illustrated in Figure 17–46.

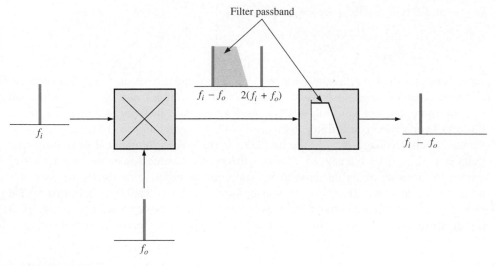

▲ FIGURE 17–46

Basic phase detector/filter operation.

EXAMPLE 17–7

A PLL is locked onto an incoming signal with a frequency of 1 MHz at a phase angle of 50°. The VCO signal is at a phase angle of 20°. The peak amplitude of the incoming signal is 0.5 V and that of the VCO output signal is 0.7 V.

(a) What is the VCO frequency?

(b) What is the value of the control voltage being fed back to the VCO at this point?

Solution **(a)** Since the PLL is in lock, $f_i = f_o = $ **1 MHz**.

(b) $\theta_e = \theta_i - \theta_o = 50° - 20° = 30°$

$$V_c = \frac{V_i V_o}{2}\cos \theta_e = \frac{(0.5 \text{ V})(0.7 \text{ V})}{2}\cos 30° = (0.175 \text{ V})\cos 30° = \textbf{0.152 V}$$

Related Problem If the phase angle of the incoming signal changes instantaneously to 30°, indicating a change in frequency, what is the instantaneous VCO control voltage?

The Voltage-Controlled Oscillator (VCO)

Voltage-controlled oscillators can take many forms. A VCO can be some type of *LC* or crystal oscillator, as you learned in Section 17–7, or it can be some type of *RC* oscillator or multivibrator, as you learned in Chapter 16. No matter the exact type, most VCOs employed in PLLs operate on the principle of *variable reactance* using the varactor diode as a voltage-variable capacitor.

The capacitance of a varactor diode varies inversely with reverse-bias voltage. The capacitance decreases as reverse voltage increases and vice versa.

In a PLL, the control voltage fed back to the VCO is applied as a reverse-bias voltage to the varactor diode within the VCO. The frequency of oscillation is inversely related to capacitance for an *RC* type oscillator by the formula

$$f_o = \frac{1}{2\pi RC}$$

and for an *LC* type oscillator by the formula

$$f_o = \frac{1}{2\pi\sqrt{LC}}$$

These formulas show that frequency increases as capacitance decreases and vice versa.

Capacitance decreases as reverse voltage (control voltage) increases. Therefore, an increase in control voltage to the VCO causes an increase in frequency and vice versa. Basic VCO operation is illustrated in Figure 17–47. The graph in part (b) shows that at the nominal control voltage, $V_{c(nom)}$, the oscillator is running at its nominal or free-running frequency, $f_{o(nom)}$. An increase in V_c above the nominal value forces the oscillator frequency to increase, and a decrease in V_c below the nominal value forces the oscillator frequency to decrease. There are, of course, limits on the operation as indicated by the minimum and maximum points. The transfer function or conversion gain, K, of the VCO is normally expressed as a certain frequency deviation per unit change in control voltage.

$$K = \frac{\Delta f_o}{\Delta V_c}$$

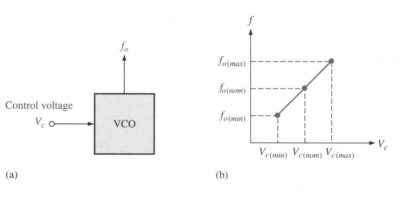

(a) (b)

▲ **FIGURE 17–47**

Basic VCO operation.

EXAMPLE 17–8

The output frequency of a certain VCO changes from 50 kHz to 65 kHz when the control voltage increases from 0.5 V to 1 V. What is the conversion gain, K?

Solution

$$K = \frac{\Delta f_o}{\Delta V_c} = \frac{65\ \text{kHz} - 50\ \text{kHz}}{1\ \text{V} - 0.5\ \text{V}} = \frac{15\ \text{kHz}}{0.5\ \text{V}} = \mathbf{30\ kHz/V}$$

Related Problem

If the conversion gain of a certain VCO is 20 kHz/V, how much frequency deviation does a change in control voltage from 0.8 V to 0.5 V produce? If the VCO frequency is 250 kHz at 0.8 V, what is the frequency at 0.5 V?

Basic PLL Operation

When the PLL is locked, the incoming frequency, f_i, and the VCO frequency, f_o, are equal. However, there is always a phase difference between them called the *static phase error*. The phase error, θ_e, is the parameter that keeps the PLL locked in. As you have seen, the filtered voltage from the phase detector is proportional to θ_e (Equation 17–4). This voltage controls the VCO frequency and is always just enough to keep $f_o = f_i$.

Figure 17–48 shows the PLL and two sinusoidal signals of the same frequency but with a phase difference, θ_e. For this condition the PLL is in lock and the VCO control

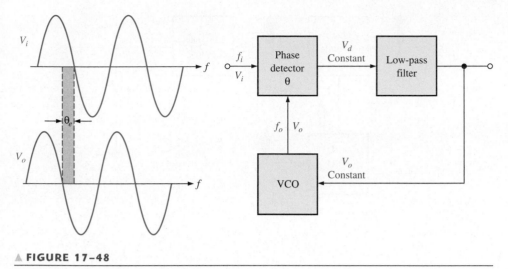

▲ FIGURE 17–48

PLL in lock under static condition ($f_o = f_i$ and constant θ_e).

voltage is constant. If f_i decreases, θ_e increases to θ_{e1} as illustrated in Figure 17–49. This increase in θ_e is sensed by the phase detector causing the VCO control voltage to decrease, thus decreasing f_o until $f_o = f_i$ and keeping the PLL in lock. If f_i increases, θ_e decreases to θ_{e1} as illustrated in Figure 17–50. This decrease in θ_e causes the VCO control voltage to increase, thus increasing f_o until $f_o = f_i$ and keeping the PLL in lock.

▲ FIGURE 17–49

PLL action when f_i decreases.

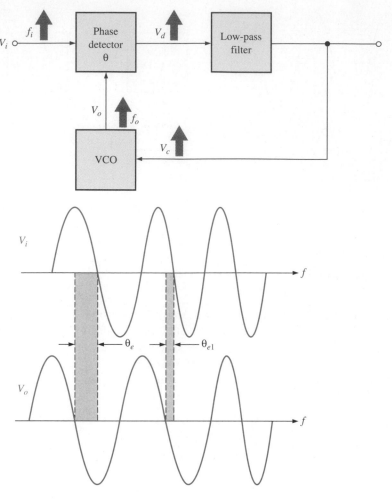

▲ **FIGURE 17–50**

PLL action when f_i increases.

Lock Range   Once the PLL is locked, it will track frequency changes in the incoming signal. The range of frequencies over which the PLL can maintain lock is called the **lock range** or tracking range. Limitations on the hold-in range are the maximum frequency deviations of the VCO and the output limits of the phase detector. The hold-in range is independent of the bandwidth of the low-pass filter because, when the PLL is in lock, the difference frequency $(f_i - f_o)$ is zero or a very low instantaneous value that falls well within the bandwidth. The hold-in range is usually expressed as a percentage of the VCO frequency.

Capture Range   Assuming the PLL is not in lock, the range of frequencies over which it can acquire lock with an incoming signal is called the **capture range.** Two basic conditions are required for a PLL to acquire lock. First, the difference frequency $(f_o - f_i)$ must be low enough to fall within the filter's bandwidth. This means that the incoming frequency must not be separated from the nominal or free-running frequency of the VCO by more than the bandwidth of the low-pass filter. Second, the maximum deviation, Δf_{max}, of the VCO frequency must be sufficient to allow f_o to increase or decrease to a value equal to f_i. These conditions are illustrated in Figure 17–51; and when they exist, the PLL will "pull" the VCO frequency toward the incoming frequency until $f_o = f_i$.

The LM565 Phase-Locked Loop

The LM565 is a good example of an integrated circuit PLL. The circuit consists of a VCO, phase detector, a low-pass filter formed by an internal resistor and an external

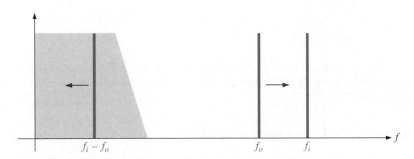

(a)

(b) $f_i - f_o$ decreases as f_o deviates towards f_i.

▲ **FIGURE 17–51**

Illustration of the conditions for a PLL to acquire lock.

capacitor, and an amplifier. The free-running VCO frequency can be set with external components. A block diagram is shown in Figure 17–52. The LM565 can be used for the frequency range from 0.001 Hz to 500 kHz.

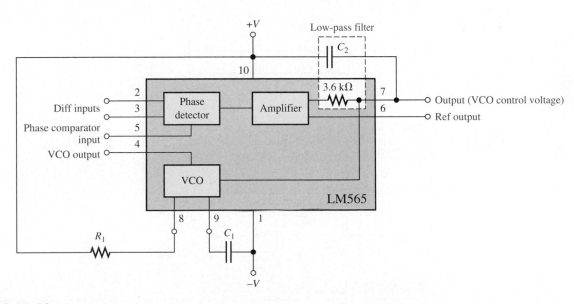

▲ **FIGURE 17–52**

Block diagram of the LM565 PLL.

The free-running frequency of the VCO is set by the values of R_1 and C_1 in Figure 17–52 according to the following formula. The frequency is in hertz when the resistance is in ohms and the capacitance is in farads.

Equation 17–5

$$f_o \cong \frac{1.2}{4R_1C_1}$$

The lock range is given by

Equation 17–6

$$f_{lock} = \pm\frac{8f_o}{V_{CC}}$$

where V_{CC} is the total voltage between the positive and negative dc supply voltage terminals.

The capture range is given by

Equation 17–7

$$f_{cap} \cong \pm\frac{1}{2\pi}\sqrt{\frac{2\pi f_{lock}}{(3600)C_2}}$$

The 3600 is the value of the internal filter resistor in ohms. You can see that the capture range is dependent on the filter bandwidth as determined by the internal resistor and the external capacitor C_2.

The PLL as an FM Demodulator

As you have seen, the VCO control voltage in a PLL depends on the deviation of the incoming frequency. The PLL will produce a voltage proportional to the frequency of the incoming signal which, in the case of FM, is the original modulating signal.

Figure 17–53 shows a typical connection for the LM565 as an FM demodulator. If the IF input is frequency modulated by a sinusoidal signal, you get a sinusoidal signal on the output as indicated. Since the maximum operating frequency is 500 kHz, this device must be used in double-conversion FM receivers. A double-conversion FM receiver is one in which essentially two mixers are used to first convert the RF to a 10.7 MHz IF and then convert this to a 455 kHz IF.

The free-running frequency of the VCO is adjusted to approximately 455 kHz, which is the center of the modulated IF range. C_1 can be any value, but R_1 should be in the range

▲ **FIGURE 17–53**

The LM565 as an FM demodulator.

from 2 kΩ to 20 kΩ. The input can be directly coupled as long as there is no dc voltage difference between pins 2 and 3. The VCO is connected to the phase detector by an external wire between pins 4 and 5.

EXAMPLE 17–9

Determine the values for R_1, C_1, and C_2 for the LM565 in Figure 17–53 for a free-running frequency of 455 kHz and a capture range of ± 10 kHz. The dc supply voltages are ± 6 V.

Solution   Use Equation 17–5 to calculate C_1. Choose $R_1 = $ **4.7 kΩ**.

$$f_o \cong \frac{1.2}{4R_1C_1}$$

$$C_1 \cong \frac{1.2}{4R_1f_o} = \frac{1.2}{4(4700\ \Omega)(455 \times 10^3\ \text{Hz})} = 140 \times 10^{-12}\ \text{F} = \textbf{140 pF}$$

The lock range and capture range must be determined before C_2 can be calculated. The lock range is

$$f_{lock} = \pm \frac{8f_o}{V_{CC}} = \pm \frac{8(455\ \text{kHz})}{12\ \text{V}} = \pm 303\ \text{kHz}$$

Use Equation 17–7 to calcualte C_2.

$$f_{cap} \cong \pm \frac{1}{2\pi} \sqrt{\frac{2\pi f_{lock}}{(3600)C_2}}$$

$$f_{cap}^2 \cong \left(\frac{1}{2\pi}\right)^2 \frac{2\pi f_{lock}}{(3600)C_2}$$

Therefore,

$$C_2 \cong \left(\frac{1}{2\pi}\right)^2 \frac{2\pi f_{lock}}{(3600)f_{cap}^2} = \left(\frac{1}{2\pi}\right)^2 \frac{2\pi(303 \times 10^3\ \text{Hz})}{(3600)(10 \times 10^3\ \text{Hz})^2}$$

$$= 0.134 \times 10^{-6}\ \text{F} = \textbf{0.134 } \boldsymbol{\mu}\textbf{F}$$

Related Problem   What can you do to increase the capture range from ± 10 kHz to ± 15 kHz?

**SECTION 17–8
REVIEW**

1. List the three basic components in a phase-locked loop.
2. What is another circuit used in some PLLs other than the three listed in Question 1?
3. What is the basic function of a PLL?
4. What is the difference between the lock range and the capture range of a PLL?
5. Basically, how does a PLL track the incoming frequency?

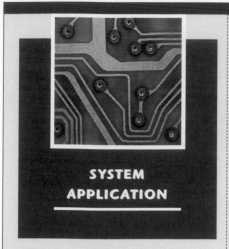

SYSTEM APPLICATION

The DCE (data communications equipment) system introduced at the opening of this chapter includes an FSK (frequency shift keying) modem (modulator/demodulator). FSK is one method for modulating digital data for transmission over voice phone lines and is basically a form of frequency modulation. In this system application, the focus is on the low-speed modulator/demodulator (modem) board, which is implemented with a VCO for transmitting FSK signals and a PLL for receiving FSK signals.

A Brief Description of the System

The FSK modem interfaces a computer with the telephone network so that digital data, which are incompatible with the standard phone system because of bandwidth limitations, can be transmitted and received over regular phone lines, thus allowing computers to communicate with each other. Figure 17–54 shows a diagram of a simple data communications system in which a modem at each end of the phone line provides interfacing for a computer.

The modem (DCE) consists of three basic functional blocks as shown in Figure 17–55: the FSK modem circuits, the phone line interface circuits, and the timing and control circuits. The dual-polarity power supply is not shown. Although the focus of this system application is the FSK modem board, we will briefly look at each of the other parts to give you a basic idea of the overall system function.

The Phone Line Interface The main purposes of this circuitry are to couple the phone line to the modem by proper impedance matching, to provide necessary filtering, and to accommodate full-duplex transmission of data. *Full-duplex* means essentially that information can be going both ways on a single phone line at the same time. This allows a computer, connected to a modem, to be sending data and receiving data simultaneously without the transmitted data interfering with the received data. Full-duplexing is implemented by assigning the transmitted data one bandwidth and the received data another separate bandwidth within the 300 Hz to 3 kHz overall bandwidth of the phone network.

Timing and Control One basic function of the timing and control circuits is to determine the proper mode of operation for the modem. The two modes are the originate mode and the answer mode. Another function is to provide a standard interface (such as RS-232C) with the DTE (computer). The RS-232C standard requires certain defined command and control signals, data signals, and voltage levels for each signal.

Digital Data Before we get into FSK, let's briefly review digital data. A detailed knowledge of binary numbers is not necessary for this system application. Information is represented in digital form by 1s and 0s, which are the binary digits or bits. In terms of voltage waveforms, a 1 is generally represented by a high level and a 0 by a low level. A stream of serial data consists of a sequence of bits as illustrated by an example in Figure 17–56(a).

Baud Rate The band rate is the number of bits/s of the transmitted data. This can be determined by taking the reciprocal of the time of the shortest pulse transmitted. A low-speed modem, such as the one we are focusing on, sends and receives digital data at a rate of 300 bits/s or 300 baud. For example, if we have an alternating sequence of 1s and 0s (highs and lows), as indicated in Figure 17–56(b), each bit takes 3.33 ms. This is the minimum time for a pulse in the data stream, so the baud rate is 1/3.33 ms = 300 baud. Since it takes two bits, a 1 and a 0, to make up the period of this particular waveform, the fundamental frequency of this format is 1/6.67 ms = 150 Hz. This is the maximum frequency of a 300 baud data stream because normally there may be several consecutive 1s and/or several consecutive 0s in a sequence, thus reducing the frequency. As mentioned earlier, the telephone network has a 300 Hz minimum frequency response, so the fundamental frequency of the 300 baud data stream will fall outside of the telephone bandwidth. This prevents sending digital data in its pure form over the phone lines.

Frequency-Shift Keying (FSK) FSK is one method used to overcome the bandwidth limitation of the telephone system so that digital data can be sent over the

| DTE | | DCE | | | DCE | | DTE |
| Computer | ⟷ | Modem | Phone line | Modem | | ⟷ | Computer |

▲ **FIGURE 17–54**

A data communications system.

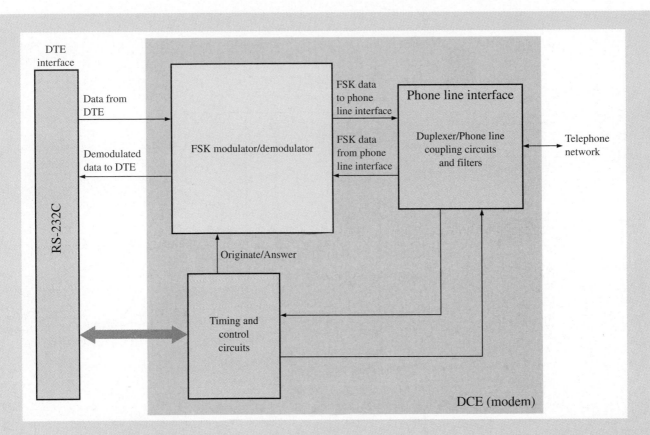

▲ **FIGURE 17–55**

Basic block diagram of a modem.

(a) Typical serial data stream

(b) Period for 300 baud square wave

▲ **FIGURE 17–56**

A serial stream of digital data.

phone lines. The basic idea of FSK is to represent 1s and 0s by two different frequencies within the telephone bandwidth. By the way, any frequency within the telephone bandwidth is an audible tone. The standard frequencies for a full-duplex 300 baud modem in the originate mode are 1070 Hz for a 0 (called a space) and 1270 Hz for a 1 (called a mark). In the answer mode, 2025 Hz is a 0 and 2225 Hz is a 1. The relationship of these FSK frequencies and the telephone bandwidth is illustrated in Figure 17–57. Signals in both the originate and answer bands can exist at the same time on the phone line and not interfere with each other because of the frequency separation.

An example of a digital data stream converted to FSK by a modem is shown in Figure 17–58.

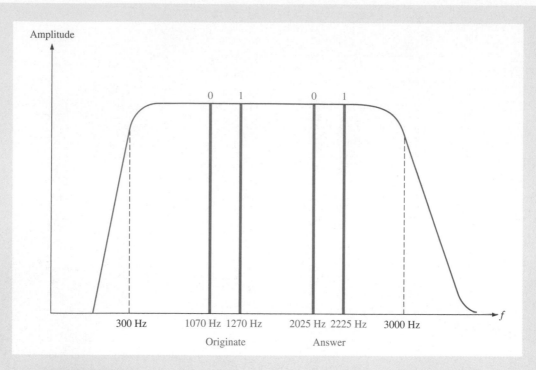

▲ **FIGURE 17–57**

Frequencies for 300 baud, full-duplex data transmission.

(a) Data stream

(b) Corresponding FSK signal (frequency relationships are not exact)

▲ **FIGURE 17–58**

Example of FSK data.

Modem Circuit Operation

The FSK modem circuits, shown in Figure 17–59, contain an LM565 PLL and a VCO integrated circuit. The VCO can be a device such as the 4046 (not covered specifically in this chapter), which is a PLL device in which the VCO portion can be used by itself because all of the necessary inputs and outputs are available. The VCO in the LM565 cannot be used independ- ently of the PLL because there is no input pin for the control voltage.

The function of the VCO is to accept digital data from a DTE and provide FSK modulation. The VCO is always the trans- mitting device. The digital data come in on the control voltage input (pin 9) of the VCO via a level-shifting circuit formed by Q_3 and Q_4. This circuit is used because the data from the RS-232C interface are dual polarity with a positive voltage representing a 0 and a negative voltage representing a 1. Potentiometer R_8 is for adjusting the high level of the control voltage and R_{11} is for adjusting the low level for the purpose of fine- tuning the frequency. Transistor Q_5 pro- vides for originate/answer mode fre- quency selection by changing the value of the frequency-selection resistance

▲ **FIGURE 17–59**

FSK modulator/demodulator circuit.

from pin 11 to ground. Transistors Q_1 and Q_2 perform a similar function for the PLL.

When the digital data are at high levels, corresponding to logic 0s, the VCO oscillates at 1070 Hz in the originate mode and 2025 Hz in the answer mode. When the digital data are at low levels, corresponding to logic 1s, the VCO oscillates at 1270 Hz in the originate mode and 2225 Hz in the answer mode. An example of the originate mode is when a DTE issues a request for data and transmits that request to another DTE. An example of the answer mode is when the receiving DTE

responds to a request and sends data back to the originating DTE.

The function of the PLL is to accept incoming FSK-modulated data and convert it to a digital data format for use by the DTE. The PLL is always a receiving device. When the modem is in the originate mode, the PLL is receiving answer-mode data from the other modem. When the modem is in the answer mode, the PLL is receiving originate-mode data from the other modem. The 741 op-amp is connected as a comparator that changes the data levels from the PLL to a dual-polarity format for compatibility with the RS-232C interface.

The PC Board

Locate and identify each component and each input/output pin on the PC board in Figure 17–60 using the schematic in Figure 17–59. Verify that the board and the schematic agree. If the PC board and the schematic do not agree, indicate the problem. Backside traces are shown as darker lines.

The Circuits

For this application, the free-running frequencies of both the PLL and the VCO circuits are determined by the formula in Equation 17–5.

▲ **FIGURE 17–60**

- Verify that the free-running frequency for the PLL IC is approximately 1070 Hz in the originate mode and approximately 1270 Hz in the answer mode.

- Repeat the previous step for the VCO.

- Determine the approximate minimum and maximum output voltages for the 741 comparator.

- Determine the maximum high-level voltage on pin 9 of the VCO.

- If a 300 Hz square wave that varies from +5 V to −5 V is applied to the data from the DTE input, what should you observe on pin 4 of the VCO?

- When the data from the DTE are low, pin 9 of the VCO is at approxi-

mately 0 V. At this level, the VCO oscillates at 1070 Hz or 2025 Hz. When the data from the DTE go high, to what value should the voltage at pin 9 be adjusted to produce a 1270 Hz or 2225 Hz frequency if the transfer function of the VCO is 50 Hz/V?

Troubleshooting

- There is no demodulated data output voltage when there are verified FSK data from the phone line interface.

- The LM565 properly demodulates 1070 Hz and 1270 Hz FSK data but

does not properly demodulate 2025 Hz and 2225 Hz data.

- The VCO produces no FSK output.

- The VCO produces a continuous 1070 Hz tone in the originate mode and a continuous 2025 Hz tone in the answer mode when there are proper data from the DTE.

Final Report (Optional)

Describe the overall operation of the FSK modem board. Specify how each circuit works and what its purpose is. Identify the function of each component. Use the results previously found as appropriate.

■ In amplitude modulation (AM), the amplitude of a higher-frequency carrier signal is varied by a lower-frequency modulating signal (usually an audio signal).

■ A basic superheterodyne AM receiver consists of an RF amplifier (not always), a mixer, a local oscillator, an IF (intermediate frequency) amplifier, an AM detector, and audio and power amplifiers.

■ The IF in a standard AM receiver is 455 kHz.

■ The AGC (automatic gain control) in a receiver tends to keep the signal strength constant within the receiver to compensate for variations in the received signal.

■ In frequency modulation (FM), the frequency of a carrier signal is varied by a modulating signal.

■ A superheterodyne FM receiver is basically the same as an AM receiver except that it requires a limiter to keep the IF amplitude constant, a different kind of detector or discriminator, and a de-emphasis network. The IF is 10.7 MHz.

■ A four-quadrant linear multiplier can handle any combination of voltage polarities on its inputs.

■ Amplitude modulation is basically a multiplication process.

■ The multiplication of sinusoidal signals produces sum and difference frequencies.

■ The output spectrum of a balanced modulator includes upper-side and lower-side frequencies, but no carrier frequency.

■ The output spectrum of a standard amplitude modulator includes upper-side and lower-side frequencies and the carrier frequency.

■ A linear multiplier is used as the mixer in receiver systems.

■ A mixer converts the RF signal down to the IF signal. The radio frequency varies over the AM or FM band. The intermediate frequency is constant.

■ One type of AM demodulator consists of a multiplier followed by a low-pass filter.

■ The audio and power amplifiers boost the output of the detector or discriminator and drive the speaker.

■ A voltage-controlled oscillator (VCO) produces an output frequency that can be varied by a control voltage. Its operation is based on a variable reactance.

■ A VCO is a basic frequency modulator when the modulating signal is applied to the control voltage input.

■ A phase-locked loop (PLL) is a feedback circuit consisting of a phase detector, a low-pass filter, a VCO, and sometimes an amplifier.

■ The purpose of a PLL is to lock onto and track incoming frequencies.

■ A linear multiplier can be used as a phase detector.

■ A modem is a modulator/demodulator.

■ DTE stands for digital terminal equipment.

■ DCE stands for digital communications equipment.

Key terms and other bold terms in the chapter are defined in the end-of-book glossary.

Amplitude modulation (AM)   A communication method in which a lower-frequency signal modulates (varies) the amplitude of a higher-frequency signal (carrier).

Balanced modulation   A form of amplitude modulation in which the carrier is suppressed; sometimes known as *suppressed-carrier modulation.*

Capture range   The range of frequencies over which a PLL can acquire lock.

Four-quadrant multiplier   A linear device that produces an output voltage proportional to the product of two input voltages.

Freqency modulation (FM)   A communication method in which a lower-frequency intelligence-carrying signal modulates (varies) the frequency of a higher-frequency signal.

Lock range   The range of frequencies over which a PLL can maintain lock.

Mixer A device for down-converting frequencies in a receiver system.

Phase-locked loop (PLL) A device for locking onto and tracking the frequency of an incoming signal.

KEY FORMULAS

17–1 $V_{OUT} = KV_XV_Y$ Multiplier output voltage

17–2 $v_1v_2 = \dfrac{V_{1(p)}V_{2(p)}}{2}\cos 2\pi(f_1 - f_2)t$ Sum and difference frequencies

$$-\frac{V_{1(p)}V_{2(p)}}{2}\cos 2\pi(f_1 + f_2)t$$

17–3 $V_{out} = V_{c(p)}^2 \sin 2\pi f_c t$ Standard AM

$$+\frac{V_{c(p)}V_{m(p)}}{2}\cos 2\pi(f_c - f_m)t$$

$$-\frac{V_{c(p)}V_{m(p)}}{2}\cos 2\pi(f_c + f_m)t$$

17–4 $V_c = \dfrac{V_iV_o}{2}\cos \theta_e$ PLL control voltage

17–5 $f_o \cong \dfrac{1.2}{4R_1C_1}$ Output frequency LM565

17–6 $f_{lock} = \pm\dfrac{8f_o}{V_{CC}}$ Lock range LM565

17–7 $f_{cap} \cong \pm\dfrac{1}{2\pi}\sqrt{\dfrac{2\pi f_{lock}}{(3600)C_2}}$ Capture range LM565

SELF-TEST

Answers are at the end of the chapter.

1. In amplitude modulation, the pattern produced by the peaks of the carrier signal is called the
   - (a) index
   - (b) envelope
   - (c) audio signal
   - (d) upper-side frequency

2. Which of the following is not a part of an AM superheterodyne receiver?
   - (a) Mixer
   - (b) IF amplifier
   - (c) DC restorer
   - (d) Detector
   - (e) Audio amplifier
   - (f) Local oscillator

3. In an AM receiver, the local oscillator always produces a frequency that is above the incoming RF by
   - (a) 10.7 kHz
   - (b) 455 MHz
   - (c) 10.7 MHz
   - (d) 455 kHz

4. An FM receiver has an IF frequency that is
   - (a) in the 88 MHz to 108 MHz range
   - (b) in the 540 kHz to 1640 kHz range
   - (c) 455 kHz
   - (d) greater than the IF in an AM receiver

5. The detector or discriminator in an AM or an FM receiver
   - (a) detects the difference frequency from the mixer
   - (b) changes the RF to IF
   - (c) recovers the audio signal
   - (d) maintains a constant IF amplitude

6. In order to handle all combinations of input voltage polarities, a multiplier must have

(a) four-quadrant capability   (b) three-quadrant capability

(c) four inputs   (d) dual-supply voltages

7. The internal attenuation of a multiplier is called the

(a) transconductance   (b) scale factor   (c) reduction factor

8. When the two inputs of a multiplier are connected together, the device operates as a

(a) voltage doubler   (b) square root circuit

(c) squaring circuit   (d) averaging circuit

9. Amplitude modulation is basically a

(a) summing of two signals   (b) multiplication of two signals

(c) subtraction of two signals   (d) nonlinear process

10. The frequency spectrum of a balanced modulator contains

(a) a sum frequency   (b) a difference frequency   (c) a carrier frequency

(d) answers (a), (b), and (c)   (e) answers (a) and (b)   (f) answers (b) and (c)

11. The IF in a receiver is the

(a) sum of the local oscillator frequency and the RF carrier frequency

(b) local oscillator frequency

(c) difference of the local oscillator frequency and the carrier RF frequency

(d) difference of the carrier frequency and the audio frequency

12. When a receiver is tuned from one RF frequency to another,

(a) the IF changes by an amount equal to the LO (local oscillator) frequency

(b) the IF stays the same

(c) the LO frequency changes by an amount equal to the audio frequency

(d) both the LO and the IF frequencies change

13. The output of the AM detector goes directly to the

(a) IF amplifier   (b) mixer   (c) audio amplifier   (d) speaker

14. If the control voltage to a VCO increases, the output frequency

(a) decreases   (b) does not change   (c) increases

15. A PLL maintains lock by comparing

(a) the phase of two signals

(b) the frequency of two signals

(c) the amplitude of two signals

PROBLEMS Answers to all odd-numbered problems are at the end of the book.

BASIC PROBLEMS

SECTION 17–1 Basic Receivers

1. Label each block in the AM receiver in Figure 17–61.

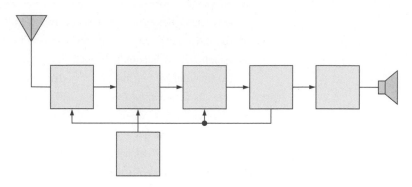

▲ **FIGURE 17–61**

2. Label each block in the FM receiver in Figure 17–62.

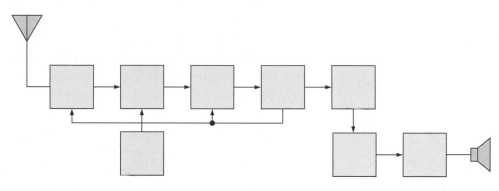

▲ **FIGURE 17–62**

3. An AM receiver is tuned to a transmitted frequency of 680 kHz. What is the local oscillator (LO) frequency?

4. An FM receiver is tuned to a transmitted frequency of 97.2 MHz. What is the LO frequency?

5. The LO in an FM receiver is running at 101.9 MHz. What is the incoming RF? What is the IF?

SECTION 17–2 The Linear Multiplier

6. From the graph in Figure 17–63, determine the multiplier output voltage for each of the following pairs of input voltages.

   (a) $V_X = -4$ V, $V_Y = +6$ V   (b) $V_X = +8$ V, $V_Y = -2$ V

   (c) $V_X = -5$ V, $V_Y = -2$ V   (d) $V_X = +10$ V, $V_Y = +10$ V

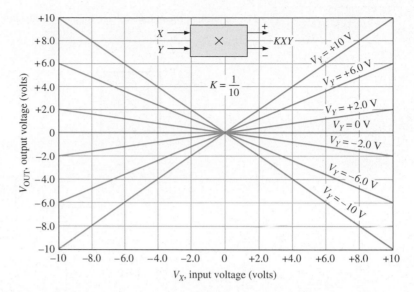

▲ FIGURE 17–63

7. How much pin 3 current is there for the multiplier in Figure 17–64? The potentiometer is set at 2.8 kΩ.

▲ FIGURE 17–64

8. Determine the scale factor for the multiplier in Figure 17–64.

9. If a certain multiplier has a scale factor of 0.8 and the inputs are $+3.5$ V and -2.9 V, what is the output voltage?

10. Show the connections for the multiplier in Figure 17–64 in order to implement a squaring circuit.

11. Determine the output voltage for each circuit in Figure 17–65.

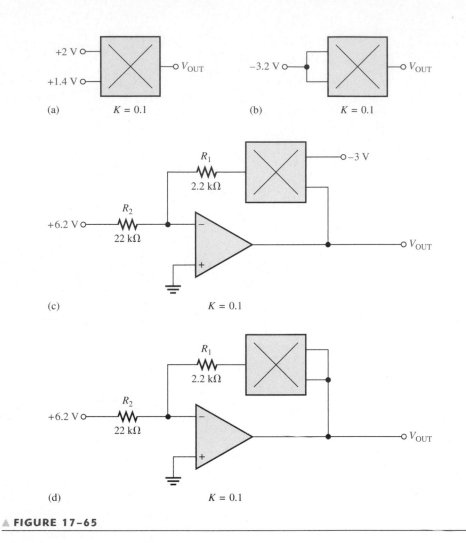

▲ **FIGURE 17–65**

SECTION 17–3   Amplitude Modulation

12. If a 100 kHz signal and a 30 kHz signal are applied to a balanced modulator, what frequencies will appear on the output?

13. What are the frequencies on the output of the balanced modulator in Figure 17–66?

▶ **FIGURE 17–66**

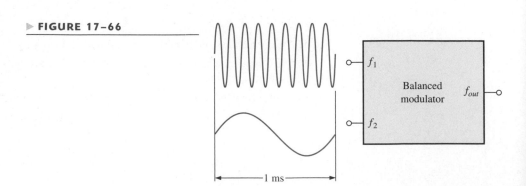

14. If a 1000 kHz signal and a 3 kHz signal are applied to a standard amplitude modulator, what frequencies will appear on the output?

15. What are the frequencies on the output of the standard amplitude modulator in Figure 17–67?

▶ FIGURE 17–67

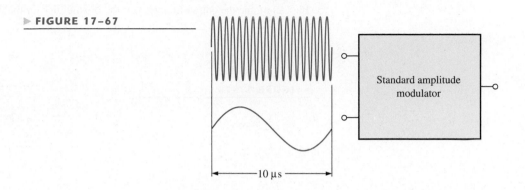

16. The frequency spectrum in Figure 17–68 is for the output of a standard amplitude modulator. Determine the carrier frequency and the modulating frequency.

1.1955 MHz   1.2 MHz   1.2045 MHz

▲ FIGURE 17–68

17. The frequency spectrum in Figure 17–69 is for the output of a balanced modulator. Determine the carrier frequency and the modulating frequency.

▶ FIGURE 17–69

847 kHz   853 kHz

18. A voice signal ranging from 300 Hz to 3 kHz amplitude modulates a 600 kHz carrier. Develop the frequency spectrum.

SECTION 17–4   The Mixer

19. Determine the output expression for a multiplier with one sinusoidal input having a peak voltage of 0.2 V and a frequency of 2200 kHz and the other input having a peak voltage of 0.15 V and a frequency of 3300 kHz.

20. Determine the output frequency of the IF amplifier for the frequencies shown in Figure 17–70.

▶ **FIGURE 17–70**

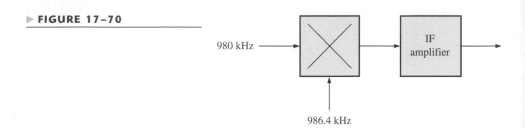

SECTION 17–5   AM Demodulation

21. The input to a certain AM receiver consists of a 1500 kHz carrier and two side frequencies separated from the carrier by 20 kHz. Determine the frequency spectrum at the output of the mixer amplifier.

22. For the same conditions stated in Problem 21, determine the frequency spectrum at the output of the IF amplifier.

23. For the same conditions stated in Problem 21, determine the frequency spectrum at the output of the AM detector (demodulator).

SECTION 17–6   IF and Audio Amplifiers

24. For a carrier frequency of 1.2 MHz and a modulating frequency of 8.5 kHz, list all of the frequencies on the output of the mixer in an AM receiver.

25. In a certain AM receiver, one amplifier has a passband from 450 kHz to 460 kHz and another has a passband from 10 Hz to 5 kHz. Identify these amplifiers.

26. Determine the maximum and minimum output voltages for the audio power amplifier in Figure 17–71.

▲ **FIGURE 17–71**

SECTION 17–7 **Frequency Modulation**

    27. Explain how a VCO is used as a frequency modulator.

    28. How does an FM signal differ from an AM signal?

    29. What is the variable reactance element shown in the MC2833D diagram in Figure 17–43?

SECTION 17–8 **The Phase-Locked Loop (PLL)**

    30. Label each block in the PLL diagram of Figure 17–72.

    31. A PLL is locked onto an incoming signal with a peak amplitude of 250 mV and a frequency of 10 MHz at a phase angle of 30°. The 400 mV peak VCO signal is at a phase angle of 15°.

        (a) What is the VCO frequency?

        (b) What is the value of the control voltage being fed back to the VCO at this point?

▶ **FIGURE 17–72**

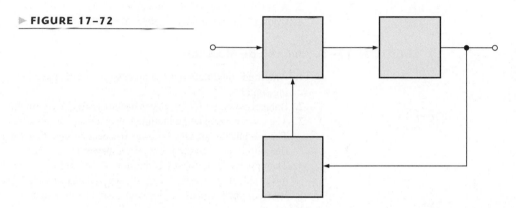

    32. What is the conversion gain of a VCO if a 0.5 V increase in the control voltage causes the output frequency to increase by 3.6 kHz?

    33. If the conversion gain of a certain VCO is 1.5 kHz per volt, how much does the frequency change if the control voltage increases 0.67 V?

    34. Name two conditions for a PLL to acquire lock.

    35. Determine the free-running frequency, the lock range, and the capture range for the PLL in Figure 17–73.

▲ **FIGURE 17–73**

ANSWERS

SECTION REVIEWS

SECTION 17–1 **Basic Receivers**

1. AM is amplitude modulation. FM is frequency modulation.
2. In AM, the modulating signal varies the amplitude of a carrier. In FM, the modulating signal varies the frequency of a carrier.
3. AM: 540 kHz to 1640 kHz; FM: 88 MHz to 108 MHz

SECTION 17–2 **The Linear Multiplier**

1. A four-quadrant multiplier can handle any combination (4) of positive and negative inputs. A one-quadrant multiplier can only handle two positive inputs, for example.
2. $K = 0.1$. K must be 1 for an output of 5 V.
3. Connect the two inputs together and apply a single input variable.

SECTION 17–3 **Amplitude Modulation**

1. Amplitude modulation is the process of varying the amplitude of a carrier signal with a modulating signal.
2. Balanced modulation produces no carrier frequency on the output, whereas standard AM does.
3. The carrier signal is the modulated signal and has a sufficiently high frequency for transmission. The modulating signal is a lower-frequency signal that contains information and varies the carrier amplitude according to its waveshape.
4. The upper-side frequency is the sum of the carrier frequency and the modulating frequency. The lower-side frequency is the difference of the carrier frequency and the modulating frequency.
5. By summing the peak carrier voltage and the modulating signal before mixing with the carrier signal

SECTION 17–4 **The Mixer**

1. The mixer produces (among other frequencies) a signal representing the difference between the incoming carrier frequency and the local oscillator frequency. This is called the intermediate frequency.
2. The mixer multiplies the carrier and the local oscillator signals.
3. 1000 kHz + 350 kHz = 1350 kHz, 1000 kHz − 350 kHz = 650 kHz

SECTION 17–5 **AM Demodulation**

1. The filter removes all frequencies except the audio.
2. Only the 1 kHz

SECTION 17–6 **IF and Audio Amplifiers**

1. To amplify the 455 kHz amplitude modulated IF coming from the mixer
2. The IF center frequency is 455 kHz.
3. The 10 kHz bandwidth allows the upper-side and lower-side frequencies that contain the information to pass.
4. The audio amplifier follows the detector because the detector is the circuit that recovers the audio from the modulated IF.
5. The IF has a response of approximately 455 kHz ± 5 kHz. The typical audio amplifier has a maximum bandwidth from tens of hertz up to about 15 kHz although for many amplifiers, the bandwidth can be much less than this typical maximum.

SECTION 17–7 **Frequency Modulation**

1. The frequency variation of an FM signal bears the information.
2. VCOs are based on the principle of voltage-variable reactance.

SECTION 17-8   **The Phase-Locked Loop (PLL)**

1. Phase detector, low-pass filter, and VCO

2. Sometimes a PLL uses an amplifier in the loop.

3. A PLL locks onto and tracks a variable incoming frequency.

4. The lock range specifies how much a lock-on frequency can deviate without the PLL losing lock. The capture range specifies how close the incoming frequency must be from the free-running VCO frequency in order for the PLL to lock.

5. The PLL detects a change in the phase of the incoming signal compared to the VCO signal that indicates a change in frequency. The positive feedback then causes the VCO frequency to change along with the incoming frequency.

RELATED PROBLEMS FOR EXAMPLES

17-1 -3.6 V from the graph in Figure 17-9

17-2   0.728 V

17-3   Modulate the carrier with a higher-frequency signal.

17-4   The balanced modulator output has the same side frequencies but does not have a carrier frequency.

17-5 $V_p = 0.025$ mV, $f = 455$ kHz

17-6   455 kHz

17-7   0.172 V

17-8   A decrease of 6 kHz; 244 kHz

17-9   Decrease C_2 to 0.0595 μF.

SELF-TEST

1. (b)   2. (c)   3. (d)   4. (d)   5. (c)   6. (a)   7. (b)   8. (c)

9. (b)   10. (e)   11. (c)   12. (b)   13. (c)   14. (c)   15. (a)

18

VOLTAGE REGULATORS

INTRODUCTION

A voltage regulator provides a constant dc output voltage that is essentially independent of the input voltage, output load current, and temperature. The voltage regulator is one part of a power supply. Its input voltage comes from the filtered output of a rectifier derived from an ac voltage or from a battery in the case of portable systems.

Most voltage regulators fall into two broad categories: linear regulators and switching regulators. In the linear regulator category, two general types are the series regulator and the shunt regulator. These are normally available for either positive or negative output voltages. A dual regulator provides both positive and negative outputs. In the switching regulator category, three general configurations are step-down, step-up, and inverting.

Many types of integrated circuit (IC) regulators are available. The most popular types of linear regulator are the three-terminal fixed voltage regulator and the three-terminal adjustable voltage regulator. Switching regulators are also widely used. In this chapter, specific IC devices are introduced as representative of the wide range of available devices.

- Describe the basic concept of voltage regulation
- Discuss the principles of series voltage regulators
- Discuss the principles of shunt voltage regulators
- Discuss the principles of switching regulators
- Discuss integrated circuit voltage regulators
- Discuss applications of IC voltage regulators

KEY TERMS

- Regulator
- Line regulation
- Load regulation
- Linear regulator
- Switching regulator
- Thermal overload

■■■ SYSTEM APPLICATION PREVIEW

A dual-polarity power supply is to be used for the FM receiver from Chapter 15. Two regulators, one positive and the other negative, provide the positive voltage required for the receiver circuits and the dual-polarity voltages for the op-amp circuits.

WWW. VISIT THE COMPANION WEBSITE

Study aids for this chapter are available at
http://www.prenhall.com/floyd

18–1 VOLTAGE REGULATION

Two basic categories of voltage regulation are line regulation and load regulation. The purpose of line regulation is to maintain a nearly constant output voltage when the input voltage varies. The purpose of load regulation is to maintain a nearly constant output voltage when the load varies.

After completing this section, you should be able to

- ▪ **Describe the basic concept of voltage regulation**
- ▪ Explain line regulation
- ▪ Calculate line regulation
- ▪ Explain load regulation
- ▪ Calculate load regulation

Line Regulation

When the dc input (line) voltage changes, the voltage **regulator** must maintain a nearly constant output voltage, as illustrated in Figure 18–1. **Line regulation** can be defined as the percentage change in the output voltage for a given change in the input (line) voltage. When taken over a range of input voltage values, line regulation is expressed as a percentage by the following formula:

Equation 18–1

$$\text{Line regulation} = \left(\frac{\Delta V_{\text{OUT}}}{\Delta V_{\text{IN}}}\right)100\%$$

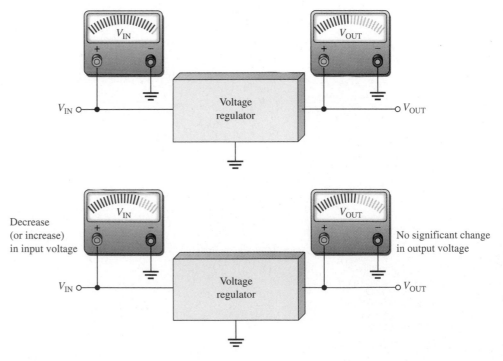

▲ **FIGURE 18–1**

Line regulation. A change in input (line) voltage does not significantly affect the output voltage of a regulator (within certain limits).

Line regulation can also be expressed in units of %/V. For example, a line regulation of 0.05%/V means that the output voltage changes 0.05 percent when the input voltage increases or decreases by one volt. Line regulation can be calculated using the following formula (Δ means "a change in"):

$$\text{Line regulation} = \frac{(\Delta V_{OUT}/V_{OUT})100\%}{\Delta V_{IN}}$$

Equation 18–2

EXAMPLE 18–1

When the input to a particular voltage regulator decreases by 5 V, the output decreases by 0.25 V. The nominal output is 15 V. Determine the line regulation in %/V.

Solution   The line regulation as a percentage change per volt is

$$\text{line regulation} = \frac{(\Delta V_{OUT}/V_{OUT})100\%}{\Delta V_{IN}} = \frac{(0.25\ \text{V}/15\ \text{V})100\%}{5\ \text{V}} = \mathbf{0.333\%/V}$$

Related Problem   The input of a certain regulator increases by 3.5 V. As a result, the output voltage increases by 0.42 V. The nominal output is 20 V. Determine the regulation in %/V.

Answers are at the end of the chapter.

Load Regulation

When the amount of current through a load changes due to a varying load resistance, the voltage regulator must maintain a nearly constant output voltage across the load, as illustrated in Figure 18–2.

▲ **FIGURE 18–2**

Load regulation. A change in load current has practically no effect on the output voltage of a regulator (within certain limits).

Load regulation can be defined as the percentage change in output voltage for a given change in load current. One way to express load regulation is as a percentage change in output voltage from no-load (NL) to full-load (FL).

Equation 18–3

$$\text{Load regulation} = \left(\frac{V_{\text{NL}} - V_{\text{FL}}}{V_{\text{FL}}}\right)100\%$$

Alternately, the load regulation can be expressed as a percentage change in output voltage for each mA change in load current. For example, a load regulation of 0.01%/mA means that the output voltage changes 0.01 percent when the load current increases or decreases 1 mA.

EXAMPLE 18–2

A certain voltage regulator has a 12 V output when there is no load ($I_L = 0$). When there is a full-load current of 10 mA, the output voltage is 11.9 V. Express the voltage regulation as a percentage change from no-load to full-load and also as a percentage change for each mA change in load current.

Solution The no-load output voltage is

$$V_{\text{NL}} = 12 \text{ V}$$

The full-load output voltage is

$$V_{\text{FL}} = 11.9 \text{ V}$$

The load regulation as a percentage change from no-load to full-load is

$$\text{load regulation} = \left(\frac{V_{\text{NL}} - V_{\text{FL}}}{V_{\text{FL}}}\right)100\% = \left(\frac{12 \text{ V} - 11.9 \text{ V}}{11.9 \text{ V}}\right)100\% = \mathbf{0.840\%}$$

The load regulation can also be expressed as a percentage change per milliamp as

$$\text{load regulation} = \frac{0.840\%}{10 \text{ mA}} = \mathbf{0.084\%/mA}$$

where the change in load current from no-load to full-load is 10 mA.

Related Problem A regulator has a no-load output voltage of 18 V and a full-load output of 17.8 V at a load current of 50 mA. Determine the voltage regulation as a percentage change from no-load to full-load and also as a percentage change for each mA change in load current.

Sometimes power supply manufacturers specify the equivalent output resistance of a power supply (R_{OUT}) instead of its load regulation. Recall that an equivalent Thevenin circuit can be drawn for any two-terminal linear circuit. Figure 18–3 shows the equivalent Thevenin circuit for a power supply with a load resistor. The Thevenin voltage is the voltage from the supply with no load (V_{NL}), and the Thevenin resistance is the specified output resistance, R_{OUT}. Ideally, R_{OUT} is zero, corresponding to 0% load regulation, but in

▶ **FIGURE 18–3**

Thevenin equivalent circuit for a power supply with a load resistor.

practical power supplies R_{OUT} is a small value. With the load resistor in place, the output voltage is found by applying the voltage-divider rule:

$$V_{OUT} = V_{NL}\left(\frac{R_L}{R_{OUT} + R_L}\right)$$

If we let R_{FL} equal the smallest-rated load resistance (largest-rated current), then the full-load output voltage (V_{FL}) is

$$V_{FL} = V_{NL}\left(\frac{R_{FL}}{R_{OUT} + R_{FL}}\right)$$

By rearranging and substituting into Equation 18–3,

$$V_{NL} = V_{FL}\left(\frac{R_{OUT} + R_{FL}}{R_{FL}}\right)$$

$$\text{Load regulation} = \frac{V_{FL}\left(\dfrac{R_{OUT} + R_{FL}}{R_{FL}}\right) - V_{FL}}{V_{FL}} \times 100\% = \left(\frac{R_{OUT} + R_{FL}}{R_{FL}} - 1\right)100\%$$

$$\text{Load regulation} = \left(\frac{R_{OUT}}{R_{FL}}\right)100\%$$

Equation 18–4

Equation 18–4 is a useful way of finding the percent load regulation when the output resistance and minimum load resistance are specified.

SECTION 18–1 REVIEW

Answers are at the end of the chapter.

1. Define *line regulation*.
2. Define *load regulation*.
3. The input of a certain regulator increases by 3.5 V. As a result, the output voltage increases by 0.042 V. The nominal output is 20 V. Determine the line regulation in both % and in %/V.
4. If a 5.0 V power supply has an output resistance of 80 mΩ and a specified maximum output current of 1.0 A, what is the load regulation? Give the result as a % and as a %/mA.

18–2 BASIC SERIES REGULATORS

The fundamental classes of voltage regulators are linear regulators and switching regulators. Both of these are available in integrated circuit form. There are two basic types of linear regulator. One is the series regulator and the other is the shunt regulator. In this section, we will look at the series regulator. The shunt and switching regulators are covered in the next two sections.

After completing this section, you should be able to

- **Discuss the principles of series voltage regulators**
- Explain regulating action
- Calculate output voltage of an op-amp series regulator
- Discuss overload protection and explain how to use current limiting
- Describe a regulator with fold-back current limiting

A simple representation of a series type of **linear regulator** is shown in Figure 18–4(a), and the basic components are shown in the block diagram in Figure 18–4(b). The control element is in series with the load between the input and output. The output sample circuit senses a change in the output voltage. The error detector compares the sample voltage with a reference voltage and causes the control element to compensate in order to maintain a constant output voltage.

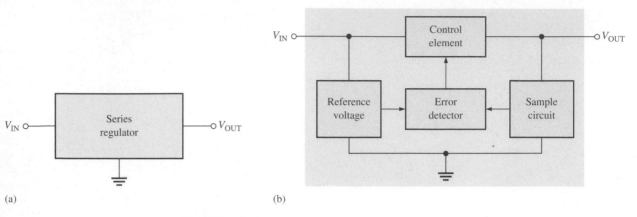

(a)   (b)

▲ FIGURE 18–4

Simple series voltage regulator and block diagram.

Regulating Action

A basic op-amp series regulator circuit is shown in Figure 18–5. The operation of the series regulator is illustrated in Figure 18–6 and is as follows. The resistive voltage divider formed by R_2 and R_3 senses any change in the output voltage. When the output tries to decrease, as indicated in Figure 18–6(a), because of a decrease in V_{IN} or because of an increase in I_L caused by a decrease in R_L, a proportional voltage decrease is applied to the op-amp's inverting input by the voltage divider. Since the zener diode (D_1) holds the other op-amp input at a nearly constant reference voltage, V_{REF}, a small difference voltage (error voltage) is developed across the op-amp's inputs. This difference voltage is amplified, and the op-amp's output voltage, V_B, increases. This increase is applied to the base of Q_1, causing the emitter voltage V_{OUT} to increase until the voltage to the inverting input again equals the reference (zener) voltage. This action offsets the attempted decrease in output voltage, thus keeping it nearly constant, as shown in part (b). The

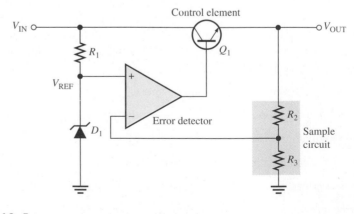

▲ FIGURE 18–5

Basic op-amp series regulator.

(a) When V_{IN} or R_L decreases, V_{OUT} attempts to decrease. The feedback voltage, V_{FB}, also attempts to decrease, and as a result, the op-amp's output voltage V_B attempts to increase, thus compensating for the attempted decrease in V_{OUT} by increasing the Q_1 emitter voltage. Changes in V_{OUT} are exaggerated for illustration.

(b) When V_{IN} (or R_L) stabilizes at its new lower value, the voltages return to their original values, thus keeping V_{OUT} constant as a result of the negative feedback.

(c) When V_{IN} or R_L increases, V_{OUT} attempts to increase. The feedback voltage, V_{FB}, also attempts to increase, and as a result, V_B, applied to the base of the control transistor, attempts to decrease, thus compensating for the attempted increase in V_{OUT} by decreasing the Q_1 emitter voltage.

(d) When V_{IN} (or R_L) stabilizes at its new higher value, the voltages return to their original values, thus keeping V_{OUT} constant as a result of the negative feedback.

▲ **FIGURE 18–6**

Illustration of series regulator action that keeps V_{OUT} constant when V_{IN} or R_L changes.

power transistor, Q_1, is usually used with a heat sink because it must handle all of the load current.

The opposite action occurs when the output tries to increase, as indicated in Figure 18–6(c) and (d). The op-amp in the series regulator is actually connected as a noninverting amplifier where the reference voltage V_{REF} is the input at the noninverting terminal, and the R_2/R_3 voltage divider forms the negative feedback circuit. The closed-loop voltage gain is

$$A_{cl} = 1 + \frac{R_2}{R_3}$$

Therefore, the regulated output voltage of the series regulator (neglecting the base-emitter voltage of Q_1) is

Equation 18–5

$$V_{OUT} \cong \left(1 + \frac{R_2}{R_3}\right)V_{REF}$$

From this analysis, you can see that the output voltage is determined by the zener voltage and the resistors R_2 and R_3. It is relatively independent of the input voltage, and therefore, regulation is achieved (as long as the input voltage and load current are within specified limits).

EXAMPLE 18–3

Determine the output voltage for the regulator in Figure 18–7.

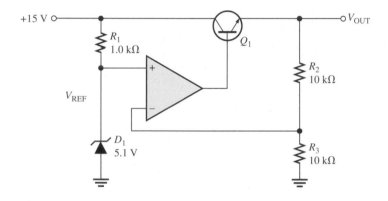

▲ **FIGURE 18–7**

Solution $V_{REF} = 5.1$ V, the zener voltage. The regulated output voltage is therefore

$$V_{OUT} = \left(1 + \frac{R_2}{R_3}\right)V_{REF} = \left(1 + \frac{10\ k\Omega}{10\ k\Omega}\right)5.1\ V = (2)5.1\ V = \textbf{10.2 V}$$

Related Problem   The following changes are made in the circuit in Figure 18–7: A 3.3 V zener replaces the 5.1 V zener, $R_1 = 1.8$ kΩ, $R_2 = 22$ kΩ, and $R_3 = 18$ kΩ. What is the output voltage?

> Open file E18-03 in the Examples folder on your CD-ROM. Measure the output voltage with 15 V dc applied to the input. Compare to the calculated value.

Short-Circuit or Overload Protection

If an excessive amount of load current is drawn, the series-pass transistor can be quickly damaged or destroyed. Most regulators use some type of excess current protection in the form of a current-limiting mechanism. Figure 18–8 shows one method of current limiting to prevent overloads called *constant-current limiting*. The current-limiting circuit consists of transistor Q_2 and resistor R_4.

The load current through R_4 produces a voltage from base to emitter of Q_2. When I_L reaches a predetermined maximum value, the voltage drop across R_4 is sufficient to forward-bias the base-emitter junction of Q_2, thus causing it to conduct. Enough Q_1 base

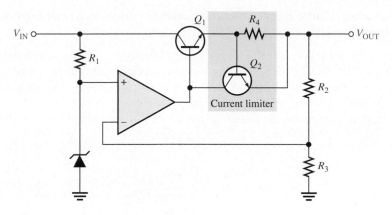

▲ FIGURE 18–8

Series regulator with constant–current limiting.

current is diverted into the collector of Q_2 so that I_L is limited to its maximum value, $I_{L(max)}$. Since the base-to-emitter voltage of Q_2 cannot exceed about 0.7 V for a silicon transistor, the voltage across R_4 is held to this value, and the load current is limited to

$$I_{L(max)} = \frac{0.7 \text{ V}}{R_4}$$

Equation 18–6

EXAMPLE 18–4

Determine the maximum current that the regulator in Figure 18–9 can provide to a load.

▲ FIGURE 18–9

Solution

$$I_{L(max)} = \frac{0.7 \text{ V}}{R_4} = \frac{0.7 \text{ V}}{1.0 \text{ }\Omega} = \textbf{0.7 A}$$

Related Problem

If the output of the regulator in Figure 18–9 is shorted, what is the current?

Regulator with Fold-Back Current Limiting

In the previous current-limiting technique, the current is restricted to a maximum constant value. **Fold-back current limiting** is a method used particularly in high-current regulators whereby the output current under overload conditions drops to a value well below the peak load current capability to prevent excessive power dissipation.

Basic Idea   The basic concept of fold-back current limiting is as follows, with reference to Figure 18–10. The circuit is similar to the constant current-limiting arrangement in Figure 18–8, with the exception of resistors R_5 and R_6. The voltage drop developed across R_4 by the load current must not only overcome the base-emitter voltage required to turn on Q_2, but it must also overcome the voltage across R_5. That is, the voltage across R_4 must be

$$V_{R4} = V_{R5} + V_{BE}$$

▲ **FIGURE 18–10**

Series regulator with fold-back current limiting.

In an overload or short-circuit condition, the load current increases to a value, $I_{L(max)}$, that is sufficient to cause Q_2 to conduct. At this point the current can increase no further. The decrease in output voltage results in a proportional decrease in the voltage across R_5; thus, less current through R_4 is required to maintain the forward-biased condition of Q_1. So, as V_{OUT} decreases, I_L decreases, as shown in the graph of Figure 18–11.

▶ **FIGURE 18–11**

Fold-back current limiting (output voltage versus load current).

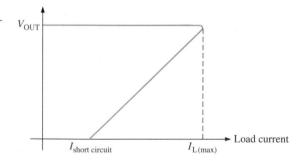

The advantage of this technique is that the regulator is allowed to operate with peak load current up to $I_{L(max)}$; but when the output becomes shorted, the current drops to a lower value to prevent overheating of the device.

SECTION 18–2 REVIEW

1. What are the basic components in a series regulator?

2. A certain series regulator has an output voltage of 8 V. If the op-amp's closed loop gain is 4, what is the value of the reference voltage?

18–3 BASIC SHUNT REGULATORS

The second basic type of linear voltage regulator is the shunt regulator. As you have learned, the control element in the series regulator is the series-pass transistor. In the shunt regulator, the control element is a transistor in parallel (shunt) with the load.

After completing this section, you should be able to

- **Discuss the principles of shunt voltage regulators**
- Describe the operation of a basic op-amp shunt regulator
- Compare series and shunt regulators

A simple representation of a shunt type of linear regulator is shown in Figure 18–12(a), and the basic components are shown in the block diagram in part (b).

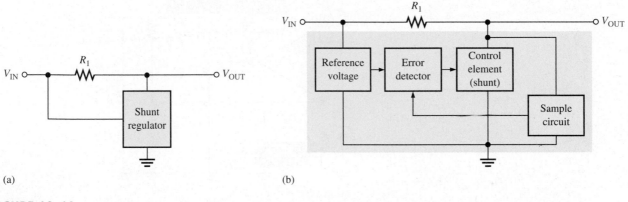

(a)   (b)

▲ FIGURE 18–12

Simple shunt regulator and block diagram.

In the basic shunt regulator, the control element is a transistor, Q_1, in parallel with the load, as shown in Figure 18–13. A resistor, R_1, is in series with the load. The operation of the circuit is similar to that of the series regulator, except that regulation is achieved by controlling the current through the parallel transistor Q_1.

▲ FIGURE 18–13

Basic op-amp shunt regulator with load resistor.

When the output voltage tries to decrease due to a change in input voltage or load current caused by a change in load resistance, as shown in Figure 18–14(a), the attempted decrease is sensed by R_3 and R_4 and applied to the op-amp's noninverting input. The resulting difference voltage reduces the op-amp's output (V_B), driving Q_1 less, thus reducing its collector current (shunt current) and increasing its effective collector-to-emitter resistance, r'_{CE}. Since r'_{CE} acts as a voltage divider with R_1, this action offsets the attempted decrease in V_{OUT} and maintains it at an almost constant level.

(a) Response to a decrease in V_{IN} or R_L

(b) Response to an increase in V_{IN} or R_L

▲ FIGURE 18–14

Sequence of responses when V_{OUT} tries to decrease as a result of a decrease in R_L or V_{IN} (opposite responses for an attempted increase).

The opposite action occurs when the output tries to increase, as indicated in Figure 18–14(b). With I_L and V_{OUT} constant, a change in the input voltage produces a change in shunt current (I_S) as follows (Δ means "a change in"):

$$\Delta I_S = \frac{\Delta V_{IN}}{R_1}$$

With a constant V_{IN} and V_{OUT}, a change in load current causes an opposite change in shunt current.

$$\Delta I_S = -\Delta I_L$$

This formula says that if I_L increases, I_S decreases, and vice versa.

The shunt regulator is less efficient than the series type but offers inherent short-circuit protection. If the output is shorted ($V_{OUT} = 0$), the load current is limited by the series resistor R_1 to a maximum value as follows ($I_S = 0$).

$$I_{L(max)} = \frac{V_{IN}}{R_1}$$

Equation 18–7

EXAMPLE 18–5

In Figure 18–15, what power rating must R_1 have if the maximum input voltage is 12.5 V?

▲ **FIGURE 18–15**

Solution The worst-case power dissipation in R_1 occurs when the output is short-circuited and $V_{OUT} = 0$. When $V_{IN} = 12.5$ V, the voltage dropped across R_1 is

$$V_{R1} = V_{IN} - V_{OUT} = 12.5 \text{ V}$$

The power dissipation in R_1 is

$$P_{R1} = \frac{V_{R1}^2}{R_1} = \frac{(12.5 \text{ V})^2}{22 \text{ }\Omega} = \textbf{7.10 W}$$

Therefore, a resistor with a rating of at least 10 W should be used.

Related Problem   In Figure 18–15, R_1 is changed to 33 Ω. What must be the power rating of R_1 if the maximum input voltage is 24 V?

Open file E18-05 in the Examples folder on your CD-ROM. Measure the output voltage with 15 V applied to the input.

SECTION 18–3 REVIEW

1. How does the control element in a shunt regulator differ from that in a series regulator?
2. What is one advantage of a shunt regulator over a series type? What is a disadvantage?

18–4 BASIC SWITCHING REGULATORS

The two types of linear regulators, series and shunt, have control elements (transistors) that are conducting all the time, with the amount of conduction varied as demanded by changes in the output voltage or current. The switching regulator is different; the control element operates as a switch. A greater efficiency can be realized with this type of voltage regulator than with the linear types because the transistor is not always conducting. Therefore, switching regulators can provide greater load currents at low voltage than linear regulators because the control transistor doesn't dissipate as much power. Three basic configurations of switching regulators are step-down, step-up, and inverting.

After completing this section, you should be able to

- ■ **Discuss the principles of switching regulators**
- ■ Describe the step-down configuration of a switching regulator
- ■ Determine the output voltage of the step-down configuration
- ■ Describe the step-up configuration of a switching regulator
- ■ Describe the voltage-inverter configuration

Step-Down Configuration

In the step-down configuration, the output voltage is always less than the input voltage. A basic step-down **switching regulator** is shown in Figure 18–16(a), and its simplified equivalent is shown in Figure 18–16(b). Transistor Q_1 is used to switch the input voltage at a duty cycle that is based on the regulator's load requirement. The *LC* filter is then used to average the switched voltage. Since Q_1 is either *on* (saturated) or *off,* the power lost in the control element is relatively small. Therefore, the switching regulator is useful primarily in higher power applications or in applications where efficiency is of utmost concern.

The *on* and *off* intervals of Q_1 are shown in the waveform of Figure 18–17(a). The capacitor charges during the on-time (t_{on}) and discharges during the off-time (t_{off}). When the on-time is increased relative to the off-time, the capacitor charges more, thus increasing

(a) Typical circuit

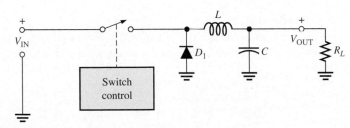

(b) Simplified equivalent circuit

▲ **FIGURE 18–16**

Basic step-down switching regulator.

(a) V_{OUT} depends on the duty cycle.

(b) Increase the duty cycle and V_{OUT} increases.

(c) Decrease the duty cycle and V_{OUT} decreases.

▲ **FIGURE 18–17**

Switching regulator waveforms. The V_C waveform is shown for no inductive filtering to illustrate the charge and discharge action (ripple). L and C smooth V_C to a nearly constant level, as indicated by the dashed line for V_{OUT}.

the output voltage, as indicated in Figure 18–17(b). When the on-time is decreased relative to the off-time, the capacitor discharges more, thus decreasing the output voltage, as in Figure 18–17(c). Therefore, by adjusting the duty cycle $t_{on}/(t_{on} + t_{off})$ of Q_1, the output voltage can be varied. The inductor further smooths the fluctuations of the output voltage caused by the charging and discharging action.

Ideally, the output voltage is expressed as

Equation 18–8

$$V_{OUT} = \left(\frac{t_{on}}{T}\right)V_{IN}$$

T is the period of the on-off cycle of Q_1 and is related to the frequency by $T = 1/f$. The period is the sum of the on-time and the off-time.

$$T = t_{on} + t_{off}$$

The ratio t_{on}/T is called the *duty cycle*.

The regulating action is as follows and is illustrated in Figure 18–18. When V_{OUT} tries to decrease, the on-time of Q_1 is increased, causing an additional charge on C to offset the

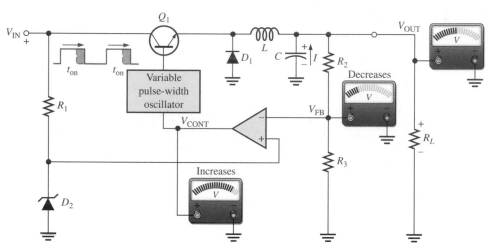

(a) When V_{OUT} attempts to decrease, the on-time of Q_1 increases.

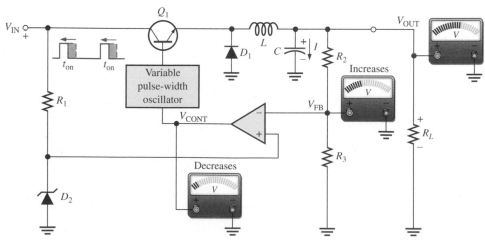

(b) When V_{OUT} attempts to increase, the on-time of Q_1 decreases.

▲ FIGURE 18–18

Basic regulating action of a step-down switching regulator.

attempted decrease. When V_{OUT} tries to increase, the on-time of Q_1 is decreased, causing the capacitor to discharge enough to offset the attempted increase.

Step-Up Configuration

A basic step-up type of switching regulator is shown in Figure 18–19, where transistor Q_1 operates as a switch to ground.

▲ **FIGURE 18–19**

Basic step-up switching regulator.

The switching action is illustrated in Figures 18–20 and 18–21. When Q_1 turns on, a voltage equal to approximately V_{IN} is induced across the inductor with a polarity as indicated in Figure 18–20. During the on-time (t_{on}) of Q_1, and inductor voltage, V_L, decreases from its initial maximum and diode D_1 is reverse-biased. The longer Q_1 is on, the smaller

▲ **FIGURE 18–20**

Basic action of a step-up regulator when Q_1 is on.

▲ FIGURE 18–21

Basic switching action of a step–up regulator when Q_1 turns off.

V_L becomes. During the on-time, the capacitor only discharges an extremely small amount through the load.

When Q_1 turns off, as indicated in Figure 18–21, the inductor voltage suddenly reverses polarity and adds to V_{IN}, forward-biasing diode D_1 and allowing the capacitor to charge. The output voltage is equal to the capacitor voltage and can be larger than V_{IN} because the capacitor is charged to V_{IN} plus the voltage induced across the inductor during the off-time of Q_1.

The longer the on-time of Q_1, the more the inductor voltage will decrease and the greater the magnitude of the voltage when the inductor reverses polarity at the instant Q_1 turns off. As you have seen, this reverse polarity voltage is what charges the capacitor above V_{IN}. The output voltage is dependent on both the inductor's magnetic field action (determined by t_{on}) and the charging of the capacitor (determined by t_{off}).

Voltage regulation is achieved by the variation of the on-time of Q_1 (within certain limits) as related to changes in V_{OUT} due to changing load or input voltage. If V_{OUT} tries to increase, the on-time of Q_1 will decrease, resulting in a decrease in the amount that C will charge. If V_{OUT} tries to decrease, the on-time of Q_1 will increase, resulting in an increase in the amount that C will charge. This regulating action maintains V_{OUT} at an essentially constant level.

Voltage-Inverter Configuration

A third type of switching regulator produces an output voltage that is opposite in polarity to the input. A basic diagram is shown in Figure 18–22.

When Q_1 turns on, the inductor voltage jumps to approximately $V_{IN} - V_{CE(sat)}$ and the magnetic field rapidly expands, as shown in Figure 18–23(a). While Q_1 is on, the diode is reverse-biased and the inductor voltage decreases from its initial maximum. When Q_1 turns off, the magnetic field collapses and the inductor's polarity reverses, as shown in Figure 18–23(b). This forward-biases the diode, charges C, and produces a negative out-

▲ **FIGURE 18–22**

Basic inverting switching regulator.

(a) When Q_1 is on, D_1 is reverse-biased.

(b) When Q_1 turns off, D_1 forward biases.

▲ **FIGURE 18–23**

Basic inverting action of an inverting switching regulator.

put voltage, as indicated. The repetitive on-off action of Q_1 produces a repetitive charging and discharging that is smoothed by the LC filter action.

As with the step-up regulator, the less time Q_1 is on, the greater the output voltage is, and vice versa. This regulating action is illustrated in Figure 18–24. Switching regulator efficiencies can be greater than 90 percent.

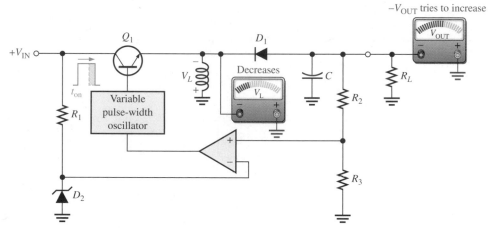

(a) When $-V_{OUT}$ tries to decrease, t_{on} decreases, causing V_L to increase. This compensates for the attempted decrease in $-V_{OUT}$.

(b) When $-V_{OUT}$ tries to increase, t_{on} increases, causing V_L to decrease. This compensates for the attempted increase in $-V_{OUT}$.

▲ **FIGURE 18–24**

Basic regulating action of an inverting switching regulator.

SECTION 18–4 REVIEW

1. What are three types of switching regulators?
2. What is the primary advantage of switching regulators over linear regulators?
3. How are changes in output voltage compensated for in the switching regulator?

18–5 INTEGRATED CIRCUIT VOLTAGE REGULATORS

In the previous sections, the basic voltage regulator configurations were presented. Several types of both linear and switching regulators are available in integrated circuit (IC) form. Generally, the linear regulators are three-terminal devices that provide either positive or negative output voltages that can be either fixed or adjustable. In this section, typical linear and switching IC regulators are introduced.

After completing this section, you should be able to

- **Discuss integrated circuit voltage regulators**
- Describe the 7800 series of positive regulators
- Describe the 7900 series of negative regulators
- Describe the LM317 adjustable positive regulator
- Describe the LM337 adjustable negative regulator
- Describe IC switching regulators

Fixed Positive Linear Voltage Regulators

Although many types of IC regulators are available, the 7800 series of IC regulators is representative of three-terminal devices that provide a fixed positive output voltage. The three terminals are input, output, and ground as indicated in the standard fixed voltage configuration in Figure 18–25(a). The last two digits in the part number designate the output voltage. For example, the 7805 is a +5.0 V regulator. Other available output voltages are given in Figure 18–25(b) and common packages are shown in part (c).

(a) Standard configuration

Type number	Output voltage
7805	+5.0 V
7806	+6.0 V
7808	+8.0 V
7809	+9.0 V
7812	+12.0 V
7815	+15.0 V
7818	+18.0 V
7824	+24.0 V

(b) The 7800 series

TO–220
T SUFFIX
CASE 221A

Heatsink surface
connected to Pin 2.

Pin 1. Input
2. Ground
3. Output

D²PAK
D2T SUFFIX
CASE 936

Heatsink surface (shown as terminal 4 in
case outline drawing) is connected to Pin 2.

(c) Typical packages

▲ **FIGURE 18–25**

The 7800 series three-terminal fixed positive voltage regulators.

Capacitors, although not always necessary, are sometimes used on the input and output as indicated in Figure 18–25(a). The output capacitor acts basically as a line filter to improve transient response. The input capacitor is used to prevent unwanted oscillations when the regulator is some distance from the power supply filter such that the line has a significant inductance.

The 7800 series can produce output currents up to in excess of 1 A when used with an adequate heat sink. The input voltage must be at least 2 V above the output voltage in order to maintain regulation. The circuits have internal thermal overload protection and short-circuit current-limiting features. **Thermal overload** occurs when the internal power dissipation becomes excessive and the temperature of the device exceeds a certain value. Almost all applications of regulators require that the device be secured to a heat sink to prevent thermal overload.

Fixed Negative Linear Voltage Regulators

The 7900 series is typical of three-terminal IC regulators that provide a fixed negative output voltage. This series is the negative-voltage counterpart of the 7800 series and shares most of the same features and characteristics. Figure 18–26 indicates the standard configuration and part numbers with corresponding output voltages that are available.

(a) Standard configuration

Type number	Output voltage
7905	−5.0 V
7905.2	−5.2 V
7906	−6.0 V
7908	−8.0 V
7912	−12.0 V
7915	−15.0 V
7918	−18.0 V
7924	−24.0 V

(b) The 7900 series

▲ FIGURE 18–26

The 7900 series three-terminal fixed negative voltage regulators.

Adjustable Positive Linear Voltage Regulators

The LM317 is an example of a three-terminal positive regulator with an adjustable output voltage. A data sheet for this device is given in Appendix C. The standard configuration is shown in Figure 18–27. The capacitors are for decoupling and do not affect the dc operation. Notice that there is an input, an output, and an adjustment terminal. The external fixed resistor R_1 and the external variable resistor R_2 provide the output voltage adjustment. V_{OUT} can be varied from 1.2 V to 37 V depending on the resistor values. The LM317 can provide over 1.5 A of output current to a load.

The LM317 is operated as a "floating" regulator because the adjustment terminal is not connected to ground, but floats to whatever voltage is across R_2. This allows the output voltage to be much higher than that of a fixed-voltage regulator.

▲ FIGURE 18–27

The LM317 three-terminal adjustable positive voltage regulator.

Basic Operation   As indicated in Figure 18–28, a constant 1.25 V reference voltage (V_{REF}) is maintained by the regulator between the output terminal and the adjustment terminal. This constant reference voltage produces a constant current (I_{REF}) through R_1, regardless of the value of R_2. I_{REF} is also through R_2.

$$I_{REF} = \frac{V_{REF}}{R_1} = \frac{1.25\ V}{R_1}$$

Also, there is a very small constant current at the adjustment terminal of approximately 50 μA called I_{ADJ}, which is through R_2. A formula for the output voltage is developed as follows.

$$V_{OUT} = V_{R1} + V_{R2} = I_{REF}R_1 + I_{REF}R_2 + I_{ADJ}R_2$$

$$= I_{REF}(R_1 + R_2) + I_{ADJ}R_2 = \frac{V_{REF}}{R_1}(R_1 + R_2) + I_{ADJ}R_2$$

$$V_{OUT} = V_{REF}\left(1 + \frac{R_2}{R_1}\right) + I_{ADJ}R_2$$

Equation 18–9

As you can see, the output voltage is a function of both R_1 and R_2. Once the value of R_1 is set, the output voltage is adjusted by varying R_2.

▲ **FIGURE 18–28**

Operation of the LM317 adjustable voltage regulator.

EXAMPLE 18–6

Determine the minimum and maximum output voltages for the voltage regulator in Figure 18–29. Assume $I_{ADJ} = 50\ \mu$A.

▲ **FIGURE 18–29**

Solution
$$V_{R1} = V_{REF} = 1.25 \text{ V}$$

When R_2 is set at its minimum of $0 \text{ }\Omega$,

$$V_{OUT(min)} = V_{REF}\left(1 + \frac{R_2}{R_1}\right) + I_{ADJ}R_2 = 1.25 \text{ V}(1) = \mathbf{1.25 \text{ V}}$$

When R_2 is set at its maximum of $5 \text{ k}\Omega$,

$$V_{OUT(max)} = V_{REF}\left(1 + \frac{R_2}{R_1}\right) + I_{ADJ}R_2 = 1.25 \text{ V}\left(1 + \frac{5 \text{ k}\Omega}{220 \text{ }\Omega}\right) + (50 \text{ }\mu\text{A})5 \text{ k}\Omega$$

$$= 29.66 \text{ V} + 0.25 \text{ V} = \mathbf{29.9 \text{ V}}$$

Related Problem What is the output voltage of the regulator if R_2 is set at $2 \text{ k}\Omega$?

Adjustable Negative Linear Voltage Regulators

The LM337 is the negative output counterpart of the LM317 and is a good example of this type of IC regulator. Like the LM317, the LM337 requires two external resistors for output voltage adjustment as shown in Figure 18–30. The output voltage can be adjusted from -1.2 V to -37 V, depending on the external resistor values. The capacitors are for decoupling and do not affect the dc operation.

▲ **FIGURE 18–30**

The LM337 three-terminal adjustable negative voltage regulator.

Switching Voltage Regulators

As an example of an IC switching voltage regulator, let's look at the 78S40. This is a universal device that can be used with external components to provide step-up, step-down, and inverting operation.

The internal circuitry of the 78S40 is shown in Figure 18–31. This circuit can be compared to the basic switching regulators that were covered in Section 18–4. For example, look back at Figure 18–16(a). The oscillator and comparator functions are directly comparable. The gate and flip-flop in the 78S40 were not included in the basic circuit of Figure 18–16(a), but they provide additional regulating action. Transistors Q_1 and Q_2 effectively perform the same function as Q_1 in the basic circuit. The 1.25 V reference block in the 78S40 has the same purpose as the zener diode in the basic circuit, and diode D_1 in the 78S40 corresponds to D_1 in the basic circuit.

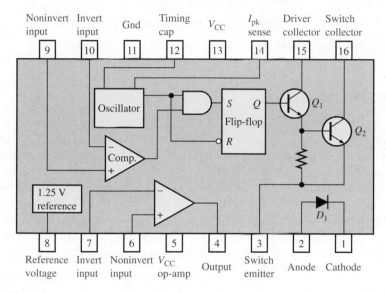

▲ **FIGURE 18–31**

The 78S40 switching regulator.

The 78S40 also has an "uncommited" op-amp thrown in for good measure. It is not used in any of the regulator configurations. External circuitry is required to make this device operate as a regulator, as you will see in Section 18–6.

SECTION 18–5 REVIEW

1. What are the three terminals of a fixed-voltage regulator?
2. What is the output voltage of a 7809? Of a 7915?
3. What are the three terminals of an adjustable-voltage regulator?
4. What external components are required for a basic LM317 configuration?

18–6 APPLICATIONS OF IC VOLTAGE REGULATORS

In the last section, you saw several devices that are representative of the general types of IC voltage regulators. Now, several different ways these devices can be modified with external circuitry to improve or alter their performance are examined.

After completing this section, you should be able to

■ **Discuss applications of IC voltage regulators**

■ Explain the use of an external pass transistor

■ Explain the use of current limiting

■ Explain how to use a voltage regulator as a constant-current source

■ Discuss some application considerations for switching regulators

The External Pass Transistor

As you know, an IC voltage regulator is capable of delivering only a certain amount of output current to a load. For example, the 7800 series regulators can handle a peak output current of 1.3 A (more under certain conditions). If the load current exceeds the maximum allowable value, there will be thermal overload and the regulator will shut down. A thermal overload condition means that there is excessive power dissipation inside the device.

If an application requires more than the maximum current that the regulator can deliver, an external pass transistor can be used. Figure 18–32 illustrates a three-terminal regulator with an external pass transistor for handling currents in excess of the output current capability of the basic regulator.

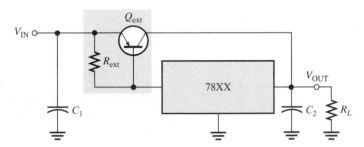

▲ **FIGURE 18–32**

A 7800-series three-terminal regulator with an external pass transistor to increase power dissipation.

The value of the external current-sensing resistor, R_{ext}, determines the value of current at which Q_{ext} begins to conduct because it sets the base-to-emitter voltage of the transistor. As long as the current is less than the value set by R_{ext}, the transistor Q_{ext} is off, and the regulator operates normally as shown in Figure 18–33(a). This is because the voltage drop across R_{ext} is less than the 0.7 V base-to-emitter voltage required to turn Q_{ext} on. R_{ext} is determined by the following formula, where I_{max} is the highest current that the voltage regulator is to handle internally.

Equation 18–10

$$R_{ext} = \frac{0.7 \text{ V}}{I_{max}}$$

When the current is sufficient to produce at least a 0.7 V drop across R_{ext}, the external pass transistor Q_{ext} turns on and conducts any current in excess of I_{max}, as indicated in Figure 18–33(b). Q_{ext} will conduct more or less, depending on the load requirements.

(a) When the regulator current is less than I_{max}, the external pass transistor is off and the regulator is handling all of the current.

(b) When the load current exceeds I_{max}, the drop across R_{ext} turns Q_{ext} on and it conducts the excess current.

▲ **FIGURE 18–33**

Operation of the regulator with an external pass transistor.

For example, if the total load current is 3 A and I_{max} was selected to be 1 A, the external pass transistor will conduct 2 A, which is the excess over the internal voltage regulator current I_{max}.

EXAMPLE 18–7

What value is R_{ext} if the maximum current to be handled internally by the voltage regulator in Figure 18–32 is set at 700 mA?

Solution

$$R_{ext} = \frac{0.7\ \text{V}}{I_{max}} = \frac{0.7\ \text{V}}{0.7\ \text{A}} = \mathbf{1\ \Omega}$$

Related Problem If R_{ext} is changed to 1.5 Ω, at what current value will Q_{ext} turn on?

The external pass transistor is typically a power transistor with a heat sink that must be capable of handling a maximum power of

$$P_{ext} = I_{ext}(V_{IN} - V_{OUT})$$

EXAMPLE 18–8

What must be the minimum power rating for the external pass transistor used with a 7824 regulator in a circuit such as that shown in Figure 18–32? The input voltage is 30 V and the load resistance is 10 Ω. The maximum internal current is to be 700 mA. Assume that there is no heat sink for this calculation. Keep in mind that the use of a heat sink increases the effective power rating of the transistor and you can use a lower rated transistor.

Solution The load current is

$$I_L = \frac{V_{OUT}}{R_L} = \frac{24\ \text{V}}{10\ \Omega} = 2.4\ \text{A}$$

The current through Q_{ext} is

$$I_{ext} = I_L - I_{max} = 2.4\ \text{A} - 0.7\ \text{A} = 1.7\ \text{A}$$

The power dissipated by Q_{ext} is

$$P_{ext(min)} = I_{ext}(V_{IN} - V_{OUT}) = (1.7 \text{ A})(30 \text{ V} - 24 \text{ V}) = (1.7 \text{ A})(6 \text{ V}) = \textbf{10.2 W}$$

For a safety margin, choose a power transistor with a rating greater than 10.2 W, say at least 15 W.

Related Problem   Rework this example using a 7815 regulator.

Current Limiting

A drawback of the circuit in Figure 18–32 is that the external transistor is not protected from excessive current, such as would result from a shorted output. An additional current-limiting circuit (Q_{lim} and R_{lim}) can be added as shown in Figure 18–34 to protect Q_{ext} from excessive current and possible burn out.

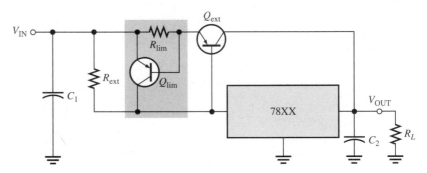

▲ **FIGURE 18–34**

Regulator with current limiting.

The following describes the way the current-limiting circuit works. The current-sensing resistor R_{lim} sets the V_{BE} of transistor Q_{lim}. The base-to-emitter voltage of Q_{ext} is now determined by $V_{Rext} - V_{Rlim}$ because they have opposite polarities. So, for normal operation, the drop across R_{ext} must be sufficient to overcome the opposing drop across R_{lim}. If the current through Q_{ext} exceeds a certain maximum ($I_{ext(max)}$) because of a shorted output or a faulty load, the voltage across R_{lim} reaches 0.7 V and turns Q_{lim} on. Q_{lim} now conducts current away from Q_{ext} and through the regulator, forcing a thermal overload to occur and shut down the regulator. Remember, the IC regulator is internally protected from thermal overload as part of its design.

This action is illustrated in Figure 18–35. In part (a), the circuit is operating normally with Q_{ext} conducting less than the maximum current that it can handle with Q_{lim} off. Part (b) shows what happens when there is a short across the load. The current through Q_{ext} suddenly increases and causes the voltage drop across R_{lim} to increase, which turns Q_{lim} on. The current is now diverted through the regulator, which causes it to shut down due to thermal overload.

A Current Regulator

The three-terminal regulator can be used as a current source when an application requires that a constant current be supplied to a variable load. The basic circuit is shown in Figure 18–36 where R_1 is the current-setting resistor. The regulator provides a fixed constant

(a) During normal operation, when the load current is not excessive, Q_{lim} is off.

(b) When short occurs ①, the external current becomes excessive and the voltage across R_{lim} increases ② and turns on Q_{lim} ③, which then conducts current away from Q_{ext} and routes it through the regulator, causing the internal regulator current to become excessive ④ which forces the regulator into thermal shut down.

▲ **FIGURE 18–35**

The current-limiting action of the regulator circuit.

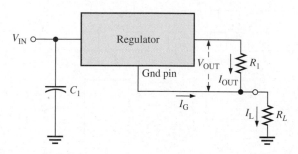

▲ **FIGURE 18–36**

The three-terminal regulator as a current source.

voltage, V_{OUT}, between the ground terminal (not connected to ground in this case) and the output terminal. This determines the constant current supplied to the load.

$$I_L = \frac{V_{OUT}}{R_1} + I_G$$

Equation 18–11

The current, I_G, from the ground terminal is very small compared to the output current and can often be neglected.

EXAMPLE 18–9

What value of R_1 is necessary in a 7805 regulator to provide a constant current of 1 A to a variable load that can be adjusted from 1 Ω to 10 Ω?

Solution

First, 1 A is within the limits of the 7805's capability (remember, it can handle at least 1.3 A without an external pass transistor).

The 7805 produces 5 V between its ground terminal and its output terminal. Therefore, if you want 1 A of current, the current-setting resistor must be (neglecting I_G)

$$R_1 = \frac{V_{OUT}}{I_L} = \frac{5\ V}{1\ A} = \mathbf{5\ \Omega}$$

The circuit is shown in Figure 18–37.

▲ **FIGURE 18–37**

A constant–current source of 1 A.

Related Problem

If a 7808 regulator is used instead of the 7805, to what value would you change R_1 to maintain a constant current of 1 A?

▲ **FIGURE 18–38**

The step–down configuration of the 78S40 switching regulator.

Switching Regulator Configurations

In Section 18–5, the 78S40 was introduced as an example of an IC switching voltage regulator. Figure 18–38 shows the external connections for a step-down configuration where the output voltage is less than the input voltage, and Figure 18–39 shows a step-up configuration in which the output voltage is greater than the input voltage. An inverting configuration is also possible, but it is not shown here.

The timing capacitor, C_T, controls the pulse width and frequency of the oscillator and thus establishes the on-time of transistor Q_2. The voltage across the current-sensing resistor, R_{CS}, is used internally by the oscillator to vary the duty cycle based on the desired peak load current. The voltage divider, made up of R_1 and R_2, reduces the output voltage to a nominal value equal to the reference voltage. If V_{OUT} exceeds its set value, the output of the comparator switches to its low state, disabling the gate to turn Q_2 off until the output decreases. This regulating action is in addition to that produced by the duty cycle variation of the oscillator as described in Section 18–4 in relation to the basic switching regulator.

▲ **FIGURE 18–39**

The step-up configuration of the 78S40 switching regulator.

SECTION 18–6 REVIEW

1. What is the purpose of using an external pass transistor with an IC voltage regulator?
2. What is the advantage of current limiting in a voltage regulator?
3. What does *thermal overload* mean?

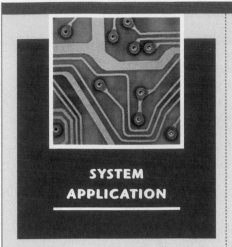

SYSTEM APPLICATION

The regulated power supply in this application provides dual-polarity dc voltages of ±12 V to the FM receiver system which you worked with in Chapter 15. The power supply consists of a transformer-coupled full-wave bridge rectifier and filter with positive and negative three-terminal voltage regulators.

The block diagram of the dual-polarity power supply is shown in Figure 18–40, and the circuit board is shown in Figure 18–41. The large vertically mounted capacitors are the 100 μF filter capacitors. The 0.33 μF and 1 μF capacitors, although not necessary in all applications, are recommended by the manufacturer for stability and improved transient response.

The Dual–Polarity Power Supply Board

■ Make sure that the circuit board shown in Figure 18–41 is correct by checking it against the schematic in Figure 18–42 and by referring to the the 7800 and 7900 series data sheets in Appendix C. Backside interconections are shown as darker traces.

■ Label a copy of the board with component and input/output designations in agreement with the schematic.

The Power Supply Circuit

■ Determine the approximate voltage at each of the four "corners" of the bridge. The transformer is rated at 24 V rms.

■ Calculate the peak inverse voltage of the rectifier diodes.

■ Determine the voltage at the input of each voltage regulator.

■ In the FM receiver from Chapter 15, assume that op-amps are used only in the channel separation circuits. If all the other circuits in the receiver, excluding the channel separation circuits, used +12 V only and draw an average dc current of 100 mA, determine how much total current each regulator must supply.

■ Based on the amount of dc current required by the receiver, do the regulators need to be attached to the heat sink?

Test Procedure

Develop a basic procedure for thoroughly testing the dual-polarity power supply.

Troubleshooting

Based on the results indicated in Figure 18–43 for four faulty power supply boards, determine the most likely fault or faults in each case.

Final Report (Optional)

Submit a final written report on the dual-polarity power supply using an organized format that includes the following:

1. A physical description of the circuit.

2. A discussion of the operation of the circuit.

3. A list of the specifications.

4. A list of parts with part numbers if available.

5. A list of the types of problems in the four faulty boards.

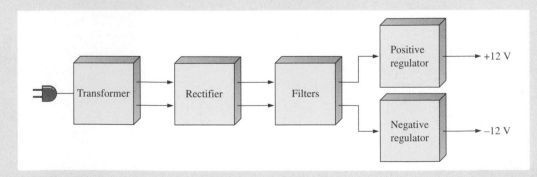

▲ FIGURE 18–40

Block diagram of the dual-polarity power supply.

▲ **FIGURE 18–41**

The dual-polarity power supply circuit board.

▲ **FIGURE 18–42**

The dual-polarity power supply schematic.

▲ **FIGURE 18–43**

Results of tests on four faulty power supply boards.

- Voltage regulators keep a constant dc output voltage when the input or load varies within limits.
- A basic voltage regulator consists of a reference voltage source, an error detector, a sampling element, and a control device. Protection circuitry is also found in most regulators.
- Two basic categories of voltage regulators are linear and switching.
- Two basic types of linear regulators are series and shunt.
- In a series linear regulator, the control element is a transistor in series with the load.
- In a shunt linear regulator, the control element is a transistor in parallel with the load.
- Three configurations for switching regulators are step-down, step-up, and inverting.
- Switching regulators are more efficient than linear regulators and are particularly useful in low-voltage, high-current applications.
- Three-terminal linear IC regulators are available for either fixed output or variable output voltages of positive or negative polarities.
- An external pass transistor increases the current capability of a regulator.
- The 7800 series are three-terminal IC regulators with fixed positive output voltage.
- The 7900 series are three-terminal IC regulators with fixed negative output voltage.
- The LM317 is a three-terminal IC regulator with a positive variable output voltage.
- The LM337 is a three-terminal IC regulator with a negative variable output voltage.
- The 78S40 is a switching voltage regulator.

KEY TERMS

Key terms and other bold terms in the chapter are defined in the end-of-book glossary.

Linear regulator   A voltage regulator in which the control element operates in the linear region.

Line regulation   The percentage change in output voltage for a given change in input (line) voltage.

Load regulation   The percentage change in output voltage for a given change in load current.

Regulator   An electronic circuit that maintains an essentially constant output voltage with a changing input voltage or load current.

Switching regulator   A voltage regulator in which the control element operates as a switch.

Thermal overload   A condition in a rectifier where the internal power dissipation of the circuit exceeds a certain maximum due to excessive current.

KEY FORMULAS

Voltage Regulation

18–1 $\text{Line regulation} = \left(\dfrac{\Delta V_{\text{OUT}}}{\Delta V_{\text{IN}}}\right)100\%$ Line regulation as a percentage

18–2 $\text{Load regulation} = \dfrac{(\Delta V_{\text{OUT}}/V_{\text{OUT}})100\%}{\Delta V_{\text{IN}}}$ Line regulation in %/V

18–3 $\text{Load regulation} = \left(\dfrac{V_{\text{NL}} - V_{\text{FL}}}{V_{\text{FL}}}\right)100\%$ Percent load regulation

18–4 $\text{Load regulation} = \left(\dfrac{R_{\text{OUT}}}{R_{\text{FL}}}\right)100\%$ Load regulation in terms of output resistance and full-load resistance

Basic Series Regulator

18–5 $V_{\text{OUT}} \cong \left(1 + \dfrac{R_2}{R_3}\right)V_{\text{REF}}$ Regulator output

18–6 $\qquad I_{L(max)} = \dfrac{0.7 \text{ V}}{R_4}$ $\qquad\qquad$ For constant-current limiting (silicon)

Basic Shunt Regulator

18–7 $\qquad I_{L(max)} = \dfrac{V_{IN}}{R_1}$ $\qquad\qquad$ Maximum load current

Basic Switching Regulators

18–8 $\qquad V_{OUT} = \left(\dfrac{t_{on}}{T}\right)V_{IN}$ $\qquad\qquad$ For step-down switching regulator

IC Voltage Regulators

18–9 $\qquad V_{OUT} = V_{REF}\left(1 + \dfrac{R_2}{R_1}\right) + I_{ADJ}R_2$ $\qquad$ IC regulator

18–10 $\qquad R_{ext} = \dfrac{0.7 \text{ V}}{I_{max}}$ $\qquad\qquad$ For external pass circuit

18–11 $\qquad I_L = \dfrac{V_{OUT}}{R_1} + I_G$ $\qquad\qquad$ Regulator as a current source

SELF-TEST

Answers are at the end of the chapter.

1. In the case of line regulation,
   (a) when the temperature varies, the output voltage stays constant
   (b) when the output voltage changes, the load current stays constant
   (c) when the input voltage changes, the output voltage stays constant
   (d) when the load changes, the output voltage stays constant

2. In the case of load regulation,
   (a) when the temperature varies, the output voltage stays constant
   (b) when the input voltage changes, the load current stays constant
   (c) when the load changes, the load current stays constant
   (d) when the load changes, the output voltage stays constant

3. All of the following are parts of a basic voltage regulator *except*
   (a) control element $\qquad$ (b) sampling circuit
   (c) voltage-follower $\qquad$ (d) error detector
   (e) reference voltage

4. The basic difference between a series regulator and a shunt regulator is
   (a) the amount of current that can be handled
   (b) the position of the control element
   (c) the type of sample circuit
   (d) the type of error detector

5. In a basic series regulator, V_{OUT} is determined by
   (a) the control element $\qquad$ (b) the sample circuit
   (c) the reference voltage $\qquad$ (d) answers (b) and (c)

6. The main purpose of current limiting in a regulator is
   (a) protection of the regulator from excessive current
   (b) protection of the load from excessive current
   (c) to keep the power supply transformer from burning up
   (d) to maintain a constant output voltage

7. In a linear regulator, the control transistor is conducting

   (a) a small part of the time   (b) half the time

   (c) all of the time   (d) only when the load current is excessive

8. In a switching regulator, the control transistor is conducting

   (a) part of the time

   (b) all of the time

   (c) only when the input voltage exceeds a set limit

   (d) only when there is an overload

9. The LM317 is an example of an IC

   (a) three-terminal negative voltage regulator

   (b) fixed positive voltage regulator

   (c) switching regulator

   (d) linear regulator

   (e) variable positive voltage regulator

   (f) answers (b) and (d) only

   (g) answers (d) and (e) only

10. An external pass transistor is used for

    (a) increasing the output voltage

    (b) improving the regulation

    (c) increasing the current that the regulator can handle

    (d) short-circuit protection

PROBLEMS

Answers to all odd-numbered problems are at the end of the book.

BASIC PROBLEMS

SECTION 18–1   **Voltage Regulation**

1. The nominal output voltage of a certain regulator is 8 V. The output changes 2 mV when the input voltage goes from 12 V to 18 V. Determine the line regulation and express it as a percentage change over the entire range of V_{IN}.

2. Express the line regulation found in Problem 1 in units of %/V.

3. A certain regulator has a no-load output voltage of 10 V and a full-load output voltage of 9.90 V. What is the percent load regulation?

4. In Problem 3, if the full-load current is 250 mA, express the load regulation in %/mA.

SECTION 18–2   **Basic Series Regulators**

5. Label the functional blocks for the voltage regulator in Figure 18–44.

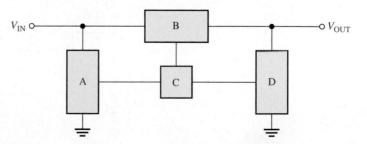

▲ **FIGURE 18–44**

6. Determine the output voltage for the regulator in Figure 18–45.

► **FIGURE 18–45**

File circuits are identified with a CD logo and are in the Problems folder on your CD-ROM. Filenames correspond to figure numbers (e.g., F18–45).

7. Determine the output voltage for the series regulator in Figure 18–46.
8. If R_3 in Figure 18–46 is increased to 4.7 kΩ, what happens to the output voltage?
9. If the zener voltage is 2.7 V instead of 2.4 V in Figure 18–46, what is the output voltage?

▲ **FIGURE 18–46**

10. A series voltage regulator with constant-current limiting is shown in Figure 18–47. Determine the value of R_4 if the load current is to be limited to a maximum value of 250 mA. What power rating must R_4 have?
11. If the R_4 determined in Problem 10 is halved, what is the maximum load current?

▲ **FIGURE 18–47**

SECTION 18–3   Basic Shunt Regulators

12. In the shunt regulator of Figure 18–48, when the load current increases, does Q_1 conduct more or less? Why?

13. Assume I_L remains constant and V_{IN} changes by 1 V in Figure 18–48. What is the change in the collector current of Q_1?

14. With a constant input voltage of 17 V, the load resistance in Figure 18–48 is varied from 1 kΩ to 1.2 kΩ. Neglecting any change in output voltage, how much does the shunt current through Q_1 change?

15. If the maximum allowable input voltage in Figure 18–48 is 25 V, what is the maximum possible output current when the output is short-circuited? What power rating should R_1 have?

▲ **FIGURE 18–48**

SECTION 18–4   Basic Switching Regulators

16. A basic switching regulator is shown in Figure 18–49. If the switching frequency of the transistor is 100 Hz with an off-time of 6 ms, what is the output voltage?

17. What is the duty cycle of the transistor in Problem 16?

▲ **FIGURE 18–49**

18. When does the diode D_1 in Figure 18–50 become forward-biased?

19. If the on-time of Q_1 in Figure 18–50 is decreased, does the output voltage increase or decrease?

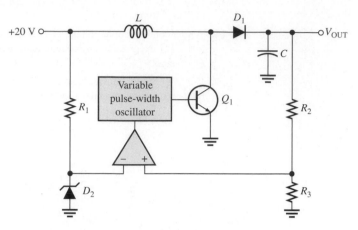

▲ FIGURE 18–50

SECTION 18–5 Integrated Circuit Voltage Regulators

20. What is the output voltage of each of the following IC regulators?

    (a) 7806   (b) 7905.2   (c) 7818   (d) 7924

21. Determine the output voltage of the regulator in Figure 18–51. $I_{ADJ} = 50 \ \mu A$.

22. Determine the minimum and maximum output voltages for the circuit in Figure 18–52. $I_{ADJ} = 50 \ \mu A$.

▲ FIGURE 18–51

▲ FIGURE 18–52

23. With no load connected, how much current is there through the regulator in Figure 18–51? Neglect the adjustment terminal current.

24. Select the values for the external resistors to be used in an LM317 circuit that is required to produce an output voltage of 12 V with an input of 18 V. The maximum regulator current with no load is to be 2 mA. There is no external pass transistor.

SECTION 18–6 Applications of IC Voltage Regulators

25. In the regulator circuit of Figure 18–53, determine R_{ext} if the maximum internal regulator current is to be 250 mA.

▲ **FIGURE 18–53**

26. Using a 7812 voltage regulator and a 10 Ω load in Figure 18–53, how much power will the external pass transistor have to dissipate? The maximum internal regulator current is set at 500 mA by R_{ext}.

27. Show how to include current limiting in the circuit of Figure 18–53. What should the value of the limiting resistor be if the external current is to be limited to 2 A?

28. Using an LM317, design a circuit that will provide a constant current of 500 mA to a load.

29. Repeat Problem 28 using a 7908.

30. If a 78S40 switching regulator is to be used to regulate a 12 V input down to a 6 V output, calculate the values of the external voltage-divider resistors.

EWB/MULTISIM TROUBLESHOOTING PROBLEMS

These file circuits are in the Troubleshooting Problems folder on your CD-ROM.

31. Open file TSP18-31 and determine the fault.

32. Open file TSP18-32 and determine the fault.

33. Open file TSP18-33 and determine the fault.

34. Open file TSP18-34 and determine the fault.

ANSWERS

SECTION REVIEWS

SECTION 18–1 Voltage Regulation

1. The percentage change in the output voltage for a given change in input voltage.

2. The percentage change in output voltage for a given change in load current.

3. 1.2%; 0.06%/V

4. 1.6%; 0.0016%/mA

SECTION 18–2 Basic Series Regulators

1. Control element, error detector, sampling element, reference voltage

2. 2 V

SECTION 18–3 **Basic Shunt Regulators**

1. In a shunt regulator, the control element is in parallel with the load rather than in series.
2. A shunt regulator has inherent current limiting. A disadvantage is that a shunt regulator is less efficient than a series regulator.

SECTION 18–4 **Basic Switching Regulators**

1. Step-down, step-up, inverting
2. Switching regulators operate at a higher efficiency.
3. The duty cycle varies to regulate the output.

SECTION 18–5 **Integrated Circuit Voltage Regulators**

1. Input, output, and ground
2. A 7809 has a $+9$ V output; A 7915 has a -15 V output.
3. Input, output, adjustment
4. A two-resistor voltage divider

SECTION 18–6 **Applications of IC Voltage Regulators**

1. A pass transistor increases the current that can be handled.
2. Current limiting prevents excessive current and prevents damage to the regulator.
3. Thermal overload occurs when the internal power dissipation becomes excessive.

RELATED PROBLEMS FOR EXAMPLES

18–1 0.6%/V
18–2 1.12%, 0.0224%/mA
18–3 7.33 V
18–4 0.7 A
18–5 17.5 W
18–6 12.7 V
18–7 467 mA
18–8 12 W
18–9 8 Ω

SELF-TEST

1. (c)   2. (d)   3. (c)   4. (b)   5. (d)
6. (a)   7. (c)   8. (a)   9. (g)   10. (c)

A: Table of Standard Resistor Values

Resistance Tolerance ($\pm\%$)

0.1% 0.25% 0.5%	1% 5%	2%	10%	0.1% 0.25% 0.5%	1% 5%	2%	10%	0.1% 0.25% 0.5%	1% 5%	2%	10%	0.1% 0.25% 0.5%	1% 5%	2%	10%	0.1% 0.25% 0.5%	1% 5%	2%	10%	0.1% 0.25% 0.5%	1% 5%	2%	10%
10.0	10.0	10	10	14.7	14.7	—	—	21.5	21.5	—	—	31.6	31.6	—	—	46.4	46.4	—	—	68.1	68.1	68	68
10.1	—	—	—	14.9	—	—	—	21.8	—	—	—	32.0	—	—	—	47.0	—	47	47	69.0	—	—	—
10.2	10.2	—	—	15.0	15.0	15	15	22.1	22.1	22	22	32.4	32.4	—	—	47.5	47.5	—	—	69.8	69.8	—	—
10.4	—	—	—	15.2	—	—	—	22.3	—	—	—	32.8	—	—	—	48.1	—	—	—	70.6	—	—	—
10.5	10.5	—	—	15.4	15.4	—	—	22.6	22.6	—	—	33.2	33.2	33	33	48.7	48.7	—	—	71.5	71.5	—	—
10.6	—	—	—	15.6	—	—	—	22.9	—	—	—	33.6	—	—	—	49.3	—	—	—	72.3	—	—	—
10.7	10.7	—	—	15.8	15.8	—	—	23.2	23.2	—	—	34.0	34.0	—	—	49.9	49.9	—	—	73.2	73.2	—	—
10.9	—	—	—	16.0	—	16	—	23.4	—	—	—	34.4	—	—	—	50.5	—	—	—	74.1	—	—	—
11.0	11.0	11	—	16.2	16.2	—	—	23.7	23.7	—	—	34.8	34.8	—	—	51.1	51.1	51	—	75.0	75.0	75	—
11.1	—	—	—	16.4	—	—	—	24.0	—	24	—	35.2	—	—	—	51.7	—	—	—	75.9	—	—	—
11.3	11.3	—	—	16.5	16.5	—	—	24.3	24.3	—	—	35.7	35.7	—	—	52.3	52.3	—	—	76.8	76.8	—	—
11.4	—	—	—	16.7	—	—	—	24.6	—	—	—	36.1	—	36	—	53.0	—	—	—	77.7	—	—	—
11.5	11.5	—	—	16.9	16.9	—	—	24.9	24.9	—	—	36.5	36.5	—	—	53.6	53.6	—	—	78.7	78.7	—	—
11.7	—	—	—	17.2	—	—	—	25.2	—	—	—	37.0	—	—	—	54.2	—	—	—	79.6	—	—	—
11.8	11.8	—	—	17.4	17.4	—	—	25.5	25.5	—	—	37.4	37.4	—	—	54.9	54.9	—	—	80.6	80.6	—	—
12.0	—	12	12	17.6	—	—	—	25.8	—	—	—	37.9	—	—	—	56.2	—	—	—	81.6	—	—	—
12.1	12.1	—	—	17.8	17.8	—	—	26.1	26.1	—	—	38.3	38.3	—	—	56.6	56.6	56	56	82.5	82.5	82	82
12.3	—	—	—	18.0	—	18	18	26.4	—	—	—	38.8	—	—	—	56.9	—	—	—	83.5	—	—	—
12.4	12.4	—	—	18.2	18.2	—	—	26.7	26.7	—	—	39.2	39.2	39	39	57.6	57.6	—	—	84.5	84.5	—	—
12.6	—	—	—	18.4	—	—	—	27.1	—	27	27	39.7	—	—	—	58.3	—	—	—	85.6	—	—	—
12.7	12.7	—	—	18.7	18.7	—	—	27.4	27.4	—	—	40.2	40.2	—	—	59.0	59.0	—	—	86.6	86.6	—	—
12.9	—	—	—	18.9	—	—	—	27.7	—	—	—	40.7	—	—	—	59.7	—	—	—	87.6	—	—	—
13.0	13.0	13	—	19.1	19.1	—	—	28.0	28.0	—	—	41.2	41.2	—	—	60.4	60.4	—	—	88.7	88.7	—	—
13.2	—	—	—	19.3	—	—	—	28.4	—	—	—	41.7	—	—	—	61.2	—	—	—	89.8	—	—	—
13.3	13.3	—	—	19.6	19.6	—	—	28.7	28.7	—	—	42.2	42.2	—	—	61.9	61.9	62	—	90.9	90.9	91	—
13.5	—	—	—	19.8	—	—	—	29.1	—	—	—	42.7	—	—	—	62.6	—	—	—	92.0	—	—	—
13.7	13.7	—	—	20.0	20.0	20	—	29.4	29.4	—	—	43.2	43.2	43	—	63.4	63.4	—	—	93.1	93.1	—	—
13.8	—	—	—	20.3	—	—	—	29.8	—	—	—	43.7	—	—	—	64.2	—	—	—	94.2	—	—	—
14.0	14.0	—	—	20.5	20.5	—	—	30.1	30.1	30	—	44.2	44.2	—	—	64.9	64.9	—	—	95.3	95.3	—	—
14.2	—	—	—	20.8	—	—	—	30.5	—	—	—	44.8	—	—	—	65.7	—	—	—	96.5	—	—	—
14.3	14.3	—	—	21.0	21.0	—	—	30.9	30.9	—	—	45.3	45.3	—	—	66.5	66.5	—	—	97.6	97.6	—	—
14.5	—	—	—	21.3	—	—	—	31.2	—	—	—	45.9	—	—	—	67.3	—	—	—	98.8	—	—	—

NOTE: These values are generally available in multiples of 0.1, 1, 10, 100, 1 k, and 1 M.

B: Derivations of Selected Equations

EQUATION 2–1

The average value of a half-wave rectified sine wave is the area under the curve divided by the period (2π). The equation for a sine wave is

$$v = V_p\sin\theta$$

$$V_{AVG} = \frac{area}{2\pi} = \frac{1}{2\pi}\int_0^\pi V_p\sin\theta\,d\theta = \frac{V_p}{2\pi}(-\cos\theta)\Big|_0^\pi$$

$$= \frac{V_p}{2\pi}[-\cos\pi - (-\cos 0)] = \frac{V_p}{2\pi}[-(-1) - (-1)] = \frac{V_p}{2\pi}(2)$$

$$V_{AVG} = \frac{V_p}{\pi}$$

EQUATION 2–11

Refer to Figure B–1.

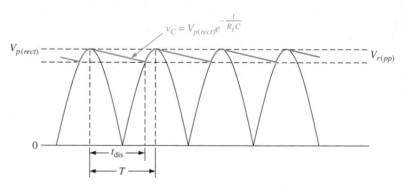

▲ **FIGURE B–1**

When the filter capacitor discharges through R_L, the voltage is

$$v_C = V_{p(rect)}e^{-t/R_LC}$$

Since the discharge time of the capacitor is from one peak to approximately the next peak, $t_{dis} \cong T$ when v_C reaches its minimum value.

$$v_{C(min)} = V_{p(rect)}e^{-T/R_LC}$$

Since $RC \gg T$, T/R_LC becomes much less than 1 (which is usually the case); e^{-T/R_LC} approaches 1 and can be expressed as

$$e^{-T/R_LC} \cong 1 - \frac{T}{R_LC}$$

Therefore,

$$v_{C(min)} = V_{p(rect)}\left(1 - \frac{T}{R_LC}\right)$$

The peak-to-peak ripple voltage is

$$V_{r(pp)} = V_{p(rect)} - V_{C(min)} = V_{p(rect)} - V_{p(rect)} + \frac{V_{p(rect)}T}{R_LC} = \frac{V_{p(rect)}T}{R_LC}$$

$$V_{r(pp)} \cong \left(\frac{1}{fR_LC}\right)V_{p(rect)}$$

EQUATION 2–12

To obtain the dc value, one-half of the peak-to-peak ripple is subtracted from the peak value.

$$V_{DC} = V_{p(rect)} - \frac{V_{r(pp)}}{2} = V_{p(rect)} - \left(\frac{1}{2fR_LC}\right)V_{p(rect)}$$

$$V_{DC} = \left(1 - \frac{1}{2fR_LC}\right)V_{p(rect)}$$

EQUATION 6–1

The Shockley equation for the base-emitter pn junction is

$$I_E = I_R(e^{VQ/kT} - 1)$$

where I_E = the total forward current across the base-emitter junction
I_R = the reverse saturation current
V = the voltage across the depletion layer
Q = the charge on an electron
k = a number known as Boltzmann's constant
T = the absolute temperature

At ambient temperature, $Q/kT \cong 40$, so

$$I_E = I_R(e^{40V} - 1)$$

Differentiating yields

$$\frac{dI_E}{dV} = 40I_Re^{40V}$$

Since $I_Re^{40V} = I_E + I_R$,

$$\frac{dI_E}{dV} = 40(I_E + I_R)$$

Assuming $I_R << I_E$,

$$\frac{dI_E}{dV} \cong 40I_E$$

The ac resistance r'_e of the base-emitter junction can be expressed as dV/dI_E.

$$r'_e = \frac{dV}{dI_E} \cong \frac{1}{40I_E} \cong \frac{25 \text{ mV}}{I_E}$$

EQUATION 6–14

The emitter-follower is represented by the r parameter ac equivalent circuit in Figure B–2(a).

▲ **FIGURE B–2**

By thevenizing from the base back to the source, the circuit is simplified to the form shown in Figure B–2(b).

$$V_{out} = V_e, I_{out} = I_e, \text{ and } I_{in} = I_b$$

$$R_{out} = \frac{V_e}{I_e}$$

$$I_e \cong \beta_{ac} I_b$$

With $V_s = 0$ and with I_b produced by V_{out}, and neglecting the base-to-emitter voltage drop (and therefore r'_e),

$$I_b \cong \frac{V_e}{R_1 \| R_2 \| R_s}$$

Assuming that $R_1 \gg R_s$ and $R_2 \gg R_s$,

$$I_b \cong \frac{V_e}{R_s}$$

$$I_{out} = I_e = \frac{\beta_{ac} V_e}{R_s}$$

$$\frac{V_{out}}{I_{out}} = \frac{V_e}{I_e} = \frac{V_e}{\beta_{ac} V_e / R_s} = \frac{R_s}{\beta_{ac}}$$

Looking into the emitter, R_E appears in parallel with R_s/β_{ac}. Therefore,

$$R_{out} = \left(\frac{R_s}{\beta_{ac}} \right) \| R_E$$

MIDPOINT BIAS (CHAPTER 7):
Proof that $I_D \cong 0.5I_{DSS}$ when $V_{GS} = \dfrac{V_{GS(off)}}{3.4}$

Start with Equation 7–1:

$$I_D = I_{DSS}\left(1 - \frac{V_{GS}}{V_{GS(off)}}\right)^2$$

Let $I_D = 0.5I_{DSS}$.

$$0.5I_{DSS} = I_{DSS}\left(1 - \frac{V_{GS}}{V_{GS(off)}}\right)^2$$

Cancelling I_{DSS} on each side,

$$0.5 = \left(1 - \frac{V_{GS}}{V_{GS(off)}}\right)^2$$

We want a factor (call it F) by which $V_{GS(off)}$ can be divided to give a value of V_{GS} that will produce a drain current that is $0.5I_{DSS}$.

$$0.5 = \left[1 - \frac{\left(\dfrac{V_{GS(off)}}{F}\right)}{V_{GS(off)}}\right]^2$$

Solving for F,

$$\sqrt{0.5} = 1 - \frac{\left(\dfrac{V_{GS(off)}}{F}\right)}{V_{GS(off)}} = 1 - \frac{1}{F}$$

$$\sqrt{0.5} - 1 = -\frac{1}{F}$$

$$\frac{1}{F} = 1 - \sqrt{0.5}$$

$$F = \frac{1}{1 - \sqrt{0.5}} \cong 3.4$$

Therefore, $I_D \cong 0.5I_{DSS}$ when $V_{GS} = \dfrac{V_{GS(off)}}{3.4}$.

EQUATION 8–5

$$I_D = I_{DSS}\left(1 - \frac{I_D R_S}{V_{GS(off)}}\right)^2 = I_{DSS}\left(1 - \frac{I_D R_S}{V_{GS(off)}}\right)\left(1 - \frac{I_D R_S}{V_{GS(off)}}\right)$$

$$= I_{DSS}\left(1 - \frac{2I_D R_S}{V_{GS(off)}} + \frac{I_D^2 R_S^2}{V_{GS(off)}^2}\right) = I_{DSS} - \frac{2I_{DSS}R_S}{V_{GS(off)}}I_D + \frac{I_{DSS}R_S^2}{V_{GS(off)}^2}I_D^2$$

Rearranging into a standard quadratic equation form,

$$\left(\frac{I_{DSS}R_S^2}{V_{GS(off)}^2}\right)I_D^2 - \left(1 + \frac{2I_{DSS}R_S}{V_{GS(off)}}\right)I_D + I_{DSS} = 0$$

The coefficients and constant are

$$A = \frac{R_S^2 I_{DSS}}{V_{GS(off)}^2}$$

$$B = -\left(1 + \frac{2R_S I_{DSS}}{V_{GS(off)}}\right)$$

$$C = I_{DSS}$$

In simplified notation, the equation is

$$AI_D^2 + BI_D + C = 0$$

The solutions to this quadratic equation are

$$I_D = \frac{-B \pm \sqrt{B^2 - 4AC}}{2A}$$

EQUATION 10–1

An inverting amplifier with feedback capacitance is shown in Figure B–3. For the input,

$$I_1 = \frac{V_1 - V_2}{X_C}$$

Factoring V_1 out,

$$I_1 = \frac{V_1(1 - V_2/V_1)}{X_C}$$

The ratio V_2/V_1 is the voltage gain, $-A_v$.

$$I_1 = \frac{V_1(1 + A_v)}{X_C} = \frac{V_1}{X_C/(1 + A_v)}$$

▶ FIGURE B–3

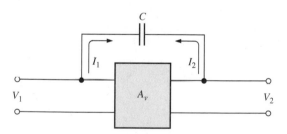

The effective reactance as seen from the input terminals is

$$X_{C_{in(Miller)}} = \frac{X_C}{1 + A_v}$$

or

$$\frac{1}{2\pi f C_{in(Miller)}} = \frac{1}{2\pi f C(1 + A_v)}$$

Cancelling and inverting,

$$C_{in(Miller)} = C(A_v + 1)$$

EQUATION 10–2

For the output in Figure B–3,

$$I_2 = \frac{V_2 - V_1}{X_C} = \frac{V_2(1 - V_1/V_2)}{X_C}$$

Since $V_1/V_2 = -1/A_v$,

$$I_2 = \frac{V_2(1 + 1/A_v)}{X_C} = \frac{V_2}{X_C/(1 + 1/A_v)} = \frac{V_2}{X_C/[(A_v + 1)/A_v]}$$

The effective reactance as seen from the output is

$$X_{C_{out(Miller)}} = \frac{X_C}{(A_v + 1)/A_v}$$

$$\frac{1}{2\pi f C_{out(Miller)}} = \frac{1}{2\pi f C[(A_v + 1)/A_v]}$$

Cancelling and inverting yields

$$C_{out(Miller)} = C\left(\frac{A_v + 1}{A_v}\right)$$

EQUATIONS 10–29 AND 10–30

The total gain, $A_{v(tot)}$, of an individual amplifier stage at the lower critical frequency equals the midrange gain, $A_{v(mid)}$, times the attenuation of the high-pass RC circuit.

$$A_{v(tot)} = A_{v(mid)}\left(\frac{R}{\sqrt{R^2 + X_C^2}}\right) = A_{v(mid)}\left(\frac{1}{\sqrt{1 + X_C^2/R^2}}\right)$$

$$f_{cl} = \frac{1}{2\pi RC}$$

Dividing both sides by any frequency f,

$$\frac{f_{cl}}{f} = \frac{1}{(2\pi f C)R}$$

Since $X_C = 1/2\pi f C$,

$$\frac{f_{cl}}{f} = \frac{X_C}{R}$$

Substitution in the gain formula gives

$$A_{v(tot)} = A_{v(mid)}\left(\frac{1}{\sqrt{1 + (f_{cl}/f)^2}}\right)$$

The gain ratio is

$$\frac{A_{v(tot)}}{A_{v(mid)}} = \frac{1}{\sqrt{1 + (f_{cl}/f)^2}}$$

For a multistage amplifier with n stages, each with the same f_{cl} and gain ratio, the product of the gain ratios is

$$\left(\frac{1}{\sqrt{1 + (f_{cl}/f)^2}}\right)^n$$

The critical frequency f'_{cl} of the multistage amplifier is the frequency at which $A_{v(tot)} = 0.707A_{v(mid)}$, so the gain ratio at f'_{cl} is

$$\frac{A_{v(tot)}}{A_{v(mid)}} = 0.707 = \frac{1}{1.414} = \frac{1}{\sqrt{2}}$$

Therefore, for a multistage amplifier,

$$\frac{1}{\sqrt{2}} = \left[\frac{1}{\sqrt{1 + (f_{cl}/f'_{cl})^2}}\right]^n = \frac{1}{\left(\sqrt{1 + (f_{cl}/f'_{cl})^2}\right)^n}$$

So

$$2^{1/2} = \left(\sqrt{1 + (f_{cl}/f'_{cl})^2}\right)^n$$

Squaring both sides,

$$2 = (1 + (f_{cl}/f'_{cl})^2)^n$$

Taking the nth root of both sides,

$$2^{1/n} = 1 + (f_{cl}/f'_{cl})^2$$

$$\left(\frac{f_{cl}}{f'_{cl}}\right)^2 = 2^{1/n} - 1$$

$$\left(\frac{f_{cl}}{f'_{cl}}\right) = \sqrt{2^{1/n} - 1}$$

$$f'_{cl} = \frac{f_{cl}}{\sqrt{2^{1/n} - 1}}$$

A similar process will give Equation 10–30:

$$f'_{cu} = f_{cu}\sqrt{2^{1/n} - 1}$$

EQUATIONS 10–31 AND 10–32

The *rise time* is defined as the time required for the voltage to increase from 10 percent of its final value to 90 percent of its final value, as indicated in Figure B–4.

Expressing the curve in its exponential form gives

$$v = V_{final}(1 - e^{-t/RC})$$

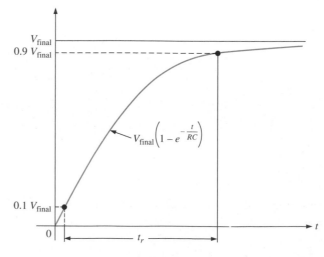

▲ FIGURE B–4

When $v = 0.1V_{final}$,

$$0.1V_{final} = V_{final}(1 - e^{-t/RC}) = V_{final} - V_{final}e^{-t/RC}$$

$$V_{final}e^{-t/RC} = 0.9V_{final}$$

$$e^{-t/RC} = 0.9$$

$$\ln e^{-t/RC} = \ln(0.9)$$

$$-\frac{t}{RC} = -0.1$$

$$t = 0.1RC$$

When $v = 0.9V_{final}$,

$$0.9V_{final} = V_{final}(1 - e^{-t/RC}) = V_{final} - V_{final}e^{-t/RC}$$

$$V_{final}e^{-t/RC} = 0.1V_{final}$$

$$\ln e^{-t/RC} = \ln(0.1)$$

$$-\frac{t}{RC} = -2.3$$

$$t = 2.3RC$$

The difference is the rise time.

$$t_r = 2.3RC - 0.1RC = 2.2RC$$

The critical frequency of an RC circuit is

$$f_c = \frac{1}{2\pi RC}$$

$$RC = \frac{1}{2\pi f_c}$$

Substituting,

$$t_r = \frac{2.2}{2\pi f_{cu}} = \frac{0.35}{f_{cu}}$$

$$f_{cu} = \frac{0.35}{t_r}$$

In a similar way, it can be shown that

$$f_{cl} = \frac{0.35}{t_f}$$

EQUATION 12–21

The formula for open-loop gain in Equation 12–19 can be expressed in complex notation as

$$A_{ol} = \frac{A_{ol(mid)}}{1 + jf/f_{c(ol)}}$$

Substituting the above expression into the equation $A_{cl} = A_{ol}/(1 + BA_{ol})$ gives a formula for the total closed-loop gain.

$$A_{cl} = \frac{A_{ol(mid)}/(1 + jf/f_{c(ol)})}{1 + BA_{ol(mid)}/(1 + jf/f_{c(ol)})}$$

Multiplying the numerator and denominator by $1 + jf/f_{c(ol)}$ yields

$$A_{cl} = \frac{A_{ol(mid)}}{1 + BA_{ol(mid)} + jf/f_{c(ol)}}$$

Dividing the numerator and denominator by $1 + BA_{ol(mid)}$ gives

$$A_{cl} = \frac{A_{ol(mid)}/(1 + BA_{ol(mid)})}{1 + j[f/(f_{c(ol)}(1 + BA_{ol(mid)}))]}$$

The above expression is of the form of the first equation

$$A_{cl} = \frac{A_{cl(mid)}}{1 + jf/f_{c(cl)}}$$

where $f_{c(cl)}$ is the closed-loop critical frequency. Thus,

$$f_{c(cl)} = f_{c(ol)}(1 + BA_{ol(mid)})$$

EQUATION 16–1

$$\frac{V_{out}}{V_{in}} = \frac{R(-jX)/(R - jX)}{(R - jX) + R(-jX)/(R - jX)} = \frac{R(-jX)}{(R - jX)^2 - jRX}$$

Multiplying the numerator and denominator by j,

$$\frac{V_{out}}{V_{in}} = \frac{RX}{j(R - jX)^2 + RX} = \frac{RX}{RX + j(R^2 - j2RX - X^2)}$$

$$= \frac{RX}{RX + jR^2 + 2RX - jX^2} = \frac{RX}{3RX + j(R^2 - X^2)}$$

For a 0° phase angle there can be no j term. Recall from complex numbers in ac theory that a *nonzero* angle is associated with a complex number having a j term. Therefore, at f_r the j term is 0.

$$R^2 - X^2 = 0$$

Thus,

$$\frac{V_{out}}{V_{in}} = \frac{RX}{3RX}$$

Cancelling yields

$$\frac{V_{out}}{V_{in}} = \frac{1}{3}$$

EQUATION 16–2

From the derivation of Equation 16–1,

$$R^2 - X^2 = 0$$
$$R^2 = X^2$$
$$R = X$$

Since $X = \dfrac{1}{2\pi f_r C}$,

$$R = \frac{1}{2\pi f_r C}$$

$$f_r = \frac{1}{2\pi RC}$$

EQUATIONS 16-3 AND 16-4

The feedback circuit in the phase-shift oscillator consists of three RC stages, as shown in Figure B–5. An expression for the attenuation is derived using the mesh analysis method for the loop assignment shown. All Rs are equal in value, and all Cs are equal in value.

$$(R - j1/2\pi fC)I_1 - RI_2 + 0I_3 = V_{in}$$
$$-RI_1 + (2R - j1/2\pi fC)I_2 - RI_3 = 0$$
$$0I_1 - RI_2 + (2R - j1/2\pi fC)I_3 = 0$$

▲ **FIGURE B–5**

In order to get V_{out}, we must solve for I_3 using determinants:

$$I_3 = \frac{\begin{vmatrix} (R - j1/2\pi fC) & -R & V_{in} \\ -R & (2R - j1/2\pi fC) & 0 \\ 0 & -R & 0 \end{vmatrix}}{\begin{vmatrix} (R - j1/2\pi fC) & -R & 0 \\ -R & (2R - j1/2\pi fC) & -R \\ 0 & -R & (2R - j1/2\pi fC) \end{vmatrix}}$$

$$I_3 = \frac{R^2 V_{in}}{(R - j1/2\pi fC)(2R - j1/2\pi fC)^2 - R^2(2R - j1/2\pi fC) - R^2(R - 1/2\pi fC)}$$

$$\frac{V_{out}}{V_{in}} = \frac{RI_3}{V_{in}}$$

$$= \frac{R^3}{(R - j1/2\pi fC)(2R - j1/2\pi fC)^2 - R^3(2 - j1/2\pi fRC) - R^3(1 - 1/2\pi fRC)}$$

$$= \frac{R^3}{R^3(1 - j1/2\pi fRC)(2 - j1/2\pi fRC)^2 - R^3[(2 - j1/2\pi fRC) - (1 - j1/2\pi fRC)]}$$

$$= \frac{R^3}{R^3(1 - j1/2\pi fRC)(2 - j1/2\pi fRC)^2 - R^3(3 - j1/2\pi fRC)}$$

$$\frac{V_{out}}{V_{in}} = \frac{1}{(1 - j1/2\pi fRC)(2 - j1/2\pi fRC)^2 - (3 - j1/2\pi fRC)}$$

Expanding and combining the real terms and the j terms separately.

$$\frac{V_{out}}{V_{in}} = \frac{1}{\left(1 - \frac{5}{4\pi^2 f^2 R^2 C^2}\right) - j\left(\frac{6}{2\pi f RC} - \frac{1}{(2\pi f)^3 R^3 C^3}\right)}$$

For oscillation in the phase-shift amplifier, the phase shift through the RC circuit must equal 180°. For this condition to exist, the j term must be 0 at the frequency of oscillation f_r.

$$\frac{6}{2\pi f_r RC} - \frac{1}{(2\pi f_r)^3 R^3 C^3} = 0$$

$$\frac{6(2\pi)^2 f_r^2 R^2 C^2 - 1}{(2\pi)^3 f_r^3 R^3 C^3} = 0$$

$$6(2\pi)^2 f_r^2 R^2 C^2 - 1 = 0$$

$$f_r^2 = \frac{1}{6(2\pi)^2 R^2 C^2}$$

$$f_r = \frac{1}{2\pi \sqrt{6} RC}$$

Since the j term is 0,

$$\frac{V_{out}}{V_{in}} = \frac{1}{1 - \frac{5}{4\pi^2 f_r^2 R^2 C^2}} = \frac{1}{1 - \frac{5}{\left(\frac{1}{\sqrt{6}RC}\right)^2 R^2 C^2}} = \frac{1}{1 - 30} = -\frac{1}{29}$$

The negative sign results from the 180° inversion. Thus, the value of attenuation for the feedback circuit is

$$B = \frac{1}{29}$$

C: Data Sheets

The data sheets in this appendix are copyrighted as follows: Motorola, Inc.: copyright of Semiconductor Components Industries, LLC. Used by permission. National Semiconductor: Reprinted with permission of National Semiconductor Corporation.

MOTOROLA

Designers'Data Sheet

"SURMETIC"▲ RECTIFIERS

. . . subminiature size, axial lead mounted rectifiers for general-purpose low-power applications.

Designers Data for "Worst Case" Conditions

The Designers▲ Data Sheets permit the design of most circuits entirely from the information presented. Limit curves — representing boundaries on device characteristics — are given to facilitate "worst case" design.

1N4001
thru
1N4007

**LEAD MOUNTED
SILICON RECTIFIERS**

**50-1000 VOLTS
DIFFUSED JUNCTION**

*MAXIMUM RATINGS

Rating	Symbol	1N4001	1N4002	1N4003	1N4004	1N4005	1N4006	1N4007	Unit
Peak Repetitive Reverse Voltage Working Peak Reverse Voltage DC Blocking Voltage	V_{RRM} V_{RWM} V_R	50	100	200	400	600	800	1000	Volts
Non-Repetitive Peak Reverse Voltage (halfwave, single phase, 60 Hz)	V_{RSM}	60	120	240	480	720	1000	1200	Volts
RMS Reverse Voltage	$V_{R(RMS)}$	35	70	140	280	420	560	700	Volts
Average Rectified Forward Current (single phase, resistive load, 60 Hz, see Figure 8, T_A = 75°C)	I_O	1.0							Amp
Non-Repetitive Peak Surge Current (surge applied at rated load conditions, see Figure 2)	I_{FSM}	30 (for 1 cycle)							Amp
Operating and Storage Junction Temperature Range	T_J, T_{stg}	–65 to +175							°C

*ELECTRICAL CHARACTERISTICS

Characteristic and Conditions	Symbol	Typ	Max	Unit
Maximum Instantaneous Forward Voltage Drop (i_F = 1.0 Amp, T_J = 25°C) Figure 1	v_F	0.93	1.1	Volts
Maximum Full-Cycle Average Forward Voltage Drop (I_O = 1.0 Amp, T_L = 75°C, 1 inch leads)	$V_{F(AV)}$	–	0.8	Volts

Maximum Reverse Current (rated dc voltage)
    T_J = 25°C
    T_J = 100°C
	I_R	0.05
		1.0
			10
			50
				μA

Maximum Full-Cycle Average Reverse Current (I_O = 1.0 Amp, T_L = 75°C, 1 inch leads)	$I_{R(AV)}$	–	30	μA

*Indicates JEDEC Registered Data.

MECHANICAL CHARACTERISTICS

CASE: Transfer Molded Plastic
MAXIMUM LEAD TEMPERATURE FOR SOLDERING PURPOSES: 350°C, 3/8" from case for 10 seconds at 5 lbs. tension
FINISH: All external surfaces are corrosion-resistant, leads are readily solderable
POLARITY: Cathode indicated by color band
WEIGHT: 0.40 Grams (approximately)

▲Trademark of Motorola Inc.

CATHODE
BAND

	MILLIMETERS		INCHES	
DIM	MIN	MAX	MIN	MAX
A	5.97	6.60	0.235	0.260
B	2.79	3.05	0.110	0.120
D	0.76	0.86	0.030	0.034
K	27.94	–	1.100	–

CASE 59-04
Does Not Conform to DO-41 Outline.

FIGURE 1 — FORWARD VOLTAGE

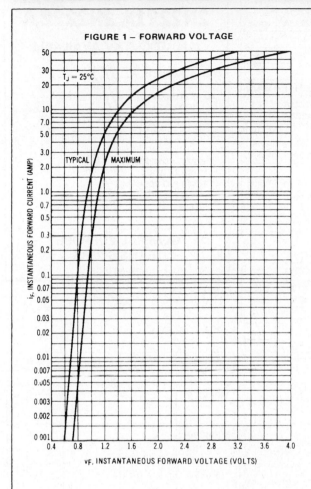

$T_J = 25°C$

TYPICAL   MAXIMUM

i_F, INSTANTANEOUS FORWARD CURRENT (AMP)

v_F, INSTANTANEOUS FORWARD VOLTAGE (VOLTS)

FIGURE 2 — NON-REPETITIVE SURGE CAPABILITY

SURGE APPLIED AT NO LOAD CONDITIONS $T_J = 25°C$

$f = 60$ Hz

1 CYCLE

V_{RRM} APPLIED AFTER SURGE
— — TYPICAL FAILURES
——— DESIGN LIMITS

SURGE APPLIED AT RATED LOAD CONDITIONS $T_J = 175°C$

I_{FSM}, PEAK HALF SINE WAVE CURRENT (AMP)

NUMBER OF CYCLES

FIGURE 3 — FORWARD VOLTAGE TEMPERATURE COEFFICIENT

TYPICAL RANGE

COEFFICIENT (mV/°C)

i_F, INSTANTANEOUS FORWARD CURRENT (AMP)

FIGURE 4 — TYPICAL TRANSIENT THERMAL RESISTANCE

P_{pk} P_{pk}
t_p TIME
t_1

DUTY CYCLE, $D = t_p/t_1$
PEAK POWER, P_{pk} is peak of an equivalent square power pulse.

$\Delta T_{JL} = P_{pk} [R_{\theta JL(\infty)} \cdot D + (1 - D) \cdot R_{\theta JL}(t_1 + t_p) + R_{\theta JL}(t_p) - R_{\theta JL}(t_1)]$
where ΔT_{JL} = increase in junction temperature above the lead temperature.

$R_{\theta JL}(t)$ = value of transient thermal resistance at time t, i.e.
$R_{\theta JL}(t_1 + t_p)$ = value of $R_{\theta JL}(t)$ at time $t_1 + t_p$
$R_{\theta JL}(t_p)$ = value of $R_{\theta JL}(t)$ at end of pulse width t_p
$R_{\theta JL}(t_1)$ = value of $R_{\theta JL}(t)$ at time t_1

$R_{\theta JL}$, JUNCTION-TO-LEAD TRANSIENT THERMAL RESISTANCE (°C/W)

$L = 1''$
$L = 1/2''$
$L = 1/32''$

t, TIME (ms)

The temperature of the lead should be measured using a thermocouple placed on the lead as close as possible to the tie point. The thermal mass connected to the tie point is normally large enough so that it will not significantly respond to heat surges generated in the diode as a result of pulsed operation once steady-state conditions are achieved. Using the measured value of T_L, the junction temperature may be determined by:

$$T_J = T_L + \Delta T_{JL}.$$

 MOTOROLA *Semiconductor Products Inc.*

MAXIMUM RATINGS

Rating	Symbol	2N2219 2N2222	2N2218A 2N2219A 2N2221A 2N2222A	Unit
Collector-Emitter Voltage	V_{CEO}	30	40	Vdc
Collector-Base Voltage	V_{CBO}	60	75	Vdc
Emitter-Base Voltage	V_{EBO}	5.0	6.0	Vdc
Collector Current — Continuous	I_C	800	800	mAdc
		2N2218A 2N2219,A	2N2221A 2N2222,A	
Total Device Dissipation @ T_A = 25°C Derate above 25°C	P_D	0.8 4.57	0.4 2.28	Watt mW/°C
Total Device Dissipation @ T_C = 25°C Derate above 25°C	P_D	3.0 17.1	1.2 6.85	Watts mW/°C
Operating and Storage Junction Temperature Range	T_J, T_{stg}	−65 to +200		°C

2N2218A,2N2219,A 2N2221A,2N2222,A

JAN, JTX, JTXV AVAILABLE

2N2218, A/2N2219, A
CASE 79-04
TO-39 (TO-205AD)
STYLE 1

2N2221, A/2N2222, A
CASE 22-03
TO-18 (TO-206AA)
STYLE 1

GENERAL PURPOSE TRANSISTORS
NPN SILICON

THERMAL CHARACTERISTICS

Characteristic	Symbol	2N2218A 2N2219,A	2N2221A 2N2222,A	Unit
Thermal Resistance, Junction to Ambient	$R_{\theta JA}$	219	145.8	°C/W
Thermal Resistance, Junction to Case	$R_{\theta JC}$	58	437.5	°C/W

ELECTRICAL CHARACTERISTICS (T_A = 25°C unless otherwise noted.)

Characteristic		Symbol	Min	Max	Unit
OFF CHARACTERISTICS					
Collector-Emitter Breakdown Voltage (I_C = 10 mAdc, I_B = 0)	Non-A Suffix A-Suffix	$V_{(BR)CEO}$	30 40	— —	Vdc
Collector-Base Breakdown Voltage (I_C = 10 μAdc, I_E = 0)	Non-A Suffix A-Suffix	$V_{(BR)CBO}$	60 75	— —	Vdc
Emitter-Base Breakdown Voltage (I_E = 10 μAdc, I_C = 0)	Non-A Suffix A-Suffix	$V_{(BR)EBO}$	5.0 6.0	— —	Vdc
Collector Cutoff Current (V_{CE} = 60 Vdc, $V_{EB(off)}$ = 3.0 Vdc)	A-Suffix	I_{CEX}	—	10	nAdc
Collector Cutoff Current (V_{CB} = 50 Vdc, I_E = 0) (V_{CB} = 60 Vdc, I_E = 0) (V_{CB} = 50 Vdc, I_E = 0, T_A = 150°C) (V_{CB} = 60 Vdc, I_E = 0, T_A = 150°C)	Non-A Suffix A-Suffix Non-A Suffix A-Suffix	I_{CBO}	— — — —	0.01 0.01 10 10	μAdc
Emitter Cutoff Current (V_{EB} = 3.0 Vdc, I_C = 0)	A-Suffix	I_{EBO}	—	10	nAdc
Base Cutoff Current (V_{CE} = 60 Vdc, $V_{EB(off)}$ = 3.0 Vdc)	A-Suffix	I_{BL}	—	20	nAdc
ON CHARACTERISTICS					
DC Current Gain (I_C = 0.1 mAdc, V_{CE} = 10 Vdc)	2N2218A, 2N2221A(1) 2N2219,A, 2N2222,A(1)	h_{FE}	20 35	— —	—
(I_C = 1.0 mAdc, V_{CE} = 10 Vdc)	2N2218A, 2N2221A 2N2219,A, 2N2222,A		25 50	— —	
(I_C = 10 mAdc, V_{CE} = 10 Vdc)	2N2218A, 2N2221A(1) 2N2219,A, 2N2222,A(1)		35 75	— —	
(I_C = 10 mAdc, V_{CE} = 10 Vdc, T_A = −55°C)	2N2218A, 2N2221A 2N2219,A, 2N2222,A		15 35	— —	
(I_C = 150 mAdc, V_{CE} = 10 Vdc)(1)	2N2218A, 2N2221A 2N2219,A, 2N2222,A		40 100	120 300	

2N2218A/19/19A/21A/22/22A

ELECTRICAL CHARACTERISTICS (continued) (T_A = 25°C unless otherwise noted.)

Characteristic		Symbol	Min	Max	Unit
(I_C = 150 mAdc, V_{CE} = 1.0 Vdc)(1)	2N2218A, 2N2221A		20	—	
	2N2219,A, 2N2222,A		50	—	
(I_C = 500 mAdc, V_{CE} = 10 Vdc)(1)	2N2219, 2N2222		30	—	
	2N2218A, 2N2221A,		25	—	
	2N2219A, 2N2222A		40	—	
Collector-Emitter Saturation Voltage(1)		$V_{CE(sat)}$			Vdc
(I_C = 150 mAdc, I_B = 15 mAdc)	Non-A Suffix		—	0.4	
	A-Suffix		—	0.3	
(I_C = 500 mAdc, I_B = 50 mAdc)	Non-A Suffix		—	1.6	
	A-Suffix		—	1.0	
Base-Emitter Saturation Voltage(1)		$V_{BE(sat)}$			Vdc
(I_C = 150 mAdc, I_B = 15 mAdc)	Non-A Suffix		0.6	1.3	
	A-Suffix		0.6	1.2	
(I_C = 500 mAdc, I_B = 50 mAdc)	Non-A Suffix		—	2.6	
	A-Suffix		—	2.0	

SMALL-SIGNAL CHARACTERISTICS

Characteristic		Symbol	Min	Max	Unit
Current Gain — Bandwidth Product(2)		f_T			MHz
(I_C = 20 mAdc, V_{CE} = 20 Vdc, f = 100 MHz)	All Types, Except		250	—	
	2N2219A, 2N2222A		300	—	
Output Capacitance(3)		C_{obo}	—	8.0	pF
(V_{CB} = 10 Vdc, I_E = 0, f = 1.0 MHz)					
Input Capacitance(3)		C_{ibo}			pF
(V_{EB} = 0.5 Vdc, I_C = 0, f = 1.0 MHz)	Non-A Suffix		—	30	
	A-Suffix		—	25	
Input Impedance		h_{ie}			kohms
(I_C = 1.0 mAdc, V_{CE} = 10 Vdc, f = 1.0 kHz)	2N2218A, 2N2221A		1.0	3.5	
	2N2219A, 2N2222A		2.0	8.0	
(I_C = 10 mAdc, V_{CE} = 10 Vdc, f = 1.0 kHz)	2N2218A, 2N2221A		0.2	1.0	
	2N2219A, 2N2222A		0.25	1.25	
Voltage Feedback Ratio		h_{re}			X 10⁻⁴
(I_C = 1.0 mAdc, V_{CE} = 10 Vdc, f = 1.0 kHz)	2N2218A, 2N2221A		—	5.0	
	2N2219A, 2N2222A		—	8.0	
(I_C = 10 mAdc, V_{CE} = 10 Vdc, f = 1.0 kHz)	2N2218A, 2N2221A		—	2.5	
	2N2219A, 2N2222A		—	4.0	
Small-Signal Current Gain		h_{fe}			—
(I_C = 1.0 mAdc, V_{CE} = 10 Vdc, f = 1.0 kHz)	2N2218A, 2N2221A		30	150	
	2N2219A, 2N2222A		50	300	
(I_C = 10 mAdc, V_{CE} = 10 Vdc, f = 1.0 kHz)	2N2218A, 2N2221A		50	300	
	2N2219A, 2N2222A		75	375	
Output Admittance		h_{oe}			μmhos
(I_C = 1.0 mAdc, V_{CE} = 10 Vdc, f = 1.0 kHz)	2N2218A, 2N2221A		3.0	15	
	2N2219A, 2N2222A		5.0	35	
(I_C = 10 mAdc, V_{CE} = 10 Vdc, f = 1.0 kHz)	2N2218A, 2N2221A		10	100	
	2N2219A, 2N2222A		15	200	
Collector Base Time Constant		$r_{b'}C_c$	—	150	ps
(I_E = 20 mAdc, V_{CB} = 20Vdc, f = 31.8 MHz)	A-Suffix				
Noise Figure		NF	—	4.0	dB
(I_C = 100 μAdc, V_{CE} = 10 Vdc,					
R_S = 1.0 kohm, f = 1.0 kHz)	2N2222A				
Real Part of Common-Emitter		Re(h_{ie})	—	60	Ohms
High Frequency Input Impedance					
(I_C = 20 mAdc, V_{CE} = 20 Vdc, f = 300 MHz)	2N2218A, 2N2219A				
	2N2221A, 2N2222A				

(1) Pulse Test: Pulse Width ≤ 300 μs, Duty Cycle ≤ 2.0%.
(2) f_T is defined as the frequency at which $|h_{fe}|$ extrapolates to unity.
(3) 2N5581 and 2N5582 are Listed C_{cb} and C_{eb} for these conditions and values.

2N2218A/19/19A/21A/22/22A

ELECTRICAL CHARACTERISTICS (continued) (T_A = 25°C unless otherwise noted.)

Characteristic		Symbol	Min	Max	Unit
SWITCHING CHARACTERISTICS					
Delay Time	(V_{CC} = 30 Vdc, $V_{BE(off)}$ = 0.5 Vdc,	t_d	—	10	ns
Rise Time	I_C = 150 mAdc, I_{B1} = 15 mAdc) (Figure 14)	t_r	—	25	ns
Storage Time	(V_{CC} = 30 Vdc, I_C = 150 mAdc,	t_s	—	225	ns
Fall Time	I_{B1} = I_{B2} = 15 mAdc) (Figure15)	t_f	—	60	ns
Active Region Time Constant (I_C = 150 mAdc, V_{CE} = 30 Vdc) (See Figure 12 for 2N2218A, 2N2219A, 2N2221A, 2N2222A)		T_A	—	2.5	ns

FIGURE 1 – NORMALIZED DC CURRENT GAIN

FIGURE 2 – COLLECTOR CHARACTERISTICS IN SATURATION REGION

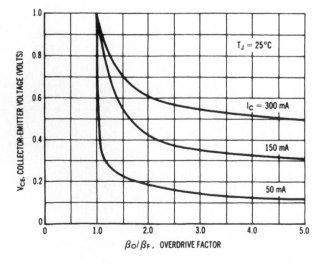

This graph shows the effect of base current on collector current. β_o (current gain at the edge of saturation) is the current gain of the transistor at 1 volt, and β_f (forced gain) is the ratio of I_c/I_{bf} in a circuit.

EXAMPLE: For type 2N2219, estimate a base current (I_{bf}) to insure saturation at a temperature of 25°C and a collector current of 150 mA. Observe that at I_c = 150 mA an overdrive factor of at least 2.5 is required to drive the transistor well into the saturation region. From Figure 1, it is seen that h_{FE} @ 1 volt is approximately 0.62 of h_{FE} @ 10 volts. Using the guaranteed minimum gain of 100 @ 150 mA and 10 V, β_o = 62 and substituting values in the overdrive equation, we find:

$$\frac{\beta_o}{\beta_f} = \frac{h_{FE} @ 1.0 \text{ V}}{I_c/I_{bf}} \qquad 2.5 = \frac{62}{150/I_{bf}} \qquad I_{bf} \approx 6.0 \text{ mA}$$

2N2218A/19/19A/21A/22/22A

FIGURE 3 — "ON" VOLTAGES

FIGURE 4 — TEMPERATURE COEFFICIENTS

h PARAMETERS

$V_{CE} = 10$ Vdc, f = 1.0 kHz, $T_A = 25°C$

This group of graphs illustrates the relationship between h_{fe} and other "h" parameters for this series of transistors. To obtain these curves, a high-gain and a low-gain unit were selected and the same units were used to develop the correspondingly numbered curves on each graph.

FIGURE 5 — INPUT IMPEDANCE

FIGURE 6 — VOLTAGE FEEDBACK RATIO

FIGURE 7 — CURRENT GAIN

FIGURE 8 — OUTPUT ADMITTANCE

2N2218A/19/19A/21A/22/22A

SWITCHING TIME CHARACTERISTICS

FIGURE 9 — TURN-ON TIME

FIGURE 10 — CHARGE DATA

FIGURE 11 — TURN-OFF BEHAVIOR

FIGURE 12 — DELAY AND RISE TIME
EQUIVALENT TEST CIRCUIT

GENERATOR RISE TIME ≤ 2.0 ns
PW ≤ 200 ns
DUTY CYCLE = 2.0%

OSCILLOSCOPE
R$_{in}$ > 100 k ohms
C$_{in}$ ≤ 12 pF
RISE TIME ≤ 5.0 ns

FIGURE 13 — STORAGE TIME AND FALL
TIME EQUIVALENT TEST CIRCUIT

DUTY CYCLE = 2.0%

SCOPE
R$_{in}$ > 100 k ohms
C$_{in}$ ≤ 12 pF
RISE TIME ≤ 5.0 ns

MAXIMUM RATINGS

Rating	Symbol	Value	Unit
Collector-Emitter Voltage	V_{CEO}	40	Vdc
Collector-Base Voltage	V_{CBO}	60	Vdc
Emitter-Base Voltage	V_{EBO}	6.0	Vdc
Collector Current — Continuous	I_C	200	mAdc
Total Device Dissipation @ T_A = 25°C Derate above 25°C	P_D	0.36 2.06	Watt mW/°C
Total Device Dissipation @ T_C = 25°C Derate above 25°C	P_D	1.2 6.9	Watts mW/°C
Operating and Storage Junction Temperature Range	T_J, T_{stg}	−65 to +200	°C

THERMAL CHARACTERISTICS

Characteristic	Symbol	Max	Unit
Thermal Resistance, Junction to Case	$R_{\theta JC}$	0.15	°C/mW
Thermal Resistance, Junction to Ambient	$R_{\theta JA}$	0.49	°C/mW

2N3946
2N3947

CASE 22-03, STYLE 1
TO-18 (TO-206AA)

3 Collector

2 Base

1 Emitter

3 2 1

GENERAL PURPOSE TRANSISTORS

NPN SILICON

ELECTRICAL CHARACTERISTICS (T_A = 25°C unless otherwise noted.)

Characteristic		Symbol	Min	Max	Unit
OFF CHARACTERISTICS					
Collector-Emitter Breakdown Voltage(1) (I_C = 10 mAdc)		$V_{(BR)CEO}$	40	—	Vdc
Collector-Base Breakdown Voltage (I_C = 10 µAdc, I_E = 0)		$V_{(BR)CBO}$	60	—	Vdc
Emitter-Base Breakdown Voltage (I_E = 10 µAdc, I_C = 0)		$V_{(BR)EBO}$	6.0	—	Vdc
Collector Cutoff Current (V_{CE} = 40 Vdc, V_{EB} = 3.0 Vdc) (V_{CE} = 40 Vdc, V_{EB} = 3.0 Vdc, T_A = 150°C)		I_{CEX}	— —	0.010 15	µAdc
Base Cutoff Current (V_{CE} = 40 Vdc, V_{EB} = 3.0 Vdc)		I_{BL}	—	.025	µAdc
ON CHARACTERISTICS					
DC Current Gain(1) (I_C = 0.1 mAdc, V_{CE} = 1.0 Vdc)	2N3946 2N3947	h_{FE}	30 60	— —	—
(I_C = 1.0 mAdc, V_{CE} = 1.0 Vdc)	2N3946 2N3947		45 90	— —	
(I_C = 10 mAdc, V_{CE} = 1.0 Vdc)	2N3946 2N3947		50 100	150 300	
(I_C = 50 mAdc, V_{CE} = 1.0 Vdc)	2N3946 2N3947		20 40	— —	
Collector-Emitter Saturation Voltage(1) (I_C = 10 mAdc, I_B = 1.0 mAdc) (I_C = 50 mAdc, I_B = 5.0 mAdc)		$V_{CE(sat)}$	— —	0.2 0.3	Vdc
Base-Emitter Saturation Voltage(1) (I_C = 10 mAdc, I_B = 1.0 mAdc) (I_C = 50 mAdc, I_B = 5.0 mAdc)		$V_{BE(sat)}$	0.6 —	0.9 1.0	Vdc
SMALL-SIGNAL CHARACTERISTICS					
Current-Gain — Bandwidth Product (I_C = 10 mAdc, V_{CE} = 20 Vdc, f = 100 MHz)	2N3946 2N3947	f_T	250 300	— —	MHz
Output Capacitance (V_{CB} = 10 Vdc, I_E = 0, f = 1.0 MHz)		C_{obo}	—	4.0	pF

2N3946, 2N3947

ELECTRICAL CHARACTERISTICS (continued) (T_A = 25°C unless otherwise noted.)

Characteristic		Symbol	Min	Max	Unit
Input Capacitance (V_{EB} = 1.0 Vdc, I_C = 0, f = 1.0 MHz)		C_{ibo}	—	8.0	pF
Input Impedance (I_C = 1.0 mA, V_{CE} = 10 V, f = 1.0 kHz)	2N3946 2N3947	h_{ie}	0.5 2.0	6.0 12	kohms
Voltage Feedback Ratio (I_C = 1.0 mA, V_{CE} = 10 V, f = 1.0 kHz)	2N3946 2N3947	h_{re}	— —	10 20	X 10^{-4}
Small Signal Current Gain (I_C = 1.0 mA, V_{CE} = 10 V, f = 1.0 kHz)	2N3946 2N3947	h_{fe}	50 100	250 700	—
Output Admittance (I_C = 1.0 mA, V_{CE} = 10 V, f = 1.0 kHz)	2N3946 2N3947	h_{oe}	1.0 5.0	30 50	μmhos
Collector Base Time Constant (I_C = 10 mA, V_{CE} = 20 V, f = 31.8 MHz)		$rb'C_c$	—	200	ps
Noise Figure (I_C = 100 μA, V_{CE} = 5.0 V, R_g = 1.0 kΩ, f = 1.0 kHz)		NF	—	5.0	dB

SWITCHING CHARACTERISTICS

Delay Time	V_{CC} = 3.0 Vdc, V_{OB} = 0.5 Vdc,		t_d	—	35	ns
Rise Time	I_C = 10 mAdc, I_{B1} = 1.0 mA		t_r	—	35	ns
Storage Time	V_{CC} = 3.0 V, I_C = 10 mA,	2N3946 2N3947	t_s	— —	300 375	ns
Fall Time	I_{B1} = I_{B2} = 1.0 mAdc		t_f	—	75	ns

(1) Pulse Test: PW ≤ 300 μs, Duty Cycle ≤ 2%.

TYPICAL SWITCHING CHARACTERISTICS
(T_A = 25°C unless otherwise noted)

FIGURE 1 — DELAY AND RISE TIME

FIGURE 2 — RISE TIME

2N3946, 2N3947

FIGURE 3 — STORAGE AND FALL TIMES

2N3946

2N3947

FIGURE 4 — TURN-ON TIME EQUIVALENT TEST CIRCUIT   FIGURE 5 — TURN-OFF TIME EQUIVALENT TEST CIRCUIT

*TOTAL SHUNT CAPACITANCE OF TEST JIG AND CONNECTORS

2N3946, 2N3947

AUDIO SMALL-SIGNAL CHARACTERISTICS

FIGURE 6 — NOISE FIGURE VARIATIONS
V_{CE} = 5.0 V, T_A = 25°C

h PARAMETERS
V_{CE} = 10 V, T_A = 25°C, f = 1.0 kc

FIGURE 7 — CURRENT GAIN

FIGURE 8 — OUTPUT CAPACITANCE

FIGURE 9 — INPUT IMPEDANCE

FIGURE 10 — VOLTAGE FEEDBACK RATIO

2N3946, 2N3947

FIGURE 11 — CURRENT GAIN CHARACTERISTICS
2N3946

2N3947

FIGURE 12 — CAPACITANCE

FIGURE 13 — CHARGE DATA

2N3946, 2N3947

FIGURE 14 — COLLECTOR SATURATION REGION
2N3946

2N3947

FIGURE 15 — "ON" VOLTAGES

FIGURE 16 — TEMPERATURE COEFFICIENTS

2N3796
2N3797

CASE 22-03, STYLE 2
TO-18 (TO-206AA)

MOSFETs
LOW POWER AUDIO

N-CHANNEL — DEPLETION

MAXIMUM RATINGS

Rating	Symbol	Value	Unit
Drain-Source Voltage 2N3796 2N3797	V_{DS}	 25 20	Vdc
Gate-Source Voltage	V_{GS}	± 10	Vdc
Drain Current	I_D	20	mAdc
Total Device Dissipation @ T_A = 25°C Derate above 25°C	P_D	200 1.14	mW mW/°C
Junction Temperature Range	T_J	+ 175	°C
Storage Channel Temperature Range	T_{stg}	− 65 to + 200	°C

ELECTRICAL CHARACTERISTICS (T_A = 25°C unless otherwise noted.)

Characteristic	Symbol	Min	Typ	Max	Unit
OFF CHARACTERISTICS					
Drain-Source Breakdown Voltage (V_{GS} = −4.0 V, I_D = 5.0 μA) 2N3796 (V_{GS} = −7.0 V, I_D = 5.0 μA) 2N3797	$V_{(BR)DSX}$	 25 20	 30 25	 — —	Vdc
Gate Reverse Current(1) (V_{GS} = −10 V, V_{DS} = 0) (V_{GS} = −10 V, V_{DS} = 0, T_A = 150°C)	I_{GSS}	 — —	 — —	 1.0 200	pAdc
Gate Source Cutoff Voltage (I_D = 0.5 μA, V_{DS} = 10 V) 2N3796 (I_D = 2.0 μA, V_{DS} = 10 V) 2N3797	$V_{GS(off)}$	 — —	 − 3.0 − 5.0	 − 4.0 − 7.0	Vdc
Drain-Gate Reverse Current(1) (V_{DG} = 10 V, I_S = 0)	I_{DGO}	—	—	1.0	pAdc
ON CHARACTERISTICS					
Zero-Gate-Voltage Drain Current (V_{DS} = 10 V, V_{GS} = 0) 2N3796 2N3797	I_{DSS}	 0.5 2.0	 1.5 2.9	 3.0 6.0	mAdc
On-State Drain Current (V_{DS} = 10 V, V_{GS} = +3.5 V) 2N3796 2N3797	$I_{D(on)}$	 7.0 9.0	 8.3 14	 14 18	mAdc
SMALL-SIGNAL CHARACTERISTICS					
Forward Transfer Admittance (V_{DS} = 10 V, V_{GS} = 0, f = 1.0 kHz) 2N3796 2N3797	$|Y_{fs}|$	 900 1500	 1200 2300	 1800 3000	μmhos
(V_{DS} = 10 V, V_{GS} = 0, f = 1.0 MHz) 2N3796 2N3797		900 1500	— —	— —	
Output Admittance (V_{DS} = 10 V, V_{GS} = 0, f = 1.0 kHz) 2N3796 2N3797	$|Y_{os}|$	 — —	 12 27	 25 60	μmhos
Input Capacitance (V_{DS} = 10 V, V_{GS} = 0, f = 1.0 MHz) 2N3796 2N3797	C_{iss}	 — —	 5.0 6.0	 7.0 8.0	pF
Reverse Transfer Capacitance (V_{DS} = 10 V, V_{GS} = 0, f = 1.0 MHz)	C_{rss}	—	0.5	0.8	pF
FUNCTIONAL CHARACTERISTICS					
Noise Figure (V_{DS} = 10 V, V_{GS} = 0, f = 1.0 kHz, R_S = 3 megohms)	NF	—	3.8	—	dB

(1) This value of current includes both the FET leakage current as well as the leakage current associated with the test socket and fixture when measured under best attainable conditions.

2N3796, 2N3797

TYPICAL DRAIN CHARACTERISTICS

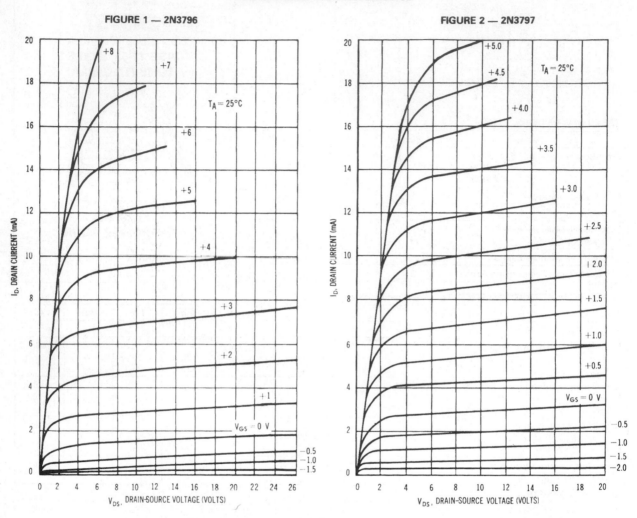

FIGURE 1 — 2N3796

FIGURE 2 — 2N3797

COMMON SOURCE TRANSFER CHARACTERISTICS

FIGURE 3 — 2N3796

FIGURE 4 — 2N3797

2N3796, 2N3797

FIGURE 5 — FORWARD TRANSFER ADMITTANCE

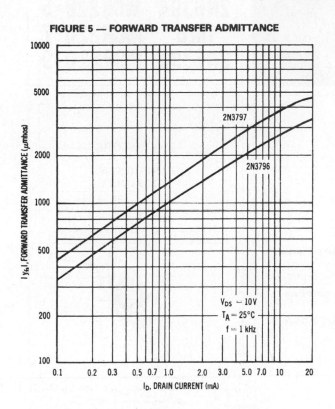

FIGURE 6 — OUTPUT ADMITTANCE

FIGURE 7 — NOISE FIGURE

2N6394 MCR220-5
thru MCR220-7
2N6399 MCR220-9

THYRISTORS

12 AMPERES RMS
50-800 VOLTS

SILICON CONTROLLED RECTIFIERS

. . . designed primarily for half-wave ac control applications, such as motor controls, heating controls and power supplies; or wherever half-wave silicon gate-controlled, solid-state devices are needed.

● Glass Passivated Junctions and Center Gate Fire for Greater Parameter Uniformity and Stability

● Small, Rugged, Thermowatt▲ Construction for Low Thermal Resistance, High Heat Dissipation and Durability

● Blocking Voltage to 800 Volts

*MAXIMUM RATINGS

Rating	Symbol	Value	Unit
Peak Reverse Voltage (1)	V_{RRM}		Volts
2N6394		50	
2N6395		100	
2N6396		200	
MCR220-5		300	
2N6397		400	
MCR220-7		500	
2N6398		600	
MCR220-9		700	
2N6399		800	
Forward Current RMS $T_J = 125^oC$ (All Conduction Angles)	$I_{T(RMS)}$	12	Amps
Peak Forward Surge Current (1/2 cycle, Sine Wave, 60 Hz, $T_J = 125^oC$)	I_{TSM}	100	Amps
Circuit Fusing Considerations ($T_J = -40$ to $+125^oC$, t = 1.0 to 8.3 ms)	I^2t	40	A^2s
Forward Peak Gate Power	P_{GM}	20	Watts
Forward Average Gate Power	$P_{G(AV)}$	0.5	Watt
Forward Peak Gate Current	I_{GM}	2.0	Amps
Operating Junction Temperature Range	T_J	-40 to +125	^oC
Storage Temperature Range	T_{stg}	-40 to +150	^oC

THERMAL CHARACTERISTICS

Characteristic	Symbol	Max	Unit
Thermal Resistance, Junction to Case	$R_{\theta JC}$	2.0	$^oC/W$

(1) V_{RRM} for all types can be applied on a continuous dc basis without incurring damage. Ratings apply for zero or negative gate voltage. Devices should not be tested for blocking capability in a manner such that the voltage supplied exceeds the rated blocking voltage.

*Indicates JEDEC Registered Data.
▲Trademark of Motorola Inc.

PIN 1. CATHODE
2. ANODE
3. GATE
4. ANODE

All JEDEC dimensions and notes apply

DIM	MILLIMETERS MIN	MAX	INCHES MIN	MAX
A	14.23	15.87	0.560	0.625
B	9.66	10.66	0.380	0.420
C	3.56	4.82	0.140	0.190
D	0.51	1.14	0.020	0.045
F	3.531	3.733	0.139	0.147
G	2.29	2.79	0.090	0.110
H	–	6.35	–	0.250
J	0.31	1.14	0.012	0.045
K	12.70	14.27	0.500	0.562
L	1.14	1.77	0.045	0.070
N	4.83	5.33	0.190	0.210
Q	2.54	3.04	0.100	0.120
R	2.04	2.92	0.080	0.115
S	0.51	1.39	0.020	0.055
T	5.85	6.85	0.230	0.270

CASE 221-02
TO 220 AB

2N6394 thru 2N6399 ● MCR220-5 ● MCR220-7 ● MCR220-9

ELECTRICAL CHARACTERISTICS (T_C = 25°C unless otherwise noted.)

Characteristic	Symbol	Min	Typ	Max	Unit
*Peak Forward Blocking Voltage (T_J = 125°C)	V_DRM				Volts
2N6394		50	–	–	
2N6395		100	–	–	
2N6396		200	–	–	
MCR220-5		300	–	–	
2N6397		400	–	–	
MCR220-7		500	–	–	
2N6398		600	–	–	
MCR220-9		700	–	–	
2N6399		800	–	–	
* Peak Forward Blocking Current (Rated V_DRM @ T_J = 125°C)	I_DRM	–	–	2.0	mA
* Peak Reverse Blocking Current (Rated V_RRM @ T_J = 125°C)	I_RRM	–	–	2.0	mA
* Forward "On" Voltage (I_TM = 24 A Peak)	V_TM	–	1.7	2.2	Volts
* Gate Trigger Current (Continuous dc) (Anode Voltage = 12 Vdc, R_L = 100 Ohms)	I_GT	–	5.0	30	mA
* Gate Trigger Voltage (Continuous dc) (Anode Voltage = 12 Vdc, R_L = 100 Ohms)	V_GT	–	0.7	1.5	Volts
* Gate Non-Trigger Voltage (Anode Voltage = Rated V_DRM, R_L = 100 Ohms, T_J = 125°C)	V_GD	0.2	–	–	Volts
* Holding Current (Anode Voltage = 12 Vdc)	I_H	–	6.0	40	mA
Turn-On Time (I_TM = 12 A, I_GT = 40 mAdc)	t_gt	–	1.0	2.0	µs
Turn-Off Time (V_DRM = rated voltage) (I_TM = 12 A, I_R = 12 A) (I_TM = 12 A, I_R = 12 A, T_J = 125°C)	tq	– –	15 35	– –	µs
Forward Voltage Application Rate (T_J = 125°C)	dv/dt	–	50	–	V/µs

*Indicates JEDEC Registered Data.

FIGURE 1 – AVERAGE CURRENT DERATING

FIGURE 2 – MAXIMUM ON-STATE POWER DISSIPATION

 MOTOROLA *Semiconductor Products Inc.*

MOTOROLA Semiconductors
BOX 20912 • PHOENIX, ARIZONA 85036

2N6027
2N6028

SILICON PROGRAMMABLE UNIJUNCTION TRANSISTORS

40 VOLTS
375 mW

DS 2520

SILICON PROGRAMMABLE UNIJUNCTION TRANSISTORS

. . . designed to enable the engineer to "program" unijunction characteristics such as R_{BB}, η, I_V, and I_P by merely selecting two resistor values. Application includes thyristor-trigger, oscillator, pulse and timing circuits. These devices may also be used in special thyristor applications due to the availability of an anode gate. Supplied in an inexpensive TO-92 plastic package for high-volume requirements, this package is readily adaptable for use in automatic insertion equipment.

- Programmable — R_{BB}, η, I_V and I_P.

- Low On-State Voltage — 1.5 Volts Maximum @ I_F = 50 mA

- Low Gate to Anode Leakage Current — 10 nA Maximum

- High Peak Output Voltage — 11 Volts Typical

- Low Offset Voltage — 0.35 Volt Typical (R_G = 10 k ohms)

MAXIMUM RATINGS

Rating	Symbol	Value	Unit
Power Dissipation (1)	P_F	375	mW
Derate Above 25°C	$1/\theta_{JA}$	5.0	mW/°C
DC Forward Anode Current (2)	I_T	200	mA
Derate Above 25°C		2.67	mA/°C
•DC Gate Current	I_G	±50	mA
Repetitive Peak Forward Current	I_{TRM}		
100 μs Pulse Width, 1.0% Duty Cycle		1.0	Amp
• 20 μs Pulse Width, 1.0% Duty Cycle		2.0	Amp
Non-Repetitive Peak Forward Current	I_{TSM}	5.0	Amp
10 μs Pulse Width			
• Gate to Cathode Forward Voltage	V_{GKF}	40	Volt
• Gate to Cathode Reverse Voltage	V_{GKR}	–5.0	Volt
• Gate to Anode Reverse Voltage	V_{GAR}	40	Volt
• Anode to Cathode Voltage	V_{AK}	±40	Volt
Operating Junction Temperature Range	T_J	–50 to +100	°C
• Storage Temperature Range	T_{stg}	–55 to +150	°C

*Indicates JEDEC Registered Data
(1) JEDEC Registered Data is 300 mW, derating at 4.0 mW/°C.
(2) JEDEC Registered Data is 150 mA.

PIN 1. ANODE
2. GATE
3. CATHODE

All JEDEC dimensions and notes apply

CASE 29-03
TO-92
PLASTIC

2N6027 ● 2N6028

ELECTRICAL CHARACTERISTICS (T_A = 25°C unless otherwise noted)

Characteristic		Figure	Symbol	Min	Typ	Max	Unit
● Peak Current		2,9,11	I_p				μA
(V_S = 10 Vdc, R_G = 1.0 MΩ)	2N6027			–	1.25	2.0	
	2N6028			–	0.08	0.15	
(V_S = 10 Vdc, R_G = 10 k ohms)	2N6027			–	4.0	5.0	
	2N6028			–	0.70	1.0	
● Offset Voltage		1	V_T				Volts
(V_S = 10 Vdc, R_G = 1.0 MΩ)	2N6027			0.2	0.70	1.6	
	2N6028			0.2	0.50	0.6	
(V_S = 10 Vdc, R_G = 10 k ohms)	(Both Types)			0.2	0.35	0.6	
● Valley Current		1,4,5,	I_V				μA
(V_S = 10 Vdc, R_G = 1.0 MΩ)	2N6027			–	18	50	
	2N6028			–	18	25	
(V_S = 10 Vdc, R_G = 10 k ohms)	2N6027			70	270	–	
	2N6028			25	270	–	
(V_S = 10 Vdc, R_G = 200 Ohms)	2N6027			1.5	–	–	mA
	2N6028			1.0	–	–	
● Gate to Anode Leakage Current		–	I_{GAO}				nAdc
(V_S = 40 Vdc, T_A = 25°C, Cathode Open)				–	1.0	10	
(V_S = 40 Vdc, T_A = 75°C, Cathode Open)				–	3.0	–	
Gate to Cathode Leakage Current		–	I_{GKS}				nAdc
(V_S = 40 Vdc, Anode to Cathode Shorted)				–	5.0	50	
● Forward Voltage (I_F = 50 mA Peak)		1,6	V_F	–	0.8	1.5	Volts
● Peak Output Voltage		3,7	V_O	6.0	11	–	Volts
(V_B = 20 Vdc, C_C = 0.2 μF)							
Pulse Voltage Rise Time		3	t_r	–	40	80	ns
(V_B = 20 Vdc, C_C = 0.2 μF)							

*Indicates JEDEC Registered Data

FIGURE 1 – ELECTRICAL CHARACTERIZATION

1A – PROGRAMMABLE UNIJUNCTION
WITH "PROGRAM" RESISTORS
R1 and R2

1B – EQUIVALENT TEST CIRCUIT FOR
FIGURE 1A USED FOR ELECTRICAL
CHARACTERISTICS TESTING
(ALSO SEE FIGURE 2)

1C – ELECTRICAL CHARACTERISTICS

FIGURE 2 – PEAK CURRENT (Ip) TEST CIRCUIT

FIGURE 3 – V_O AND t_r TEST CIRCUIT

MOTOROLA *Semiconductor Products Inc.*

N *National Semiconductor*

August 2000

LM741
Operational Amplifier

General Description

The LM741 series are general purpose operational amplifiers which feature improved performance over industry standards like the LM709. They are direct, plug-in replacements for the 709C, LM201, MC1439 and 748 in most applications.

The amplifiers offer many features which make their application nearly foolproof: overload protection on the input and output, no latch-up when the common mode range is exceeded, as well as freedom from oscillations.

The LM741C is identical to the LM741/LM741A except that the LM741C has their performance guaranteed over a 0°C to +70°C temperature range, instead of −55°C to +125°C.

Connection Diagrams

Metal Can Package

DS009341-2

Note 1: LM741H is available per JM38510/10101

**Order Number LM741H, LM741H/883 (Note 1),
LM741AH/883 or LM741CH
See NS Package Number H08C**

Dual-In-Line or S.O. Package

DS009341-3

**Order Number LM741J, LM741J/883, LM741CN
See NS Package Number J08A, M08A or N08E**

Ceramic Flatpak

DS009341-6

**Order Number LM741W/883
See NS Package Number W10A**

Typical Application

Offset Nulling Circuit

DS009341-7

© 2000 National Semiconductor Corporation DS009341

Absolute Maximum Ratings (Note 2)

If Military/Aerospace specified devices are required, please contact the National Semiconductor Sales Office/ Distributors for availability and specifications.

(Note 7)

	LM741A	LM741	LM741C
Supply Voltage	±22V	±22V	±18V
Power Dissipation (Note 3)	500 mW	500 mW	500 mW
Differential Input Voltage	±30V	±30V	±30V
Input Voltage (Note 4)	±15V	±15V	±15V
Output Short Circuit Duration	Continuous	Continuous	Continuous
Operating Temperature Range	−55°C to +125°C	−55°C to +125°C	0°C to +70°C
Storage Temperature Range	−65°C to +150°C	−65°C to +150°C	−65°C to +150°C
Junction Temperature	150°C	150°C	100°C
Soldering Information			
N-Package (10 seconds)	260°C	260°C	260°C
J- or H-Package (10 seconds)	300°C	300°C	300°C
M-Package			
Vapor Phase (60 seconds)	215°C	215°C	215°C
Infrared (15 seconds)	215°C	215°C	215°C

See AN-450 "Surface Mounting Methods and Their Effect on Product Reliability" for other methods of soldering surface mount devices.

ESD Tolerance (Note 8)	400V	400V	400V

Electrical Characteristics (Note 5)

Parameter	Conditions	LM741A			LM741			LM741C			Units
		Min	Typ	Max	Min	Typ	Max	Min	Typ	Max	
Input Offset Voltage	$T_A = 25°C$										
	$R_S \leq 10\ k\Omega$					1.0	5.0		2.0	6.0	mV
	$R_S \leq 50\Omega$		0.8	3.0							mV
	$T_{AMIN} \leq T_A \leq T_{AMAX}$										
	$R_S \leq 50\Omega$			4.0							mV
	$R_S \leq 10\ k\Omega$						6.0			7.5	mV
Average Input Offset Voltage Drift				15							μV/°C
Input Offset Voltage Adjustment Range	$T_A = 25°C$, $V_S = ±20V$	±10				±15			±15		mV
Input Offset Current	$T_A = 25°C$		3.0	30		20	200		20	200	nA
	$T_{AMIN} \leq T_A \leq T_{AMAX}$			70		85	500			300	nA
Average Input Offset Current Drift				0.5							nA/°C
Input Bias Current	$T_A = 25°C$		30	80		80	500		80	500	nA
	$T_{AMIN} \leq T_A \leq T_{AMAX}$			0.210			1.5			0.8	μA
Input Resistance	$T_A = 25°C$, $V_S = ±20V$	1.0	6.0		0.3	2.0		0.3	2.0		MΩ
	$T_{AMIN} \leq T_A \leq T_{AMAX}$, $V_S = ±20V$	0.5									MΩ
Input Voltage Range	$T_A = 25°C$							±12	±13		V
	$T_{AMIN} \leq T_A \leq T_{AMAX}$				±12	±13					V

Electrical Characteristics (Note 5) (Continued)

Parameter	Conditions	LM741A			LM741			LM741C			Units
		Min	Typ	Max	Min	Typ	Max	Min	Typ	Max	
Large Signal Voltage Gain	$T_A = 25°C$, $R_L \geq 2$ kΩ										
	$V_S = \pm20V$, $V_O = \pm15V$	50									V/mV
	$V_S = \pm15V$, $V_O = \pm10V$				50	200		20	200		V/mV
	$T_{AMIN} \leq T_A \leq T_{AMAX}$, $R_L \geq 2$ kΩ,										
	$V_S = \pm20V$, $V_O = \pm15V$	32									V/mV
	$V_S = \pm15V$, $V_O = \pm10V$				25			15			V/mV
	$V_S = \pm5V$, $V_O = \pm2V$	10									V/mV
Output Voltage Swing	$V_S = \pm20V$										
	$R_L \geq 10$ kΩ	±16									V
	$R_L \geq 2$ kΩ	±15									V
	$V_S = \pm15V$										
	$R_L \geq 10$ kΩ				±12	±14		±12	±14		V
	$R_L \geq 2$ kΩ				±10	±13		±10	±13		V
Output Short Circuit Current	$T_A = 25°C$	10	25	35		25			25		mA
	$T_{AMIN} \leq T_A \leq T_{AMAX}$	10		40							mA
Common-Mode Rejection Ratio	$T_{AMIN} \leq T_A \leq T_{AMAX}$										
	$R_S \leq 10$ kΩ, $V_{CM} = \pm12V$				70	90		70	90		dB
	$R_S \leq 50\Omega$, $V_{CM} = \pm12V$	80	95								dB
Supply Voltage Rejection Ratio	$T_{AMIN} \leq T_A \leq T_{AMAX}$, $V_S = \pm20V$ to $V_S = \pm5V$										
	$R_S \leq 50\Omega$	86	96								dB
	$R_S \leq 10$ kΩ				77	96		77	96		dB
Transient Response	$T_A = 25°C$, Unity Gain										
Rise Time			0.25	0.8		0.3			0.3		μs
Overshoot			6.0	20		5			5		%
Bandwidth (Note 6)	$T_A = 25°C$	0.437	1.5								MHz
Slew Rate	$T_A = 25°C$, Unity Gain	0.3	0.7			0.5			0.5		V/μs
Supply Current	$T_A = 25°C$					1.7	2.8		1.7	2.8	mA
Power Consumption	$T_A = 25°C$										
	$V_S = \pm20V$		80	150							mW
	$V_S = \pm15V$					50	85		50	85	mW
LM741A	$V_S = \pm20V$										
	$T_A = T_{AMIN}$			165							mW
	$T_A = T_{AMAX}$			135							mW
LM741	$V_S = \pm15V$										
	$T_A = T_{AMIN}$					60	100				mW
	$T_A = T_{AMAX}$					45	75				mW

Note 2: "Absolute Maximum Ratings" indicate limits beyond which damage to the device may occur. Operating Ratings indicate conditions for which the device is functional, but do not guarantee specific performance limits.

LM741

Electrical Characteristics (Note 5) (Continued)

Note 3: For operation at elevated temperatures, these devices must be derated based on thermal resistance, and T_j max. (listed under "Absolute Maximum Ratings"). $T_j = T_A + (\theta_{jA} P_D)$.

Thermal Resistance	Cerdip (J)	DIP (N)	HO8 (H)	SO-8 (M)
θ_{jA} (Junction to Ambient)	100˚C/W	100˚C/W	170˚C/W	195˚C/W
θ_{jC} (Junction to Case)	N/A	N/A	25˚C/W	N/A

Note 4: For supply voltages less than ±15V, the absolute maximum input voltage is equal to the supply voltage.

Note 5: Unless otherwise specified, these specifications apply for $V_S = ±15V$, $-55˚C ≤ T_A ≤ +125˚C$ (LM741/LM741A). For the LM741C/LM741E, these specifications are limited to $0˚C ≤ T_A ≤ +70˚C$.

Note 6: Calculated value from: BW (MHz) = 0.35/Rise Time(μs).

Note 7: For military specifications see RETS741X for LM741 and RETS741AX for LM741A.

Note 8: Human body model, 1.5 kΩ in series with 100 pF.

Schematic Diagram

DS009341-1

MOTOROLA ■ SEMICONDUCTOR ■
TECHNICAL DATA

**LM117
LM217
LM317**

**THREE-TERMINAL
ADJUSTABLE POSITIVE
VOLTAGE REGULATORS**

SILICON MONOLITHIC
INTEGRATED CIRCUIT

THREE-TERMINAL ADJUSTABLE
OUTPUT POSITIVE VOLTAGE REGULATORS

The LM117/217/317 are adjustable 3-terminal positive voltage regulators capable of supplying in excess of 1.5 A over an output voltage range of 1.2 V to 37 V. These voltage regulators are exceptionally easy to use and require only two external resistors to set the output voltage. Further, they employ internal current limiting, thermal shutdown and safe area compensation, making them essentially blow-out proof.

The LM117 series serve a wide variety of applications including local, on card regulation. This device can also be used to make a programmable output regulator, or by connecting a fixed resistor between the adjustment and output, the LM117 series can be used as a precision current regulator.

- Output Current in Excess of 1.5 Ampere in K and T Suffix Packages
- Output Current in Excess of 0.5 Ampere in H Suffix Package
- Output Adjustable between 1.2 V and 37 V
- Internal Thermal Overload Protectiion
- Internal Short-Circuit Current Limiting Constant with Temperature
- Output Transistor Safe-Area Compensation
- Floating Operation for High Voltage Applications
- Standard 3-lead Transistor Packages
- Eliminates Stocking Many Fixed Voltages

K SUFFIX
METAL PACKAGE
CASE 1

Adjust ① ② V_{in}

V_{out}

(Bottom View)
CASE
IS OUTPUT

Pins 1 and 2 electrically isolated from case.
Case is third electrical connection.

T SUFFIX
PLASTIC PACKAGE
CASE 221A

PIN 1. ADJUST
2. V_{out}
3. V_{in}

Heatsink surface connected to Pin 2

H SUFFIX
METAL PACKAGE
CASE 79

CASE
IS OUTPUT

(Bottom View)

PIN 1. V_{in}
2. ADJUST
3. V_{out}

STANDARD APPLICATION

V_{in} — LM117 — V_{out}

$I_{Adj} \downarrow$ Adjust

R_1 240

* C_{in} 0.1 μF

** C_O 1 μF

R_2

* = C_{in} is required if regulator is located an appreciable distance from power supply filter.

** = C_O is not needed for stability, however it does improve transient response.

$$V_{out} = 1.25 \text{ V} \left(1 + \frac{R_2}{R_1}\right) + I_{Adj} R_2$$

Since I_{Adj} is controlled to less than 100 μA, the error associated with this term is negligible in most applications.

ORDERING INFORMATION

Device	Tested Operating Temperature Range	Package
LM117H LM117K	$T_J = -55°C$ to $+150°C$	Metal Can Metal Power
LM217H LM217K	$T_J = -25°C$ to $+150°C$	Metal Can Metal Power
LM317H LM317K LM317T	$T_J = 0°C$ to $+125°C$	Metal Can Metal Power Plastic Power
LM317BT#	$T_J = -40°C$ to $+125°C$	Plastic Power

#Automotive temperature range selections are available with special test conditions and additional tests.
Contact your local Motorola sales office for information.

LM117, LM217, LM317

MAXIMUM RATINGS

Rating	Symbol	Value	Unit
Input-Output Voltage Differential	V_I-V_O	40	Vdc
Power Dissipation	P_D	Internally Limited	
Operating Junction Temperature Range LM117 LM217 LM317	T_J	 -55 to $+150$ -25 to $+150$ 0 to $+150$	°C
Storage Temperature Range	T_{stg}	-65 to $+150$	°C

ELECTRICAL CHARACTERISTICS (V_I-V_O = 5.0 V; I_O = 0.5 A for K and T packages; I_O = 0.1 A for H package; T_J = T_{low} to T_{high} [see Note 1]; I_{max} and P_{max} per Note 2; unless otherwise specified.)

Characteristic	Figure	Symbol	LM117/217 Min	LM117/217 Typ	LM117/217 Max	LM317 Min	LM317 Typ	LM317 Max	Unit
Line Regulation (Note 3) T_A = 25°C, 3.0 V ≤ V_I-V_O ≤ 40 V	1	Reg_{line}	—	0.01	0.02	—	0.01	0.04	%/V
Load Regulation (Note 3) T_A = 25°C, 10 mA ≤ I_O ≤ I_{max} V_O ≤ 5.0 V V_O ≥ 5.0 V	2	Reg_{load}	 — —	 5.0 0.1	 15 0.3	 — —	 5.0 0.1	 25 0.5	 mV %/V_O
Thermal Regulation (T_A = +25°C) 20 ms Pulse	—		—	0.02	0.07	—	0.03	0.07	%/W
Adjustment Pin Current	3	I_{Adj}	—	50	100	—	50	100	μA
Adjustment Pin Current Change 2.5 V ≤ V_I-V_O ≤ 40 V 10 mA ≤ I_L ≤ I_{max}, P_D ≤ P_{max}	1,2	ΔI_{Adj}	—	0.2	5.0	—	0.2	5.0	μA
Reference Voltage (Note 4) 3.0 V ≤ V_I-V_O ≤ 40 V 10 mA ≤ I_O ≤ I_{max}, P_D ≤ P_{max}	3	V_{ref}	1.2	1.25	1.3	1.2	1.25	1.3	V
Line Regulation (Note 3) 3.0 V ≤ V_I-V_O ≤ 40 V	1	Reg_{line}	—	0.02	0.05	—	0.02	0.07	%/V
Load Regulation (Note 3) 10 mA ≤ I_O ≤ I_{max} V_O ≤ 5.0 V V_O ≥ 5.0 V	2	Reg_{load}	 — —	 20 0.3	 50 1.0	 — —	 20 0.3	 70 1.5	 mV %/V_O
Temperature Stability (T_{low} ≤ T_J ≤ T_{high})	3	T_S	—	0.7	—	—	0.7	—	%/V_O
Minimum Load Current to Maintain Regulation (V_I-V_O = 40 V)	3	I_{Lmin}	—	3.5	5.0	—	3.5	10	mA
Maximum Output Current V_I-V_O ≤ 15 V, P_D ≤ P_{max} K and T Packages H Package V_I-V_O = 40 V, P_D ≤ P_{max}, T_A = 25°C K and T Packages H Package	3	I_{max}	 1.5 0.5 0.25 —	 2.2 0.8 0.4 0.07	 — — — —	 1.5 0.5 0.15 —	 2.2 0.8 0.4 0.07	 — — — —	A
RMS Noise, % of V_O T_A = 25°C, 10 Hz ≤ f ≤ 10 kHz	—	N	—	0.003	—	—	0.003	—	%/V_O
Ripple Rejection, V_O = 10 V, f = 120 Hz (Note 5) Without C_{Adj} C_{Adj} = 10 μF	4	RR	 — 66	 65 80	 — —	 — 66	 65 80	 — —	dB
Long-Term Stability, T_J = T_{high} (Note 6) T_A = 25°C for Endpoint Measurements	3	S	—	0.3	1.0	—	0.3	1.0	%/1.0 k Hrs.
Thermal Resistance Junction to Case H Package K Package T Package	—	$R_{\theta JC}$	 — — —	 12 2.3 —	 15 3.0 —	 — — —	 12 2.3 5.0	 15 3.0 —	°C/W

NOTES: (1) T_{low} = -55°C for LM117 T_{high} = $+150$°C for LM117
      = -25°C for LM217   = $+150$°C for LM217
      = 0°C for LM317   = $+125$°C for LM317
(2) I_{max} = 1.5 A for K and T Packages
      = 0.5 A for H Package
   P_{max} = 20 W for K Package
      = 20 W for T Package
      = 2.0 W for H Package
(3) Load and line regulation are specified at constant junction temperature. Changes in V_O due to heating effects must be taken into account separately. Pulse testing with low duty cycle is used.

(4) Selected devices with tightened tolerance reference voltage available.
(5) C_{ADJ}, when used, is connected between the adjustment pin and ground.
(6) Since Long-Term Stability cannot be measured on each device before shipment, this specification is an engineering estimate of average stability from lot to lot.

MC7800, MC7800A, LM340, LM340A Series

Three-Terminal Positive Voltage Regulators

These voltage regulators are monolithic integrated circuits designed as fixed–voltage regulators for a wide variety of applications including local, on–card regulation. These regulators employ internal current limiting, thermal shutdown, and safe–area compensation. With adequate heatsinking they can deliver output currents in excess of 1.0 A. Although designed primarily as a fixed voltage regulator, these devices can be used with external components to obtain adjustable voltages and currents.

- Output Current in Excess of 1.0 A
- No External Components Required
- Internal Thermal Overload Protection
- Internal Short Circuit Current Limiting
- Output Transistor Safe–Area Compensation
- Output Voltage Offered in 2% and 4% Tolerance
- Available in Surface Mount D^{2}PAK and Standard 3–Lead Transistor Packages

MAXIMUM RATINGS (T$_A$ = 25°C, unless otherwise noted.)

Rating	Symbol	Value	Unit
Input Voltage (5.0 – 18 V)	V$_I$	35	Vdc
(24 V)		40	
Power Dissipation Case 221A T$_A$ = 25°C	P$_D$	Internally Limited	W
Thermal Resistance, Junction–to–Ambient	R$_{\theta JA}$	65	°C/W
Thermal Resistance, Junction–to–Case	R$_{\theta JC}$	5.0	°C/W
Case 936 (D^{2}PAK) T$_A$ = 25°C	P$_D$	Internally Limited	W
Thermal Resistance, Junction–to–Ambient	R$_{\theta JA}$	See Figure 13	°C/W
Thermal Resistance, Junction–to–Case	R$_{\theta JA}$	5.0	°C/W
Storage Junction Temperature Range	T$_{stg}$	–65 to +150	°C
Operating Junction Temperature	T$_J$	+150	°C

NOTE: ESD data available upon request.

ON Semiconductor

http://onsemi.com

TO–220
T SUFFIX
CASE 221A

Heatsink surface connected to Pin 2.

D^{2}PAK
D2T SUFFIX
CASE 936

Pin 1. Input
2. Ground
3. Output

Heatsink surface (shown as terminal 4 in case outline drawing) is connected to Pin 2.

STANDARD APPLICATION

A common ground is required between the input and the output voltages. The input voltage must remain typically 2.0 V above the output voltage even during the low point on the input ripple voltage.

XX, These two digits of the type number indicate nominal voltage.

  * C$_{in}$ is required if regulator is located an appreciable distance from power supply filter.

 ** C$_O$ is not needed for stability; however, it does improve transient response. Values of less than 0.1 µF could cause instability.

ORDERING INFORMATION

See detailed ordering and shipping information in the package dimensions section on page 16 of this data sheet.

DEVICE MARKING INFORMATION

See general marking information in the device marking section on page 18 of this data sheet.

© Semiconductor Components Industries, LLC, 2000
April, 2000 – Rev. 6

Publication Order Number:
MC7800/D

MC7800, MC7800A, LM340, LM340A Series

Representative Schematic Diagram

This device contains 22 active transistors.

MC7800, MC7800A, LM340, LM340A Series

ELECTRICAL CHARACTERISTICS (V_{in} = 10 V, I_O = 500 mA, $T_J = T_{low}$ to T_{high} [Note 1.], unless otherwise noted.)

Characteristic	Symbol	MC7805B			MC7805C/LM340T–5			Unit
		Min	Typ	Max	Min	Typ	Max	
Output Voltage (T_J = 25°C)	V_O	4.8	5.0	5.2	4.8	5.0	5.2	Vdc
Output Voltage (5.0 mA ≤ I_O ≤ 1.0 A, P_D ≤ 15 W)	V_O							Vdc
7.0 Vdc ≤ V_{in} ≤ 20 Vdc		–	–	–	4.75	5.0	5.25	
8.0 Vdc ≤ V_{in} ≤ 20 Vdc		4.75	5.0	5.25	–	–	–	
Line Regulation (Note 2.)	Reg_{line}							mV
7.5 Vdc ≤ V_{in} ≤ 20 Vdc, 1.0 A		–	5.0	100	–	0.5	20	
8.0 Vdc ≤ V_{in} ≤ 12 Vdc		–	1.3	50	–	0.8	10	
Load Regulation (Note 2.)	Reg_{load}							mV
5.0 mA ≤ I_O ≤ 1.0 A		–	1.3	100	–	1.3	25	
5.0 mA ≤ I_O ≤ 1.5 A (T_A = 25°C)		–	0.15	50	–	1.3	25	
Quiescent Current	I_B	–	3.2	8.0	–	3.2	6.5	mA
Quiescent Current Change	ΔI_B							mA
7.0 Vdc ≤ V_{in} ≤ 25 Vdc		–	–	–	–	0.3	1.0	
5.0 mA ≤ I_O ≤ 1.0 A (T_A = 25°C)		–	–	0.5	–	0.08	0.8	
Ripple Rejection 8.0 Vdc ≤ V_{in} ≤ 18 Vdc, f = 120 Hz	RR	–	68	–	62	83	–	dB
Dropout Voltage (I_O = 1.0 A, T_J = 25°C)	$V_I - V_O$	–	2.0	–	–	2.0	–	Vdc
Output Noise Voltage (T_A = 25°C) 10 Hz ≤ f ≤ 100 kHz	V_n	–	10	–	–	10	–	μV/V_O
Output Resistance f = 1.0 kHz	r_O	–	0.9	–	–	0.9	–	mΩ
Short Circuit Current Limit (T_A = 25°C) V_{in} = 35 Vdc	I_{SC}	–	0.2	–	–	0.6	–	A
Peak Output Current (T_J = 25°C)	I_{max}	–	2.2	–	–	2.2	–	A
Average Temperature Coefficient of Output Voltage	TCV_O	–	–0.3	–	–	–0.3	–	mV/°C

ELECTRICAL CHARACTERISTICS (V_{in} = 10 V, I_O = 1.0 A, $T_J = T_{low}$ to T_{high} [Note 1.], unless otherwise noted.)

Characteristic	Symbol	MC7805AC/LM340AT–5			Unit
		Min	Typ	Max	
Output Voltage (T_J = 25°C)	V_O	4.9	5.0	5.1	Vdc
Output Voltage (5.0 mA ≤ I_O ≤ 1.0 A, P_D ≤ 15 W) 7.5 Vdc ≤ V_{in} ≤ 20 Vdc	V_O	4.8	5.0	5.2	Vdc
Line Regulation (Note 2.)	Reg_{line}				mV
7.5 Vdc ≤ V_{in} ≤ 25 Vdc, I_O = 500 mA		–	0.5	10	
8.0 Vdc ≤ V_{in} ≤ 12 Vdc, I_O = 1.0 A		–	0.8	12	
8.0 Vdc ≤ V_{in} ≤ 12 Vdc, I_O = 1.0 A, T_J = 25°C		–	1.3	4.0	
7.3 Vdc ≤ V_{in} ≤ 20 Vdc, I_O = 1.0 A, T_J = 25°C		–	4.5	10	
Load Regulation (Note 2.)	Reg_{load}				mV
5.0 mA ≤ I_O ≤ 1.5 A, T_J = 25°C		–	1.3	25	
5.0 mA ≤ I_O ≤ 1.0 A		–	0.8	25	
250 mA ≤ I_O ≤ 750 mA		–	0.53	15	
Quiescent Current	I_B	–	3.2	6.0	mA
Quiescent Current Change	ΔI_B				mA
8.0 Vdc ≤ V_{in} ≤ 25 Vdc, I_O = 500 mA		–	0.3	0.8	
7.5 Vdc ≤ V_{in} ≤ 20 Vdc, T_J = 25°C		–	–	0.8	
5.0 mA ≤ I_O ≤ 1.0 A		–	0.08	0.5	

1. T_{low} = 0°C for MC78XXAC, C, LM340AT–XX, LM340T–XX T_{high} = +125°C for MC78XXAC, C, LM340AT–XX, LM340T–XX
   = –40°C for MC78XXB
2. Load and line regulation are specified at constant junction temperature. Changes in V_O due to heating effects must be taken into account separately. Pulse testing with low duty cycle is used.

MC7800, MC7800A, LM340, LM340A Series

ELECTRICAL CHARACTERISTICS (continued) (V_{in} = 10 V, I_O = 1.0 A, T_J = T_{low} to T_{high} [Note 1.], unless otherwise noted.)

Characteristic	Symbol	MC7805AC/LM340AT–5			Unit
		Min	Typ	Max	
Ripple Rejection 8.0 Vdc ≤ V_{in} ≤ 18 Vdc, f = 120 Hz, I_O = 500 mA	RR	68	83	–	dB
Dropout Voltage (I_O = 1.0 A, T_J = 25°C)	$V_I - V_O$	–	2.0	–	Vdc
Output Noise Voltage (T_A = 25°C) 10 Hz ≤ f ≤ 100 kHz	V_n	–	10	–	µV/V_O
Output Resistance (f = 1.0 kHz)	r_O	–	0.9	–	mΩ
Short Circuit Current Limit (T_A = 25°C) V_{in} = 35 Vdc	I_{SC}	–	0.2	–	A
Peak Output Current (T_J = 25°C)	I_{max}	–	2.2	–	A
Average Temperature Coefficient of Output Voltage	TCV_O	–	–0.3	–	mV/°C

ELECTRICAL CHARACTERISTICS (V_{in} = 11 V, I_O = 500 mA, T_J = T_{low} to T_{high} [Note 1.], unless otherwise noted.)

Characteristic	Symbol	MC7806B			MC7806C			Unit
		Min	Typ	Max	Min	Typ	Max	
Output Voltage (T_J = 25°C)	V_O	5.75	6.0	6.25	5.75	6.0	6.25	Vdc
Output Voltage (5.0 mA ≤ I_O ≤ 1.0 A, P_D ≤ 15 W) 8.0 Vdc ≤ V_{in} ≤ 21 Vdc 9.0 Vdc ≤ V_{in} ≤ 21 Vdc	V_O	 – 5.7	 – 6.0	 – 6.3	 5.7 –	 6.0 –	 6.3 –	Vdc
Line Regulation, T_J = 25°C (Note 2.) 8.0 Vdc ≤ V_{in} ≤ 25 Vdc 9.0 Vdc ≤ V_{in} ≤ 13 Vdc	Reg_{line}	 – –	 5.5 1.4	 120 60	 – –	 0.5 0.8	 24 12	mV
Load Regulation, T_J = 25°C (Note 2.) 5.0 mA ≤ I_O ≤ 1.5 A	Reg_{load}	–	1.3	120	–	1.3	30	mV
Quiescent Current (T_J = 25°C)	I_B	–	3.3	8.0	–	3.3	8.0	mA
Quiescent Current Change 8.0 Vdc ≤ V_{in} ≤ 25 Vdc 5.0 mA ≤ I_O ≤ 1.0 A	ΔI_B	 – –	 – –	 – 0.5	 – –	 0.3 0.08	 1.3 0.5	mA
Ripple Rejection 9.0 Vdc ≤ V_{in} ≤ 19 Vdc, f = 120 Hz	RR	–	65	–	58	65	–	dB
Dropout Voltage (I_O = 1.0 A, T_J = 25°C)	$V_I - V_O$	–	2.0	–	–	2.0	–	Vdc
Output Noise Voltage (T_A = 25°C) 10 Hz ≤ f ≤ 100 kHz	V_n	–	10	–	–	10	–	µV/V_O
Output Resistance f = 1.0 kHz	r_O	–	0.9	–	–	0.9	–	mΩ
Short Circuit Current Limit (T_A = 25°C) V_{in} = 35 Vdc	I_{SC}	–	0.2	–	–	0.2	–	A
Peak Output Current (T_J = 25°C)	I_{max}	–	2.2	–	–	2.2	–	A
Average Temperature Coefficient of Output Voltage	TCV_O	–	–0.3	–	–	–0.3	–	mV/°C

1. T_{low} = 0°C for MC78XXAC, C, LM340AT–XX, LM340T–XX T_{high} = +125°C for MC78XXAC, C, LM340AT–XX, LM340T–XX
   = –40°C for MC78XXB
2. Load and line regulation are specified at constant junction temperature. Changes in V_O due to heating effects must be taken into account separately. Pulse testing with low duty cycle is used.

MOTOROLA

Three-Terminal Negative Voltage Regulators

MC7900 Series

The MC7900 series of fixed output negative voltage regulators are intended as complements to the popular MC7800 series devices. These negative regulators are available in the same seven–voltage options as the MC7800 devices. In addition, one extra voltage option commonly employed in MECL systems is also available in the negative MC7900 series.

Available in fixed output voltage options from −5.0 V to −24 V, these regulators employ current limiting, thermal shutdown, and safe–area compensation – making them remarkably rugged under most operating conditions. With adequate heatsinking they can deliver output currents in excess of 1.0 A.

- No External Components Required
- Internal Thermal Overload Protection
- Internal Short Circuit Current Limiting
- Output Transistor Safe–Area Compensation
- Available in 2% Voltage Tolerance (See Ordering Information)

THREE–TERMINAL NEGATIVE FIXED VOLTAGE REGULATORS

T SUFFIX
PLASTIC PACKAGE
CASE 221A

Heatsink surface connected to Pin 2.

Pin 1. Ground
2. Input
3. Output

D2T SUFFIX
PLASTIC PACKAGE
CASE 936
(D²PAK)

Heatsink surface (shown as terminal 4 in case outline drawing) is connected to Pin 2.

Representative Schematic Diagram

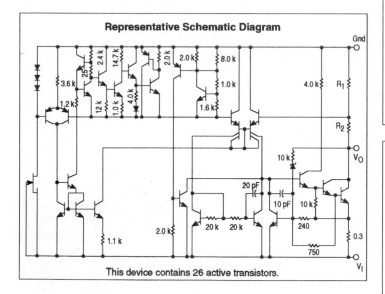

This device contains 26 active transistors.

STANDARD APPLICATION

A common ground is required between the input and the output voltages. The input voltage must remain typically 2.0 V above more negative even during the high point of the input ripple voltage.

XX, These two digits of the type number indicate nominal voltage.
  * C_{in} is required if regulator is located an appreciable distance from power supply filter.
  ** C_O improve stability and transient response.

ORDERING INFORMATION

Device	Output Voltage Tolerance	Operating Temperature Range	Package
MC79XXACD2T	2%		Surface Mount
MC79XXCD2T	4%	$T_J = 0°$ to +125°C	
MC79XXACT	2%		Insertion Mount
MC79XXCT	4%		
MC79XXBD2T	4%	$T_J = -40°$ to +125°C	Surface Mount
MC79XXBT			Insertion Mount

XX indicates nominal voltage.

DEVICE TYPE/NOMINAL OUTPUT VOLTAGE

MC7905	5.0 V	MC7912	12 V
MC7905.2	5.2 V	MC7915	15 V
MC7906	6.0 V	MC7918	28 V
MC7908	8.0 V	MC7924	24 V

© Motorola, Inc. 1996

Rev 6

MC7900

MAXIMUM RATINGS (T_A = +25°C, unless otherwise noted.)

Rating	Symbol	Value	Unit
Input Voltage (−5.0 V ≥ V_O ≥ −18 V) (24 V)	V_I	−35 −40	Vdc

Power Dissipation
Case 221A
   T_A = +25°C
   Thermal Resistance, Junction–to–Ambient
   Thermal Resistance, Junction–to–Case
Case 936 (D2PAK)
   T_A = +25°C
   Thermal Resistance, Junction–to–Ambient
   Thermal Resistance, Junction–to–Case
	
	P_D
	θ_{JA}
	θ_{JC}
	
	P_D
	θ_{JA}
	θ_{JC}
		
		Internally Limited
		65
		5.0
		
		Internally Limited
		70
		5.0
			
			W
			°C/W
			°C/W
			
			W
			°C/W
			°C/W

Storage Junction Temperature Range	T_{stg}	−65 to +150	°C
Junction Temperature	T_J	+150	°C

THERMAL CHARACTERISTICS

Characteristics	Symbol	Max	Unit
Thermal Resistance, Junction–to–Ambient	$R_{\theta JA}$	65	°C/W
Thermal Resistance, Junction–to–Case	$R_{\theta JC}$	5.0	°C/W

MC7905C
ELECTRICAL CHARACTERISTICS (V_I = −10 V, I_O = 500 mA, 0°C < T_J < +125°C, unless otherwise noted.)

Characteristics	Symbol	Min	Typ	Max	Unit
Output Voltage (T_J = +25°C)	V_O	−4.8	−5.0	−5.2	Vdc

Line Regulation (Note 1)
  (T_J = +25°C, I_O = 100 mA)
    −7.0 Vdc ≥ V_I ≥ −25 Vdc
    −8.0 Vdc ≥ V_I ≥ −12 Vdc
  (T_J = +25°C, I_O = 500 mA)
    −7.0 Vdc ≥ V_I ≥ −25 Vdc
    −8.0 Vdc ≥ V_I ≥ −12 Vdc
	Regline	
		
		−
		−
		
		−
		−
			
			7.0
			2.0
			
			35
			8.0
				
				50
				25
				
				100
				50
					mV

Load Regulation, T_J = +25°C (Note 1)
  5.0 mA ≤ I_O ≤ 1.5 A
  250 mA ≤ I_O ≤ 750 mA
	Regload	
		−
		−
			
			11
			4.0
				
				100
				50
					mV

Output Voltage −7.0 Vdc ≥ V_I ≥ −20 Vdc, 5.0 mA ≤ I_O ≤ 1.0 A, P ≤ 15 W	V_O	−4.75	−	−5.25	Vdc
Input Bias Current (T_J = +25°C)	I_{IB}	−	4.3	8.0	mA

Input Bias Current Change
  −7.0 Vdc ≥ V_I ≥ −25 Vdc
  5.0 mA ≤ I_O ≤ 1.5 A
	ΔI_{IB}	
		−
		−
			
			−
			−
				
				1.3
				0.5
					mA

Output Noise Voltage (T_A = +25°C, 10 Hz ≤ f ≤ 100 kHz)	V_n	−	40	−	μV
Ripple Rejection (I_O = 20 mA, f = 120 Hz)	RR	−	70	−	dB
Dropout Voltage I_O = 1.0 A, T_J = +25°C	V_I–V_O	 −	 2.0	 −	Vdc
Average Temperature Coefficient of Output Voltage I_O = 5.0 mA, 0°C ≤ T_J ≤ +125°C	$\Delta V_O/\Delta T$	 −	 −1.0	 −	mV/°C

NOTE: 1. Load and line regulation are specified at constant junction temperature. Changes in V_O due to heating effects must be taken into account separately. Pulse testing with low duty cycle is used.

MC7900

MC7905AC
ELECTRICAL CHARACTERISTICS ($V_I = -10$ V, $I_O = 500$ mA, $0°C < T_J < +125°C$, unless otherwise noted.)

Characteristics	Symbol	Min	Typ	Max	Unit
Output Voltage ($T_J = +25°C$)	V_O	−4.9	−5.0	−5.1	Vdc
Line Regulation (Note 1) −8.0 Vdc ≥ V_I ≥ −12 Vdc; I_O = 1.0 A, T_J = +25°C −8.0 Vdc ≥ V_I ≥ −12 Vdc; I_O = 1.0 A −7.5 Vdc ≥ V_I ≥ −25 Vdc; I_O = 500 mA −7.0 Vdc ≥ V_I ≥ −20 Vdc; I_O = 1.0 A, T_J = +25°C	Reg$_{line}$	 − − − −	 2.0 7.0 7.0 6.0	 25 50 50 50	mV
Load Regulation (Note 1) 5.0 mA ≤ I_O ≤ 1.5 A, T_J = +25°C 250 mA ≤ I_O ≤ 750 mA 5.0 mA ≤ I_O ≤ 1.0 A	Reg$_{load}$	 	 11 4.0 9.0	 100 50 100	mV
Output Voltage −7.5 Vdc ≥ V_I ≥ −20 Vdc, 5.0 mA ≤ I_O ≤ 1.0 A, P ≤ 15 W	V_O	−4.80	−	−5.20	Vdc
Input Bias Current	I_{IB}	−	4.4	8.0	mA
Input Bias Current Change −7.5 Vdc ≥ V_I ≥ −25 Vdc 5.0 mA ≤ I_O ≤ 1.0 A 5.0 mA ≤ I_O ≤ 1.5 A, T_J = +25°C	ΔI_{IB}	 − − −	 − − −	 1.3 0.5 0.5	mA
Output Noise Voltage ($T_A = +25°C$, 10 Hz ≤ f ≤ 100 kHz)	V_n	−	40	−	μV
Ripple Rejection (I_O = mA, f = 120 Hz)	RR	−	70	−	dB
Dropout Voltage I_O = 1.0 A, T_J = +25°C	V_I–V_O	 −	 2.0	 −	Vdc
Average Temperature Coefficient of Output Voltage I_O = 5.0 A, 0°C ≤ T_J ≤ +125°C	$\Delta V_O/\Delta T$	 −	 −1.0	 −	mV/°C

MC7905.2C
ELECTRICAL CHARACTERISTICS ($V_I = -10$ V, $I_O = 500$ mA, $0°C < T_J < +125°C$, unless otherwise noted.)

Characteristics	Symbol	Min	Typ	Max	Unit
Output Voltage ($T_J = +25°C$)	V_O	−5.0	−5.2	−5.4	Vdc
Line Regulation (Note 1) (T_J = +25°C, I_O = 100 mA) −7.2 Vdc ≥ V_I ≥ −25 Vdc −8.0 Vdc ≥ V_I ≥ −12 Vdc (T_J = +25°C, I_O = 500 mA) −7.2 Vdc ≥ V_I ≥ −25 Vdc −8.0 Vdc ≥ V_I ≥ −12 Vdc	Reg$_{line}$	 − − − −	 8.0 2.2 37 8.5	 52 27 105 52	mV
Load Regulation, T_J = +25°C (Note 1) 5.0 mA ≤ I_O ≤ 1.5 A 250 mA ≤ I_O ≤ 750 mA	Reg$_{load}$	 	 12 4.5	 105 52	mV
Output Voltage −7.2 Vdc ≥ V_I ≥ −20 Vdc, 5.0 mA ≤ I_O ≤ 1.0 A, P ≤ 15 W	V_O	−4.95	−	−5.45	Vdc
Input Bias Current (T_J = +25°C)	I_{IB}	−	4.3	8.0	mA
Input Bias Current Change −7.2 Vdc ≥ V_I ≥ −25 Vdc 5.0 mA ≤ I_O ≤ 1.5 A	ΔI_{IB}	 − −	 − −	 1.3 0.5	mA
Output Noise Voltage ($T_A = +25°C$, 10 Hz ≤ f ≤ 100 kHz)	V_n	−	42	−	μV
Ripple Rejection (I_O = 20 mA, f = 120 Hz)	RR	−	68	−	dB
Dropout Voltage I_O = 1.0 A, T_J = +25°C	V_I–V_O	 −	 2.0	 −	Vdc
Average Temperature Coefficient of Output Voltage I_O = 5.0 mA, 0°C ≤ T_J ≤ +125°C	$\Delta V_O/\Delta T$	 −	 −1.0	 −	mV/°C

NOTE: 1. Load and line regulation are specified at constant junction temperature. Changes in V_O due to heating effects must be taken into account separately. Pulse testing with low duty cycle is used.

MC7900

MC7906C
ELECTRICAL CHARACTERISTICS ($V_I = -11$ V, $I_O = 500$ mA, $0°C < T_J < +125°C$, unless otherwise noted.)

Characteristics	Symbol	Min	Typ	Max	Unit
Output Voltage ($T_J = +25°C$)	V_O	−5.75	−6.0	−6.25	Vdc

Line Regulation (Note 1)
 ($T_J = +25°C$, $I_O = 100$ mA)
   −8.0 Vdc ≥ V_I ≥ −25 Vdc
   −9.0 Vdc ≥ V_I ≥ −13 Vdc
 ($T_J = +25°C$, $I_O = 500$ mA)
   −8.0 Vdc ≥ V_I ≥ −25 Vdc
   −9.0 Vdc ≥ V_I ≥ −13 Vdc
	Reg$_{line}$	
		
		−
		−
		
		−
		−
			
			9.0
			3.0
			
			43
			10
				
				60
				30
				
				120
				60
					mV

Load Regulation, $T_J = +25°C$ (Note 1) 5.0 mA ≤ I_O ≤ 1.5 A 250 mA ≤ I_O ≤ 750 mA	Reg$_{load}$	 − −	 13 5.0	 120 60	mV
Output Voltage −8.0 Vdc ≥ V_I ≥ −21 Vdc, 5.0 mA ≤ I_O ≤ 1.0 A, P ≤ 15 W	V_O	 −5.7	 −	 −6.3	Vdc
Input Bias Current ($T_J = +25°C$)	I_{IB}	−	4.3	8.0	mA
Input Bias Current Change −8.0 Vdc ≥ V_I ≥ −25 Vdc 5.0 mA ≤ I_O ≤ 1.5 A	ΔI_{IB}	 − −	 − −	 1.3 0.5	mA
Output Noise Voltage ($T_A = +25°C$, 10 Hz ≤ f ≤ 100 kHz)	V_n	−	45	−	μV
Ripple Rejection ($I_O = 20$ mA, f = 120 Hz)	RR	−	65	−	dB
Dropout Voltage $I_O = 1.0$ A, $T_J = +25°C$	V_I–V_O	 −	 2.0	 −	Vdc
Average Temperature Coefficient of Output Voltage $I_O = 5.0$ A, $0°C ≤ T_J ≤ +125°C$	$\Delta V_O/\Delta T$	 −	 −1.0	 −	mV/°C

MC7908C
ELECTRICAL CHARACTERISTICS ($V_I = -14$ V, $I_O = 500$ mA, $0°C < T_J < +125°C$, unless otherwise noted.)

Characteristics	Symbol	Min	Typ	Max	Unit
Output Voltage ($T_J = +25°C$)	V_O	−7.7	−8.0	−8.3	Vdc

Line Regulation (Note 1)
 ($T_J = +25°C$, $I_O = 100$ mA)
   −10.5 Vdc ≥ V_I ≥ −25 Vdc
   −11 Vdc ≥ V_I ≥ −17 Vdc
 ($T_J = +25°C$, $I_O = 500$ mA)
   −10.5 Vdc ≥ V_I ≥ −25 Vdc
   −11 Vdc ≥ V_I ≥ −17 Vdc
	Reg$_{line}$	
		
		−
		−
		
		−
		−
			
			12
			5.0
			
			50
			22
				
				80
				40
				
				160
				80
					mV

Load Regulation, $T_J = +25°C$ (Note 1) 5.0 mA ≤ I_O ≤ 1.5 A 250 mA ≤ I_O ≤ 750 mA	Reg$_{load}$	 − −	 26 9.0	 160 80	mV
Output Voltage −10.5 Vdc ≥ V_I ≥ −23 Vdc, 5.0 mA ≤ I_O ≤ 1.0 A, P ≤ 15 W	V_O	 −7.6	 −	 −8.4	Vdc
Input Bias Current ($T_J = +25°C$)	I_{IB}	−	4.3	8.0	mA
Input Bias Current Change −10.5 Vdc ≥ V_I ≥ −25 Vdc 5.0 mA ≤ I_O ≤ 1.5 A	ΔI_{IB}	 − −	 − −	 1.0 0.5	mA
Output Noise Voltage ($T_A = +25°C$, 10 Hz ≤ f ≤ 100 kHz)	V_n	−	52	−	μV
Ripple Rejection ($I_O = 20$ mA, f = 120 Hz)	RR	−	62	−	dB
Dropout Voltage $I_O = 1.0$ A, $T_J = +25°C$	V_I–V_O	 −	 2.0	 −	Vdc
Average Temperature Coefficient of Output Voltage $I_O = 5.0$ mA, $0°C ≤ T_J ≤ +125°C$	$\Delta V_O/\Delta T$	 −	 −1.0	 −	mV/°C

NOTE: 1. Load and line regulation are specified at constant junction temperature. Changes in V_O due to heating effects must be taken into account separately. Pulse testing with low duty cycle is used.

Answers To Odd-Numbered Problems

CHAPTER 1

1. 6 electrons; 6 protons

3. **(a)** insulator **(b)** semiconductor **(c)** conductor

5. Four

7. Conduction band and valence band

9. Antimony is a pentavalent material. Boron is a trivalent material. Both are used for doping.

11. No. The barrier potential is a voltage drop.

13. To prevent excessive forward current.

15. A temperature increase.

17. (a) -3 V **(b)** 0.7 V **(c)** 0.7 V **(d)** 0.7 V

19. $V_A = 25$ V; $V_B = 24.3$ V; $V_C = 8.7$ V; $V_D = 8$ V

21. Diode open

23. Diode shorted

25. Diode shorted

27. Diode open

CHAPTER 2

1. See Figure ANS–1.

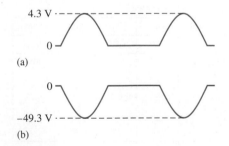

▲ **FIGURE ANS–1**

3. 23 V rms

5. **(a)** 1.59 V **(b)** 63.7 V
   (c) 16.4 V **(d)** 10.5 V

7. 173 V

9. 78.5 V

11. See Figure ANS–2.

▲ **FIGURE ANS–2**

13. $V_r = 8.33$ V; $V_{DC} = 25.8$ V

15. 556 μF

17. $V_{r(pp)} = 1.25$ V; $V_{DC} = 48.9$ V

19. 4%.

21. See Figure ANS–3.

▲ **FIGURE ANS–3**

23. See Figure ANS–4.

▲ **FIGURE ANS–4**

25. See Figure ANS–5.

(a) (b)

▲ **FIGURE ANS–5**

27. **(a)** A sine wave with a positive peak at +0.7 V, a negative peak at -7.3 V, and a dc value of -3.3 V.

   (b) A sine wave with a positive peak at +29.3 V, a negative peak at -0.7 V, and a dc value of +14.3 V.

   (c) A square wave varying from +0.7 V down to -15.3 V, with a dc value of -7.3 V.

   (d) A square wave varying from +1.3 V down to -0.7 V, with a dc value of +0.3 V.

29. 56.6 V

All 1N4001

▲ FIGURE ANS–6

31. 50 V

33. 62.5 mΩ

35. R_{surge} is open. Capacitor is shorted.

37. The circuit should not fail because the diode ratings exceed the actual PIV and maximum current.

39. Excessive PIV or surge current causes diode to open each time power is turned on. It could be a faulty transformer or maybe the diodes do not have sufficient PIV rating.

41. 177 μF

43. 651 mΩ (nearest standard 0.68 Ω)

45. See Figure ANS–6.

47. $V_{C1} = 155$ V; $V_{C2} = 310$ V

49. Diode leaky

51. Bottom diode open

53. Open filter capacitor

55. D_1 open

CHAPTER 3

1. See Figure ANS–7.

Zener equivalent

▲ FIGURE ANS–7

3. 5 Ω

5. 6.92 V

7. 14.3 V

9. See Figure ANS–8.

▲ FIGURE ANS–8

11. 14.3%

13. 3.13%

15. 5.88%

17. 3 V

19. ≈ 2.5 V

21. See Figure ANS–9.

▲ FIGURE ANS–9

23. (a) 30 kΩ   (b) 8.57 kΩ   (c) 5.88 kΩ

25. −750 Ω

27. The reflective ends cause the light to bounce back and forth, thus increasing the intensity of the light. The partially reflective end allows a portion of the reflected light to be emitted.

29. (a) ≈ 30 V dc

(b) 0 V

(c) Excessive 120 Hz ripple limited to 12 V by zener

(d) Full-wave rectified waveform limited at 12 V by zener

(e) 60 Hz ripple limited to 12 V

(f) 60 Hz ripple limited to 12 V

(g) 0 V

(h) 0 V

31. Could be fuse, transformer, limiting resistor, or surge resistor open. Also, capacitor or zener could be shorted. Cannot isolate further with given measurements.

33. D_1 open, R_1 open, no dc voltage, short in threshold circuit

35. (a) 60 V   (b) 307 mW

(c) 1.27 W   (d) 21 pF

(e) 1N5139   (f) 4.82 pF

37. (a) Reverse-bias voltage   (b) 940 nm

(c) 40 nA   (d) 940 nm

(e) 40 μA/mW/cm^2 (f) 104 μA

39. $V_{OUT(1)} = 6.8$ V; $V_{OUT(2)} = 24$ V

41. See Figure ANS–10.

▲ **FIGURE ANS–10**

43. See Figure ANS–11.

45. Zener diode open

47. Zener diode shorted

CHAPTER 4

1. Holes

3. The base is narrow and lightly doped so that a small recombination (base) current is generated compared to the collector current.

5. Negative, positive

7. 0.947

9. 102.5

11. 8.98 mA

13. 0.99

15. (a) $V_{BE} = 0.7$ V, $V_{CE} = 5.10$ V, $V_{BC} = -4.40$ V

(b) $V_{BE} = -0.7$ V, $V_{CE} = -3.83$ V, $V_{BC} = 3.13$ V

17. $I_B = 30$ μA, $I_E = 1.3$ mA, $I_C = 1.27$ mA

19. 3 μA

21. 425 mW

23. 33.3

25. 500 μA, 3.33 μA, 4.03 V

27. See Figure ANS–12.

29. Open, low resistance

31. (a) 27.8   (b) 109

33. Q_2 or Q_4 shorted collector to emitter, short from pin 2 of relay to ground, Q_1 or Q_3 open collector to emitter

35. (a) 40 V   (b) 200 mA dc   (c) 625 mW

(d) 1.5 W   (e) 35

37. 1.26 W

39. (a) Saturated   (b) Not saturated

▲ **FIGURE ANS–11**

▲ **FIGURE ANS–12**

41. **(a)** No parameters are exceeded.

　　(b) No parameters are exceeded.

43. Yes, marginally; $V_{CE} = 0.8$ V; $I_C = 75$ mA

45. See Figure ANS–13.

▲ **FIGURE ANS–13**

47. R_B shorted

49. Collector-emitter shorted

51. R_E leaky

53. R_B open

CHAPTER 5

1. Saturation

3. 18 mA

5. $V_{CE} = 20$ V; $I_{C(sat)} = 2$ mA

7. See Figure ANS–14.

▲ **FIGURE ANS–14**

9. 69.1

11. $I_C \cong 809$ μA; $V_{CE} = 13.2$ V

13. See Figure ANS–15.

▲ **FIGURE ANS–15**

15. **(a)** -1.41 mA, -8.67 V

　　(b) 12.2 mW

17. $I_{CQ} = 92.5$ mA; $V_{CEQ} = 2.75$ V

19. 27.7 mA to 69.2 mA; 6.23 V to 2.08 V; Yes

21. $V_B = -391$ mV; $V_E = -1.10$ V; $V_C = 3.22$ V

23. 0.09 mA

25. $I_C = 16.3$ mA; $V_{CE} = -6.95$ V

27. 2.53 kΩ

29. 7.87 mA; 2.56 V

31. **(a)** Open collector

　　(b) No problems

　　(c) Transistor shorted collector-to-emitter

　　(d) Open emitter

33. **(a)** 1: 10 V,　2: float,　　3: -3.59 V,　4: 10 V

　　(b) 1: 10 V,　2: 4.05 V,　3:　4.75 V,　4: 4.05 V

　　(c) 1: 10 V,　2: 0 V,　　3:　0 V,　　4: 10 V

　　(d) 1: 10 V,　2: 570 mV,　3:　1.27 V,　4: float

　　(e) 1: 10 V,　2: 0 V,　　3:　0.7 V,　4: 0 V

　　(f) 1: 10 V,　2: 0 V,　　3:　3.59 V,　4: 10 V

35. R_1 open, R_2 shorted, BE junction open

37. $V_C = V_{CC} = 9.1$ V, V_B normal, $V_E = 0$ V

39. None are exceeded.

41. 457 mW

43. See Figure ANS–16.

▲ **FIGURE ANS–16**

45. See Figure ANS–17.

▲ **FIGURE ANS–17**

47. Yes

49. V_{CEQ} will be less, causing the transistor to saturate at a slightly higher temperature, thus limiting the low temperature response.

51. R_C open

53. R_2 open

55. R_C shorted

CHAPTER 6

1. Slightly greater than 1 mA min.

3. 8.33 Ω

5. $r'_e \cong 19$ Ω

7. See Figure ANS–18.

DC equivalent

AC equivalent

▲ **FIGURE ANS–18**

9. (a) 1.29 kΩ   (b) 968 Ω   (c) 171

11. (a) $V_B = 3.25$ V   (b) $V_E = 2.55$ V
    (c) $I_E = 2.55$ mA   (d) $I_C \cong 2.55$ mA
    (e) $V_C = 9.59$ V   (f) $V_{CE} = 7.04$ V

13. $A'_v = 131$; $\theta = 180°$

15. $A_{v(max)} = 65.5$, $A_{v(min)} = 2.06$

17. A_v is reduced to approximately 30. See Figure ANS–19.

19. $R_{in(tot)} = 3.1$ kΩ; $V_{OUT} = 1.06$ V

21. 270 Ω

23. 8.8

25. $R_{in(emitter)} = 2.28$ Ω; $A_v = 526$; $A_i \cong 1$; $A_p = 526$

27. 400

▲ **FIGURE ANS–19**

29. (a) $A_{v1} = 93.6$, $A_{v2} = 302$
    (b) $A'_v = 28,267$
    (c) $A_{v1(dB)} = 39.4$ dB, $A_{v2(dB)} = 49.6$ dB,
    $A'_{v(dB)} = 89.0$ dB

31. $V_{B1} = 2.16$ V, $V_{E1} = 1.46$ V, $V_{C1} \cong 5.16$ V,
    $V_{B2} = 5.16$ V, $V_{E2} = 4.46$ V, $V_{C2} \cong 7.54$ V,
    $A_{v1} = 66$, $A_{v2} = 179$, $A'_v = 11,814$

33. (a) 1.41   (b) 2.00   (c) 3.16   (d) 10.0   (e) 100

35. Cutoff, 10 V

37.

TEST POINT	DC VOLTS	AC VOLTS (rms)
Input	0 V	25 μV
Q_1 base	2.99 V	20.8 μV
Q_1 emitter	2.29 V	0 V
Q_1 collector	7.44 V	1.95 mV
Q_2 base	2.99 V	1.95 mV
Q_2 emitter	2.29 V	0 V
Q_2 collector	7.44 V	589 mV
Output	0 V	589 mV

39. (a) $V_C = 5.87$ V, $V_c = 850$ mV
    (b) $V_C = 5.87$ V, $V_c = 0$ V
    (c) $V_C = 5.87$ V, $V_c = 0$ V
    (d) $V_C = 5.87$ V, $V_c = 203$ mV
    (e) $V_C = 5.87$ V, $V_c = 0$ V
    (f) $V_C \cong 0$ V, $V_c = 0$ V

41. (a) Q_1 is off   (b) 9 V   (c) 5.87 V

43. (a) 700   (b) 40 Ω   (c) 20 kΩ

45. A leaky coupling capacitor affects the bias voltages and attenuates the ac voltage.

47. See Figure ANS–20.

49. See Figure ANS–21.

51. See Figure ANS–22.

53. $A_v = R_C/r'_e$
    $A_v \cong (V_{RC}/I_C)/(0.025 \text{ V}/I_C) = V_{RC}/0.025 = 40V_{RC}$

▲ **FIGURE ANS–20**

▲ **FIGURE ANS–21**

▲ **FIGURE ANS–22**

55. C_2 shorted

57. C_1 open

59. C_3 open

CHAPTER 7

1. (a) Narrows   (b) Increases

3. See Figure ANS–23.

5. 5 V

7. 10 mA

9. 4 V

▲ **FIGURE ANS–23**

11. −2.63 V

13. $g_m = 1429\ \mu S$, $y_{fs} = 1429\ \mu S$

15. $V_{GS} = 0\ V$, $I_D = 8\ mA$

   $V_{GS} = -1\ V$, $I_D = 5.12\ mA$

   $V_{GS} = -2\ V$, $I_D = 2.88\ mA$

   $V_{GS} = -3\ V$, $I_D = 1.28\ mA$

   $V_{GS} = -4\ V$, $I_D = 0.320\ mA$

   $V_{GS} = -5\ V$, $I_D = 0\ mA$

17. 800 Ω

19. (a) 20 mA   (b) 0 A   (c) Increases

21. 211 Ω

23. 9.80 MΩ

25. $I_D \cong 5.3\ mA$, $V_{GS} \cong 2.1\ V$

27. $I_D \cong 1.9\ mA$, $V_{GS} \cong -1.5\ V$

29. The enhancement mode

31. The gate is insulated from the channel.

33. 4.69 mA

35. (a) Depletion   (b) Enhancement

   (c) Zero bias   (d) Depletion

37. (a) 4 V   (b) 5.4 V   (c) −4.52 V

39. (a) 5 V, 3.18 mA   (b) 3.2 V, 1.02 mA

41. R_D or R_S open, JFET open D-to-S, $V_{DD} = 0\ V$, or ground connection open.

43. No change

45. The 1.0 MΩ bias resistor is open.

47. V_{OUT} = 300 mV for pH = 5; V_{OUT} = −400 mV for pH = 9

49. V_{OUT} = +12.1 V assuming typical values.

51. (a) −0.5 V   (b) 25 V   (c) 310 mW

    (d) −25 V

53. 2000 μS

55. 1 V

57. I_D ≅ 13 mA when V_{GS} = +3 V, I_D ≅ 0.4 mA when V_{GS} = −2 V.

59. −3.0 V

61. I_D = 3.58 mA; V_{GS} = −4.21 V

63. 6.01 V

65. See Figure ANS−24.

+12 V

R_1 15 kΩ

R_D 200 Ω

R_2 10 kΩ

R_S 100 Ω

▲ **FIGURE ANS−24**

67. R_D shorted

69. R_1 open

71. R_D open

73. Drain-source shorted

CHAPTER 8

1. (a) 60 μA   (b) 900 μA

    (c) 3.6 mA   (d) 6 mA

3. 14.2

5. (a) *n*-channel D-MOSFET with zero-bias; V_{GS} = 0

    (b) *p*-channel JFET with self-bias; V_{GS} = −0.99 V

    (c) *n*-channel E-MOSFET with voltage-divider bias; V_{GS} = 3.84 V

7. (a) *n*-channel D-MOSFET   (b) *n*-channel JFET

    (c) *p*-channel E-MOSFET

9. Figure 8–18(b): approximately 4 mA

    Figure 8–18(c): approximately 3.2 mA

11. 920 mV

13. (a) 4.32   (b) 9.92

15. ≈7.5 mA

17. 2.54

19. 33.6 mV rms

21. 9.84 MΩ

23. V_{GS} = 9 V; I_D = 3.13 mA; V_{DS} = 13.3 V; V_{ds} = 675 mV

25. R_{in} ≅ 10 MΩ; A_v = 0.783

27. (a) 0.906   (b) 0.299

29. 250 Ω

31. (a) V_{D1} = V_{DD}; no Q_1 drain signal; no output signal

    (b) V_{D1} ≅ 0 V (floating); no Q_1 drain signal; no output signal

    (c) V_{GS1} = 0 V; V_S = 0 V; V_{D1} less than normal; clipped output signal

    (d) Correct signal at Q_1 drain; no Q_1 gate signal; no output signal

    (e) V_{D2} = V_{DD}; correct signal at Q_2 gate; no Q_2 drain signal or output signal

33. The 10 μF capacitor between Q_1 drain and Q_2 gate is open.

35. V_{DC} = 4.35 V; V_{ac} = 1.29 V rms

37. (a) −3.0 V   (b) 20 V dc   (c) 200 mW

    (d) ±10 V dc

39. 900 μS

41. 1.5 mA

43. 2.0; 6.82

45. See Figure ANS−25.

+24 V

2.2 kΩ

5 kΩ

2N3797

▲ **FIGURE ANS−25**

47. C_2 open

49. R_S shorted

51. R_1 open

53. R_2 open

CHAPTER 9

1. (a) I_{CQ} = 68.4 mA; V_{CEQ} = 5.14 V

    (b) A_v = 11.7; A_p = 263

3. The changes are shown on Figure ANS–26. The advantage of this arrangement is that the load resistor is referenced to ground.

▲ **FIGURE ANS–26**

▲ **FIGURE ANS–27**

5. **(a)** 39.8 mA, 1.99 V **(b)** 12.3 mA, 2.88 V

7. 265 mW

9. **(a)** $V_{B(Q1)} = +0.7$ V; $V_{B(Q2)} = -0.7$ V; $V_E = 0$ V;
  $V_{CEQ(Q1)} = +9$ V; $V_{CEQ(Q2)} = -9$ V; $I_{CQ} = 8.3$ mA
  (b) $P_L = 0.5$ W

11. **(a)** $V_{B(Q1)} = +8.2$ V; $V_{B(Q2)} = +6.8$ V
  $V_E = +7.5$ V; $I_{CQ} = 6.8$ mA;
  $V_{CEQ(Q1)} = +7.5$ V;
  $V_{CEQ(Q2)} = -7.5$ V
  (b) $P_L = 167$ mW

13. **(a)** C_2 open or Q_2 open
  (b) Power supply off, open R_1, Q_1 base shorted to ground
  (c) Q_1 has collector-to-emitter short
  (d) One or both diodes shorted

15. 450 μW

17. 24 V

19. Negative half of input cycle

21. **(a)** No dc supply voltage or R_1 open
  (b) D_1 or D_2 open
  (c) No fault
  (d) Q_1 shorted C-to-E

23. 6 V dc, positive alternation of input signal

25. C_1 is connected backwards.

27. 51 W

29. Gain decreases.

31. T_C is much closer to the actual junction temperature than T_A. In a given operating environment, T_A is always less than T_C.

33. See Figure ANS–27.

35. C_{in} open

37. Q_1 collector-emitter open

39. Q_2 drain-source open

CHAPTER 10

1. If $C_1 = C_2$, the critical frequencies are equal, and they will both cause the gain to drop at 40 dB/decade below f_c.

3. BJT: C_{be}, C_{bc}, C_{ce}; FET: C_{gs}, C_{gd}, C_{ds}

5. 812 pF

7. $C_{in(miller)} = 6.95$ pF; $C_{out(miller)} \cong 5.28$ pF

9. 24 mV rms; 34 dB

11. **(a)** 3.01 dBm **(b)** 0 dBm **(c)** 6.02 dBm
  (d) -6.02 dBm

13. **(a)** 318 Hz **(b)** 1.59 kHz

15. At $0.1f_c$: $A_v = 18.7$ dB
  At f_c: $A_v = 35.7$ dB
  At $10f_c$: $A_v = 38.7$ dB

17. Input RC circuit: $f_c = 3.34$ Hz
  Output RC circuit: $f_c = 3.01$ kHz
  Output f_c is dominant.

19. Input circuit: $f_c = 4.32$ MHz
  Output circuit: $f_c = 94.9$ MHz
  Input f_c is dominant.

21. Input circuit: $f_c = 12.9$ MHz
  Output circuit: $f_c = 54.5$ MHz
  Input f_c is dominant.

23. $f_{cl} = 136$ Hz, $f_{cu} = 8$ kHz

25. $BW = 5.26$ MHz, $f_{cu} \cong 5.26$ MHz

27. 230 Hz; 1.2 MHz

29. 514 kHz

31. ≈ 2.5 MHz

33. Increase the frequency until the output voltage drops to 3.54 V rms. This is f_{cu}.

35. 23.1 Hz

37. No effect

39. 112 pF

41. $C_{gd} = 1.3$ pF; $C_{gs} = 3.7$ pF; $C_{ds} = 3.7$ pF

43. ≈10.5 MHz

45. R_C open

47. R_2 open

CHAPTER 11

1. $I_A = 24.1$ mA

3. See "Turning the SCR On" in Section 11–2.

5. Add a transistor to provide inversion of negative half-cycle in order to obtain a positive gate trigger.

7. See Figure ANS–28.

▲ **FIGURE ANS–28**

9. See Figure ANS–29.

▲ **FIGURE ANS–29**

11. Anode, cathode, anode gate, cathode gate

13. 6.48 V

15. (a) 9.79 V   (b) 5.2 V

17. See Figure ANS–30.

▲ **FIGURE ANS–30**

19. (a) 12 V   (b) 0 V

21. When the switch is closed, the battery V_2 causes illumination of the lamp. The light energy causes the LASCR to conduct and thus energize the relay. When the relay is energized, the contacts close and 115 V ac are applied to the motor.

23. 30 mA

25. 0 V

27. As the PUT gate voltage increases, the PUT triggers on later to the ac cycle causing the SCR to fire later in the cycle, conduct for a shorter time, and decrease power to the motor.

29. See Figure ANS–31.

31. 4-layer (Shockley diode) shorted (EWB only)

33. R_1 shorted

CHAPTER 12

1. *Practical op-amp:* High open-loop gain, high input impedance, low output impedance, high CMRR.

Ideal op-amp: Infinite open-loop gain, infinite input impedance, zero output impedance, infinite CMRR.

▲ **FIGURE ANS–31**

3. **(a)** Single-ended input

 (b) Differential input

 (c) Common-mode input

5. 120 dB

7. 8.1 μA

9. 1.6 V/μs

11. **(a)** Voltage-follower

 (b) Noninverting

 (c) Inverting

13. **(a)** $A_{cl(NI)} = 374$

 (b) $V_{out} = 3.74$ V rms

 (c) $V_f = 9.99$ mV rms

15. **(a)** 49 kΩ **(b)** 3 MΩ

 (c) 84 kΩ **(d)** 165 kΩ

17. **(a)** 10 mV, in phase

 (b) -10 mV, 180° out of phase

 (c) 223 mV, in phase

 (d) -100 mV, 180° out of phase

19. **(a)** $Z_{in(NI)} = 8.41$ GΩ; $Z_{out(NI)} = 89.2$ mΩ

 (b) $Z_{in(NI)} = 6.20$ GΩ; $Z_{out(NI)} = 4.04$ mΩ

 (c) $Z_{in(NI)} = 5.30$ GΩ; $Z_{out(NI)} = 19.0$ mΩ

21. **(a)** $Z_{in(I)} = 10$ kΩ; $Z_{out(I)} = 5.12$ mΩ

 (b) $Z_{in(I)} = 100$ kΩ; $Z_{out(I)} = 7.41$ mΩ

 (c) $Z_{in(I)} = 470$ Ω; $Z_{out(I)} = 6.22$ mΩ

23. **(a)** 2.69 kΩ **(b)** 1.45 kΩ **(c)** 53 kΩ

 R_c is placed between V_{in} and the $+$ input

25. 175 nV

27. $A_v = 125{,}892$; $BW_{ol} = 200$ Hz

29. **(a)** 0.997   **(b)** 0.923   **(c)** 0.707

 (d) 0.515   **(e)** 0.119

31. **(a)** $-51.5°$ **(b)** $-7.17°$ **(c)** $-85.5°$

33. **(a)** 90 dB   **(b)** $-281°$

35. **(a)** 29.8 dB   **(b)** 23.9 dB   **(c)** 0 dB

 All are closed-loop gains.

37. 71.7 dB

39. **(a)** $A_{cl(VF)} = 1$; $BW = 2.8$ MHz

 (b) $A_{cl(I)} = -45.5$; $BW = 61.6$ kHz

 (c) $A_{cl(NI)} = 13$; $BW = 215$ kHz

 (d) $A_{cl(I)} = -179$; $BW = 15.7$ kHz

41. **(a)** Faulty op-amp or R_1 open

 (b) R_2 open, forcing open-loop operation

43. The gain becomes a fixed -100, with no effect as the gain potentiometer is adjusted.

45. Op-amp gain will be ten times too high.

47. $Z_{in(NI)} = 3.96$ GΩ

49. 50,000

51. See Figure ANS–32.

▲ **FIGURE ANS–32**

53. See Figure ANS–33.

▲ **FIGURE ANS–33**

55. See Figure ANS–34.

▲ **FIGURE ANS–34**

57. R_f open

59. R_f leaky

61. R_f shorted

63. R_f leaky

65. R_i shorted

67. R_f open

69. R_f open

71. R_i open

CHAPTER 13

1. 24 V, with distortion
3. $V_{UTP} = +2.77$ V, $V_{LTP} = -2.77$ V
5. See Figure ANS–35.

(a)

(b)

▲ FIGURE ANS–35

7. $+8.57$ V and -0.968 V
9. (a) -2.5 V   (b) -3.52 V
11. 110 kΩ
13. $V_{OUT} = -3.57$ V, $I_f = 357$ μA
15. -4.46 mV/μs
17. 1 mA
19. See Figure ANS–36.

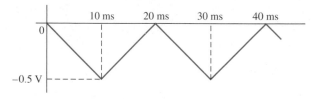

▲ FIGURE ANS–36

21. Output not correct; R_2 is open.
23. 50 kΩ resistor open
25. The output of IC2 will saturate positively.
27. $+8$ V
29. $f_{in} = 100$ kHz. See Figure ANS–37.
31. Op-amp inputs shorted together
33. D_1 shorted
35. Middle 10 kΩ resistor shorted
37. R_f open
39. C open

▲ FIGURE ANS–37

CHAPTER 14

1. $A_{v(1)} = A_{v(2)} = 101$
3. 1.005 V
5. 51.5
7. Change R_G to 2.2 kΩ.
9. 300
11. Change the 18 kΩ resistor to 23.2 kΩ.
13. Connect pin 6 directly to pin 10, and connect pin 14 directly to pin 15 to make $R_f = 0$.
15. 500 μA, 5 V
17. $A_v \cong 11.6$
19. See Figure ANS–38.
21. See Figure ANS–39.
23. (a) -0.301 (b) 0.301   (c) 1.70   (d) 2.11
25. The output of a log amplifier is limited to 0.7 V because of the transistor's *pn* junction.
27. -157 mV
29. $V_{out(max)} = -147$ mV, $V_{out(min)} = -89.2$ mV; the 1 V input peak is reduced 85% whereas the 100 mV input peak is reduced only 10%.
31. See Figure ANS–40.
33. TP1: ≈ 0 V
    TP2: ≈ 0 V
    TP3: 21.2 mV @ 1 kHz
    TP4: $+15$ V
    TP5: 0 V
35. R open
37. Zener diode open

CHAPTER 15

1. (a) Band-pass   (b) High-pass   (c) Low-pass
   (d) Band-stop
3. 48.2 kHz, No
5. 700 Hz, 5.04

▲ FIGURE ANS–38

▲ FIGURE ANS–39

▲ FIGURE ANS–40

7. (a) 1, not Butterworth

   (b) 1.44, approximate Butterworth

   (c) 1st stage: 1.67; 2nd stage: 1.67; Not Butterworth

9. (a) Chebyshev   **(b)** Butterworth

   (c) Bessel   **(d)** Butterworth

11. 190 Hz

13. Add another identical stage and change the ratio of the feedback resistors to 0.068 for first stage, 0.586 for second stage, and 1.482 for third stage.

15. Exchange positions of resistors and capacitors.

17. (a) Decrease R_1 and R_2 or C_1 and C_2.

   (b) Increase R_3 or decrease R_4.

19. (a) $f_0 = 4.95$ kHz, $BW = 3.84$ kHz

   (b) $f_0 = 449$ Hz, $BW = 96.4$ Hz

   (c) $f_0 = 15.9$ kHz, $BW = 838$ Hz

21. Sum the low-pass and high-pass outputs with a two-input adder.

23. R_4 shorted

25. C_3 shorted

27. R_1 open

29. R_1 open

31. R_7 open

CHAPTER 16

1. An oscillator requires no input (other than dc power).

3. $1/75 = 0.0133$

5. 733 mV

7. 50 kΩ

9. 2.34 kΩ

11. 136 kΩ, 691 Hz

13. 9.4

15. Change R_1 to 3.54 kΩ

17. $R_4 = 65.8$ kΩ, $R_5 = 47$ kΩ

19. 3.33 V, 6.67 V

21. 0.0076 μF

23. Drain-to-source shorted

25. Collector-to-emitter shorted

27. R_2 open

CHAPTER 17

1. See Figure ANS 41.

3. 1135 kHz

5. $f_{RF} = 91.2$ MHz; $f_{IF} = 10.7$ MHz

7. 739 μA

9. -8.12 V

11. **(a)** $+0.28$ V   **(b)** $+1.024$ V

     (c) $+2.07$ V   **(d)** $+2.49$ V

13. $f_{diff} = 8$ kHz; $f_{sum} = 10$ kHz

15. $f_{diff} = 1.7$ MHz; $f_{sum} = 1.9$ MHz; $f_c = 1.8$ MHz

17. $f_c = 850$ kHz; $f_m = 3$ kHz

19. $V_{out} = 15$ mV $\cos[2\pi (1100$ kHz$)t] -$
       15 mV $\cos[2\pi (5500$ kHz$)t]$

21. See Figure ANS–42.

23. See Figure ANS–43.

25. 450 kHz–460 kHz: IF amplifier; 10 Hz–5 kHz: Audio/Power amplifiers

27. The modulating input signal is applied to the control voltage terminal of the VCO. As the input signal amplitude varies, the output frequency of the VCO varies proportionately.

29. Varactor

▲ **FIGURE ANS–43**

31. **(a)** 10 MHz   **(b)** 48.3 mV

33. 1005 Hz

35. $f_o = 233$ kHz; $f_{lock} = \pm 104$ kHz; $f_{cap} = \pm 4.56$ kHz

CHAPTER 18

1. 0.0333%

3. 1.01%

5. A: reference voltage, B: Control element, C: Error detector, D: Sampling circuit

7. 8.51 V

9. 9.57 V

11. 500 mA

▲ **FIGURE ANS–41**

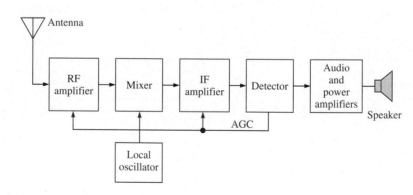

▲ **FIGURE ANS–42**

13. 10 mA
15. $I_{L(max)} = 250$ mA, $P_{R1} = 6.25$ W
17. 40%
19. V_{OUT} decreases
21. 14.3 V
23. 1.3 mA
25. 2.8 Ω
27. $R_{limit} = 0.35$ Ω
29. See Figure ANS–44.
31. R_2 leaky
33. Q_2 collector-to-emitter open

▲ **FIGURE ANS–44**

Glossary

ac ground   A point in a circuit that appears as ground to ac signals only.

active filter   A frequency-selective circuit consisting of active devices such as transistors or op-amps coupled with reactive components.

A/D conversion   A process whereby information in analog form is converted into digital form.

alpha (α)   The ratio of dc collector current to dc emitter current in a bipolar junction transistor.

amplification   The process of increasing the power, voltage, or current by electronic means.

amplifier   An electronic circuit having the capability to amplify power, voltage, or current.

amplitude modulation (AM)   A communication method in which a lower-frequency signal modulates (varies) the amplitude of a higher-frequency signal (carrier).

analog   Characterized by a linear process in which a variable takes on a continuous set of values.

anode   The p region of a diode.

antilogarithm   The result obtained when the base of a number is raised to a power equal to the logarithm of that number.

astable   Characterized by having no stable states.

atom   The smallest particle of an element that possesses the unique characteristics of that element.

atomic number   The number of protons in an atom.

attenuation   The reduction in the level of power, current, or voltage.

audio   Related to the frequency range of sound waves that can be heard by the human ear and generally considered to be in the 20 Hz to 20 kHz range.

avalanche   The rapid buildup of conduction electrons due to excessive reverse-bias voltage.

avalanche breakdown   The higher voltage breakdown in a zener diode.

balanced modulation   A form of amplitude modulation in which the carrier is suppressed; sometimes known as *suppressed-carrier modulation.*

band-pass filter   A type of filter that passes a range of frequencies lying between a certain lower frequency and a certain higher frequency.

band-stop filter   A type of filter that blocks or rejects a range of frequencies lying between a certain lower frequency and a certain higher frequency.

bandwidth   The characteristic of certain types of electronic circuits that specifies the usable range of frequencies that pass from input to output.

barrier potential   The amount of energy required to produce full conduction across the *pn* junction in forward bias.

base   One of the semiconductor regions in a BJT. The base is very thin and lightly doped compared to the other regions.

Bessel   A type of filter response having a linear phase characteristic and less than -20 dB/decade/ pole roll-off.

beta (β)   The ratio of dc collector current to dc base current in a BJT; current gain from base to collector.

bias   The application of a dc voltage to a diode, transistor, or other device to produce a desired mode of operation.

bipolar   Characterized by both free electrons and holes as current carriers.

BJT   Bipolar junction transistor; a transistor constructed with three doped semiconductor regions separated by two *pn* junctions.

Bode plot   An idealized graph of the gain in dB versus frequency used to graphically illustrate the response of an amplifier or filter.

bounding   The process of limiting the output range of an amplifier or other circuit.

breakdown   The phenomenon of a sudden and drastic increase when a certain voltage is reached across a device.

bridge rectifier   A type of full-wave rectifier consisting of diodes arranged in a four-cornered configuration.

Butterworth   A type of filter response characterized by flatness in the passband and a -20 dB/ decade/pole roll-off.

bypass capacitor   A capacitor placed across the emitter resistor of an amplifier.

capture range   The range of frequencies over which a PLL can acquire lock.

carbon   A semiconductive material.

carrier   The high radio frequency (RF) signal that carries modulated information in AM, FM, or other systems.

cascade   An arrangement of circuits in which the output of one circuit becomes the input to the next.

cathode   The *n* region of a diode.

center-tapped rectifier   A type of full-wave rectifier consisting of a center-tapped transformer and two diodes.

channel   The conductive path between the drain and source in a FET.

Chebyshev   A type of filter response characterized by ripples in the passband and a greater than -20 dB/decade/pole roll-off.

clamper   A circuit that adds a dc level to an ac voltage using a diode and a capacitor.

class A   A type of amplifier that operates entirely in its linear (active) region.

class AB   A type of amplifier that is biased into slight conduction.

class B   A type of amplifier that operates in the linear region for 180° of the input cycle because it is biased at cutoff.

class C   A type of amplifier that operates only for a small portion of the input cycle.

clipper   See Limiter.

closed-loop   An op-amp configuration in which the output is connected back to the input through a feedback circuit.

closed-loop voltage gain (A_{cl})   The voltage gain of an op-amp with external feedback.

CMRR   Common-mode rejection ratio; the ratio of open-loop gain to common-mode gain; a measure of an op-amp's ability to reject common-mode signals.

coherent light   Light having only one wavelength.

collector   The largest of the three semiconductor regions of a BJT.

common-base (CB)   A BJT amplifier configuration in which the base is the common terminal to an ac signal or ground.

common-collector (CC)   A BJT amplifier configuration in which the collector is the common terminal to an ac signal or ground.

common-drain (CD)   A FET amplifier configuration in which the drain is the grounded terminal.

common-emitter (CE)   A BJT amplifier configuration in which the emitter is the common terminal to an ac signal or ground.

common-gate (CG)   A FET amplifier configuration in which the gate is the grounded terminal.

common mode   A condition characterized by the presence of the same signal on both op-amp inputs.

common-source (CS)   A FET amplifier configuration in which the source is the grounded terminal.

comparator   A circuit which compares two input voltages and produces an output in either of two states indicating the greater or less than relationship of the inputs.

complementary symmetry transistors   Two transistors, one *npn* and one *pnp,* having matched characteristics.

conduction electron   A free electron.

conductor   A material that conducts electrical current very well.

core   The central part of an atom, includes the nucleus and all but the valence electrons.

covalent   Related to the bonding of two or more atoms by the interaction of their valence electrons.

critical frequency   The frequency at which the response of an amplifier or filter is 3 dB less than at midrange.

crossover distortion   Distortion in the output of a class B push-pull amplifier at the point where each transistor changes from the cutoff state to the *on* state.

crystal   A solid material in which the atoms are arranged in a symmetrical pattern.

current   The rate of flow of electrical charge.

current mirror   A circuit that uses matching diode junctions to form a current source. The current in a diode junction is reflected as a matching current in the other junction (which is typically the base-emitter junction of a transistor). Current mirrors are commonly used to bias a push-pull amplifier.

cutoff   The nonconducting state of a transistor.

cutoff frequency   Another term for critical frequency.

cutoff voltage   The value of the gate-to-source voltage that makes the drain current approximately zero.

D/A conversion   The process of converting a sequence of digital codes to an analog form.

damping factor A filter characteristic that determines the type of response.

dark current The amount of thermally generated reverse current in a photodiode in the absence of light.

darlington pair A configuration of two transistors in which the collectors are connected and the emitter of the first drives the base of the second to achieve beta multiplication.

dc load line A straight line plot of I_C and V_{CE} for a transistor circuit.

decade A ten-times increase or decrease in the value of a quantity such as frequency.

decibel (dB) A logarithmic measure of the ratio of one power to another or one voltage to another.

demodulation The process in which the information signal is recovered from the IF carrier signal; the reverse of modulation.

depletion In a MOSFET, the process of removing or depleting the channel of charge carriers and thus decreasing the channel conductivity.

depletion region The area near a *pn* junction on both sides that has no majority carriers.

derivative The instantaneous rate of change of a function, determined mathematically.

diac A two-terminal four-layer semiconductor device (thyristor) that can conduct current in either direction when properly activated.

differential A mode of op-amp operation in which two opposite polarity signal voltages are applied to two inputs.

differential amplifier (diff-amp) An amplifier that produces an output voltage proportional to the difference of the two input voltages.

differentiator A circuit that produces an output which approximates the instantaneous rate of change of the input function.

digital Characterized by a process in which a variable takes on either of two values.

diode A semiconductor device with a single *pn* junction that conducts current in only one direction.

diode drop The voltage across the diode when it is forward-biased; approximately the same as the barrier potential and typically 0.7 V for silicon.

discriminator A type of FM demodulator.

doping The process of imparting impurities to an intrinsic semiconductive material in order to control its conduction characteristics.

drain One of the three terminals of a FET analogous to the collector of a BJT.

dynamic resistance The nonlinear internal resistance of a semiconductive material.

efficiency The ratio of the signal power delivered to a load to the power from the power supply of an amplifier.

electroluminescence The process of releasing light energy by the recombination of electrons in a semiconductor.

electrostatic discharge (ESD) The discharge of a high voltage through an insulating path that can destroy an electronic device.

electron The basic particle of negative electrical charge.

electron-hole pair The conduction electron and the hole created when the electron leaves the valence band.

emitter The most heavily doped of the three semiconductor regions of a BJT.

emitter-follower A popular term for a common-collector amplifier.

enhancement In a MOSFET, the process of creating a channel or increasing the conductivity of the channel by the addition of charge carriers.

feedback The process of returning a portion of a circuit's output back to the input in such a way as to oppose or aid a change in the output.

feedback oscillator An electronic circuit that operates with positive feedback and produces a time-varying output signal without an external input signal.

FET Field-effect transistor; a type of unipolar, voltage-controlled transistor that uses an induced electric field to control current.

filter A capacitor in a power supply used to reduce the variation of the output voltage from a rectifier; a type of circuit that passes or blocks certain frequencies to the exclusion of all others.

floating point A point in the circuit that is not electrically connected to ground or a "solid" voltage.

fold-back current limiting A method of current limiting in voltage regulators.

forced commutation A method of turning off an SCR.

forward bias The condition in which a diode conducts current.

forward-breakover voltage ($V_{BR(F)}$) The voltage at which a device enters the forward-blocking region.

4-layer diode The type of two-terminal thyristor that conducts current when the anode-to-cathode voltage reaches a specified "breakover" value.

four-quadrant multiplier A linear device that produces an output voltage proportional to the product of two input voltages.

free electron An electron that has acquired enough energy to break away from the valance band of the parent atom; also called a conduction electron.

frequency modulation (FM) A communication method in which a lower frequency intelligence-carrying signal modulates (varies) the frequency of a higher frequency signal.

full-wave rectifier A circuit that converts an ac sinusoidal input voltage into a pulsating dc voltage with two output pulses occurring for each input cycle.

fuse A protective device that burns open when the current exceeds a rated limit.

gain The amount by which an electrical signal is increased or amplified.

gain-bandwidth product A constant parameter which is always equal to the frequency at which the op-amp's open-loop gain is unity (1).

gate One of the three terminals of a FET analogous to the base of a BJT.

germanium A semiconductive material.

half-wave rectifier A circuit that converts an ac sinusoidal input voltage into a pulsating dc voltage with one output pulse occurring for each input cycle.

high-pass filter A type of filter that passes frequencies above a certain frequency while rejecting lower frequencies.

holding current (I_H) The value of the anode current below which a device switches from the forward-conduction region to the forward-blocking region.

hole The absence of an electron in the valence band of an atom.

hysteresis Characteristic of a circuit in which two different trigger levels create an offset or lag in the switching action.

index of refraction A property of light-conducting materials that specifies how much a light ray will bend when passing from one material to another.

infrared (IR) Light that has a range of wavelengths greater than visible light.

input resistance The resistance looking in at the transistor base.

instrumentation amplifier An amplifier used for amplifying small signals riding on large common-mode voltages.

insulator A material that does not conduct current.

integrated circuit (IC) A type of circuit in which all the components are constructed on a single tiny chip of silicon.

integrator A circuit that produces an output which approximates the area under the curve of the input function.

intrinsic The pure or natural state of a material.

inverting amplifier An op-amp closed-loop configuration in which the input signal is applied to the inverting input.

ionization The removal or addition of an electron from or to a neutral atom so that the resulting atom (called an ion) has a net positive or negative charge.

irradiance (H) The power per unit area at a specified distance for the LED; the light intensity.

isolation amplifier An amplifier with electrically isolated internal stages.

JFET Junction field-effect transistor; one of two major types of field-effect transistors.

large-signal A signal that operates an amplifier over a significant portion of its load line.

LASCR Light-activated silicon-controlled rectifier; a four-layer semiconductor device (thyristor) that conducts current in one direction when activated by a sufficient amount of light and continues to conduct until the current falls below a specified value.

laser Light amplification by stimulated emission of radiation.

light-emitting diode (LED) A type of diode that emits light when there is forward current.

limiter A diode circuit that clips off or removes part of a waveform above and/or below a specified level.

linear Characterized by a straight-line relationship.

linear region The region of operation along the load line between saturation and cutoff.

linear regulator A voltage regulator in which the control element operates in the linear region.

line regulation The change in output voltage for a given change in input (line) voltage, normally expressed as a percentage.

load The amount of current drawn from the output of a circuit through a load resistance.

load regulation The change in output voltage for a given change in load current, normally expressed as a percentage.

lock range   The range of frequencies over which a PLL can maintain lock.

logarithm   An exponent; the logarithm of a quantity is the exponent or power to which a given number called the base must be raised in order to equal the quantity.

loop gain   An op-amp's open-loop gain times the attenuation.

lower-side frequency   In balanced modulation, the difference of the carrier frequency and the modulation frequency.

low-pass filter   A type of filter that passes frequencies below a certain frequency while rejecting higher frequencies.

majority carrier   The most numerous charge carrier in a doped semiconductive material (either free electrons or holes).

midrange   The portion of a response curve between the lower and upper critical frequencies.

minority carrier   The least numerous charge carrier in a doped semiconductive material (either free electrons or holes).

mixer   A device for down-converting frequencies in a receiver system.

modem   A device that converts signals produced by one type of device to a form compatible with another; *mo*dulator/*dem*odulator.

modulation   The process in which a signal containing information is used to modify the amplitude, frequency, or phase of a much higher-frequency signal called the carrier.

monochromatic   Related to light of a single frequency; one color.

MOSFET   Metal oxide semiconductor field-effect transistor; one of two major types of FETs; sometimes called IGFET for insulated-gate FET.

multistage   Characterized by having more than one stage; a cascaded arrangement of two or more amplifiers.

natural logarithm   The exponent to which the base e ($e = 2.71828$) must be raised in order to equal a given quantity.

negative feedback   The process of returning a portion of the output signal to the input of an amplifier such that it is out of phase with the input signal.

neutron   An uncharged particle found in the nucleus of an atom.

noise   An unwanted signal.

noninverting amplifier   An op-amp closed-loop configuration in which the input signal is applied to the noninverting input.

nucleus   The central part of an atom containing protons and neutrons.

octave   A two-times increase or decrease in the value of a quantity such as frequency.

open-loop voltage gain (A_{ol})   The voltage gain of an op-amp without external feedback.

operational amplifier (op-amp)   A type of amplifier that has a very high voltage gain, very high input impedance, very low output impedance, and good rejection of common-mode signals.

operational transconductance amplifier (OTA)   A voltage-to-current amplifier.

orbit   The path an electron takes as it circles around the nucleus of an atom.

order   The number of poles in a filter.

oscillator   A circuit that produces a periodic waveform on its output with only the dc supply voltage as its input.

output resistance   The resistance looking in at the transistor collector.

overall voltage gain   The product of the attenuation and the gain from base to collector of an amplifier.

passband   The range of frequencies that are allowed to pass through a filter with minimum attenuation.

peak inverse voltage (PIV)   The maximum value of reverse voltage which occurs at the peak of the input cycle when the diode is reversed-biased.

pentavalent atom   An atom with five valence electrons.

phase-locked loop (PLL)   A device for locking onto and tracking the frequency of an incoming signal.

phase margin   The difference between the total phase shift through an amplifier and 180 degrees; the additional amount of phase shift that can be allowed before instability occurs.

phase shift   The relative angular displacement of a time-varying function relative to a reference.

phase-shift oscillator   A type of feedback oscillator that is characterized by three RC circuits in the positive feedback loop that produces a phase shift of 180°.

photodiode   A diode in which the reverse current varies directly with the amount of light.

photon   A particle of light energy.

phototransistor   A transistor in which base current is produced when light strikes the photosensitive semiconductor base region.

piezoelectric effect   The property of a crystal whereby a changing mechanical stress produces a voltage across the crystal.

pinch-off voltage   The value of the drain-to-source voltage of a FET at which the drain current becomes constant when the gate-to-source voltage is zero.

***pn* junction**   The boundary between two different types of semiconductive materials.

pole   A circuit containing one resistor and one capacitor that contributes -20 dB/decade to a filter's roll-off.

positive feedback   The return of a portion of the output signal to the input such that it reinforces and sustains the output. This output signal is in phase with the input signal.

power gain   The ratio of output power to input power of an amplifier.

power supply   A circuit that converts ac line voltage to dc voltage and supplies constant power to operate a circuit or system.

proton   The basic particle of positive charge.

push-pull   A type of class B amplifier with two transistors in which one transistor conducts for one half-cycle and the other conducts for the other half-cycle.

PUT   Programmable unijunction transistor; a type of three-terminal thyristor (more like an SCR than a UJT) that is triggered into conduction when the voltage at the anode exceeds the voltage at the gate.

Q-point   The dc operating (bias) point of an amplifier specified by voltage and current values.

quality factor (Q)   For a reactive component, a figure of merit which is the ratio of energy stored and returned by the component to the energy dissipated; for a band-pass filter, the ratio of the center frequency to its bandwidth.

radiant intensity (I_e)   The output power of an LED per steradian in units of mW/sr.

radiation   The process of emitting electromagnetic or light energy.

recombination   The process of a free (conduction band) electron falling into a hole in the valence band of an atom.

rectifier   An electronic circuit that converts ac into pulsating dc; one part of a power supply.

regulator   An electronic device or circuit that maintains an essentially constant output voltage for a range of input voltage or load values; one part of a power supply.

relaxation oscillator   An electronic circuit that uses an *RC* timing circuit to generate a nonsinusoidal waveform without an external input signal.

reverse bias   The condition in which a diode prevents current.

RF   Radio frequency.

ripple factor   A measure of effectiveness of a power supply filter in reducing the ripple voltage; ratio of the ripple voltage to the dc output voltage.

ripple voltage   The small variation in the dc output voltage of a filtered rectifier caused by the charging and discharging of the filter capacitor.

***r* parameter**   One of a set of BJT characteristic parameters that include α_{DC}, β_{DC}, r'_e, r'_b, and r'_c.

roll-off   The rate of decrease in the gain above or below the critical frequencies of a filter.

saturation   The state of a BJT in which the collector current has reached a maximum and is independent of the base current.

schematic   A symbolized diagram representing an electrical or electronic circuit.

Schmitt trigger   A comparator with hysteresis.

Schottky diode   A diode using only majority carriers and intended for high-frequency operation.

SCR   Silicon-controlled rectifier; a type of three-terminal thyristor that conducts current when triggered on by a voltage at the single gate terminal and remains on until the anode current falls below a specified value.

SCS   Silicon-controlled switch; a type of four-terminal thyristor that has two gate terminals that are used to trigger the device on and off.

semiconductor   A material that lies between conductors and insulators in its conductive properties.

shell   An energy band in which electrons orbit the nucleus of an atom.

signal compression   The process of scaling down the amplitude of a signal voltage.

silicon   A semiconductive material.

single-ended   A mode of op-amp operation in which a signal voltage is applied to only one input.

slew rate   The rate of change of the output voltage of an op-amp in response to a step input.

source   One of the three terminals of a FET analogous to the emitter of a BJT.

source-follower   The common-drain amplifier.

spectral   Pertaining to a range of frequencies.

stability   A measure of how well an amplifier maintains its design values (Q-point, gain, etc.) over changes in beta and temperature.

stage One of the amplifier circuits in a multistage configuration.

standoff ratio The characteristic of a UJT that determines its turn-on point.

summing amplifier An op-amp configuration with two or more inputs that produces an output equal to the sum of the inputs.

suppressed-carrier modulation A form of amplitude modulation in which the carrier is suppressed; also called *balanced modulation*.

switching current (I_S) The value of anode current at the point where the device switches from the forward-blocking region to the forward-conduction region.

switching regulator A voltage regulator in which the control element operates as a switch.

thermal overload A condition in a rectifier where the internal power dissipation of the circuit exceeds a certain maximum due to excessive current.

thermistor A temperature-sensitive resistor with a negative temperature coefficient.

thyristor A class of four-layer (*pnpn*) semiconductor devices.

transconductance (g_m) The ratio of a change in drain current for a change in gate-to-source voltage in a FET; in general, the ratio of the output current to the input voltage.

transistor A semiconductive device used for amplification and switching applications.

triac A three-terminal thyristor that can conduct current in either direction when properly activated.

trigger The activating input of some electronic devices and circuits.

trivalent atom An atom with three valence electrons.

troubleshooting The process and technique of identifying and locating faults in an electronic circuit or system.

tuning ratio The ratio of varactor capacitances at minimum and at maximum reverse voltages.

tunnel diode A diode exhibiting a negative resistance characteristic.

UJT Unijunction transistor; a three-terminal single *pn* junction device that exhibits a negative resistance characteristic.

upper-side frequency In balanced modulation, the sum of the carrier frequency and the modulated frequency.

valence Related to the outer shell of an atom.

varactor A variable capacitance diode.

V-I characteristic A curve showing the relationship of diode voltage and current.

voltage-controlled oscillator (VCO) A type of relaxation oscillator whose frequency can be varied by a dc control voltage; an oscillator for which the output frequency is dependent on a controlling input voltage.

voltage-follower A closed-loop, noninverting op-amp with a voltage gain of 1.

voltage multiplier A circuit using diodes and capacitors that increases the input voltage by two, three, or four times.

wavelength The distance in space occupied by one cycle of an electromagnetic or light wave.

Wien bridge oscillator A type of feedback oscillator that is characterized by an *RC* lead-lag circuit in the positive feedback loop.

zener breakdown The lower voltage breakdown in a zener diode.

zener diode A diode designed for limiting the voltage across its terminals in reverse bias.

Index